Proceedings of the Sixth International Conference on Computer Supported Cooperative Work in Design

July 12–14, 2001
London, Ontario, Canada

Edited by

Weiming Shen

Zongkai Lin

Jean-Paul Barthès

Mohamed Kamel

Sponsored by

NRC·CNRC
National Research Council Canada

IEEE Networking the World™
IEEE Canada and IEEE London/Canada Section

Western
University of Western Ontario

University of Waterloo
University of Waterloo

NRC·CNRC
NRC Research Press
Ottawa 2001

Published by
NRC Research Press
National Research Council Canada
1200 Montreal Road, Building M-55
Ottawa, Ontario, Canada K1A 0R6
www.monographs.nrc.ca

Printed in Canada on acid-free paper. ♾

Publisher's note: This book has been prepared from camera-ready copy provided by the authors.

ISBN 0-660-18493-1
IEEE Catalog No. 01EX472
NRC No. 44458

National Library of Canada cataloguing in publication data

Conference on CSCW in Design (6th : 2001: London, Ont.)

Proceedings of the Sixth International Conference on CSCW in Design

"July 12–14, 2001, London, Ontario, Canada"
"Sponsors: National Research Council Canada, IEEE Canada and IEEE London/Canada Section, University of Western Ontario, and University of Waterloo."
Includes bibliographical references.
ISBN 0-660-18493-1

1. Engineering design — Data processing — Congresses.
2. Teams in the workplace — Data processing — Congresses.
3. Design, Industrial — Congresses.
4. System design — Congresses.

I. Weiming, Shen.
II. National Research Council Canada.
III. IEEE Canada.
IV. University of Western Ontario.
V. University of Waterloo.

PA174.C66 2001 620'.0042'0285 C2001-980190-4

Preface

Welcome to the Sixth International Conference on Computer Supported Cooperative Work in Design (CSCWD 2001).

Design of complex artifacts and systems requires the cooperation of multidisciplinary design teams. CSCWD 2001 provides a forum for researchers and practitioners involved in different but related domains to confront research results and discuss key problems. The major topics of CSCWD 2001 include the following:

- Techniques, methods, and tools for CSCW in design;
- Social organization of the CSC process;
- Integration of methods and tools within the work organization;
- Cooperation in virtual enterprises and e-businesses;
- CSCW in Design and Manufacturing;
- Interaction between the CSC approach and knowledge reuse as found in knowledge management;
- Intelligent agent and multi-agent systems;
- Internet/Web and CSCW in design;
- Applications and testbeds.

The Technical Program Committee has developed an exciting and provocative technical program that includes 29 technical sessions (contributed and invited), with papers from over 18 countries.

We would like to thank all members of the Program Committee for their effort in promoting the conference and carefully reviewing the submitted papers. Special thanks are due to the authors who contributed to this conference.

We would also like to thank the chairs and members of the organizing committee for taking care of all the details to make this conference successful. We acknowledge the sponsorship of National Research Council Canada, IEEE Canada, and IEEE London/Canada through seed funding. We gratefully appreciate the support we received from the staff of the Conference Services at the University of Western Ontario during the conference preparation and organization.

Special thanks to Leah Paterson who provided a great help during the preparation of the conference and the editing of the conference proceedings.

Weiming Shen
Integrated Manufacturing Technologies Institute
National Research Council Canada
London, Ontario, Canada

Zongkai Lin
Institute of Computing Technology
Chinese Academy of Sciences
Beijing, P.R. China

Jean-Paul. A. Barthès
CNRS UMR 6599 HEUDIASYC
Université de Technologie de Compiègne
France

Mohamed Kamel
Dept. of Systems Design Engineering
University of Waterloo
Waterloo, Ontario, Canada

Contents

Part III: Collaborative Design **161**

Part IX: Applications 523

Author Index 581

International Steering Committee

Chair: Prof. Zongkai Lin (China)

Co-Chair: Prof. Jean-Paul Barthès (France)

Secretary: Dr. Minzhe Zhao (Singapore)

Members:
Prof. Igor Hawryszkiewycz (Australia)
Dr. Karl R. Lang (Hong Kong, China)
Prof. Mary Lou Maher (Australia)
Prof. Jay Nunamaker (USA)
Prof. Hwa Gyoo Park (Korea)
Dr. Weiming Shen (Canada)
Prof. Gilbert Young (Hong Kong, China)

Organizing Committees

General Conference Chair:
Dr. Mohamed Kamel (Univ. of Waterloo, Canada)

Vice Conference Chairs:
Dr. Sherman Lang (NRC-IMTI, Canada)
Dr. Rajni Patel (UWO, Canada)
Prof. Haicheng Yang (NWPU, P.R. China)
Mr. Ashfaq Husain (IEEE London/Canada)

Program Committee Chairs:
Dr. Weiming Shen (NRC-IMTI, Canada)
Prof. Zongkai Lin (Chinese Academy of Sciences, P.R. China)
Prof. Jean-Paul Barthès (UTC, France)

Financial Committee Chair:
Dr. Lihui (Ian) Wang (NRC-IMTI, Canada)

Workshops Chairs:
Dr. Hamada Ghenniwa (UWO, Canada)
Dr. Paul Woodard (NRC-IMTI, Canada)

Publication Chair:
Mr. Vibhor Gupta (NRC-IMTI, Canada)

Publicity Chair:
Dr. Zakaria Maamar (Zayed U., UAE)

Local Arrangement Chair:
Dr. Ralph Buchal (UWO, Canada)

Local Arrangement Committee Members:
Brian Corrie (NRC-IMTI, Canada)
John Dickinson (NRC-IMTI, Canada)
Costantino Mongelli (NRC-IMTI, Canada)
Ajit Pardasani (NRC-IMTI, Canada)
Gail Percy (NRC-IMTI, Canada)
Helen Xie (NRC-IMTI, Canada)

Program Committee Members

Wil van der Aalst (NL)
Marcos Borges (BR)
Carlos Bremer (BR)
Ralph Buchal (CA)
Catherine Burns (CA)
Paul Calamai (CA)
Luis Camarinha-Matos (PT)
Alvin Chan (HK)
Stephen Chan (HK)
Li Chen (CA)
Chin-Jung Chen (US)
To-Yat Cheung (HK)
Moon Jung Chung (US)
Jiati Deng (CN)
Roberto DiRrado (CA)
Jinxiang Dong (CN)
Zuming Dong (CA)
Shaw Feng (US)
Rajit Gadh (US)
Hamada Ghenniwa (CA)
Dan Grecu (US)
Ning Gu (CN)
Yuchai Guo (CN)
Igor Hawryszkiewycz (AU)
Yeh-Liang Hsu (TW)
George Huang (HK)
Anne James (UK)
Jano Moreira de Souza (BR)
Michael Johnson (AU)
Catholijn Jonker (NL)
Mohamed Kamel (CA)
Mark Klein (US)
Laryy Korba (CA)
Rob Kremer (CA)
Ernest Lam (HK)
Karl Lang (HK)
W.B. Lee (HK)
Tae-Eog Lee (KR)
Tangqiu Li (CN)
Shouxin Lin (CN)
Hong Liu (CN)
Kecheng Liu (UK)
Jian Lu (FR)
Junzhou Luo (CN)
Carolyn MacGregor (CA)
Mary Lou Maher (AU)
K.L. Mak (HK)
Zakaria Maamar (AE)
Tsutomu Matsumoto (JP)
Vincent Ng (HK)
Jay Nunamaker (US)
Eugenio Oliveira (PT)
Vaggelis Ouzounis (DE)
Zhigeng Pan (CN)
Hwa Gyoo Park (KR)
Yun Peng (US)
Bruno Ramond (FR)
Ramana Reddy (US)
Paul Rodgers (UK)
Edson Scalabrin (BR)
Meilin Shi (CN)
Yuanchun Shi (CN)
Haitao Song (CN)
Shunlin Song (CN)
Ram D. Sriram (US)
Yong Tang (CN)
Michael Terk (US)
Francis Thibault (CA)
Huaglory Tianfield (UK)
Tetsuo Tomiyama (JP)
Brigitte Trousse (FR)
Patri Venuvinod (HK)
Haiyang Wang (CN)
Shaomei Wang (CN)
Hong Xu (CN)
Deyi Xue (CA)
Chenhui Yang (CN)
Gilbert Young (HK)
K.M. Yu (HK)
Jyh-Cheng Yu (TW)
Matthew Yuen (HK)
Chris Zhang (CA)
Fuyan Zhang (CN)
Minzhe Zhao (SG)
Diji Zhou (CN)
Yu Zhou (CN)
Khaldoun Zreik (FR)

Part I: Keynote Papers

Design Repositories and Product Representation for Collaborative Product Development
(Extended Abstract)

Ram D. Sriram (sriram@cme.nist.gov)
Leader, Design and Process Group

Simon Szykman (szykman@cme.nist.gov)
Leader, Design Repositories and Knowledge-based Interoperability Projects

National Institute of Standards and Technology
100 Bureau Drive, MS 8263
Gaithersburg, MD 20899
USA

Design of complex engineering systems is increasingly becoming a collaborative task among designers or design teams that are physically, geographically, and temporally distributed. The complexity of modern products means that a single designer or design team can no longer manage the complete product development effort. Developing products without sufficient expertise in a broad set of disciplines can result in extended product development cycles, higher development costs, and quality problems. Designers are no longer merely exchanging geometric data, but more general knowledge about design and design process, including specifications, design rules, constraints, rationale, etc.

As design becomes increasingly knowledge-intensive and collaborative, the need for computational design frameworks to support the representation and use of knowledge among distributed designers becomes more critical. The significant barrier to product development is not the problem of providing distributed access to distributed information, but of finding the information that is needed. Industry's need for rapid retrieval and subsequent reuse of knowledge, driven by pressure to reduce product development times in industry, has resulted in an increased focus on methods for representing and storing engineering artifact knowledge. Traditional design databases, which merely provide access to schematics, computer-aided design (CAD) models, and documentation, are inadequate for this purpose. The emerging research area of design repositories is aimed at addressing these industry needs.

One focus of ongoing research activities at the National Institute of Standards and Technology (NIST) is the development of representations of information that are unavailable in traditional computer-aided engineering tools to support the exchange of product information in a distributed product development environment. Traditional CAD systems are limited to representation of geometric data and other types of information relating to geometry that may include constraints, parametric information, features, and so on. Because of industry's increasing dependence on other types of knowledge in the design process, new classes of tools to support knowledge-based design, product data management (PDM), and concurrent engineering have begun emerging in the engineering software marketplace.

When contrasted with traditional CAD tools, these new systems are making progress toward the next generation of engineering design support tools. However, these systems have focused primarily on database-related issues and do not place a primary emphasis on information models for artifact representation. Furthermore, although these systems can often represent non-geometric knowledge (e.g. information about the design process, manufacturing process, bills of materials, etc.), representation of the *artifact* itself is still generally limited to geometry. Work at NIST has involved the development of a core artifact representation that encompasses broader engineering design content. This content includes representation not only of geometry, but also of function, behavior, physical and functional decompositions, mappings between physical structures and function, and various kinds of relationships among these entities.

In parallel with this work, the NIST Design Repository Project involves research toward providing a technical foundation for the creation of design repositories—repositories of heterogeneous knowledge and data that are designed to support representation, capture, sharing, and reuse of corporate product knowledge. The NIST Design Repository Project is engaged in developing a prototype suite of software tools built upon the core artifact representation. The various facets of this project include the development of a distributed web-based communications architecture, interfaces for browsing and authoring design repositories, and search and retrieval of information from these bodies of knowledge. The project also involves the creation of taxonomies of engineering terminology to support product modeling activities as well as to facilitate knowledge retrieval.

This presentation will discuss the role of the concept of design repositories in next generation collaborative product development systems. An overview of the core artifact representation will be provided, followed by a discussion of various aspects of the NIST Design Repository Project. In addition, the core product representation used in a related project—the Knowledge-based Interoperability project—will be discussed. An interactive scenario being developed in conjunction with researchers at Carnegie Mellon University (CMU) will be described. The objective of this scenario is to demonstrate the integration of composable simulation research at CMU and design repository research at NIST. The scenario involves the addition of behavior model information to design repositories, and the retrieval of individual models fragments associated with different parts of a product. These model fragments are then assembled into a simulation of the overall product behavior in order to evaluate the designed product against desired function and performance requirements.

Acknowledgments: Current funding for the Design Repository and the Knowledge-based Interoperability projects comes from the Systems Integration for Manufacturing Applications Program (http://www.mel.nist.gov/div826/msid/sima/sima.htm). Previous funding was provided by the NIST Advanced Technology Program's intramural research program. The current team members are: Simon Szykman (Project Lead), Ram D. Sriram, Steven J. Fenves, Robert Allen, Khanh Trieu, Guilhem Assant, Sylvain Archaimbault, Jean-François Heintz, and Fujun Wang.

What Complex Systems Research Can Teach Us About Collaborative Design

Mark Klein
Massachusetts Institute of Technology
m_klein@mit.edu

Hiroki Sayama
New England Complex Systems Institute
sayama@necsi.org

Peyman Faratin
Massachusetts Institute of Technology
peyman@mit.edu

Yaneer Bar-Yam
New England Complex Systems Institute
yaneer@necsi.org

Abstract

Collaborative design is challenging because strong interdependencies between design issues make it difficult to converge on a single design that satisfies these dependencies and is acceptable to all participants. Complex systems research has much to offer to the understanding of these dynamics. This paper describes some insights from the complex systems perspective.

1. The Challenge: Collaborative Design Dynamics

Collaborative design is challenging because strong interdependencies between design issues make it difficult to converge on a single design that satisfies these dependencies and is acceptable to all participants. The dynamics of collaborative design are thus typically characterized by (1) multiple iterations and/or heavy reliance on multi-functional design reviews, both of which expensive and time-consuming, (2) poor incorporation of some important design concerns, typically later life-cycle issues such as environmental impact, as well as (3) reduced creativity due to the tendency to incrementally modify known successful designs rather than explore radically different and potentially superior ones.

Complex systems research is devoted to understanding, at a fundamental level, the dynamics of systems made up of interdependent components, and has we argue much to offer to our understanding of the dynamics of collaborative design. Previous research on design dynamics has focused on *routine* design [1] where the design space is well-understood (e.g. as in brake or transmission design), and the goal is to optimize a design via incremental changes for requirements similar to those that have been encountered many times before [2] [3]. Rapid technological and other changes have made it increasingly clear, however, that many of the most important collaborative design problems (e.g. concerning software, biotechnology, or electronic commerce) involve *innovative* design, radically new requirements, and unfamiliar design spaces. In this paper we explore some of what complex systems research can contribute to this important challenge. We will begin by defining a simple model of collaborative design, review the strengths and weaknesses of current collaborative design approaches, discuss some of the insights a complex systems perspective has to offer, and suggest ways to better support innovative collaborative design building on these insights.

2. Defining Collaborative Design

A design (of physical artifacts such as cars and planes as well as behavioral ones such as plans, schedules, production processes or software) can be represented as a set of *issues* (sometimes also known as *parameters*) each with a unique value. A complete design for an artifact includes issues that capture the *requirements* for the artifact, the *specification* of the artifact itself (e.g. the geometry and materials), the *process* for creating the artifact (e.g. the manufacturing process) and so on through the artifacts' entire life cycle. If we imagine that the possible values for every issue are each laid along their own orthogonal axis, then the resulting multi-dimensional space can be called the *design space*, wherein every point represents a distinct (though not necessarily good or even physically possible) design. The choices for each design issue are typically highly *interdependent*. Typical sources of inter-dependency include shared resource (e.g. weight, cost) limits, geometric fit, spatial separation requirements, I/O interface conventions, timing constraints etc.

Collaborative design is performed by multiple participants (representing individuals, teams or even entire organizations), each potentially capable of proposing values for design issues and/or evaluating these choices from their own particular perspective (e.g. manufacturability). Figure 1 below illustrates this model: the small black circles represent design issues, the links between the issues represent design issue inter-dependencies, and the large ovals represent the design *subspace* (i.e. subset of design issues) associated with each design participant.

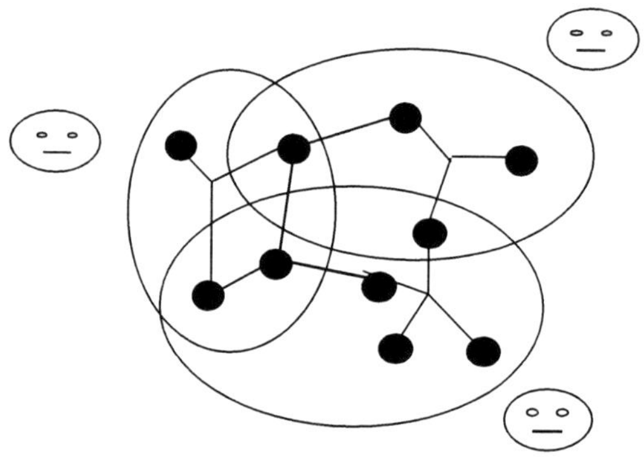

Figure 1: A Model for Collaborative Design.

In a large artifact like a commercial jet there may be millions of components and design issues, hundreds to thousands of participants, working on hundreds of distinct design subspaces, all collaborating to produce a complete design.

Some designs are better than others. We can in principle assign a *utility* value to each design and thereby define a *utility function* that represents the utility for every point in the design space (though in practice we may only be able to assess *comparative* as opposed to *absolute* utility values). A simple utility function might look like the following:

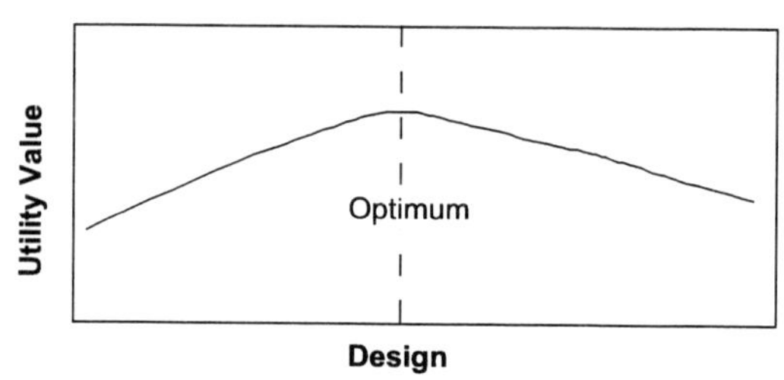

Figure 2. A simple utility function.

The *goal* of the design process can thus be viewed as trying find the design with the optimal (maximal) utility value, though often optimality is abandoned in favor of 'good enough'.

The key challenge raised by the collaborative design of complex artifacts is that the design spaces are typically huge, and concurrent search by the many participants through the different design subspaces can be expensive and time-consuming because design issue interdependencies lead to conflicts (when the design solutions for different subspaces are not consistent with each other). Such conflicts severely impact design utility and lead to the need for expensive and time-consuming design rework.

3. Strengths and Limitations of Current Approaches

Traditionally, collaborative design has been carried out using a serialized process, wherein for example a complete requirement set would be generated, then given to design engineers who would completely specify the product geometry, which in turn would then be given to the manufacturing engineers to create a manufacturing plan, and so on. This has the problem that if an earlier decision turns out to be sub-optimal from the perspective of someone making dependent decisions later on in the design process (e.g. if a requirement is impossible to achieve, or a particular design geometry is very expensive to manufacture): the process of revising the design is slow and expensive, and often only the highest priority changes are made. The result is designs that tend to be poor from the standpoint of later life-cycle perspectives, including for example environmental concerns such as recyclability that are becoming increasingly important.

More recently, several strategies have emerged for better accounting for the interdependencies among collaborative design participants. These include concurrent engineering and least-commitment design:

Concurrent engineering involves the creation of design teams with representatives of all important design perspectives, for each distinct design subspace. Design decisions can be reviewed by all affected design perspectives when they are initially being considered, so bad decisions can be caught and revised relatively quickly and cheaply. While this approach has proven superior in some ways to traditional serial design, it often incurs an overwhelming burden on engineers as they attend many hours of design meetings and review hundreds of proposed changes per week [4].

Least-commitment design is a complimentary approach that attempts to address the same challenges by allowing engineers to specify a design incompletely, for example as a rough sketch or set of alternatives, and then gradually make the design more specific, for example by pruning some alternatives [5] [6]. This has the advantage that bad design decisions can be eliminated before a lot of effort has been invested in making them fully specific, and engineers are not forced to make arbitrary commitments that lead to needless conflicts.

While the adoption of approaches has been helpful, major challenges remain. Consider for example the Boeing 767-F redesign program [4]. Some conflicts were not detected until long (days to months) after they had occurred, resulting in wasted design time, design rework, and often even scrapped tools and parts. It was estimated that roughly half of the labor budget was consumed dealing with changes and rework, and that roughly 25-30% of design decisions had to be changed. Since maintaining scheduled commitments was a priority, design rework often had to be done on a short flow-time basis that typically cost much more (estimates ranged as high as 50 times more) and sometimes resulted in reduced product quality. Conflict cascades that required as many as 15 iterations to finally produce a consistent design were not uncommon for some kinds of design changes. All this in the context of Boeing's industry-leading concurrent engineering practices. The dynamics of current collaborative design processes are thus daunting, and have led to reduced design creativity, a tendency to incrementally modify known successful designs rather than explore radically different and potentially superior ones.

Improving the efficiency, quality and creativity of the collaborative innovative design process requires, we believe, a much better understanding of the dynamics of such processes and how they can be managed. In the next section we will review of the some key insights that can be derived from complex systems research for this purpose.

4. Insights from Complex Systems Research

A central focus of complex systems research is the dynamics of distributed networks, i.e. networks in which there is no centralized controller, so global behavior emerges solely as a result of concurrent local actions. Such networks are typically modeled as multiple nodes, each node representing a state variable with a given value. Each node in a network tries to select the value that optimizes its own utility while maximizing its consistency with the influences from the other nodes. The global utility of the network state is simply the sum of local utilities plus the degree to which all the influences are satisfied. The dynamics of such networks emerge as follows: since all nodes update their local state concurrently based on their current context (at time T), the choices they make may no longer be the best ones in the new context of node states (at time T+1), leading to the need for further changes.

Is this a useful model for understanding the dynamics of collaborative design? We believe that it is. It is straightforward to map the model of collaborative design presented above onto a network. We can map design participants onto nodes, where each participant is trying to maximize the utility of the choices it makes for the design subspace (e.g. subsystem) it is responsible for, while ensuring its decisions will satisfy the relevant dependencies (represented as the links between nodes). As we shall see, to understand network dynamics, the links between nodes need capture only quite abstract properties of the dependencies. As a first approximation, it is reasonable to model the utility of a design as the local utility achieved by each participant plus a measure of how well all the decisions fit together. Even though real-world collaborative design clearly has top-down elements, the sheer complexity of many design artifacts means that no one person is capable of keeping the whole design in his/her head and centralized control of the design decisions becomes impractical, so the design process is dominated by concurrent local activities. The remainder of this paper will be based on this view of the collaborative design process.

How do such distributed networks behave? Let us consider the following simple example, a network consisting of binary-valued nodes where each node is influenced to have the same value as the nodes it is linked to (Figure 3):

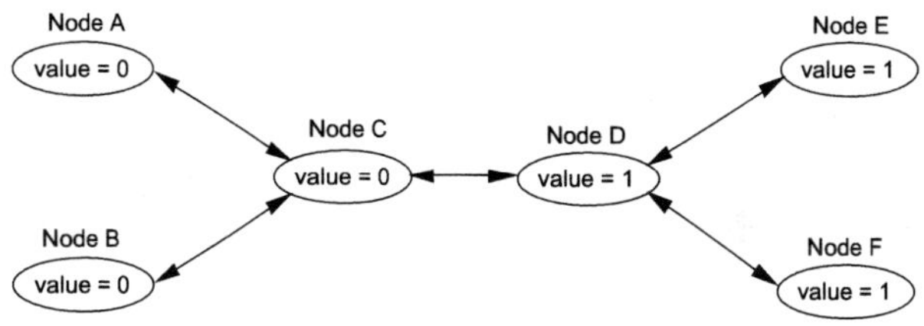

Figure 3: A simple network.

We can imagine using this network to model a real-world situation wherein six subsystems are being designed and we want them to use matching interfaces. The network has converged, using the concurrent update procedure described above, onto a *local* optimum (no node can increase the number of influences it satisfies by a local change), so it will not reach as a result a *global* optimum (where all the nodes have the same value). Generally speaking, networks may not always converge upon the global optimum, and in some cases (as we shall see with *dynamic attractors*), a network may not converge at all. *Insights into whether and how global optima can be found in networks represent the heart of what complex systems research offers to the understanding of collaborative design.*

We will discuss these insights in the remainder of this section. The key factor determining network dynamics is the nature of the influences between nodes. We will first consider how such influences can be defined. We will then consider two important distinctions: whether the influences are *linear* or not, and whether they are *symmetric* or not. We will finally discuss subdivided network topologies, and the role of learning. Unless indicated otherwise, the material on complex systems presented below is drawn from [8].

4.1. How Are Influences Defined?

It is, in principle, straightforward to compute what the inter-node influences should be in order to create a network that implements a given global utility function. In design practice, however, we almost invariably do *not* know the global utility function up front; it is revealed incrementally, rather, by the process of defining and evaluating different candidate designs. Utility evaluations are apt in any case to be approximate at best, because among other things of uncertainties about the context the artifact will exist in. Imagine for example that our goal is to design the most profitable airplane possible: so many imponderable factors heavily influence this (e.g. oil prices, wars, government subsidies for competitors) that the only way to *really* know the utility of a design is to build it and see what happens! It is usually much easier, as a result, to define the influences directly based on our knowledge of design decision dependencies. We know for example that parts need to have non-overlapping physical geometries, that electrical interfaces for connected systems must be compatible, that weight limits must be met, and so on.

Care must be taken in defining these influences, however. We face the risk of neglecting to give sufficient prominence to important concerns. Traditionally, influences from later stages of the life cycle (e.g. the manufacturing or recycling of the product) tend to be the ones most neglected, and the consequences are only encountered when that life cycle stage has been reached and it is typically much more difficult, time-consuming and expensive to do anything about it. Another concern is that, while there is always a direct mapping from a utility function to a set of influences, the opposite is *not* true. Asymmetric influences, in particular, do not have a corresponding utility function, and the network they define does not converge to any final result. This will be discussed below further in the section on asymmetric networks.

4.2. Linear vs. Non-Linear Networks

If the value of nodes is a linear function of the influences from the nodes linked to it, then the system is linear, otherwise it is non-linear. Linear networks have a single *attractor*, i.e. a single configuration of node states that the network converges towards no matter what the starting point, corresponding to the global optimum. Their utility function thus looks like that shown in Figure 2 above. This means we can use a 'hill-climbing' approach (where each node always moves directly towards increased local utility) because local utility increases always move the network towards the global optimum.

Non-linear networks, by contrast, are characterized by having multiple attractors and *ultrametric* (bumpy) utility functions, like that shown in Figure 4:

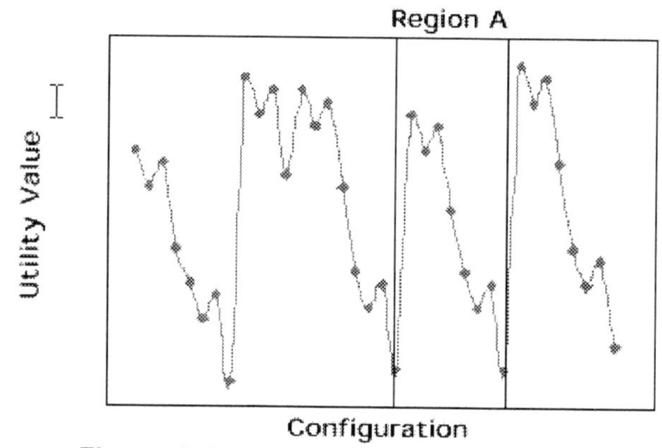

Figure 4. An ultrametric utility function.

Ultrametric utility functions have a *fractal* structure (i.e. self-similar at different scales) so they are bumpy at all scales, the highest peaks also tend to be the widest ones, and high peaks tend to be widely separated from each other. The total number of peaks is an exponential function of the number of nodes in the network.

A key property of non-linear networks is that search for the global optima can *not* be performed successfully by pure hill-climbing algorithms, because they can get stuck in local optima that are globally sub-optimal. Consider, for example, what would happen if the system started searching in region A in Figure 4 above. Hill-climbing (if it started from the left of region A) would take it to the top of the local optimum, which is substantially lower than optima in other regions of the utility function. Hill-climbing would do even more poorly if it started at the right of region A.

One consequence of this reality is a tendency to stick near well-known designs. When a utility function has widely separated optima, once a satisfactory optimum is found the temptation is to stick to it. This design conservatism is exacerbated by the fact that it is often difficult to compare the utilities for radically different designs. We can expect this effect to be especially prevalent in industries, such as commercial airlines and power plants, which are capital-intensive and risk-averse, since in such contexts the cost of exploring new designs, and the risk of getting it wrong, can be prohibitive.

A range of techniques have emerged that are appropriate for finding optima in ultrametric utility functions, all relying on the ability to search past valleys in the utility function. Stochastic approaches such as simulated annealing have proven quite effective. Simulated annealing endows the search procedure with a tolerance for moving in the direction of lower utility that varies as a function of a virtual 'temperature'. At first the temperature is high, so the system is as apt to move towards lower utilities as higher ones. This allows it to range widely over the utility function and possibly find new higher peaks. Since higher peaks are also wider ones, the system will tend to spend most of its time in the region of high peaks. Over time the temperature decreases, so the algorithm increasingly tends towards pure hill-climbing. While this technique is not provably optimal, it has been shown to get close to optimal results in most cases.

Annealing, however, runs into a dilemma when applied to systems with multiple actors. Let us assume that at least some actors are self-interested 'hill-climbers', concerned only with directly maximizing their local utilities, while others are 'annealers', willing to accept, at least temporarily, lower local utilities in order to increase the utility in other nodes. Simulation reveals that while the presence of annealers always increases *global* utility, annealers always fare *individually* worse than hill-climbers when both are present [9]. The result is that globally beneficial behavior is not individually incented.

How do these insights apply to collaborative design? Linear networks represent a special case and we would expect because of this that most collaborative design contexts are non-linear. There is a particular class of collaborative design, however, that has been successfully modeled as linear networks: *routine* design [1]. Routine design involves highly familiar requirements and design options, as for example in automobile brake or transmission design. Designers can usually start the design process near enough to the final optimum, as a result, to be able to model the design space as having a single attractor. Linear network models of collaborative design have generated many useful results, including approaches for identifying design process bottlenecks [2] and for fine-tuning the lead times for design subtasks [3] in routine design domains.

As we argued above, however, today's most challenging and important collaborative design problems are *not* instances of routine design. The requirements and design options for such *innovative* design challenges are typically relatively unfamiliar, and it is unclear as a result where to start to achieve a given set of requirements. There may be multiple very different good solutions, and the best solution may be radically different than any that have been tried before. For such cases non-linear networks seem to represent a more accurate model of the collaborative design process.

This has important consequences. Simply instructing each design participant to optimize its own design subspace as much as possible (i.e. 'hill-climbing') can lead to the design process getting stuck in local optima that may be significantly worse than radically different alternatives. Design participants must be willing to explore alternatives that, at least initially, may appear much worse from their individual perspective than alternatives currently on the table. Designers often show greater loyalty to producing a good design for the subsystem they are responsible for, than to conceding to make someone else's job easier, so we need to find solutions for the dilemma identified above concerning the lack of individual incentives for such globally helpful behavior. We will discuss possible solutions in the section below on "How We Can Help".

4.3. Symmetric vs. Asymmetric Networks

Symmetric networks are ones in which influences between nodes are mutual (i.e. if node A influences node B by amount X then the reverse is also true), while asymmetric networks do not have this property. Asymmetric networks (with an exception to be discussed below) add the complication of *dynamic* attractors, which means that the network does not converge on a *single* configuration of node states but rather cycles indefinitely around a relatively small *set* of configurations. Let us consider the simplest possible asymmetric network: the 'odd loop' (Figure 5):

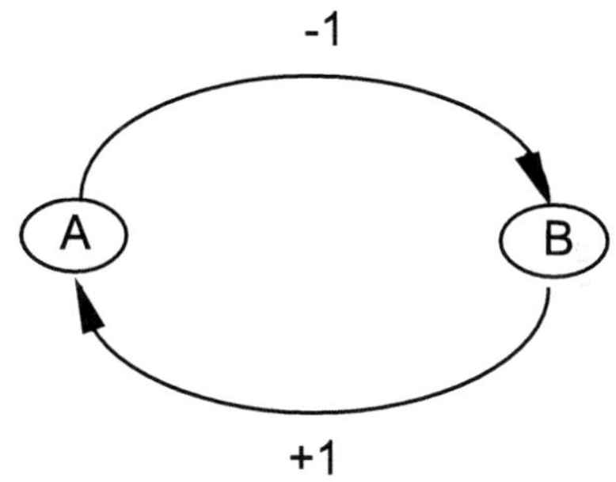

Figure 5. The simplest possible asymmetric network – an 'odd loop'.

This network has two links: one which influences the nodes to have the same value, the other which influences them to have opposite values. Imagine we start with node A having the value 1. This will influence node B to have the value –1, which will in turn influence node A towards the value –1, which will in turn cause node B to flip values again, and so on *ad infinitum*. If we plot the state space that results we get the following simple dynamic attractor (Figure 6):

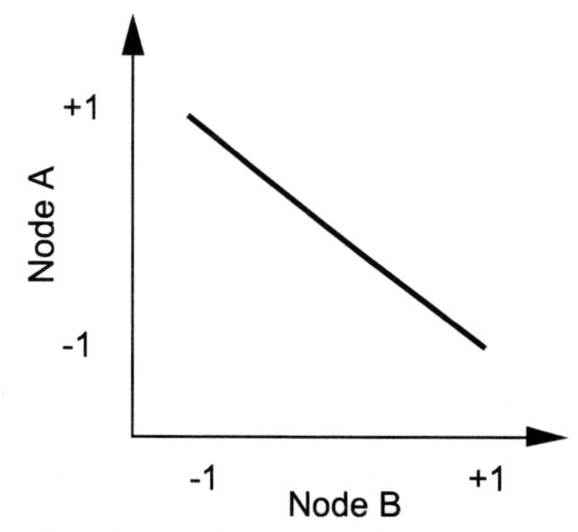

Figure 6. The dynamic attractor for the odd loop.

More complicated asymmetric networks will produce dynamic attractors with more complicated shapes, but the upshot is the same: the only way to get a definite solution (i.e. configuration of node states) with a dynamic attractor is to arbitrarily pick one point along its length. There is one important special case, however: feed-forward networks. The influences in feed-forward networks are *acyclic,* which means that a node never is able to directly or indirectly influence its own value (there are in other words no *loops*). Feed-forward networks do not have dynamic attractors.

How does this apply in collaborative design settings? Traditional serialized collaborative design is an example of an asymmetric feed-forward network, since the influences all flow uni-directionally from the earlier product life cycle stages (e.g. design) to later ones (e.g. manufacturing) with only weak feedback loops if at all. In such settings we may not expect particularly optimal designs but the attractors should be static and convergence should always occur, given sufficient time. 'Pure' concurrent engineering, where all design disciplines are represented on multi-functional design teams, encourage roughly symmetric influences between the participants and thus can also be expected to have convergent dynamics with static attractors. Current collaborative design practice, however, is a hybrid of these two approaches, and thus is likely to have the combination of asymmetric influences and influence loops that produces dynamic attractors and therefore non-convergent dynamics. Dynamic attractors were found to not to have a significant effect on the dynamics of at least some routine (linear) collaborative design contexts [3], but may prove more significant in innovative (non-linear) collaborative design. It may help to explain, for example, why it sometimes takes so many iterations to account for all the consequences of changes in complex designs [4].

4.4. Subdivided Networks

Another important property of networks is whether or not they are sub-divided, i.e. whether they consist of sparsely interconnected 'clumps' of highly interconnected nodes, as for example in Figure 7:

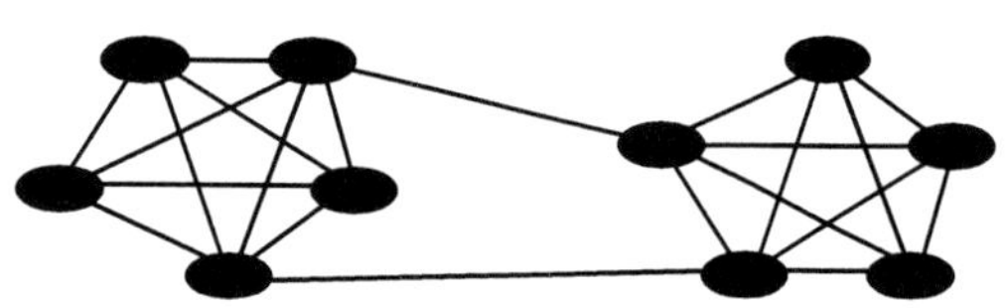

Figure 7. An example of a subdivided network.

When a network is subdivided, node state changes can occur within a given clump with only minor effects on the other clumps. This has the effect of allowing the network to explore more states more rapidly. Rather than having to wait for an entire large network to converge, we can rely instead on the much quicker convergence of a number of smaller networks, each one exploring possibilities that can be placed in differing combinations with the possibilities explored by the other sub-networks.

This effect is in fact widely exploited in design communities, where it is often known as *modularization.* This involves intentionally creating subdivided networks by dividing the design into subsystems with pre-defined standardized interfaces, so subsystem changes can be made with few or any consequences for the design of the other subsystems. The key to using this approach successfully is defining the design decomposition such that the impact of the subsystem interdependencies on the global utility is relatively low, because the standardized interfaces rarely represent an optimal way of satisfying these dependencies. In most commercial airplanes, for example, the engine and wing subsystems are designed separately, taking advantage of standardized engine mounts to allow the airplanes to use a range of different engines. This is not the optimal way of relating engines and wings, but it is good enough and simplifies the design process considerably. If the engine-wing interdependencies were crucial, for example if standard engine mounts had a drastically negative effect on the airplane's aerodynamics, then the design of these two subsystems would have to be coupled much more closely in order to produce a satisfactory design.

4.5. Imprinting

One common technique used to speed network convergence is *imprinting*, wherein the network influences are modified when a successful solution is found in order to facilitate quickly finding (similar) good solutions next time. A common imprinting technique is reinforcement learning, wherein the links representing influences that are satisfied in a successful final configuration of the network are strengthened, and those representing violated influences weakened. The effect of this is to create fewer but higher optima in the utility function, thereby increasing the likelihood of hitting such optima next time.

Imprinting is a crucial part of collaborative design. The configuration of influences between design participants represents a kind of 'social' knowledge that is generally maintained in an implicit and distributed way within design organizations, in the form of individual designer's heuristics about who should talk to whom when about what. When this knowledge is lost, for example due to high personnel turnover in an engineering organization, the ability of that organization to do complex design projects is compromised. It should be noted, however, that imprinting reinforces the tendency we have already noted for organizations in non-linear design regimes to stick to tried-and-true designs, by virtue of making the previously-found optima more prominent in the design utility function.

5. How We Can Help?

What can we do to improve our ability to do innovative collaborative design? We will briefly consider several possibilities suggested by the discussion above.

Information systems are increasingly becoming the medium by which design participants interact, and this fact can be exploited to help monitor the influence relationships between them. One could track the volume of design-related exchanges or (a more direct measure of actual influence) the frequency with which design changes proposed by one participant are accepted as is by other participants. This can be helpful in many ways. Highly asymmetric influences could represent an early warning sign of non-convergent dynamics. Detecting a low degree of influence by an important design concern, especially one such as environmental impact that has traditionally been less valued, can help avoid utility problems down the road. A record of the influence relationships in a successful design project can be used to help design future projects. Influence statistics can also be used to help avoid repetitions of a failed project. If a late high-impact problem occurred in a subsystem that had a low influence in the design process, this would suggest that the influence relationships should be modified in the future. Note that this has the effect of making a critical class of normally implicit and distributed knowledge more explicit, and therefore more amenable to being preserved over time (e.g. despite changes in personnel) and transferred between projects and even organizations.

Information systems can also potentially be used to help assess the degree to which the design participants are engaged in routine vs innovative design strategies. We could use such systems to estimate for example the number and variance of design alternatives being considered by a given design participant. This is important because, as we have seen, a premature commitment to a routine design strategy that optimizes a given design alternative can cause the design process to miss other alternatives with higher global optima. Tracking the degree of innovative exploration can be used to fine-tune the use of innovation-enhancing interventions such as incentives, competing design teams, introducing new design participants, and so on.

6. Conclusions

Existing collaborative design approaches have yielded solid but incremental design improvements, which has been acceptable because of the relatively slow pace of change in requirements and technologies. Consider for example the last 30 years of development in Boeing's commercial aircraft. While many important advances have certainly been made in such areas as engines, materials and avionics, the basic design concept has changed relatively little (Figure 8):

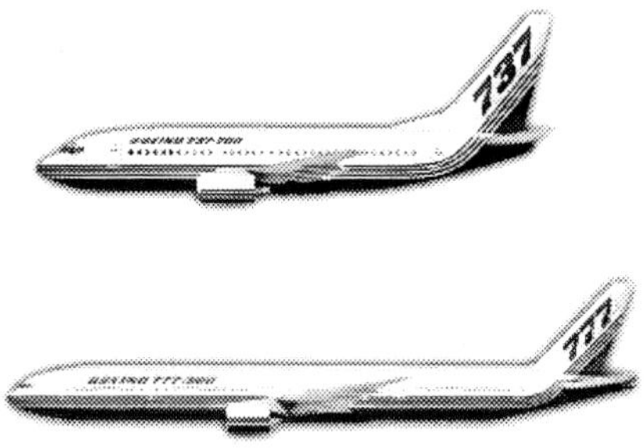

Figure 8. The Boeing 737 (inaugurated 1965) and the Boeing 777 (1995).

Future radically innovative design challenges, such as cost-effective supersonic commercial transport, will probably require, however, substantial changes in design processes:

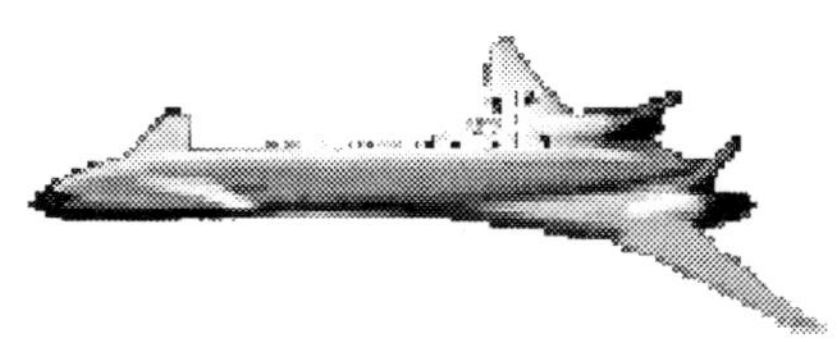

Figure 9. A concept for the Boeing supersonic commercial transport.

This paper has begun to identify what a complex systems perspective can offer in this regard. The key insight is that the dynamics of collaborative design can be understood as reflecting the fundamental properties of a very simple abstraction of that process: distributed dependency networks. This is powerful because this means that our growing understanding of such networks can be applied to help us better understand and eventually better manage collaborative design regardless of the domain (e.g. physical vs behavioral artifacts) and type of participants (e.g. human vs software-based).

This insight leads to several others. Most prominent is the suggestion that we need to embrace a change in thinking about how to manage complex collaborative design processes. It is certainly possible for design managers to have a very direct effect on the content of design decisions during preliminary design, when a relatively small number of global utility driven high-level decisions are made top-down by a small number of players. But once the design of a complex artifact has been distributed to many players, the design decisions are too complex to be made top-down, and the dominant drivers become local utility maximization plus fit between these local design decisions. In this regime encouraging the proper influence relationships and local search strategies is the primary tool available to design managers. If these are defined inappropriately, we can end up with designs that take too long to create, do not meet important requirements, and/or miss opportunities for significant utility gains through more creative (far-ranging) exploration of the design space.

7. References

[1] Brown, D.C. *Making design routine.* in *Proceedings of IFIP TC/WG on Intelligent CAD.* 1989.

[2] Smith, R.P. and S.D. Eppinger, *Identifying controlling features of engineering design iteration.* Management Science, 1997. **43**(3): p. 276-93.

[3] Eppinger, S.D., M.V. Nukala, and D.E. Whitney, *Generalized Models of Design Iteration Using Signal Flow Graphs.* Research in Engineering Design, 1997. **9**(2): p. 112-123.

[4] Klein, M., *Computer-Supported Conflict Management in Concurrent Engineering: Introduction to Special Issue.* Concurrent Engineering Research and Applications, 1994. **2**(3).

[5] Sobek, D.K., A.C. Ward, and J.K. Liker, *Toyota's Principles of Set-Based Concurrent Engineering.* Sloan Management Review, 1999. **40**(2): p. 67-83.

[6] Mitchell, T.M., L.I. Steinberg, and J.S. Shulman, *A Knowledge-Based Approach To Design.* IEEE Transactions on Pattern Analysis and Machine Intelligence, 1985. **PAMI**(7): p. 502-510.

[7] Mezard, M., G. Parisi, and M.A. Virasoro, *Spin glass theory and beyond.* 1987, Singapore ; New Jersey: World Scientific. xiii, 461.

[8] Bar-Yam, Y., *Dynamics of complex systems.* 1997, Reading, Mass.: Addison-Wesley. xvi, 848.

[9] Klein, M., et al., *Negotiating Complex Contracts.* 2001. Working paper ROMA-2001-01. Massachusetts Institute of Technology: Cambridge MA USA.

Part II: CSCW

A Method for Specification of Collaborative Interfaces through the Use of Scenarios

Anamaria Montandon Dumont
União de Escolas Superiores de Rondonópolis – FAIR/UNIR-MT
anamaria@unir-roo.br

Carlos Alberto Marques Pietrobon
Pontifícia Universidade Católica de Minas Gerais – PUC-MG
capietro@pucminas.br

Abstract

The complexity and the extension of tasks in the world today demand an extensive interaction among people with different knowledge backgrounds, who need to work in cooperation, aiming at a common goal. In this paper, we present a method to model the tasks and actions that a group of users carries out when working cooperatively through a computational system interface. For this purpose, we use scenarios for requirements gathering for user-cooperative interfaces. Along with the method, we propose the use of a notation that was developed to support interaction characteristics of users in cooperative work.

1. Introduction

Computational systems have been used to support completely different human activities, and they are essential work tools today. Recently, besides supporting the execution of individual tasks, they started becoming an important instrument of interaction and cooperation among people. However, this technology, even when used to support the activities carried out by a group, has been used to improve the individual's work, not the group's. For instance, text editors available enhanced the productivity of a person who is writing, but did not permit that two or more people write a document concurrently.

With the demand for softwares that support cooperative work, (Virtual Class, software development, etc), the challenge of specifying, designing and implementing these systems appears. One of the limiting points of these computational systems is the lack of an interface that supports appropriately groups of people that exerts its activities at the same time on the same representation or task. It is not properly an implementation problem, but rather a design one: there is a lack of method (and notation) means to express the dynamics related to using this kind of interface.

In this paper we present a method to obtain the requirements that a cooperative interface should comply with in order to support people working in a group. This method is based on the analysis of scenarios obtained from text descriptions of the tasks sent by the user.

Initially, we will approach the concept of cooperative work, the requirements of cooperative user interfaces and scenarios. Next, we will consider papers related to this area. In the following sections, we will deal with the method for obtaining cooperative interaction requirements from scenarios.

1.1. Computer-Supported Cooperative Work

In [15], Computer-Supported Cooperative Work (CSCW) "is considered a generic term that combines the understanding of the ways in which people work in groups with the available technologies (groupware) of software, hardware, services, and related techniques".

It's a research area that aims at integrating the works of several people involved in a common goal, inside a cooperative universe, through the sharing of resources in an efficient way, establishing, in this way, dynamic social and technological behaviors [24].

The requirements to support cooperative environments are [8], [15], [20], [24]: people coordination, information sharing (access concurrent control), workspace sharing, joint-work environment technical and social analysis, support to individual and joint work aspects, joint work dynamic control according to immediate necessities; group organizational memory, group dynamics, communication support, interaction transparence support, access policy establishment, recognition and dissemination of modifications.

From what was depicted above, it follows that the interface of collaborative systems is one of the vital aspects for the usability of such systems, for without it there is no communication, coordination or collaboration. And, exactly for these reasons, the design of such interfaces isn't an easy activity because it involves interaction among people.

2. Requirements Cooperative Interfaces

In this section it's presented a set of requirements that shall be complied with when people work cooperatively. These requirements will be used by designer and in the user-specified cooperative activities extraction algorithm by means of scenarios, that will be showed in section 2.

Defining U = {u1, u2, ..., un} the set of system users in a cooperative activity. Defining E = {e1, e2, ..., en} the set of workstations where the several users are working in. For simplicity sake, let's also consider that for each user ui there exists one and only one workstation ej associated. Defining A = {a1, a2, ..., an} the set of actions that can be carried out by the user.

Defining ai an action carried out by the user ui that drives the computational system and for which a response is to be given (feedback). This response must be observed locally and in all other ek-1 workstations/users where it's sensible to have feedback for that action. This response can be produced by either a user or the computational system.

If we are considering a group of users, it's necessary not only to model the feedback, but also who will receive this feedback. Thus it's necessary to model the following relationships among the several users:

a) relationships 1-1 (one user communicating with other user)
b) relationships 1-n (one user communicating with several users)
c) relationships n-m (several users communicating with several users)
d) relationships n-1 (several users communicating with one user)

Other aspect that has to be distinguished is the feedback origin: system feedback or users feedback.

When dealing with groups, it's necessary to model the synchronism or asynchronism among actions and feedbacks, that is identify if two or more actions / feedbacks may/must happen at the same time or at different times.

3. Scenarios and the Use of Scenarios in the Software Development Process

According to [18], scenarios are "narrative descriptions of how an application can be used and include the context in which they occur, its goal, and a narrative of the sequence of executed actions". They may be produced by users and designers alike, with different purposes. When produced by the user, each user specifies the scenarios that describe his tasks. The designer can make use of the tasks identified to help the users identify their scenarios.

In [19], it's seen the use of scenarios in the different stages of the software development process. In [9], scenarios are used as a technique of elicit the application requirements, through the identification of actors, entities and actions that describe it.

In [18], is presented a method for gathering requirements and specification of interaction in a process of Software Engineering. The method consists in obtain use cases from scenarios and user interaction diagrams (a notation graphical proposed) from this method. These diagrams represent the interaction between the user and the application. This method, however, just does worry about obtaining of data and process; does not worry about approach as people interact cooperatively.

[21] presents an approach of HCI design oriented to web sites which is based on scenarios which describe interface aspects. In [9], scenarios are used to elicit the application requirements from the users and to refine and validate these requirements. [12] proposes an approach based on user stories and use cases for gathering requirements.

4. Proposed Method

In [18], the authors describe a method for requirements gathering and conceptual design. The requirements gathering stage presents the following phases: identification of actors and tasks, specification of scenarios, specification of use cases from scenarios, specification of user interaction diagrams (a graphical representation for the use cases), validation of use cases and user interaction diagrams. The following stage (conceptual design) has only one phase: specification of the conceptual scheme (a model of the application domain).

That method approaches only the exchange of information between the system and the user. In this work, we propose to model the interaction among the users starting from the "Specification of Scenarios" phase.

In this way, we intend to create the phase of "Specification of cases of collaborative interaction among the users", as an alternative way to specify aspects of collaborative interaction for multi-user applications. This phase is considered the key for the success of the collaborative applications development. The result of this phase is the "User Cooperation Diagram", which, graphically, represents the cooperation among people manifested in the scenarios. These diagrams are built using an existing graphic notation, which are the Sequence Diagrams (section 3), that specific the requirements. See Figure 1.

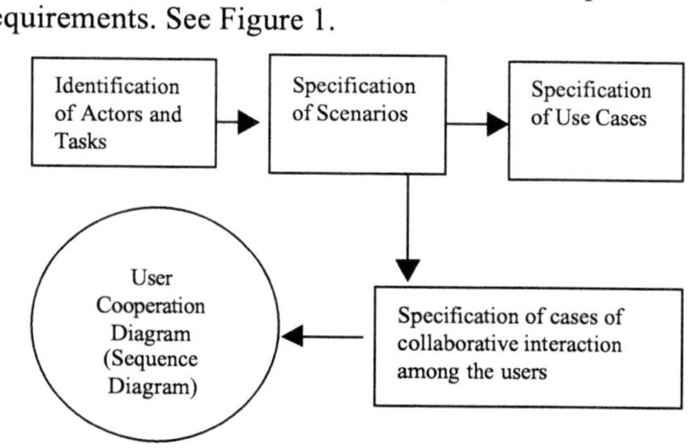

Figure 1. Phases of the requirements gathering of the proposed method.

To extract the cooperative activities from the scenarios, we propose the following steps (algorithm) to model or specify the collaborative interaction cases among the users. They are:

Identify the actors (1)
  Identify the users (2)
  Identify the hierarchies among the users (manager, leader, group, member) (3)
  Identify others situations, in which users are associated among themselves, organized as a group. These users appear in the scenarios as a group of people (4)
Identify tasks (decomposition and sequence) (5)
  Carried out individually/isolatedly (6)
  Carried out in groups (7)
    Specify collaborative interactions (8)
      Identify interaction categories (relationships 1-1; 1-N; N-M) (9)
        Identify types of interaction (10)
        Model the synchronism or assynchronism among the actions and feedbacks (11)
Identify products (12)
Identify others approaches (13)

Actors are people who exchange information among them (line 1 of the algorithm). They represent an specific type of user (line 2 of the algorithm), being a coordinator of others users, a group leader or an isolated member (line 3 of the algorithm). These actors may be organized according to a pre-established hierarchy among them. For instance, a member may be submitted to a group leader, which may be submitted to a manager. In a hierarchy it's necessary to provide "powers" to the higher levels to make them perform some kind of control on their subordinates. When the actors are at the same level of the hierarchy, they may form a group (line 4 of the algorithm). So that, it's necessary to identify and name these groups.

It's important to say that different actors can participate at the same scenario. As well as, the same actor can participate at different scenarios.

There will be variations on the vocabulary domination depending on the context. The actors identification will happen through NOUNS that represent the people involved in the context. For instance, in a company we can identify the actors through the words: president, director, manager, employee, and so on. while in a classroom we can identify: teacher, student, classmate, group leader and so on.

The tasks of each actor must be identified (line 5 of the algorithm). This identification will happen through VERBS that specify pertinent activities to the context. For example, in a company we can identify tasks through the verbs: protocol, send, count and so on, while in a classroom we can identify the verbs: ask, discuss, evaluate and so on.

A task can be decomposed in small tasks. We have to identify this decomposition and the sequence that these tasks are done (line 5 of the algorithm). The decomposition of the tasks shows that many people work at the system and, therefore, they are supposed to integrate with each other or with a coordinator who controls and/or integrates all the parts. The sequence of the tasks may also shows another way of interaction (workflow).

It's necessary to provide a short description of the tasks to identify which ones are done individually by only one user (line 6 of the algorithm), and which ones are done by a group (line 7 of the algorithm). This identification will happen through the use of the verbs in the SINGULAR or in the PLURAL. For example, when the user says "I do", "I send", we realize that he does the task alone. When the user says "we do", "we send", we realize that he does the task in group.

It's important to remind the tasks which are done individually must not be discarded at the first time, because they may be part of a cooperative activity bigger, in a decomposing structure or sequence.

For the tasks done by a group (line 7 of the algorithm) we need to identify the collaborative interactions (line 8 of the algorithm), in other words, we need to consider every relevant aspect in the interactions among the members of this group.

One of these aspects is the identification of the interaction categories (line 9 of the algorithm). We can classify three types of relationships among these members: a) relationships 1-1, b) relationships 1-N, c) relationships N-M.

In the first case (relationships 1-1) we are looking for expressions that identify that a member of the group communicates with only one person. For example, if the sentence "...if there is any problem, I send a message to the coordinator..." is described in a scenario, we realize that there is a communication among the elements from the group (send a message). This interaction happens between two people (the member and the coordinator).

In the relationships 1-N we are looking for expressions that identify that a member of the group communicates with several members. For example, the sentence "...I send the test for the students..."

In the relationships N-M we are looking for expressions that identify that several members of the group communicates with several members. For example, the sentence "...we discuss the question in the group..."

There are several situations that permit to identify the interactions, like: safety aspects (messages/data public or private and access permission) and alert situations such as errors of communication and warnings.

It's necessary to identify types of interaction (line 10 of the algorithm) such as the exchange of message/data and the exchange of social signals or non-techniques.

An example of synchronism (line 11 of the algorithm) is "...the teacher send the test at the same time for all the students..." and an example of assynchronism is "...the students send the test for the teacher at different times...".

The identification of the products (line 12 of the algorithm) which complement, and, in many cases, overcomes the identification of the tasks and actors suggests collaborative activities and the actors interaction, from manipulated products. So that, we ask about how each product was made (in group or not), if the end is in it or if it is part of a "big" product and who is going to use it.

There are others approaches (line 13 of the algorithm) that represent several kinds of situations the people have to get in touch with each others. It's impossible to enumerate all these situations; however, we must be awake to their existence. A good designer must have a list of them, got from the experience resulted by the development of systems. The designer, using its sensibility, may also notice others more complex situations do not related in this paper.

With these activities, we bring the requirements gathering for the support of a group working cooperatively. To put in documents these interactions, we propose the User Cooperation Diagrams. This is a Sequence Diagram [28],[29] specialized for our propose. The activities described in [18] only deal with data (model of data); we need as well to treat with the interaction (communication or interaction model) and with the interface (interface model), seen in [2]. The detailed description of the diagrams can be seen in the master's dissertation [27].

In the next section, it is presented an example of the application of the algorithm through the use of scenarios.

3. Example

The figure 2 below shows a simple example of an "User cooperation diagram" (Sequence Diagram [28], [29]) of the scenario "Employee receives a report, makes the analysis and send its opinion to the manager". Next we will comment about this example showing how we gather the requirements to make the diagram based on the scenario and using the algorithm proposed on section 2. During the explanation we will mention the meaning of the symbols used on the diagram.

At the scenario, we identify two users (line 2 of the algorithm): manager and employee (represented by the retangulars at the top of the Sequence Diagram in the figure 2).

To represent "Employee receives a report...", we realize that the manager creates the report which is represented at the Sequence Diagram by the arrow named "To Create Report" and send it to an employee, represented by the arrow named "To Send Report".

Note that several employees (relationships 1-N) (line 9 of the algorithm) might be receiving this report at the same time (synchronism) (line 11 of the algorithm). If we would like to represent this situation, we would use the notation in [27].

To represent "...makes the analysis...", we realize that the employee makes the analysis individually (line 6 of the algorithm), represented in the Sequence Diagram by the self-delegation symbol (an arrow that starts and finishes at the object, named "To Analyse Report"). This analysis could be done by a group (line 7 of the algorithm).

Note that the time the employee has spent to make the analysis, may be placed like a parameter, in brackets, for example (2 minutes), to represent time aspects.

To represent "...send its opinion to the manager" (arrow named "To Send Opinion"), we realize that the opinion sent to the manager is private (safety aspect) and that the relationship between the employee and the manager is 1-1 (line 9 of the algorithm).

Other aspects which were not modelled in this diagram can be analysed such as:

- The manager may send commentaries to the employee or store questions/answers in a repository so that the user has access all the time (assyncronism);
- The repository may work like a group memory to store tasks versions.

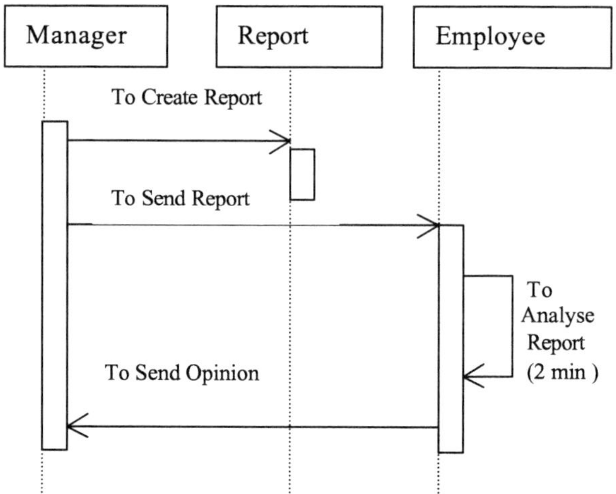

Figure 2. Scenario: "Employee receives a report, makes the analysis and send its opinion to the manager" and derived Sequence Diagram.

4. Conclusion

In this work, a method (a collection of activities) was proposed to specify collaborative interaction cases among the users from scenarios and using an existing notation (Sequence Diagram) to model these interactions.

There are several approaches of requirements gathering using scenarios. In general, these approaches use scenarios to determine data and functions modelled in different ways to which we have proposed in this paper. There is no knowledge of a method which uses scenarios to specify cooperation among users.

As we see, to support this group cooperative work, we need, among others aspects, to an interface adequated to this kind of work. The modelling of cooperative interfaces requires incremental methodologies which are not found in the literature. For this, we propose the method seen in section 2.

This work did not worry about implement aspects of the interface which the groups will support collaborating, but with the gathering and documentation of the requirements that have to be satisfied to support the activities of this group.

As well as scenarios, use cases can also provide information to this specification. In future works, use cases will be able to be used (explored) to provide more subsidy to reach an specification more completed. Collaboration diagrams will also be considered like an alternative documentation diagram.

We believe that our method and the notation that we use helps designers in the modelling of cooperative interactions. This method and notation can be seen much more detailed in the master's dissertation [27].

5. References

[1] Souza, C. S., Leite, J. C., Prates, R.O., and Barbosa, S. D. J., "Projeto de Interfaces de Usuário: Perspectivas Cognitivas e Semióticas"; National Congress of Society Brasilian of Computer, vol. II, 1999, pp. 425-476.

[2] Silva, E. J., "Projeto de Interfaces Homem-Máquina", 1999. IEC-PUC-MG.

[3] Maia, A.C., "A model for Cooperative Software Design"; 1996. http://www-nt.inf.puc-rio.br/cgilua/cgilua.exe/html

[4] Oliveira, O.L., "Interface extendida como um espaço de comunicação", 1999. IHC99.

http://www.unicamp.br/~ihc99/artigo_aceito.html

[5] Furtado, E., "Integrando fatores humanos no processo de desenvolvimento de interfaces homem-computador adaptativas", 1999, IHC99. http://www.unicamp.br/~ihc99/artigo_aceito.html

[6] Prates, R. O., and Souza, C. S., "Um modelo de apoio à expressão de projetistas de interfaces multi-usuário", 1999. IHC99. http://www.unicamp.br/~ihc99/artigo_aceito.html

[7] Barroso, G., "Teste de usabilidade de software" , 1998. Technical Report.

http://www-nt.inf.puc-rio.br/cgilua/cgilua.exe/html

[8] Grudin, J., "CSCW – History and Focus", 1994.

http://www.ics.uci.edu/~grudin/Papers/IEEE94/IEEEComplastsub.html

[9] Weidenhaupt, K., Klaus, P., Jarkem, M., and Haumer, P., "Scenarios in System Development : Current Practice" , IEEE Software , March/April, 1998.

[10] Bacelo, A. P., and Becker, K., "Uma ferramenta de apoio à discussão e deliberação em grupo", April, 1997. São Carlos – SP.

[11] Leite, J. C., and Souza, C. S., "Uma linguagem de especificação para Engenharia Semiótica de Interfaces de usuário", 1999. IHC99.

http://www.unicamp.br/~ihc99/artigo_aceito.html

[12] Shneiderman, B., "Designing the User Interface: Strategies for Effective Human-Computer Interaction", 3rd ed. , Addison-Wesley Longman, Reading Mass., 1998.

[13] Valaer, L. A., "Choosing a User Interface Development Tool", IEEE Software, July/August 1997.

[14] Pereira, A. C., and Sica, F.C., "Modelo para Construção de Serviços Cooperativos Baseado na Visão Vygotskyana e Formalizado pelos Conceitos Estruturalistas" , Technical Report (001/1999) – January, 1999 – Ouro Preto – MG.

[15] Pietrobon, C.A.M., Dumont, A.M., and Netto, O.A.M., "Extensão da Linguagem UAN para Especificação de Interfaces Cooperativas", Article, May, 2000.

[16] Souza, C. S., Prates, R. O., and Varejão, F. M., "Multimodal Communication between Software Designers and Software Users", Anais do III WorkshopHCI, May, 1997.

[17] Maia, A.C. P., Fuks, H., and Lucena, C. J. P., "A model of the communicative Behavior of Designers Involved in Cooperative Software Design", PUC-Rio, 1995.

[18] Vilain, P., Schwabe, D., and Souza, C. S., "Use Cases and Scenarios in the Conceptual Design of Web Applications", PUC-Rio, 2000.

[19] Breitman, K. K., and Leite, J. C. S. P., "Processo de Software Baseado em Cenários", PUC-Rio, 1999.

[20] Bock, G. E. and Marca, D. A., "Designing Groupware", August, 1995.

[21] Prates, R. O., Souza, C. S., "On the Rationale of Interface Semiotics for Multi-User Applications", 1999.

http://www.unicamp.br/~ihc99/artigo_aceito.html

[22] Christiansen, E., and Dirckinck-Holmfeld, L., "Making Distance Learning Collaborative", 1995.

http://www-cscl95.indiana.edu/cscl95/christia.html

[23] Collings, P., and Walker, D., "Applications to Support Student Group Work", 1995.

http://www-cscl95.indiana.edu/cscl95/collings.html

[24] Sica, F. C., "Camada Groupware: Uma camada de Suporte a Aplicações Cooperativas em Ambientes Distribuídos Abertos", Master's dissertation, 1996, Campinas – SP.

[25] Yourdon, E., "Análise Moderna Estruturada", 1989.

[26] Erikson, H.; Penker, M. – "UML Toolkit", September, 1997.

[27] Dumont, A. M., "Utilização de Cenários para Especificação de Interfaces Colaborativas", Master's dissertation, 2001, Pontifícia Universidade Católica de Minas Gerais – PUC-MG Brazil, Eletrical Engineering Departament.

[28] Fowler, M.; Scott, K. – "UML Distilled – Applying the Standard Object Modeling Language", Addison-Wesley, 1997.

[29] Harmon, P.; Watson, M. – "Understanding UML - The Developer's Guide", Morgan Kaufmann Publishers, 1997.

An Event and Service Interacting Model and Event Detection Based on the Broker/Service Model

Ping-peng Yuan, Gang Chen, Jin-xiang Dong, Wei-li Han
State Key Lab of CAD&CG , Zhejiang University , Hangzhou, 310027, P.R. China
peepeng@sina.com, cg@cs.zju.edu.cn, djx@cs.zju.edu.cn, olivine@sina.com

Abstract

In distributed systems based on B/SM, events and event-condition-action rules are the fundamental metaphors for defining and enforcing system. Processing entities execute behavior by reacting to and generating new events. It is important that event semantics is explicit. Although a lot of research has concentrated on event semantics, there still exists potentially semantic ambiguity which is unacceptable in some circumstances.

In this paper, we propose an event model and incorporate an interaction group identifier into it. We emphasize that the semantics is a key to capture the actual meaning of events in real life, in which the same composite event patterns may imply totally different semantic meanings. Based on this observation, a graph-ESIG(Event and Service Interacting Graph) and ESIC(Event and Service Interacting Chain), which are constructed according to ECA-Rules, are proposed to trace events. ESIG greatly enhanced the expressive power to allow event detection correctly, which matches situations in reality. Finally, we develop a event detection algorithm that uses ESIG and ESIC.

1. Introduction

The functionality of complex information systems is often achieved through the cooperation of collections of distributed heterogeneous software systems and humans. This kind of information systems is always refered to as Cooperative Information Systems(CIS). CIS-construction should support the integration of existing heterogeneous applications, the definition of the overall behavior in terms of processes, distribution, and maintenance of the autonomy of component systems. Andreas Geppert, etc[2] presented the Broker/Services Model(B/SM, which is described in detail in [1]), a system architecture model for Cooperative Process-oriented Enviroments (CPE)[1]. In our research, we extended B/SM to address heterogeneity of component systems and distribution within CIS. In CIS based on the B/SM, the basic constructs are *brokers* representing the actors in a CIS and *services* defining the functionality of these actors. events and event-condition-action rules are the fundamental metaphors for defining and enforcing the systems. Brokers are *reactive* in the sense that they respond to *events* occurring in the CIS or its environment, and their behavior is expressed by ECA-rules. Brokers execute behavior by reacting to and generating new events.

In the CIS, events are used in ECA rules to specify the time to check the conditions of the rules, therefore events play a important role in the CIS. Although much research is dedicated to event specification, owing to applying the event operators to the space of events instead of the space of event instances introduces semantic ambiguity and one problem still exists. The problem is that events' multiple occurrences affect the semantics of events. The answer to the question is the motivation to develop an event model, *ESIG*(Event and Service Interacting Graph) and *ESIC* (Event and Service Interacting Chain) presented in this paper.

The remainder of this paper is organized as follows: in section 2, we consider related work. In section 3 ,we propose extended ECA-Rule and an event and service interacting model. In section 4, we develop a detection algorithm that uses the event and service interacting model to efficiently monitor composite events. In section 5, we conclude the paper.

2. Related Work

Rule-based systems are a commonly accepted solution in the area of nonstandard applications in order to express an event-driven and constraint-driven system environment. Many attempts have been made to gain more and more expressive power and flexibility in rule-based systems. The specification of the events is one of focuses in study of rule-based systems. Research efforts studying event specification are forced in the design of specification languages, event semantics, expressive power of the language and event evaluation model. In the following, we introduce some aspects briefly.

Specification languages deal with systems that describe events, primitive or composite. Most of research is forced to enhance expressive power and flexibility of specification languages. R.J. Zhang and E.A. Unger[2] incorporated a calendar algebra to conveniently allow complex temporal event descriptions. G. Liu, et al[3] proposed a formalism to specify a wider range of composite events with formal semantics in the logic RTL which is especially amenable for specifying timing constraints in real-time systems. G.Liu, A.K.Mok, and E.J.Yang[4] presented a composite event specification approach that can precisely express complex timing constraints among correlated event instances.

Event evaluation models study systems that detect the occurrences of events. Basically, there are three types of composite event detection mechanisms seen in literature. Snoop[5,11] and TriGS[6] use syntax trees and syntax graphs. Completely different approaches for the detection of composite events are followed within the systems Ode[7] and SAMOS[8]. Ode uses finite state automata which are based on the equivalence of event expressions and regular expressions. Since a finite automata can remember only the state, but not the input symbols, some sentences from a context free grammar cannot be detected. SAMOS uses colored petri nets for the detection of composite events. petri nets are best suited for detecting events in a chronological context, since the transition takes place as soon as all required tokens are available.

All research mentioned above has proposed various event operators to form abundant composite event types. However, because composite events specified with these event operators often have different interpretations in terms of their occurrences, semantics from intuition for event specification language is not sufficient due to its ambiguity in tracing reasons for event occurrence. Therefore the problem that events' multiple occurrences affect the semantics of events still exists.

3. Interacting Model of Brokers

3.1 ECA-Rule

The reaction of a broker to events (the *behavior* of the broker) is defined by one or more event-condition-action rules (ECA-rules), as proposed for active database management systems. ECA-rules are used to define how brokers implement services. Brokers provide services for other brokers by replying events. Generally, services refered to a software module in a device or behavior. But with diverse requirements, services become more and more complex, and are composed of not only a single software module on a single device, but many software modules running on different devices. When services become much more complex, it is necessary to organize services clearly. There are many ways to organize services. Our way is as tree, whose leaf nodes represent atomic services and nontrminals represent the operators by which child services can be combined into more complex services. The relation of action list in the action part of general ECA-Rule is specified as and-relation. Although action list is different from services composed of single action, it does not fit for diverse requirements. We extend ECA-Rule with logical constraints among actions, so it is possible to provide Client Brokers with complex service. The syntax of extended ECA-rules is given as follows:

```
< ECA-rule> ::= on [name =] event [if
condition] do <Complex_Action>
<Complex_Action>::={Atomic_Action | <
Complex_Action > } { { AND | XOR | OR } {
Atomic_Action | < Complex_Action >}}*.
```

3.2 Event

Event is the occurrence of a particular situation or condition[9]. In this research, events are roughly divided into three categories, temporal, primitive and composite. Temporal events are event specification expression only containing pure times. Primitive events cover all events detectable by a system (not including the time). And composite events are those formed by composing primitive, temporal or other composite events. Much research has proposed composite event specifiction. For instance, *Andreas Geppert, et al*[10] have proposed a event model in which composite event types include *conjunctive* event type, *exclusive or disjunctive* event type, *sequential* event type, *concurrent* event type, *negative* event type, and *repetitive* event type. Although there are plenty of composite event types in Andreas Geppert's event model, the event model does not satisfy diverse constraints in practical cases. Moreover, the model adds timing constraints to composite event specification, it thus requires globle ordering of event timestamps to be implemented. It is the most challenging in distributed enviroments to Implement the event Broker which has all those functions. We propose an event model which only concerns with logical constraints among events. In this model, the condition parts of ECA-Rules and Brokers deal with temporal constraints and other constraints, it simplify event detection. A formal definition of event occurrence is given below.

Definition 1 Event Occurrence. A Event Occurrence is a 8-tuple(*Geid, EId, B, S, T, GroupId, Priority, D*).

- *Geid* is the global unique identifier, which is assigned by *ERC*;
- *EId* is the event identifier, and the unique identifier for site where the event occured;
- *B* is the source of the event;
- *S* is the site where the event occured;
- *T is* the timestamp of event;
- *GroupId* is the identifiers of interacting group, interacting group is defined in section 3.3;
- *Priority* is assigned to events by suppliers, and used to order the delivery of notifications.

- *D* is the additional data of the event.

We simplify the composite event model. In our model, there are three composite event types: AND event type, OR event type, XOR event type. We define composite event below. For the convenience, we define function symbols *occurrence* and *duration*. the function symbol *occurrence* test whether a event occurred. The function *duration* returns the time interval between two timepoints. lowercase *and, or* and *xor* are logical operators.

Definition 2 AND Event. (*defAND(?e1 ?e2) := (and (occurrence ?e1) (occurrence ?e2)))*.

Definition 3 OR Event. (*def OR (?e1 ?e2) := (or (occurrence ?e1) (occurrence ?e2)))*.

Definition 4 XOR Event. (*def XOR (?e1 ?e2) := (xor (occurrence ?e1) (occurrence ?e2)))*.

3.3 Event and Service Interacting Model

We argue that semantics from intuition for event specification language is not sufficient due to its ambiguity in tracing reasons for event occurrence and over-triggering in some application circumstances. Based on this observation, we put forward an *ESIG* and *ESIC* to provide a modification for general semantics. The *ESIG* and *ESIC* keep track of all event occurrences, on the general semantics that impose additional constraints on the event when an event is viewed as triggered. The *ESIG* and *ESIC* set up relationship among occurrences of different events. This kind of relationship can not be expressed by most event specification languages in literature.

An *ESIG*(Event and Service Interacting Graph) is a directed graph, which keeps information about how brokers interact with other Brokers by events. Every event and service, which are respectively the event part and action part of ECA-Rules, connect each other in an *ESIG*. Therefore an *ESIG* includes all possible interaction among all brokers in system, and correctness of behavior of Brokers thus can be verified with the help of *ESIG*.

Definition 5 *ESIG* (*Event and Service Interacting Graph*). *ESIG* is a 6-tuple($E, S, C; EC, SC; A$).

- E is a finite set of events, $E=PE \cup CE$, where PE, CE are respectively the set of Primitive Events and the set of Composite Events;
- $S= AS \cup CS$, where AS, CS are respectively the set of atomic services and the set of complex services of Brokers.
- C is the set of constraints;
- $EC \subseteq (E \times C)$;
- $SC \subseteq (S \times C)$;
- $A \subseteq (EC \times SC) \cup (SC \times EC) \cup (EC \times EC) \cup (SC \times SC)$;

And satisfies the following properties:

- $\forall\ a \in AS,\ Pre(a) \subseteq CS \cup E,\ Suc(a) \subseteq E;$
- $\forall\ p \in PE,\ Pre(p) \subseteq CE \cup AS,\ Suc(p) \subseteq S;$
- $\forall\ c \in CS,\ Pre(c) \subseteq CS \cup E,\ Suc(c) \subseteq E \cup S;$
- $\forall\ c \in CE,\ Pre(c) \subseteq S \cup CE,\ Suc(c) \subseteq E \cup S;$

Note: *Pre, Suc are respective function symbols to get predecessors and successors.*

If there are circles in an *ESIG*, the circles are easy to be found according to the adjacent matrix representation of the *ESIG*, because the arcs of a circle should set a flag in the corresponding places of the upper triangle and the lower triangle of adjacent matrix. Therefore we can determine whether there are circles in an ESIG only by searching the upper triangle or lower triangle of the adjacent matrix.

Every node of an *ESIG* is attached a constraint. According to the constraints, the reachability of every node of an *ESIG* can be determined, so behavior of Brokers is verified.

Definition 6 Logical Connection. A path $P(v_1, v_2, \ldots, v_n)$ of *ESIG* is called logical connection iff $c_1 \wedge c_2 \wedge \ldots \wedge c_n$ is satisfaction or tautology, where $c_1, c_2, \ldots, c_n$ are constraints corresponding to nodes $v_1, v_2, \ldots, v_n$.

Definition 7 Logically Reachable. There exists a path P where v_s is the source node and v_e is the sink node. If P is logical connection, then node v_s, v_e is logically reachable.

Definition 8 *ESIC* (Event and Service Interacting Chain). *ESIC* is a set of correlated logical connection pathes in *ESIG* .

An *ESIC* may be viewed as a dynamically executed *ESIG*. It represents a series of interaction among Brokers. According to the above definition, an *ESIC* may have one or more source nodes or sink nodes. Owing to the difficulty to coordinate among brokers in an *ESIC* with multiple source nodes, so our focus is on the ESIC with only one source nodes. We call it regular *ESIC*. The definition of regular *ESIC* is given below.

Definition 9 Regular. An *ESIC* is regular if and only if:

- *ESIC* has a source node e_{start} and one or more sink nodes e_{final};
- Every node $n \in N$ ($N=E \cup S$) is on a path from e_{start} to e_{final}s.

A regular *ESIC* represents an interacting group, which our composite event detection approach is based on. Here we briefly describe the algorithm for regularizing *ESIC*: start with a source node of an *ESIC*, find all input nodes whose immediate successors are nodes of type *AND,* and mark all those nodes as $v_1, v_2, \ldots, v_n$. For every node $v_i(i$=1,n), retrace the predecessors of v_i in *ESIG*, and add all predecessors of v_i into queue q. For every $v_p \in q$, if v_p is empty, then finish. If v_p satisfies the following conditions: v_p is automatic services or services which are provided by persons or data, then mark the node as a source node; else retrace the predecessors of v_p; finally construct a fake node and connect the fake node with those source nodes.

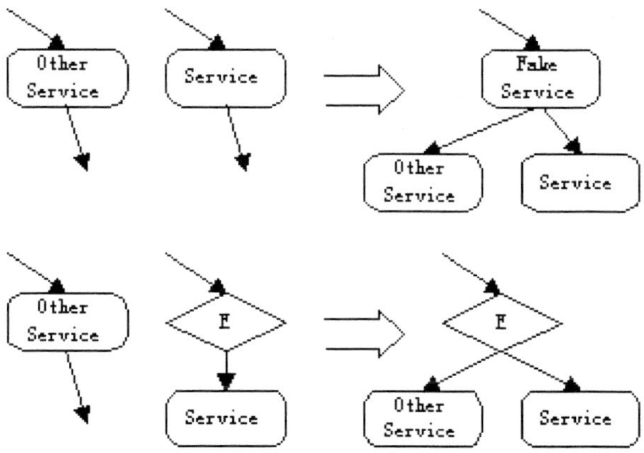

figure 1 Non-regular ESIC mapped to regular ESIC

Fig 1 shows two cases in which nonregular *ESICs* are converted to regular *ESICs*. The top of Fig.1 illustrates the case in which the source node is service. This case is complex, and regularizing the *ESIC* need to replace the original service with a fake complex service. The fake complex service composed of all source nodes which denote service. The foot of Fig.1 shows the case where the source nodes represent event, in that case, only a fake connection need to be constructed.

4. Composite Event Detection

The rule-based systems need facility to support brokers by providing event management, storage, and notification functionality. In the present paper, the facility is called as *ERC* (*Event Relay Center*). The principle by which brokers use *ERC* is the following: brokers' ECA-rules are translated to rules stored and executed in *ERC*. The *ERC* then performs event detection and determines brokers that have registered a notification interest for such event occurrences which may result from this event. These brokers are then appropriately informed and react as defined by their ECA-rules. We propose a framework of *ERC*, which is composed of ES(Event Sender), ER(Event Receiver), EA(Event Analyzer), ESYN(Event Synthesizer), EM(Event Matcher), ECA Register, and Interaction History Management. When a broker generates an event, it notifies the ER of the *ERC* about the occurrence. The ER inserts the event into HEQ(Handling Event Queue), EM gets event from HEQ and sets visited flag, determines brokers that have registered a notification interest for such event occurrences, then adds the pair of action and event into SEQ(Sending Event Queue). The ESYN also gets events from HEQ, and checks whether new composite events can be detected. The EA decomposes composite events into primitive events. The ES posts events to respective brokers. We will not consider in detail the technical aspects of *ERC* and focus on event detection. Composite event detection is more difficult than primitive event detection, so we introduce how to detect composite events elaborately.

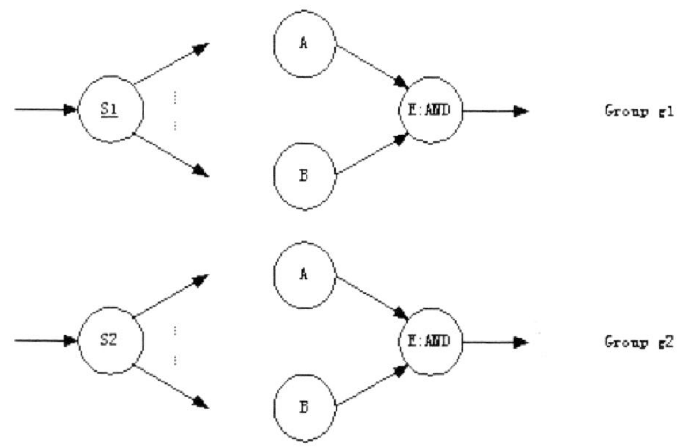

Figure 2 two interacting groups

As mentioned in section 1, an event which may have multiple instances during a computation often incurs semantic ambiguity, thus limiting the expressiveness of the event specification languages. We propose an approach where events are assigned a group identifier. When the source node of an *ESIC* is fired, an interacting group will be created, and assigned a group identifier by *ERC*. The *GroupId* of the events following the source node will assigned the same group identifier as the source node. In our approach, events with different group identifiers do not form new composite events. Fig.2 illustrates two interacting groups g_1, g_2, where events *A* and *B* will be generated and a composite event defined with *AND* operator, *E=A* AND *B*, will also be generated. When interaction reaches the nodes *A* and *B* of groups g_1, g_2, there will be four event instance occurrences a_1, a_2, b_1, b_2 of *A* and *B* (we use a small letter with a subscript index as the event identifier), where a_1.*GroupId*= b_2.*GroupId* and a_2.*GroupId*= b_1.*GroupId*. So event occurrences a_1 and b_2, a_2 and b_1 will form two occurrences of event *E*. Other potential triggering combinations of events, such as a_1 and b_1, a_2 And b_2 will not be tested, because the occurrence of E caused by (a_1, b_1) or (a_2 , b_2) is not possible. An algorithm for composite event detection is given below.

Algorithm 1 Composite Event Detection
Step 1 Get a event from *HEQ*, Mark the event as *e*;

If EM has visited event *e*, then delete e from *HEQ*, and write it into *History*;
Else set the visited flag of event *e*.
Step 2 search *ESIG*,
If event *e* is one of child events of a event, then add a flag indicating the occurrence of event *e* in *SEQ*
If the parent event type of event *e* is *AND*,
then check whether the other child events of the parent event have occured,
If *TRUE*, add the parent event into *HEQ*;
If the parent event type of event *e* is *OR*, add the parent event into *HEQ*;
If the parent event type of event *e* is *XOR*,
then check whether the other child events of the parent event have occurred
If *FALSE*, add the parent event into *HEQ*;
If corresponding parent event has fired, then clear the occurrence flag of child events;
Step 3 goto Step 1;

5. Conclusion

In this paper, we study an event and service intracting model and event detection. We propose an event model. In the event model we separate timing constraints on event semantics from logical constraints. Moreover, we incorporate an interaction group identifier into the event model It is argued that it's necessary and convenient to have such facility in the event model. We emphasize that the semantics is a key to capture the actual meaning of events in real life, in which the same composite event patterns may imply totally different semantic meanings. Based on this observation, a graph-*ESIG*, which is constructed according to ECA-Rules, and a graph-*ESIC*, which keep track of Brokers' interaction, are proposed to detect event. The *ESIG* and *ESIC* greatly enhanced the expressive power to allow event detection correctly, which matches situations in reality. The *ESIG* and *ESIC* solve the problems that multiple event occurrences incurs semantic ambiguity. Event detection is one of topic addressed in this paper. We develop a detection algorithm that uses the *ESIG* and *ESIC*.

6. References

[1] Dimitrios Tombros, Andreas Geppert, Klaus R. Dittrich. The Broker/Services Model for the Design of Cooperative Process-Oriented Environments. ftp://ftp.ifi.unizh.ch/pub/techreports/TR-97/ifi-97.06.ps.gz. Department of Computer Science, University of Zurich.

[2] Robert J. Zhang, Elizabeth A. Unger. Event Specification and Detection. Department of Computing and Information Sciences, Kansas State University. TR CS-96-8, June, 1996.

[3] Guangtian Liu, Aloysius K.Mok and Prabhudey Konana. A Unified Approach for Specifying Timing Constraints and Composite Events in Active Real-Time Database Systems. In Real-Time Technology and Applications Symposium, June 1998.

[4] G.Liu, A.K.Mok, and E.J.Yang. Composite Events for Network Event Correlation. http://www.cs.utexas.edu/users/liugt/ publications/im99.ps.gz. Department of Computer Sciences, The University of Texas at Austin.

[5] Sharma Chakravarthy and Deepak Mishra. Snoop: An Expressive Event Specification Language For Active Databases. Technical Report UF-CIS-TR-93-007, University of Florida, Computer and Information Sciences Department, 1993.

[6] Werner Retschitzegger. Composite Event Management in TriGS-Concepts and Implementation. In Proceedings of the 9th International Conference on Database and Expert Systems Applications (DEXA'98), Vienna, August, 1998, pp. 1-15.

[7] Narain Gehani, H.V.Jagadish and Oded Shmueli. COMPOSE: A System For Composite Event Specification and Detection. In Book chapter in Advanced Database Concepts and Research Issues (N. R. Adam and B. Bhargava, eds.), Lecture Notes in Computer Science. Springer Verlag, 1994.

[8] Stella Gatziu, Klaus R. Dittrich. Detecting Composite Events in Active Database Systems Using Petri Nets. In Proc. 4. Intl. Workshop on Research Issues in Data Engineering, pages 2--9, Houston, USA, 1994.

[9] WfMC: Workflow Management Coalition White Paper – Events 1999. http://www.aiim.org/wfmc/standards/docs/eventsdh1.pdf.

[10] Andreas Geppert, Dimitrios Tombros and Klaus R.Dittrich. Defining the semantics of reactive components in event-driven workflow execution with event histories. Information Systems 23(3/4):pp235-252.

[11] S.Chakravarthy, V.Krishnaprasad, E.Anwar, S.-K Kim. Anatomy of a Composite Event Detector. Technical Report UF-CIS-TR-93-039, CIS Department, University of Florida, December 1993.

The Design and Implementation of a Real-time Walkthrough System Supporting Cooperative Manipulation[⁂]

Hongmei Sun, Shenquan Liu
CAD Lab, Institute of Computing Tecnology,
Chinese Academy of Sciences, P.O. Box 2704, Beijing, 100080, P.R. China
hmsun@ict.ac.cn, liushq@ict.ac.cn

Abstract

In DIS, simulated scenes with extreme complexity become much more common, which will definitely advance much higher requirement on both real-time interaction and cooperative manipulation among group members. Aiming at this, we propose a kind of schedule based parallel processing model, in which the parallel processing of the whole rendering pipeline and cooperation among members are combined together effectively. The constitution of our model and its implementation are discussed in detail in our paper. This model has been successfully applied in the development of OADS, and its satisfying results, including the fine image, real-time performance, well workload balance at the server, friendship cooperative interface and so on, illustrate its efficiency.

1. Introduction

Distributed Interactive Simulation (DIS) is a hotspot in system simulation at all times. And that real-time rendering of the virtual scene and well support to cooperative manipulation are two sticking points of the DIS supported software environment [1-2]. Along with the extensive application of DIS, especially in military field, virtual scenes containing many millions of polygons become common. Complex scene makes visibility culling, collision detection and geometric transformation in computing stage before rendering more perplexing and time consuming, and that more and more objects are included in the virtual scene makes the cooperative control between different group members much more complex and more difficult to implement.

Lots of system for distributed real-time rendering and navigation are running on Graphics Workstation before, as the effective Message Passing Interface(MPI) and Parall Virtual Machine(MVP) technology are provided under the Unix Operation System, which make the implementation of the network communication much easier [4], while very little method are based on Personal Computer. In addition, although peer-to-peer model and Client/Server model are representative communication model and often adopted by CSCW application, they both have disadvantage. Systems using peer-to-peer model must process more communication packages and thus have bad scalability, while systems using Client/Server model must endure the boring message transmitting delay, which is not helpful for real-time interaction. Moreover, lots of work just focus on how to provide the cooperative support for multi-user system, but seldom very little can integrate the parallel rendering and cooperative manipulation together as the data communication will then become much more complex [3, 5].

Aiming at these aspects, a kind of schedule based distributed parallel processing model is put forward in our paper. Firstly, the distributed computing is introduced into dynamic rendering of the large-scale complex scene. By re-partitioning the whole graphics rendering flow, the computational part and the rendering part are separated from each other and performed at different computer nodes (computing node and rendering node) in parallel. Thus, we decreased the rendering time by shorten the length of the whole rendering streamline in each frame. Secondly, our model is a kind of hybrid one of peer-to-peer model and Client/Server model. A control entity named task schedule agent is used to control the cooperative manipulation among different members and manage the communication between computing nodes and rendering nodes. Thus communication quantity decreased effectively and workload balance among different servers is well achieved, which makes our system easy to be extended in the future. Thirdly, the hardware environment of our system is based on personal computer, which is helpful for DIS system popularization.

A general introduction of our idea is given in section 2. The implementation of the main parts of our model, which includes the implementation of the schedule agent, the server and the client, is discussed in detail in section 3 and the experiment result is given in section 4. Finally, conclusion and our future work are arranged in section 5.

⁂ This work is supported by 95 National Defense Pre-research Project

2. Overview of Our System Structure

2.1. System structure

Aiming at the real-time rendering of large scale complex scene and the cooperative interaction between different group members and the scene, a kind of schedule based parallel processing model is presented in our paper. Its basic structure is shown in Figure 1. Mainly three parts are included in it: server set, client set and a task schedule agent. The server nodes and client nodes are used to accomplish the total rendering process and the task schedule agent is used to manage the cooperative manipulation.

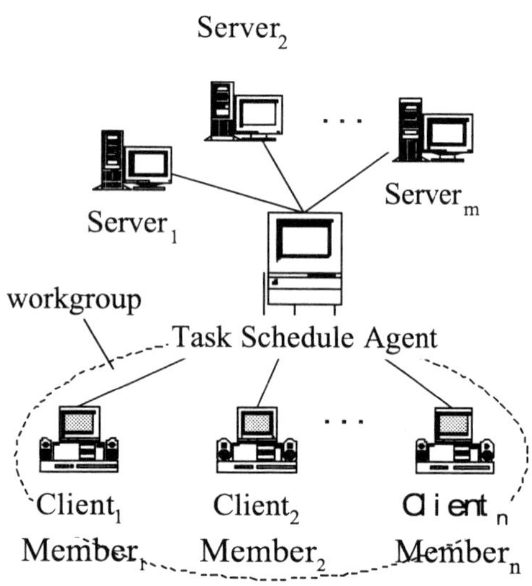

Figure 1. Schedule based parallel processing model

Firstly, in order to shorten the rendering time, the whole rendering process is divided into 2 parts, which are named data process part and rendering part and are executed in parallel at different computer.

(1) Client node: also called as romance-computer, is equipped with high-speed graphics accelerator and possesses strong graph interaction ability and display ability. It is mainly responsible for real-time rendering.

(2) Server, also called control-computer, is equipped with great capacity memory and owns strong data processing ability. It mainly takes charge of the source data (including terrain data, object model data) storage, and computation before rendering, such as visible determination, hidden elimination, automatic generation of LOD models and so on.

So each group member can send its computing request from its Client node as well as the associated parameters to Server according to the granularity of task while rendering. Then Server can process the data used for rendering future visible region and return the result to the Client if being requested. Therefore, the computing and the rendering can be accomplished in parallel.

During implementation, the server maintains only the initial model data of terrain and the detail boundary information of the objects in the scene, no need to save the texture data. While the client node must save the data used for rendering the scene, but no need to save middle data, such as the bounding box data.

Secondly, in order to support the cooperation manipulation of the simulation environment for our multi-user system effectively, a control entity named task schedule agent is added in our system. Its responsibility is to manage the communication between the client node and server nodes in our system, assign the data process task coming from client node to server, and so on.

Thus, client and server are transparent to each other as both the client send request and the server return the data process result only to the schedule agent, and no need to care for who will process the task, and whose data it processes. Then the server and client are liberated from the more and more complex task schedule and communication management in multi-user system and can thus focus on their own process task and high process efficiency can be achieved.

2.2. Cooperative mechanism and concurrent control

Cooperative mechanism is one of the kernel problems to be solved in shared-object cooperative manipulation system. Some individuals may be separated from each other by time or space, but they are interdependent on each other, thus their activities can be united together well through effective cooperative mechanism. In distributed virtual environment, cooperation among different group members primarily means the cooperative control toward the scene, as well as the dynamic objects in the scene. When user applies to participate in the workgroup, different authority will be give according to its role by the system. There are totally 3 kinds of role in the system, which are system administrator, super-member and common-member. System administrator is equivalent to the chief of staff, and is responsible for set up a cooperative workgroup, formulate the total scheme and has the right to invite other member to attend. Super-member can control the scene interactively in command of the administrator. Common-member can only walkthrough in the scene, but has no right to change the scene or control the moving object in the scene.

The simulation task such as battle and drilling are completed through the cooperation among group members. Thus synchronization control mechnism is adopted in our system, and the concurrent control to the virtual scene is achieved by lock.

3. Model implementation

3.1. The implementation of the task schedule agent

The function of the schedule agent is to receive the request from clients and assign it to a server according to the granularity of the current requested task and the total task quantity assigned to each server, then decide the return type according to the type of the task. Its structure is shown in Figure 2. The User Log In Module is responsible for recording user information, such as IP address, port number, etc., inform other members and delete related information when the user exits the system. Task schedule module is responsible for receiving and assigning task from client, receiving and returning the result from the server, and so on.

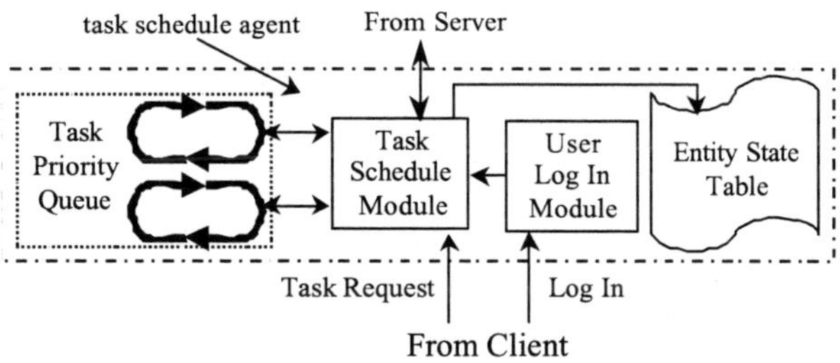

Figure 2. The structure of task schedule agent.

3.1.1. Data structure. There are totally two kinds of task, broadcast type and non-broadcast type.

Broadcast Type: This type related with the shared data, such as: the state change, the interaction and collision of the objects in the scene. The result will be send to all members existed in the system.

Non-broadcast Type: This type related with the private data, such as the change of the view, the multi-degree computation of the view, the interaction between the view and the objects in the scene, and so on. The result is only returned to the member who sends out the request.

According to this kind of classification, two task lists are maintained at the task schedule agent, used for saving the data for broadcast task and non-broadcast task respectively. There are totally 5 fields in the task list, which are named FromAddr, FromContent, ToAddr, ToContent and PortNumber, as shown in Figure 3.

FromAddr: records the address of the client sending the request;

FromContent: saves the pointer to the buffer saving the request content;

ToAddr: records the address of the server processing the request;

ToContent: saves the pointer to the buffer saving the process result;

PortNumber: records the port number between the client and the server while using UDP protocol to transmit data.

When the task put forward by group member arrived at task schedule agent, all the entities are filled except the entity named "**ToContent**". Only when the task finished and the task schedule agent received the corresponding result from server, this entity is filled.

In order to guarantee that the whole scene be the same to all group members at all time, a scene object state list is maintained in the system, which including 5 entries: object_type, object_index, object_visibility, object_stance and Object-LOD. Thus, any change to them made by any group member will be broadcast to all other members in the system. However the navigation of different members and the interaction between the scene and different members is independent to others. The concurrent control on the shared objects in the virtual scene is implemented by share-exclusive lock-mechanism.

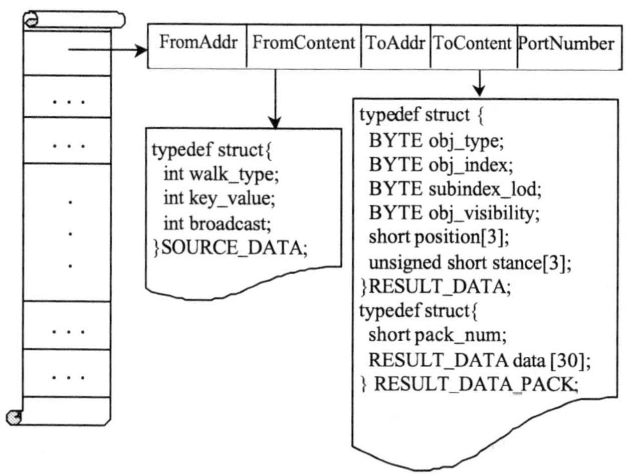

Figure 3. Task list.

3.1.2 Task schedule. There are two steps for the agent to assign the task coming from different client nodes:

(1) **Sort task:** all the tasks received will be sorted in order of requesting time, which guaranteed that the earliest task will be processed first.

(2) **Assign task:** compute the workload of each server, and choose the server with the least workload to process the new task. If several servers exist, the one with the highest data processing capabilities will be chosen.

To ensure the smooth transition of each frame image and give group members a real simulation effect, the synchronization between the client and the server is indispensability. In fact, the client node, which is responsible for rendering the scene, is so busy that it seldom has time to receive the data from network, so lots of packages of data will be lost if the schedule agent sends data package very fast. In implementation, we set the server's receiving thread with the highest priority, which makes it receive the data safely and doesn't lay heavy influence on its data process as the server isn't so busy as client. But if we use this method on the rendering thread, then the frame rate will be decreased greatly and the image quality is influenced seriously. So, the schedule agent uses the system timer to achieve a steady-going speed and let the data arrives in a fixed speed acceptable by the client. And we find that it is effective in practice.

3.2. The Implementation of the Server Node

We can regard the data processing happened at the server as a 3-stage pipeline, which are receiving data, processing data and sending the result. We use the window's socket and thread synchronization technology to implement the server, which make the three stages of the whole pipeline running in parallel. We call the three threads running at the server as Receive Thread, Data Process Thread and Send Thread respectively. Their relations are well shown in Figure 4.

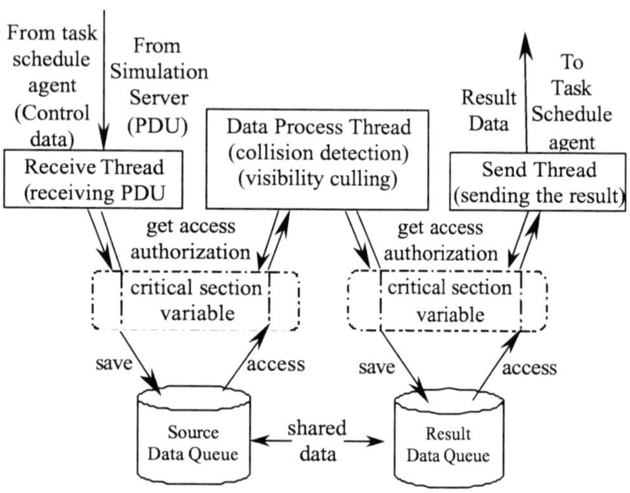

Figure 4. Pipeline structure of the server.

Receive Thread: Receive processing task from the task schedule agent and PDU(Protocol Data Unit) data from the simulation server called simulator. The data from the simulator includes the route of the plane, the position of the tank, when will the tank fire and so on. PDU is a criterion defined by the military organization for simulation. We adopt it in our system. Of course, if no simulation server exists, we can still use keyboard to simulate the PDU data. In addition, in our system, we use keyboard and mouse to change the parameters related with the navigation and viewpoint, such as the walkthrough type, speed, walk direction, view direction, and so on.

Data Processing Thread: It mainly includes 3 parts: multi-freedom view movement computation, visibility culling, and collision detection. All the view movement type implemented in our system are listed below:

(1) Terrain following: means flying over a terrain at a constant height. Thus, the viewpoint may glide over the surface at a constant height.,which makes flying easier and also more pleasant. Bilinear interpolation to get the height of viewpoint at different position;

(2) Overlook: that is the user can lift the view point up highly and observe the whole scene in upper air;

(3) Turn left and right: change the move direction of the view in X-Z plane when walkthrough;

(4) Look up and down: change the view direction in Y-Z or X-Y plane when walkthrough.

(5) Viewpoint switch: the position of the viewpoint can be changed randomly through mouse click.

(6) Locked Observation: the viewpoint can be locked on a object, thus the change of the viewpoint is controlled by the movement of the object.

Multi-degree observation and walkthrough are supported in this system by using multi-view window, as shown in Figure 5. Thus members can observe the virtual scene much more accurately and clearly, which is helpful to achieve cooperative control and interactive manipulation toward the scene effectively.

We use a kind of fast visibility-culling algorithm in [6] in our system, please refer this paper for the algorithm details as this is out of the scope of our paper.

Send Thread: In this thread, the result data are taken out from the Result Data Queue and sent to the task schedule agent.

main-view window

back-view mirror

survey-view window

Figure 5. A part of the scene in OADS.

3.3. The Implementation of the Client Node

There are also 3 threads in the program running at the client, which are Sending Thread, Receiving Thread and Rendering Thread. The implementation of the Sending

Thread and Receiving Thread is similar to the corresponding threads running at the server. In Rendering Thread, WTK(World Tool Kit) is used to render the scene. The reason is that the virtual scene can be described in an object-oriented style in WTK, which makes the implementation easy to understand and implement.

4. Experimental Results

We applied our method to the development of OADS, a subsystem of RSSE. In OADS, various kinds of entities are included in the whole scene. The terrain size is about 40 km^2, and its triangle number is 2, 875, 202. Nearly 30 kind of weapons and objects are arranged on this terrain. The total entity number exceeds 200, and their total triangle number in model data is 2, 262, 305. It is obvious that such large quantities of data will definitely make the rendering and cooperative manipulation much more complex. Our system is implemented under the Window NT 4.0 and Visual C++6.0 environment, using win-sock and thread synchronization technology, and running on computer with the PII 400 CPU. The frame rate can obtain over 18 frames per seconds, while only 2~3 frames can be obtained without the distributed computation. And through the schedule agent, the cooperative manipulation between different group member is well managed. The satisfying result illustrates the efficiency of our method.

5. Conclusion and Future Work

Shared object cooperative manipulating system is an important branch of CSCW and has a wide application field[7]. In our paper, a kind of schedule based parallel processing model are presented and has achieved satisfying result in the development of OADS. On the one hand, our method short the stream length of the whole rendering stage, not only shorten just one stage, so it is more effective than the former ones. And this advantage will become much more obvious when the scene becomes more complex. However it doesn't conflict with the real-time technique whose purpose is just to shorten one stage or several stages of the whole rendering. These two kinds of method can be integrated together. On the other hand, in our model, all the work associated with communication are accomplished by a task schedule agent, which makes the server and the client transparent to each other and the data quantity transmitted decrease effectively. Our future work is to propose a new concurrent control model to improve the concurrence of the cooperative manipulation.

6. References

[1] Li Bohu, etc., The Research of Synthetic Simulation System, Journal of System Simulation, 2000(9), Vol.12, No.5, pp. 429-434.

[2] Hong Xu, etc., the Application of Virtual Reality in Military Simulation. Proceeding of 10th Chinese CAD&CG Conference, Guilin City, Guangxi Province, China, 1998.10, pp. 169-173.

[3] Wen Ye, etc., An Architecture of Web-based Distributed CAD system, Proceedings of Second International Conference on Computer-Aided Industrial Design and Conceptual Design, Thailand, 1999, Nov26-28, pp. 267-272.

[4] Ren Jicheng, etc., Computer Aided Collaborative and Parallel Visualization System Model, Proceedings of Second International Conference on Computer-Aided Industrial Design and Conceptual Design, Thailand, 1999, Nov26-28, pp. 250-254.

[5] Thomas A. Funkhouser, Ring: A Client-Server System for Multi-user Virtual Environment. http://www.cs.princeton.edu/~funk/ring.html.

[6] Sun Hongmei and Liu Shenquan, A Fast Visibility Culling Algorithm for Outside Complex Terrain Scene, Computer Applications, 2000, vol.20, Suppl, pp. 1-3.

[7] Ma Xianlin, Research on Interface for Human-human Interactions and Application, Proceedings of Second International Workshop on CSCW in Design, Thailand, Nov.26-28, 1997. Pp. 321-325.

Supporting Cooperation with Off-the-Shelf Applications in Wired and Wireless Networks

Federico Bergenti, Agostino Poggi, Matteo Somacher
Dipartimento di Ingegneria dell'Informazione
Università degli Studi di Parma
Parco Area delle Scienze 181/A, 43100 Parma, Italy
bergenti@ce.unipr.it, poggi@ce.unipr.it, somacher@ce.unipr.it

Abstract

This paper presents a cooperative platform that allows the sharing of information as well as the interactive use of different kinds of application in a synchronous mode. The access to the collaborative sessions is possible from different kinds of terminal and network (fixed or mobile). In particular, we address Web users and mobile users with sufficient bandwidth. The platform is based on the centralised sharing of an off-the-shelf application that relies on a collaborative version of the graphics libraries used in Java and Microsoft Windows. A first prototype of the platform has been experimented by Ph.D. students during a seminary distributed across Italy.

1. Introduction

Increasingly more people need to access the enterprise information remotely as well as to collaborate with others from different places, not only from their offices. The Web is assuming a central role in the way people share the information in the local and geographic networks, mainly because browsers are available everywhere and they integrate different services into a common, easily accessible, platform-independent user interface. For this reason, the Web has already been adopted as one of the principal media capable of supporting the collaboration between people [1]. Nevertheless, the basic communication support offered by the Web is not sufficient to support an interactive approach to collaboration. The communication needs for which the Web was designed was about consulting structured documents and was not about supporting an interactive discussion between virtual teams.

The migration of Web technologies toward mobile networks allows porting consolidated Web services and facilities to mobile users. The widespread availability of high-bandwidth and mobile infrastructures allows people sharing information with heterogeneous hardware (desktop and laptop computers, PDAs, wearable computers) using different operating systems and different kinds of communication systems (LAN, Internet, wireless networks). The integration of these technologies to create flexible and effective virtual teams has to take into account the heterogeneous environments and the constraints that could limit the performances and the quality of the works. We exploit the availability of the same technologies for fixed and mobile networks to build an integrated, collaborative platform that can be accessed seamlessly from mobile and fixed users.

C/WebTop, the collaborative platform described in this paper, enhances the classic Web communication mechanisms to support synchronous sharing of off-the-shelf applications. The platform is designed to support a collaboration model in which people can take full benefit of the emerging hardware and software technologies such as: third generation mobile networks and terminals, new generation home appliances, multimedia contents and Java technologies. A set of interactive services is provided to deal with the most common activities of a virtual team and to adapt the platform automatically according to each specific user profile. The system allows accessing the shared applications from fixed or wireless network seamlessly, providing a high level of flexibility to users. Moreover, C/Webtop can adapt the user interface of applications to the capacity of the specific terminal. For example, it can shape the user interface to the available bandwidth managing the resolution and the colour depth of images. Users from different terminals simply receive different displays according to the capability of their connections.

This paper is organised as follows: section 2 introduces the collaboration model supported that *C/WebTop* implements. Section 3 describes the platform architecture and the services it offers for virtual teams using Java or Microsoft Windows applications. Section 4 presents the related work found in the literature on platforms supporting virtual teams. Finally, section 5 discusses the main features of the presented platform and outlines some future research directions.

2. Collaboration Model

A collaborative session is the activity of a communication group in which the participants exchange information among them [3]. Collaborators may play different roles in a session and the roles can be changed dynamically, moreover collaborators may also join and leave a running session. A collaborative platform is required to provide all the facilities needed to support the dynamic nature of the collaboration while guaranteeing the availability of suitable media for information exchange. In fact, the exchanged information essentially depends on the media provided by the collaboration support. A collaborative activity can be roughly classified into two categories, depending on the information exchange dynamism: *synchronous* or *asynchronous* [3]. Synchronous collaboration is characterised by a high level of interaction within the group: all collaborators share a single view of the discussion and the information is exchanged when it becomes available. Conversely, in an asynchronous collaboration the information is transferred only when requested, thus lowering the degree of interaction in the group. The classic Web communication facility support only an asynchronous collaboration, mainly because HTTP protocols rely on a communication model in which the browser needs to request the information, the HTML pages, from the server.

Any collaboration support needs to provide *consistency-guarantee mechanisms* [2] to correctly manage the shared information. In synchronous collaborative environment, where the collaborators share a single view of the shared information, consistency is typically managed by a *floor-control policy* [4]. This policy guarantees that when a collaborator is free of modifying the shared data, no other collaborator has the same possibility. This modifying privilege is commonly described in terms of possessing the *modification token*, and the floor-control policies can be roughly classified into *explicit* and *implicit* depending on how the collaboration support assign the modification token. Explicit floor control states that the collaboration support explicitly assigns the modification token; while in implicit policies the token is assigned without the collaborators being aware of this.

C/WebTop is designed to provide Web and mobile users with a collaboration tool implementing a synchronous collaboration model with an explicit floor control policy. The platform is mainly aimed at supporting collaborative activities using one or more off-the-shelf applications. This virtual team is characterised by a dynamic life cycle involving both the users and the state of the applications. The system supports *unanticipated sharing* [5] of the application and the latecomers can decide to enter into the discussion in a *synchronous* way. This means that the latecomer is immediately accommodated in the virtual team with a view of the shared document. At any time during the conversation, a user can leave the discussion simply closing its terminal application. Users are immediately informed of this event.

The explicit floor control policy enables only one user at a time to control the state of the collaborative session. When she finishes its work, the floor passes automatically to another user. This token-moving scheme implies that users can hold the floor only after receiving it by someone else. The order in which users acquire the floor follows a FIFO policy. At any time, users can ask to have the floor. Their requests are queued and when the floor is released the system give control of the applications to the user with the older request. To prevent discussion starvation caused by indefinitely possessing of the floor, the team manager can also decide to take the floor away from a user, giving it automatically to the user on the top of the waiting queue. If nobody held the floor, the applications cannot be modified.

3. The Architecture of C/WebTop

Collaboration platform can be roughly classified into *centralized* and *replicated* [6]. In a centralized architecture, the shared application is maintained in a single physical location and participants to the virtual team are supplied only with the output of this application. In a replicated architecture, each user owns an instance of the application and the platform provides the mechanisms to synchronize the instances. The validity of the replicated approach is lost when taking into account users accessing the virtual meeting through small devices. The replicated approach is based on the possibility of installing an instance of the shared application on each terminal and to synchronize the various instances by means of events broadcasting. We cannot rely on installing a real-world application on a device with small processing power and memory, and therefore, C/Webtop is based on a centralised architecture.

Figure 1 depicts the architecture of C/Webtop. Users interact with the shared applications through their terminals. The client application running on the terminal receives the graphic output of the shared application (solid lines) and captures all mouse and keyboard events. Such events are the redirected to the server through dedicate channels where they are recreated and passed to the application (dashed lines). Only the events arriving from the member that currently has the floor are passed to the shared application, while the others are ignored. The result of this mechanism is allowing only the user having the floor to work actively on the shared application. Whenever a new member joins the collaborative session, the server synchronizes the state of the session requiring all applications to completely redraw their user interfaces.

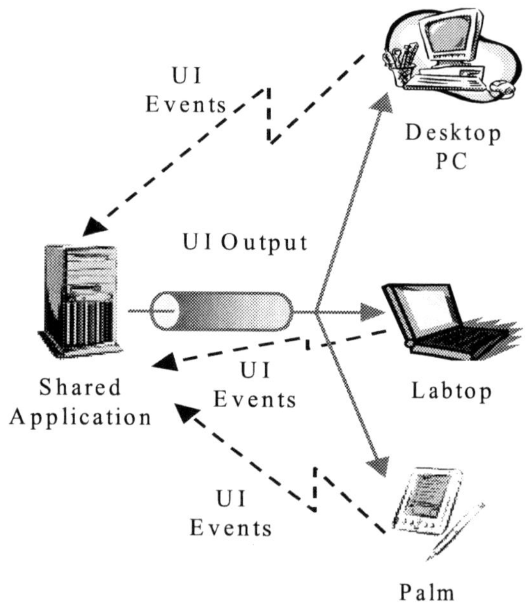

Figure 1. The Architecture of C/WebTop.

The platform is not limited to a single application and the virtual team is provided with a shared desktop capable of containing many applications concurrently. To start a collaborative session, the team manager launches C/WebTop on the server with one or more off-the shelf applications. She also fixes the authorization policies and the permissions each user have. Then, she opens C/WebTop user-interface, takes the floor and begins working with the shared applications.

In addition to the sharing of applications, the desktop that C/Webtop provides acts as a transparent white board to allows users drawing geometric shapes on the desktop. This supports the creation of logical links between different applications and it enhances the quality of communication among the virtual teams. This is important in particular when the desktop contains more than one application and users want to express relationships between them.

A videoconferencing service, based on the VIC [18] tool developed by the Lawrence Berkeley Laboratory, is integrated in the platform in order to increase the level of interaction of the virtual team. We patched VIC providing the mechanisms to synchronize the videoconference with the other services of the shared desktop. The videoconferencing service is independent from the shared applications running on the server side and is managed among the client interfaces. A Java AWT Canvas controls VIC, so each applet or Java interface can display the output. If the client terminal has enough processing power and memory, the client interface adds the Canvas and launches VIC using a multicast channel. At any time during the conversation a user can enter or leave the videoconference without interfere with the collaborative session. It is also possible for people outside the virtual group to participate only to the videoconference. All the images of the participants to videoconference are displayed in 50x40 format and one of them is displayed in 175x140. The user can choose which image to maximize double clicking on the corresponding small image.

The broadcasting mechanism that C/Webtop uses, differs for Windows and Java applications. Since Java technology is platform independent we implemented clients as Java applets. This eases the broadcasting of the user interface of Java applications because we can send the calls to the graphic primitives, such as `DrawLine`, and have them replicated on the clients, see Figure 2(a). This is not the case for Windows applications because we cannot rely on having the Windows graphic engine on terminals with small capabilities. Therefore, we share Windows applications broadcasting bitmaps representing the user interface of application as shown in Figure 2(b).

J/WebTop is the C/WebTop subsystem designed for Java applications. Our approach to make Java off-the-shelf applications collaborative relies on a collaborative version of Swing that we plug into the server-side Java virtual machine.

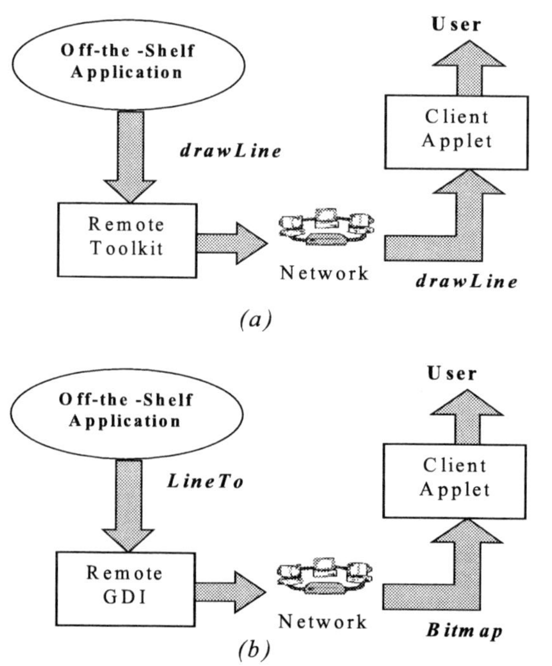

Figure 2. The architectures of (a) J/WebTop and (b) W/WebTop.

As long as Swing 1.x is concerned, our problem simply requires implementing a substitute of the AWT library supporting remote users because Swing 1.x completely relies on AWT for its input/output operations. Implementing a remote AWT is straightforward because AWT was designed to be easily substitutable. It relies completely on the Toolkit singleton class that acts as an abstract factory and its substitution allows controlling the whole AWT. This abstract factory allows creating the peers that AWT components use for performing input/output operations.

We implement remote peers in terms of the existing peer interface and we provide an active version of the Toolkit, that we call RemoteToolkit, used on the server side instead of the one provided by default. This Toolkit manages the connections between the shared application and the user-interfaces and it is responsible for the state of the collaborative session. The introduction of Java 2D as part of Java 2 makes the remotisation of Swing 2 much more complex. Java 2D is a big library that exploits native code intensively to achieve reasonable performances and many of the services that it provides are not mediated by an abstract factory. In particular, one of the problems that we have to face is that Java 2D allows creating images simply instantiating the class BufferedImage. This is a problem for the buffering of images and we solved it implementing a substitution of some of Java 2D classes.

Users join a running collaborative session simply starting the client application setting the IP address of the server where the shared application runs. The client captures the mouse and the keyboard events generated by the users and it sends them to the RemoteToolkit on the server side through a dedicated channel. Only the events received by the user holding the modification token are posted to the application by the dispatchEvent() method, while the others events are ignored. Consequently, the application will produce the output generated by the user currently having the floor. The class RemoteGraphics, that extends Graphics, intercepts all the graphics primitives that the application calls sending them to all the terminals involved in the collaborative session.

In order to optimise the performances, the graphic events are sent with a reliable multicast protocol implemented in Java by INRIA [7]. The user interface of the shared application is sent to a multicast address and packet loss can be recovered. When a new group member joins the session it begins to receive the data packets available in the network. A non-reliable solution, i.e. simple IP multicast, cannot be adopted because we cannot determine if the eventual loss of information is important or not for the usage of the application.

A virtual team composed by Ph.D. students tested the platform during a virtual seminary distributed across Italy. We used a simple editor and a Web browser written in Java to discuss topics about the Web technology. The professor played the role of team manager and presented the course using HTML pages to the students that joined the virtual team from different cities. Figure 3 shows the prototypal Java user-interface experimented by the Ph.D. students during the seminary.

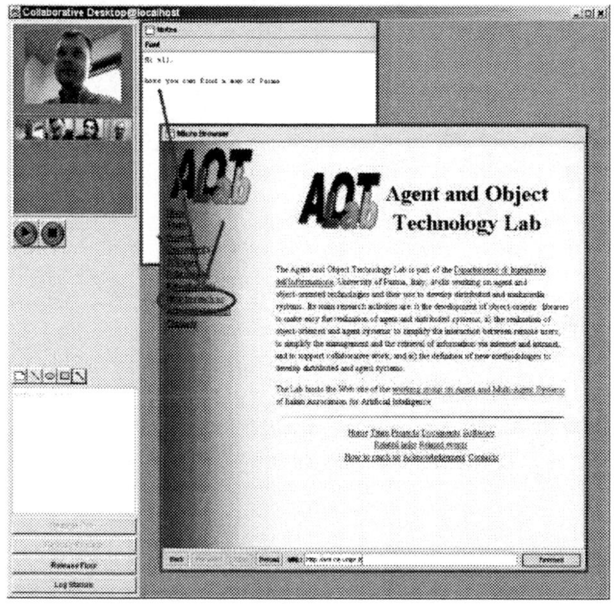

Figure 3. J/WebTop snapshot on a PC.

W/WebTop is the C/WebTop subsystem designed for sharing Windows applications. On the server side, the off-the-shelf application graphic calls are directed to an off-screen bitmap. This is achieved hooking the graphic primitives exported by the GDI library. This approach allows redirecting most of the graphic output of the shared application to the off-screen bitmap. This bitmap is then forwarded to the virtual team. In order to reduce the network load, the colour depth of the bitmap is reduced to 8bpp and some of the compression techniques used by VNC [13] are implemented. The client application is the same of J/Webtop and it uses the same mechanism to communicate with the shared application.

4. Related Work

A number of collaborative platform designed to work in the Web are found in the literature as the interest in both the Internet and collaboration has grown in the past few years [8]. Some systems provide special-purpose shared applications that are intended to support virtual meetings in particular domains. This is the case, for example, of the MASH [9] and MERCI [10] projects: they provide a shared whiteboard, a shared text editor and a videoconference system. Another category of collaborative systems supports collaboration unawareness allowing the sharing off-the-shelf applications. We are more interested in such systems as C/WebTop falls in that category.

The principal method employed to implement centralized systems is display broadcasting. This method is based on the assumption that participants can access a client application able to show the output of the centralized application. These systems can be classified into *heavyweight* and *lightweight* clients on the basis of the information the centralized application transmit to the client applications. The heavyweight-client approach is mainly used for collaborative systems based on the X-Windows system, such as Hewlet-Packard SharedX [11], Old Domino University XTV [12] where the client is supplied with a description of the user-interface event occurred on the centralized application. On the contrary, the lightweight-client approach, also known as *thin clients*, is used when the capabilities required to the client are only the possibility to display the bitmap of the actual output of the centralized application. Such a method is used by Microsoft NetMeeting and by the collaborative version of VNC [13][14] and TridiaVNC[15]. This approach is the basis of the ITU T.128 protocol.

The centralized approach seems not well suited for the integration in the Web because some of the basic requirements of a Web-centric application are not satisfied. The heavyweight-client approach is not platform-agnostic, as it requires some sort of X-Server to be installed on the client platform. The lightweight-client approach is not fully satisfying because it tends to overload the network: studies on X-based applications shows a 10 to 1 ratio between the network load caused by graphics bitmaps and X events [16].

The replicated approach seems to suit better the Web assumption as it can be implemented using Java clients integrated with the most common browsers by means of events broadcasting. A number of critics are moved against this approach [6]:

- The environments where the multiple instances of the application are necessarily different;
- In order to support latecomers, application instances must be able to externalise their states;
- The identification of the user-interface components receiving the events is not a trivial task and relies on the capabilities of the user interface.

Recently it has been noted that constraining the application to be purely-reactive and replicating the entire environment in which the application runs can be sufficient to give a positive answer to the previous three critics [16]. At the moment, the most interesting Java-based collaboration frameworks supporting both collaboration transparency and replicated architecture with event broadcasting are the Java Collaborator Toolset [12] from Old Domino University and JAMM [16] from Virginia Tech. Both the Java Collaborator Toolkit and JAMM implement a collaboration model similar to C/WebTop, but at present they do not provide the same set of system services. In particular, those systems implement a collaborative version of the Java AWT, without integrating it with a smart decision policy. Moreover, the JAMM relies on a hidden-floor collaboration model and requires a modified version of the Java Virtual Machine in order to keep related events together and to be able to send them to all application instances. While this approach can be considered more general than the collaboration model previously described, it prevents the use of JAMM applets within a Web browser. However, the validity of the replicated approach is lost when taking into account participants accessing the virtual meeting through small devices

TridiaVNC and jVNC are two products that provide seamless access to shared applications exploiting display broadcasting. They are designed to work with thin clients possibly on small devices and they offer some of the features that C/WebTop provides. Nevertheless, they are not an integrated solution implementing a virtual meeting room as:

- They do not provide any support for multimedia communication;
- They do not integrate any tool to manage virtual meetings;
- They offer a very weak synchronization mechanism to prevent multiple modifications on the shared documents.

Basically, TridiaVNC and jVNC can be considered enabling technology for the implementation of collaboration solutions.

As far as application sharing is concerned, C/WebTop exploits the display broadcasting approach optimising it to allow the use of thin clients. In particular, we exploit the availability of Java on the client systems and rely on its user-interface services to render the output of the shared application. This approach allows using J/WebTop seamlessly on any device and it significantly reduces the development effort because we can exploit the availability of a common environment, i.e., Java.

The use of mobile devices in supporting virtual teams is nowadays limited to telephone conferences. This is mainly due to the limited processing power of current mobile devices and to the limited bandwidth. These limitations will be overcome in next generation mobile devices as the previsions of companies in this field guarantee [17] and companies in the market of next generation mobile devices are still looking for applications capable of exploiting these new features.

5. Conclusions

This paper presents *C/WebTop*, a collaborative platform that enhances the classic Web communication mechanisms to support synchronous sharing of off-the-shelf applications from fixed and wireless networks. The platform has a centralized architecture based on the broadcasting of the graphic output towards the users and provides an explicit floor-control policy. The platform is composed of two modules for different kinds of application: *W/WebTop*, based on Windows technologies and designed for Windows applications, and *J/WebTop*, based on Java technologies and supporting Java applications. The integration of mobile networks in the system allows the users to access the collaborative sessions from different environments with different hardware.

6. Acknowledgements

This work is partially supported by CNR Applied Research Project 5% "Multimedialità" and by Microsoft Research that provided detailed documentation and development tools.

7. References

[1] R. Bentley, W. Appelt, U. Busbach, E. Hinrichs, D. Kerr, K. Sikkel, J. Trevor and G. Woetzel, "Basic support for cooperative work on the World Wide Web", *Int. Journal of Human Computer Studies*, 1997.

[2] P. Dourish, "Consistency guarantees: Exploiting application semantics for consistency management in a collaboration toolkit", in *Proceedings of the ACM, Conference on Computer-Supported Collaborative Work*, 1996.

[3] U. Gall and F.J. Hauck, "Promondia: a java-based framework for real-time group communication in the Web", in *Proceedings of the 6th Int. WWW Conference*, 1997.

[4] A. Poggi and G. Golinelli, "Automatic Storing and Retrieval of Large Collections of Images", In *Proceedings of the 11th International Conference on Industrial & Engineering Applications of Artificial Intelligence & Expert Systems (IEA/AIE-98)*, Spain, 1998.

[5] J. Begole, A. Struble, C.A. Shaffer and R.B. Smith, "Transparent sharing of Java applets: A replicated approach", in *Proceedings of the Symposium on User Interface and Technology*, 1997.

[6] W. Minenko, "The application sharing technology", *The X Advisor*, 1995.

[7] T. Liao, INRIA, "Light-weight Reliable Multicast Protocol", 1997. *Available at http://webcanal.inria.fr/lrmp/index.html*

[8] D. Ramduny and A. Dix, "Why, What, Where, When: Architectures for cooperative work on the World Wide Web" In *Proceedings of the Conference on Human Factors in Computing Systems*, H. Thimbley, B. O'Connail and P. Thomas editors, pages 283–301, 1997.

[9] T. Hodes and R.H. Katz "A Document-based Framework for Internet Application Control". In *Proceedings of the Second USENIX Symposium on Internet Technologies and Systems (USITS '99)*, Boulder, CO, 1999.

[10] R. Bennett and P.T. Kirstein "Technical Innovations deployed by the MERCI Project". In *Proceedings of Networkshop 25*, Belfast, pages 181-189, 1997.

[11] D. Garfinkel, B.C. Welti and T.W. Yip, "HP SharedX: A tool for real-time collaboration", *HP Journal*, 1994

[12] H. Abdel-Wahab, B. Kvande, S. Nanjangud, O. Kim and J.P. Favreau, "Using Java for multimedia collaborative applications", In *Proceedings of the 3rd Workshop on Protocols for Multimedia Systems*, 1996.

[13] K.R. Wood, T. Richardson, F. Bennett, A. Harter and A. Hopper, "Global teleporting with Java: towards ubiquitous personalized computing", In *Proceedings of Nomadics '96*, 1996.

[14] S.F. Li and A. Hopper, "A framework to integrate synchronous and asynchronous collaboration", In *IEEE Seventh International Workshop on Enabling Technologies: Infrastructure for Collaborative Enterprises*, 1998.

[15] D. MacVittie, AT&T Labs', "Virtual Network Computing Is Virtually Perfect", 2000. *Available at* http://www.networkcomputing.com.

[16] J. Begole, A. Struble and C.A. Shaffer, "Collaboration transparency in Java through event broadcasting", In *Proceeding of the Symposium on User Interface Software and Technology*, pages 55–64, 1997.

[17] Ericsson prevision about the processing power of next generation mobile devices at http://www.ericsson.com.au/ThirdGeneration

[18] User Guide for VIC v2.8, Lawrence Berkeley Laboratory, 1998. *Available at* http://www-mice.cs.ucl.ac.uk/multimedia/software

Supporting Group Awareness in Collaborative Design

Yan Liu, Yuanchun Shi, Guangyou Xu
Dept. of Computer Science and Technology
Tsinghua University, Beijing, China
liuyan98@mails.tsinghua.edu.cn

Abstract

Different from multi-user database management systems, CSCW system must support group awareness explicitly, namely that participants should perceive the presence of each other during the process of cooperating, which is essential to effective collaboration. In this paper we firstly analyze and compare several means of supporting sensibility, then investigate the model of collaborative design and suggest achieving group awareness by product data which forms the foundation of relationship among designers. The information model and the shared workspace organization of CECAD (Collaborative Environment for Computer-Aided Design, a prototype system we developed) are presented and its group awareness supporting mechanism is described in succession.

1. Introduction

As a sort of multi-user system, CSCW faces issues such as data sharing and multi-user access. In contrast with multi-user database management systems, in which users are isolated by a variety of transparency provided by system and think they are accessing system resources independently, CSCW system must present group awareness explicitly [1], to achieve the second C in CSCW, i.e. "Cooperative".

The simplest understanding of group awareness is that one could realize others' presence in the course of cooperation. This is relatively simple to implement in synchronous systems. However, different means may cause dramatically different effect due to the involvement of social and psychological factors. Generally, Synchronous systems support group cooperation through three categories of media: audio, video and text.

➢ *Audio.* Ochsman& Chapanis[4] in a series of studies investigated the effect of different communication modes in a range of tasks and found that audio is the most essential factor in synchronous communication. In fact, the communication-rich mode, such as face-to-face mode, does not have significantly enhanced communication effect compared with the voice-only mode. On the other hand, from the perspective of social psychology, audio communication is more personal, which is identification of persons could be more easily recognized through audio. Studies on tele-presence found that audio telecommunication could reinforce greatly the sense of being in the same physical space.

➢ *Video.* Although a series of studies showed that video communication does not have apparent improvement on collaboration efficiency than voice-only communication, it could provide more awareness of the presence of others for the participants, especially when new users join in and old users drop out. Also video could transfer more meaningful information, which is obviously observed at communication interval. When participants are contemplating or interacting with computers and do not want to speak, the video channel can deliver the scenery through face expression or gesture, instead of the "silence anxiety" raised by lack of interaction in audio communication.

➢ *Text.* Text-based communication could remove traditional social barrier, the inequity on communication caused by disparity on social status [6]. But an obvious deficiency in text-based communication is the difficulty on text interpretation or even misunderstanding due to the lack of video and audio clues, such as face expression, tone, etc. Besides, studies of Ellis&Gibbs [7] showed that for the absence of direct feedback in video and audio communication, text-based communication will be more demanding on coordination, and consensus is usually difficult to be achieved.

Similarly, in asynchronous systems, the implementation of group awareness requires certain forms of feedback. Multi-user database systems enforce the illusion of single-user activities for the absence of feedback. In collaborative system, two mechanisms could enhance the group awareness:

➢ *Identified locking.* In database system, lock mechanism ensures the integrity and consistency of data. But you don't know who is accessing the locked data and which sort of operation is being executed. Identified locking allows access to the information about the locked object, i.e. who is currently locking the objects and why he got the authorization. Given this kind of information, a user may enter into a discussion with the lock holder, and this gives basis for negotiation and possible collaboration.

➢ *Event notification*. In database system, data change is invisible to other users. To improve the visibility in collaborative system, event notification mechanism could be used. The initiation of events could be caused by operation to shared object, e.g. the creation or modification of shared object. Response to events could be messages sent to people who will be interested in the event.

This article discusses in detail how to support group awareness in cooperative design systems. In section two, we describe a collaborative model in cooperative design and analyze how group awareness is represented in this model; in reference to the model, we describe how CECAD, a prototype system we developed, supports the implementation of group awareness in its shared workspace. In the final section, we make a brief summary to our work.

2. Collaboration Model of Cooperative Design

In cooperative design, the design process consists of a series of discrete activities. Designers do not always work together and in quite long period they work independently with their own expertise. From the collaborative model in Figure 1, we can see that interaction arise only at the phase of negotiation and evaluation, and the span is very short.

During this very short span, designers collaborated in three patterns [2]:

➢ *Mutual collaboration*, in which participants are ''busy working with the other''.

➢ *Exclusive collaboration*, in which the participants "work on separate parts of the problem, negotiating occasionally by asking for advice from the other''.

➢ *Dictator collaboration*, where the participants decide who is "in charge" and that person leads the process.

Studies show that the "exclusive collaboration" is the most effective pattern and in which they observed most productive results. "Mutual collaboration" led to no result at the end of very busy exchange between the participants, whereas "dictator collaboration" came to a conclusion as soon as the leader made up his mind. So, although "Mutual collaboration" exists in cooperative design, such as drafting with whiteboard, we will put the focus on "exclusive collaboration".

"Exclusive collaboration" generally takes place in asynchronous environments, whose interactive pattern is illustrated in Figure 2. Designers browse, operate and store data through access tools, and in many cases, not only does he need know the current value of the data, but also he need know if the value have been changed and the reason of the change. This is associated with the feature of cooperative design. When several people co-design a product, each will be responsible for different parts or different aspects of one part (appearance, structure, material, etc.), and will set the value of parameters of those parts. Usually these parts are not isolated, but connected in a variety of ways.

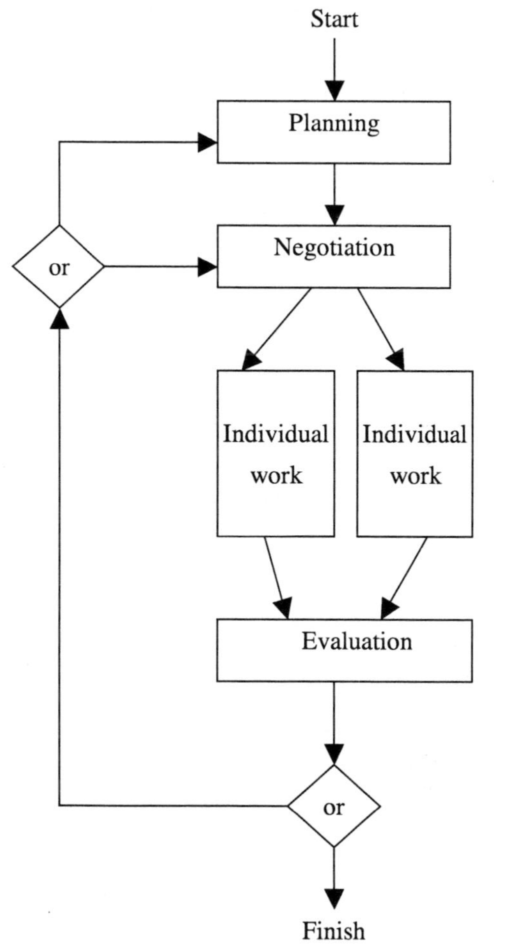

Figure 1. Collaborative model of cooperative design.

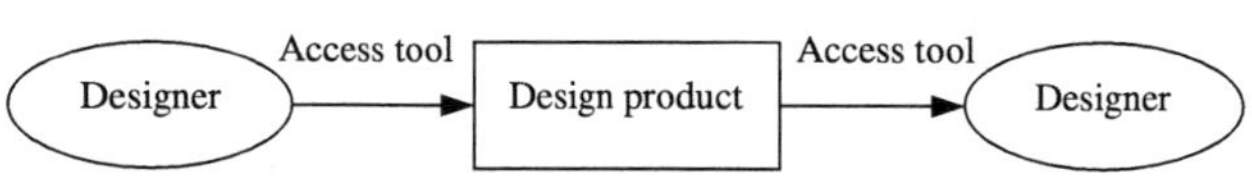

Figure 2. Exclusive collaboration pattern.

Thus, in cooperative design, group awareness is presented by medium of product data and guaranteed by the following functions of the system:

➢ Designers could access the scheme of product parameters and their underlying rationale, for effective evaluation.

➢ Designers could access different proposals of the scheme and their rationale, so as to facilitate the reference and coordination.
➢ In access to shared data, designers can know its current status, e.g. if the value contradicts with the requirement.
➢ When some changes take place to one parameter, designers could sense in some way so as to response appropriately.

3. Organization of shared workspace

Shared workspace is a virtual space in which several collaborators work together. In cooperative design it is a platform established on the basis of information model, and designers interact with shared workspace through a range of access tools, including hypertext tool, event management tool, session management tool, negotiation tool, and etc. Figure 3 displayed the organization of shared workspace in CECAD.

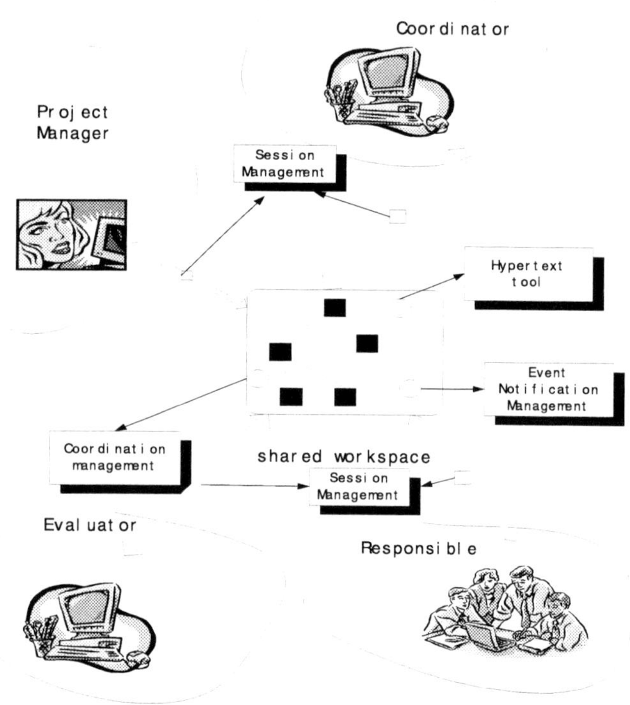

Figure 3. Organization of shared workspace in CECAD.

Information model

There exists a problem of understanding when product data is taken as medium of collaboration in cooperative design. So an appropriate description of information is the foundation of effective evaluation and negotiation. The following layers of information should be described in the information space:

➢ Design object, including function, goal and requirement of the design object
➢ Decision, including proposals or positions such as positive or negative position.
➢ Rationale, including design principals, cases, rules and etc.

As mentioned in section one, to implement group awareness, certain identification mechanism should be provided when data is accessed and, most importantly, the visibility of data change should be improved. This could be implemented by event and message mechanism. In multi-agent system[3], designers distribute their design interests through agent interfaces, and each agent maintains a record of interest list. When data changes, the agent will search in its interest list if the data is contained and found, an event will be initiated. In another pattern, the initiation of events and sending of messages are determined by the interdependency among product data. Due to the adjacency or assembly relationship, certain constraint should be satisfied. If one parameter changes, system would judge if it leads to status change of one constraint ('satisfied' or 'not satisfied'), and determine whether to send this event to designers associated with the constraint. We adopt the second pattern in our system, for it represents the relation among designers more directly.

Figure 4 is the information model of CECAD. This model describes involved information during design procedure and the diverse and complicated linkage among them. Label 1 indicates the relation of design products, mainly the 'consist-of' relation between parent parts and children parts; label 2 indicates the relation of design parameters; Label 3 indicates the 'is-required-by' relation between parameters and their requirements; Label 4 indicates the relations of different proposals, which could be 'support', 'contradict', or 'is-alternative-of'; Label 5 indicates the 'is-referenced-by' relation between proposals and their rationale, which could be principals, rules, cases, constraint networks, catalogs and design prototypes. When contradiction occurs and negotiation is required, every involved person would see the underlying rationale in order to make further choice.

Hypertext tool

Hypertext permits the browse and generation of non-linear text. Links in hypertext can express rich connection of information. CECAD uses hypertext as a uniform representation of structure and function of shared information as well as medium of collaboration. We take advantage of hypertext tool to create connection among information conveniently and also browse and update them, e.g. the pertinence of geometric data of design objects. In the phase of negotiation and evaluation, hypertext provides effective mechanism for group awareness. By hypertext, we could view if a design object has different schemes and rationale of these schemes, or if there are different opinions toward a design result, so as to make an effective judgment.

On account of the rich information maintained in links, interpretation of links is accomplished by linking server in CECAD.

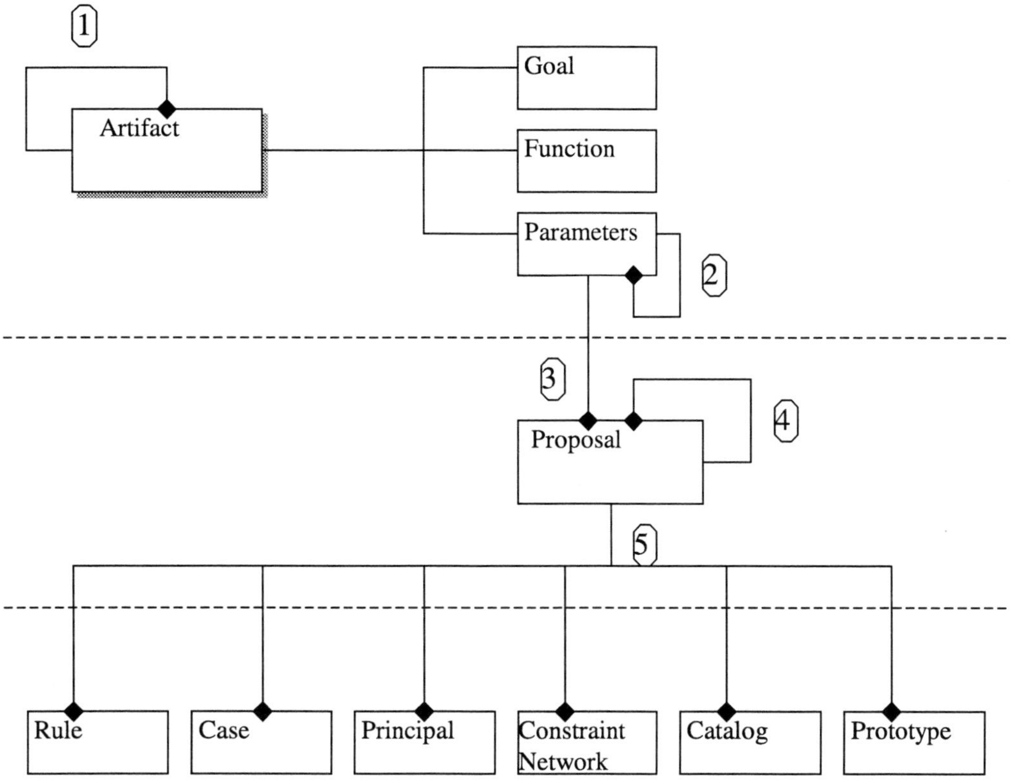

Figure 4. Information model in CECAD.

Role control

Database management systems assign access right to each user, such as view, create, update, delete, etc. However, the mapping of users to rights in this access mode is too simple, and is especially inconvenient to cooperative design system in which staffs are always dynamic and their rights are also altered frequently. So, in CECAD, a concept of role is provided, which means rights will not be directly assigned to users but to roles. Users are mapped to roles and gain rights of access to data through roles. So far, roles defined in CECAD are:

- Project Manager (PM): to create a project
- Responsible (RB): to be responsible for the content of certain objects
- Valuator (VA): to be responsible for the evaluation of design result
- Coordinator (CO): to be responsible for the procedure of coordination when contradiction takes place.

This access mode shows a high degree of flexibility and scalability. When a design system is upgraded, definition of rights will be supplemented or adapted, but the access infrastructure need not be overhauled as long as we redefine the content of role rights or add new roles. When a user changes his work, system would just re-map him and the corresponding role. In addition, roles convey more social information and thus will improve the group awareness during interaction.

Session management

In CECAD, Web is used as a platform for collaboration. Web provides a good environment for cooperation: as a form of data representation, it facilitates browse and operation; as a kind of user interface, its update could be noticed by all users and does not require repeated distribution.

In session management, session objects associate workspace with user interface. CECAD will record all the users participating one session and they will see the same interface. As one user operates an object, session object will record the identity of both the object and the user, and make a mapping. When other users operate the same object, they will know how it is being used.

Coordination management

When contradiction occurs, e.g. data value does not satisfy the requirement, or several values do not satisfy certain constraint, notification will be sent to designers by message mechanism and designers may coordinate with coordination management tool. Coordination might be carried out in synchronous or asynchronous way.

So designers use CAD or other drawing software to work independently on their own task. When negotiation is needed, they enter into the shared workspace and interact with shared data with the tools provided by shared workspace. The information model, which is the kernel of workspace, and a range of tools guarantee the group awareness during the collaboration.

4. Conclusion

This article aims at providing group awareness in collaborative design system. We analyze collaboration mode of cooperative design and achieving group awareness by product data that forms the foundation of relationship among designers. This article investigates the pertinence of design information, and gives the information model in cooperative design, then on its basis a collaboration platform, i.e. a shared workspace, is constructed. In the workspace, designers use hypertext tool to browse and operate information and, most importantly, through the hyperlink, designers could understand the relation among information which implicates the relation among designers. This provides prerequisite for further collaboration. Session management associates users with shared workspace, and enforces the group awareness using role control and message mechanism. When users operate shared object, according to the relation among information, system will send message or mail to corresponding designers, who can give timely response in reference to the nature of events.

Group awareness is crucial for effective collaboration of multi-users. We do some research in our CECAD. But so far there is few evaluating standard toward group awareness.

5. Reference

[1] John A. Martin and Tom Rodden, Cooperative Information Sharing: Developing a Shared Object Service, The Computer Journal, Vol 39.No.6, 1996

[2] Maher.M.L., Cicognani.A.,&Simoff.S.J., An experimental study of computer mediated collaborative design, International Journal of Design Computing.,Vol.6, June 1998

[3] Michael P.C. and Stephen C-Y Lu, Discourse Model for collaborative design, Computer-Aided Design,Vol.28, No.5. 1996□333-345

[4] Ochsman.R.B.,& Chapanis.A. The effects of 10 communication modes on the behavior of teams during co-operative problem-solving. International Journal of Man-Machine Studies,Vol 6, 1974, 579-619

[5] Böcker.M., & Mühlbach,L. Communicative presence in video-communication. In Proceedings of the Human Factors and Ergonomics Society 37th Annual Meeting, 1993.Santa Monica, CA: Human Factors and Ergonomics Society.

[6] Kiesler.S.,Siegel.J., & McGuire.T.W. Social Psychological aspects of computer-mediated communication. American Psychologist, 39(10),1984, 1123-1134.

[7] Ellis, C.A., Gibbs.S.J. & Rein. G.L. Groupware: Some issues and experiences. Communications of the ACM. 34(1), 1991, 39-58

[8] Thomas Kvan, Collaborative Design: what is it? Automation in Construction, Vol. 9, 2000, 409-415.

Improving the Selectivity of Awareness Information in Groupware Applications

José Maria N. David
COPPE-PESC / UFRJ &
Faculdade Ruy Barbosa / BA
david@cos.ufrj.br

Marcos R. S. Borges
NCE & IM
Federal University of Rio de Janeiro
mborges@nce.ufrj.br

Abstract

Many awareness mechanisms in groupware systems have been searching to reduce the amount of information found in the workspace where activities develop. Several approaches and proposals found in literature are based on filter usage for best identifying and presenting information according to the users' choices or previous ways of working. The goal of this paper is to present a strategy to both guide the selection of information from the Group Memory and support coordination actions in groupware applications. This proposal is based on filters and awareness profiles established by the participants in groupware activities.

1. Introduction

An asynchronous way of work implies that users share the same object for a large period of time and need appropriate mechanisms to support that work. Only occasionally are the users interested in all current events. In most circumstances, they are interested only in events relevant to their work context, i.e. events that direct them make a better decision and/or widen their capacity to decide on their work aspects. Filtering mechanisms (or "*user profiles*") can be used to select relevant information and to reduce an eventual overload of information, usually imposed on group participants.

Many awareness mechanisms are projected and shown to participants according to the planned results of the application, or to the roles that they have been playing in the activity [1]. However, if the applications and the roles change from time to time, mechanisms should also change in an evolutionary way to assist the participants' needs without overloading the workspace.

In these environments, the resources used by the coordination should be configurable - and complemented - according to the participants' abilities, and to the activities indicated by their dynamically established filter [3]. The strategy used in this paper suggests that the Organizational Memory (OM) should be built up gradually, improved and enriched as the system is being used, supplying support to participants and coordinators throughout the construction of new awareness components.

In Section 2 we discuss the issues related to awareness and selectivity in asynchronous environments, soon some concepts related to the selectivity and the density of information in the workspace have been presented. The filters used either by the users or in application as a form of reducing the distance between the overload and the absence of awareness information are described in Section 3. Section 4 presents the group memory. Section 5 describes an example of a prototype built with the objective to study and to validate the topics presented in the previous sections. Finally, Section 6 concludes the paper.

2. Selectivity and Granularity in the Workspace

The selectivity of information in a groupware environment is directly associated with the amount of information available to the users and to the mechanisms to support workspace awareness. These are the mechanisms that will give the users necessary support to perceive the overall state of the application [8]. Thus, both the coordinator and other participants will efficiently be able to understand not only their activities, but also the dynamic of the CSCW environment.

Referring to awareness, we are talking about of a vast group of information and resources that can be used with the same objective. These resources, presented in an inadequate way to the users, can trigger an information overload for the events that happen in the workspace. They could also shift the attention to inadequate information, and eventual dispersions could be the result of the amount of actions that the user should take to obtain the wanted information. Thus, we should have a good interface project, capable of taking us to a point to either the lack of or the overload of information, which is no easy task.

It is not enough to build interfaces and awareness mechanisms that meet the needs and desires of the group at a certain moment. Due to the fact that their interests change throughout the time and during activities, many times these mechanisms become ineffective. Only through information, and directions supplied by the users, can the coordinators perceive what the participants are doing. As a result, new awareness information can become available; others omitted.

Additionally, we need mechanisms and objects able to accomplish a filtering of awareness information. We also need to present them to the participants of the group in a denser way, always trying to reduce the cognitive overload, without omitting important information. More available information doesn't necessarily mean larger perception. Besides the consequences already presented, it also means larger potential of privacy violation for the participants and consumption of resources of the system [6]. Then, it is important to ask: which is the ideal amount of information for the perception, and consequently for the selectivity in a group memory.

Users must not only understand their own compositions, but also the successes and failures, strategic goals, positions and the directions taken by other participants. They should also suggest them solutions for similar situations, so that, he/she can select and make the appropriate information available. A useful metaphor for the perception of the knowledge is presented by Tucker and Frappaolo [9], where the authors divide the existent knowledge in an organization in four possible states:

- **KK** *know that you know*;
- **KDK** *know that you don't know*;
- **DKK** *don't know that you know*;
- **DKDK** *don't know that you don't know*.

In a group activity, it is also necessary to pay attention to these subjects. They have to be quickly discovered, or lots of time will be wasted on group activities. Participants always work in the sense of increasing the percentage of the first two items, and of reducing the percentage of the last two. However, this will only happen through an appropriate support of awareness, supplying better visualization techniques and, consequently, increasing the capacity to select important and relevant items in group memory.

An important aspect to be considered in the process of information selection concerns the nature of the integrated information that is presented to the users. The data granularity has serious implications in the quality of the information, mainly when we are considering big geographically dispersed groups. During the activities in which they participate, those groups select, interpret the available data, and build the knowledge producing information in different places and contexts. Thus, the more dispersed the information is, the harder for the participants to accomplish their activities.

The technological aspects involved in environments with high granularity are also factors that can hinder the establishment of awareness mechanisms able to contribute to the information selectivity. Additionally, we can consider work groups with habits, cultures and different values, constituting one more subject, which deserves special attention.

We can affirm that the smaller the granularity and the volume of information in the workspace are, the higher the probability of increasing the percentage of KK and KDK is. We also need techniques to process and to classify information. They then can be accessed, reused, organized and integrated into other important information, building a cohesive, significant and necessary set to support group activities [7].

Recent works have been developing more general theories on the selection and presentation of the awareness information, and their mechanisms [1, 7]. This information is usually related to events in real time. Nevertheless, many environments have been exploring awareness in activities that happened during a certain period. We then need specific tools to guide them properly. This paper focuses especially on this issue.

3. Input and Output Filters

3.1. The Application Filters

We can relate the input application filters to the persistence of the information in the group memory. Through these filters we define the quality of the information and the appropriate time for them to stay in the group memory. In asynchronous systems we need to maintain the awareness information at least until all the involved participants are aware of it. The time that the information is kept in the group memory is also related to the information relevance for the group activities. However, how can we determine the relevance of this information? A possible indication could be the filters defined by the users. After analysis of the user's profile the coordinator can opt to discard this information or to transfer it to the Organizational Memory (see Figure 1).

The output filters are directly related to the roles that the participants are playing at a given moment. These roles indicate to the application the type of awareness information needed, so that the role can be performed successfully. This may change over time, as users become more skilled and require a different set of information to perform their activities. Thus, the system should allow users to adjust their awareness profile accordingly.

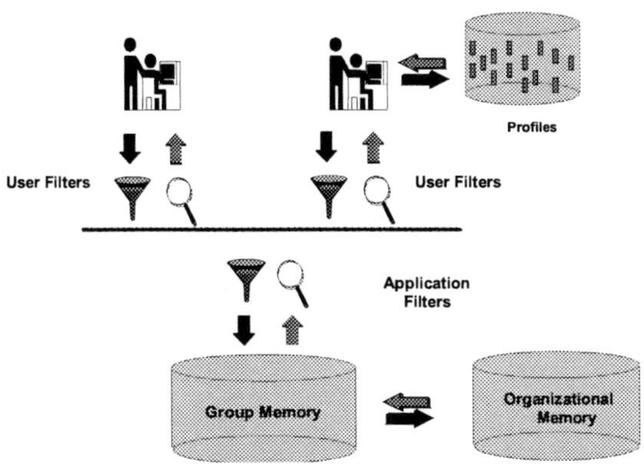

Figure 1. Filters of awareness and memories.

3.2. Filters for Users

There are several ways to notify the users according to their needs [4], which may change in the course of their activities. The input filters established by users could also be seen as part of awareness information. Many times the users filter certain information because it is meaningless for them. Not for the coordinator though, for him/her the profiles can be meaningful. For example, in a cooperative edition, if one of the users filters the list of updated documents or the list of people who have contributed to the edition, it may seen to the coordinator low participation, or lack of involvement of the user's part.

The coordinator must be informed about the participants' activities, their interests, needs, as well as their participation and/or contribution rates. This will increase the KK ("know that you know") and KDK ("know that you don't know") rates of the group. Otherwise, the possibility of increasing DKK ("don't know that you know") and DKDK ("don't know that you don't know") rates would still exist.

Also, visualizing some details of group activities and interactions coordination could be more effective. Only thus, group participant's dispersion could be reduced, or on the other hand the contribution rate could be raised. The faster he/she receives — and perceives — the users' attitudes, the quicker he/she will be able to react by altering the management of the group activities towards best results. Without enough interest, it is likely that a certain item of the group activity does not develop as expected. The users can also infer this conclusion from the analysis of the established awareness profiles. Frequently, the coordinator needs to know, for example, why a certain task has not been concluded? Through profile analysis the coordinator can also conclude that some awareness components in OM should be redesigned.

The output filters established by the users are related to the degree of desired information privacy and their interest, which is related to the information available in the workspace area. The participants, for some reason, may not have the same interest in presenting to the rest of the group all the information generated by them. Nor does all available information have the same importance and relevance to all the participants. For example, some participants can be interested in some comments on certain ideas while others are not. Additionally, many times users also refuse to supply information if they don't perceive the benefits that derive from it. If this participant does not filter this awareness information, others probably may use it as input filter.

Thus, we can assume that the privacy is not related to the desires and the participants' wills, but to the context in which they are inserted and to the course of the activities they participate. This context can be inferred through an analysis of the filters by either the system (automatic) or the coordinator. Boyle et al. [2] examine the use of filters in video scenes and present the results obtained in different situations. They also discuss the impact of filters in the workspace perception and privacy.

That is, a user who does not filter certain awareness information ought to be sure of the usefulness of this information in a certain activity. Without an explicit attitude from the participants of the group about the usefulness of awareness information, as for example, through an event produced by the system [4], a participant will only have access to this information through the analysis of the filters established by others users. Consequently, they will increase the KK and KDK rates of the group. Otherwise, the possibility of increasing DKK and DKDK rates would still exist. Of course, this is what we don't want in a cooperative activity.

Filters established either by the system or by the users are not unique. As long as the users familiarize with the information in groupware application, the previous established filters may become obsolete. They must allow adjustments capable of reflecting the real changes of the group. They are "filter objects"; artifacts that can be inserted or removed by the application or by the users who can also modify or even reuse them.

4. Group Memory

In its initial state, DB of awareness components should be capable to supply support for relationships between coordination actions, ***Coord(*Δt*)***, and user awareness profiles, $P_i(\Delta t)$, established by the participants at time interval Δt. Gutwin and Greenberg [5] consider that coordination activities can be accomplished in two ways in a shared workspace: (i) through explicit communication at work; (ii) through perception in workspace ("workspace awareness"). As already mentioned, we are particularly interested in activities that happen in shared workspace.

For the coordinator it is important to have past event and activity reports from each participant. Based on the report of historical events, next activities can be planned and necessary resources allocated for them. If the coordinator is aware of the activities accomplished by each participant in the past, it is also possible to focus on both the next activities and the skilled participants to carry them out.

Through filter analysis established in the past, the coordinator can infer that if a user filtered certain awareness information, he probably was not aware of it. The coordinator could then: try to avoid unnecessary conflicts; help in the division of work and roles; plan new subsequent activities; motivate the participants or take actions to effective task accomplishment. On the other hand, if the participants did not filter the awareness information, the coordinator can infer that they had access to that information. Thus, awareness information used in the past, or in the present can be used to take specific coordination actions, while awareness information in the future can only be used to foresee actions.

Moreover, coordination actions are related to eventual support to the participants for the accomplishment of their activities. The faster the coordinator identifies this need, the better results can be obtained.

Through appropriate awareness information the coordinator could be able to foresee user actions, expectations, and in a recurrent way the use and construction of new awareness components. Each participant could then anticipate his/her needs, resources for the next activities and prevent misunderstandings.

One of the objectives of this model, that we will still have to evaluate, is exactly this: predict the use of awareness components, and thus make them available in group memory for future use, minimizing eventual conflicts, not necessary for the accomplishment of activities.

Then we can consider that group memory based on awareness components, which are used by each participant $\underline{i}$ at a certain interval $\underline{\Delta t}$, could be composed by the following artifacts: (i) a set of awareness components, $Aw_i(\Delta t)$; (ii) a set of each participant's profiles, $P_i(\Delta t)$; and (iii) a set of past activities accomplished by each participant, $Ha(\Delta t)$ (see Figure 2). In other words,

$$GM: \{ Aw(\Delta t), P_i(\Delta t), Ha(\Delta t) \}.$$

Workspace awareness is related to both user and coordination perception in group activities that happen in a time interval Δt. We are not interested in explicit synchronous communication during work, but on those that occur through group memory.

Considering that this approach is addresses coordination issues of CSCW, we aim at relating them with the awareness framework proposed by Gutwin and Greenberg [5]. According to the authors, coordination actions must assure that cooperative activities will happen in the right order and time, and satisfy the imposed restrictions by the system.

Anticipation and assistance concepts proposed by the same authors are considered here through the awareness profiles established by group participants'. Through these profiles, coordinator could anticipate both the knowledge about activities accomplishment and past technology usage (*anticipation*).

For the accomplishment of some activities mandatory awareness components are established by the coordinator, participants establish others as the activities occur. The analysis of these profiles can identify that some components, that are not being used, could increase the perception and help the accomplishment of next activities (*assistance*).

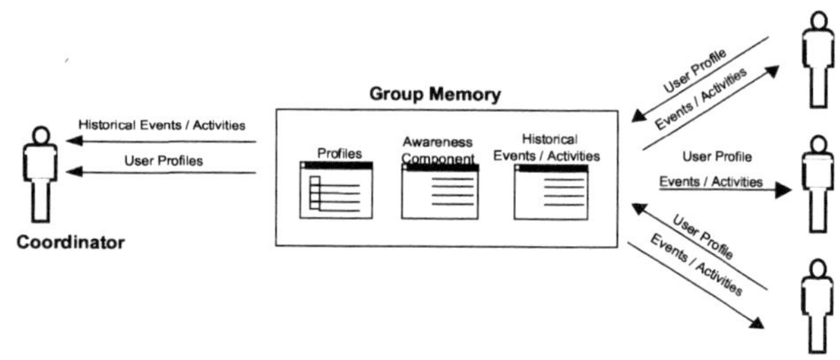

Figure 2. Group memory components.

Thus, DB of awareness components supporting coordination actions maps:

$$GM: \{ P_i(Aw_i(Act(\Delta t)), \ldots , P_I(Aw_i(Act(\Delta t)) \} \text{->} Coord(t);$$

where, $Aw_i(Act(\Delta t))$ is the set of awareness components necessary for the accomplishment of activities $Act(\Delta t)$.

5. An Example on the Use of Filters

Aiming at further analysis for the issues discussed in the previous sections, a prototype is being conceived based on a typical asynchronous discussion activity. During this activity it can be necessary to determine the time a certain item, or the associated document, will be kept in the group memory. After this period has expired, a decision should be made whether the discussion items should or not be stored in the Organizational Memory.

A strategy frequently used during the discussion is the attribution of roles and responsibilities to the participants. Depending on these roles, the system can establish a group of awareness mechanisms for specific discussion items and for documents associated by them. Examples of elements that can be filtered, or made available by the system are: historical reports of occurred events; time of resource usage; event tracking; list of participants; list of document readers; and people who contributed to the discussion.

Awareness components can become available to participants as they are built. Each participant can choose which filters will be used in the workspace. For example, we can mention filters set to prevent annotations or references to external documents to be published for the whole group.

Eventually, as a result of some updating performed in a discussion item, events can be triggered by notifications, for example. As they are broadcasted, the participants, who may not be interested in them, can filter these events [4]. Furthermore, they can filter reports generated during their discussions.

It is also important to implement priority policies when establishing awareness filters. For example, the filters established by the coordinator will have priority over those established by the users. The information filtered by the coordinator will not be available to users, independently of the filters established by them.

The user can select the awareness components in such way that these selections suggest possible coordination actions. Thus, these same actions can result in new awareness component usage suggestion, or in building other newer components.

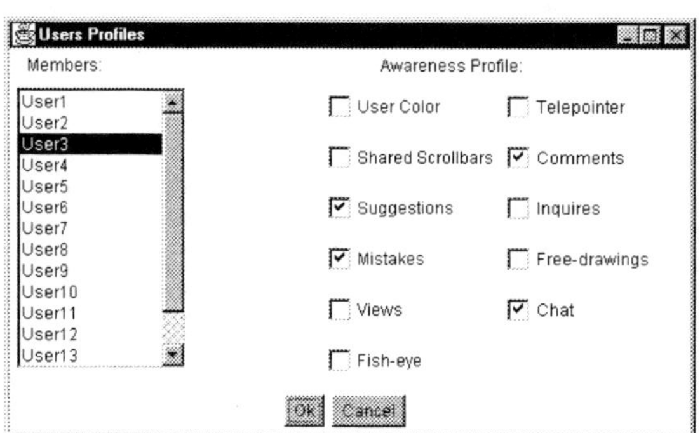

Figure 3. Awareness profile of 'User3' viewed by the coordinator

Initially, in this prototype the coordinator could establish some mandatory awareness components that cannot be filtered by the users. Other mechanisms are established as the participants begin their activities, or when their interests change, generating new awareness profiles. Eventually, the coordinator will access the profiles (Figure 3) and decide on the appropriate actions. For example, the User5 in Figure 3 is not aware of what other participants are doing in real time because the user filtered the Telepointer and the Shared Scrollbar awareness information.

Besides reducing the workspace overload, awareness profile could indicate to coordinators the level of awareness the participants have. We are also investigating a suitable way for the coordinator to analyze the generated profiles.

6. Conclusions

This paper showed a strategy to both guide the selection of information from the Group Memory, and support coordination actions in groupware applications. The approach was based on filter usage and awareness profiles established by the participants in CSCW activities.

We aimed at avoiding that participants interact with a set of available awareness information in a submissive way, i.e., dependent on the mechanisms established by the system designers. Group participants do not cooperate and interact according to the initial system design.

Alternatively, they could choose to participate in the selection process actively, building and rebuilding ideas, interacting with what was previously proposed. Thus, the participants and the coordinator work in a cooperative way to develop the use of the OM proposed by the system.

This process certainly depends on the previous knowledge accumulated by the participants of the group. It also depends on the beliefs and values that will contribute to define their attitudes toward the judgements of the intentions (presumed) of those who recommended or presented the information to the members of the group for selection.

The functionality of the profiles both for the coordinator and the participants still needs extensive testing. We are also trying out ways to test the overall usability of the system.

7. Acknowledgement

This work has been partially supported by grants from Faculdade Ruy Barbosa (Brazil – Salvador / BA).

8. References

[1] M. R. S. Borges and J. A. Pino, "Awareness Mechanisms for Coordination in Asynchronous CSCW", *Proceedings of 9th Workshop on Information Technologies and Systems (WITS 99)*, Charlotte, NC, USA, Dec. 1999, pp. 69-74.

[2] M. Boyle, C. Edwards and S. Greenberg, "The Effects of Filtered Video on Awareness and Privacy", *Proceedings of ACM Conference on Computer Supported Cooperative Work (CSCW'00),* Philadelphia, USA, Dec. 2000, pp. 241-250.

[3] H. J. L. Domingos, N. M. Preguiça and J. L. Martins, "Coordination and Awareness Support for Adaptive CSCW Sessions", *Proceedings of Forth International Workshop on Groupware (CRIWG'98)*, Armação de Búzios, RJ, Brazil, Sept. 1998, pp.21-37.

[4] L. Fuchs, "AREA: A Cross-Application Notification Service for Groupware", *Proceedings of the Sixth European Conference on Computer-Supported Cooperative Work (ECSCW'99)*, Kluwer Academic Publishers, Copenhagen, Denmark, Sept. 1999, pp. 61-80.

[5] C. Gutwin and S. Greenberg, "A Framework of Awareness for Small Groups in Shared-Workspace Groupware", *Technical Report 99-1*, Department of Computer Science, University of Saskatchewan, Canada, 1999.

[6] S. E. Hudson and I. Smith, "Techniques for Addressing Fundamental Privacy and Disruption Tradeoffs in Awareness Support Systems", *Proceedings of ACM Conference on Computer Supported Cooperative Work (CSCW'96)*, Boston, MA, USA, Nov. 1996, pp. 248-257.

[7] J. A. Mariani, "SISCO: Providing a Cooperation Filter for a Shared Information Space", *Proceedings of the International ACM SIGGROUP Conference on Supporting Group Work (Group'97)*, Phoenix, Arizona, USA, Nov. 1997, pp. 376-384.

[8] M. Sohlenkamp, W. Prinz and L. Fuchs, "PoliAwac: Design and Evaluation of an Awareness Enhanced Groupware Client", *AI & Society Journal*, vol. 14, 2000, pp. 31-47.

[9] M. Tucker and C. Frappaolo, "Search Engines in the Age of Knowledge", *Intelligent Enterprise*, vol. 2(18), Dec. 1999.

Consistency Control in Synchronous Cooperative Work

Lei Yu, Zongkai Lin, Yuchai Guo, Shouxun Lin
Cooperative Design Group, CAD Lab., Institute of Computing Technology
Chinese Academy of Sciences, Beijing, 100080, P.R. China
happyu@263.net

Abstract

*Consistency control is an important research field in CSCW. In this paper, we present an improved algorithm of concurrency control based on operation transformation. Our fastAdOPT algorithm improves other algorithms in two main aspects: first, we improve the transforming efficiency through saving original form of operations in request log and through the special process of "**intimate preceding**" operations. Secondly, it can realize natural interactive mode through not using Undo/Redo operation.*

1. Introduction

Consistency control is an important research field in CSCW, on which people have done lots of studies and got much achievement, such as locking, serialization, operational transformation, and so on. Currently, most research works are concentrated on the synchronous cooperative work, and none of them alone could satisfy the requirements of CSCW very well.

The locking method can solve concurrency control in a simple way. However, there may be three main problems with locking: the overhead in requesting and obtaining lock; the question of granularity, people must make the tradeoff between concurrency level and overhead; it is difficult to determine when lock should be requested and released.

The discussion of other approaches' problems can be found in [1], we mainly focus on the transformation-Oriented approaches.

The transformation-Oriented concurrency control approaches are rather popular in real-time cscw systems with fully replicated architecture, especially in text-based group editors. When used in multi-media environment, the transformation-Oriented method will be infeasible because the transformation of operations is difficult to realize. In this paper, we mainly focus on the consistency control in text-based group editors.

2. Related Work

The main mature examples of transformation-Oriented approaches include: dOPT algorithm of Ellis; forward/backward transposition of M. Suleiman; adopted dOPT algorithm of M. Ressal; and reversible inclusion and exclusion algorithm of C. Sun [1][2][5][6].

Ellis proposed the dOPT algorithm in 1989, in which he firstly presented the transformation-Oriented approaches. The dOPT algorithm has the properties like good responsiveness and fully distributed. The main principle of dOPT algorithm is: local operations are performed immediately on their originating site, and remote operations are transformed with the performed concurrent operations to include the effect of them before they are performed on other sites.

Unfortunately, there is a "Partial concurrency problem" with the dOPT algorithm. As shown in figure 1, operation O2 are executed at site 2 before operation O3 is generated, we call O2 is **causal preceding** O3 (or O3 is **dependent** on O2). The operation O1 generated on site1 and the operation O2 generated on site 2 are **concurrent** (**independent**) because O1 not causal preceding O2 and O2 not causal preceding O1. O1 and O3 are **partial concurrent** because they are generated in different state. When O3 reached site1, it must be transformed according to O1 to include the effect of O1. In dOPT algorithm, however, O3 can't be transformed correctly because they are generated in different state. So, the transformed result will be wrong.

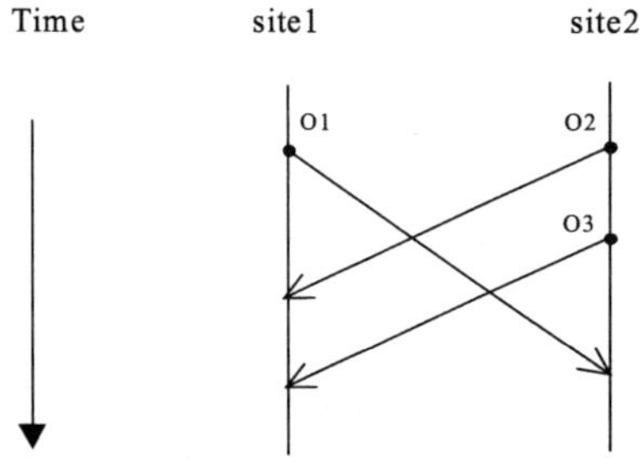

Figure 1. Partial concurrency problem.

All other algorithms have the similar principle in preserving the precedence property (all operations which are causal preceding other operation must be executed before it on all sites) using state vectors. They have been proposed to solve the "Partial concurrency problem" in different ways. [2]'s AdOPT algorithm used an N-dimensional interaction graph containing all operations of the original, intermediate, and executed forms. It is inefficient to maintain and search a dynamically growing and potentially large N-dimensional graph at run time, especially when N become big.

C. Sun uses exclusion transformation to solve the partial concurrency problem. As dOPT algorithm, the inclusion transformation is used to include the effect of executed concurrent operations, while the exclusion is used to exclude the effects of executed preceding operations to achieve the same state with partial concurrent operation. In this paper, we'll show that the exclusion transformation can be simplified with other approach.

To achieving convergence property (which ensures the consistency of the final results at the end of a cooperative editing session), C. Sun adopted Undo/Redo to ensure the total order of performed operations, which may induce the unnatural leap or twinkle. Others allowed the different performing order of operations on different sites, however, some special process should be done to achieve the consistency of the final results. For example, the priorities of sites are used to decide the correct order when concurrent operations insert or delete character in the same position.

When a site is processing remote operations from other sites, it needs finishing the transformation work before it can perform the operation. During this process, the site can't perform local operations for the following reasons: (1) all the operation requests are placed into the queue waiting for processing, the remote operation being processed by the site is "causal preceding" to the local ones generated after performing it. So it must be performed before those local requests; (2) local requests should "include" the effects of all it's "causal preceding" operations. So local member should first see the preceding operation's result, then generate the new request. Therefore, how to improve the transformation efficiency as much as possible is the key factor of algorithm's feasibility.

We present an fast transforming algorithm, which solves above problems well both in transforming efficiency and natural interactive mode.

3. The Improved Algorithm: fastdOPT

3.1 State Vectors

The State Vector (a kind of timestamp) is useful to ensure the causal relationship and the total order among all operations, which provide the basis to maintain the causality preservation and convergence property.

Definition: Let N be the number of sites in the cooperative system, which are numbered from 0 to N-1; On every site S, a vector SV composed of N elements is defined where: SV[i] = number of operations generated by site i, received by the site S and executed by S on its copy of the share data. For every site S, initialize SV[i] = 0□i = 0,1,...,N-1); when site S executed an operation generated by site i, SV[i] = SV[i] + 1.

When an operation is generated on a site S, the current value of the state vector SV is associated with it. The operation request is broadcasted to other sites in a triplet: <op, S_{op}, V_{op} >, where op is the operation, S_{op} is the generator site and V_{op} the state vector associated with the operation.

Cite 1: Let O be an operation generated at site i and timestamped by SV_o. O is *causally-ready* for execution at site j ($j \neq i$) with a state vector SV_j only if: $SV_o[i] \leq SV_j[i]$, $i \in \{0,1,...,N-1\}$.

Cite 2: For all pair of operations O_a and O_b, if O_a is $\rightarrow$ (causal preceding) O_b, then O_a is executed before O_b at all sites. Then the causality-preservation of system is ensured.

There are also some other basic cites or definitions, we'll use them directly and won't explain them for length limitation.

3.2 Data Structure of Algorithm

Most data structures are same with that of the dOPT algorithm [1]:

State Vector: S;

Operation request: r = <i, s, o, p>, where i is the generator site, s is the state vector of site i, o is the operation and p the priority of site i.

Request Queue of site i: Q_i, which contains the requests waiting to be executed by site i. The element of Q_i is request r, and will be deleted form Q_i when r is being executed at site i.

Request Log of site i: L_i, which contains all executed operations at site i. The element of L_i is Er = <i, s, o, o', p>, where o is the original operation, o' is the execution form of o. o' may be different with o after necessary transformations. Saving original form of operations in L_i can provide great optimization of algorithm, which is also the key difference from other algorithms.

3.3 Basic Ideas of Improved Algorithm

We improved dOPT and other algorithms in two main aspects, which are to be discussed in detail in the following sections.

3.3.1 Resolve "Partial Concurrency" Problem

When a remote operation O_p is "causal ready" at local site, it can be processed at local site. First, O_p will be transformed according to executed concurrent operations, and then be executed in the transformed form. The process is as followed:

Scan the request log (operation history) of local site in left-to-right order to find all the executed operations which are concurrent with O_p. Let them be $O'_1, O'_2 \dots O'_t$, then transform O_p with them in turn.

For each O'_i, scan after it in request log to find all operations which are causally preceding to O_p. If there are none of these operations, then O_p and O'_i are not of partial concurrent relationship. So we can transform O_p with O'_i directly, just like what be done in dOPT algorithm, to include the effect of O'_i into O_p.

If there are these operations, let them be $O_{p1}, O_{p2} \dots O_{ps}$, then O_p and O'_i are of partial concurrent relationship because O_p have included the effects of $O_{p1} \dots O_{ps}$, while O'_i not. O_p and O'_i are generated in different state. So we can't transform O_p with O'_i directly.

To solve this problem, two choices can be made: first, include the effects of $O_{p1} \dots O_{ps}$ into O'_i, like M. Suleiman's "Backward" transformation[6]; secondly, exclude the effects of $O_{p1} \dots O_{ps}$ form O_p, like C.Sun's "Exclusion" transformation. For "Backward" transformation need to reorder the request log which will induce additional overheads, so we adopt the idea of exclusion transformation.

Differed from C.Sun, however, we can exclude the effects of preceding operations from O_p in rather easy and high efficient way, because we save operation's original form in request log. We needn't do complex work like C.Sun, who first do exclusion transformation on O_p's all preceding operations with O_p's all concurrent operations, then do exclusion transformation on O_p with the above transformed results. When the preceding operations are from the same site with O_p, all his above work can be proved equivalent with the original form of O_p's preceding operations.

The main idea is that first do exclusion transformation on O_p with O_p's preceding operations $O_{p1}, O_{p2} \dots O_{ps}$ (original form), then do inclusion transformation on O_p with O'_1 (execution form) and those after O'_1 in request log (including $O'_{p1}, O'_{p2} \dots O'_{ps}$).

In this way, the transformation times are decreased and algorithm efficiency is increased considerably.

3.3.2 Use the transformation result of preceding operation to realize the current transformation

The operating speed of a member is much lower than machine's proceeding speed, so in a short period of time, the operations generated by a member are concentrative, limiting in a small position space. Consider the following two examples:

(1) Member A (at site i) inserts a series of characters continuously in a text line, producing a series of continuous operations. Assume that these operations are OP_0: Insert(p, X^0), OP_1: Insert(p + 1,) ... OP_t: Insert(p + t, X^t), where p is the insertion position, and X^i is the inserted char. However, when these continuous operations are broadcasted to site j, they may be separated by other concurrent operations. The fact may be an overlapped array.

When OP_t is causal ready on site j, before executing it, we need do transformation on it with all executed concurrent operations. Assume these operations are $O'_1 \dots O'_{k-1}, O'_k \dots O'_n$, and the execution order of OP'_{t-1} is between O'_{k-1} and O'_k.

If we can utilize the transformation result of OP_{t-1} (which already includes the effect of $O'_1 \dots O'_{k-1}$) and the special relationship between OP_{t-1} and OP_t, to educe directly the transformation result of OP_t against $O'_1 \dots O'_{k-1}$, we'll skip at least k-1 times of transformation. Then, all we need to do is to transform OP'_t, the above intermediate result of OP_t, with $O'_k \dots O'_n$ to get the final execution form.

In fact, if we ignore the effect of $O'_k \dots O'_n$, then OP_t

only need to transform with $O'_1 ... O'_{k-1}$, and the final result must be insert char in the adjacent position compared with OP'_{t-1}. For example, if OP'_{t-1} = Insert (s, X_1), then OP'_t = Insert (s + 1, X_2).

Because we save both the original and the execution forms of operations, we can judge whether the relationship of OP_t and OP_{t-1} belong to the above situation.

(2) Similar thing happens when a member deletes chars continuously. The operations generated are Delete (p, X^0)...Delete (p, X^t), we can adopt the similar process with (1).

(3) In most editors, "Modify" command is provided. In current concurrency control algorithms, however, the "Modify" command is considered as the combination of "Delete" and "Insert" commands. In this case, the separated two commands are also of the same position.

For the above three cases, we call OP_t and OP_{t-1} are of "***intimate preceding***" relationship and the final execution forms of them must keep continuous (Insert) or same (Delete/Modify) in position space.

3.4 The fastdOPT Algorithm

Most parts of our improved fast algorithm are same or similar with those of dOPT, including Initialization, Generate local Operations, Receive remote Operations, and Concrete Transformation Functions. The realization of Inclusion and Exclusion Transformation Functions can be found in **[5]**. For the limitation of length, we don't list all of them. We'll focus on the difference part: Execute Operations.

The following is the fastdOPT algorithm as executed by the site i.

Initialization:

$Q_i \leftarrow$ empty; $L_i \leftarrow$ empty; $S_i \leftarrow$ <0, 0, ..., 0>

Generate local Operations:

receive operation o from the user interface
calculate the priority, p, of site i
$Q_i \leftarrow Q_i$ + <i, S_i, o, p>
broadcast <i, S_i, o, p> to all other sites

Receive remote Operations

receive <j, S_j, o_j, p_j> from the network
$Q_i \leftarrow Q_i$ + <j, S_j, o_j, p_j>

Execute Operations:

for each <j, S_j, o_j, p_j> $\in Q_i$ where $S_j \le S_i$ begin
$Q_i \leftarrow Q_i$ - <j, S_j, o_j, p_j>
if $S_j < S_i$
o'_j = Integrate (j, S_j, o_j, p_j)
Perform operation o'_j on site i
$L_i \leftarrow L_i$ + <j, S_j, o_j, o'_j , p_j>
$S_i \leftarrow S_i$ with j'th component incremented by 1
end for

The function of Integrate (j, S_j, o_j, p_j) is to transform "causally ready" remote operation o_j to get the proper form o'_j of o_j. It's realization is presented below:

Function Integrate (j, S_j, o_j, p_j)

Begin

Assume there are m elements in L_i, i.e., m operations have been executed on site i.

(1) Scan L_i from left to right to find the first operation O'_{c1} which is concurrent with o_j. If no such an operation is found, then o'_j = o_j, GOTO (7). Otherwise, assume O'_{c1} = L_i [k], and all o_j's concurrent operations in L_i compose EOC, EOC = [O'_{c1}, $O'_{c2} ... O'_{ct}$];

(2) Search L_i [K+1, m] to find all operations which are causally preceding o_j and let EOL denote these operations. If EOL = [ ], then o_{tmp} = o_j, GoTo (6);

Otherwise, suppose EOL = [O_{p1}, $O_{p2} ... O_{ps}$], EOL' = [$O'_{p1}, O'_{p2} ... O'_{ps}$] where O_{ps} is the original form of O'_{ps}, and O'_{ps} is the execution form of O_{ps};

(3)Consider whether O_{ps} and o_j belong to "***intimate preceding***" relationship. If so, get o'_j according to section 3.3.2, then continue (4). Otherwise, GOTO (5);

(4) If O_{p1}, $O_{p2} ... O_{ps}$ were not originated from the same site with o_j, then get o'_j through the same proceeding as C.Sun has done. Else, Let EOC' = [$O'_{cu} ... O'_{ct}$], where O'_{cu} is the first one in EOC which is executed after O'_{ps}. If $EOC' \ne$ [], then o'_j = LIT (o'_j , EOC'). GOTO (7);

(5) o_{tmp} = LET (o_j, EOL^{-1});

(6) o'_j = LIT (o_{tmp} , L_i [K, m]);

(7) Return o'_j .

end

Explanations:

- $EOL^{-1} = [O_{ps}, O_{ps-1} \ldots O_{p1}]$;
- LIT (O, OL) is defined to include the effects of a list of operations OL into operation O, and LET (O, OL) is defined to exclude the effects of a list of operations OL from operation O. The realization of functions LIT and LET can be seen in [5].
- In step (6), we do inclusion transformation on o_{tmp} with all operations (the execution form) in L_i [K, m] from left to right, which also include all o_j's preceding operations.

The example of our fastAdOPT algorithm is omitted for limitation of paper length.

3.5 Performance Analysis

For the operation sequence is of various changes, we just give the performance comparison of fastAdOPT and C.Sun [5] based on process of one special remote operation. When a remote operation O_p is causal ready at local site, it must be transformed before it can be performed, which is the most time-consuming step in both algorithms.

Assume O_p's first concurrent operation in L is L[k], and the sub list(from k till the end of L) is named L_S. Assuming the length of L_S is m, and the number of concurrent operations is t, so the number of proceeding operations is (m-t). To transform O_p into O'_p, some times of ET (exclude transformation) and IT (include transformation) need to performed, which cab be used to analyze the complexity of both algorithm. The comparison result is shown in the following table. We have shown the results in two opposite situations: the best and the worst situations. The best case is shown in Figure 2, and the worst case is shown in Figure 3.

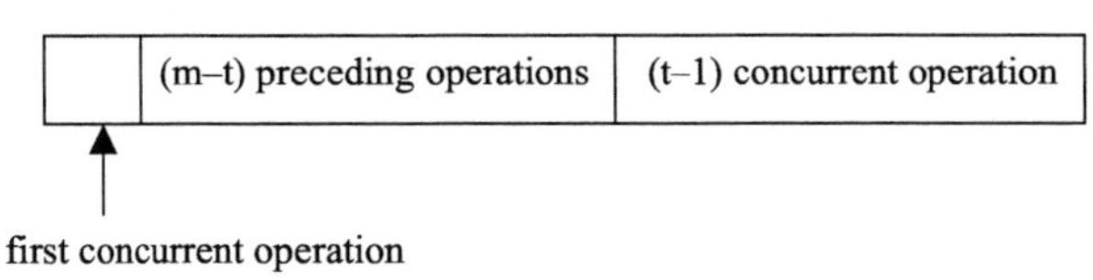

Figure 2. The best case.

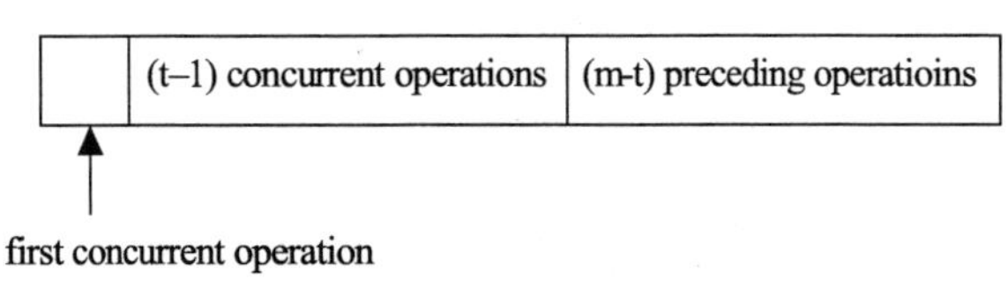

Figure 3. The worst case.

When all concurrent operations are originated from the same site with o_j, the fastAdOPT algorithm always do the same times of ET and IT transformation. In this case, the "best and worst situations" is of no sense.

Table 1 has shown the performance analysis of the fastAdOPT algorithm and the algorithm of C.Sun.

Table 1. Performance Comparison.

	[Sun97]	FastAdopt	
		Intimate preceding	Same Site
BEST	$\frac{(m-t)\times(m-t-1)}{2}+(m-t)$ ET $\frac{(m-t)\times(m-t-1)}{2}+m$ IT	1 Add	(m-t) ET m IT
WORST	$\frac{(m-t)\times(m+t-1)}{2}+(m-t)$ ET $\frac{(m-t)\times(m-t-1)}{2}+m$ IT	1 Add (t-1) IT	

4. Conclusion

Analyzed the various conflict probabilities in CSCW, and based on the study of other people's research fruits and our own previous work, we present an improved algorithm of concurrency control based on operation transformation in this paper, which solves other algorithms' problem in transforming efficiency, and natural interactive mode.

5. References

[1] C. A. Ellis and S. J. Gibbs: "Concurrency Control in groupware systems", In Proc. Of ACM SIGMOD Conference on Management of Data, pp. 399-407, 1989

[2] M. Ressel, D. Nitsche-Ruhland, and R. Gunenbauser: "An integrating, transformation-oriented approach to concurrency control and undo in group editors", In Proc. Of ACM Conference on Computer Supported Cooperative Work, pp.288-297, 1996

[3] Muson J, Dewan P: "A concurrency Control Framework For Collaborative Systems", In Proc. Of ACM Conference on Computer Supported Cooperative Work, pp.278-287, 1996

[4] Robert S, Guruduth B, Kevan M, etl: "Concurrency Control and View Notification Algorithms for Collaborative Replicated Objects", In Proc. Of the 17th IEEE International Conference on Distributed Computing Systems, pp. 194-203, 1997

[5] C.Sun, X. Jia, Y.Yang, and Y.Zhang: "Generic Operation Transformation Scheme for Consistency Maintenance in Real-time Cooperative Editing Systems", In Proc. Of ACM SIGGROUP Conference on Supporting Group Work, pp.425-434, 1997

[6] Suleiman M, Cart M, Ferrie J: "Serialization of Concurrent Operations in A Distributed Collaborative Environment", In Proc. Of ACM SIGGROUP Conference on Supporting Group Work, pp.435-445, 1997

Research on Cooperative Template Design

Xiao-ping Liu
Institute of Plasma Physics
Chinese Academy of Sciences
Hefei, 230031, P.R. China
lxp@mail.ipp.ac.cn

Xiao-yun Liu, Zheng-feng Hou, Xiao-rong Du
Faculty of Computer Science
Hefei University of Technology
Hefei, 230009, P.R. China

Abstract

The design method based on template used way of decomposing, combination, replacement and merger of template for agility and variation of model in entirety and local design. About template authors had a deep research in the fields of mechanical CAD and engineering CAD systems. Based on these works a concept of cooperative template is presented in order to control and manage design task on the net. It can inherit components which have been designed and give a nimble response for new project if template technology is introduced into the public area of cooperative design. This paper discusses work process of template, its elements based on cooperative mechanism and combines lock and semaphore function and expands TDL (Template Design Language) using cooperating primitive. Users are allowed to compete about design elements in the framework of visualization. Some technical problems about XML and JSP+Java Servlet are also discussed.

1. Introduction

Engineering design is a problem solved process under constraints. Some advanced developing strategy and produce model,just like FMS(Flexible Manufacturing System), Virtual Manufacturing, Agile Manufacturing, Lean Production, Reorganization, Concurrent Engineering and so on, it can be found from their develop and change process [1]□no one is out of support of CAD□CAPP□CAM□MIS techniques. Several commercial CAD package used variation modeling based on features, and made design convenient in certain degree. The localization is appear also. The first it not support individuation modification of agile. Something lack bout components management and working mechanism. The second one is the range of constraints. The most parts of research is about geometry constraints. Less about engineering constraints and fuzzy constraints. The third one is the lack of cooperative mechanism on the net. With direction of market internationalization and make study about model of cooperative and algorithms become a hot point in the world, it also the base point of concurrent engineering. [1].

"Template" is not new concept in the fields of information. In china <Guangyun> say :" Template is rules shapes and standards". This concept has been used in the OA software and improved efficiency of management and design. Features and constraints is something about standard, rule and shape in some meaning. Because there are many model and method in engineering design. The task of us is to discovery and abstract templates and give them grammar of TDL(Template Design Language).

Srivastava in Berkeley is the first one introduce template into the field of ICCAD(1991□, but it was about hardware template design. Sallade, J.S.; Philpott, M.L. (USA Illinois Univ.) [2] presented Image templates for CAD directed inspection and control in the machine vision domain. Chang, Hyunman; Ong, Soohwan; Sunwoo, Myung H. From Ajou Univ, Suwon, South Korea describes[3] template matching in architectures and design of a general purpose parallel image processor chip called a SliM-II Image Processor. In paper[4] Seo, Seungwan; Sunwoo, Myung H, Ajou Univ. presents an enhanced template matching algorithm and its chip implementation. Princeton Univ. Zhong, Peixin; Martonosi, Margaret; Ashar, Pranav; Malik, Sharad [5] describes and evaluates a formula-specific method for implementing Boolean satisfiability solver circuits in configurable hardware. That is, using a template generator, they create circuits specific to the problem instance to be solved.

USA Defense put forward HLA (High Level Architecture□in 1996. There is a important concept in HLA is Object Model Templates, OMT in short form. The aim of this is to make process of modeling standard. [7]. Xiaojun Shen. Hage R. Georganas N. From Multimedia Commun. Res. Lab., Ottawa Univ., Ont., Canada introduces a collaborative virtual environment (CVE) system over the Internet. A Virtual Reality Modeling Language (VRML) 97-based user interface is developed to permit users to navigate around the virtual world. [6] Buckle DF. From UK [8] introduced SIMTEC as both a large model with a wide range of applications and an initiative in software reuse. It has the modelling functionality to simulate in C3I (Command, Control, Communication and Information) hierarchies. Page EH. Griffin SP. Rother SL[9].from The MITRE Corp., McLean, VA, USA and Lutz R. Scrudder R. Graffagnini J[10].from Johns Hopkins Univ. Discussed tools and system frameworks for HLA also.

China researchers have a lot of work about template[11] also. With global competition is forcing a reduction in design cycle time. Thus, the cost and time associated with building and testing prototypes often cannot be tolerated. Reforming methodology of tradition is need. Papers [12] discusses some problems about CSCW. But no any paper is about ccoperative template mechanism and design. Authors think it is very important for the future CAD system.

2. Definition Of Template

The basic idea of template is for the similarity of everything in the world. The modes or patterns of framework should be abstracted from the similar objects, and something else which is correspond to it may be thought as an instance derived by the template. Template is a constraints graph of relationship among the objects.

[Definition 1]Template is a group of four elements, Template = <ID,O,S,R>, ID is the only code of template, O is the set of objects in template, S is the serves set of object operation, R is constrains rules set of object serves.

A template is a named set of slots, indentifying a particular collection of slots that has a physical or conceptual meaning described by the name. By defining new templates from existing ones, through the use of inheritance, a template hierarchy is implicitly defined.

[Definition 2] Cooperative Template not only includes all elements of general template about design and process but also contain a agent program which is develop in Java AWT package in order to produce a event and follow up design process of client user. Cooperative template is corresponding with the nodes in PG graph.

[Definition 3] Precedence Graph (PG): $PG =< N, E >$, $N \in (p_1, p_2, \cdots p_m), E \in (e, e_2 \cdots e_k)$, p_i is node of PG, it describes a task of design or a sub-task if the task can be divided into small one. e_j describes a edge which links two nodes in PG. The direction indicated sequence relation of design task. The relation can be described in Figure 1.

In Figure 1. describes sequence of design task is:

- P1 must be completed before any one.
- P2, p3 and p4 can running at same time. It mcans these task can be down in different places on the net.
- Only p2 and p3 are OK, p5 can be deal with.
- Only p5 and p4 are OK, P6 can be running in the last.

Based on Precedence Graph, system can control task management and design schedule. This is a important difference point between template and cooperative template in paper here.

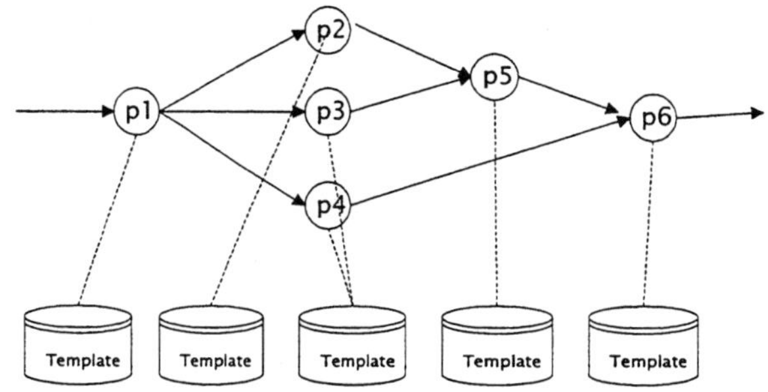

Figure 1. Precedence Graph and template library.

3. Mechanism Design

3.1. Control to the Workflow on the Net

XML, which is the most advanced language, can be used for the page layout making. XML overthrow tumbledown groundwork of HTML. According to standard, no built-in label for XML. So the developer of XML document can construct any significative label freely. Task to exchange data with other application program is simplified to make it easier to understand. In a another wise It can identify different language of mankind use its attribute of xml:lang. It give a great support for collaborative work on the net in global range. Here is a example wrote in dialect of XML:

<socialtalk xml:lang="EN">
<greeting>Hi</greeting>
<closing xml:lang="CH">□□</closing>
</socialtalk>

WWW server program not only use XML introduced above writing client file of page layout, but also need background support program in order to save date about cooperative template on the net for communication between different designer. Up to now there are many CGI languages which can be used for software developing. But JSP+Java Servlet+XML is the best combination tools to develop communication program between clients and servers and implement access in phase from clients to server and complete date processing in server. Because JSP technique is easy be moved to many application architectures for using some tools and skills known and extend distributing application supported for enterprises. As a part of using Java technique family and a part of composing of Java 2 ,which is version of enterprise, JSP technique is competent deeply for a complex application based on

Web. Otherwise , inner script language of JSP page layout is based on Java programming language and all JSP page layout are compiled into Java Servlet. So JSP page layout have all strongpoints of Java technique, including vigorous store management and security. As a part of Java platform, JSP have the characteristics of Java programming language of "edit once and run everywhere". With more and more commercial provider put JSP into their product, no any influence about current application you can use server and tools you selected and change them. Shown in Figure 2.

3.2. Implement knowledge base share of experts on the net

- The first, store all experts knowledge base to a server.
- Each expert can update their knowledge base at any moment and can use whole knowledge base in the server.
- Checking CLF (Common Log Format) file, session and cookies of server program and follow up the scent of user operation. Search useful message from share base and provide on-line help for user. Shown in Figure 2.

4. Conclusions

Scientific feasibility has been substantiated for developing cooperative design based on template. In this international environment, it will set the pace to discover feasibility of engineering design. The generic algorithm ideas here provides a flexible framework for designing and testing . But application of engineering algorithms in CAD system is not a simply thing. Multiformity was appeared in different application fields. It contains abstraction of engineering design process and some patterns of design. Authors have completed a series of template research in CAD system. The results make us sure that combination of template technique and CSCW will be challenging and significative.

5. Acknowledgement

This paper is supported by "100-People Project" of Chinese Academy of Sciences, National Nature Science Fund of China (No:19985004 and No:19785003). Natural Science Fund of Anhui Province China (No:98334724).

6. References

[1] Lin Zongkai, "Recent Developments In CAD", *Proceedings of 2nd International Conference On CAID&CD'99*, Nov,1999, Bangkok, P333.

[2] Sallade, J.S.; Philpott, M.L., "Synthetic template methodology for CAD directed robot vision", *International Journal of Machine Tools & Manufacture* v 37 n 12 Dec 1997. p 1733-1744.

[3] Chang, Hyunman; Ong, Soohwan; Sunwoo, Myung H.," Linear array parallel image processor: SliM-II", Proceedings of the International Conference on Application-Specific Systems,

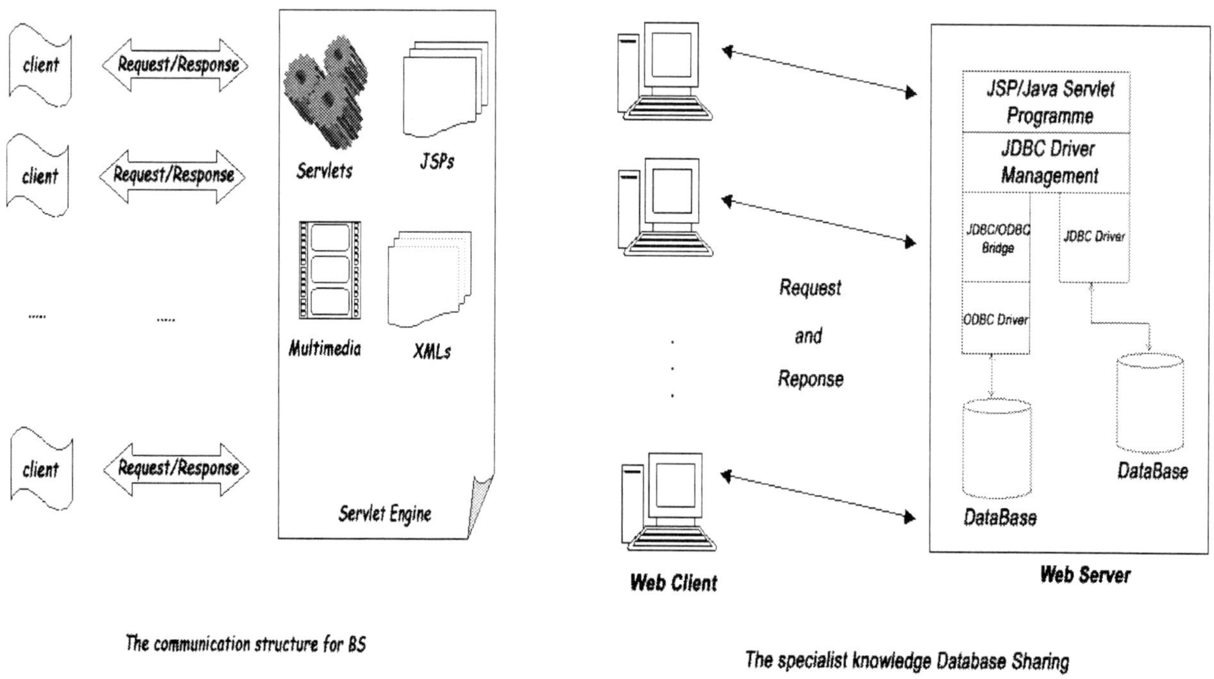

Figure 2. Mechanism of Cooperative Template.

Architectures and Processors 1997. IEEE, Piscataway, NJ, USA. p 34-41

[4] Seo, Seungwan; Sunwoo, Myung H., "Enhanced template matching algorithm and its chip implementation", *Design and Implementation IEEE Workshop on Signal Processing Systems, SiPS: Design and Implementation* 1998. IEEE, Piscataway, NJ, USA,98TH8374. p 162-171.

[5] Zhong, Peixin; Martonosi, Margaret; Ashar, Pranav; Malik, Sharad, "Using configurable computing to accelerate Boolean satisfiability", *IEEE Transactions on Computer-Aided Design of Integrated Circuits and Systems* v 18 n 6 1999. p 861-868.

[6] Xiaojun Shen. Hage R. Georganas N. "Agent-aided collaborative virtual environments over HLA/RTI". *Proceedings 3rd IEEE International Workshop on Distributed Interactive Simulation and Real-Time Applications*. IEEE Comput. Soc. 1999, pp.128-35. Los Alamitos, CA, USA.

[7] Dahmann JS. "The High Level Architecture and beyond: technology challenges". *Proceedings Thirteenth Workshop on Parallel and Distributed Simulation. PADS99.* (Cat. No.PR00155). IEEE Comput. Soc. 1999, pp.64-70. Los Alamitos, CA,USA.

[8] Buckle DF. "The SIMTEC simulation framework". *GEC Review*, vol.14, no.1, 1999, pp.28-36. Publisher: GEC, UK.

[9] Page EH. Griffin SP. Rother SL. "Providing conceptual framework support for distributed Web-based simulation within the High Level Architecture". *SPIE-Int. Soc. Opt. Eng. Proceedings of Spie - the International Society for Optical Engineering,* vol.3369, 1998, pp.287-92. USA.

[10] Lutz R. Scrudder R. Graffagnini J. "High Level Architecture object model development and supporting tools". *Simulation*, vol.71, no.6, Dec. 1998, pp.401-9. Publisher: Simulation Councils, USA.

[11] Fazhi He et al, "Cooperative Design System for Mechanical Drawings", *Proceedings of 6th International Conference On CAD&CG*, Dec,1999,Shanghai,P303-307.

[12] Jiyuan Liang et al, "Development of Collaborative CAD/CAM System for CIMS", *Proceedings of 6th International Conference On CAD&CG*, Dec,1999,Shanghai,P344-348.

X-Arc Spatial Data Integration in the SPeCS Collaborative Design Framework

Gustavo da R. B. Pinto
COPPE/UFRJ
PO BOX 68511
Rio de Janeiro, Brazil
gpinto@cos.ufrj.br

Sergio Palma J. Medeiros
COPPE/UFRJ
PO BOX 68511
Rio de Janeiro, Brazil
palma@cos.ufrj.br

Julia Celia M. Strauch
EMBRAPA/Solos
R. Jardim Botânico, 1024
Rio de Janeiro, Brazil
julia@cnps. embrapa.br

Jano Moreira de Souza
COPPE/UFRJ
IM/UFRJ
PO BOX 68511
Rio de Janeiro, Brazil
jano@cos.ufrj.br

Carlete R. F. Marques
COPPE/UFRJ
PO BOX 68511
Rio de Janeiro, Brazil
carlete@cos.ufrj.br

Abstract

X-Arc is a web-based architecture for the integration and sharing of spatial data sources in XML (eXtensible Markup Language) format that applies metadata and mediation concepts. It is an important component of SPeCS (Spatial Decision Collaborative Support System), an architecture which intends to support the aspects of multi-criteria spatial analysis within a distributed GIS with tools that can help a group cope with all activities in this kind of project. In X-Arc Framework, the availability and computing capabilities of data sources, as well as the heterogeneity of its data (structure, format, type and metadata) are explored by the data mediation and publishing services. Through SPeCS the users can cooperate, collaborate and coordinate their integration activities during their spatial analysis. This work presents a framework that supports the georeferencing of argumentations and allows transparent access to heterogeneous data sources involving the members of the design team in a spatial collaborative context.

1. INTRODUCTION

The need to share data from several sources has been present during the evolution of Geomatic technology. This results from the existence of a great volume of spatial data stored in several GIS (Geographical Information System). Sharing spatial data is necessary due to the growing complexity of the environmental analyses and due to the high cost of the data acquisition.

During this last decade, the data interchange between organizations through the Internet became not only possible, but also essential. The largest barrier found has been the heterogeneous data structures offered by the different data models to represent the graphic and non-graphic components of the spatial data [8]. GIS, in general, uses proprietary structures that are designed to maximize the manipulation performance of the spatial data, assisting the needs of the manipulation language [4]. Therefore, the integration of these data sources is a very complex task because of the diversity of the existing applications and the fact that spatial data present themselves with different semantics, formats and representations.

Mediation is one of the database integration techniques, which provides data access for distributed and heterogeneous data sources. The mediation technique uses one global component (mediator) and multiple local components (wrappers). The mediator accesses the information shared by the local components, which in turn, manage the access to the local data sources and the wrappers store the mapping to the local data [5, 6, 16, 17].

We developed X-ARC, a data integration architecture that uses metadata, heterogeneous data access and mediation concepts in its components. It provides services to locate, share and publish data sources through the web. The metadata managed by X-Arc aids the services' execution and can be used as a catalog of the available data sources.

However, considering that GIS users belong to specific application areas that represent different competencies, political agendas and social interests and often lack the knowledge about other areas with which they may have to work with, we propose the use of SPeCS [12, 13]. SPeCS supplies a collaborative work environment increasing synergy and user cooperation during the data interchange between the GIS data sources. In this way, we used the

SPeCS architecture, which is composed of three functional layers: decision tools, knowledge tools and integration layer. X-Arc belongs to the last layer.

In the literature we find some proposals for data integration such as TSIMMIS, DISCO, GARLIC and LE SELECT [9, 15, 3, 17] and others for collaboration and group decision as, for example, the AR/GIS [7], GEO-WASA [11], SDSS [10], among others [14]. We took advantage of SPeCS' support for collaboration in X-Arc's integration, publication and access tasks.

This work presents X-Arc Spatial Data Integration in the SPeCS Collaborative Design Framework. It is organized as follows: the **next section** presents a short introduction of the SPeCS system with its architecture, functional modules, interface and its relation with X-Arc; the **third section** presents the X-Arc proposal along with its components, services and functionalities, and in the **fourth section,** this work's final considerations are brought forth.

2. THE SPECS ARCHITECTURE

SPeCS architecture offers a cooperative, common, flexible and easy to use work environment where the members of a group may be geographically distributed in heterogeneous environments, interacting during a decision making process.

2.1 Interface

The prototype herein presented brings an interface whose main screen presents the components necessary for the work group (Figure 1). It shows the areas for the discussion and allows private or shared notes, which can be seen by the rest of the team. Besides these graphic interface areas, there is a menu with the operations of creation and alteration of the model entities. For example, a user may introduce a new climatic characteristic, not foreseen up to that moment, or subdivide an area for having important details that must be studied by the group.

The system supplies the user with the ability to make textual notes, with the use of his preferred word processor and to share this information with other members of the work group, optionally indicating the region or sub region of the cartographic basis, which, in turn, refers to the argumentation. Members of the group may present proposals at any time, and a historical list may aid in the argumentation.

As in all argumentation systems, the SPeCS prototype allows positive or negative positions, with the related explanations and also the thoughts conceived as the proposal is being discussed. The system still allows open proposals, by anonymous authors because this may be

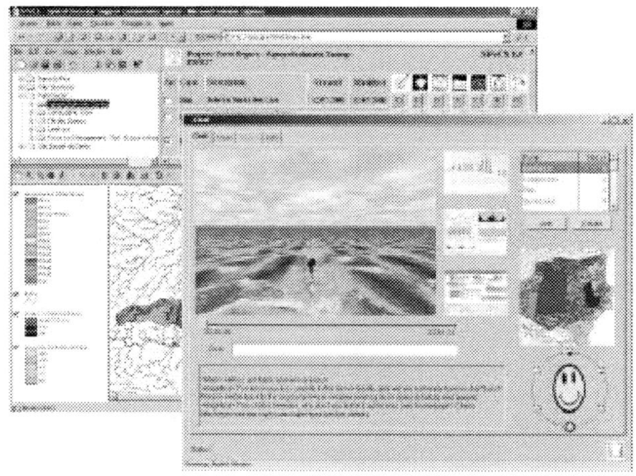

Figure 1. SPeCS interface prototype.

interesting for institutional political reasons. This feature may be turned off to attend the more rigid administrative structures, or in cases where this attribute is used unduly. The capacity for geographic expression must be made easy and the access to all kinds of heterogeneous data sources must be made feasible since this is the main characteristic of this environment.

2.2 Functional Modules

The main functional modules that compose the SPeCS solution are: Decision Tools, Knowledge Tools, Integration Layer and Knowledge Repositories. Figure 2 details SPeCS' main modules.

The Decision Tools aid the activities involved in a decision making process from the problem definition phase to the specification and documentation of all solutions created by the discussion. It also helps track the responsibilities assumed at the time the decisions were

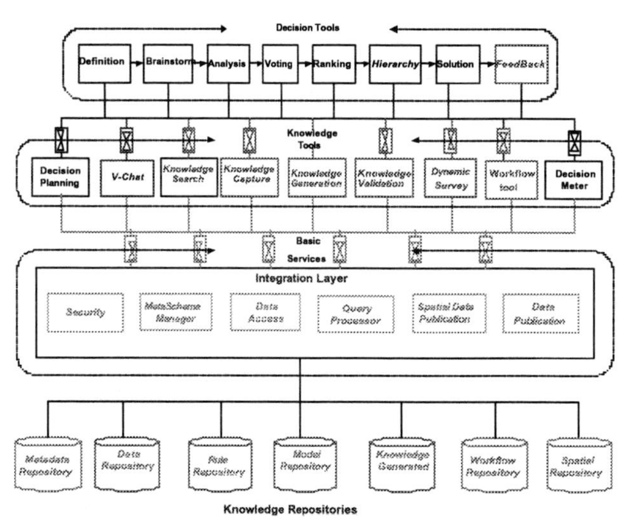

Figure 2. SpeCS functional modules.

made by the group members, which can also be all referenced to geographic elements. These tools offer an environment where members of a group can prioritize issues, perform different types of voting (yes/no, multiple choice and weighed choice), evaluate different criteria, elaborate documents in group, perform brainstorming, analyze projects, rank solutions, etc.

The Knowledge Tools includes mechanisms for decision planning and meter, visual chatting, knowledge management and survey control.

The Decision Planning tool allows the customization of the workflow decision process, disabling and introducing decision steps in accordance to the problem solving features.

The V-Chat (Visual Chat) tool encapsulates Chat, Forum and email facilities, as show in Figure 1. The basis of the conversation is a semantic framework, allowing the reuse of decision argumentation.

The knowledge management facilities include: a search tool for locating relevant information from the integrated databases or from the Internet; a knowledge capturing tool to input new or provide access to existing rules, models and data in participating data; a knowledge generation engine with an inference machine which provides the integration of rules, models and data for the generation of new knowledge based on validated ones. A knowledge validation tool helps the users, through automatic and manual mechanisms, retrieve, analyze and transfer the validated knowledge from the Knowledge Generated Repository to the appropriate Data, Rule and/or Model repositories.

A Dynamic Survey tool was implemented to help researchers cope with the constant changes in the GIS requirements enabling easy change of survey structure and attributes. This will also be used to capture the feedback from local actors to be compared with those from others regions where some kind of interference has been made in the past by group decisions, allowing a measure of decision efficiency. The Decision Meter tool can compare solutions proposed in the past and the actual results to improve the conceptual basis for future decision, so researchers can discover the success and failure factors to improve and make better decisions. A Workflow Engine is also in development to allow a smooth interaction with the GIS environment.

The Integration Layer, implemented by X-Arc, permits the integration and sharing of heterogeneous data sources using mediation techniques. This module provides basic data services to the upper level modules. X-Arc will be better discussed in the next section.

When dealing with the knowledge repositories, it is important that all group components share the same taxonomy and that they may enjoy the same standard of knowledge, thus encouraging consensus based on common premises. The absence of this type of perspective may lead to the failure of all effort put into the assembly of this argumentation structure, since the lack of common knowledge concerning the problem and the solution, makes it difficult for the group ideas and opinions to converge [1].

3. X-ARC

In the database area, there are several solutions for database integration, which are presented through different names: multi-databases systems (MDBS), mediators, federated databases and interoperable systems. These solutions goal to preserve the existing systems, maintaining their autonomy through a system with facilities for the access and sharing of the data stored by the local databases. The mediation architecture emerges as an interesting integration proposal and is formed by one global component (mediator) and multiple local components (wrappers) [16, 17, 6, 5, 2].

The eXtended Architecture (X-Arc) herein proposed intends to integrate the data sources using the mediation technology providing interoperability among the data repositories spread over the web, maintaining their autonomy. It encapsulates mediation systems in its components to provide integration, publishing and searching services to its users. Furthermore, the availability and computing capabilities of data sources, as well as the heterogeneity of its data (structure, format, type and metadata) are explored by the components and its services that provide a uniform access to the data.

This architecture is based on mediators and ontologies to provide the binding of heterogeneous data to their domain applications. To assist the identification of related data sources and their appropriate domain application, each data source is associated to one metadata, which provides a domain ontology.

X-Arc provides a uniform view of the available data repositories organized in a domain taxonomy. Domain ontologies are used to help the search for related environmental data through the representation of domain semantic concepts. Therefore, this approach promotes data integration and mechanisms to translate data requests across ontologies.

Figure 3 shows a basic configuration of the X-Arc architecture with its main components, Manager and Partner. For each main component and published data source, there is an associated metadata. In X-Arc, the Manager is an agent essentially responsible for the management of the all architecture, while the Partner is an agent responsible for the data publication.

The Manager component is responsible for the metadata maintenance of all data sources integrated by the architecture. It has a Metadata database that is constructed during the architecture startup and updated at each new Partner initialization. Due to a Query Processor Service, this component is able to rewrite and send all sub-queries to the appropriate Partners for query execution.

Nonetheless, the Manager does not have to be consulted for a Partner to have its data accessed, but the metadata provided by the Manager can help users on the localization of the shared data spread through the architecture.

The Partner component is mainly responsible for the data integration and publication services. A data repository has to be controlled by one Partner, which can publish and integrate more than one repository. Each Partner is a mediator and for each data source there is an associated wrapper and metadata, which describes the localization, platform details, data format, structure and representation, as well as, the access type available. The data sources can be descriptive or spatial and be stored in tables or files (containing structured or semi-structured data). Therefore, the Partners can be:

Descriptive – in this case, the mediation system built-in the data integration services supplies descriptive data stored in tables, texts, files, etc;

Spatial – in the same way, the mediation system built-in the data integration services supplies spatial data through maps and images.

The Metadata manipulated by the architecture consist of information collected by the Manager from the Partners and from each data source to be published by the architecture. The Metadata aids the user in the localization of the interesting data and helps the Manager during the query execution. Both Manager and Partner are responsible for metadata maintenance and data source integration (descriptive or spatial).

With X-Arc, environmental users have the autonomy and agility provided by the mediation-based integration to share theirs data with each other, as well as, to achieve cooperation and data sharing between the heterogeneous and distributed databases.

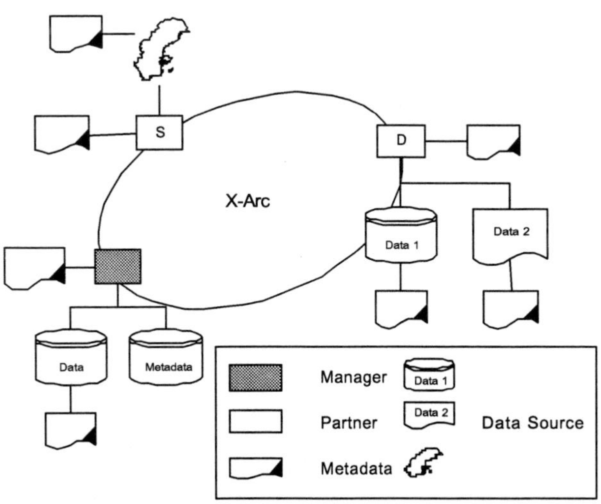

Figure 3. X-Arc main components.

3.1 X-Arc Services

The X-Arc services are responsible for the integration, publishing and searching facilities of the architecture. They provide the basic data services needed in the SPeCS architecture as showed in the figure 1. The main services are presented in the figure 1 and discussed bellow.

- Meta Schema Manager – responsible for the collection, management and publishment of metadata associated with each data source. It provides information to the Query Processor during the query rewriting and the sub-query distribution process;
- Data Publication – responsible for the descriptive data publication. All descriptive data retrieved by the Query Processor or by the Data Access are published in XML format. Through the Meta Schema Manager, the Data Publication can convert the content of the data sources from its format to XML;
- Spatial Data Publication – similar to the Data Publication service, however the data to be published in XML format is spatial;
- Data Access – responsible for the direct access to the published data in the partners and the passing of the data to the descriptive and spatial publication services. This service encapsulates descriptive and spatial mediation systems, which access the data sources through wrappers;
- Query Processor – responsible for the manipulation of each query sent to the architecture. The main operations executed by the Query Processor are: query optimization, rewriting and distribution; and query result aggregation. Thus, the sub-queries are sent only to the partners who publish the data requested in the queries; and
- Security – responsible for the data security, providing three data access levels: complete, read only and denied. Each data request to the Data Access or Query Processor is validated by the Security service before the execution.

4. FINAL CONSIDERATIONS

The requirements and specifications of a Spatial Decision Support Collaborative System in environmental projects aided to understand the complexity of spatial decision making and the need for coordination among the researchers, thus introducing new interaction perspectives within the work arena.

In this work we presented X-Arc, a SPeCS component which provides autonomy and agility in the data integration and publication phases. Through this mediation-based architecture users can share descriptive and spatial data through the web, using XML as a standard for information interchange. The metadata managed by this architecture helps the localization, access and the sharing of the spatial data. The SPeCS tools supply the collaborative work environment with several facilities for remote GIS users and through X-Arc services, a user can locate, access and retrieve all data available in the architecture in an easy, flexible and agile way.

The X-Arc architecture proposal in the SPeCS prototype opens new perspectives in studying the collaborative and cooperative interaction among users in a GIS environment to share and integrate data sources during a decision making process.

5. ACKNOWLEDGEMENTS

This work was partially financed by CNPq and CAPES agencies.

6. REFERENCES

[1] BARTHÈS, J.; "Ontologies as a Backbone for Knowledge Management in Cooperative Design". Proceedings of Fourth International Workshop in CSCW in Design. Compiègne, France. September 1999.

[2] BOUGUETTAYA, A.; BENATALLAH, B.; ELMAGARMID, A., *Interconnecting Heterogeneous Information Systems*. 1st ed., Boston, Kluwer Academic Publishers, 1998.

[3] CAREY, M. J. et al., "Towards Heterogeneous Multimedia Information Systems: The Garlic Approach". Technical Report RJ991. 1997.

[4] CLEMENT, G. et al., "OGDI: Toward Interoperability among Geospatial Databases", SIGMOD Record, v. 26, n. 3 (Set), 1997, pp. 18-23.

[5] DOMENIG, R.; DITTRICH, K. R., "An Overview and Classification of Mediated Query Systems". 1999.

[6] ELMAGARMID, A., RUSINKIEWICZ, M., SHETH, A., *Management of Heterogeneous and Autonomous Database Systems*. 1st ed., San Francisco, Morgan Kaufmann Publishers. 1999.

[7] FABER, G. W.; WALLACE, K.; CROTEAU, V. & THOMAS, L. Small, "Active Response GIS: An Architecture for Interactive Resource Modeling." Proceedings of the GIS'97 Annual Symposium on Geographic Information Systems. Vancouver, B.C, GIS World Inc., 1997.

[8] FROMENT, E. D., "GIS and Data Integration: a user's perspective". In: GIS'94 Symposium, Vancouver, February 1994, pp. 157-161.

[9] GARCIA-MOLINA, H. et al., "The TSIMMIS Approach to Mediation: Data Models and Languages", In Journal of Intelligent Information Systems, 1997.

[10] JABLONSKI, S. & BUSSLER, C., *Workflow Management Modeling Concepts, Architecture and Implementation*. International Thompson Computer Press, London, UK, 1996.

[11] MEDEIROS, CLAUDIA B.; VOSSEN, GOTTFRIED; WESK, MATHIAS, GEO-WASA – "Combining GIS Technology with Workflow Management", February 1996, pp. 1-23.

[12] MEDEIROS, S. et al., "SPeCS - A Spatial Decision Support Collaborative System For Environment Design". Proceedings of Fifth International Workshop in CSCW in Design. Hong Kong. November 2000.

[13] MEDEIROS, S. et al., "Coordination Aspects In a Spatial Group Decision Support Collaborative System". Proceedings of 16º ACM SAC 2001 Symposium on Applied Computing. Las Vegas, USA, March 2001.

[14] PARK H. G., KIM, H., OH, C., J., "A Design Support System Using Concurrent Engineering Concept", Proceedings of Second International Workshop on CSCW in Design. Bangkok, Thailand. November 1997.

[15] TOMASIC, A.; RASCHID, L.; VALDURIEZ, P., "Scaling Access do Distributed Heterogeneous Data Sources with DISCO". In IEEE Trans. Knowledge and Data Eng., v. 10, n. 5, September 1998.

[16] WIEDERHOLD, G., "Mediators in the architecture of future information systems". IEEE Computer, 25(3), 1992.

[16] WIEDERHOLD, G., "Mediation in Information Systems", ACM Computing Surveys, v. 27, n.2, 1995.

[17] XHUMARI, F.; AMZAL, M.; SIMON, E., "Le Select: a Middleware System for Publishing Autonomous and Heterogeneous Information Sources", Inria, Caravel Group, France, 1999.

Managing Dynamic Shared State in Virtual Space for Collaborative Design

Lilly Li, Chenhui Yang, Tangqiu Li
Computer Science Department, Xiamen University, Xiamen, 361005, P.R.China
xyli@yanan.xmu.edu.cn, chyang@xmu.edu.cn, tqli@xmu.edu.cn

Abstract

In building the virtual world for distributed collaborative design, the senses of sharing the world and cooperation between physically distributed hosts and users are usually achieved by maintained dynamic shared state. To make those events that happen in another physical location to be virtually "seen" by other participants, communication through network is indispensable. Because of the limited network bandwidth and limited CPU resources reserved for communication in a virtual world simulation, reducing communication traffic is a critical issue. In this paper, we present an efficient model, improved "Lazy model", to reduce communication for maintaining dynamic shared state. After that, we propose an approximate method using AOI, Area of Interest, to enhance the implementation of this model.

1. Introduction

We share this real world. If we are located in the same place, then we should feel the existence of the same objects, and also get to know the same properties and action of the same objects.

Similarly, in the virtual world, the users also need to share the dynamic virtual environment. In this case, what to be shared is usually called DSS, standing for Dynamic Shared State. The dynamic shared states deserve much effort to maintain, because violations of this rule may destroy the users' sense of reality.

Usually there are three solutions to this problem: centralized information repositories, frequent state regeneration and dead reckoning.

1.1 Centralized information repositories

This solution maintains the consistent information of all objects in a central server. It could be a file repository[1][4], a repository in server memory [2] or a virtual repository using a distributed consistency protocol[3]. The former two are relatively simple to understand. The third one is a little more complex. In this method, in fact there is no central server. A distributed consistency protocol must be established first. Each client caches all the necessary data for the dynamic shared state locally, and then communicates with each other via the distributed consistency protocol. Through this protocol, the participation hosts can exchange messages directly with each other to disseminate state updates, make sure that all host receive those updates, and determine a common global ordering for those updates.

Table 1: Evolution of Central Repository Approaches for Shared State Management.

TECHNIQUE	EXAMPLE	CHARACTERISTICS
File system – pull	NFS, AFS	Simplest to program Highest consistency Slow data updates and access Automatic persistence
Central server—pull	Networked VR (NEC)	Faster data updates and access "Eventual" consistency No persistence in FIFO queues
Central server--push	Shastra	Immediate data access Slower updates
Virtual central repository	DIVE with ISIS	Better fault tolerance (no server to crash) Immediate data access Greater protocol overhead
Client consistency selection	BrickNet	Greater complexity Better management of bandwidth resources Potential data inconsistencies

According to when to send the update information of the source objects, this solution also can be classified into two categories: Pull or Push. In the Pull method, the clients will query the information when they need it, while in the Push method, the server will decide when to send the information to the local cache of the client so that it can be ready for use when the client need it.

As to the different implementations of the idea, we use table 1-1 [1] to explain the pros and cons of them.

1.2 Frequent state regeneration

There are some conditions that users may prefer smooth motion to absolute accuracy, if the result of this temporal inconsistency is slight and temporal. The solution "Frequent state regeneration" is used to replace the distributed consistency protocol in this situation.

There is no central server in this solution. Instead, each transmitted update contains a full description of the entity's state, whether or not it has changed, so that receivers can frequently capture the entire state of the entity. The "owner" of each piece of state information transmits these current state values using blind broadcasts that are sent asynchronously and unreliably at some regular interval to everyone participating in the virtual world. Each client simply maintains a cache of the most recent update received of reach piece of shared state.

It's possible that some updates might not be received (or may arrive only after some significant delay), but in this solution it is assumed that a high transmission frequency will make such inconsistencies relatively unnoticeable to the users.

One of the limits of the frequent state regeneration is that it not only sacrifices absolute consistency, but also reduces some of the flexibility that centralized repositories. In the centralized repositories, any host could modify information about any entity in the virtual environment, which will ensure that every one will see the updates reliably and in the same order. However, in the frequent state regeneration systems, at any time only one host could update an entity state (This is enforced by the state ownership), since it can't be guaranteed that those updates will be received and process at each host.

Another limit of frequent state regeneration is the network bandwidth utilization and network latency; It assumes that all hosts are broadcasting at the same rate, equal to their display frame rates. However, remote send rate may be various and the speed may be not the desired one of the local display rate, which will make it difficult to mask or conceal the distinction between the local or the remote hosts.

1.3 Dead reckoning

In the dead reckoning method, the host will receive the update information from source object, but less frequently. During the interval of the two updates, the host is expected to approximate the shared state of the object using the information in the previous updated information through the "prediction" algorithm. Once the host receives the updates information, it will update the local information for the object according them through "convergence" algorithm.This method will sacrifice accuracy of the shared state, but it reduces message traffic and therefore supports more users or participants. Dead reckoning is most appropriate when bandwidth is limited, computational cycles are available, or the net-VE contains many participants.

Now many ways are proposed to predict the entity's state, including Zero-order polynomial, derivative polynomials, PHBDR (position history-based dead reckoning) and more advanced method--object specified prediction which will take into account about the objects' current status and physical properties.

The convergence algorithm should not only correct predicted state quickly but with creating noticeable visual distortion to the users. And there also many methods proposed for the convergence algorithm, such as :

1) zero-order: simple to understand and implement, but not provide the best visual display.
2) linear convergence. Can give us continuity along the visual path.
3) other high order convergence.

Non-regular transmission is a variant of the dead reckoning algorithm. Rather than transmit updated information regularly, this variant will require remote host to send updated information only when the divergence between the actual state and the predicted state is greater than a certain predefined value.

The advantages of this method is obvious--if the prediction algorithm is relatively accurate, then source host prediction can greatly reduce updates values, and the overall accuracy of the remote state information will be guaranteed. Moreover, this method allows the source host to dynamically balance its network transmission load based on the current situation. For instance, the source host could temporarily increase the error threshold on entities that reside within wide open space and simultaneously reduce the error threshold on entities that are located close to remote entities in the network.

But this method is also not perfect. If the prediction is good, or there is little change of the objects state, then there may be no updates about this objects are transmitted. New users many not receive any information about them. And it's also difficult to tell whether there is little change in it or the objects have logout out. This could be solved by setting a time limits to decide whether to generate a update and broadcast it.

In short, in dead reckoning algorithm, the updates information could be sent at lower-than-frame-rate frequencies and each host use prediction and convergence to seamlessly integrate remote and local entities on the display, maintain a smooth rendering for the local user based on the previous update packets. However, this algorithm doesn't guarantee that all hosts share identical state about each entity and it is more complex to implement and evaluate.

Except the central file server, whatever method we may use, there is a problem of scope of broadcast. When a host needs to update its shared states, it needs to let other hosts receive this update. Typically and traditionally, blind broadcasts—that is, the update information is sent to all the hosts, are used. Obviously, this is not a satisfactory solution for this. We need to reduce the broadcast scope. One method is proposed by the VEOS of UW to filter the updates before they get sent to inappropriate recipients. This filtering may be done in the server, or use epidemic approach for it (U of Washington). In VEOS, they organize the host in a hierarchical structure. At the end of each entity's events loop, a communication daemon will send updated state information to the neighbors in the host hierarchy. These states information is stored in the queue of the hosts, and will be used in its next event loop.

In this paper, we will propose our own way to reduce the broadcast scope.

2. Using improved Lazy Model to reduce the communication flow

In this paper, we'd like to introduce a **"Lazy Model"** of the managing shared dynamic state. There are two principles for our model:

1) Never send updates to an objects that doesn't need the information
2) Never send update information when there is no update in the source object

The basic idea behind our model is that: every users or objects have its "Potential Interactive Set", The objects sets that are possible will interact with other objects. (This may include the telecommunication interaction, for example, one person in the building may talk with another one outside the building via the phone). Usually, we use "PIS" designated those objects or users that can feel its existence or will interact with it.

Definition of PIS: those must should be autonomic objects, or those objects with non-zero objects which could have interaction with the object.

Further, we could classify the objects into two categories: (1) Autonomic objects that can manipulate other objects or that have a velocity. (2) The passive objects that is static. For the first kind of objects, they possibly will change their own states, and therefore, they will not only need to receive the update reports of other objects, but also frequently send update information to other objects. For the second category objects, they will stay static until a certain event happens to them and change their states. This kind of objects can't manipulate other objects, and also need relatively few update information since they seldom change their states. However, once they need to update their states to other objects, they should send the updates to all the objects that could feel the existence of them. Is it necessary for them to receive the update information of other objects? Since this object can't feel the existence of other objects, it seems that they need not to know the update information of the other objects.

You may ask, if, for example, one static object is bumped by another object, should it need to know the update information of the other objects? Not necessary. The moving objects will of course trigger a certain event of the object to respond to the other object, but not need to receive the other information of it.

Therefore, for this kind of object, we could subdivide the PIS into two categories: one is the SPIS (Sending PIS); the other is RPIS (Receiving PIS). The SPIS refers to the objects list to which the object needs to send update information. The RPIS designates those objects from which the object needs to receive the update information. For example, for the static objects we've mentioned above, once the states of them change, they need to send to all the object of the first category their update information. Therefore, their SPIS will be the link to those objects. However, their RPIS will be void since they need not to know the existence of the other objects.

This is similar to the consciousness of the real world. That is, those that have consciousness need to get their consciousness through the update information of other objects, while those have no consciousness need not to know the existence of the other objects. Here our idea is a little different in order to deal with the interaction between two unconscious objects. And we deem the one that has velocity conscious.

In this way, we could easily achieve our goals that lie in the two principles in our "Lazy Model".

3. Implementation of improved Lazy Model

3.1 Effort-saving solution

The PIS is a little different from the area of interest in the 3D simulation. The area of interest refers to those objects and terrain that must be rendered in the users virtual world display. This may include those objects that are outside the PIS. But to save effort, we could deem the PIS the same as AOI in order not to impose an extra computation load on the computer.

In this solution, we can have the following equation:

AOI=RPIS=SPIS

Therefore, we only need to copy the object list in the AOI, and make the RPIS and SPIS the same as the AOI. And we then add the two links to the objects:

Structure Object

```
{
// other properties
..................

// other events
.................
Object &SPIS; // the object linked list that need to send update information to them
Object &RPIS; // the object linked list that the object need to know the update information of them.
}
```

One thing deserve notice is that the server must be in the RPIS of all objects to maintain the persistency and at the same time, consistency of the world for the new users. Once a new user enters the virtual world, it needs to query the central server to obtain its PIS. The serve responds by send its PIS back to the user. The user then sends notification to the objects of the objects in the PIS to tell them to add it to they SPIS. Since only these updates are sent to the server, and almost no other condition (except the one mentioned below) will the users query the server, therefore, the server will not be the bottleneck of the application.

The process goes as the following:

```
void Object.updateinfo( )
{
    Object Update it information;
    Read A.SPIS,
    For each object in A.SPIS
      send the update information to each object in the A.SPIS
    End for
}
```

An object/ user's action or motion will possibly change its SPIS and RPIS. On this condition, the object will respond by the following procedure: then it notify the old objects that are not currently lies in the areaofiterest of it that they should discard it form their sendingPIS. And it will notify those new objects that lies in the new PIS of the objects that they should add the object into their sendingPIS.

```
void Object.ChangePIS( NewAOI)
{
  OldSPIS= Object.SPIS;
  OldRPIS=Object.RPIS;
  Object.SPIS=NewAOI;
  Object.RPIS=NewAOI;
  For each object A that is in OldRPIS but not in the new RPIS
    Notify A (object, DSPIS) // send notification to A to inform A that the object will not need its update information any more, so that the object could delete the object from its SPIS
  End for

  For each object A that is not in OldRPIS but in the new RPIS
      Notify A (object, ASPIS) // send notification to A to inform A that the object will need to send its update information to the object
  End for

  For each object B that is in the OldSPIS but not in the NewSPIS
   Notify B( object, DRPIS); // the object will delete the object from its RPIS;
  End For

  For each object B that is not in the OldSPIS but in the NewSPIS
   Notify B( object, ARPIS); // the object will delete the object from its RPIS;
  End For

}
```

As to the object A's response, it goes as following:

```
void uponnotification()
{
 RemotObject=object;
 Actionclass=event;
  Switch case Actionclass
    Case "ASPIS"
        Add object into the SPIS of the local object
    Case "DSPIS"
      Delete object into the SPIS of the local object
    Case "ARPIS"
      Add object into the RPIS of the local object
    Case "DRPIS"
      Delete object from the RPIS of the local object
End case
}
```

Once we build those structures, we've implemented the "Lazy Model" of the shared dynamic state.

3.2 More efficient solution

As we can see from the above, the solution doesn't make good use of the difference between the AOI and PIS, also SPIS and RPIS, as a result , the SPIS and RPIS is very loose. Next we will propose another more efficient solution, which make good use of the convenience provided by the SPIS and RPIS.

First, let's make it clear the difference between the SPIS and RPIS. For an object, the SPIS indicates which objects the local object needs to send its update information, while the RPIS indicates that it need which object's update information. For example, If in a virtual environment area where object A is located. There are totally N objects in the environment, M of them are static with no velocity. Then object A's SPIS could exclude these objects, making the broadcasting of the update information of objects limited to the other (N-M) objects, While A will receive the update information from all the objects (including the static objects with no velocity). For a static object, say B, when its RPIS will be void, while its SPIS is full: that is, it needs to send its update information to all the other (N-M) objects.

From the above analysis, we could see much communication saving if we use the above "tight" PIS. To construct this "tight" or "economic", PIS is also not a formidable task. Again, we could start with AOI of the users.

Once we get the AOI of the object, we could then check properties of the objects in it. The properties that could affect the objects' SPIS and RPIS include the category (users, animals, or just common objects) of the object, and its velocity. We then could classify those objects into two categories and then decide further their SPIS and RPIS. This process will be performed in the beginning of the application, or upon the change of the AOI.

```
void TightenPIS(AOI)
{
 Subdivide(AOI, Set1, Set2); //Set1 designates the first
category Set2 designates the second category;
 For each object A in Set1
       A.RPIS=AOI
       A.SPIS=Set1
 End for

 For each object A in Set2
    A.SPIS=Set1;
    A.RPIS=NULL;
   End for
}

void Subdivide(AOI, Set1, Set2)
{For each object A in AOI
  If (A.category<>user ) & (A.category<>animal) &
(A.velocity <>0)
      Add A into Set1
  Else
   Add A into Set2
  Endif
Endfor}
```

The above computation will certainly impose the some computation load on the server computer. But since different users may not log in at the same time, this may not cause a bottleneck in the server.

4. Conclusion

In this paper, we introduce an improved "Lazy Model" to reduce the communication load on the network and also the computation load on the local computer. Of course, there are still some problems are left here. For example, we assume that the AOI of the users include the PIS of all kinds of objects. this is reasonable in some conditions, but in other condition, this may not be true. Then how to deal with the problem? Again, we could use the virtual grid we present in another paper. And we then make the objects in a certain peripheral area as the PIS. But this also leave many problem to be explored later.

5. Acknowledgements

Most of ideas in this paper come from the work of FL group in MCS division of Argonne National Laboratory in United States. It is with great pleasure for me, the second author of this paper, to thank my supervisor, professor Rick Stevens, director of MCS division and team leader of Grid Alliance, who gave me lots of innovative ideas and kind help when I was a visiting faculty in his group. I am also very grateful to all the colleagues in FL group who put tremendous amount of efforts to help me. These individuals include Mike Papka, Mark Hereld, Lisa Childers, Tom Brown, Terry Disz, Ivan Judson, Tushdar Udeshi, Bob Olson and Randy Hudson.

6. References

[1] Sandeep Singhal and Michael Zyda, Networked Virtual Environment: Design and Implementation. ACM Press, 1998

[2] Defense Modeling and Simulation Office, U.S. Department of Defense High Level Architecture Interface Specification, Version 1.3, April 1998.

[3] Campbell, R. Managing AFS: The File System, Upper Saddle River, NJ: Prentice-Hall, 1998

[4] Anupam, V. and C. Bajaj, Distributed and collaborative visualization, IEEE multimedia 1(2): 39-49, Summer 1994.

Collaborative Technique Integration in Knowledge-Based System

Kuo-Wei Su
Dept. of Accounting & Statistics
Tak Ming College
Taipei, Taiwan, R.O.C., 114.
kwsu@mail.takming.edu.tw

Thu-Hua Liu
College of Management
Chang Gung University
Taoyuan, Taiwan, R.O.C., 259.

Sheue-Ling Hwang
Institute of Industrial Eng. &
Eng. Management
National Tsing Hua University
Hsinchu, Taiwan, R.O.C., 300.

Abstract

In this research, a methodology of collaborative technique integration (CTI) is developed to implement the proposed strategy for people working in the maintenance domain. This conceptual model CTI includes system hazard assessment, knowledge-based expert system, and web-based collaborative environment. The system enables administrators to share the relative maintenance information and then may prevent or mitigate human errors. Such a hybrid approach could serve as a guide for other similar logistic systems to prevent maintenance errors and for the different kinds of high technological industrial that needs to manage a domain-specific knowledge.

1. Introduction

For an operational transportation system, the logistic maintenance tasks must commit the responsibility of maintaining the quality and safety of the system for carefully attending to dependability (i.e. reliability) [2][14]. Up to now, the automation and modulus degrees of the transportation system are getting higher, the more complex technique and professional skill of diagnosis of the maintenance engineers should be well equipped for the maintenance service.

Moreover, a vast amount of maintenance data that included lots of implicit information could indicate critical incidents for the policymaker by availability assessment and hazard analyses in the maintenance domain, but the important regular maintenance data is often suffering from neglect. Designing a structured and well-implemented model could help to reveal the implicit information. This study proposes an availability model (including reliability, maintainability, and supportability) in the hazard assessment process.

At the beginning, by analyzing the current problem-solving process in the maintenance domain, we identified that the main problem is lack of knowledge sharing and consistent direction that is referable. As for maintenance experience, it is as valuable as textbook knowledge. Therefore, all knowledge related to the problem-solving process must be accumulated and shared in an integrated system [5]. In the sense, comprehensive maintenance support for the maintenance engineers in critical events to offset experience lack and reduce human errors is strongly recommended. This is an ideal case for the application of knowledge-based expert system (KBES) technology.

With a view to sharing experts' domain-specific knowledge and utilizing well-preserved documents, this research introduces a conceptual model (CTI) for achieving an effective knowledge management. The CTI includes system hazard assessment, knowledge-based expert system, and web-based collaborative environment.

2. Assumptions and domain specification

For fault diagnosis and recovery, this study constructs some basic assumptions as below,

(1) The experienced knowledge is very wide and valuable because of the design and maintenance of various buses.
(2) The regular documents are not easy to be preserved or appeared because of vast amount of maintenance data.
(3) Cognitive type can be investigated through systematic method and could affect directly mental model; namely, cognitive type and cognitive process are consistent [7]. Mental models are task-specific. Therefore, the same maintenance engineer is likely to employ different models for different tasks.
(4) For the same task, different maintenance engineers may think in terms of different models.
(5) Even the same maintenance engineers may use different models for the same task depending on his/her actual condition (e.g. degree of skill, fatigue...).
(6) To allow for these differences as above (4), (5), and (6) [3][9], KA process and KBES interface should be adaptable to different mental models.
(7) To Choose and interview the senior maintenance engineers who can express in words clearly. This will let the KA process be more exact.

To acquire experienced knowledge, knowledge about

fault diagnosis tasks and management problems were elicited from senior maintenance engineers and managers (8 people), and the seniority of the 8 experts is over 20 years. To analyzing document, the data about 2 years and the data relative to the system availability (reliability, maintainability, and supportability) data were collected for a suitable assessment.

The research tasks are supported and implemented in the hall and the maintenance department of the Taipei City Bus Administration.

3. Research framework

According to this research scope and method, this skeleton of the paper is exhibited in Figure 1. Management and technique aspects are primarily considered into the maintenance knowledge management mechanism.

4. Collaborative technique integration (CTI)

As mentioned above, an operational system performance and safety rely on logistic maintenance and support quality; therefore, an effective maintenance management mechanism is necessary. This research proposed a practical knowledge management mechanism named collaborative technique integration (CTI) in a maintenance domain.

There are two key features in the CTI system. One is to standardize the development of the expert system by the process of maintenance protocol development based on a cognitive-driven approach; the other is to apply the availability analysis for hazard assessment based on dependability data. Based on human experts' domain-specific knowledge sharing and well-preserved documents utilizing, a CTI model (both the KBES and the HA techniques are considered simultaneously which were very few applied for practical examples) can achieve an

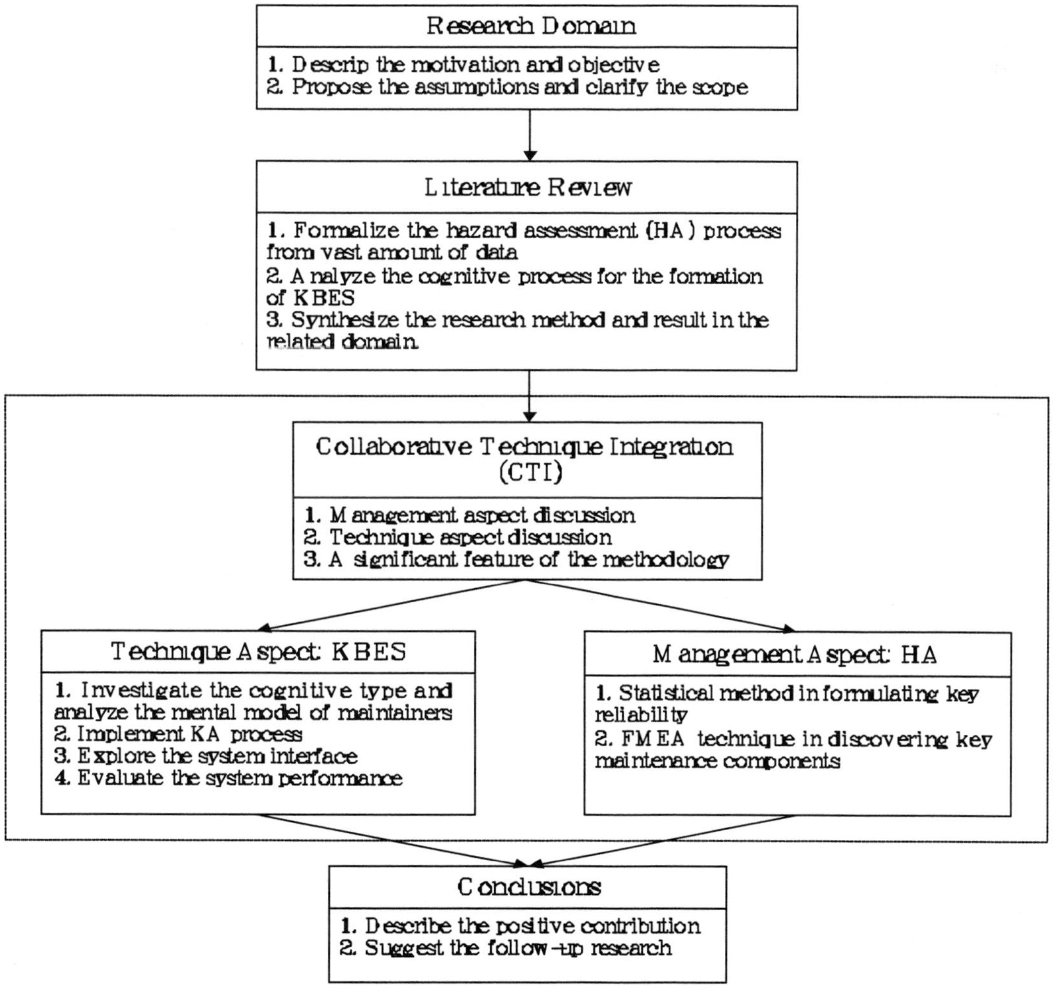

Figure 1. Research framework.

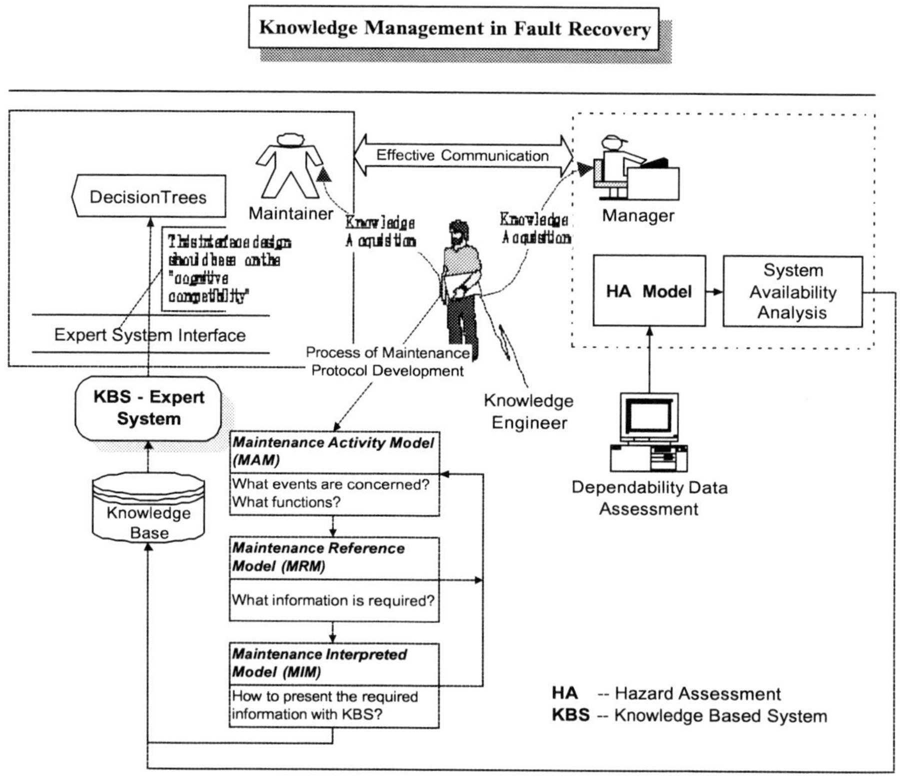

Figure 2. The formation of CTI model [11].

effective maintenance management. Figure 2 formulates the system structure of CTI.

CTI emphasizes information and knowledge integration and implements in Web-based system, which enables maintenance professionals to share and reuse knowledge/information in active and latent human failures. It combines the structured system dependability reports on each event with a list of related fault recovery records [4] and the use of knowledge based system about human experts' behavior that has accumulated plentiful experience for problem solving. Moreover, the role of the knowledge engineer in the diagram (Figure 2) is a multi-skilled, multi-talented individual with experience and expertise in maintenance area including KBS technology, human and interpersonal skills, requirements gathering, and task analysis [6][8] as follows,

- KBS technology is to support a process known as knowledge acquisition, knowledge representation, software design, and implementation in an appropriate computer language.
- Human and interpersonal skills aid knowledge engineers to elicit knowledge of experts through interviews and other techniques.
- Requirements gathering or analysis is the process of finding out what a maintenance engineer requires from an expert system. General techniques such as interviewing, observation and document analysis are used in requirements gathering.
- Task analysis (TA) is a generic term for a rather bewildering range of techniques. There are some techniques aimed at eliciting descriptions of what people do, representing those descriptions, predicting difficulties and evaluating systems against usability of functional requirements.

Moreover, the result of system availability analysis can be established into knowledge base for technique tasks.

5. Hazard assessment (HA)

About system availability, three aspects (reliability, maintainability, and supportability) are considered. System availability names operational availability in this study. It is the probability that a device can operate normally in any moment under the given supporting environment and operating conditions. The formulation is presented below:

$$\textit{Availability} = \textit{reliability} \div (\textit{reliability} + \textit{maintainability} + \textit{supportability}) \quad (1)$$

Where the definitions are the mean times to failure (MTTF), the mean time to repair (MTTR), and the mean time to support (MTTS) for reliability, maintainability, and supportability respectively [14].

Reliability is the measure of system quality performance, which defines the probability that a device will satisfactorily perform a specified function for a specified period of time under the given operating conditions. In this study, the anchorage rate and the failure rate are defined as the major reliability factors; therefore, the maintenance rate is regarded as cross-reference in reliability analysis.

Maintainability is the measure of system maintenance quality performance, which defines the time that a maintenance task will promote a device to retain or retrieve a normal operational function by a specified procedure and resource under the indicated conditions. In this study, the mean time to repair (MTTR) is defined as the major maintenance factor.

Supportability is the ability of supporting maintenance material, which defines the time when the request unit proposes the requirement by indicated repair policy under the specified conditions. In this study, the mean time to support (MTTS) is defined as the major supportability factor.

This study can infer the operational availability from the analysis of reliability, maintainability, and supportability, which can be taken to compare with other bus systems or different periods for performance improvement and cost control [1][15].

6. Knowledge-based expert system (KBES)

An interface of an artificial intelligence system shell is provided to facilitate the task of knowledge acquisition, knowledge refinement, knowledge base edition/modification, knowledge base maintenance, and information management relating to specific expert system project [13]. In other words, the structure of the interface should match the cognitive structures of the users (i.e. the maintenance engineers). For considering cognitive compatibility, such an interface was developed from knowledge acquisition (KA) process based on experienced maintenance engineers' cognitive formation.

The key to the efficient implementation of a KBES is finding a suitable representation for the real problem-solving knowledge that is to be incorporated in the logistic system. It is therefore very important that knowledge engineer represents the knowledge in the best way possible [8]. There are quite a few competitive candidate representations and it is up to the maintenance engineer to select the most appropriate one.

The forms of 'Decision Trees' representation chosen for the knowledge could match the structure of the tasks in hand. It captures analysis information in hierarchy. The decision tree represents failure modes, possible causes, and analysis processing of underlying problems in the given event.

A graphical display of the user interface provides maintenance's users with a flexible mechanism for viewing knowledge base structure and is shown in Figure 3 [11]. The interface explanations is part of the interface environment that utilizes production rules, procedures and data in order to simulate the intelligent behavior of a competent maintenance engineers. It executes the following tasks:

- Forward and backward chaining of rules
- Start up of procedures
- Debug mode of rules

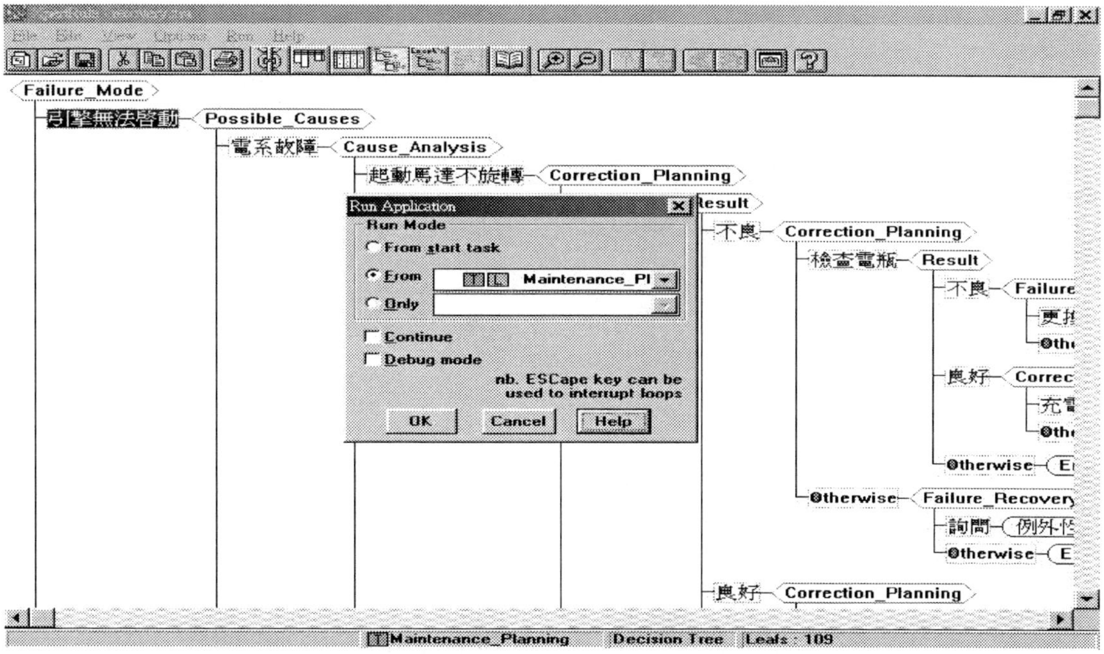

Figure 3. Decision tree and inference explanations [11]

The three tasks are common to the majority of inference engines. The explanations offered by the user interface are not limited to the conventional 'why' or 'how' explanation; deep explanation about a specific task can be provided through manual or senior maintenance engineers.

The core of KBES is domain knowledge base. An optimum maintenance program should be based on actual field experiences. A cognitive-driven KBES interface achieves a successful performance and interaction with users through an effective KA technique that has approved in this case study.

7. Conclusions

To evaluate the operational availability of a system could help one to understand whether the old system can cope with the need of operation conditions or not and ensure its quality. For the actual service aspect, it will help to identify the major inspection point for later maintenance service to accumulate decision-making information for implementing subassembly renewal project [10].

Moreover, to construct an expert system, which could assist engineering and management for decision-making and analysis for diagnosing failure modes, maintaining system components, and preserving system function needs to enhance system reliability and stability.

The CTI is designed as a Web-base system, which enables maintenance professionals to share and reuse knowledge/information in active and latent human failures.

The main contributions of this study can be generalized as follows,

(1) The critical items of a system can be identified and prioritized for treatment such as preventive maintenance or renewal;
(2) During the study process, the field knowledge and experiences of the maintenance department are served as the heuristic rules for the future service;
(3) The purchase of component decisions can be made according to the analysis;
(4) The most important point is that it could improve the quality and safety standards of a public transportation system;
(5) Such a counterexample discussed by Su *et al.* [12] indeed encourages knowledge engineering to develop a suitable or specific human-machine interface corresponding to specific user's cognitive types and mental model;
(6) The study as reported here is to contribute the feasible tool for the KBES interface development, which is originally considered the consistency among the cognitive types of the users and KA process [12].

The developed CTI conceptual model could not only redeem the previous theory but also serve as a guide for other similar logistic systems to prevent maintenance engineer errors or the different kinds of high technological industrial that needs to manage a domain-specific knowledge.

8. Acknowledgments

The authors would like to express their gratitude to the manager, and engineers of the maintenance department of the Taipei City Bus Administration for their crucial help and many useful comments in regard to the development of the system.

9. References

[1] Blanchard, B.S. (1967). Cost effectiveness, system effectiveness, integrated logistics support, and maintainability. *IEEE Transaction on Reliability, 16* (3), 117-126.

[2] Dhillon, B.S. and Reiche, H. (1985). Reliability and Maintainability Management. Van Nostrand Reinhold, Inc., New York, U.S.A.

[3] Fuchs-Frohnhofen, P., Hartmann, E.A., Brandt, D., and Weydandt, D. (1996). Designing human-machine interfaces to match the user's mental models. *Control Eng. Practice, 4,* 13-18.

[4] Lee, M. R., Wong, W. Y., & Zhang, D. M. (1999). A knowledge-based framework for clinical incident management. *Expert Systems with Applications, 17,* 315-325.

[5] Park, M. K., Lee, I., & Shon, K. M. (1998). Using case based reasoning for problem solving in a complex production process. *Expert Systems With Applications. 15,* 69-75.

[6] Preece, J. (1998). *Human-computer interaction.* England: Addison-Wesley.

[7] She, Y. M., & Hwang, S. L, (1996). *The approach of user's mental model and cognitive type in process control system*, Taiwan, ROC: National Tsing-Hua University Press.

[8] Smith, P. (1996). *An introduction to knowledge engineering.* Oxford: International Thomson Publishing Press.

[9] Streufert, S.,& Swezey, R. W. (1986). *Complesity, managers, and organizations.* Orlando, FL: Academic Press.

[10] Su, K. W., Hwang, S. L., & Liu, T. H. (2000). Reliability-centered maintenance management: a case study on a public bus system. *International Journal of Industrial Engineering, 7* (3), 232-241.

[11] Su, K. W., Hwang, S. L., & Liu, T. H. (2000). Knowledge architecture and framework design for preventing human error in maintenance tasks. *Expert Systems With Applications, 19* (3), 219-228.

[12] Su, K. W., Liu, T. H., & Hwang, S. L. (2001). A developed model of expert system interface (DMESI). *Expert Systems With Applications, 20* (4), arranged.

[13] Usha, S., & Strikanth, S. (1998). *Flexible system for knowledge acquisition in expert system development,* Ft Collins, CO, UNITED STATES Assigned to Hewlett-Packard Company.

[14] Villemeur, A., Cartier, A., & Lartisien, M-C. (1992). Reliability, Availability, Maintainability and Safety Assessment. John Wiley & Sons, Ltd., England.

[15] Winlund, E. S. (1964). Cost-effectiveness analysis for optimal reliability and maintainability. *Proceedings, 11th National Symposium on Reliability and Quality Control, IEEE,* (pp.107-114).

On Sharing of Objects in Concurrent Design

Aiko Frank
Universität Stuttgart, Breitwiesenstr. 20-22
D-70565 Stuttgart, Germany
Aiko.Frank@informatik.uni-stuttgart.de

Bernhard Mitschang
Universität Stuttgart, Breitwiesenstr. 20-22
D-70565 Stuttgart, Germany
mitsch@informatik.uni-stuttgart.de

Abstract

Sharing data is an important aspect in distributed design environments and should be supported by an underlying system. Any synchronous access to data is conflict prone. Applying concurrency control and two phase commit are an option to be considered. But design processes also demand cooperation between the designers. Negotiation about actions on the product under design and the early exchange of preliminary results are crucial issues. Controlled data access by itself doesn't fullfil all the needs for cooperation. We will present a new approach that relies on a concept and system model which integrates concurrent activities by a common information space offering flexible protocols for cooperation on the shared objects. We will describe the customizability of the protocols to allow the approach to be adapted to different cooperative scenarios.

1. Introduction

Processes in concurrent engineering and design rely on cooperation between the different participants. This cooperation is necessary to establish and coordinate the tasks needed to fullfil the goal of the design process. Those tasks (also called design steps) - in which the complete process can be subdivided into - rely on the results, states and progress of concurrent, previous, and future tasks. Typically it can not be exactly planned beforehand which tasks are necessary, how the tasks depend on each other, and how and when those tasks have to interact. Thus it is necessary to offer mechanisms to support cooperation at runtime.

We already presented our activity model for coordinating design tasks in [3]. In this paper we will discuss data sharing as the second pillar of the ASCEND project (*Activity Support in Co-operative ENvironments for Design issues*). The situation we started from is given by non-integrated systems which should share data in order to coordinate tasks in a joint process (Figure 1). The difficulty is that the data to be exchanged is not under a common system control. That might cause inconsistent views, conflicts and errors. So our answer is to transfer the data to be shared under a common and adequately customized system control.

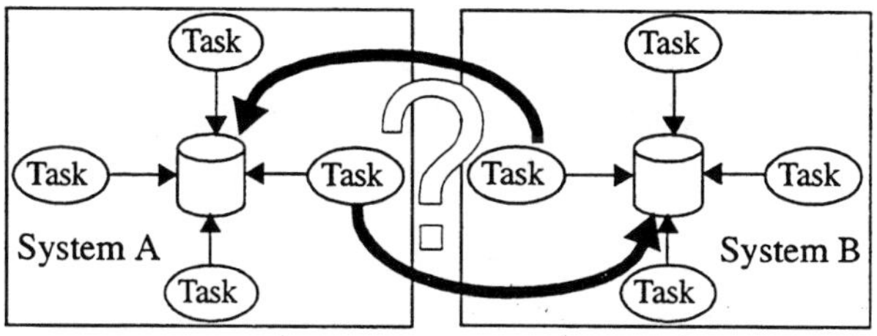

Figure 1. Data sharing for heterogenous systems

1.1. Process example

A new technical term for activities that integrate both, groupware as well as workflow is called *designflow*. To give an idea of our activity model, we describe the design of a new computer printer as a designflow (Figure 2).

First the general specifications for the printer have to be established: type (laser, ink), processor and chipsets, page format, OS platforms, pricing, and so on. This results into a specification document, which is stored in the shared information space (SIS). Also a preliminary list of hardware components and software modules (generic driver files, ...) to be used is collected and stored in the SIS.

We didn't detail the activity *Determine Specs* which might contain group meetings, and analysis of vendor documents. The first printer specification is used by the groupware activity *Design Coordination* to delegate the design of hard- and software. Those activities are further detailed, containing specific sub-activities. The hardware design activities have to work closely together, since they share e.g. the circuit layout, hardware specs and components. Changes in the specs or layout can be propagated and when necessary negotiated. Similar, the software design activities have to have access to the current hardware design and specification in order to correctly write software for the printer. The software activities have also to interact on shared data (i.e. files, specs) as is typical for software processes. Tests are executed to validate the software and simulate the hardware (e.g. electro magnetic compatibility). The hardware simulation is described as a workflow since it is a automatable process. Finally the *Design Coordination* is responsible to decide if the design is finished.

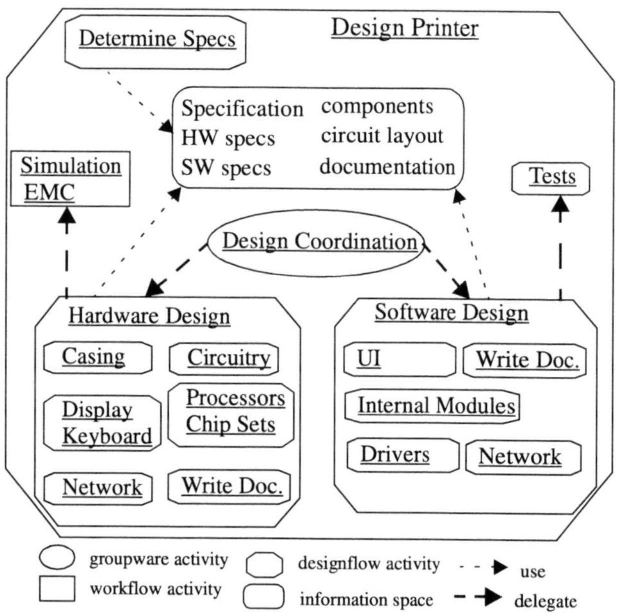

Figure 2. Designflow for printer design

The SIS in the example is important for the activities to exchange current design and preliminary results. Thus all activities can be aware of the work of others which they depend upon. For example, if the display is altered, the *UI* activity gets notified and can react to those changes. The notification is possible since the *UI* activity is registered on the (output) components.

1.2. ASCEND approach

A straightforward approach to support concurrent tasks is synchronizing them, like DB transactions or workflow systems do [5]. The processes we look at can be dynamic and require interaction and sharing of preliminary results. Thus they shouldn't be synchronized nor can be completely defined in advance. We actually don't want to serialize tasks anyway, since this prohibits any kind of parallelism. Instead we want to encourage cooperation, awareness and early data exchange which is typical for design and engineering processes. Additionally, we want to have system support for executing a process and individual actvities.

First we have a look at the activity model as part of the *ASCEND Designflow Model* (ADM). Our approach is combining groupware and workflow technology to comprehensively support structured and less structured parts of processes. The complete integration of existing groupware and workflow systems is not readily possible. We suggest an integration by encapsulating corresponding processes into ASCEND groupware and workflow activities, which can then be combined in ASCEND designflow activities (see [3] and Figure 3).

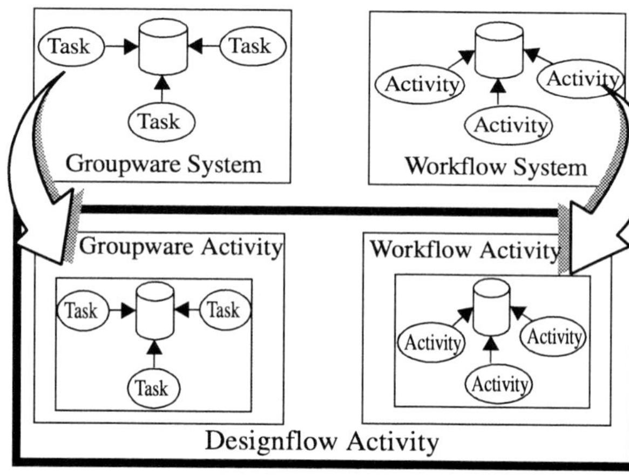

Figure 3. Combining different systems in a common activity model

The second part of the ADM regards data handling. The control of data access is a very well suited mechanism for coordination between concurrent tasks since data dependencies also translate into task interdependency. Those dependencies can be solved by concurrency control or additionally by cooperative interaction (e.g. negotiation) between tasks, respectively their actors (human users of activities).

The wrapping of the activities, as pictured in the above figure, realizes the integration into our designflow system. Our system also offers cooperation functionality like data sharing. For this the shared data has to be moved into the SIS (Figure 4). Logically the usage of those shared objects in the SIS is handled through the designflow system. I.e. the system guarantees that access to the objects is executed like the objects are natively stored through the designflow system. Resulting problems are mentioned in Section 3.

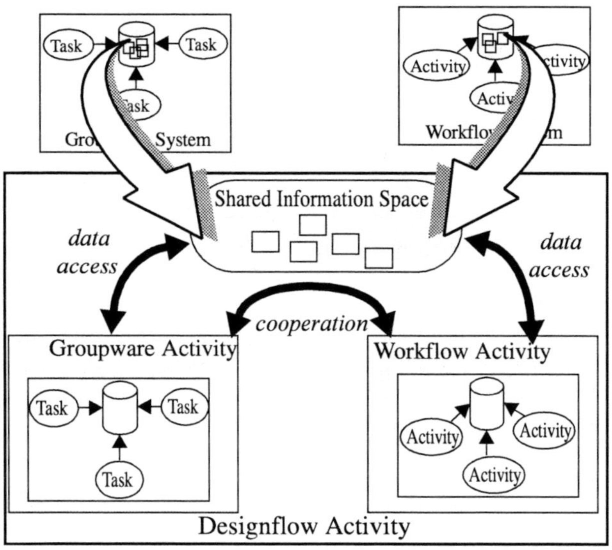

Figure 4. Transferring data into SIS

When examining objetcs to be used, one has to be aware that the objects often have dependencies between themselves, i.e. we have *complex object structures*. Sometimes complex objects are summarized as documents. Using this abstraction, one can think of a document being simultaneously edited by two users A and B. If the document is a single object, all actions on it are potentially in conflict. When the document can be "looked into"—i.e. it is separated into different objects like paragraphs or chapters—those sub-objects can be individually assigned and coordinated (e.g. [1]). Thus interference occurs less, when each user can only write to certain objects, which are distinct from the other user's ones. On the other hand if sub-objects are shared, awareness can be reached by making this fact known (e.g. displaying locked paragraphs). Another advantage is using object decomposition. Sub-objects can then be delegated to other tasks. This sub-object is exactly the part that the task has to work on, like delegating only the printer's case instead of all components. Of course it is possible to query any other sub-object of the complex object if it is necessary to access their information to complete a task, as would be necessary to know the dimension of keyboard and display to correctly define the printer's case. The sub-objects also have dependencies amongst them, like the printer's case has to be large enough to fit in the corresponding toner cartridge. We propose to use dependencies for objects in Section 2.2 for building complex objects to support such scenarios.

The last aspect discussed is the customization of objects. This is advantageous on two levels. First it is important to support requirement of different applications with different semantics of objects and even different object types. Secondly, as already discussed in [4], the granularity of objects is a crucial issue, which determines the degree of cooperation (and coupling). By using an object-oriented approach this should be fairly easy (Section 3). To finally use the objects with varying granularity and semantics, the access protocol also has to be flexible to adapt to different application semantics. Our answer to this is the use of conversation protocols in Section 2.3.

2. Design of the shared information space

In this chapter we discuss the concepts of the SIS. The information space comprises many aspects to support data exchange and cooperation. Since we use protocols to share objects, we can easily establish awareness and cooperation patterns. Also direct cooperation on the basis of interaction through protocols is possible.

2.1. Basic concept

In the ADM we call objects and activities entities, since they are somehow self-contained and can interact with each other. The basic aspects of the SIS comprise the information space itself containing objects and their dependencies, a communication layer for exchanging messages, and finally the protocols for communication between the entities (Figure 5).

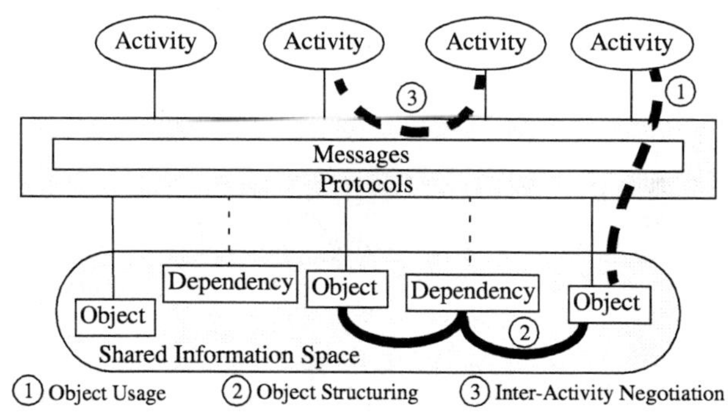

Figure 5. Communication and cooperation infrastructure for ASCEND

2.2. Objects and dependencies

The SIS is basically a simple name service, where all available objects and dependencies are being referenced. Thus objects to be shared have to be registered into this name service by some tool or application. Removal from the SIS is realized by unregistering the corresponding object. The object has to make sure that all connected activities and dependencies are then also disconnected.

Our approach to SIS can be characterized as being object-oriented. An object of the SIS implements at least the following structures and functions:

- user register
  Activities intending to use an object have to register on it. Thus depending on the access protocols implemented and applied to the object, users can be e.g. notified on object modifications caused by others.
- communication
  In order to communicate with an object, like modification, registration and so on, a receive function for messages has to be offered. Also the object itself has to have some kind of send function to communicate with other entities.
- privileges
  Typically the different users may have also different access rights to an object. Those have to be managed by the objects. Negotiation for privileges is executed via according protocols.
- dependencies
  Dependencies among objects are recognized.
- identifier
  An object has an unique identifier, and is stored by this in a name service.

All objects derive from the same interface, allowing dynamic handling of different objects types. Communication with an object relies on a known protocol (as opposed to fully dynamic conversations). This independency from the encapsulated resource eases the handling of different objects.

As explained before, dependencies are introduced for establishing complex object structures, e.g. to support task delegation scenarios with decomposed design items. The semantics of dependencies can be manifold. Currently we only look at "part of" and "is a" as dependencies, but whenever necessary, additional dependencies can be introduced. Of course the handling of dependencies can get in-efficient, when complex interactions between the objects have to be handled. It is up to specific implementations to find a compromise between functionality and efficiency. A dependency has to fullfil the following:

- object list
  Here all dependent objects are referenced. In case of e.g. a "part of" dependency, they can be distinguished regarding super-object(s) and sub-objects. By querying this list a sub-object may find its siblings and its super-object (vice versa for super-object).
- communication
  To realize interaction for dependencies and objects (e.g. notification) we also use the same communication interface like entities.
- identifier
  Like an object, a dependency is also stored in the name service by this identifier.

Typical scenarios involving dependencies include bill of materials, hierarchically structured objects, or objects with specific semantical relationships. Regarding our previous example, we may have a structure like the one shown in Figure 6. Further part-of relationships as for the chip set or the circuit board are existent, but have not been detailed. As one can see, the is-a relationship can be also very useful to express semantics for scenarios.

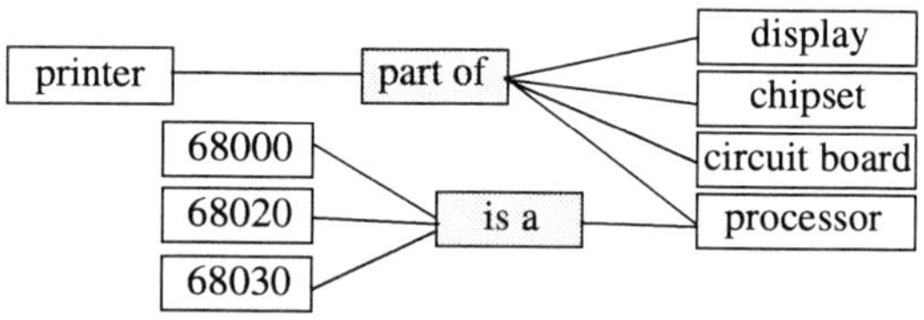

Figure 6. Object relationships for printer structure

2.3. Protocols

We picked aspects from conversation nets, speech acts [7] and agent technology to build powerful but simple protocols for interaction among the entities in our model. The advantage of combining those technolgies is the simple definition of protocols, comprehensibility and the possible deployment of a generic protocol engine, which can be customized to support different protocols. Our protocols realize the access, interaction and negotiation between the ADM entities. Thus cooperation patterns can be established and group awareness (e.g. querying all users of an object) achieved.

Figure 7 depicts a typical usage protocol, from the object's view, as a simple example. First the object receives the registration request of an actvity (`usage`). Now the object has to decide, whether to accept the request or reject it (`reject/accept`). This decision can be delayed when the object decides to negotiate parameters of the request (`propose`), e.g. by proposing a read only privilege instead of an exclusive one. This negotiation can take some iterations until a final agreement or denial is reached.

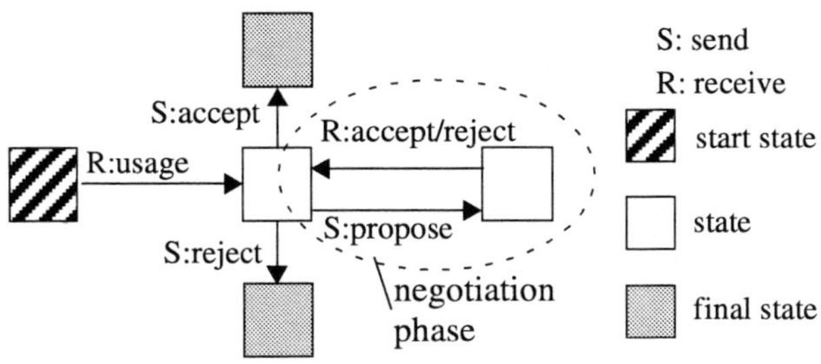

Figure 7. Usage request for SIS object

An object access protocol can look similar to the usage protocol. Instead of `usage`, the conversation topic would be for example `read`. But depending on the object type, an access protocol can become a lot more complex (Figure 8):

The activity E tries to modify object y (1). Since object y knows that it's in a tight coupled part-of relationship, it has to ask its super-object x for a permit to be modified (2). Object x's implementation requires registered activities to agree on changes, so it has to ask these activities for a permit itself (3). The result is transferred back to object y (4) and also made known to activity E (5). If the requested modification has been successful, the part-of dependency can then actively notify its other users object y about the modification (6) to ensure consistent views.

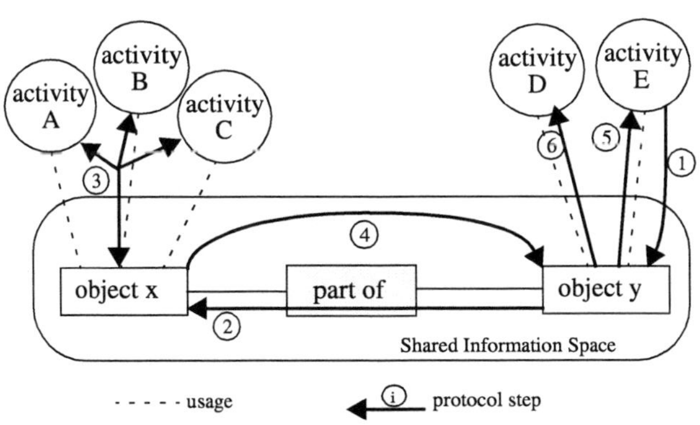

Figure 8. Complex access protocol

The above scenario is kind of a worst case of an access request (esp. if you take into account that step 3 could additionally require a further negotiation between activities A, B, and C to reach a common decision). But it illustrates what can be achieved through the combination of customizable protocols, dependencies, and objects. Generally one will not always implement dependencies and objects (and their corresponding protocols) as suggested by this scenario, but instead as is sensible for a specific application scenario.

3. Realization

We will now briefly describe how to realize the presented concepts and where customization is possible.

3.1. Object centered concept

The objects are a central concept to our approach since they are used to coordinate the work, raise group awareness, and support cooperation. Our implementation CASSY (*Cooperative Activity Support SYstem*) uses CORBA as middleware. This allows wrapping of proprietary systems and resources as well as the exploitation of existing services (events, workflow facility, ...). The groupware facility has been implemented by ourselves using CORBA. It offers typical groupware functionality like tasks, workspaces, groups, resources. Problems arise from the API supported by legacy systems, if they don't offer access to internal structures and functionality (e.g. IBM's FlowMark). In that case they can't be fully integrated by CASSY. E.g. if objects in the workflow system can not be locked over the API, CASSY has to make sure to lock the object for modification access outside of the workflow activity while the corresponding workflow is running.

ADM objects and activities are implemented as CORBA objects. Adaptibility of objects and dependencies is fairly easy by exploiting object-oriented concepts like inheritance. E.g. a simple part-of dependency implementation can just realize the storing of the referenced objects and querying them. When a tighter coupled approach is required, a dependency can be derived and additionally implement mechanisms like propagation of changes from one object to the other referenced objects.

3.2. Protocol realization

An adaptive object model alone is not sufficient for supporting manifold scenarios in design and concurrent engineering. Flexible protocols are as well needed to use the object model as suggested by the application scenario.

Being CORBA objects, the objects of the ADM have already some characteristics of active objects, which led us to look into agent technology. We found the approach of agent communication promising, especially since it is very adaptable and supports arbitrary interactions. For our prototype implementation we used the JATLite package [6] as platform for agent communication. The message format has been defined in KQML [2]. KQML is very flexible and thus lets us define very sophisticated protocols. As can be seen in the screenshot (Figure 9) we implemented some protocols and corresponding entities to test the definition of protocols and their usability. Currently we want to implement this approach into CASSY, because of the positive results from the prototype.

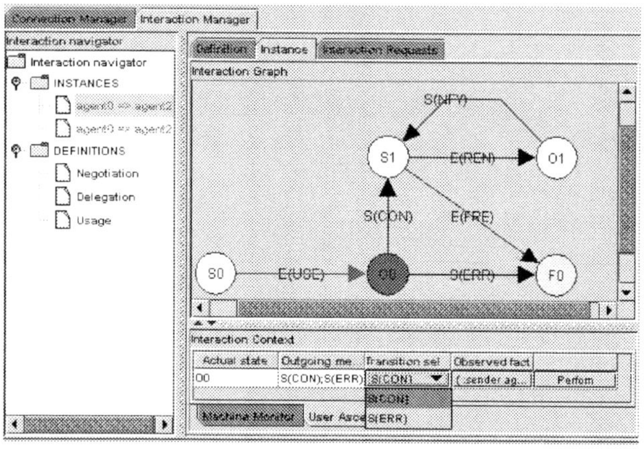

Figure 9. Screenshot of protocol agent

A further advantage of this implementation is that actors can also be integrated into protocols. The implementation can be used to offer an actor choices of answers (or requests) valid in a protocol (see list of transitions in Figure 9). Thus the proposed approach can allow direct interaction between actors and/or entities. If a protocol between entities reaches a point where human interaction is required a corresponding choice of messages can be presented to the actor. Through this we can integrate the actors more tight into the process.

4. Conclusion

Efficient design processes are characterized by the high degree of concurrency and interdependence of the single tasks or design steps. Those tasks are based on the structuring and execution of activities, and the usage of common data. We described a model to integrate shared objects in a common information space for concurrent activities. Our approach promotes cooperative techniques like awareness and negotiation through the actvity model and SIS combined with the protocols (e.g. delegation). This allows cooperation to take place between the actors and entities in our system model.

The adaptability enables one to customize the model to specific needs. The object protocols can be designed for many scenarios, even very sophisticated cooperation patterns. Object implementations can be suited to specific object types and exploit their full functionality through the generic message interface. Finally by the exploitation of dependencies, complex object structures can be established and functionally used. The degree of cooperation can be adapted to the scenario by accordingly adapting protocols and object granularity. Already the protocols we established realize a high degree of dynamic communication. We apply an implementation approach that is built on agent technology. In doing so the model can be even extended to dynamic cooperation interactions and ad hoc adaptability for newly arising situations.

Further work requires the implementation of exemplary protocols and object structures into CASSY to further validate the proposed concepts. Also it seems worthwhile to investigate whether those protocols could also be advantageous for communication between system services in CASSY, like the activity management or user client.

5. Acknowledgements

This work is partly sponsored by DFG grant MI 311/8. We also thank Erol Bozak for his work on the prototype implementation.

6. References

[1] Borghoff, U. M., Schlichter , J. H.: Computer-Supported Cooperative Work: Introduction to Distributed Applications. Springer, Berlin, 2000

[2] Finin, T., et al.: KQML: A Language and Protocol for Knowledge and Information Exchange. In: Fuchi, K., Yokoi, T.: Knowledge Building and Knowledge Sharing. Ohmsha and IOS Press, 1994

[3] Frank, A.: Towards an Activity Model for Design Applications. ISCA 14th Intern. Conference on Computers and Their Applications, April 7-9, Cancun, Mexico, 1999

[4] Frank, A., Sellentin J., Mitschang, B.: TOGA - A Customizable Service for Data-Centric Collaboration. Information Systems Journal, Vol. 25, No. 2, Elsevier Science Ltd., UK, 2000

[5] Gray, J., Reuter A.: Transaction Processing: Concepts and Techniques. Morgan Kaufmann Publ., San Mateo, CA, 1993

[6] Jeon, H., Petrie, C., Cutosky, M. R.: JATLite: A Java Agent Infrastructure with Message Routing. IEEE Internet Computing, IEEE, 200

[7] Winogard, T., Flores, F.: Understanding Computers and Cognition: A New Foundation for Design. Norwood, NJ: Ablex, 1986

A Web-based Collaborative Workspace Using Java 3D

Lihui Wang, Brian Wong, Weiming Shen, and Sherman Lang
Integrated Manufacturing Technologies Institute
National Research Council of Canada
800 Collip Circle, London, Ontario N6G 4X8, Canada
[firstname].[lastname]@nrc.ca

Abstract

This paper presents a framework for building web-based collaborative workspaces using the latest Java technologies – Java 3D, JavaServer Page (JSP), and Java Servlet. This web-based approach allows designers, engineers and production managers to share a common workspace that can be used for design review, production monitoring, remote control, and trouble-shooting, based on a set of interactive Java 3D models that represent the physical world with common interests. Following a brief overview of the related research work, this paper discusses the Java 3D concept from its scene graph structure to behavior control, and explains our approach to building web-based collaborative workspaces using Java 3D. The proposed framework uses the popular client-server architecture and view-control-model design pattern with a secured session control. Control logic and the interfaces, which interact with the real world, are handled by an application server through servlets. The benefits enabled by the framework include reduced network traffic, increased flexibility of remote monitoring, interactive control, web-based synchronous collaboration and quick response. It also shows a large potential for various web-based real-time and distributed applications.

1. Introduction

Looking at today's manufacturing, the decentralization of organizations has become a major success factor. Outsourcing, joint ventures, and global collaborations have led to work environments that are geographically distributed across organizational and national boundaries. During the last decade, the World Wide Web (WWW) has gained broad acceptance in academic and commercial areas. Today, it is widely used for development of collaborative applications to support dispersed working groups and organizations because of its platform, network and operating system transparency, and its easy-to-use user interface. In addition to the Web technology, Java has brought about a fundamental change in the way that applications are designed and deployed. Java's "write once, run anywhere" model has reduced the complexity and cost traditionally associated with producing software on multiple distinct hardware platforms. With Java, the browser paradigm has emerged as a compelling way to produce applications for collaboration over the Internet.

The objective of this research is to develop a Web- and Java-based framework with appropriate open architecture for effective collaboration among a dispersed engineering team. Following a brief overview of the related work, this paper presents our idea and approach towards building a Web-based collaborative workspace using the latest Java technologies – Java 3D, JSP, and Servlet. This Web/Java-based framework allows designers, engineers and production managers to share a common workspace that can be used not only for design review and production monitoring, but also for remote machine control and trouble-shooting, based on a set of interactive Java 3D models. Different from VRML (Virtual Realty Modeling Language), Java 3D gives end users high-level constructs for creating and manipulating 3D geometry using a separate behavior model. Instead of camera images, the physical world with common interests (such as a product, a device, or even a production line, etc.) can be represented by a set of Java 3D models with behavioral control nodes embedded. Once being downloaded, the 3D models are rendered by the CPUs at client side, but remain alive by connecting with the real world through message passing (sensor signals and control commands). This framework uses popular client-server architecture and the VCM (view-control-model) design pattern with secured session control. Central focus is given to the open architecture that supports the framework. As a proof-of-concept prototype, a user interface demonstrates cross-platform and distributed access to Java 3D models. It is believed that the proposed framework has large application potentials. It can be utilized to such areas as design verification, remote monitoring and control, and navigation in augmented reality.

2. Related Work

In 1993 shortly after its debut, Web was used by many as a medium to share data, information, and knowledge such as *SHARE* [1] and *PACT* [2]. Since then, a number of methods and frameworks have been proposed for web-based collaborative systems [3-8], but most of them are still under proof-of-concept prototype development stage. *DOME* is one of the web-based frameworks, which provides an open product development environment for distributed, collaborative, and integrated design [9]. Another commercial software suite called *ipTeam* [http://www.NexPrise.com/] is available from NexPrise Inc. for collaborative product development. However, it is primarily focused for the use of virtual enterprise and supply chain management. Most of the frameworks are developed for collaborative design and project management, while a few are focus on conflict resolutions during collaboration such as *CONCENSUS* [10] and *NegotiationLens* [11]. A shared workspace or framework for real-time distributed manufacturing (remote control, monitoring, inspection, and trouble-shooting, etc.) is still missing from the literature.

Web-based collaborative systems use the client/server architecture. In terms of technologies used in the existing collaborative systems, HTML and Java Applets are widely adopted for developing the client side user interfaces, in addition to ActiveX and VRML [8][12], while server site technologies including JSP, XML, and Servlet are still obtaining attentions for future system developments. To facilitate a viable collaborative environment, application servers must engage users in a 3D graphical interaction in addition to the dialog-like data sharing, because remote engineers and managers need active and visual helps to coordinate their efforts in a distributed environment. Reducing network traffic and increasing system performance are the other concerns in collaborative system developments. The framework, proposed in this paper, is focused on how to facilitate users in working effectively in a collaborative workspace.

3. A Web and Java-Based Framework

As new classes of collaborative applications emerge for the Web environment, there is increasing pressure for full multimedia capabilities (audio, video, and 3D graphics) to become seamlessly integrated into the browser paradigm. Users are now demanding a collaborative workspace that is easy to follow, operate, and maintain. In addition to software innovations, today's technologies have made accelerated 3D graphics- and multimedia-capable hardware affordable for such advanced Web applications. However, the obvious conservation of the Internet bandwidth makes itself the bottleneck for most multimedia Web applications when real-time constraints are required.

Targeting at distributed, real-time monitoring and control in manufacturing sectors, a framework of high efficiency for Web-based collaboration is carefully examined in this research. The resolutions of meeting both the user requirements and the real-time constraints are: (1) using interactive Java 3D graphic models instead of camera images for visualization; (2) Choosing lightweight Swing components for client-side graphical user interface (GUI) implementation; and (3) encapsulating major control logic at server side by JSP and Servlet.

As shown in Fig.1, the proposed framework is based on the popular VCM (view-control-model) design pattern by dividing itself into three tiers: *Presentation*, *Application*, and *Data*. Java 3D is used to model the physical world that is of common interests. A Java 3D model may reside in both *Data* and *Presentation* tiers. Once downloaded to a standard Web browser, the Java 3D model is rendered by using local CPU and thus can provide a look-and-feel of its physical counterpart. The global behavior of the Java 3D model is driven by real-time sensor signals, which are the limited datagrams transmitting from server to the client machine. On demand, an end user can send requests (or commands) for other services (or instructions) to the application server, or broadcast messages to other team members who are participating in the same collaborative workspace for discussion and trouble-shooting.

Security issues, such as session control and user access control, are handled by the application server.

For the same reasons, the physical world (milling machine, robot, and turning machine, etc.) is only accessible by *SignalCollector* and *Commander* that reside in the application server. The *SignalCollector* is responsible for data collection from the physical devices. It utilizes the concept of *Pushlet* developed by Just van den Broecke [http://www.javaworld.com/jw-03-2000/jw-03-pushlet.html].

Based on the Publish-Subscribe design pattern, the *SignalCollector* maintains a list of subscribers for real-time data distribution. Authorized users can submit commands through *CyberController* to the server. The *Commander* then takes over the control for device manipulations. Another server component called *DataAccessor* is designed to separate logical and physical view of data. It encapsulates complex JDBC and SQL code and provides standard methods for accessing data (Java 3D models, historical data of the devices, or XML documents). The historical data are found helpful for trouble-shooting, while XML will be used for data communication in future extension. All transactions need to pass through a session manager on the server, who checks user credibility and assigns proper access rights. Although the global behavior of a Java 3D model is controlled by the server based on real-

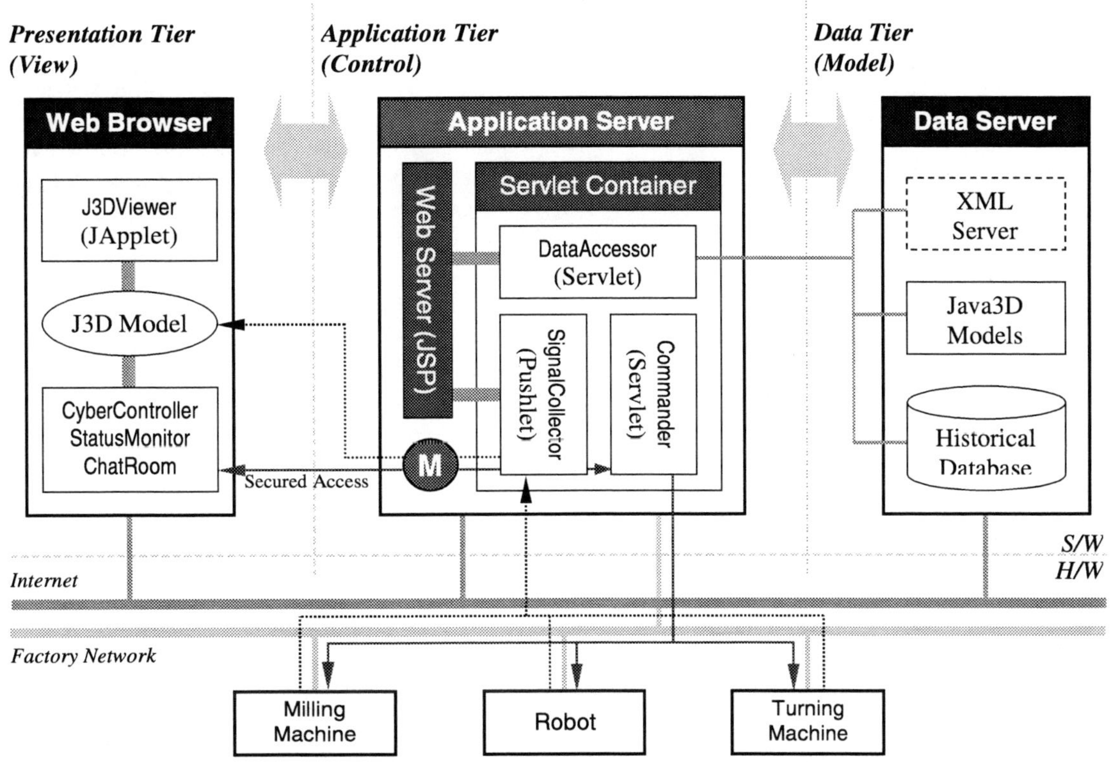

Figure 1. Web and Java-based framework.

Table 1. Modular functionality of the framework.

DataAccessor	Server-side component. Designed to separate logical and physical view of data. Providing standard methods for accessing data to/from a data server. Responsible for serving Java 3D models to clients and recording device data.
SignalCollector	Server-side component. Responsible for sensor signal collection through handshaking with physical devices. Maintaining a list of subscribers for sensor data distribution using multicasting.
Commander	Server-side component. Responsible for controlling the physical devices on behalf of a client user. Separating clients from the physical world to enhance security.
M	Server-side session control manager. Authorized to check clients' authentication and assign appropriate access rights back to the clients.
J3DViewer	Client-side component. Built by lightweight Swing. Providing means for end users to control the view of Java 3D models, individually.
CyberController	Client-side component. Providing a set of widgets for end users to manipulate the physical devices through a collaborative workspace. User authentication is required.
StatusMonitor	Client-side lightweight component (JTables). Used to monitor the run-time status of a physical device whose data are served by the application server.
ChatRoom	Client-side component. A shared space designed for negotiation and trouble-shooting. Messages are exchanged and coordinated by server-site session manager.

time sensor signals, users still have the flexibility of monitoring the 3D model from different perspectives, such as selecting model, changing viewpoint, and zooming, etc., through *J3DViewer*. Table 1 summarizes the major functionality of each component within the proposed framework.

4. Modeling in Java 3D

Java 3D is considered to be a mid- to high-level fourth-generation 3D API [13]. What sets a fourth-generation API apart from its predecessors is the use of scene-graph architecture for organizing graphical objects in the virtual 3D world. Unlike the display lists used by the third-generation APIs (such as VRML, OpenInventor, and OpenGL, etc.), scene graphs mercifully hide a lot of the rending details from the programmer while offering opportunities for more flexible and efficient rending. Enabled by the scene-graph architecture, Java 3D provides an abstract, interactive imaging model for behavior and control of 3D objects. Because Java 3D is part of the Java pantheon, it assures developers ready access to a wide array of applications and network support functionality [14]. Figure 2 illustrates a scene graph for a Direct Drive testbed at authors' lab. This testbed consists of a linear slide and a brushless rotary motor with encoder feedback.

Java 3D differs from other scene graph-based systems in that scene graphs may not contain cycles. Thus, a Java 3D scene graph is a directed acyclic graph. The individual connections between Java 3D nodes are always a direct relationship: parent to child, as shown in Fig.2. The scene graph contains a complete description of the entire scene, or virtual universe. This includes the geometric data, the attribute information, and the viewing information needed to render the scene from a particular point of view.

As shown in Fig.2, the *Virtual Universe* object provides a grounding for the entire scene graph. All Java 3D scene graphs must connect to a *Virtual Universe* object to be displayed. A scene graph itself starts with the *BranchGroup* nodes (although only one BG node in this case). A *BranchGroup* node serves as the root of a sub-graph, or branch graph, of the scene graph. The *TransformGroup* nodes inside of a branch graph specify the position, the orientation, and the scale of the geometric objects in the virtual universe. Each geometric object consists of a *Geometry* object, an *Appearance* object, or both. The *Geometry* object describes the geometric shape of a 3D object. The *Appearance* object describes the appearance of the geometry (color, texture, material reflection characteristics, etc.). The behavior of a Java 3D model is controlled by *Behavior* nodes, which contain user-defined Java codes and state variables. Sensor signal processing can be embedded into the Java codes of a Java 3D model, for remote monitoring of its physical counterpart in the real world. Once applied to a *TransformGroup* node, the so-defined behavior control affects all the descending nodes. In this example, the two movable objects – Linear Slide and Rotary Motor are both controlled by two behavior nodes, one for online control/ monitoring and the other for offline simulation/animation.

5. Collaborative Workspace

In the near future, distributed design, manufacturing, and facility management can easily be conducted through

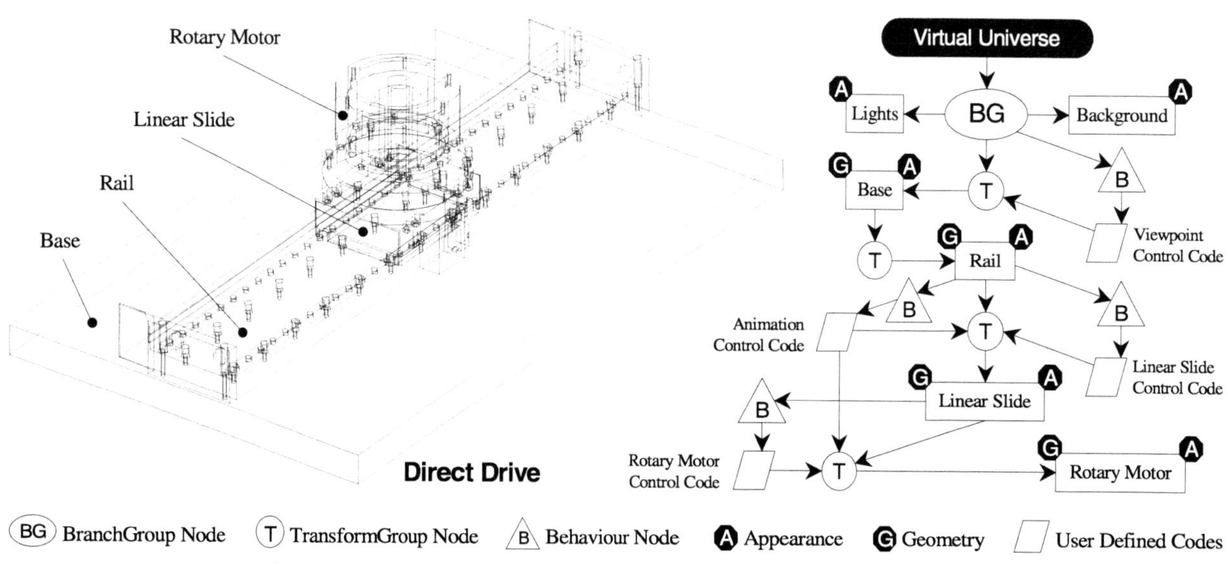

Figure 2. Scene graph architecture of Java 3D through an example.

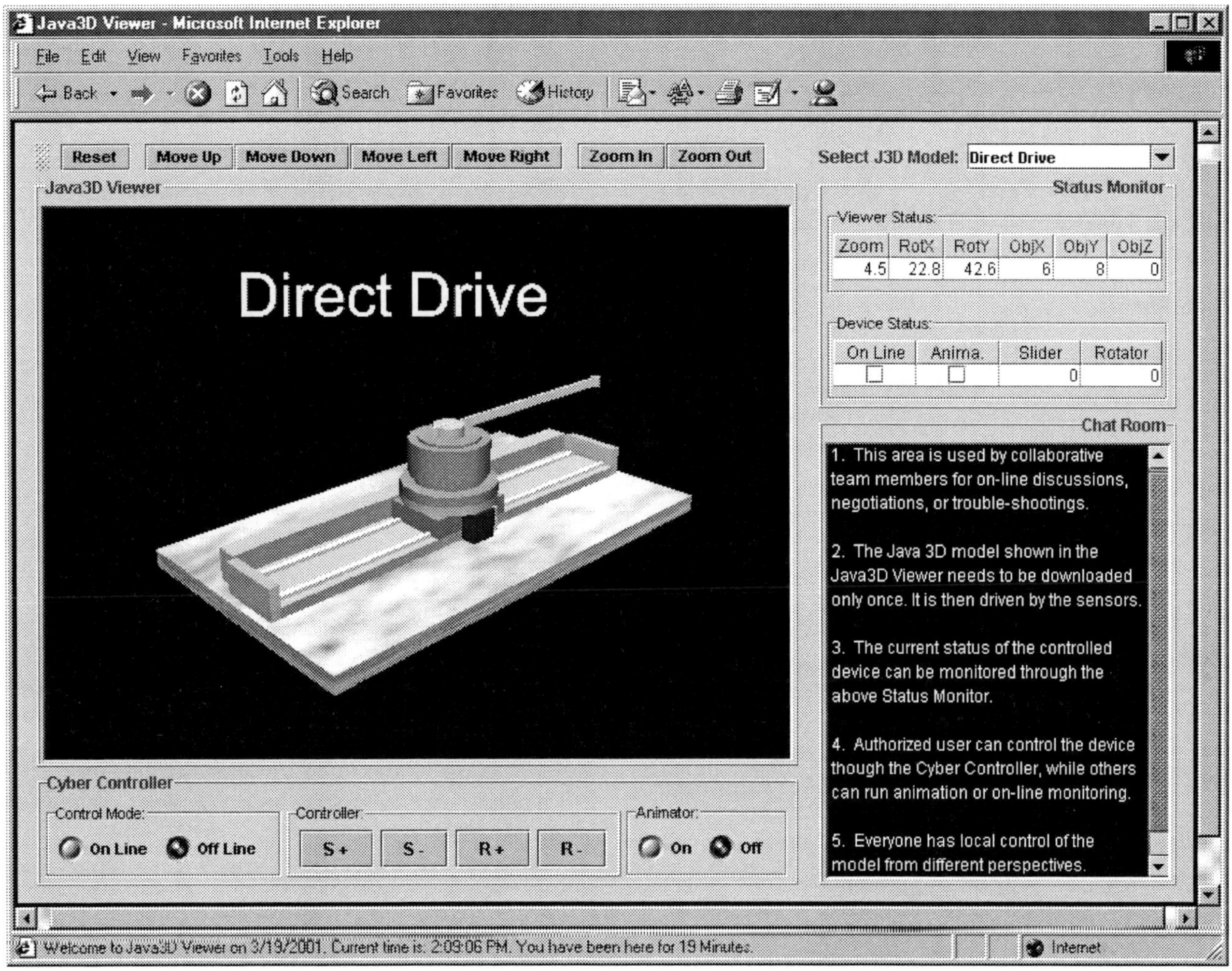

Figure 3. Collaborative workspace from users' perspectives.

an effective collaborative workspace. As shown in Fig.3, such a collaborative workspace prototype has been implemented on top of the proposed framework, through a Java applet-based graphical user interface. A collaborative team member can share real-time data with others while performing his or her own duty through this portal.

The shared workspace eliminates network traffic by replacing camera images with Java 3D models. By choosing a viewpoint carefully, an engineer can even use the "walk-through" capability of 3D models to perform internal inspections or trouble shootings of physical devices that are impossible by using optical cameras. The largely reduced network traffic also makes real-time monitoring, control, inspection, and trouble-shooting practical for users on relatively slow hock-ups such as modem and wireless connections. This is especially useful for field engineers or shop floor managers who want to access rich information through cellular phone lines. Figure 4 illustrates the data flow via the workspace.

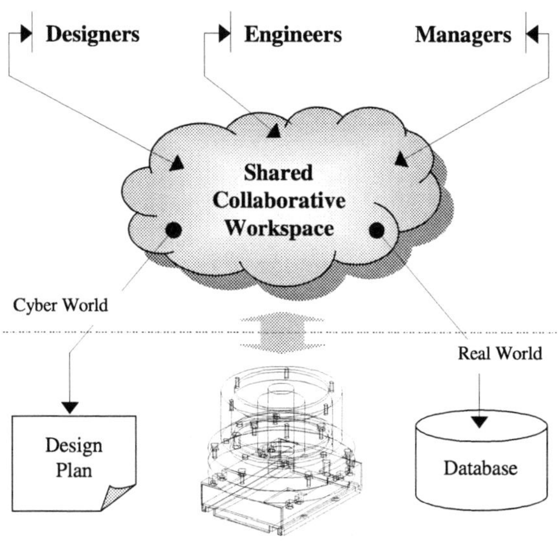

Figure 4. Data flow via a shared workspace.

The functionality of the shared workspace is largely determined by the control and decision-making logics that are embedded in the application server by means of servlets. Responding to end users' requests, these servlets should be able to search and provide solutions effectively. Permissions of user accessibility to the physical devices via the workspace are granted by a session manager (see Fig.1), based on the authentication of each participating member. Complete server-side functionality of the shared workspace is currently being developed at authors' lab.

6. Discussions

As an extension to the underlying framework, XML (eXtensible Markup Language) may slash the efforts needed to define the control logics and to communicate between various physical devices, software modules, and Web browsers.

In the near future, open-architecture devices (like OpenPLCs and CNC Controllers, etc.) will have web servers and Java VMs (virtual machines) embedded. This will make the proposed framework and the workspace function more efficiently for raw data collection, real-time data transmission, and synchronous communication.

The proposed web and Java-based framework shows promise for applications and extensions to other areas. For example, if combining Java 3D models properly with camera images, the so-called augmented reality can be applied to web-based multi-party collaborations, such as in construction where a virtual structure is merged into a real scene of landscape for design verification. Once it is recognized as a promising technology, Java 3D will be supported by major CAD system venders. It is quite likely that independent software suppliers will build authoring systems that generate Java 3D applications as output [15].

7. Conclusions

This paper proposes a Web and Java-based framework for efficient and distributed collaborations. Java 3D is used as one of the enabling technologies in developing the framework. A shared collaborative workspace is implemented on top of the framework for use of remote monitoring, control, and trouble-shooting. Participating in the workspace, a collaborative team member not only can feel reduced network traffic by real-time interactions with quick response, but also can obtain more flexible control while still enjoying intuitive visualization of the real world. A large application potential of this work is anticipated. It can be extended for remote inspection even if the physical counterpart of the model is invisible or non-accessible (inside of a nuclear power plant or outside of a space station). The benefits enabled by the framework include reduced network traffic, increased flexibility of remote monitoring, interactive control, quick response, and Web-based synchronous collaboration.

8. References

[1] G. Toye, M. Cutkosky, L. Leifer, J. Tenenbaum and J. Glicksman, SHARE: A Methodology and Environment for Collaborative Product Development, *Proceedings of 2nd Workshop on Enabling Technologies: Infrastructure for Collaborative Enterprises*, pp.33-47, 1993.

[2] M. Cutkosky, R. S. Engelmore, R. E. Fikes, M. R. Genesereth, T. R. Gruber, W. S. Mark, J. M. Tenenbaum and J. C. Weber, PACT: An Experiment in Integrating Concurrent Engineering Systems, *IEEE Computer*, **26/1**, pp.28-37, 1993.

[3] J. Numata, Knowledge Amplification: An Information System for Engineering Management, *Sony's Innovation in Management Series*, Vol.7, Sony Corporation, Japan, 1996.

[4] S. F. Li and A. Hopper, A Framework to Integrate Synchronous and Asynchrinous Collaboration, *Proceedings of the IEEE Workshop on Enabling Technologies Infrastructure for Collaborative Enterprises*, pp.96-101, 1998.

[5] N. H. M. Caldwell and P. A. Rodgers, WebCADET: Facilitating Distributed Design Support, *IEE Colloquium on Web-based Knowledge Servers*, London, U.K., pp.9/1-9/4, 1998.

[6] G. Q. Huang and K. L. Mak, Web-based Collaborative Conceptual Design, *Journal of Engineering Design*, **10/2**, pp.183-194, 1999.

[7] J. L. Rodgers and A. O. Salas, Toward a More Flexible Web-based Framework for Multidisciplinary Design, *Advances in Engineering Software*, **30/7**, pp.439-444, 1999.

[8] U. Roy and S. S. Kodkani, Product Modeling within the Framework of the World Wide Web, *IIE Transactions*, **31/7**, pp.667-677, 1999.

[9] G. D. F. Pahng, S. Bae and D. Wallace, A Web-based Collaborative Design Modeling Environment, *Proceedings of the IEEE Workshop on Enabling Technologies Infrastructure for Collaborative Enterprises*, pp.161-167, 1998.

[10] S. Cooper and A. Taleb-Bendiab, CONCENSUS: Multi-Party Negotiation Support for Conflict Resolution in Concurrent Engineering Design, *Journal of Intelligent Manufacturing*, **9**, pp.155-159, 1998.

[11] B. Adelson, Developing Strategic Alliances: A Framework for Collaborative Negotiation in Design, *Research in Engineering Design*, **11**, pp.133-144, 1999.

[12] G. Q. Huang, S. W. Lee and K. L. Mak, Web-based Product and Process Data Modeling in Concurrent "Design for X", *Robotics and Computer-Integrated Manufacturing*, **15/1**, pp.53-63, 1999.

[13] J. Barrilleaux, *3D User Interfaces with Java 3D*, Manning Publications Co., 2001.

[14] H. Sowizral, K. Rushforth and M. Deering, *The Java 3D API Specification, 2nd Edition*, Addison-Wesley, 2001.

[15] Sun Microsystems, *The Java 3D API*, Technical White Paper, July 1997.

Multi-user Shared Virtual Reality in the Exhibition of Chinese Nationalities
—*Virtual Museum of Chinese Nationalities*

Jun Zhang, Fengsen Li, Hua Li
Lab of Intelligent Information Processing, Institute of Computing Technology,
Chinese Academy of Sciences, P.O. Box 2704, Beijing, 100080, P.R. China
zhangjun@ict.ac.cn, lfs@ict.ac.cn, lihua@ict.ac.cn

Abstract

Virtual Museum of Chinese Nationalities is a multi-user shared virtual reality system developed to efficiently popularize the cultures and folk-customs of various ethnic groups for the public by means of cooperative work. We regard the project as long-term research, in that there are so many problems to solve. Unlike existing systems, we should maintain scalable and interactive performance on a wide variety of computing platforms, not only high-end graphics workstations, and distribute the virtual world via a bandwidth-limited network. In this paper we will introduce some architectures and algorithms realized in our current prototype, e.g., shared event mechanism, area of interest algorithm, distributed server architecture, etc.

1. Introduction

As is well known, China is a great country with about 5000 years of resplendent civilization. All of her 56 nationalities have their own brilliant cultures. How to conserve those cultural treasures, and popularize them for the public to provide more opportunities for the research and study in the history, folk-custom, and geographical distribution of various ethnic groups using advanced technologies, has attracted more and more attention of the society.

On the other hand, with the development of computing technology, virtual reality (VR) and computer-supported cooperative work (CSCW) have gradually become the focus of people's attention, and the combination of the two technologies has become a trend. More and more architectures and systems in these fields are developed to solve some practical problems, such as the distributed virtual environment (DVE).

The well-known SIMNET [1] is the original DVE system developed by some research groups of DoD background, which is often designed for military purpose, to develop a networked simulation environment for training small units to fight as a team.

Nevertheless, they often provide limited support for the collaboration between multiple users and can hardly be used for arbitrary location independent application environments.

Some academic research groups presented the multi-user environment (MUE) and did many works in multi-user collaboration such as W. Broll [2], who described a framework for scene sharing and in the meanwhile guaranteeing the consistency. These methods bring me some enlightenment, but they are usually based on dedicated and expensive hardware and I/O devices, which makes those technologies difficult to be popularly accepted by the public.

So our work should be done to realize a multi-user shared virtual environment to content actual requirements and apply it on commonly low-end hardware platforms and through a bandwidth-limited network.

In those backgrounds, we developed the *Virtual Museum of Chinese Nationalities* (or more appropriately, a Virtual Park) based on our initial prototype *VRBuilder* (a small platform used to construct multi-user virtual environments on the Internet). Our virtual museum consists of many distributed ethical communities of various ethnic groups, and each community is composed of several ethical constructions, a virtual guide (a smart agent) who can interact with the visitors, several virtual performers exhibiting their folk-customs, showrooms for history and folk-custom, surrounded sound effect, and finally, the avatars of other visitors who you can communicate with. You can login as a tourist, to visit anywhere you like to go and communicate with anybody connected. Figure 1 shows a scene on the client side.

In brief, our system has elements of a CSCW environment, a virtual reality system and an on-line Schmoozing forum. And the main idea of our approaches is to leave as many as possible roles taken over by the client to reduce the burden on the server and the network transfer.

2. System architecture

This section describes some essential pieces of our system architecture (Figure 2).

Figure 1. Screenshot of *Virtual Museum of Chinese Nationalities.*

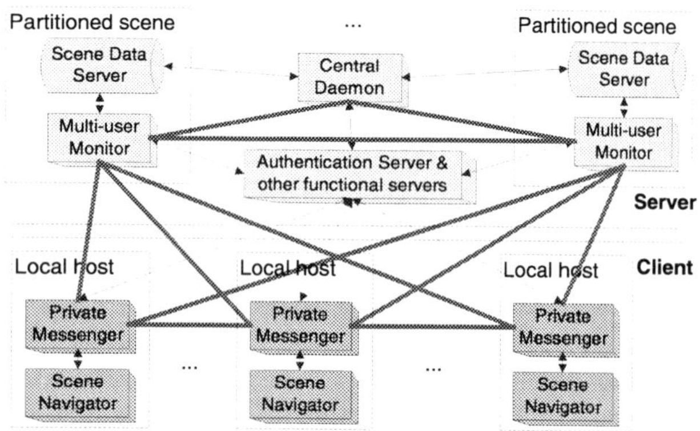

Figure 2. System network architecture.

The following items are basic elements of system architecture:

- **Central Daemon:** the control center of our system. It is responsible for the management of the other server components, (e.g. partitioning of subdivided scenes, etc.).
- **Multi-user Monitor:** the communication mechanism of the server, which is responsible for monitoring the user connections (whether a visitor join the world or leave it), and forwarding all user updates to other participants.
- **Scene Data Server:** the database of the distributed virtual worlds.
- **Other functional servers:** such as the Authentication Server, etc.
- **Scene Navigator:** the 3D browser of our system that renders the 3D scene and allows users to navigate through it. In other words, it is a local process to control the user's own avatar.
- **Private Messenger:** the communication mechanism of the client designed to receive the modifications of the remote avatars and subsequently update the position of their local representatives (we name them "ghosts"), and some other server information. On the other hand, it sends all local updates to the server and forwards them to all other current visitors.

2.1 Transmission Protocols

First of all, we should employ an appropriate protocol strategy to guarantee a maximum of interactivity as well as reliability.

In order to increase interactivity, we take some connectionless network protocols into account and employ them in data transmission while employing some connection-oriented communication mechanisms such as TCP/IP to establish a connection.

In our system, we mainly use IP Multicasting groups [3] to address a large number of users. Multicast communication allows a single message send to be delivered to a group of receivers. In hardware supported multicast environments, e.g. Ethernet, this allows for very efficient messaging. In the Internet, an experimental multicast layer built above IP has been used.

But a general problem of using multicasting is, that the IP Multicast protocol is neither reliable nor order preserving. Thus packets might get lost, be duplicated or arrive in different orders. The reason is, that multicasting is a connectionless service similar to UDP/IP.

Several approaches exist to build reliable protocols on top of IP multicast. Although some of them (e.g. ISIS/HORUS [4]) can even provide a total ordering of all messages within a distributed system, they do not fit very well for the task. The reasons are:

- High interactivity can not be provided, if all messages have to be acknowledged by all visitors
- Realizations do not scale very well to large numbers of visitors, especially for frequently changing sets of visitors
- Solutions require additional UDP/IP transmission of messages between all visitors

That is why we use our slightly different approach. In general, reliable transmission can be realized either by sending acknowledge message when receiving a message or by sending negative acknowledges, when not receiving a message or receiving a corrupted message. Similar to ISIS and some other approaches we use positive acknowledge messages.

2.2 Consistency

Because the distributed virtual space is shared, all participants of that space must see it consistently. To sharing a virtual space, we mean not only independent interactions, e.g. navigating through the scene, but also the modification of the scene itself by mutually exclusive interactions (e.g. moving scene objects), and furthermore, we also allow for concurrent multi-user interactions (involving two or more participants who concurrently aim at the same object).

So, some synchronization mechanisms are needed to handle access-control, i.e. read / write access to distributed objects.

Independent interactions do not require any synchronization, whereas mutually exclusive interactions require absolute synchronization, in addition to which, multi-user interaction requires synchronization of multiple users. There are some existing approaches provided to address synchronization issues.

And we use an optimistic approach relying on social locks rather than on synchronization mechanisms provided by the system — requiring fewer messages to be transferred. Additionally our approach uses locks on a per-interaction basis instead of a per-object basis used by traditional approaches, increasing the interactivity of users and enables us to support advanced features such as multi-user interactions.

In our model, interactions that require synchronization try to acquire a lock on activation. The lock is released on deactivation of the specific interaction or after a certain time-out.

Participants usually cannot acquire a lock on an interaction already locked. The multi-user monitor as well as each client keeps track of all locks. If the multi-user monitor receives a lock message for an interaction already locked, the lock is rejected, by sending an appropriate message. This message also will correct the locks on any clients, which might have received the lock requests in a different order as the multi-user monitor. However if a lock can be realized, there are actually two possible approaches. Either the lock is acknowledged, or all locks that are not rejected within a certain period are estimated to be acquired. However the second alternative might fail due to network delays. Locks that are not released by a participant within a certain time are released by the multi-user monitor. This is realized by sending an appropriate message and resetting the lock. This locking mechanism allows us to support multi-user locks. It can be acquired several times by different participants before the lock is rejected. This can easily be supported, by adding a counter to each lockable interaction. Multi-user locks are especially useful for concurrent multi-user interactions.

2.3 Server

The server of our system consists of the central daemon, the multi-user monitor, the scene data server, and other functional servers, but the most important component on such side is the multi-user monitor, which is the spirit of multi-user environment.

- **Multi-user Monitor**

In our approach we use IP Multicasting as the default communication protocol between all visitors of a particular virtual world. Although all visitors of a virtual world are equal and can send events directly to other visitors by this mechanism, this architecture is not completely symmetrical. The reason for that is, that new visitors or visitors separated (disconnected) for a certain time (e.g. due to network failure) need a certain (fixed) dial-in point to connect to the virtual world. This requires a location, where the current state of the virtual world is kept or even updated continuously, even if no visitors are connected to this particular world at that time. In our model this is realized by the multi-user monitor. The multi-user monitor is a kind of home of the virtual world. Similar to files located on an HTTP Server, the virtual worlds managed by a multi-user monitor are specified by a URL.

Besides providing a unique identifier for accessing a shared world, the multi-user monitor is responsible for

- The transmission of virtual world contents including the remote avatars' local representatives (ghost)
- Assigning multicast groups and ports to virtual worlds or parts of them
- Sending appropriate messages to support reliability for these multicasting groups
- Supervising the presence of connected participants
- Supporting visitors to recover from temporary disconnections
- Supporting consistency preserving mechanisms, such as locking

Although the multi-user monitor has to perform a number of tasks in our model, its average load to support a single shared virtual world is even lower than that of an individual visitor that has to realize the visualization of the virtual world in addition to the network communication. Thus our approach is much closer to a symmetrical solution than to a traditional client/server model. Multicast groups and port numbers are assigned by applying a hash function on the URL of the virtual world, if not defined explicitly within the virtual world.

2.4 Client

The client supports the VRML97 standard and uses Java as its scripting language, because VRML, whose initial version was based on SGI's Open Inventor file format and has been closely related to the WWW, has become the standard description language for the distribution of 3D models or virtual worlds on the Internet, and on the other hand, only pure Java solutions (such as Java3D) are unable to be an unaided basis for mass participation at distributed virtual events.

When a VRML file is downloaded to the local 3D browser that renders its contents, the script nodes in the VRML scene point either to local Java scripts, or to Java scripts on an HTTP server. In the latter case, our client uses the associated HTML browser to subsequently download these scripts. Scripts are able to manipulate scene graph nodes by generating events that are delivered to the node and change one or more of its properties, for example, its position in the scene, its shape or one of its material attributes. Obviously, since the scripts are fully functional Java code, they are not restricted to just changing the scene graph. They can, for example, dynamically generate additional VRML nodes, or locate and add existing VRML to the base scene downloaded in the original VRML file. This may be carried out using a call to an HTTP server or by a request to another network machine. Further, they can also interact with other applications, for example mining data from the scene database that can subsequently be turned into VRML and added to the shared scene.

2.5 Joining & Leaving a World

When new visitors join a virtual world, specified by an URL, they first connect to the multi-user monitor via a reliable unicast connection (see figure 3). The virtual world description is then transmitted to the new visitor. In our prototype this is realized by transmitting VRML code. Since the world's contents continuously changes, this description cannot be transferred from a file but has to be generated on the fly by the multi-user monitor.

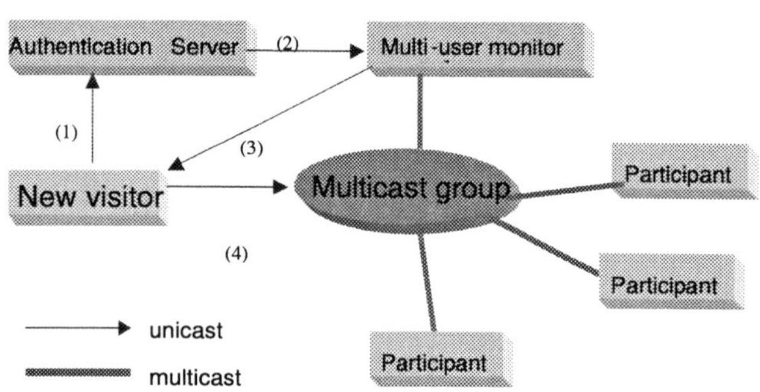

Figure 3. Joining a new world.

Nevertheless new messages, modifying the virtual world contents, might arrive during this transfer. While messages are usually delivered (incorporated into the virtual world) as soon as they have been completely received, the incoming message queue is frozen during these initial transfers. This means, that new messages will still be received and acknowledged, but do not modify the world as stored within the multi-user monitor immediately. As soon as the transfer is completed, the multi-user monitor continues to modify the current virtual world according to the received messages. Finally the messages that have not been included in the transmitted world are sent to the new visitor. Additionally it receives the multicast group address and port to connect to for further messages. It also receives one or several addresses for unicast (UDP/IP) connections that are used for recovery as well as for non-multicast capable clients. These connections will usually be located at a different host than the one running the multi-user monitor. If the client is not multicast capable, is has to send a connection message to one of these addresses to receive all subsequent messages by this connection. As soon as the new visitor has received its first message as well as the corresponding acknowledge message at a UDP or multicast port, the connection is assumed to be stable and the TCP/IP connection is closed down. Since most participants are represented by avatars, the first message transferred by each new visitor usually contains a description of the user's embodiment in the virtual world. When leaving a world, the visitor is supposed to send an appropriate message. The multi-user monitor however, will detect inactivity of visitors and remove them after a certain time out in order to prevent virtual corpses.

3. Key issues

We have two important things to do in this work.

First, client-oriented retractility should be realized, which means that scalable and interactive performance should be maintained on a wide variety of computing platforms, not only high-end graphics workstations. In order to solve this problem, we should do at least two things:

- Probing pre-transmission
- Retractility of the entity description (include that of the scene and avatars)

Second, we should break through the bottlenecks of the network transfer bandwidth. There are so many algorithms we can submit to address this issue, like shared event, area of interest algorithm, and distributed server architecture, which are described in the following subsections, and some other methods, such as using the dead reckoning techniques [5] to predict the state of each entity in a scene, and thus to reduce network traffic, and the combination of image based rendering (IBR) [6] and traditional geometry based rendering, etc.

3.1 Shared Event Mechanism

In our multi-user shared system, each participant has at least one avatar, and each avatar can act on not only each other, but also their environment. It means that, any modifications that occur in the shared scene must be duly propagated to all participants in that space to inform them of updating.

So one of the most important things is to decide what does the data transmission contain on different condition. If in most case it were the updates on current scene contents that overwhelmed network transfer, the server would suffer from tremendous data transmission.

Other than world descriptions, there are some

information types, which consume significantly less bandwidth, to be shared by other participants and make the scene consistent, such as events, instructions, parameters, etc. And sharing events by sending those events to other participants via the server is one of the most efficient ways.

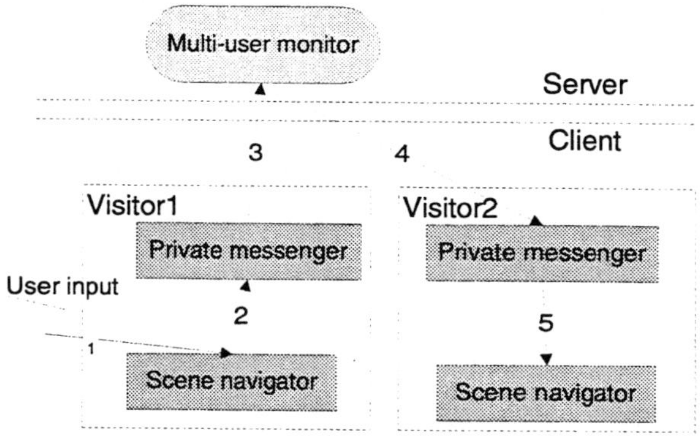

Figure 4. Flow chart of shared event mechanism.

In figure 4 we can see message flows as a result of a user interaction (e.g. a mouse click) in the shared event model. A user interaction (1) causes a local script to run (2). This in turn converts the event into a message and sends it through its private messenger to the multi-user monitor (3). The multi-user monitor sends the message to all other private messengers (4) who then convert the message to an event that causes execution of the local script (5).

3.2 Areas Of Interest (AOI) Algorithm

In previous experiments, we have observed that visitors form sub-groups where activities occur in clusters or peer-to-peer within the global session. This mimics the way we use the spatial model in the real world. The observation can be exploited to decrease overall message passing if one can deliver packets only to the recipients they are intended for, i.e. those within the sub-group. In this way, the amount of global traffic is limited, and the number of incoming messages to each user is reduced.

Using the three dimensions of space is a well known approach to partition virtual worlds into several more or less disjoint AOIs. Static geographical regions are used in applications based on natural terrains such as in DIS based systems [7].

It is this model that we seek to use to drive our consistency mechanism and to reduce the number of visitors in any consistency algorithm.

To achieve this, the server is structured as shown in figure 5. An AOI manager (AOIM) is responsible for tracking the spatial location of any user and for determining if two user's AOIs have collided. If they have, the AOIM causes those two avatars to join a consistency group, which is defined as a set of objects who have shared data that must be maintained consistent. For example, in figure 5, user 1 and user 2 are in each other's AOI, but user 1 is not in user 3's AOI. Thus, any updates to user 3, e.g. a position update, will be sent to user 2 but not user 1. Ghosts, i.e. local representatives of the remote avatars, denote the actual replicas. In the case where the avatars are all local to one server, these ghosts are generally pointers to the master avatar.

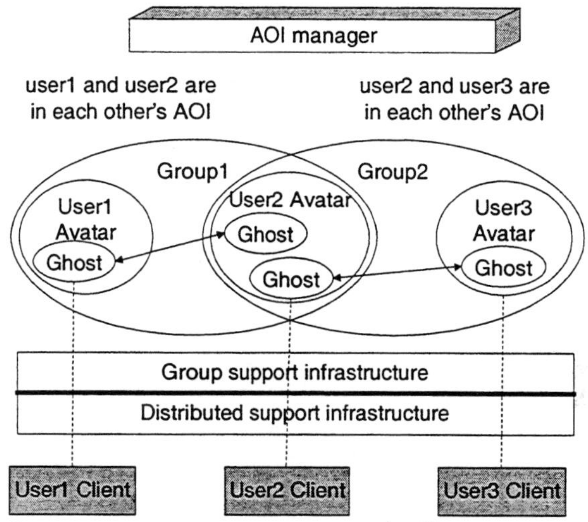

Figure 5. Areas of interest and groups.

In essence, the AOIM is responsible for defining groups of spatially co-located avatars that need to maintain a degree of consistency. As it decreases the degree of sharing, this mechanism is used to reduce the amount of information that has to be sent out from the server as a result of any state changes.

3.3 Distributed Server Architecture

Moreover, as the scale of the virtual worlds that consists of various ethical communities (especially those geographically far away from each other) increase, the server eventually becomes a bottleneck as it manages state on behalf of these visitors, and forwards messages between visitors, and between scenes and visitors. A major mechanism to reduce the required network traffic among visitors of shared virtual worlds is replicating the scene data server at which the virtual worlds are located and partitioning such worlds into several zones (distributed scene data servers). This mechanism is also called Zoning. In our system, to address this issue, we need to replicate the scene data server as well as the multi-user monitor, which constitute a server pair. By this means, we are able to spread the processing and communication loads between several server nodes and so scale the entire system.

The model described above deals gracefully with

server replication. Our ghosts become real remote avatars and are responsible for updating their replicas.

The basic architecture of the distributed version of the server is shown in figure 6. The AOI manager (not shown) now becomes a replicated entity that we organize in a hierarchy. Each server pair has an associated AOIM that is responsible for periodically sending updates to a master AOIM (i.e. the next AOIM in the hierarchy) who then calculates AOI collisions between avatars managed by different server pairs. In the case of a collision, the master AOIM informs the other AOIMs, who then add the colliding avatar to the respective consistency groups.

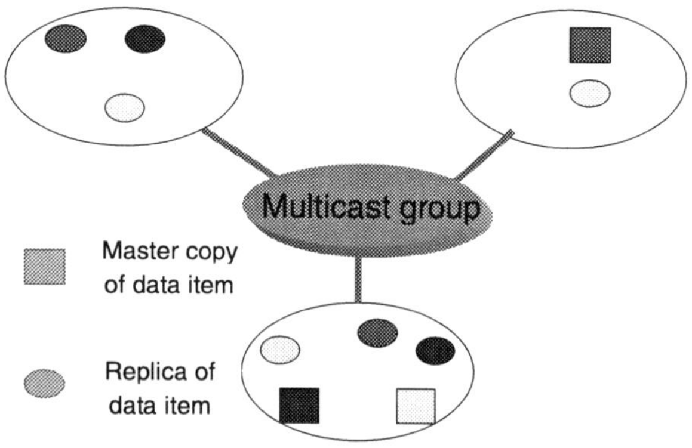

Figure 6. Distributed architecture.

4. Results and future work

Virtual Museum of Chinese Nationalities is a multi-user distributed virtual reality system used to efficiently popularize the cultures and folk customs of various ethnic groups for the public. We have implemented a prototype using VRML as the scene description language and Java as the scripting language. We temporarily choose two ethnic groups (the Hans and the Mongolians) to realize and test our prototype on a local area network environment constructed with several common PC computers. Figure 7 shows the different view of two independent users when they are looking at each other. We may notice their respective avatars.

From the implementation of our prototype, we can draw a conclusion that the system infrastructure is feasible. The further step is to extend the scale of ethnic group and geographical distribution (nation-wide database rather than small-scaled communities), and consider the increasing number of participants and some security mechanisms.

Figure 7. Screenshots of the views of two different Users in a shared scene.

5. References

[1] J. Locke. An Introduction to the Internet Networking Environment and SIMNET/DIS [Technical Report], Computer Science Department, Naval Postgraduate School, 1993

[2] W. Broll. Distributed Virtual Reality for Everyone — a Framework for Networked VR on the Internet. Proceedings of the IEEE Virtual Reality Annual International Symposium 1997 (VRAIS'97), IEEE Computer Society Press, 121-128

[3] V. Kumar. MBone: Interactive Multimedia on the Internet. New Riders, Indianapolis, Indiana, 1995.

[4] R. v. Renesse, K. P. Birman and S. Maffeis. Horus, a flexible Group Communication System. Communi -cations of the ACM, April 1996.

[5] Singhal, S. and Cheriton, D. Exploiting position history for efficient remote rendering in networked virtual reality. Presence. Vol. 4 No. 2. Spring 1995. pp. 169-193. MIT Press.

[6] S. Eric Chen and L. Williams. View interpolation for image synthesis. In Computer Graphics Proceedings, Annual Conference Series, pages 279–288, ACM SIGGRAPH, August 1993.

[7] DIS ANSI/IEEE std 1978-1993. Standard for information technology, Protocols for distributed interactive simulation. March 1993.

Integrity Validation for Object Versions in a Co-operative Design Environment

J.S.Goonetillake , T.W.Carnduff
University of Wales Institute, Cardiff, Cardiff, UK
goonetilla,TCarnduff@uwic.ac.uk

W.A. Gray
University of Wales Cardiff, Cardiff, UK

Abstract

Due to the evolutionary nature of the engineering design process it is necessary to provide support for version management. The version model should be provided with an integrity checking mechanism to assure the consistency of object versions with the design constraints. Managing integrity constraints is complex due to their evolving nature and raises number of issues that need to be dealt with. In this paper we present a conceptual framework for providing an integrity mechanism for object versions. The framework is based on the concept of the Constraint Version Object (CVO).

1. Introduction

Due to the evolutionary nature of the engineering design process it is necessary to provide support for version management. Katz [12] defines a version as a semantically meaningful snapshot of a design object (artifact) at a point in time. There can be a number of solutions for a design problem. Version management facilitates a return to previous stages of work if a particular line of evolution doesn't work out [15]. Moreover, it enables the reusability of a design since a new design version can be derived from a previous design version [14]. Several version models have been proposed in the literature [3,4,6,9,12,13,18], only some of which [6,12,18] consider collaboration in the design environment.

2. The Constraint Evolution Problem and Integrity Validation Requirements

It is important to augment a version model with an integrity mechanism to check each object version for data correctness and consistency. We consider integrity here as the accuracy or adherence of data to a given set of design constraints. A common problem encountered during the design process is that of frequent constraint changes. This can be due to changes in product design specification, sub component interdependency or constraint interdependency (e.g. a cost reduction may have an impact on other constraints) [8,16,22]. It is not possible to predict how frequently constraints change. This may depend on how successfully the design solution is achieved using the initial constraints. As design is an evolutionary process where a solution is sought by trial and error, it is necessary to recognise that integrity constraints in engineering applications will not remain fixed but will tend to evolve throughout the design process. Consequently, the following issues should be considered in providing an integrity mechanism for object versions.

(i) Changes in constraints may require the creation of new object versions which satisfy new constraints. This requires a more flexible environment.
(ii) It is unlikely that all of the constraints for an object will be changed in a modification. Most of the previous constraints imposed on a parent object version are re-applicable to the new object version and will not require re-definition and re-validation. This raises the importance of constraint reusability and the need for an efficient constraint validation mechanism.
(iii) Even though a new set of constraints is currently applicable it is recognised that the consistency of previous versions is still valid against the set of constraints imposed on them. The maintenance of a constraint evolution graph is an important feature in this regard, particularly when considered against object version evolution.
(iv) The ability to retrieve the set of constraints applicable to a particular version is important.

It is evident that the management of integrity constraints in an object versioning environment is non-trivial due to the evolutionary nature of design. Moreover, in a co-operative design environment a complex object is broken down to its component sub objects which are assigned to different design groups or individuals [19]. The design requirements of artifacts are represented as constraints at component and sub component levels. These constraints specify the nature of individual components and the way in which they fit together [16]. Therefore integrity validation is required not only at configuration level but also at sub component level. This may be achieved by distributing constraints amongst the designers of each of the configurations and sub-components. As a result the ability to manage the consistency of constraint changes within this environment is required.

Constraints are normally defined either as active rules [5,7] attached to classes or within class definitions [5]. The mechanism of defining/calling constraints inside a class definition has the disadvantage that a new class has to be introduced for each group of versions that share common constraints. This type of schema/class evolution is undesirable as it imposes significant complexity on the versioning environment.

For the reasons given earlier, the provision of integrity mechanisms for object versions in a co-operative design environment cannot be achieved without a well-developed conceptual framework. We have been unable to find a version model that deals with the consistency of version objects through handling constraint evolution in collaborative design environment. In this paper we propose a framework that supports design integrity of object versions in a co-operative design environment considering the evolutionary nature of design constraints. For reasons of space we have not addressed the integrity mechanism for configurations separately but present a general framework for both sub components and configurations. Section 3 reviews the related work. Section 4 briefly describes the version model and the constraint types that are to be used in the proposed framework. Sections 5,6,7,8 and 9 explain the proposed framework addressing each of the required features from section 2. Section 9 also outlines our data validation mechanisms. Section 10 and 11 provide the implementation and conclusion respectively.

3. Related Work

It does not appear that existing version models [3,4,9,12,14,18] provide any integrity checking for sub component versions or configurations to check their design integrity. This imposes a high burden on the designer, since ensuring design consistency of individual versions and finding the versions that go together to form a consistent configuration are designer's responsibilities. Doucet & Monties [10] propose a framework to deal with both versions and integrity constraints. The version model of this framework is based on the Database Version (DBV) approach [4]. A constraint version is associated with a DBV. The constraint reusability aspect has not been considered under this framework. Keeping track of constraint evolution with respect to a particular design object is also not possible. It is also not clear whether this model can be applied to a co-operative and distributed design environment. We believe that in a co-operative design environment, the system may end up with a large number of database versions which will impose a considerable overhead.

Systems proposed in [23,25] are concerned with managing design consistencies of configuration objects as well as sub component objects in a design environment which is co-operative and distributed. The integrity constraints are categorised as local and global constraints related to sub objects and configuration objects respectively considering the co-operative and distributed nature. However, these approaches do not consider the connection between design constraints and object versions.

4. Version Model & Constraint Categories

We have defined our framework around the version model described in [11]. In this version model we have two types of artifact version, namely design versions (alternatives) and design equivalences (revisions) as in [17]. *Logical-version clusters* [9] are used to group versions that are revisions of the same alternative. We define the default version as the version which is closest to completion at any given time. Unlike other version models which only allow the derivation of a new version from the default version [3,12], we believe that designers should be provided with the flexibility to derive a new version by experimenting with other active versions. A new version can be made the default version at the time of its creation if the designer wishes. Sub component versions that go together, form a configuration version which is identified by a configuration number. A hierarchical database architecture [6,12,18] is adopted which supports the collaborative nature of design. It separates private and shared work by means of global and local workspaces/databases and provides enough flexibility for individual designers to work in their local databases while co-operating with other designers.

The constraint categories that are going to be considered in our framework are value based. Value-based constraints assure accuracy of the values assigned to corresponding attributes of the object. These constraints are mainly divided into following categories [2]: **Range Constraints** (e.g. 35 cm $\leq$ frame_size $\leq$ 58 mm) which check the validity of an attribute value specified by the designer by means of returning true if the value is within the specified range and false otherwise.
Enumeration Constraints (e.g. {steel, stainless steel, titanium, aluminium} $\supseteq$ material) which may also return a boolean after validation.
Relationship constraints are used to establish a relationship between two or more attributes that may belong to the same object or different objects. The validation of this constraint may either return a derived value (e.g. tube_surface = π* tube_diameter *tube_length) or a boolean (e.g. checking whether frame_size > 40 cm => wheel_diameter > 60 cm).

5. The Conceptual Solution

In proposing our framework, we take the view that evolving integrity constraints in an engineering design environment exhibit similar characteristics to that of object version data. Embedding this type of dynamic constraint information in a class definition clearly requires a redefinition of that class whenever a change occurs. It is also necessary to support the features discussed in section 2. Our framework is based on the concept of the Constraint Version Object (CVO). Each CVO contains a set of integrity constraints that needs to be satisfied for a version of a particular artefact. If the artefact is a bicycle frame, the CVO may contain constraints for its size, tube diameter, material and seat tube length. An attribute of a design object has a constraint in the corresponding CVO only if the values of that attribute are drawn from a particular domain. The relevant constraint provides a specification of the domain.

Definition: O is an object class with attributes ($A_1 \ldots\ldots A_n$) where Aj, $1 \leq j \leq n$. x is an instance of O and x has the form x: ($A_1(x)$, $A_2(x) \ldots\ldots A_n(x)$) where $A_j(x)$ denotes the value of attribute A_j for instance x. Let R be the CVO with a set of constraints ($r_1(d_1)..r_m(d_m)$), which defines a set of domains D = ($d_1 \ldots d_m$). If $A_j(x)$ is drawn from domain d_i, $d_i \in D$, $1 \leq i \leq m$ then $\forall\ A_j(x)$ $Dom(A_j(x)) = d_i \Rightarrow \exists\ (r_i(d_i) \in R)$ where $r_i(d_i)$ is the constraint that validates the value $A_j(x)$ of attribute A_j against the domain d_i.

Maintaining CVOs enables the changed constraints to be captured independently from the artifact class definition. Constraint versions govern the validity of data in object versions. Hence each version object should keep track of the CVO which has been used to validate its data. What is required here is an association between an object version and its corresponding constraint version. The resultant initial framework is depicted in Figure 1. Any number of object versions can be associated with one CVO. Grouping a set of constraints in a CVO also facilitates the retrieval of the set of constraints that apply to a particular version. It is necessary to employ a naming scheme for CVOs for their unique identification.

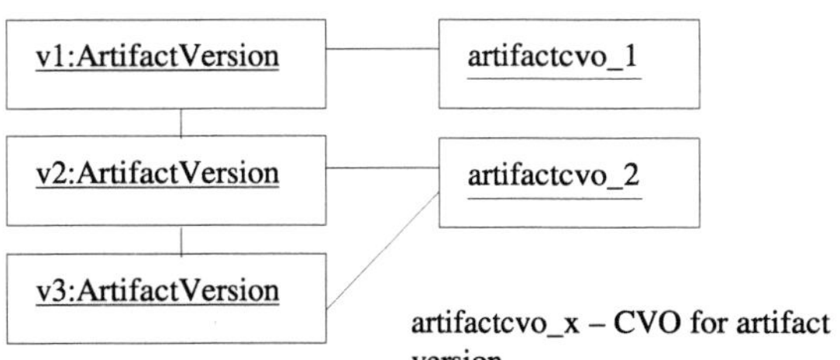

Figure 1. Initial Framework.

5.1 Managing Constraint Evolution

We consider that the CVOs are affected due to:

(i) a change in the specified domain from which attribute values are drawn. For example the tube diameter constraint of bicycle frame may change from (25 mm ≤ tube_diameter ≤ 28 mm) to (29 mm ≤ tube_diameter ≤ 31 mm).

(ii) introduction of new domains to validate values of some attributes that were not validated before (e.g. the head angle of a bicycle frame which was not restricted before can be constrained to a particular set of values).

(iii) omission of previously used constraints. For example moving on to free size bicycle frames means the size constraint will not be applicable, anymore. or

(iv) any combination of (i)-(iii).

A new CVO has to be created in a situation like this. It is important to consider how the relationship between constraint versions are maintained. The important aspects that should be taken into account at this stage are constraint evolution representation and constraint reuse (i.e. avoiding redefinition). As a result a new CVO is defined only with changes to the parent CVO providing ability to refer to the parent CVO for the unchanged rules. This may be achieved within a child CVO if we provide the means to:

- delegate constraints to the parent object if not declared within the child object
- override constraints in the parent when they are redefined in the child object.

We propose *instance inheritance* [24] as the mechanism to provide those features in CVOs. In this way a child CVO only contains the refined constraints and inherits the unchanged ones from the parent. For reason of simplicity it is considered that the CVO evolution graph has the form of a tree. The CVO containing the full set of constraints for an artifact is called the *generic CVO.* The first CVO created for a design object is the initial *generic CVO.*

6. Constraint Version Object (CVO) Categories

As a result of the co-operative nature and the resultant hierarchical database architecture that separates private and shared work, CVOs can be divided into two categories: local and global. Local CVOs contain local constraints [23,25] that entirely belong to the local object and are valid only in that particular object space. The global CVOs contain global constraints [23,25] whose validity span multiple objects. For example inter-dependencies between the attribute values of sub components of an artifact can be considered in global

CVOs. These constraints check the relationship between attributes of the sub component objects and ensure that the valid components are selected in forming the complex object (e.g. if frame_size > 40 cm then the diameter of the wheel should be > 60 cm). However, the terms local and global are relative in relation to CVOs and are dependent on the number of levels in the design database hierarchy. A global object at one level may be a local object at the next level up in the hierarchy.

7. Default CVO Concept

The design environment is normally collaborative and distributed. Hence, designers may be dispersed at different locations designing the artifacts assigned to them. The default concept with respect to CVOs is important in this regard to maintain the consistency within the design environment. The default CVO enables designers to uniformly identify the currently active CVO for a particular artifact. Unlike the default object version the default CVO should be decided by the project leader and be communicated to the team members. Usually, the last CVO created in the system will become the default object. The default object version should generally be linked with the default CVO. The current default object version can be changed to another object version only if the latter object version is associated with the default CVO. Hence, the default object version depends on the default CVO.

8. Management of CVO Creation

Ideally, the proposed framework should be managed in a form which allows the CVOs to be easily created and read by the design engineers themselves without the necessity for expert programmers' involvement. As a step towards this ideal, we have identified two alternative mechanisms for simplifying the process. As a result of design decomposition designers are only required to deal with the CVO applicable to their design component. Effective communication in the co-operative design environment results in a good information flow among the designers, especially when a change occurs [18,1]. Changes are governed by project leaders with the consent of the subordinate design members in the project group. Consequently,

(i) It can be assumed that the project leader is responsible for sending the new constraints objects relevant to each designer with the instruction to add them to their environment. or
(ii) Individual designers should be able to create new CVOs by themselves in accordance with the changes sent by their project leader. This mechanism requires a user friendly environment to automate the production of CVOs.

9. Data Validation

CVOs are used to check the consistency or adherence of a particular object version to a given set of design requirements before making them persistent in the database. The application is responsible for communicating with the corresponding (normally default) CVO and invoking the relevant constraints to validate each of the changed attribute values. Normally data validation is associated with a transaction when a change or update in attribute values directly affect the consistent state of the database [10]. In our versioning framework changing the values of a parent version does not affect the database state until a new version is formed which incorporates these changes and is made persistent in the database. Forming a new version after successful completion of the validation process ensures the design integrity of object versions stored in the database. In our framework there are two main situations in which design object version data is validated. The first situation is at the time of a new version derivation. This is on-line validation. The second situation is the validation of an existing version against a newly created CVO.

9.1 Validation of a New Object Version

The important aspects that have to be taken into account in data validation are: reduction of evaluation cost and prevention of inadvertent insertion of any illegal data due to designers' freedom to change attribute values. We believe that the data validation should thus consist of

(i) Checking relevant attribute values against corresponding changed constraints, which we call *delta validation*. However, this is only applicable when parent and child object versions are created under two different CVOs,
(ii) Checking any other attribute values changed by the designer against their existing integrity constraints.

This will reduce the validation cost since only the changed values and affected values (due to constraint modifications) are checked. However, delta validation (which checks the attribute values relevant to the modified constraints) becomes tedious when the two CVOs concerned are not directly connected as parent and child as in the situation depicted by Figure 2(a) (i.e. deriving v3 from v2 considering CVOs cvo_1 and cvo_2 respectively).

An improvement to this would be to: (i) Derive the new version from one of the versions associated with the same default CVO (Figure 2(b)), or (ii) If the default CVO is a newly created one with no object versions created under it, then it would be reasonable to derive the new object version, from object versions belonging to the parent CVO of the default CVO (Figure 2(c)). Hence it is easier to make V3 derive from V1 in Figure 2(c). V1 is a version created under the parent CVO of *CVO_2*. The designer is notified when a constraint is violated. At the sub component level the user should be given the option to proceed by changing the attribute value in order to satisfy the constraint. At the configuration level, validation is concerned with selecting valid components and should be carried out with global CVOs.

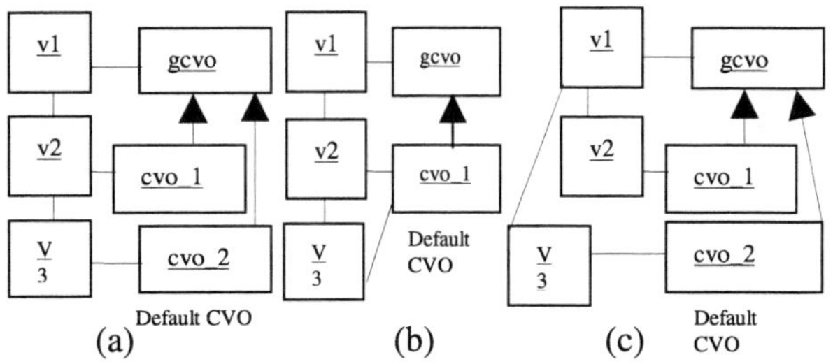

Figure 2. Delta Validation.

9.2 Validation of an Existing Object Version

When a new CVO is created it is reasonable to check some of the existing versions against the integrity constraints within the newly created CVO (which is default). This approach is beneficial from the point of view that if an existing version (or versions) is validated successfully the necessity for creation of a new object version can be eliminated. The object versions used or qualified for this type of validation are the versions created under the parent CVO of the newly created CVO. The designer will be notified on the outcome of the validation. A successful validation is feasible when constraints are relaxed in the child CVO. The successfully validated object version can be set to the default version. This facility requires that an object version is associated with more than one CVO version as depicted in Figure 3.

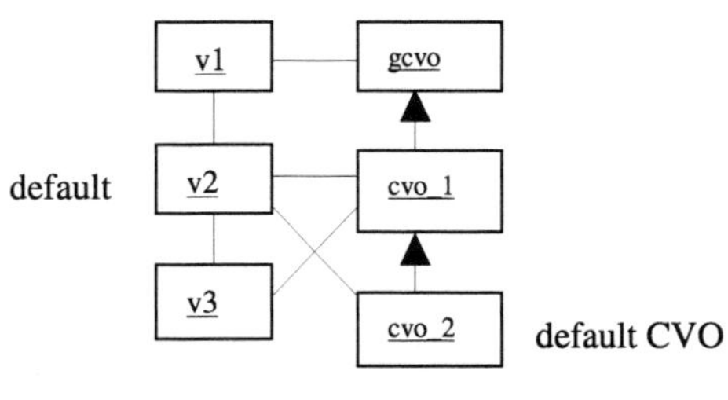

Figure 3.Validation of an Existing object Version.

10. Implementation

The implementation of our prototype is in progress and will be carried out in two phases. The first phase is to develop the version model. This will be enhanced with the integrity mechanism from the second phase. We have chosen to implement our model in Java2. Objectivity has been chosen as the database system to add persistence to design objects and versioning data since it is an object-oriented database with rich OO features. Prolog is well suited for defining CVOs since the integrity constraints can be handled in a straight forward manner as user defined predicates in Prolog. SICStus Prolog has the advantage that it provides SICStus objects with named collection of predicates with a facility for inheritance. Furthermore, the package called Jasper provides a bi-directional interface between Java and SICStus [21]. For reasons of space, we have not considered implementation details any further here.

11. Conclusion

In this paper we have described the complex nature of the management of evolving integrity constraints to support integrity validation for design artifact versions in a co-operative design environment. A framework is presented where an integrity management mechanism is described which addresses these issues. We base our framework on the concept of CVOs. CVOs provide a flexible environment in which evolving constraints and object versions are managed in a co-operative design environment. The components of the artifact and associated CVOs are considered to be distributed across the system. The CVOs were categorised as local to ensure consistency of sub components and global to ensure the consistency of configurations that is complex objects. Our approach is novel since it both maintains the consistency of versions whilst catering for constraint evolutions within a co-operative and distributed design environment using efficient data validation mechanisms.

12. References

[1] Branki, C.,1995, The Acts of Cooperative Design, Concurrent Engineering Research and Applications, Vol 3(3), September, pg: 237-244.

[2] Buchmann, A.P., Carrera, R.S. and Vazquez-Galindo, M.A.,1992, Handling Constraints and their Exceptions: An Atached Constraint Handler for Object-Oriented CAD Databases,On Object-Oriented Database Systems, eds.,K.R.Dittrich,U.Dayal and P.Buchmann, Springer-Verlag, pg: 65-83.

[3] Carnduff, T.W., 1993, Supporting Engineering Design with Object-Oriented Databases, PhD thesis, Department of Computer Science, University of Wales Cardiff, UK.

[4] Cellary, W., Jomier, G.,1990, Consistency of versions in object oriented databases,16th International Conference on Very Large Databases (VLDB), pg: 432-441, Australia.

[5] Ceri, S., Fraternali, P.,1997, Database Applications with Objects and Rules, Addison-Wesley.

[6] Chou, H., Kim, W.,1986, A unifying Framework for Version Control in a CAD Environment ,12th International Conference on Very Large Databases (VLDB), pg: 336-344. Kyoto,Japan.

[7] Diaz, O., Paton, N. and Gray, P.,1991, Rule Management in Object Oriented Databases: A Uniform Approach, 17th International Conference on Very Large Databases (VLDB), pg: 317-326.,Barcelona.

[8] Dieter, E.G.,1991, Engineering Design, A Materials and Processing Approach, Second Edition, McGraw-Hill,Inc.

[9] Dittrich, K.R. and Lorie, R.,1988, Version Support for Engineering Database Systems, IEEE Transactions on Software Engineering, Vol.14(4), pg:429-437.

[10] Doucet, A and Monties, S.,1997, Versions of Integrity Constraints in Multiversion Databases. 8th International Conference on Database and Expert System Applications (DEXA'97), pg. 252-261, Toulouse,France, Springer.

[11] Goonetillake, J.S., 2001, Managing Evolving Design Data in an OODB, under review for CONCORD Journal, University of Wales Institute, Cardiff, UK.

[12] Katz, R.H.,1990, Towards a Unifying Framework for Version Modeling in Engineering Databases, ACM Computing Surveys, Vol.22(4), pg: 376-408.

[13] Kim, W., Ballou, N., Garza, J.F. and Woelk, D.,1991, A Distributed Object-Oriented Database System Supporting Shared Private Databases, ACM Transactions on Information Systems, January, Vol.9(1), pg:31-51.

[14] Krishanmurthy, K. and Law, K., 1995, A Data Management Model for Design Change Control, Concurrent Engineering: Research and Applications, Vol. 3(4):pg:329-343.

[15] Loomis, M.E.S.,1992, Object Versioning, In Journal of Object-Oriented Programming, January, pg:40-43.

[16] Pugh, S.,1991, Total Design – Integrated Methods for Successful Product Engineering, Addison-Wesley Publishing company.

[17] Ramakrishnan.R, and Janaki Ram.D, 1996, Modeling Design Versions. 22nd International Conference on Very Large Databases (VLDB), pg: 556-566,India.

[18] Santoyridis, I., Carnduff, T.W., Gray, W.A and Miles, C.J. ,1997,An object versioning system to support collaborative design within a concurrent Engineering context. 15th British National Conference on Databases (BNCOD 15), Advances in Databases, pg: 184-199. London, UK, Springer.

[19] Sauce, R., Martini, K. and Powell, G.,1992, Object Oriented Approaches for Integrated Engineering Design Systems, ASCE Journal of Computing in Civil Engineering, Vol. 6(3), pg:248-265.

[20] Sciore, E.,1994, Versioning and Configuration Management in an Object-Oriented Data Model, VLDB Journal, Vol.3(1), pg: 77-106.

[21] SICStus Prolog User's Manual, 2000, Release 3.8.4, Swedish Institute of Computer Science,Sweden.

[22] Thompson, D.R., Tomski, T., Ellacott, S.W., Kuczora, P., 1993, An expert system for preliminary design of timber roofs: Information Technology for Civil and Structural Engineers, eds. B.H.Topping and I.Khan,Civil-Comp Press, pg: 187-196.

[23] Twari, S. and Franklin, H.,1994, Automated Configuration Management in Concurrent Engineering Projects, Concurrent Engineering Research and Applications, Vol 2(3), pg: 149-161.

[24] Wilkes, W.,1988, Instance Inheritance Mechanism for OODBS, Object Oriented Database Systems (OODBS 88), pg:274-279.

[25] Yoo, S.B., Suh, H.W., 1999, Integrity Validation of Product Data in a Distributed Concurrent Engineering Environment in Concurrent Engineering Projects, Concurrent Engineering Research and Applications, Vol 7(3), pg: 201-213.

Towards Distributed Privacy for CSCW[1]

Larry Korba
National Research Council of Canada, Ottawa, Canada
Larry.Korba@nrc.ca

Abstract

Tools for computer supported collaborative work offer many advantages for their users. Some developments in this field have included agent-based approaches for CSCW. Many challenges face the development and application of distributed, agent-based solutions for CSCW. A concern often overlooked is the acceptability and social impact of these technologies. In particular, the management of the privacy of collaborators is a paramount issue that must be accommodated. Included and related to the privacy of user information and interactions are assurances of integrity, certification, validation, ownership, non-repudiation, and authentication of individuals and software systems operating on disparate computers and operating systems, communicating over heterogeneous networks. In this paper we outline our early work on an approach for accountable privacy that may be applied to distributed computer supported cooperative work (CSCW) environments. Particular challenges for the issue of privacy associated with the CSCW application scenario are discussed.

1. Introduction

Recent developments in Computer Supported Cooperative Work (CSCW) explore the application of cooperating, distributed software rather than single, monolithic applications. These approaches offer the advantage of problem partitioning and the potential for improving application scalability. These distributed approaches include agent-based systems.

A number of challenges face the developers of distributed agent-based solutions in this domain. Key challenges include secure and efficient software agent deployment, secure communications and the difficulties of developing and monitoring agent systems deployed across a network. While there have been many approaches developed for dealing with these challenges, still other challenges persist. One of them involves protecting privacy of personal information in CSCW operation.

Protection of privacy and policies describing how organizations handle personal information have broad impact on how networked applications are used and whether or indeed whether or not a company can do business. The openness of group communication may be seriously impaired in cases where personal privacy is not upheld. The way in which technology is used is affected by how an organization deals with privacy issues. In a restrictive environment, individuals would be less likely to use tools like CSCW with abandon [1].

Over the last few years privacy laws have been put in place in different jurisdictions. The European Union's 1995 adoption of the "Data Directive" ushered in a new era of data privacy regulated by government [2]. Other countries including Australia, Argentina and Canada, have more recently enacted privacy legislation. In the United States, though there is a consensus on the need for data privacy legislation, movement has been very slow in that direction. The disparity of regulations for privacy between different countries may make it difficult or impossible for them to exchange data. For instance, the EU could sanction against data exchange with the U.S. using Article 25 of the directive. Even when all nations are operating on a level legislative playing field, technology implementing and proving adherence to privacy regulations will be required.

In this paper we examine recent developments related to the social and technical aspects related to privacy in CSCW systems. This examination underscores why privacy management is important in CSCW environments. We also describe our current work in the development of privacy enhancing technologies for distributed applications.

2. Problem Statement

Distributed object or multi-agent technology offer many potential advantages for CSCW [3] and other domains [4], [5]. Some of the benefits include:

[1] NRC publication number: NRC 44197

1) Decrease network traffic by performing computational intensive processing near the server that is the source of information.
2) Greater autonomy due to the asynchronous nature of operation of an agent.
3) Ease transportability of services across providers.
4) Increased network availability by autonomy and asynchronous agent operations.
5) Reduce time and effort for installation, operation and management.
6) Enable "on demand" provision of special services.
7) Allow a more decentralized realization of management and service control thereby reducing the dependence on network availability, resulting in a more robust system.

Distributed CSCW systems are dynamic in nature. Ideally, users may come together to collaborate in distributed work spaces in an ad hoc fashion. Relationships between individuals may be established and maintained only through virtual interactions. During the course of interactions, direct and indirect information related to the tasks in which the users participate are gathered or shared. In addition, subtler, yet personal and potentially valuable information about users and organizations may be collected. Three types of the information that may be collected or exchanged during the course of CSCW interactions include:

1) Technical or business-related information. In order for individuals to interact with each other, information related to assigned tasks must be exchanged. Depending on the nature of the collaboration, this information may be quite detailed and sensitive. How do collaborators share information in a selective manner with other participants? For instance, different collaborators either through their roles, personal or corporate preferences, may sanction different levels of privacy related to information exchanged or collected during CSCW interactions.

2) Information about the way in which people interact with others. Interacting with others in distributed CSCW environments leaves traces of information relating to the manner in which one interacts with others. Information such as the speed with which a response is made, the number of interactions, the quantity of information exchanged with others and the appropriateness and content of the information exchanged all have an important bearing upon how one is perceived to operate within a CSCW team environment. This information may be collected by direct and indirect means. It may be mined from logs relating information exchange between users. Generally, collaborators would like to limit the collection of this type of data. While companies might have policies that determine how and what type of information is captured and how it is used, individuals would like to have clear policies enunciated regarding data mining and the applications to which the data may be put.

3) Information about the technologies used to interact with others. This information is more functional in that it relates for instance to how well CSCW applications operate. It could also include information concerning most often used helper applications. It might be used to improve the performance or to correct the operation of CSCW applications. Just as easily, such information would be useful for vendors of competing products to target marketing and application development.

In general, an agent-based approach provides the advantage of problem partitioning and the potential for application scalability. Meeting the technical challenges for securing distributed system deployment for e-commerce and other applications like CSCW are a challenge [6]. In addition, as described above, privacy issues can impede widespread, accepted use of distributed CSCW solutions either from a sociological or legal point of view. To meet and to manage the challenges for privacy in a distributed environment, we are developing a distributed policy-based approach. The next section describes the building blocks of the system.

3. Distributed Privacy System

Our approach attempts to meet the following requirements for privacy implementation:

1) Policy-based privacy. Policy is considered to be information which can modify the behavior of a system [7]. All users and providers of services may specify policies indicating the type of information considered to be private and how it will or should be dealt with.
2) Accountability. The system audits activities of participants, keeping track of compliance with individual and agreed upon privacy policies. Participating entities may receive audit logs of activities pertaining to triggered, scheduled or negotiated privacy exceptions.
3) Accessibility/Flexibility. Participants may access and adapt policies to specify privacy requirements in detail.
4) Automated policy negotiation. To determine the classification, handling and disposition of information considered to be private, a policy negotiation engine deals with reconciling differences between collaborating parties.
5) Consensual assignment. Participants may be notified in situations wherein their informed consent is required to release information requested by any party.

In meeting these requirements, the system complies actively with the policies expressed by the participants while providing a means for automatically negotiating and resolving disputes. Currently privacy and security for

CSCW environments are provided only in a rather ad-hoc manner. In our work we add a policy-based approach for specifying, negotiating and managing privacy and security policies for users, managers, and system administrators. The privacy policy describes how and what type of information may be shared with other users and the triggers and actions related to privacy compliance. Security policies specify the security details regarding the operation of the system.

Figure 1 illustrates the overall architecture in terms of the core services supplied for privacy management.

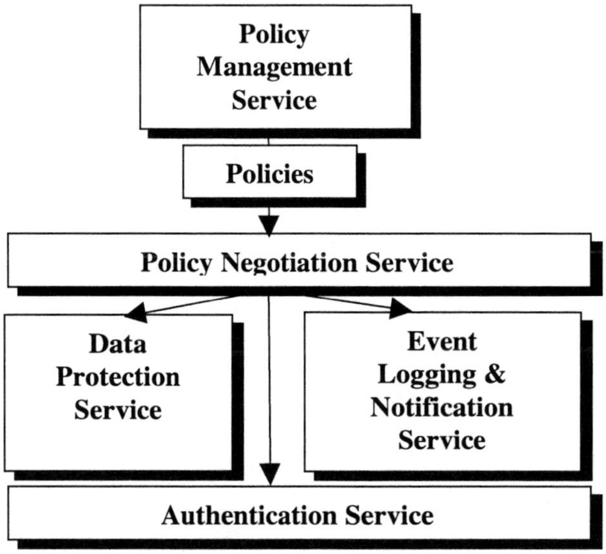

Figure 1: Policy-based Privacy and Security Services

Services may be attached to the end user, terminal device or other agents in the CSCW system, depending on policy requirements. Below is a description of each of the privacy and security services.

3.1. Policy Management Service

This service manages the development and maintenance of policies for system privacy and security. Policies are embodied as XML documents. Users develop and maintain policies using a web-based interface. Authorized users may manage different policy options for handling the detailed privacy and security options for system operation. For instance, in the case of privacy policies, a variety of options for users (supervisors, peers and administration) privacy are possible. Typical options include:

- The level of protection required for potentially sensitive information.
- The types of events that will be logged (e.g. changes and access to profiles or student records, access to services, changes, etc.).
- Event filtering for user notification. For instance, a user may choose to be notified, when his or her information is accessed by someone outside of the core team.
- An organization would have a general policy for privacy. Overall options presented to CSCW participants would be derived from this general policy. The overall policy may however offer flexibility and individual discretion for some privacy options, depending upon the role of the individual. In these cases, personal preferences may predominate in the negotiated privacy behavior of the system.

In a similar fashion, policies for system security are adjusted. In this case however, only the system administrator is authorized to change these policies.

3.2. Policy Negotiation Service

This service has three functions regarding policy: interpretation, negotiation and resolution. Policy actions are triggered by the occurrence of events for the execution of which, policy has been defined. Policy expressed in policy documents must be must be first interpreted. Where the privacy policy event is triggered for the first time, the respective policy engines for the two parties must negotiate how the event will be handled in a mutually agreeable fashion. Negotiation involves a process, the outcome of which will assure that both parties will understand why and how a privacy event will be handled. In most cases negotiation will result in an automated resolution. In other words, a policy may be amended automatically to create a new policy. The outcome will depend upon the degree of flexibility expressed by administration and users. In cases where the user or institution has expressed inflexibility and there is a conflict in resolution, users are notified and allowed to select options as to how to proceed.

Figure 2 illustrates the schema for policy negotiation. Policy negotiation begins with the selection of an appropriate trial policy. By analyzing the context of the interaction, the policy engine chooses a trial policy from a set of existing policies. A set of core policies is created from the privacy policy for the organization. Different policy documents are created for different categories of operation. For instance, two higher level categories are Internet or Intranet interactions. Within both of these two broad categories, there are further sub-categories of policies. These sub-categories are determined from the core privacy policy of the organization but as well from the policies that are negotiated and settled upon through negotiation with CSCW partners. As shown in Figure 3, successful policy negotiation may result in the production of an amended version of a core policy. Some amendments may be made automatically; others may require the intervention of the participants.

The approach taken in policy negotiation is one that accommodates the privacy policy for the organization while attempting to handle the wide range of variation of privacy policy between individuals and organizations interacting in a CSCW setting.

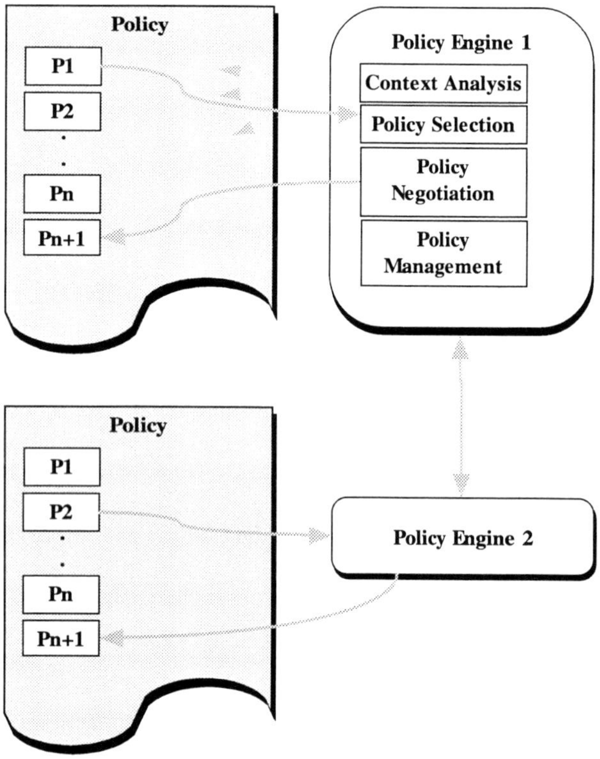

Figure 2. Policy negotiation and management details.

3.3. Data Protection Service

Depending on policy requirements, data relating to privacy and security (including policy documents themselves) may be stored and protected. Some institutions for instance may have a policy indicating that all records of CSCW interactions be stored encrypted in a particular fashion (e.g. DES, triple DES, Rijndael, etc.), in a particular location (locally, or at a trusted hosted), as well as archiving requirements. This service handles the encryption and decryption process, and in concert with the authentication service for the generation and access to the encryption keys.

3.4. Event Logging and Notification Service

This service implements the logging and filtering of events for notification. The objective of this service is to provide users an indication of compliance to expressed policy through a notification mechanism. Under the control of policy, the event logging service may invoke a process to monitor and filter privacy-related events in the system. These processes may be located at the user terminal, or at one of the host servers for the system. Events may include changes to profiles, alterations or other accesses to collaborator records etc. This service may also filter the logged events (based upon expressions in user policy) to determine when particular types of events occur.

3.5. Authentication Service

The Authentication Service assures that the identity of the users any of the system services. This service also manages user authorization. For the different types of users, there are different levels of authorization. Normally, the authentication service acts as a trusted third party for determining access to services or other resources. In the situation where a user is operating remotely, without Internet access, the authentication service provides for distributed authentication, asserting only authorized access local project-related files.

3.6. General Issues

Our system is based upon a framework that provides some of the essential requirements for implementing secure agent-based applications in a variety of areas. In the development of our architecture, the main goals in the security and privacy area are:

- Secure channels for communications between agents or operators;
- Scalable design;
- Secure delivery of software agents,;
- Secure operation environment for the agents;
- Flexible security provisions;

Developments to support these requirements have been reported previously [6]. An important issue is scalability. This is important especially when many hundreds of agents may interact with each other. The approach of having a central authority for managing policy would be a bottleneck. To alleviate this bottleneck we use a distributed policy approach [8]. A policy engine for handles and maintaining user policies exists for each user.

Figure 3 illustrates conceptually the central role the policy engine plays in system operation. The various processes in Figure 3 may be distributed over different computers. The Policy Engine implements the policy negotiation service. Policy negotiation involves an exchange of policy entities between negotiating agents. Depending on the resolution of policy negotiation, the engine mediates privacy exception-based actions. The policy engine modulates what an authenticated task or individual is allowed to do. Privacy exceptions may trigger an event and transaction-logging mechanism to provide a means for recording the occurrence of specific policy-triggered activities. This approach offers a means for tracking and monitoring that the implemented policies are appropriately carried out for all privacy exception initiated actions.

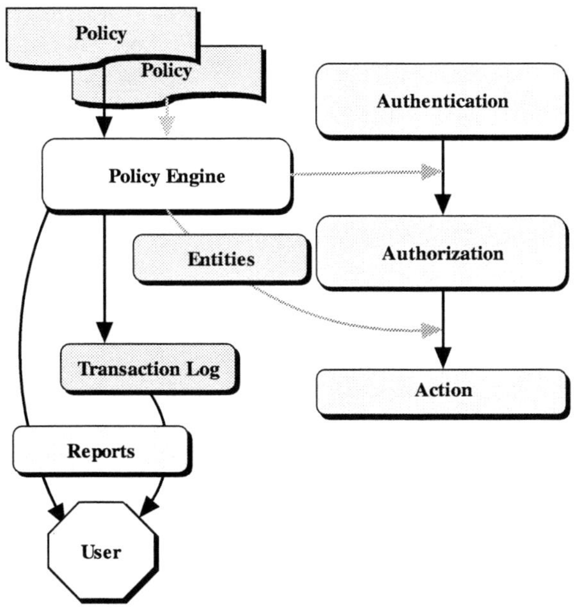

Figure 3. The policy engine plays a central role approach for managing privacy.

4. Related Work

With the development of the wired society it has become clear that technology is not privacy neutral [8]. In his book, Garfinkel suggests that current developments leave us with only two choices: 1) allow our personal data to rest in the public domain or 2) become hermits (no network/telephone use, no credit cards, no web surfing, etc.). The latter option is not much of an option considering the advantages Internet technologies, particularly CSCW bring to workplaces.

Security for interaction within CSCW systems has been examined more from the point of view of protecting shared information from access by others outside of particular working groups, i.e. via network snooping. Foley and Jacob describe a technique for specifying security in the development of secure electronic examinations for courseware [9]. Based upon activity analysis this approach provides some preliminary techniques for assessing the requirements for confidentiality in dealing with CSCW systems as applied to the courseware development application domain. There does not appear to be further development of this approach. While it does not describe a system for implementing a policy-based approach for dealing with confidentiality requirements in CSCW, it does offer promise for analysis for the purpose of policy generation.

Security and privacy are considered to be a basic requirement in multi-agent applications. Considering the development and general acceptance of multi-agent systems for electronic commerce as well as computer supported cooperative work, it is clear that agent systems must have a certification process [10]. There are several multi-agent environments that provide different levels of security [11] [12]. The approach for most of these is to secure communication channels [6], [13] with protocols like Secure Sockets Layer (SSL) version 3. This approach does not address the need for flexibility in specifying application issues for privacy. For instance, it is often important to specify which communications should be protected and what information may be exchanged with third parties not only for privacy purposes but also to reduce computational overhead.

Prevelakis et al. [14] describe an agent-based framework for the exchange of electronic documents over open networks. This approach builds upon the low level security functionality offered in many agent platforms to offer some security regarding data exchange over networks. The authors state that the system eliminates unauthorized copying, redistribution or modification of the documents. The system provides authentication of the sender, recipient, and the document. It implements a uniform policy and a trusted third party approach for authentication.

Agent-based applications have been developed for electronic commerce [15], [16], [17], supply chain management [18] and for CSCW [3], [19], [20]. The agent model provides system characteristics of cooperation, distribution, interaction, concurrence and autonomy. There has been relatively little attention paid to the security and privacy of interactions between participants in CSCW systems.

The impact of a lack of a well-defined structure for privacy and security of interactions can be broad on both the acceptance and the use of any technology. User perceptions of privacy provide an example of how policies can differ across organizational boundaries [1]. Different organizations have different values regarding interactions and actions of their employees. For instance, in universities freedom of speech is considered of paramount importance, whereas in a military establishment controls may be in place to ensure the protection of a nation's security. Depending on the organizational setting, there can be a marked contrast as to how much personal and sensitive information individuals are willing to share through computer mediated communications. In effect, the privacy offered by a technology affects how individuals use that technology. User trust can be enhanced when organizations recognize the expectations of employees, provide policies expressing the guidelines and implement processes to carry out those policies.

Enormous amounts of information about web site visitors are being collected by thousands of web sites. The information collected may be aggregated, filtered and used for a variety of purposes. This may be done with or without the permission of the web site visitors. In a movement toward developing in the World Wide Web consortium (W3C) has developed the platform for privacy preferences (P3P) [21], [22]. P3P offers a way for users to reach an agreement with services such as web sites or applications that request data and offer a privacy practice. P3P offers a machine-readable means for specifying the privacy practice for a service. A user defines his or her privacy options in a user agent (potentially a plug-in for a web browser) and may be alerted when there is disagreement between a service proposal and the selected privacy options. Issues not covered by P3P include:

1) a means for negotiating a contract especially in a peer-to-peer exchange,
2) a means for proving privacy policies of all parties have been upheld,
3) a specification of how private information is protected,
4) a non-web specific approach providing interactions beyond the simple exchanges,
5) a more detailed specification of the requirement for the private information. A description of why it is needed and the other organizations that will share the information.

Interestingly, early versions of P3P included some negotiation capability. Compromise in the its development led to the current specification. The Privacy Server Protocol (PSP) [23] is an effort to incorporate simple privacy negotiation between web client and server. PSP offers a simple peer-to-peer privacy policy negotiation, using http protocol.

5. Discussion and Conclusions

In this paper we have described early work in the development of an approach for managing privacy in distributed applications such as CSCW. In a discussion of the background to this work we have drawn from experience in the legal, sociological and technical developments of technologies for CSCW.

Issues associated with creating and operating CSCW applications in a secure fashion on open computing platforms form a major challenge not addressed in this paper. Indeed, they are very difficult to address at all. While secure communication channels and encryption technologies may be used to assure privacy of information exchange, safeguarding against denial of service attacks is difficult. In addition, it is difficult, or impossible to build applications that would protect information from unauthorized access on open computing platforms. Software and hardware in-circuit emulation techniques and reverse engineering may be used to launch an attack on privacy information. Reverse engineering an application to determine secrets though illegal in some jurisdictions can be trivial with applications that are not designed to safeguard against this intrusion. For the most part, attempts at protecting against this sort of exposure involve protection by obfuscation. Designing systems to operate with a high degree of security in open computing environments is a challenging area for research.

Currently we are further developing our work on several fronts. We are collaborators in the Privacy Incorporated Software Agent (PISA) project. PISA is a European Union 5th framework project involving researchers from the Netherlands, France, Italy and Canada [24]. The objective of the work is develop privacy enhancing technologies for agent-based electronic commerce applications. The project involves designing an architecture that will enable privacy for distributed applications, and to build several demonstrators of the technology in action.

In addition, we are examining the application of privacy technologies in other domains. For instance, we are investigating approaches and technologies for privacy enhancement is distributed manufacturing [25]. Here, we are developing approaches that would offer privacy and security controls for Internet-based manufacturing. In this application domain, different organizations may have different policies regarding the type and the amount of information that may be shared with other organizations. A flexible approach for managing and for negotiating privacy and security policies automatically would be a vital enabling technology for this area.

We are also investigating the applicability of distributed privacy approaches in the area of electronic learning. One of the challenges in this work is assuring that the privacy transaction logs are original, i.e. unaltered. While there have been some developments in this area [26] they have not been applied in distributed applications.

6. References

[1] S.P. Weisband, B.A. Reinig, Managing user perceptions of email privacy, Communications of the ACM, vol. 38, no. 12, 1995, pp. 40-47.

[2] Information on the EU Directive on Data Protection http://www.privacy.org/pi/intl_orgs/ec/eudp.html

[3] A.P. Kosoresow, G.E Kaiser, Using agents to enable collaborative work, IEEE Internet Computing, Vol. 2, No. 4, July-Aug. 1998, pp. 85-87.

[4] Breugst, L. Hagen, and T. Magedanz, Impacts of Mobile Agent Technology on Mobile Communications System

Evolution, *IEEE Personal Communications*, vol. 5, no. 4, (1998), pp. 56-69.

[5] S. Krause and T. Magendanz, Mobile service agents enabling "Intelligence on Demand: in telecommunications, *IEEE Global Telecommunications Conf.*, London, UK, Nov, 1996, pp. 78-85.

[6] Korba, L. Issues in Agent-Based Electronic Commerce. Proc. of the SSGRR 2000: International Conference on Advances in Infrastructure for Electronic Business, Science, and Education on the Internet, l'Aquila, Rome, July 31-August 6, 2000.

[7] M. Sloman, Policy driven management for distributed systems, J. Network and Systems Management, vol. 2, no. 4, pp. 333-360, Plenum Press, 1994.

[8] M. Blaze, J. Feigenbaum, J. Lacy. *Decentralized Trust Management*. In *Proc. of the 17th Symposium on Security and Privacy*, pp. 164-173. IEEE Computer Society Press, Los Alamitos, 1996.

[9] S. Garfinkel, *Database Nation: The Death of Privacy in the 21st Century*, O'Reilley & Associates, 2001.

[9] S.N. Foley, J. Jacob, Specifying security for CSCW systems, Proc. 8th IEEE Computer Security Foundations Workshop, June 13-15, 1995, pp. 136-145.

[10] H.S. Nwana, J. Rosenschein, T. Sandholm, C. Sierra, P. Maes, R. Guttman, Agent-mediated electronic commerce: issues, challenges and some viewpoints, Proc. of the 2nd Int. Conf. on Autonomous Agents, May 10-13, 1998, pp. 189-196.

[11] Links to other agent tools collected by AgentBuilder. http://www.agentbuilder.com/AgentTools/

[12] P. Bellavista, A. Corradi, C. Stefanelli, A secure and open mobile agent programming environment, Proc. 4th Int. *Symp. On Autonomous Decentralized Systems*, March 21-23, 1999, pp. 238-245.

[13] Mitsubishi Concordia Agents Information Page, http://www.meitca.com/HSL/Projects/Concordia/Welcome.html

[14] V. Prevelakis, J-H. Morin. And D. Konstantas, Controlling the dissemination of electronic documents, Proc. Tenth Int. Workshop on Database and Expert Systems Applications, Sept. 1-3, 1999, pp. 869-873.

[15] J.J. Jung, D.Y. Hwang, and S.B. Jeon, Agents-based framework for brokerage between buyers and sellers on electronic commerce, *Proc. Int. Conf. on Electronic Commerce*, 1998, pp. 17-22.

[16] H.S. Yoon and J.K. Lee, Intelligent agents based virtually-defaultless check system: safecheck system, *Proc. Int. Conf. on Electronic Commerce*, 1998, pp. 224-231.

[17] J.K. Lee, and W. Lee, Intelligent Agent Based Contract Process in Electronic Commerce: UNIK-AGENT Approach, *Proc. 30th Hawaii Int. Conf. on System Science*, 1997, pp. 230-241.

[18] M. Barbuceanu, M.S. Fox, Coordinating multiple agents in the supply chain, Proc. of the 5th Workshop on Enabling Technologies: Infrastructure for Collaborative Enterprises, 1996, June 19-21, 1996, pp. 134-141.

[19] J. Dong, J.C. Zheng, S.Y. Xiao, Software Models in CSCW, Proc. IEEE Int. Conf. on Intelligent Processing Systems, Oct. 28-31, 1997, pp. 881-885.

[20]L. Hongchen, S. Meilin, Application of agents in workflow management system, APCC/OECC '99 5th Asia-Pacific conference on Communications & 4th Optoelectronics and Communications Conf., Oct. 18-22, 1999, pp. 1068-1072.

[21] Platform for Privacy Preferences http://www.w3c.org/p3p

[22] J. Reagle and L. F. Cranor, The platform for privacy preferences, Communications of the ACM, Vol. 42, No. 2, 1999, pp. 48-55

[23] Robert Thibideau, Privacy Server Protocol (PSP), http://yuan.ecom.cmu.edu/psp/

[24] Privacy Incorporated Software Agent (PISA) Project, http:// www.tno.nl/instit/fel/pisa

[25] Weiming Shen, Sherman Lang, Larry Korba, Lihui Wang and Brian Wong, "Reference Architecture for Internet Based Intelligent Shop Floors", in Network Intelligence: Internet-Based Manufacturing, Proceedings of SPIE Vol. 4208, pp. 63-71 (2000).

[26] B. Schneier, J. Kelsey, Secure audit logs to support computer forensics, ACM Trans. on Information and System Security, Vol. 2, No. 2, 1999, pp. 159-176.

A New Algorithm of Collision Detection in Shared Space for Collaborative Design

Chenhui Yang, Lilly Li, Tangqiu Li
Computer Science Department, Xiamen University, Xiamen, 361005, P.R.China
chyang@xmu.edu.cn, xyli@yanan.xmu.edu.cn, tqli@xmu.edu.cn

Abstract

Fast and accurate collision detection is critical for the performance of the simulation of a virtual space where a great number of objects are presented and groups of people are working collaboratively. Most of previous algorithms consume much memory resource or CPU resource. In this paper, we present a new algorithm that can greatly reduce the number of pairs needed to check for collision detection in a distributed large-scale virtual space for collaborative design. Based on the idea of object oriented modeling, it makes good use of the concept of virtual grid which will consume no unnecessary memory resource and CPU resource, and at the same time, fully utilize the distributed resource to boost up the speed of collision detection.

1. Introduction

One of the important feature and goal of virtual environment is the sense of immersion. The result of geometric modeling and rendering of 3D objects may give users a sense of immersion. However, once the users find that they can fly through a wall, or two cars can pass through each other with no damage, their sense of immersion in the virtual world will be fully destroyed. The reason for this is that the behaviors of the object are against to what the users expect. Users, according to their common sense of the real world, expect the virtual world to realistically reproduce the real world. Therefore, they expect naturally objects to collide if they come too closer to each other. As a result, the virtual world simulation should have a collision handling mechanism. An ideal condition is that, in the virtual world simulation, there will be an automatic collision handling mechanism, which determine how the object moves, when the object collide, and how the objects respond to the collision according to the physical properties of objects and physical rules.

Another important feature of virtual world simulation is its real-time response. Virtual world simulation is a time critical application. The users in the real world can receive an immediate response or feedback from their action. So, in the virtual world, users will also expect to see the happening of the collision immediately when the collision occurs. Therefore, it is very necessary for collision detection to be finished in sharp time. This real-time requirement is an extremely important problem when dealing with collision, since in a large scale virtual world, the number of the objects is large, and detecting collisions among so many objects demands huge computational load.

In a virtual environment application, good collision detection algorithm is critical. Poor algorithm can not precisely predict the happening of the collision and may bog down the performance of real time interaction.

2. Related Work

2.1 The naïve solution

Consider a simulation involving N Object:

$$O_1, O_2, \ldots, O_n$$

Following presents a naive algorithm by Hubbard[Hub 93] that attempts to detect collisions among the agents in the period [0-t].

for t := 0 to t in steps of ?t
  for each object O_i ? { $O_1, O_2, \ldots, O_n$}
    move O_i to its position at time t
      for each Object O_j ? { $O_{1+i}, O_{2+i}, \ldots, O_n$}
        move O_j to its position at time t
        if (the geometric representations of O_i and O_j penetrate) , then a collision occurs at time t

2.2 Previous work

One bottleneck of the basic algorithm is obvious-- $O(n^2)$ pair wise tests. The algorithm loops cycle through every pair of objects at every time step, making the processing time spent by the algorithm increase quadratically with N, the number of objects. As users demand more detailed modeling of virtual space, the typical values of N will increase, hurting the performance of the algorithm.

Different methods have been proposed to overcome the bottleneck of $O(n^2)$ pair wise tests in an environment of n bodies. The basic idea behind these methods is to prune the number of pairs that the algorithm must process, that is, to avoid processing the pairs of objects that are obviously far away from each other.

A typical method is to add a time bound to the object in addition to the 3D bound (the three coordinates of the object which indicates the object's location) and make the object to be of 4D. Samet and Tamminen's [1] method falls into this category. They describe a 4D approach based on recursive subdivision of the time--cubes that contain two objects. In their method, the objects are represented by 4D(3D location + 1D time) geometry. If the objects do collide, subdivision will be performed to locate where the collision occurs. The limit of this algorithm is that it only dealt with two objects, so it could not address the pair-processing weakness. For two objects that have linear motions, they gave detailed algorithm. For more general motion, the authors suggested an approach based on interval analysis.

As to the interval analysis, it is Mudur and Koparkar [2] who first described its application to graphics, but the work of Snyder [3], Du [4] and Snyder et al. [5] addressed more to the problem of collision detection. In this context, intervals are used to represent regions in space and time. The detection algorithm takes intervals as its parameters, and recursively subdivides (the process of subdivision is boosted by the Newton-Raphson method) the argument interval to locate the interval (or region) that contains collisions. If any objects collide within these regions, the functions return intervals that contain designated values. To decide when to check for collision, Snyder et al. [5] put some constraints on each pair of objects that are met only when the pair has just started to collide. Du [4] 's method is a great improvement. He uses simpler functions and avoids the all-pairs weakness. Unfortunately, the algorithm shares the disadvantage of 4D analysis, that it must know each object's exact position throughout time.

Another common approach is to subdivide 3D space into regions. A pair of objects must intersect the same region in order to collide with each other, so the detection algorithm can use the regions to reduce the number of pairs it must consider.

One simple subdivision technique is to partition 3D space along a uniform grid. Zyda et al. [6] describes an implementation of this idea. They partition the space with grids whose cells are isothetic and axis-aligned cubes that allow the detection algorithm to ascertain whether an object belongs to a cell quickly.

Turk [7] adopts a similar idea and presents two ways to determining relationship between objects and cells (i.e., whether an object belongs to a cell) in a molecular docking simulation. One way to implement this grid is to find the box that bounds the molecular and then divide it into cells. According to Turk, there is no obvious performance improvement over a simpler approach that uses spatial hashing. This method creates axed grid of cells and then uses a hash function to map any position in R^3 to a cell in that grid. This approach has low overhead and prevents the all-pairs weakness. It can be used for part of effective detection algorithm in an interactive application. It would quickly identify pairs of objects whose bounding spheres collide and then, a pair-processing algorithm would then check those objects for actual collisions.

Octree is involved in a more sophisticated approach to space subdivision. The root of an octree is an isothetic box that bounds all the objects. The children of the root are the eight octants defined by dividing each edge of the box in half. In this way, the octree acts like a hierarchy of grids at different resolutions. The subdivision algorithm also determines which objects belong in each octant. If an octant contains only a few objects, the detection algorithm can efficiently check pairs of those objects for collisions. If an octant contains many objects, the algorithm recursively subdivides that octant into smaller octants. This approach adaptively refines the resolution of the grid, making the cells smaller only where necessary. Hahn [8], Moore and Wilhelms [9] and Shaer and Herb [10] describe variations on this approach for collision detection. These approaches address the all-pairs weakness. The disadvantage of this method is that octrees may cause high computation overhead. To rebuild or modify the octree each time it is called involves considerable computation, particularly for applications involving many moving objects.

All of the works mentioned above have some contribution to improve the naïve algorithm for the collision detection. But none of them addressed that, in a distributed interactive virtual world simulation, how to detect collision between the objects manipulated by different users. In this paper, we will present a new way to reduce the pairs of object to detect with easy extensibility in a large-scale virtual world simulation.

3. Collision detection in distributed interactive virtual world simulation

In a large-scale interactive distributed environment, the virtual world should reproduce a perfect virtual presence and reasonable behaviors of all kinds of objects as well as users for the different users scattered in the physically distributed locations. Since the number of the objects can be incredibly large, and the user's manipulation of the object can be frequent, the performance of those brute-force approaches that test all the possible pairs for collision is not tolerable in this kind of real time application.

The collision detection in the large-scale virtual environment is different from other applications. In robotics, computer animation and simple vehicle simulation, the positions and orientation of the objects may be known in advance. This is not the case in a large-scale interactive of virtual world simulation. Not only the physical properties and the external forces between the objects, but also the users' interaction decide the occurrence of the collisions. It's seldom possible to predict what kind of manipulation the users will perform to the object; therefore, the velocity or acceleration can't be prophesied.

Although the distributed environment may add some complexity to the collision detection algorithm, it also brings some benefits –it also boosts the collaborations between users.

The collaborations can be classified into two categories. One is to co-computation for communication, i.e., the collaboration will enable the users to share and interact with the virtual world. The other is the co-computation for speed: the CPU of different users can cooperate with each other to finish the one task to boost up the speed of the virtual world simulation.

With the rapid development of hardware, it's reasonable to the users that access the virtual world will have a CPU with fair good computing ability, therefore, can deal with fairly heavy computation requirements.

Collision detection is a fairly computation consuming process. We should take the best use of the CPU resources scattered in the different locations with the users and distribute processing load among a set of CPUs. In the algorithm proposed in this paper, we will fully utilized the CPUs resources to boost up the computation speed.

4. Object-driven collision detection

The virtual world consists of various objects, whose properties and the events of those objects are defined in their data structures. The properties include the shape, the volume, the velocity, the current location, or any other necessary for constructing the virtual world. The events of the objects include autonomic activities and responsive activities. Both the autonomic activities and responsive activities must act according to the physical rules and dynamic rules. Therefore we need the third class events to enforce these objective rules of our real world. We call these events arbitrate events.

In the OO collision detection in a distributed environment, the collision detection will be triggered by the objects' arbitrate events. The arbitrate events will judge the need to detect collision, and will send request to the appropriate CPU to detect the collision. And the same CPU is also responsive for the notification of the object's current's status.

4.1 Basic idea of object-driven collision detection in a large scale distributed virtual world

In the virtual world, the objects are autonomic, will perform certain actions and respond to certain stimulations (or inputs). For example, the objects will move according to their own physical properties and external forces. Once the objects collide with other objects, it's these objects that will feel the collision and respond to the collision. Therefore, the collision detection and collision response computation should only occur when two objects actually collide. In a large-scale virtual environment, this is an issue of critical importance. If there are N objects, and M of them are moving objects; then the M objects have displacement and possibly create new collisions between themselves and the static objects; and those (N-M) relatively static objects will never create new collision between each other. As a result, we only need to check those moving objects to see whether collision occurs.

If a moving object A collide with another object B (moving or static) at time t at a location L, then at time t, object B must be in the proximity of location L. Therefore, once A move to a new place, we only need to check the proximity of the new place of object A. In this way, the computation of the collision detection for the virtual environment is O(c), in which c refer to the pair numbers of those objects whose distance are smaller that a predefined valve and at least one is moving. This avoid the demand to detect those objects are very near to each other but remain static to each other.

4.2 The collision detection algorithm

4.2.1Virtual grid

We will build a virtual dynamic grid for the virtual environment space. The grids are virtual because they actually do not exist, and therefore, do not require any memory resources. But we have a way to build necessary grid quickly and can refine them when necessary.

To build the virtual grids when necessary, we first fix the origin and the three axes of the virtual world. All the virtual grids' edges will parallel to one of the axes of the virtual world, and the vertical distance two ends of the edges can be divided by a preset value. And this value may be large initially, but it can be refined when necessary.

The virtual grid in the 2D space can be shown in the Figure 1:

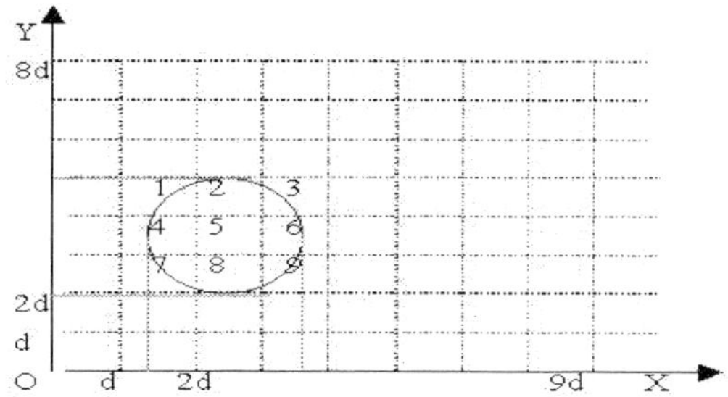

Figure 1: Virtual grid in 2D space.

When one certain area does not contain any object, we do not compute or store any grid information. However, when one object occupies certain region, then the corresponding grids are created and stored in occupied grid lists. The data structure of the grid can be as follow:

Structure Grid
{
Integer interval; // defines the power to divide the length of the grid, as d, d/2,….., $d/2^n$, is useful when it is necessary to refine the grids.
Real Xstartpoint;
Real Ystartpoint;
Real Xstartpoint;
Boolean Occupied;
Boolean Fulloccupation; //indicate that whether this grid is entirely in the objects.
String Owner ; //indicate which one occupy this grid;
}

For example, for the ellipse in the above figure, we call it A, it occupied the 9 grids indicated by the numbers. Then we create (or instantiate the structure) the nine grids according to the maximum x and minimum x, maximum y and minimum y. This could be rather easily done. The minimum starting point of the x axis will be floor of $|x|/d$ when x is located the positive axis. While the maximum ending point of the x will be ceiling of $|x|/d$. And we can easily divide the distance of the minimum and maximum x into the scale of the grid. During this division, the structure of the grid are built up, and the grid.Occupied will be set to 1(that is , Occupied), and the Owner will be set to the object's name. Building the grid could be finished in O(1) time.

4.2.2 The data structure of the objects for collision detection

The objects in the virtual world should appear and behavior as we may expect them to be in a real world, this may demand many complex properties for the object. However, we will ignore those are irrelevant to collision detection here, and define the structure of the objects as following:

Structure Object {
// properties:
String Name; // indicate the name of the object;
Shape ShapeOfObject; //the shape is also a structure, may include the vertices, edges
//and faces of the object.
Real Velocity;
CPUobject ownerCPU: // indicate which the objects presents in which CPU, or Users. This is used to sent multicast update information for the other users. Include the server
Cube Boundingshape; // the cube is also a structure, with a center and a radius.

// The events:
Public:
void checkvelocity ( ) // may add some necessary parameters here;
{
if .velocity >0 then
{call displacement(object) ;
call detectproximity(object);
}

Boolean detectproximity(Objecta)
{
Loop
Computing the starting point of the grids.
Check if there already exsits such a grid.
If there exist such a grid[I]
then
{refine the grid
If collisiondetect (Objecta, grid[I].owner, grid[I])
then return true;
}
Else
Computing the next grid.
End loop
}
void collision response(Object) {
//According to the properties of the object and the other colliding object, computing the result the collision and then call the rendering functions.
…}}

4.2.3 Contact determination

Once a new moving object wants to move to a new location, the grid occupation check thread will destroy the old occupied grids and remove it from the global list if there is no other objects indicate occupation of the grids. When the check thread intend to create a new grid structure, it will check first in the global occupied grids spool to see whether a grid at this location already exists. If all the grids are new, then there is no collision. However, If it already exists, the program will report a collision possibility. It will then refine recursively the grid that report already existence to progressively narrow the possible collision area. The recursion will stop when it find the two refinement grid actually apart from each other (i.e., there is no collision), or it find the two objects actually overlap each other.

Usually the contact will occur in the frontier area with more probability, since we know before this check they have no contact. We can use this to improve the detect velocity.

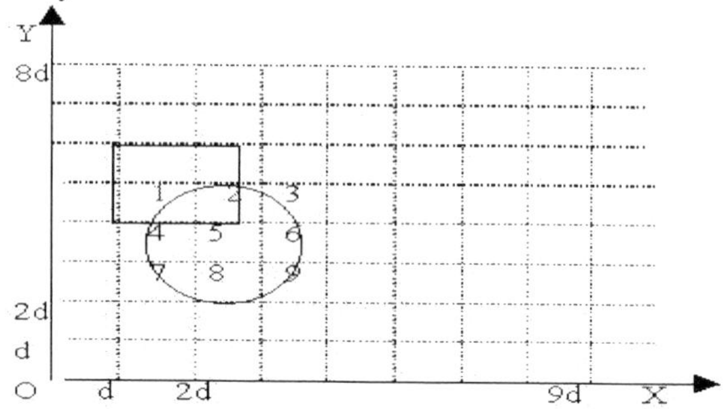

Figure 2: Contact determination.

For example, in the figure above, the rectangular and the ellipse collide at time t. Let us assume the rectangular is moving toward the static ellipse. When the rectangular attempt to create the new occupied grid1 and grid2, it will find grid1 and grid2 already exist in the global grid list. It will then find out the owner of the grid1 and grid2, that is, the ellipse.

4.2.4 refinement and levels of the virtual grid

Once the collision detect thread find the possibility of overlap, it necessary to decide weather the two objects actually overlap. In this condition, the next step is to refine the virtual grids. Supposing d is the largest predefined virtual grid length, then the virtual grid can be refined when necessary as, $d/2^2$, $d/2^4$ …, $d/2^{2n}$.

According to the length of the grid, we define different levels of grids list. For example, to a grid that is $d/2^{2n}$, we call it nth level. The grids of different levels will be stored in a different global list.

For the Occupied grids list, we've consider two ways to organize the grids; One way is by the order of the starting point, the other is to organize the grids by the group of it's occupied objects. After consideration, we choose to store the grids by the order of the starting point of the grids. If the data were stored by the order of the starting point of the grid, then we could use dichotomy retrieving method to retrieve the possible grid existence. And the object need not to store a list to the grid structure here. This could be done in ? (Len), the len is the length of the occupied grids of all the objects. To detect the collision detection for the two objects above, we then refine the two grids, grid 1 and grid 2. First, we divide the grid into d/4 and then check each refined grid whether is occupied by both objects.

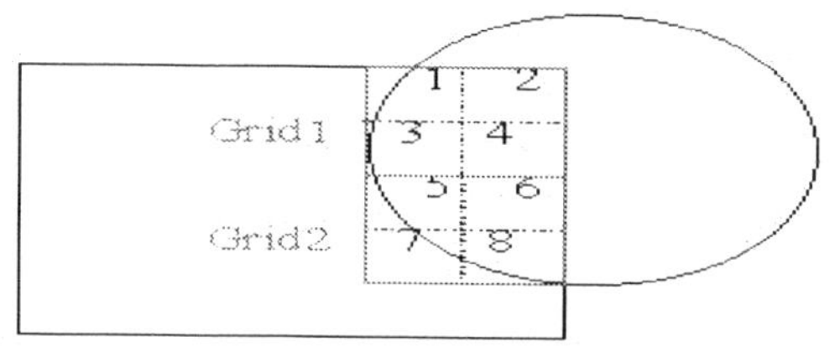

Figure 3: Level of virtual grid.

We check the full-occupied properties of the grid, once we find a grid that is both occupied by the two objects, and at the same time, either one of the full-occupied is set to true, then the two objects overlap each other. For example, we will check that grid6 are contained in both objects. To decide whether a grid is completely contained in the object can be rather easily done for convex objects. Just check maximum and minimum coordinates of the objects, and the work gets done. For this instance, since the application can check out that grid b6 is contained by both the rectangular and the ellipse, then the collision is detected and the algorithm stops. Or we can also set a valve small enough,

4.2.5 Collision detection for distributed virtual world simulation

Our proposed algorithm can be adapted for the distributed collision detection easily. The local CPU has access to the global occupied list through a pointer, and at the same time, it also maintain a local copy of existing occupied grids for effective simulation region. This could be easily done since it is not hard to decide the coordinates of the bounds of simulation region of the computer--find the staring point and ending point for the local area, and then copy the global occupied lists from the starting point to the ending point, the work gets done. And this synchronization will not consume much network time and bandwidth.

The CPUs maintain a list to the objects that is located in the CPUs effective region. But at a time, only one CPU needs to compute the object collision detection. If a user

U_2 try to manipulate one object, say O_4 then we call the U_2 local CPU the object O_4's control CPU. In the next step when there is a need to check the collision between the object and any other object, then it is the U_2's local CPU that performs this function.

The reasons for the movement of the objects may be classified into to two categories: One is that the object initially has a velocity; the other is that there is an external force on it. This external force includes the users' manipulation. It's obvious that the user's computer should deal with the collision detection. For instance, if U_2 like to manipulate the object O_i, then U_2 's local CPU should first check the possible collision between the object and other objects when the object intend to move to its destination. In this condition, we call U_2 CPU the control (or owner) CPU of the O_i. The owner CPU needs to maintain a copy of its controlling object from which the CPU could find the necessary information (such as velocity, location information) to process the collision.

There are some objects that there is no manipulator, we call them no-owner objects. These objects could be those static objects, but also could be some moving object with an original velocity. We could ignore those static objects and perform no collision detection for them. But for those that have an original velocity, we should also assign a CPU to process the collision detection for them. The solution here is to find those that have least workload to be responsive for its collision detection.

We set a flag indicate the control CPU of the object, and also set a list in the CPU indicate the object that controlled by the CPU. Therefore, we could find the appropriate CPU to deal with the collision detection. As a result, we need to add some properties to the objects' structure, and make the structure of the objects look like the following:

```
Structure Object
{
OwnerCPU
}
```

For example, if two objects, A and B, and CPU_i is the control CPU for object B. If the object B trigger the collision detection procedure, CPU_i will compute the grid of the object and to see if they already exists by checking the local copy of the global occupied grid list. We could find those objects (in this case, object A) that are located in B's proximity by access the grids' owner property. If necessary, or if we initially have no information of Object A, we could copy it from the central world server. If we already have it in the local CPU, then we can perform the collision detection immediately.

If no collision is detected, then Object B will move to its new location, and update its properties, update the new grid it will occupied, and discard the old grids it occupied if no other object occupied it now. According to the objects' presence CPU list, the local CPU may maintain a multicast list of the users and their corresponding CPU, and very important, to the central world server.

5. Performance Analysis

This algorithm to reduce the pairs that need to be checked for collision can be implemented rather easily. Here, we will give a theoretical analysis. Unlike the previous space division algorithm to divide the space to uniform girds and build structure for Octrees or BSP trees before it's necessary to detect any collision. And these reconstruction of the Octrees or BSP trees must be called each time the objects move to a new location. This is a great computing consumption.

Our algorithm will not divide any region of the space until an object occupies there. We will not maintain any $O(n^2)$matrix for the n objects in the virtual environment. We only build those necessary pairs and do not waste any memory and computing resources. Our space region division mechanism enable the quick construction of the virtual grid for the collision detection.

In an environment of N objects, M of them are moving objects. Then for the (N-M) objects, we only need to build the occupied grids once and for ever before the simulation starts. For the M moving objects, we need to update the occupied grids according to certain time step. But for each moving objects, in most time, the grids construction can be done in short constant time(find the maximum and minimum values and divide into small grids). Therefore, each time to compute collision detection, the time need for the grid construction will be O(M). When the application proceeds to process collision detection, only those very close pair of object are checked for possible collision. For example, if there are n pairs of objects are in the predefined proximity of each other, then we only need to perform collision detection n times. Therefore, here, we greatly reduce the number of pairs need to check for collision.

For the distributed algorithm, we could see the time needed by the CPU is about from O(N) to $O(N/U_k)$ for one collision detection, in which N is the number of the moving objects, and U_k is the number of the users that have the N objects presented in their simulation.

6. Conclusion

Collision detection is a very critical part if we try to improve the performance of large scale virtual world. By borrowing the ideas from OO, we greatly reduce the number of pairs needed to be checked for collision detection and greatly boost the collision detection speed.

7. Acknowledgements

Most of ideas in this paper come from the work of FL group in MCS division of Argonne National Laboratory in United States. It is with great pleasure for me, the first author of this paper, to thank my supervisor, professor Rick Stevens, director of MCS division and team leader of Grid Alliance, who gave me lots of innovative ideas and kind help when I was a visiting faculty in his group. I am also very grateful to all the colleagues in FL group who put tremendous amount of efforts to help me. These individuals include Mike Papka, Mark Hereld, Lisa Childers, Tom Brown, Terry Disz, Ivan Judson, Tushdar Udeshi, Bob Olson and Randy Hudson.

8. References

[P.M.Hubbard 93] Collision Detection for Interactive Graphics Applications, Ph.D. thesis of Department of Computer Science Brown University

[1] Hanan Samet and Markku Tamminen. Bintrees, CSG Trees, and Time. In Proceedings of SIGGRAPH '85, published as Computer Graphics, volume 19, number 3, pages 121-130, July 1985.

[2] S. P. Mudur and P. A. Koparkar. Interval Methods for Processing Geometric Objects.IEEE Computer Graphics and Applications, volume 4, number 2, pages7--17, February1984.

[3] John M. Snyder. Interval Analysis for Computer Graphics. In Proceedings of SIG GRAPH '92, published as Computer Graphics, volume 26, number 2, pages 121-130, July 1992.

[4] Tom Duff. Interval Arithmetic and Recursive Subdivision for Implicit Functions and constructive Solid Geometry. In Proceedings of SIGGRAPH '92, published as Computer Graphics, volume 26, number 2, pages 131--138, July 1992.

[5] John M. Snyder, Adam R. Woodbury, Kurt Fleischer, Bena Currin, and Alan H. Barr.Interval Methods for Multi-Point Collisions between Time-Dependent Curved Surfaces.In Proceedings of SIGGRAPH '93, published as Computer Graphics Proceedings, Annual Conference Series, pages 321-334, August 1993.

[PTVF92] William H. Press, Saul A. Teukolsky, William T. Vetterling, and Brian P. Flannery. Numerical Recipes in C. Cambridge University Press, Cambridge, England, 2nd edition,1992.

[6] Michael J. Zyda, William D. Osborne, James G. Monahan, and David R. Pratt. NPSNET: Real-Time Vehicle Collisions, Explosions and Terrain Modifications. TheJournal of Visualization and Computer Animation, volume 4, number 1, pages 13--24, 1993.

[7] Greg Turk. Interactive Collision Detection for Molecular Graphics. Technical Report TR90-014, Department of Computer Science, The University of North Carolina at Chapel Hill, January 1990.

[8] James K. Hahn. Realistic Animation of Rigid Bodies. In Proceedings of SIGGRAPH '88, published as Computer Graphics, volume 22, number 4, pages 299{308, August 1988.

[9] Matthew P. Moore and Jane Wilhelms. Collision Detection and Response for Computer Animation. In Proceedings of SIGGRAPH '88, published as Computer Graphics, volume 22, number 4, pages 289-298, August 1988.

[10] Clifford A. Shaffer and Gregory M. Herb. A Real-Time Robot Arm Collision Avoidance System. IEEE Transactions on Robotics and Automation, volume 8, number 2, pages 149-160, April 1992.

System Model and Implementation of Constraint-Based Distributed Intelligent Conceptual Design System *

Yurong Xu, Shouqian Sun
State Key Lab of CAD & CG
Dept. of Computer Science and Eng.
Zhejiang University,
Hangzhou, 310027, P.R. China
Xyr_internet@iname.com,
ssqq@mail.hz.zj.cn

Zongkai Lin
CAD Lab, Institute of
Computing Technology,
Chinese Academy of Sciences,
Beijing, 100080, P.R. China
lzk@ict.ac.cn

Yingxin Zhao
Dept. of Industrial Design,
South Campus,
Shandong University,
Jinan, 250014, P.R. China
leeyohn88@sina.com

Abstract

The objective of Computer Supported Knowledge Aided Collaborative Conceptual Design (CSKACCD) is to aid group design activities semi-automatically or automatically with multidisciplinary knowledge. This paper presents a general distributed design system model to give a global view of CSKACCD, and then designs a constraint-based distributed knowledge representation model based on different constraints abstracted from the domain specific knowledge of the designing artifact. Finally, this paper describes a robust distributed conflict detection and resolution mechanism developed for different designers to interact with each other effectively.

1. Introduction

The efforts of many designers with different expertise should be needed to design large complex artifacts. In such a fairly complex collaborative design process, we should provide a computation environment to make different terms of designers to work on it efficiently and concurrently. It is requested for us to provide such a CSKACCD environment. It should give a well defined interacting interface to the designers for system to apperceive the activities of designers to create constraints. It should represent them with constraints in knowledge representation model, which is used to stand for special expertise of the design artifact. If the conflicts are created, then they will be detected by the environment, then will be resolved by the conflict-resolution mechanism. So, it is guaranteed that whenever, wherever a designer made a design modification in the environment, the knowledge representation model, which is to express the target product, will be kept integration and unification.

In order to implement such environment, we should think about some facets as following:

1. Because of the boosting of WWW, we think about that a general design environment is recommended to be hetero-architecture. It should be built on commercially successful CAD systems and can help designer to finish design activity by provide an easy-learn, easy-use design interface.
2. Such environment offers a distributed knowledge representation model to express the different expertise, which is combined in the particular target product, and to apperceive design knowledge from individual designer's sketching activity with the aid of the design interface.
3. In case of large complex artifact design, it is impossible for a single designer to possess the entire knowledge of the artifact. Therefore, it is extremely difficult and some time impossible for a designer to determine all the possible conflictions in the design activity. Hence, to define a sophisticated mechanism is necessary to automatically or semi-automatically resolute the conflicts, and when the system can't resolute the conflicts, a man-man conflict resolution will be needed.
4. In such a distributed environment, because there will be so many 2D, 3D graphic data to be propagated. The communication will be a bottleneck, especially in a low-speed network. In order to implement rapid 3D, 2D design in low-speed environment, it is necessary to disseminate parameters among different clients in stead of images and graphs.

In the following chapters, with the sample of motorcycle conceptual design, we will introduce a general distributed design system model for CSKACCD, as described in chapter 2. A constraint-based distributed knowledge representation model is given in chapter 3.

*This research is supported by the National 863 Project (863-511-942-016) of P. R.China and YunNan Province-University Collaboration project, P. R.China.

While the creation and detection and resolution of conflicts, which is created in such collaborative environment, will be presented in chapter 4, and a robust distributed conflict detection-resolution mechanism to resolve the conflicts, is described here.

2. Relation Works & System Model

In many distributed design systems, such as AURA [1], DIVE [18], MASSIVE [19], VLNET [20] and so on, almost all of them give powerful systems to make a simulative model and a mechanism for users to interact each other synchronously. However, they pay little attention to use virtual model to apperceive the distinctive designers' activities and provide automatic and semi-automatic conflict detection and resolution by using the special knowledge. As in the [3][4],"knowledge intensive" computer aided design is advocated. Such as M. Hale and Graid [5] developed a distributed intelligent system for aircraft design based on the conception of a design integration framework, a "Virtual Design Environment" was described by R. Subbu et. Al. [6] to utilize evolutionary intelligent agents in order to resolute the conflict among the designers, and J.D'Ambrosio et. Al. [7] have presented a framework in which, the constraints of a design supervisor are distributed to design subordinates, so that they can exploit the local expertise with the context provide by global design information.

Such systems can be categorized into three major classes of [2](1) Interpreters, In such system, a virtual model environment is given for designers to simulate the design object, and they designed for interpreting the design situations. (2) Advisory Systems. In systems with special knowledge concerning of the specific design object, by using their interpretation of the design situation, the designers will be able to improve the conflicts manually. (3) Prescriptive systems. Systems with a capacity for controlling the conflict detection and correctness process in the distributed simulation environment. They can apperceive the design information of the designers and compare with the global design context or with a series of modeling experiments, and then optimize their virtual model, in this type of systems, semi-automatic and/or automatic conflict resolution in the distributed simulation system.

Although, almost the above works about distributed design system can be classified as the category of 'prescriptive system'. They thought about a kind designers with only one expertise, did not cogitate the collaborating among the distinctive designers with their unique proficiencies. Besides, in the above works, the conflicts are created only from 1. the relationship of the operation of distinctive designers (R. Subbu et. Al. [9]), or 2. the hierarchical relationship of the operating activities between the supervisor and subordinates (J.D'Ambrosio et. Al. [10]). The conflicts in such design activity don't represent the collisions in the exertion of expertise knowledge, and the resolving of the conflicts offers only an auto-harmony in the collaboration of distinctive designers, can't function as knowledge aiding design.

The following work not only considers collaboration

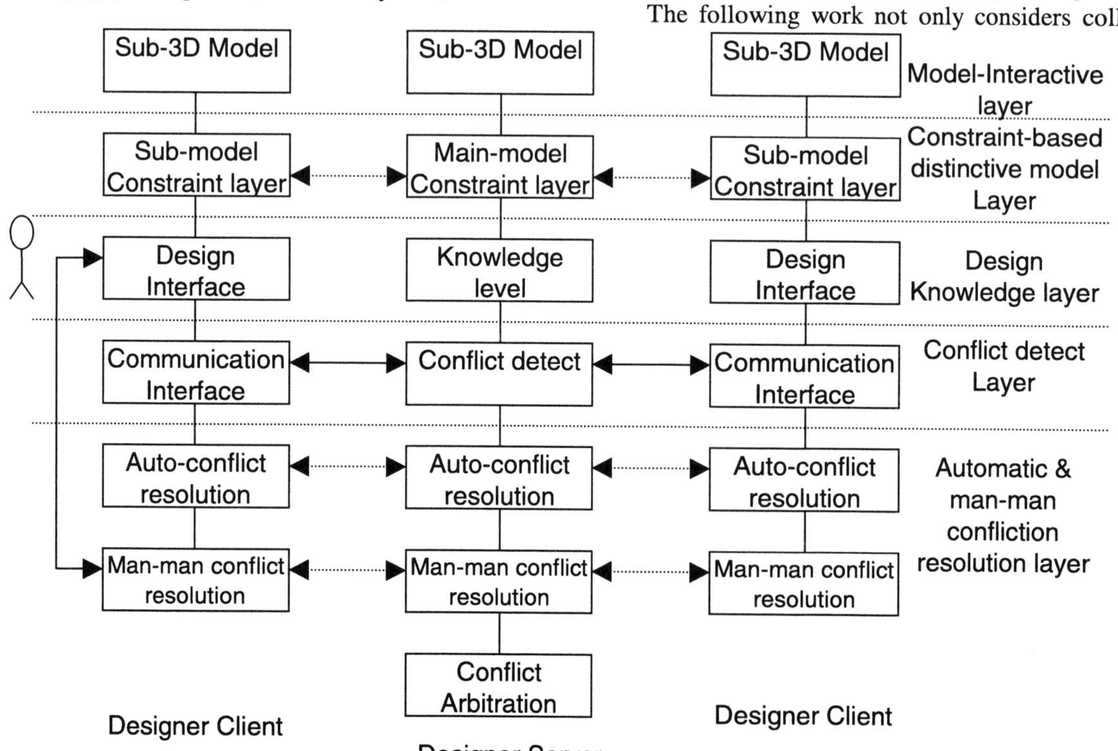

Fig.1 Distributed constraint-based 3D design system model.

among different designers with similar expertise, but also estimates cooperation among different designers with distinctive expertise. By offering different sub-model (which is based the special knowledge) to individual designer, offering main-model to express the abstract knowledge of the designing artifact, the design question solution will be archived from the resolution of conflicts which is created in the collision between the main-model and sub-model. The following work not only considers collaboration among different designers with similar expertise, but also estimates cooperation among different designers with distinctive expertise. By offering different sub-model (which is based the special knowledge) to individual designer, offering main-model to express the abstract knowledge of the designing artifact, the design question solution will be archived from the resolution of conflicts which is created in the collision between the main-model and sub-model.

The following Fig.1 is the Distributed Constrain-based Design system model. It is disported into five layers

1□ Model-interactive Layer

The first layer is model-interactive layer, which is used to give a universal interface to let designer manipulate the distributed simulation 3D model and to apperceive the designer's activity. It has a respondent between 3D model and 3D graphic display object.

This layer is the user interface of distributed 3D simulation design environment. Users immerged in the distributed 3D simulation design Environment through interactive facilities and other kinds of interactive technologies (such as VR). Its functions are (1) to perfect the action of the designer activity and the design thinking, (2) give the correct view to the specific sub-model. It decodes input events and adequately reacts, or it sends an adequate message to its corresponding sub-model. As is showed in Fig.2.

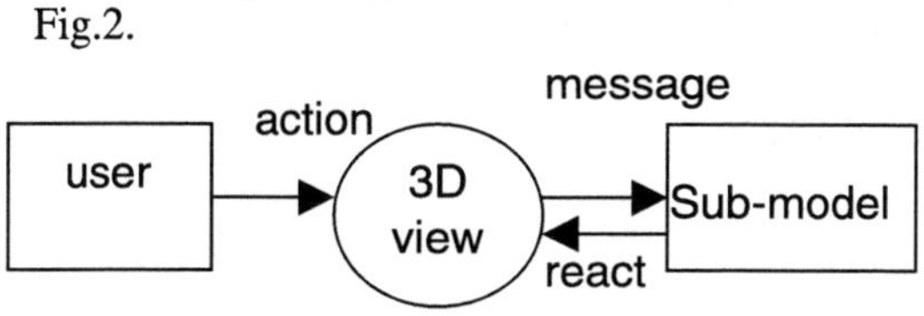

Fig.2 model-interactive layer.

2□ Constraint-based distinctive model layer

(1) Deducing sub model: A simulation main-model is constructed in advance through the constraint deduced from the designing artifact, in such a model the largely relative knowledge is inferred to make the different kinds of constraints. Based on the different expertise of distinctive designers, specific sub-models is got by aggrandizing and infirming the constraints in the main-model, so that the layer gives the virtual view to the main-model among the distinct designers, and make the designer to integrate its interesting in its special design field.

(2) Detecting conflicts in the interactive process with the designer or among distinctive designers. When a designer submitted a change in his sub-model with his expertise, the change conflicts with some kind of constraint in the sub-model, then a conflict is produced; Of source, some constraint lied in the sub-model is changed by the designer based on the expertise, then the conflict between sub-model and main-model is made out.

3□ Design knowledge layer

In this layer, the design knowledge is cataloged by the specific major field, then they are interpreted into the constraint format, so that at the initiation, the sub-model can be produced to represent the distinctive knowledge of the different designer by giving the different weight to the different constrain in the specific sub-model which is product by the model layer. When in design activity, the system can use them to understand the designer's thinking, and so that the system can gives the advice to the designer and do further work (such as adding the specific knowledge attributions) to the confliction produced in the upper layer, then for the lower layer to resolute them.

4□ Conflict Detection Layer

In this layer, we will detect the conflicts which is created in the above layers. This and the following layers will be discussed in detail.

5□ Conflict Resolution Layer

The conflicts detected in the above layer will be auto-resolved if possibly; or un-resolved conflicts in them will be sent into the lower sub-layer of human-human confliction resolution. Here we will give a general distributed mechanism to detect and resolute the conflictions in a distributed design environment.

3. Constraint-based Distributed Knowledge Representation Model

In this chapter, we will describe the definition of constraint-based distributed knowledge representation model, which is founded under the expertise of a sub-field in conceptual design – layout design. At first, we will utility special knowledge of the artifact to get an abstract model of mechanics. In the second, we will give the formal definition of the constraint-based distributed knowledge representation model. At last, a sample of the model is given.

3.1. An Model of Mechanics

Layout design, as a conceptual design, is to decide the

spatial existing mode of the layout units in a complex artifact. It is decided by several factors. Some factors are come from statics, while many more factors are raised in kinemics. Undoubtedly, there are many components in a complex product, so it is necessary to describe them with a large number of different formulations for relating the forces within a physical structure. It is lucky that all of them can be described by mechanics and statics [8] in particular as a close-in system. So many of the formulations can be reduced to a set of formulations, which only represent the place restrictions on how these components may be arranged, and on what order forces may be propagated.

The following is the abstract model of mechanics which is used to describe the artifact's and its components' statics characters, specified by mass, position (P), force and so on.

The model of general constraint (mechanics) ::
the body of module
::mass and position and force and momentum (kinematics)
mass :: mass and center of mass
position :: geometry position and layout position
force :: extern force and internal force
internal force :: gravity force and stress force
extern force :: spring force | rod force
and the system
::horizon normal and surface normal and origin point
the model of specific constraint (such as power system design, aerodynamics design, layout design)
...

As shown above, we can see that an artifact can be disparted into several modules constituted with several components (such as front-wheel module). Each module has static, kinematical characters, can be functioned as a whole. The module has mechanical and kinematical relationships with other modules, (such as the relationship between front-wheel module with front-fork module) also. In a layout design, we consider the static, kinematical characters of the modules between the whole motorcycle as strong constraints, which can't be violated in design activity.

3.2. The Constraint-based Distributed Knowledge Representation Model

It will be exciting, if it is possible to give a simple model for different types of product design, so that the components can be arranged in any desired configuration. The efficiency is contained in the models, also. Therefore, it is triable to introduce the expertise into the modeling by using constraint-based mechanism, as such CLP language [9]. Witkin et al.[10] had been used constraints to build mechanisms out of components for interactive simulation. As well as, t Lakmazaheri and Rasdorf implemented a system in CLP(R) [11][12][13]. However, they utility it only in single-user environment, and do not deal with cooperative impact. A constraint-based Distributed knowledge representation Model is needed to implement an efficient resolution in a collaborative environment.

The model can be described as following:

<LAYOUT>::=<LAYOUT_ID>+<SUB_LAYOUT$_1$>...<SUB_LAYOUT$_n$>+RELATIONSHIP($\prod$(<SUB_LAYOUT$_1$> ...<SUB_LAYOUT$_n$>))

<SUB_LAYOUT>::=<SUB_LAYOUT_ID>+<MODULE$_1$>...<MODULE$_n$>+CONSTRAINT($\prod$(<MODULE$_1$> ...<MODULE$_n$>))

<MODULE>::=<MODULE_ID>+<DISPLAY>+<REACTOR>+<STATICS_ATTRIBUTE>+<KINEMATICS_ATTRIBUTE>

<DISPLAY>::=<DISPLAY_ATTRIBUTE>+<DISPLAY_CONSTRAINT>

<REACTOR>::={When the MODULE is changed by the designer, the REACTOR will be called to manage the transactions, which is produced by the constraints come from interacting between the module and the other modules.}

RELATIONSHIP(<SUB_LAYOUT$_1$>...<SUB_LAYOUT$_i$>)::={deduces the constraints from the special field knowledge of distinctive SUB_LAYOUTs.}

CONSTRAINT(<MODULE$_1$>...<MODULE$_i$>)::={deduces the constraints between the specific MODULE and other distinctive MODULEs.}

The Constraint-based Distributed Knowledge Representation Model is composed of the set of several sub-layouts, on which designers with specific expertise focus, and a main-model combined with the sub-layouts, which have the relationships among each other. <SUB_LAYOUT> represents the sub-system in a specific product system such as power system, ergonomics system, bracing system and so on. So the sub-layout is constructed with some relative components labeled "module". Here, the specific knowledge is represented by the constraints among the modules. In a layout design, the minor component is <MODULE>, which includes the statics attribution and kinematics attributions of the module. A <REACTOR> is acted as an agent, which is independent, autonomic and mainly focus on the apperceiving the possible conflicts among the constraints. A <DISPLAY> object offers a view for end-designer to interact with the sub-model and main-model, and apperceives the character of the design activity.

The model has three functions:

<REACTOR>: is a similar agent program unit. It is corresponded to one-to-one < MODULE>. When the display object is changed, the reactor will be called to judge whether the conflict is produced between the module constraint and the display. In additionally, when a new rule is created by a designer, then new correspond constraints are formed by the reactor.

RELATIONSHIP: In such a function, series constraints are produced dynamically when some sub-layouts are changed by designer actively. In upper process the systemic expertise discipline is used, when the main layout model is contained integrity.

CONSTRAINT: In order to maintaining integrity of module between graphics and statics & kinematics, many general attributions should be used to construct constraints. The specific knowledge is used to deduce different constraints, as such 'key point' constraints for layout design.

A sample of constraint-based distributed knowledge representation model is described as Fig3.

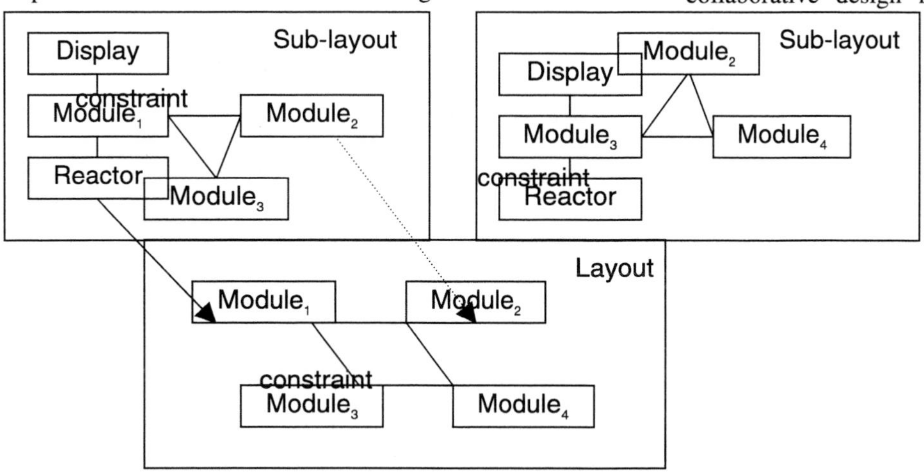

Fig.3 A sample of constraint-based distributed knowledge representation model.

In the sample of Constraint-based Distributed Knowledge Representation Model, there are four module represented four components in layout. In one sub-layout, there are three modules labeled 1,2,3, and $modules_{234}$ belong to the other sub-layout. In the sub-model, a module has a <display> and a <reactor> connected with it. The <display> is to show the 3D, 2D graph, while the <reactor> is monitoring the operation on the module. When a module is modified by a designer, the <reactor> will compute the constraints which is connect with the module in particular sub-layout, at the same time the modification will be send back to <layout> for it to keep integrity.

The Constraint-based Distributed Knowledge Representation Model, is integrated with the model interactive layer, distributed constraint-based model layer and the design knowledge layer, and some constraint detection, and focuses on the constraint-based distributed modeling. It has following properties: (1) Descriptive. The specification of a component is described easily. (2) Consistent. Giving a same mathematical model for a different mechanical system, regardless of which syntactic description or other specific discipline. (3) Efficient. In a distributed simulation environment, it is important to implement a competent communication and coordination mechanism. Now by the constraint-based model we reduce the communication quantity by exchange only different conflicts instead of many graphic or multimedia data largely [14].

4. Conflict creation, detection and resolution

In this chapter, we will describe the creation, detection and resolution of conflicts.

Conceptual design for a complicated artifact is a fairly complex cooperative design process. Many different conflicts deduced from single designer mismatches or collaborative design mismatches [15] will be created under different expertise disciplines. When the conflicts is created in collaborative environment, the different kinds of conflicts should be separated with different statuses (single user environment & multi-user environment), so that, the conflicts can be resolved efficiently.

4.1. Process of Conflict Creating &Detection

When a designer is interacting with different sub-model by specific display object, the activity of design will be apperceived by the sub-model to create possible conflicts by struggling with the constraints contained in the sub-model. All of the conflicts will be detected by the <reactor> in every sub-model, but it is not practical to resolve the conflicts in main-model in such a complicated collaborative environment. Many conflicts will be resolved in the sub-model, while some conflicts will be sent to the main-model for resolution. So the main-model will gather many conflicts and there are mobile-agents (defined in the 4.2) created to represent them. The Fig.4 describes it.

4.2 Definition of Mobile-Agent

In the above model, we use a mobile agent to represent a latency conflict source, in a distributed environment, some a latency conflict source comes from a designer of the kernel design group possibly, or comes from periphery participant. Each agent has the attributes of the time-field, knowledge-field, knowledge-depth-field, project correlative degree, corresponding to the relative attributes of each latency conflict source. As be shown in Fig.5.

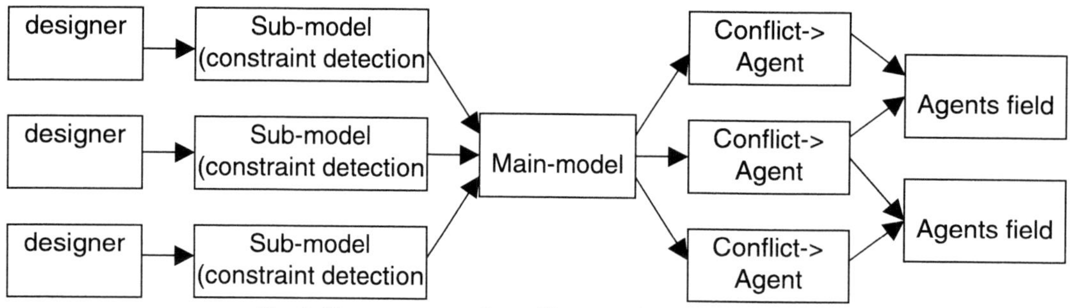

Fig 4. The process of conflict creating.

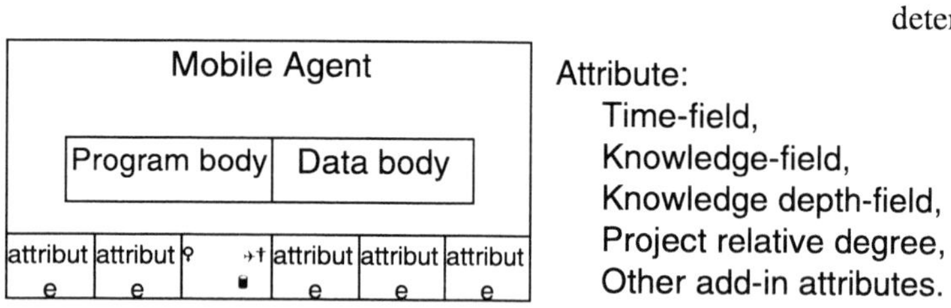

Fig. 5 An Agent with multi-attribute.

4.3. The Main Conflict Resolution Mechanism

Traditionally, two major approaches have been adopted to achieve collaboration. (1). Distribution lists [16]: In the distribution lists, a major problem is that the lists are static in nature; they are generally based on the names of the people and not their current roles. Because the organization itself is quite dynamic and so the relationship between the people and their roles is changeable, so the 'static' distribution lists fail to achieve the coordination of designers. (2). Multi-disciplinary team meetings [17]: It has the similar problem, too. It wastes people's time, because only a small fraction of all the people are involved in a particular design conflict[28]. At the same time, meetings often require physical co-location of the people, and in many times, it is often impractical for the projects in which teams are located at geographically different places.

Also, aimed at the individuating and customizing increasingly in the product, conceptual design changes from only a phrase in the industry production dissolve into full-life-time production from market investigation, to decision-making, to product design, to product making, to market vendition.

So, in collaborative conceptual design, the source of conflict is not only limited in the designers with similar time-field, similar knowledge-field, similar knowledge deepness-field, it will be expended to different designers and participants with distinct time-field, distinct knowledge-field, distinct knowledge deepness-field, distinct project relationship degree, etc.

The characters of the conflict model determine that it will have many advantages for using an opening, self-study intellectualized conflict harmony mechanism.

The main advantages of the mechanism are:

1. Elastic, the mechanism can work properly regard few of the different characters of designers such as attending time, working place and specific expertise discipline.
2. Self-study, Because the conceptual design is an intelligent design process in collaborative work environment, the constraints in the constraint-based distributed model are deduced from the specific expertise knowledge. They are changed dynamically in the design activity if necessarily, so it is unavoidable for the constraint resolution mechanism to have ability of self-study and self-précising.
3. Robust. In a simulation design environment with variable scales, there is protean number of conflicts in the system. Sometimes the agents represented the conflict will terminate illegally. A robust mechanism to administer agents and conflicts.

The mechanism for conflict-resolution is described in Fig. 6.

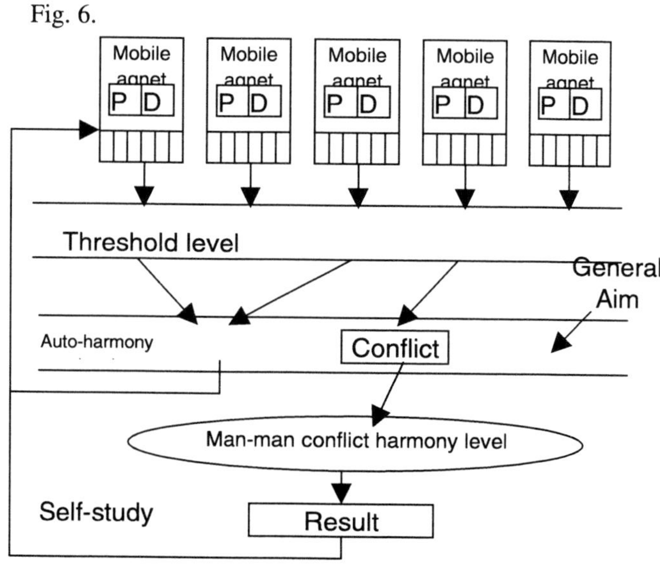

Fig.6 Self-study conflict resolution mechanism.

All of the latency conflict sources represented by the Agents are filtrated by a threshold level. The Agents whose value below to the threshold value, will be hold out the upper level of conflict resolution model. The agents which come into the threshold level, will be grouped into distinct conflict field with representing to different conflicts, then the additional restrictions will be added into the Agents. The distinct conflict fields composed with the different agent group will be put into the auto-conflict harmony level to resolve the conflicts. If the auto-conflict harmony level cannot decide the conflicts of the specific conflict fields, the unresolved conflict fields will enter the man-man conflict harmony level.

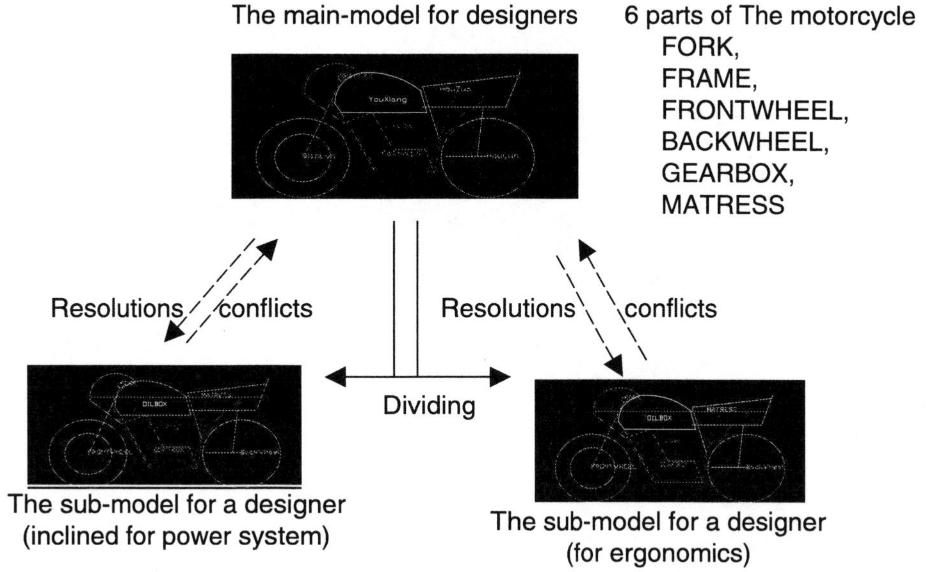

Fig.7 The example of constraint-based distributed knowledge representation model.

5. The example of system verification

After we perfectly define the system model, knowledge representation model and the conflict resolution mechanism, it is time to give our prototype system of the conceptual design for motorcycle (layout design). As the request of national 863 project, we designed a knowledge aided collaborative concept design system for motorcycle(MotorCCDS). The client software of MotorCCDS is designed under commercial software AutoCAD2000™ by added the communication partition, knowledge representation level and constraint-based level. The server software is structured under SqlServer 2000 and NT4.0.

In Fig.7, the main-model of motorcycle in server, which have general constraints of motorcycle system, is disparted into two sub-models, after a dividing-work algorithm. The first sub-model focuses on layout design of power system, whose constraints, inherited from the main-model, will be focused on power system include back-wheel, engine. The second focuses on the layout design of ergonomics, whose constraints will pay attention to ergonomics. A virtual human will be generated to evaluate the layout design of such sub-model, as Fig.8.

Each sub-model for a specific designer is integrality except by adding different constraints in that expertise field, while the other constraints which are little relative with the such sub-model is infirmed. When a designer in the layout design of power system, changed his design in his sub-model and submitted his result, sub-model would send the possible conflicts to the main-model in server. In server, there will be a large number of synchronal, asynchronous conflicts. All conflicts are expressed by diffident mobile-agents, then they are dealt with the mechanism for conflict-resolution.

Fig.8 the sub-model of layout design in ergonomics.

The fig.9 Shows the user interface of MotorCCDS, which is included two sub-models. Fig.10 gives the design output in 3D. Fig.11 is the 3D texturized motorcycle, which is generated by specific software 'Mechanical Desktop'.

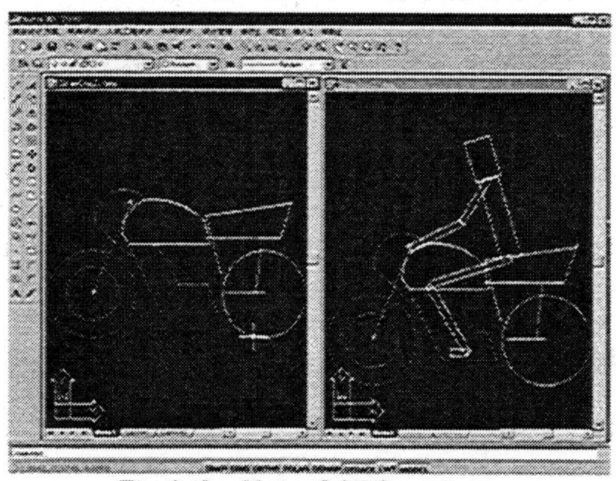

Fig.9 the MotorCCDS system.

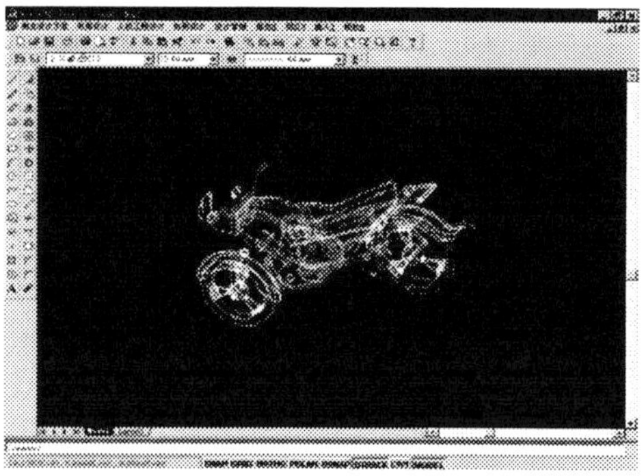

Fig.10 3D motorcycle Layout.

Fig.11 3D texturized model for motorcycle.

6. Conclusion

By giving a universal distributed design system model and then describing the details of the system mainly in constraint-based distributed knowledge representation model and conflict creation, detection and resolution, we depicted the implement of collaborative knowledge aided design in the distributed constraint-based intelligent conceptual design. We combine graphics display with the design knowledge represented by specific constraints to produce a constraint-based distributed knowledge representation model for collaborative design environment. By being experienced in our national 863 project on it, the result is exciting, it produce some kind brains for a distributed conceptual design. The system can resolve a few conflicts automatic when multi-designers are active. Such as, when a designer modified the layout of front tyre in a motorcycle model, then the model will try to keep integration of the motorcycle sub-model and main-model, by attempting to modify the front brace, which is connect with the front tyre, to resolve the tyre-brace conflict. It is only primary attempt by using distributed constraint-based model to automatically and semi-automatically resolve the conflicts in the collaborative environment.

7. References

[1]Lennart E. Fahlen, Olov Stahl, et al., "A Space Based Model for User Interaction in Shared Synthetic Environments", *ACM INTERCHI'93*, 1993,pp.43-48.

[2] Taratoukhine, V;Bechkoum, K. "Towards a Consistent Distributed Design: a Multi-Agent Approach". *Info. Visualization99 VI99* IEEE,1999,pp384-389.

[3]T. Tomiyama, "Towards knowledge intensive intelligent CAD",*JSME-ASME Workshop on Design*, 1993,pp.46-51.

[4] T. Tomiyama, T. Kiriyama, and Y. Umeda, "Towards knowledge intensive engineering", *in Knowledge Building and Knowledge Sharing*, IOS Press, 1994,pp. 308-316.

[5]M.A.Hale, and J.I.Craig, "Preliminary Development of Agent Technologies for a Design Integration Framework" *Proc.5th Sym. on Multi-disciplinary Analysis and Optimisation*, Panama City, FL USA, Sep. 7-9 1994.

[6] R. Subbu,C.Hocaoglu, and A. Sanderson, "A Virtual Design Environment using Evolutionary Agents," *Proc. of 1998 IEEE Internatoinal Conference on Robotics&Automation* Belgium, 1998, pp.247-253.

[7] J. G. D'Ambrosio, T. Darr, and W.P. Birminghum, "Hierarchical Concurrent Engineering in a Multiagent Framework," *Concurent Engineering: Research and Applications Journal*, 1996,Vol.4,pp.47-57.

[8] Loofbourrow, N "A compositional, constraint-based model of statics" *Computer Animation 98. Proceedings*, 1998,pp177 – 184.

[9] J.Jaffar and J.L. Lassez. "Constraint logic programming". *Proc. of the 14th ACM Symp. On Princ. Of Prog. Lang.(POPL)*, January 1987, pp111-119.

[10] A. Witkin, M. Gleicher, and W, Welch. "Interactive dynamics" . *In Comp. Graph.(1990 Symp. On Interact. 3D Graph.)*, Mar. 1990,pagers 11-21.

[11]S.Lakmazaheri ,W.J.Rasdorf. "Constraint logic programming for the analysis and partial synthesis of truss structures". *Artif. Intel. Eng. Design, Analysis and Manuf.*, ,1989,3(3):157-173.

[12] S.Lakmazaheri and W.J.Rasdorf. "The analysis and patial synthesis of truss structures via theorem proving". *Int. J. Eng. With Comp.* Apr.1989,6(1):31-45.

[13] S.Lakmazaheri and W.J.Rasdorf. "A formal approach to structural design using predicate logic". *Proc. First Intl. Workshop Formal Methods in Eng. Design, Manuf. And Assem.* Jan. 1990, Pages 247-259.

[14] Shigeaki Aoki et al., "Design and Control of Shared Conference Environments for Audio Telecommunication Using Individually Measured RTFs", *Presence*, 1994, Vol.3,No.1, p.60-72.

[15] *Danforth, S.*"onsistent choice narrowing-a new model for constraint based programming" *Software Track, Proceedings of the 21st Annual Hawaii International Conference on System Sciences*, Vol.II.1988, pp: 595 –604.

[16]Gupta, L."Thesis Interviews –Coordination and Concurrency" ; *Technical Report*, Enterprise Integration Laboratory, Department of Industrial Engineering, Univ. Toronto, Toronto, Canada, 1996.

[17]Klein,M. "Conflict Resolution in Cooperative Design"; *The International Journalfor Artifical Intelligence in Engineering* Vol.4 ,No.4; 1994, pp.168-180.

[18] DIVE: http://www.sics.se/dce/dive.html

[19]MASSIVE: http://www.crg.cs.mott.ac.uk/~cmg/massiv.html

[20] VLNET: http://ligwww.epfl.ch/~captin/en/vlnet.html.

Collaborative Discussion and Decision-Support in a Manufacturing Organization

André Marcos da Silva
Inst. of Computing/Unicamp
amarcos@ic.unicamp.br

Rodrigo Bonacin
Inst. of Computing/Unicamp
ra000470@ic.unicamp.br

M. Cecília C. Baranauskas
IC & Nied/Unicamp
cecilia@ic.unicamp.br

Abstract

Most work activities in organizations involve problem discussion, interpretation and decision-making. Literature has shown that an understanding of the nature of groups and their organizational work context is essential to the designs of systems to support these activities. Andon is a CSCW (Computer-Supported Cooperative Work) system, designed to support discussion during problem-solving processes in a manufacturing organization. This paper addresses the reuse of results from participatory techniques applied to the Andon interface design to propose an artificial agent structure aimed at improving decision-making in workgroups.

1. Introduction

Despite the fact that they usually have particular viewpoints and slightly different concerns, technological support for groups in organizations has received widespread attention in a variety of research areas such as: Human-Computer Interaction (HCI), Computer-Supported Cooperative Work (CSCW), Group Decision Support Systems (GDSS).

Most work activities in organizations involve sense-making, interpretation and decision-making. Systems for groups working together usually involve some support for different phases of group activity, such as idea generation and selection, and tools intended to assist in aspects of decision-making processes. Electronic Meeting Systems and computer-augmented meeting rooms are examples of this category aimed at dealing with issues of decision-making.

Decision Support Systems (DSS) include information technology applications that support people in making decisions about what actions to take in problem-solving situations. Group Decision Support Systems (GDSS) can be seen as a DSS augmented by the fact that there is more than one decision maker involved, and so some computer support for communication and the integration of multiple inputs should be provided [3]. As an extension of DSS, GDSS inherits the theoretical background of the former.

The tradition in DSS has evolved from pure rationality to naturalistic decision theory. In the first generation of decision support systems, the decision problem was formalized as a series of well-defined mathematical relationships. People were seen as inherently poor decision-makers as they were unable to maintain and manipulate the underlying relationships mentally. External decision support had to be offered as a way to compensate for this deficiency, in terms of mathematical algorithms implemented on computers [11]. The second generation of DSS was grounded on the assumption that formally optimal decisions had little benefit over sub-optimal decisions that achieved the same desired goal. The concept of bounded (rather than pure) rationality was created. A decision-maker acting under bounded rationality needed knowledge about the behavioral context of the decision (i. e. the boundaries to formal rationality). This generation emphasized an analytical approach to decision making, in which problems were systematically analyzed considering also knowledge and problem-specific data. A third generation of decision support has arisen in the last decade, which considers naturalistic settings, i. e. socially and organizationally situated contexts. Decision-makers in naturalistic settings are experts in the domain, and have extensive knowledge of the tasks in which the decision process is embedded. This fact compensates for the ill-defined problems found in natural contexts, which makes them intractable for the rational or bounded rational decision methods. By the end of the last decade naturalistic decision theory became the dominant theory about decision making.

While accepting the contributions of this field, the perspective of GDSS as an extension of DSS has its drawbacks. A first problem is the "group" concept adopted among researchers and practitioners. In most cases groups are characterized by a number of specific properties such as homogeneity, equal status, shared goals, etc., while real groups in real organizations include interactions among people with different backgrounds and different expertise about the domain in which decisions have to be taken. Shop-floor workers, supervisors, and managers are the personnel involved in decision-making, in the context of problem-solving in a manufacturing line, for example.

Some researchers argue that the studies of the decision-making genre lack concern about the social and organizational aspects of decision making and technology support [3]. Also, there is an interest in allowing for richer forms of interpretation of how people function in organizations, and how decision-making actions are taken.

As pointed out by Bannon [3], GDSS is seen by the CSCW community as a field of study that relies too heavily on a set of rationalistic assumptions about human behavior in organizations. The CSCW field, on the other hand, is seen as relying too much on either political-sociological and ethnomethodological approaches or, alternatively, of having a technocentric focus. The field of Computer Supported Cooperative Work has embraced a variety of research in overlapping areas such as groupware, group decision support systems, cooperative work support, coordination technology, etc. One of the important contributions of CSCW work has been the detailed investigation of exactly how work gets done in organizations, to provide an appropriate technology to support it. Ethnographers have pointed to the necessity of detailed studies of the circumstances in which work is conducted, at all levels of the organization, from the factory floor to the executive management staff. Transforming this kind of information in the design of effective artifacts for use by groups working together is a subject of HCI studies and techniques.

In this paper we argue that methodologies and techniques of HCI can bring the focus to issues which interest both areas of GDSS and CSCW. In particular, we are addressing participatory methodologies for the design of *Andon*: a system designed for problem discussion and decision-making in the real context of manufacturing. The paper is organized as follows: Section 2 presents de organizational scenario of the work, Section 3 presents the design rationale of the *Andon* system, Section 4 highlights the decision making model to be embedded in the system. Section 5 concludes.

2. The organizational scenario of the work

Human resources are, indeed, the greatest good of an industrial organization nowadays. History shows, however, that it was not always this way. The philosophy of "mass production", which was predominant in the USA in the twenties, had its focus on quantity, on economies of scale and on sales [12]. The power of decision-making was concentrated at the top of the hierarchy, while workers on the production line followed work routines defined by an expert.

The philosophy of "lean production", created by the pioneer work of Deming [13], in the late 1940s, argued that an organized industrial system with focus on the quality of its goods not only supported, but also enhanced productivity. The set of new techniques and strategies used to enhance quality and productivity changed the focus from the end product itself to the quality of the manufacturing process. Another important change is the role of the worker in a lean system, which is not anymore as a piece in the gears of mass production, but an intelligent element necessary to the drive for quality in the production process.

The organization we are working with is a multinational factory, which manufactures components of automotive systems. It presents the first signs of a culture tending to use of the lean production ideas. Among the main characteristics of the lean production ideas is the distribution of decision-power among the personnel, the incentive to teamwork and a more dynamic role for the shop floor workers. Together, these ideas propose that solutions to problems in the factory routine should be created by people who are closer to the problems, that the work in the organization must be more collaborative and that the operators should be more dynamic and multifunctional in the workgroup [10]. The design methodology for problem-solving and decision-making proposed in this paper is in line with the challenges imposed by the lean production paradigm in manufacturing organizations.

3. *Andon*, a CSCW system for problem-discussion

This section will briefly present *Andon* [5], a system that allows the application of participatory methodologies to model an intelligent agents' architecture for decision-making support in workgroups.

Andon is composed by *e-brainstorming* tools designed to support discussion and problem-solving processes in a lean organization. This system addresses the practices of the lean production methodology, facilitating and stimulating collaboration among different categories of workers in the organization staff (engineers, operators, and administrators), in problem discussion at work.

Instead of imposing new tasks in the organization personnel, the proposed goal of this system is to improve some existent practices in the company. In current problem-solving processes, several difficulties make almost impossible the participation of all staff in problem-discussion. These difficulties are inherent in synchronous meeting/sessions: available time away from the production line, a limited period to discuss problems, the difficulty of exposing the problems, and the difficulty of controlling the conversation. The *Andon* system aims to overcome these inconveniences through an asynchronous computational artifact, which facilitates the exhibition of the problem, the conversation among the workgroup, and the collection of contributions. Other goals involve stimulating the participation of different people in the process, to support exchange of knowledge within the workgroup, and to make the problem solution more efficient and to reach it faster than through the early processes.

Figure 1 shows a snapshot of the latest prototype of *Andon*. This image shows different problems opened to discussion, their id-numbers, the names of people who identified them, the date, and location where the problems occurred. In the early process a small work group or department knew about each problem; with *Andon* it is expected that many more people in the factory will have access to the problems being discussed.

The success of a system to support workgroup discussion is not guaranteed by the technology itself, as discussed by Grudin [1], but it is directly influenced by its acceptance in the workgroup. *Andon* was designed to stimulate the participation of the whole staff in finding solutions to problems. To reach this goal, the personnel were directly involved in the design process through participatory design techniques.

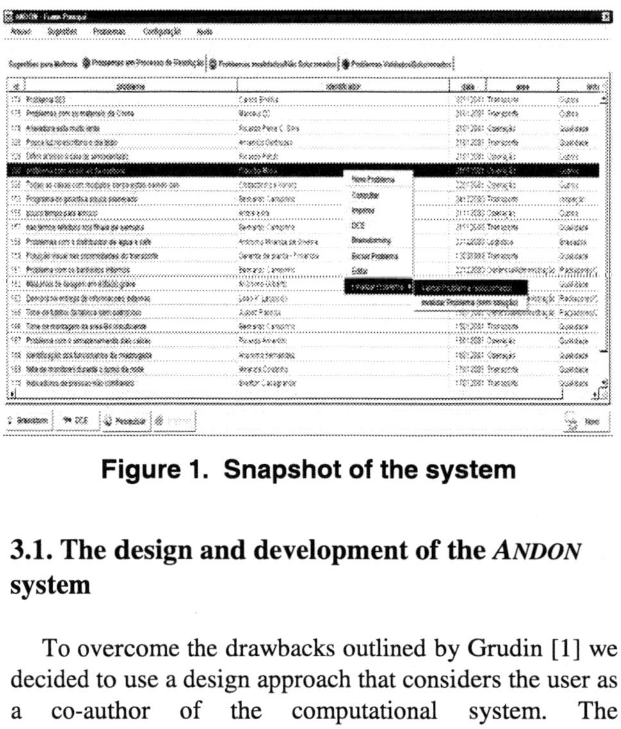

Figure 1. Snapshot of the system

3.1. The design and development of the *ANDON* system

To overcome the drawbacks outlined by Grudin [1] we decided to use a design approach that considers the user as a co-author of the computational system. The *Participatory Design* methodology was developed initially in Scandinavia and employs a variety of techniques to carry design *with* the user, rather than *for* the user [2]. In *Participatory Design*, the users interact directly with designers during the overall cycle of development, having some control over design decisions.

With this concept in mind we allowed users to participate in the development of the system, through special activities carried on during the whole development lifecycle. Aiming to elaborate an ideal prototype model of the expected system, techniques such as Ethnography [4] and Started Conference [2] were used. These techniques were embedded in a proposed lifecycle model for software development [5], which inherits some characteristics of the traditional Software Engineering Models, such as fast prototyping and evaluation phases among each development activity.

The techniques of participatory design used in the *Andon* development were carefully chosen among several techniques proposed by Muller, et. al [2]. We aimed to find a model that does not impose rules to be followed during the development of the software, but that stimulates the users' participation in the system design, through activities and techniques that really consider the needs, the demands, and the opinions of people from the work domain.

The design of the whole system was supported by the *Participatory Design* approach. The system characteristics were identified, based on results of these participatory practices. The logical concept of the system and its computational structure definition were also derived from these results. Figure 1 shows some features derived from participatory techniques, such as the functional arrangement, the identification of objects, the flow of tasks and the interface design of component with specific icons and words used in the routine of the organization.

3.2. The designers' conception of the current process

A lot of information collected through participatory activities was related to the current work process, as for instance, the identification of tasks, the tools involved in the problem-solving process, and so on. Because *Andon* proposes to uphold some activities already running in the organization, this information was very important during the *Andon* design.

The most important considerations about the early process in the organization involved the concepts of *problems* and *ideas*. The participatory activities enabled the discussion of them and the definition of their possible states in the whole process. Through these results the designers built a conceptual model centered on *problems* and *ideas*, facilitating the design of some system components, such as the menu structure, owner permissions, data base tables, and problems, ideas and flow of suggestions.

The different possible states of the problems in discussion are illustrated in Figure 2 Each of the three states has a group of functionalities to deal with the problem.

When a problem is identified, it is included in the system as a "new" problem and a person is allocated to be in charge of it. During this stage the problem is open to receiving all kinds of contribution, through ideas proposed to solve it, and trials to identify its causes. The problem stays in this state until it is considered solved or invalidated.

Solved and unsolved problem states in the system, allow only searches for information. Solved problems can be reported and unsolved problems can return to the problem-solving process state.

Every problem in the solving-process state can receive ideas proposing a solution to it. Because these ideas represent the contribution of the organization's community they are very important to the problem-solving process and to the system. As in the case of problems, the ideas are associated with states in the problem-solving process: a new proposed idea state, an idea in the implantation state, validated, invalidated, and canceled states.

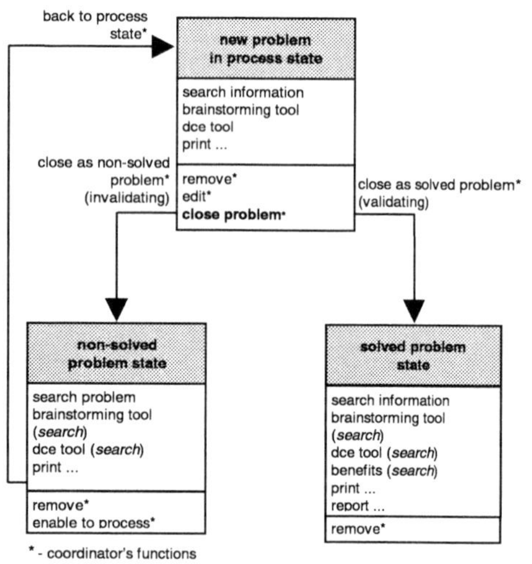

Figure 2. Flow of problem states

The **new state** contains ideas proposed by someone and not yet implemented in the factory. To put an idea in the implantation state, means that this idea will be tested as a potential solution to the problem. In this case some actions are planned and some people are assigned to control this attempt to solve the problem. During this phase an idea can be canceled, validated, or invalidated. To cancel an idea means that the attempt to solve the problem is being abandoned. If an idea reaches a solution to solve the problem, this idea is validated, and the problem is validated too. Otherwise the idea is invalidated.

The knowledge about the problem-idea relationship was elicitated as a result of the *Artifact Walkthrough* technique. It proposes activities to study the tools used in the current work, facilitating the interpretation of the real system and reducing the gap between the work context and the designers' conception of the system.

4. Decision support through the *ANDON* system

Between the problem identification and the problem being considered solved, some decisions must to be taken. When a new idea is proposed, the participants will be able to criticize it and this process must end in a decision that results in a solution to the problem. The electronic support for a standard quality process method, the *Ishikawa Diagram*, was embedded in the *Andon* System. Methods like the *Ishikawa Diagram* are helpful for the identification of the causes and for representation of the problem, but they do not eliminate the human cognitive load necessary to solve the problem. An architecture based on artificial agents is being proposed with the aim of decreasing the human cognitive load and facilitating decision-making through the *Andon* system.

4.1. The artificial agents approach

"*An agent is anything that can be viewed as perceiving its environment through sensors and acting upon that environment through effectors*" [6]. Using Russel and Norving´s classification of environments [6], the *Andon* problem-solving environment is inaccessible, nondeterministic, nonepisodic, dynamic and continuous, characterizing a very complex environment. Multiple agents are necessary to address distinct tasks. Because these agents are in a common environment, an infrastructure is necessary to support a multi-agent system [7]. Each agent's behavior and its relation with other agents, cooperatively or competitively, is determined by its task and the *design* objectives.

Before defining the infrastructure and each agent's behavior, the *design* objectives must be described precisely. Some questions such as: "How can they aid the users?", "How could they improve the human activities in the organization?" and "What is the task that the agent should carry out?" must be answered. The next sections address these questions.

4.2. Artificial Intelligence in the work context

In a complex environment artificial agents are hardly able to make correct decisions without human intervention. Considering Roth et al's [8] proposed taxonomy for intelligent interface design, some considerations have to be made concerning the work context. The traditional paradigm known as the "Greek Oracle", proposes "superhuman" reasoning to solve the user's problem. The humans provide the necessary information to a "black box", which solves the problem. After that the human serves as a *solution filter*. There are a lot of problems with decision-aids within this paradigm. Considering *Andon*, such problems include:

- in the current state-of-the-art in artificial intelligence, it is not possible to build agents capable of solving (alone) the problems in the organization context with the necessary accuracy to obtain the user's acceptance. The problems in the organization in general are associated with creative decisions/solutions;
- the "Greek Oracle" paradigm does not contribute to improvement in the human activities in the organization. This paradigm "reduces the opportunity for the user to utilize and sharpen his skill" [8], therefore the workers will not be able to promote the continuous improvement philosophy of lean production;
- in the "Greek Oracle", the task of providing the necessary information to solve the problem can restrict the vocabulary used in real problems identification/discussion. The organization staff demands flexible vocabulary and methods.

The other categorizations of the Roth's paradigm [8] of the development of intelligent interfaces considers:

- intelligent interfaces as *cognitive tools* that can be utilized by practitioners in solving their problems;
- intelligent interfaces as *members of cooperative person-machine* systems that jointly work on problems and share task responsibility (as a critic or a team player);
- intelligent interfaces as *representational aids* that dynamically structure the presentation of information to make key information perceptually prominent.

These three paradigm will overlap in the *Andon* system, whose different agents can act in three ways:

- a *personal assistant*, as suggested by Maes [9], collaborates with the user in the same work environment, aiding the users in part of their tasks, hence reducing their workload.
- a *critic* analyzing the human solutions and advising if a solution is deemed incomplete or incorrect. The critics agent could be used to promote continuos improvement in the organization's activities.
- a *team player* proposes new solutions and collaborates in decision processes.

4.3. Designing artificial agents for the organizational context

To really improve the human activities, the agents must aid in real problems found in the workplace, fitting themselves into the organizational structures, its political and strategic objectives. Our approach to reach these objectives is to design the artificial agents analyzing the structure of the organization with the users.

As in the case of the system interface design, the design and development of the agents to support group decision-making must start from knowledge about the organization. Although the participatory activities applied during *Andon* design were not concerned about the artificial agents structure project, some results from that activities do contribute to the proposal of an agent's structure for decision-making through *Andon*.

During the design of *Andon* some new functionalities were proposed by the users. These new functionalities suggest the design of rational agents to aid many distinct tasks in the organization, such as: to aid human decision making to reach the organizational objectives (making decisions in line with the organizational strategic plan), to support decisions based on expert knowledge, to support the coordination of the processes and the knowledge of past decisions.

Through the brief description of the artificial agents grouped below we can foresee how the agents could improve the human activities, and what kind of task they could carry out.

Tracking Agents – the importance of this group of agents was first noticed during an evaluation activity carried out with some people from the factory. Tracking Agents are *personal assistant* agents involved in the problems' follow-up. During a problem-solving process it is important for track the problem situation, and this activity is hard to the people in the organization. These agents propose to reduce the management workload following a specific problem discussion, their ideas, and their related actions. For example, if a problem evaluation deadline is passed and this problem has not been solved, its "owner" must track everything related to this problem. It is necessary to find the action or implementation of an idea, which is preventing the solution to the problem, the production area and line involved with it. The main goal of the Tracking Agents is to support these activities. These agents group are in line with the actual organizational objectives differentiating problems from the implementation of proposed solutions.

Backward Agents – the necessity of these agents was identified by the factory staff during the *Andon's* design. Many problems in the organization have similar characteristics to former problems, but knowledge about past decisions have not been recovered and considered, mainly when the decisions were made by other workgroups. Backward Agents are *personal assistant* agents who search for similar solved problems (and their solutions) when a new problem is identified. Besides helping to enhance the quality of decisions these agents promote the organization's objective of integrating the workgroups into the company.

Manufacturing System Agents – the importance of these agents was perceived during the course of *Andon's* design, by the quality managers. There is a special concern about knowing if the decisions/solutions adopted are in accordance with the manufacture system guidelines of the factory. These agents are *critical* agents. They analyze the human-proposed solution and show the consequences of the solutions in terms of the manufacture system reference guide. In this organization many objectives are specified from parameters/scores specified in this manufacture system documentation. As a *critic* these agents can also promote improvement in human decision-making.

Coordinator Assistant Agents – the existence of many repetitive tasks which increase very fast as problems and contributions are inserted into the system, motivated us to propose this kind of agents group. These artificial agents aim to learn about routine tasks, which is accomplished by owners or problem coordinators. When these agents perceive a list of repetitive tasks they try to intermediate and make suggestions. For instance, considering the deadline for implementation evaluation, every day someone must verify which idea implementation or specific action needs to be assessed. Once noticed by the Coordinator Assistant Agents, they start to help the user through advice assuring the success of this task.

Expert Agents – these agents act in collaboration with a *team player*. They could interact with other agents and use their knowledge to elaborate solutions, identify the best solution to implement, and discuss decisions. This group of agents can be very helpful in situations where several types of expertise are required. This necessity was suggested by the factory staff during the *Andon*'s design.

5. Conclusion

Literature has shown that an understanding of the work situation, the group members, and the task specifics is critical for designing systems to support groups in an organization. This work represents a joint effort being carried out within a manufacturing organization, aimed at articulating the design of a problem discussion and decision-making system with the understanding of the nature of group and organizational work.

This paper showed that some participatory techniques that were useful in the design the *Andon* system for problem discussion, were also helpful in the definition of an Agent-based structure to be embedded in the system to improve the decision-making processes in organizations.

The collaboration of the factory personnel in these participatory activities allowed the acquisition of knowledge about the organizational knowledge and the workgroup context that was essential for the intelligent interface design of the system. The next steps involve the detailed design of the proposed agents and its integration with the *Andon* environment to support decision-making in the factory context.

6. Acknowledgment

We would like to thank *Delphi Automotive System Organization*/Piracicaba - Brazil and Nied/Unicamp for collaboration and technological support to this work. This work is been supported by FAPESP and CNPq. We also tank Prof. Steven F. Durrant for his critical comments.

7. References

[1] Grudin, J. "Groupware and Social Dynamics: Eight Challenges for Developers", *Communications of the ACM*, 37, 1, pp. 92-105, 1994.

[2] Muller, M. J., Haslwanter, J. H. and Dayton, T. "Participatory Pratices in the Software Lifecycle" in *Handbook of Human-Computer Interaction*, M. Helander, T. K. Landauer, P. Prabhu (eds.), 2 ed., Elsevier Science B.V., pp. 255-297, 1997.

[3] Bannon, L. J. *CSCW* – A challenge to certain (G)DSS perspectives on the role decisions, information, and technology in organizations?. *Decision Support in Orgamizational Transformation.* P. Humphreys, S. Ayestaran, A. McCosh, B. Mayon-White (Eds.), 1997.

[4] Rose, A., Shneiderman, B. and Plaisant, C. "An applied ethnographic method for redesigning user interfaces," *ACM Proc. Of DIS'95 Symposium on Designing Interactive Systems: Processes, Practices, Methods & Techniques*, pp.115-122, 1995.

[5] Silva, A. M. and Baranauskas, M. C. C. The *Andon* System: Designing a CSCW Environment in a Lean Organization. *Proc. of Sixth International Conference on Groupware (CRIWG'2000).* IEEE Computer Society Press, 2000.

[6] Russell, S. J. *Artificial intelligence: a modern approach.* Ed. Prentice Hall, 1995.

[7] Huhns, M., Stephens, L. M. Multiagent Systems and Societies of Agents. *Multiagent Systems: A Modern Approach to Distributed Artifical Intelligence*, Gerhard Weiss (ed). The MIT Press, pp. 79-120, 1999.

[8] Roth, E. M., Malin T. M., Schreekenghost, D. L. Paradigms for intelligent interface design. *Handbook of human-computer interaction.* M. Helander, T. K. Landauer, P. Prabhu (eds.), 2 ed., Elsevier Science B.V., pp. 1177-1201, 1997.

[9] Maes, P. Agents that reduce work and information overload. *Communication of the ACM*, 35 (7), pp.31-41, 1994.

[10] Womack, J. P., Jones, D.T. and Roos, D. *The Machine that Changed the World*, Harper Collins Publishers, 1990.

[11] Zachary, W. W. And Ryder, J. M. Decision Support Systems: Integrating Decision Aiding and Decision Training. *Handbook of human-computer interaction.* M. Helander, T. K. Landauer, P. Prabhu (eds.), 2 ed., Elsevier Science B.V., pp. 1235-1258, 1997.

[12] Mazzone, J., "O Sistema Enxuto e a Educacao no Brasil", *in Computadores e Conhecimento: Repensando a Educacao*, J.A.Valente (ed.), Gráfica Central da Unicamp, pp.247-312, 1993.

[13] Deming, W. E. *Out of Crisis.* The MIT press, 1992.

Cooperation as a Communication process

Jean-Philippe Kotowicz and Xavier Briffault
LIMSI-CNRS, Université de Paris-Sud, Batiment 508, 91403 Orsay,
http://www.limsi.fr, {kotowicz, briffault}@limsi.fr

Abstract

The actors of a company must collaborate in an efficient way to achieve the common purposes of the company (the Project). Therefore, they must have competencies, manage and produce knowledge, perform tasks, and communicate between them; all that in an environment permanently evolving. The methods and tools of work organization can benefit from new technologies of the communication to return these more effective adaptations. This system will be a major constituent element of companies' memories. A new type of software will be then necessary to disambiguate, to facilitate, to structure and to allow complex searches in the exchanges arising among actors of a cooperative socio-technical system. We thus propose a system of communication mediation and of memories companies management, based on a multi-agent architecture. This architecture allows a use adapted to the "professional" activities of tool users associated to the various levels of language treatment.

1. Introduction

Cooperation is, according to the dictionary, "the action to participate in a common work"[1]. According to this point of view, facilitating cooperation consists in giving possibilities to act collectively. A common method is to **drive** the person in the organization of its actions: i.e. to clarify which are the permissible actions and how they get organized.

These approaches, which assist actions and their temporal progress, concern workflow management and are used very often [18] [20]. However, these approaches are structuring by definition, and they restrain the collective adaptations to the variations of the environment. A lot of work, theoretical as well as practical, has been dedicated to reengineering these processes when they turn out unsuitable ("Business Process Reengineering" -BPR- [12]). BPR is relevant when the environment doesn't change too quickly. These methods reach their limits when the frequency of the evolutions requires a permanent BPR. Then, we can wonder about the aptness of this notion of structuring process [1]. Moreover, the communications between actors are seldom correctly taken into account in the information processing system [19]. We develop in this article the following hypothesis: *information and cooperation mechanisms are going to move massively from a structured information system towards a communication system without any a priori structure*. This system will be a major constituent element of companies memories. We propose that, if this hypothesis is verified, a new type of software will be necessary to disambiguate, to facilitate, to structure, to improve, to augment, and to allow complex searches in the exchanges arising among the actors (human beings and software) of a cooperative socio-technical system.

Since these exchanges are essentially of linguistic nature in the tertiary sector, it looks natural to envisage to use natural language processing (NLP). However, we know that present-day technologies often require a preliminary structuring of information (lexical, syntactic, pragmatic), which seems incompatible with ours objectives of flexibility and of ability to react. So, we shall propose an architecture allowing the opportunist use (i.e. dynamically determined by the availability and the capacities of the tools) of various tools associated to the various levels of languages treatments.

The proposed architecture is based on the technology of multi-agent systems (as defined by FIPA [8]) and the formal methods of NLP. This architecture allows the management of the lexical, syntactical, semantic, pragmatic, interactive levels (reference, argumentation), in a system of mediation of the communications and of the management of companies memories.

2. A Multi-Agent System to support Information System and to mediate communication

2.1. How to assist cooperation?

Two solutions are conceivable to assist cooperation. The first would consist in endowing the information systems of dynamic adaptation capacities to the

[1] Petit Robert

environment evolutions. This program has been the subject of work in artificial intelligence for many years. The use of these techniques in real environments should not be envisaged before numerous years, even though interesting specifications are produced in the scope of BDI systems and Multi-Agent Systems [26] (KQML [16], FIPA ACL [7]). The other solution consists in increasing the speed of the human collective adaptation to new situations, when there are no predefined procedures. The heuristic mechanisms of human planning are undoubtedly more effective in very complex situations than those of the most powerful algorithms, provided that the necessary information for this planning is available quickly and easily. In current tertiary structures, the cooperation and the interactions are essentially of linguistic nature. So, we propose to provide users with tools to extract information in the exchanged messages between the members of a collective work. And we postulate that these tools can allow a greater ability to react, while preserving the efficiency of the predefined procedures. These tools would allow the users the easiest possible *a posteriori* reconstruction of the meaning of the interactions.

2.2. The metaphor of "virtual organization"

The efficiency of an enterprise information system thus depends on its adequacy to requirements and structure of the organization that uses it. These needs and this structure are progressive by nature, and even more so when the environment is changing or when the organization of the company must frequently adapts to its works.

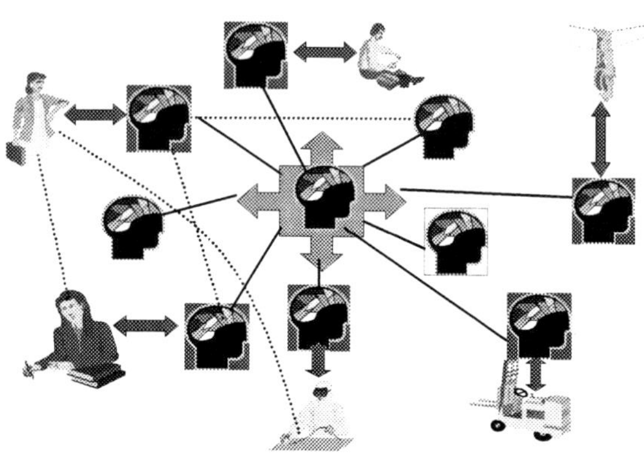

Figure 1. Metaphor of "Virtual Company"

The multi-agents approach allows an effective management of these dynamic aspects; the use of a high level communication language preserves the entire richness of the communications forwarding within a company. The objective of the EUREKA project "MERCURE" is to propose an assistance tool to the collaboration and to the communication between actors [3]. It is based on a multi-agents platform (autonomous and co-operating software agents) communicating by speech acts [21] of high level language (ACL[1] of KQML[2] type). Our platform is developed according to the standards suggested by the FIPA (FIPA ACL in XML [9], FIPA SL [11], FIPA KIF[3] [10]...). We use the metaphor known as "virtual company" (Cf. Figure 1), in which each person of the company is associated to a software agent, which represents him/her, assists him/her and to which certain tasks are deputy. The management of the interactions among users is delegated to software agents, in charge of message treatments and of exchange management (Cf. Figure 2).

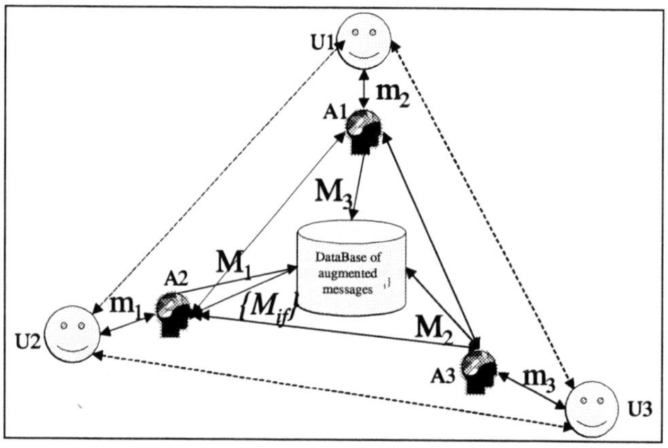

Figure 2. Man-Man Communication mediated by software agents

2.3. Transmission of augmented messages

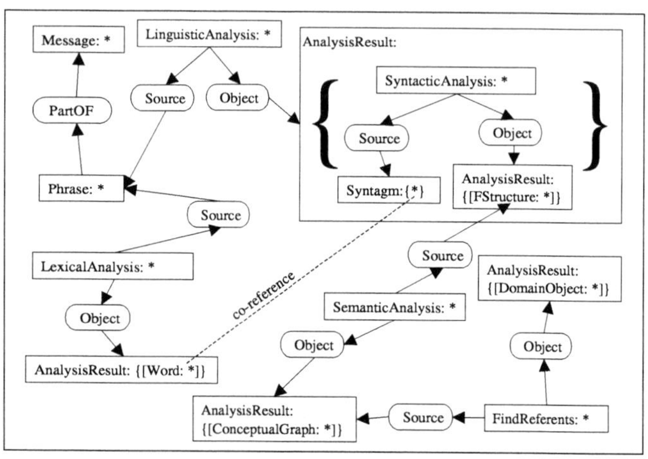

Figure 3. Conceptual graph representing an augmented message

---

[1] ACL: Agent Communication Language

[2] KQML: Knowledge Query and Manipulation Language

[3] KIF: Knowledge Interchange Format

The communication between human actors is relieved by software agents, it becomes then very well-to-do to set up treatments on messages transiting by the system (Cf. Figure 4). The purpose is to add a maximum of information which on the one hand will allow to make the messages more accessible to the receiver (direct access to the referenced objects, argumentative links), on the other hand will add criteria of search in the base of messages more relevant than of the simple keywords. The conceptual graph of the Figure 3 is the pattern of an increased message: it allows to store the raw message and all information added in a manual or automatic way.

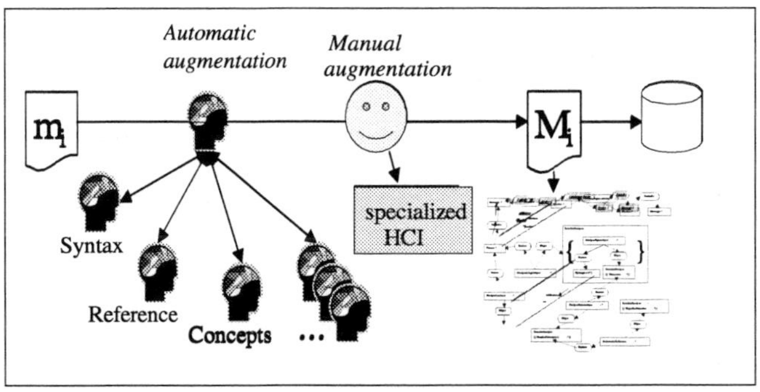

Figure 4. Process to augment a message[1]

The search for an augmented message and the navigation in the messages base correspond to search and to navigate inside a conceptual graphs base, through a hypertextual interface with semantic links (Cf. Figure 5).

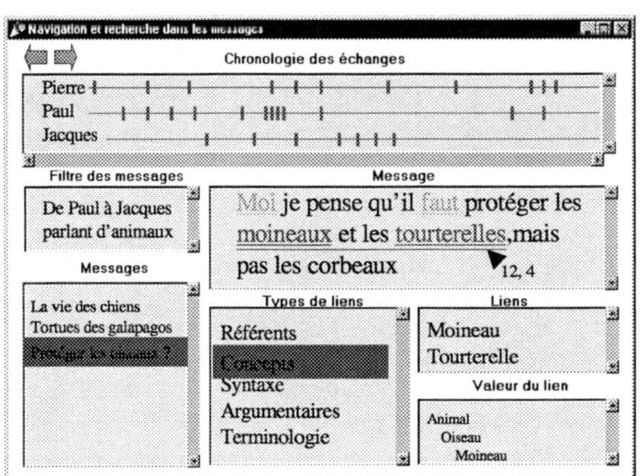

Figure 5. Navigation in a base of augmented messages by the way of an hypertextual interface

The information available for a message arise from data associated to the augmented messages. The editor of the message constructs this data by adding information in the original messages (assisted and manual augmentation) with the help of his agent (using modules of linguistic treatment associated to the various linguistic levels) (Cf. Figure 4). One can see semantic hyperlinks in the interface for augmented messages. When the mouse is moved on one of the hyperlinks, the types of available links is displayed; if one activates a hyperlink, the list of the links types appear, one can then select the one in which one is interested and see the value associated to this link (a concept, an argumentative link, an object...).

3. Tools in charge of augmenting message

3.1. Linguistic tools

Segmenting a message into words allows research by keywords. As numerous words are polysemic, there is often a high resultant noise in this sort of research. The syntactic category and/or the syntactic context of the word often allow to leave out some number of ambiguities in an automatic way. So, we propose a word labeling to identify their grammatical category. The mechanism of labeling (tagging) we developed is based on Brill's tagger[2] [5]. The obtained result after tagging is an association of words and of grammatical labels. This information can be stored in the conceptual graph of the augmented message (cf. Figure 3) in the concept called **Word**.

Such information (grammatical category of words) may be set up in a totally automatic way and has the advantage to solve a great number of simple ambiguities which remain with keywords. Knowing the grammatical category of words, we can also use higher levels of linguistic analysis. This permits to set up tools of search based on sentence patterns. Indeed, by having lexicons and grammatical rules, it becomes possible to build a grammatical representation of sentences. For it, we use the LFG[3] analyzer [13] [24] developed at LIMSI.

Our LFG analyzer allows to manage simultaneously one or more lexicons [23], so, each actor disposes of a generic lexicon (common and pre-defined) and one or more specific lexicons containing its own vocabulary (specific to its domain and defined by an ergonomic study [6]). The advantage is that it's possible to complete the vocabulary of an actor without overloading the vocabulary of actors who don't use these particular terms (domain terms, exotic word). Also, even though we dispose a general quite enough complete French grammar, it's impossible to consider specific expressions in all specialized domains. That is why our LFG analyser is conceived to manage configurations of grammatical rules.

[1] The method used by an agent to look dynamically for the relevant agents according to the declared competences and its requirements is described in [3] [14] and in [8].

[2] A detailed documentation for French tagging is available: winbrill by INALF [17]

[3] LFG: Lexical Functional Grammar

In accordance with the example of lexicons, each actor has a configuration of rules (for fluent French) and one or more configurations. Additional configurations allows to analyze the expressions in the proper domain of the actor [4]. Whatever the type of syntactic analysis used, resulting information can be stored in the conceptual graph of the augmented message (Cf. Figure 3) which is defined in a general way but it is possible to specify it according to the analyzers used.

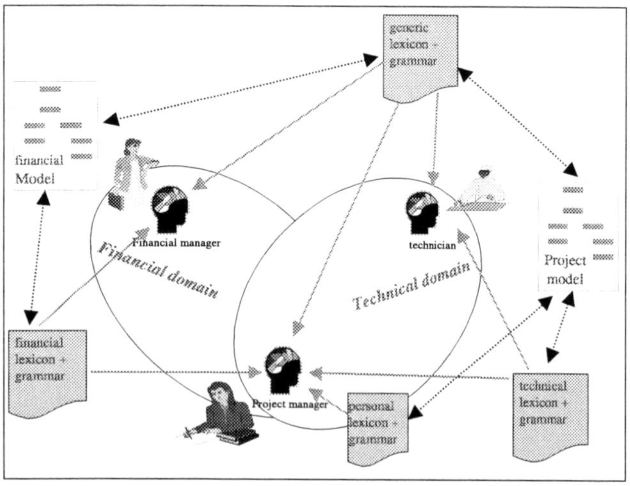

Figure 6. Schema of lexicons and ontology access by domains

Other ambiguities can be solved by using semantic information. So, we also complete the messages by adjoining them the meaning which must be conveyed. To acquire flexibility about the meaning associated to a word, and to take into account polysemic phenomena, we chose to set up an intermediary data structure linking the syntax and the semantic. It's a dictionary linking up a predicate (that characterizes a word of the lexicon) and a concept of the ontology (which defines its meaning). Once more, we can't identify in advance which will be all the meanings of a word. That's why we allow the association of several tables to the same lexicon, each table having a priority level: a table for the current meanings of words and other tables for the specific meanings in a domain. The last phase is therefore a semantic analysis which is going to supply a "meaning" to the message. For semantic analysis we use the formalism of conceptual graphs [22]. This knowledge formalism is an intermediary between logic and natural language. The semantic analyser tries to realize joint operations on concepts (denoted by lexical entries) of which canonical graphs precise the relations [2] [27]. The result of the semantic analysis will be associated to the message specifying the ontology in which it takes its meaning (that of its domain). So, each actor will have to have its own ontology which will correspond to its point of view on the objects that it handles. Each can so point out an object according to the outstanding attributes in its domain. Only the common objects are shared in an ontology, the manners to describe an object (by its attributes) are different varying from domain to domain even though the object is the same; then we speak about multiple ontology (cf. Figure 6).

All these augmentations of the message are purely linguistic. They essentially allow to structure the information and to have at disposal more complete and more definite criteria of search than simple keywords. But the use of these linguistic tools also gives us the possibility to look for the objects referenced by the messages. Indeed, in a collaborative work interaction, the exchanged messages essentially concern objects of the information system. "*At what time is the meeting of tomorrow ?*", "*I rescheduled the planning concerning the T12 task of the PVS project*"... It seems to us of the utmost importance to have a direct access to the objects as tasks, meetings, projects, resources... Having a semantic representation of each object in the ontology, we can construct a semantic representation of nominal groups of messages and so we can find referenced objects; i.e. objects with a conceptual graph corresponding to the semantic representation of part of a sentence. The link between messages and objects they reference is done by the way of an abstract middleware which encapsulates data and software.

3.2. Management of the interlocutor level

To constitute a memory of the exchanges, it is essential to represent -as precisely as possible- the structure of these exchanges, and in particular, speaking of an intervention, the reasons for an interlocutor reaction. Numerous works stemming from the linguistic pragmatics domain (Genevan model for example) propose modes of construction and of representation of dialogues structures. These propositions turn out inadequate for our objective for several reasons (refer to [25] for a detailed argumentation).

By observing real exchanges, it seems in particular that the speech acts, that are primitive elements of numerous models, represent *an insufficient level of structuring* (cf. [25] p. 255). The mechanism of exchange structuring we suggest to use arises from the PragmaMail system [25]. PragmaMail is based simultaneously on the principles of hypertexts and of mail system: Vapillon proposes a functionality of **explicit creation of links between parts of message contents** (and not only between messages). The system builds, from user's interventions, a shared database of messages under form of a hypertext with multi-sort links. That way, readers can navigate and reconstruct *a posteriori* the meaning of the interactions. In this framework, reacting to a message means realizing a *particular communicative act in reaction to a specific* *textual element* of the previous intervention. With respect to argumentative links, one will say that a *reactive intervention is **co-oriented** with the initiative if the sender of the reactive thinks that it satisfies the **favorite intentions**[1] (or the expected conclusion) of the speaker expressed in the initiative. In any other cases, the reactive is **anti-oriented**.* In this system we use symbols to mark the reactive orientation (the symbol "+" indicates the co-orientation; the symbol "-" indicates the anti-orientation).

The mark of orientation facilitates the navigation because it shows clearly the relation between the initiative and the reactive. For example, if a user sends a demand to ten different persons, he will be able to read the answers from co-oriented reactive (those being more susceptible to contain the expected answer). On the contrary, if he sends a demand of confirmation on a point which seems to him granted, he will be interested first and foremost to the anti-oriented answers. However, the orientation of a reactive is not always "an unequivocal position", it could be a subordinate negotiation from an answer. That is a way to "moderate his orientation". The intervention contains then an *anti-oriented argumentative connector*. "No, but…" can be interpreted as a proposition of another solution. On the contrary, co-oriented reactions can correspond to the addition of prerequisites. The system proposes the symbol "->" to mark the opening of a subordinate negotiation. This symbol is always used conjointly with the mark of co-orientation or of anti-orientation (+ -> or - ->). In certain cases, it can be difficult to account for an orientation, for example when the receiver is lacking information or did not understand what was asked to him. In that case, the user will use the mark of interrogation "?". It's also possible to complement or to cancel a negotiation (symbol ">>").

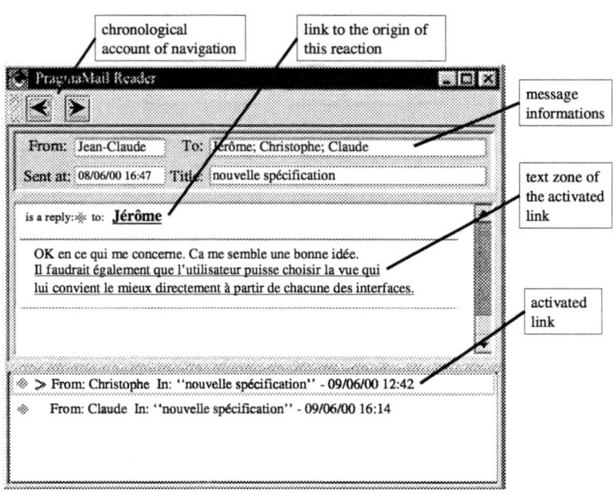

Figure 7. Interface of message manipulation and of multi-sorts links

[1] We define the favorite intention as what the interlocutor believes for that his partner waits as answer. It is so a quite subjective definition.

The hypertext links of PragmaMail are multi-destinations links so we can't use the usual display of hypertexts. For example, several links of different persons can overlap completely or partially. Besides, a message contains both links "going towards" a reactive of an another message and links "returning from" the initiatives. To deal with this problem, we propose an interface in two main parts: the top of the interface contains the message and the bottom contains the active links associated to the portion of the text containing the cursor (Cf. Figure 7).

In the same way that for the linguistic information, which augments the message, we can store the argumentative information and we can find messages according to criteria of search on types of argumentative links. For it, we defined the conceptual graph of the Figure linking two messages by way of an argumentative link (the various argumentations appear in the lattice of relations and subsumes the ArgumentativeLink relation).

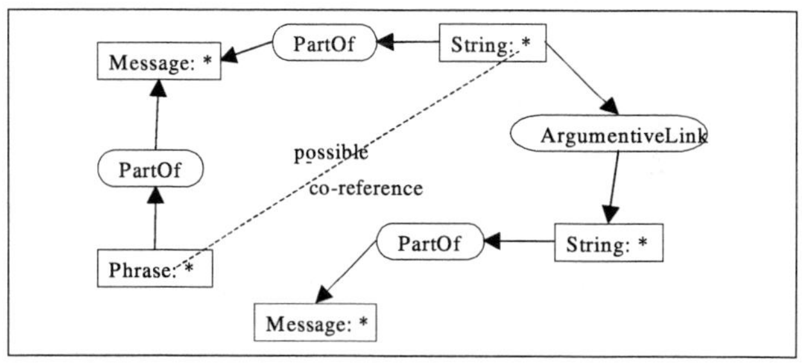

Figure 8. Conceptual graph allowing to store and to go through the argumentative links

The representation of the links associated with the marks of orientation provides discriminating information and permits **a faster selection of the relevant information.**

4. Conclusion

The setting of the described tools requires many resources. But the multi-agent system relieves user of the modules management. Indeed, the software agent (associated to a user) takes care to retrieve needed resources, and dynamic management of competences allow to realize linguistic analyses. So the software agent proposes to the user (who emits a message) the possible analyses of the message without this last one has to be engaged in the way which were obtained it. Furthermore the system just puts propositions forward and the user will accept or reject these propositions. The system just suggests to the user the best way to structure information as it is produced. So, the exchanged messages base would have more powerful functionality of search. For example, argumentative links allow to prove exactly how one managed to reach such a decision; the addition of referenced objects gives an immediate access to the striking information.

All the described modules were implemented and some were tested in real situations. The multi-agents architecture is also totally implemented, but all described modules have not yet been integrated. The lexical analysis, the tagger, the LFG analyser, the semantic analyser, the referent search, the management of lexicons and of ontology are at present totally integrated into the architecture.

On the basis of this first application, the main problems are due to the global performances of the multi-agent system (because of the dynamic organization of the agents). This problem is currently being solved by mechanisms of dynamic subscriptions. Another difficulty is due to the conception and the formalization of the competences ontology of the linguistic agents. This conception is a relatively long and boring process and it seems nevertheless difficult to improve in the current state of the technique. However, for us, the most important point is the evaluation of the effective improvements of such a system to the cooperative efficiency of a group.

5. References

[1] Bannon L. (1997) "CSCW - A challenge to certain GDSS perspectives on the role of decisions, information, and technology in organizations ?" in *Decision Support in Organizational Transformation*, P. Humphreys, A. McCosh & B. Mayon-White (Ed.), Chapman & Hall, London.

[2] Briffault X., Chibout K., Sabah G. & Vapillon J. (1997) "A Linguistic Engineering Environment based on LFG and CG" in *LFG97*, University of California, San Diego.

[3] Briffault X., Guichard N., Vapillon J. & Pierre D. (1998) "MERCURE : une plate-forme multi-agent cognitifs communiquant par KQML, appliquée à la communication d'entreprise" in *Complex Systems, Intelligent Systems & Interfaces*, Nîmes, France.

[4] Briffault X., Kotowicz J.-P. & Chibout K. (2000) "Systèmes collecticiels et communication naturelle : apports conjoints de l'Informatique Linguistique et de l'Ergonomie" in *Revue Interaction Homme Machine*, numéro spécial psychologie de l'interface (à paraître).

[5] Brill E. (1992) "A simple Rule-based Part Of speech Tagger" in *3rd Conference on Applied Computational Language (ACL) Processing*, Trento, Italy.

[6] Chibout K. (1999) "La polysémie lexicale : observations linguistiques, modélisation informatique, études ergonomique et psycholinguistique" Sciences Cognitives, Université de Paris XI, Orsay, France.

[7] FIPA. (1997) "FIPA 97 Specification Part 2 : Agent Communication Language" , FIPA, Genève, Suisse.

[8] FIPA. (1998) "FIPA 98 Agent Management" *v0.1*, http://www.fipa.org/spec/index.html, Genève, Suisse.

[9] FIPA. (2000a) "FIPA ACL Message Representation in XML", *00071*, http://www.fipa.org/specs/fipa00071/, FIPA, Geneva, Switzerland.

[10] FIPA. (2000b) "FIPA KIF Content Language" *00010*, http://www.fipa.org/specs/fipa00010/, Geneva, Switzerland.

[11] FIPA. (2000c) "FIPA SL Content Language Specification" *00008*, http://www.fipa.org/specs/fipa00008/, FIPA, Geneva, Switzerland.

[12] Hammer M. & Champy J. (1993) Reengineering the Corporation: a Manifesto for Business Revolution, Harper, New-York.

[13] Kaplan R. & Bresnan J. (1982) "LFG: a formal system for grammatical representation" in *The Mental Representation of Grammatical Relations*, MIT Press, Cambridge, pp. 173-281.

[14] Kotowicz J.-P. & Briffault X. (2000b) "Système multi-agent et médiatisation : vers une communication plus naturelle" in *JFIAD & SMA 2000*, ENS des Mines, St Jean La Vêtre, France.

[15] Kotowicz J.-P., Guichard N. & Briffault X. (2000) "Aide au travail collaboratif et système multi-agent : le cas de la gestion de projet" in *Ingénierie de systèmes et NTIC 2000*, Nîmes, France.

[16] Labrou Y. & Finin T. (1994) "A Semantic Approach for KQML - a General Purpose Communication Language for Software Agents" in *3rd International Conference on Information and Knowledge Management*, Gaithersburg, Maryland.

[17] Lecomte J. (1998) "Le cartégoriseur BRILL14-JL5 / Winbrill-0.3" , http://www.inalf.fr/WinBrill/, INALF/CNRS, Nancy, France.

[18] Levan S. K. (1999) *Le projet Workflow*, Eyrolles, Paris.

[19] Prince V. (1996) Vers une informatique cognitive dans les organisations : le rôle central du langage, Masson, Paris.

[20] Schael T. (1997) *Théorie et pratique du workflow*, Springer, Berlin.

[21] Searle J. R. (1969) *Speech Acts: an Essay in the Philosophy of Language*, University Press, Cambridge, Massachusetts.

[22] Sowa J. (1984) Conceptual Structures: Information Processing in Mind and Machine, Addison Wesley, London.

[23] Vapillon J. (1997a) "Analyseur Syntaxique LFG & Atelier Syntaxique" *version 2.5b*, LIMSI-CNRS, Orsay.

[24] Vapillon J. (1997b) "Un atelier de "Génie Linguistique" pour l'analyse syntaxique fondée sur le formalisme LFG" in *JST Francil*, Avignon, France.

[25] Vapillon J. (2000) "Contribution à l'étude de la conversation dans le cadre du travail coopératif assisté par ordinateur (TCAO)" Informatique, Paris-XI Orsay, Orsay.

[26] Wooldridge M. & Jennings N. R. (1994) "Towards a Theory of Cooperative Problem Solving" in *Modeling Autonomous Agents in a Multi-Agents World*, Odense, Denmark, pp. 15-26.

[27] Zweigenbaum P. & Bouaud J. (1997) "Construction d'une représentation sémantique en Graphes Conceptuels à partir d'une analyse LFG" in *TALN97*, Grenoble, France.

A Novel Method for Image Retrieval in Distributed Multimedia System

S.X. Sun, T.Y. Huang, C.Y. Pang
Computer Science Department, UEST of China, Chengdu, P.R. China, 610054
sxsun@uestc.edu.cn Huang_ty@263.net Pangchaoyang@263.net

Abstract

With the development of Internet and multimedia technologies, Images retrieval based on their visual features becomes an important research area in distributed multimedia system. In this paper, we propose a new method for images retrieval by their visual appearance—Multi Scale Gauss-filtering Representation (MSGF). Images are first filtered in several scales with Gauss derivative filters, and the feature vectors are constructed on filtering outputs in several scales. Next, images are matched by their feature vectors, the matching results are scored and returned to user. We develop a fast search algorithm of index structure for local similarity retrieval of images. We then generalize the idea to a wide range of image retrieval by global similarity. Experiment shows that this method is effective in feature extraction and images retrieval by their visual appearance.

1. Introduction

With the development of Internet & multimedia technologies, the need for searching and classifying a great deal of multimedia information has become more stringent. Conventional text-based search engine can't retrieve images (videos) effectively over the Internet. Recent researches in the area of content-based image retrieval (CBIR) [1,2] can retrieve images more effectively by feature extraction such as color, texture, shape, etc.

A full-functional image (video) retrieval system must have the ability to catch the features of image, to adapt to all diversities user requests. We can classify the main retrieval methods into the following types: 1) Keywords-based retrieval; 2) Query by image example (QBE); 3) Content based visual Retrieval (CBVR). The prototype QBIC[3] provides retrieval by color, shape, texture of images, but it is not fully automatic. In the remarkable work of Smith[1,4] etc., Content based visual Query of images are discussed. While building an efficient image retrieval system, one must solve the following four questions: 1) What features to choose? 2) How to represent them? 3) How to index the chosen features? 4) How to process the similarity matching?

We propose a new method for feature representation and retrieval of images—Multi-Scale Gauss Filtering Representation (MSGF). Images are first filtered by Gauss function and its derivatives in sample points with several scales. Feature vector sets are constructed by filtering outputs. Images in remote server database are filtered off-line and the feature structures are stored. While user commit requests, the feature of image template is extracted on-line and sent to the server. Therefore, the server can compare the template feature with stored feature vector sets using a fast search algorithm of distributed index structure. The comparison results are scored and returned to user as request responses.

In the following sections, we first outline the main theory framework; we then discuss how to use the method for images retrieval by their local & global similarity in section 3 and section 4. The experimental results are presented in section 5.

2. Framework of the MSGF Theory

Gauss filtering of images had been used in many areas. In earlier days, Marr and Hildreth [5] used it for noise elimination and edge detection. The Laplacian transformation [6] of Gauss function and its derivatives have been used in multi-resolution analysis. Recent researches [2,7,8] show, Gauss derivative functions can represent the structure of image in several scales. It had been used in the area of image retrieval in some papers. In this paper, We propose a new sample rule, provide a distributed index structure and its fast search algorithm, construct a space matching criteria based on *affine transformation*.

2.1. Gauss Filtering Representation of Image

Filtering of Gauss function and its derivatives can represent spatial and frequency feature of images[9]. Koenderink and Van Doorn [7] call it N-jet. Gauss filtering representation of image is as follows:

$$I_g = I * G \tag{1}$$

In which, I is the image, G is Gauss function, $G(p,\sigma)=\frac{1}{\sqrt{2\pi}\sigma}e^{-\frac{pp'}{2\sigma^2}}$, I_g is the filtering output image. $p=(x,y)\in R^2, \sigma\in R$ is the scale.

By defining $G_{i_1,i_2,\ldots,i_n}(X,\sigma)=\frac{\delta^n G(X,\sigma)}{\delta i_1 \delta i_2 \ldots \delta i_n}$, we have [10]:

$$I_{i_1 i_2 \ldots i_n}(X,\sigma)=(I*G)_{i_1 i_2 \ldots i_n}(X,\sigma)= \\ (I_{i_1 i_2 \ldots i_n}*G)(X,\sigma)=(I*G_{i_1 i_2 \ldots i_n})(X,\sigma) \tag{2}$$

$I_g(X,\sigma)$ is the smoothed result of Gauss function, which represents feature distribution of point $X=(x,y)$ in the neighboring area $\delta(\sigma)$ with scale σ.So, every point $X+\Delta X$ in $\delta(\sigma)$ can be represented with the combination of the feature of point X using Taylor expansion:

$$J^N[I](X+\Delta X)=\sum_{n=0}^{N}\frac{1}{n!}(\Delta X_{i_1}\Delta X_{i_2}\ldots\Delta X_{i_n}\frac{\delta^n I(X)}{\delta_{i_1}\delta_{i_2}\ldots\delta_{i_n}}) \tag{3}$$

From above, we believe that Gauss filtering outputs in different orders at point X represent the feature distribution in the neighboring area $\delta(\sigma)$ with scale σ. Similarly, feature of the neighboring area $\delta(\sigma)$ with scale σ can also be represented by Gauss filtering outputs with scale σ. If two images have local similarity, then their Taylor expansions must be appropriately equal in the neighboring area $\delta(\sigma)$. Therefore, we can represent image feature in the neighboring area $\delta(\sigma)$ of X by features combination of point X. Through Gauss filtering outputs in different orders, we can obtain feature vector of image at point X :

$$J^N[I](X,\sigma)=\{I(.,\sigma)*G(.,\sigma), I(.,\sigma)*G_x(.,\sigma), \\ I(.,\sigma)*G_y(.,\sigma),\cdots I(.,\sigma)*G_{x^i y^j}(.,\sigma)\cdots\} \tag{4}$$

When we choose N=2, the 2-jet of image at point X can be written as:

$$J^2[I](X,\sigma)=\{I, I_x, I_y, I_{xx}, I_{xy}, I_{yy}\} \quad (I_{xy}=I_{yx}) \tag{5}$$

Gauss filtering outputs of image 'Ferrari' (100*100) are in Fig. 1. (From top-left to bottom-right are I, $I_g, I_x, I_y, I_{xx}, I_{xy}, I_{yy}$.)

Fig 1. Gauss filtering outputs of image 'Ferrari'

2.2. Multi-Scale Representation of Image

Since Gauss function can effectively describe the properties of scale space, We chose it as the smoothing function for the neighboring area at point X. Scale space[11] describes feature change of the scene while the focus and length of camera change. Fixing scale σ, feature of the neighboring area can be well presented; while changing scale σ, description of image also changes. Therefore, in order to retrieve similar images from server database, We call the feature representation of point X at different scales, *Multi-scale vector representation*:

$$\{J^N[I](X,\sigma_1), J^N[I](X,\sigma_2),\cdots J^N[I](X,\sigma_k)\} \tag{6}$$

In which, $(\sigma_1,\sigma_2,\cdots\sigma_k)$ is scale set.

2.3. Match on Different Scales between Query Image and Database Images

The query area user chose affects query results. However, no matter which area chosen, the database images similar to query area of template image should be retrieved. These images may be different with query area in size, shape, etc. But, we think they are *affine similarity* within the allowed Errors, i.e. $I_1(r)=I_2(sr)$, S is the Scale. When we don't know S, How to match the template image and database image? In fact, because of the strong correlation of neighboring scales, we can use only finite scales, e.g. $\sigma/4, \sigma/2\sqrt{2}, \cdots, 2\sqrt{2}\sigma, 4\sigma$, etc., to construct multi-scale feature vector.

2.4. Feature Invariant and Index Structure

Now, we write following two equations:

$$d_0=I, d_1=I_x^2+I_y^2, d_2=I_{xx}+I_{yy}, \\ d_3=I_{xx}I_xI_x+2I_{xy}I_xI_y+I_{yy}I_yI_y, d_4=I_{xx}^2+2I_{xy}^2+I_{yy}^2 \tag{7}$$

$\bar{D}_\sigma=\Delta\sigma=(d_1,d_2,d_3,d_4)_\sigma$. We won't use d_0 because it is sensitive to gray-scale shift. By smoothing database images using the first and second order derivatives of Gauss function at three different scales (How many SCALES to use is controlled by user), we obtain *Multi-Scale Invariant*

$$D=(\Delta\sigma_1,\Delta\sigma_2,\Delta\sigma_3)=(\bar{D}_1,\bar{D}_2,\bar{D}_3) \tag{8}$$

The point $X=(x,y)$ of database image can be represented as: I£º£¨ $i,x,y,D)$ (9)

In (9), i is image I.D., $X=(x,y)$ are the coordinates of point X , D is multi-scale feature invariant.

Assume that the corresponding points of query image (I.D. is i) and database image (I.D. is i') are $X=(x,y)$ and $X'=(x',y')$, so the corresponding multi-scale features invariant: $D=(\vec{D}_1,\vec{D}_2,\vec{D}_3)$¡¢$D'=(\vec{D}'_1,\vec{D}'_2,\vec{D}'_3)$. We illustrate their relationship and the index structure of database images in Fig 2 and Fig 3.

$$\left\{ i \leftrightarrow [X] \begin{bmatrix} \vec{D}_1 \\ \vec{D}_2 \\ \vec{D}_3 \end{bmatrix} \Leftrightarrow \begin{bmatrix} \vec{D}'_1 \\ \vec{D}'_2 \\ \vec{D}'_3 \end{bmatrix} [X'] \leftrightarrow i' \right\} \quad (10)$$

Fig 2 Relationship of feature vectors

$$\begin{array}{ccc} & \begin{bmatrix} \vec{D}' \\ \downarrow \end{bmatrix} & \\ \left[\begin{matrix} \vec{D}'_1 \\ \downarrow \end{matrix} \right. & \begin{matrix} \vec{D}'_2 \\ \downarrow \end{matrix} & \left. \begin{matrix} \vec{D}'_3 \\ \downarrow \end{matrix} \right] \\ \begin{bmatrix} d_1\, d_2\, d_3\, d_4 \\ \downarrow\ \downarrow\ \downarrow\ \downarrow \end{bmatrix} & \begin{bmatrix} d_1\, d_2\, d_3\, d_4 \\ \downarrow\ \downarrow\ \downarrow\ \downarrow \end{bmatrix} & \begin{bmatrix} d_1\, d_2\, d_3\, d_4 \\ \downarrow\ \downarrow\ \downarrow\ \downarrow \end{bmatrix} \end{array}$$

Fig 3 Index structure of sample point

3. Local Similarity Retrieval of Images

3.1. Search of Matching Points

While using local similarity retrieval, user first chooses interesting area in template image, the area is sampled and multi-scale invariant $\vec{D}$ is constructed by Gauss filtering output at every sample point with three different scales. Query vector $(x_i, y_i, \vec{D}^i)_{i=1}^N$ (N is sum of sample points) are formed by all multi-scale invariants and their coordinates. Query vector is then transmitted to server database over network. As before, we assume the feature vector of sample point P in query image is $\vec{D}^k=\{\vec{D}_1^k,\vec{D}_2^k,\vec{D}_3^k\}$, among which, $\vec{D}_i^k$ is the ith component of feature vector $\vec{D}^k$, and the sub-components of $\vec{D}_i^k$ are $d_{ij}{}^k (j=1,2,3,4)$. In addition, we define matching threshold $t=\{t_j\}_{j=1}^4$. All above are also defined for database point P'. Now, we can say, if and only if the following equation holds, query point P and database point P' are matched between scales σ_i and $\sigma_{i'}$:

$$\forall i,i' \in \{1,2,3\}, if\ \exists i,i', \vec{D}_i^k=\{d_{ij}^k\}_{j=1}^4, \vec{D}_{i'}^{k'}= \{d_{i'j}^{k'}\}_{j=1}^4; \ni \bigcap_{j=1}^{4} | d_{ij}^k - d_{i'j}^{k'} | \le t_j, then\ P \leftrightarrow P' \quad (11)$$

Write down the corresponding images I.D. and coordinates $(i,X)\leftrightarrow(i',X')$, Sort on image I.D. i'. For points P_1,P_2,P_3 in query area, if there exist corresponding points:

$P'_1=\{i'_1,x'_1,y'_1),P'_2=(i'_2,x'_2,y'_2),P'_3=(i'_3,x'_3,y'_3)$ of database image, then we have:

$$if\ \exists i', \ni i'_1=i'_2=i'_3=i'\ then\ P_1 \leftrightarrow P'_1, P_2 \leftrightarrow P'_2, P_3 \leftrightarrow P'_3; \Rightarrow i \leftrightarrow i' \quad (12)$$

3.2. Fast Algorithm of Index Structure

We implement the fast matching algorithm of feature vector invariants of sample points between query image and database images using 4 PC CLUSTER, One as front-end N_f, three Back-end $\{N_j\}_{j=1}^3$ for storing index structure. Database images can reside on N_f.

First, put the components of feature vectors at three different scales evenly on three PCs: N_1□N_2□N_3. On every PC, we construct four binary tree indexes $\{B_{ij}\}_{i=1,2,3}^{j=1,2,3,4}$ based on sub-components $\{d'_1,d'_2,d'_3,d'_4\}$ (see Fig 3.). Images I.D. are stored with index structure. Feature vectors of query area and database images are defined as Section 3.1. Fast search algorithm of distributed index structure is listed below:

```
Nf: For j=1 to 3
{
  For i=1 to 3
  {
    Dj → Ni
    Ni: For each Ni
    {
      Create 4 threads Tj in Ni ;
      For k=1 to 3
      {
        Tj: For each thread Tj
        {
          dkj → Tj As input parameter
          Find dkj in indexed binary tree Bij with threshold tkj
          If Found, resultk (j)=True,
          Return (i', x', y')k to Ni
          Else return resultk (j)=False
        }
      }
    }
}
Ni: For k=1 to 3
{
  If ∩(j=1..3) resultk(j) = True then RESULT (k) =True,
  return all (i', x', y')k to Nf
  Else return RESULT (k)=False
}
Nf: If ∃RESULT(k) = True, sort all (i', x', y')k on number
    i and k
(i,x,y) ↔ (i',x',y')k □Go to spatial matching. (Section 3.4)
}
```

3.3. Sample Criteria

We can determine the matching points between query area and database images through fast index algorithm in section 3.2. However, a stricter assumption must be given, that is, only there exist at least three matching points between query area and database images, We then think the two images match on *point features*. So far, we give the *Sample Criteria* below:

1). While choosing query area, at least three sample points in this area must be promised; 2). Size of query area is greater than 10*10 matrix (pixels); 3). While preprocessing database images, the sum of sample points is proportional to image size (So, we can fully represent the structure of image, at the same time, decrease computation complexity).

3.4. Spatial Matching

Errors may be introduced if the match and matching degree of images are only determined by sum of matching points and the equality of their Euclid distances, because of the two reasons: 1). The size of query image and database images may be different greatly; 2). So do database images. From the perspective of *Affine Transformation*, We assume the query area and the area of target image, which contains object similar to query area, is *affine similar*.

We assume $P_1 \leftrightarrow Q_1$, $P_2 \leftrightarrow Q_2$, $P_3 \leftrightarrow Q_3$, δ is the Euclid distance of two points. Then, only the Eq.13 holds, we can say that two images are *spatial matching*, and the similarity degree is marked as sum of spatial matching points. (In Eq.13, T is the similarity *Threshold*).

$$\begin{cases} |\,\delta(P_1P_2)/\delta(P_1P_3) - \delta(Q_1Q_2)/\delta(Q_1Q_3)\,| \le T \\ |\,\delta(P_1P_3)/\delta(P_2P_3) - \delta(Q_1Q_3)/\delta(Q_2Q_3)\,| \le T \\ |\,\delta(P_1P_2)/\delta(P_2P_3) - \delta(Q_1Q_2)/\delta(Q_2Q_3)\,| \le T \end{cases} \tag{13}$$

In practical computation, we use a *heuristic algorithm*. If there exist two similar points in areas of query image and target image, the ratio of their distances is K, then we check other corresponding points, see whether their ratio in $[K\text{-}T, K\text{+}T]$ or not. If so, the similarity degree adds 1. At last, database images are sorted on their similarity degree and retrieved images and their I.D. returned to user as query results.

4. Global Similarity Retrieval

User can also retrieve database images by the global similarity of query image and database images. Global similarity retrieval is easier than local similarity retrieval, but there is still some difference between them. Global similarity is useful in video scene check, face recognition, trademark retrieval, etc.

We only mention about global similarity based on *Moment*. Readers interested in can find more in Ref. 12. Moment technology had been used in QBIC, it is very suitable to represent global feature of images. However, notice that moment invariant is so coarse that it can't represent whole features of image very well.

We believe that global similarity can be determined by comparing local features of images and the spatial distribution of these features, and this method can tolerate the change of focus and length to some extent. Schiele[14] etc., use this method to recognize gray-scale images. First, we obtain multi-scale Gauss filtering representation of images through Gauss filtering, then, we construct *histogram* of these local features. If two images are very similar in their histograms, we can conclude that the two images are global similar.

The efficiency of image retrieval depends on the choice of feature. Gauss filtering feature representation can capture the appearance similarity of images. More importantly, How to quantize the 3-D surface of images. We know that if the *curvature* of every point in the image is known, then the 3-D surface is determined uniquely. Therefore, the local curvature can be treated as a feature to represent surface of images. However, the information of surface orientation would lose if we use only curvature to construct histogram. Hence, we used both *curvature* and *phase* information of 3-D surface to construct global feature histogram. We first compute local curvatures and phases of image through Gauss filtering, then their normalized histograms are used as measurement of global similarity.

4.1. Definition of Curvature & Phase

Definition of Curvature & Phase of 3-D surface (*x,y,Intensity*) of images is as follows[14]:

$$\begin{cases} N(P,\sigma) = \dfrac{I_x^2 I_{yy} + I_y^2 I_{xx} - 2I_x I_y I_{xy}}{(I_x^2 + I_y^2)^{3/2}} \quad (P,\sigma) \\ T(P,\sigma) = \dfrac{(I_x^2 - I_y^2)I_{xy} + (I_{xx} - I_{yy})I_x I_y}{(I_x^2 + I_y^2)^{3/2}} \quad (P,\sigma) \end{cases} \tag{14}$$

In which, $I_x(P,\sigma), I_y(P,\sigma)$ are the filtering outputs of the first order derivative of Gauss function at point P in image I with scale σ. Similarly, others are the corresponding second order derivatives.

The normal curvature N and tangential curvature T are then combined to generate the shape index[15] as follows:

$$C(P,\sigma) = \text{atan}\left[\frac{N+T}{N-T}\right](P,\sigma) \tag{15}$$

The index value C is $C = \pi/2$ when N=T and is undefined when either N and T are both zero, and is therefore not computed. The very flat portions of image are eliminated and not taken into consideration.

The phase is defined as:

$$P(p,\sigma) = \operatorname{atan2}\left(I_y(p,\sigma), I_x(p,\sigma)\right) \quad (16)$$

As before, P is defined only at those locations where C is defined and ignored elsewhere. Experiment shows that the retrieval accuracy can by greatly improved while using more scales. Because which represents different details observed under different scales. Therefore, multi-scale histogram can represent the global feature of images more effectively.

4.2. Histogram Normalization

The variation of signal can be decreased by normalization when convoluting with filters. The noise in the images can also be eliminated in this way. In tradition, we can use *energy normalization* or *variance-mean normalization* [13], in which *energy normalization* is more effective as shown by experiment. The retrieval inaccuracy caused by the intensity change of surface can be smoothed out by dividing energy of the neighboring area. Therefore, the curvature & phase histogram is invariable to surface intensity after normalized by energy. This method can also eliminate additional gauss noise. Equation of *energy normalization* is as follows:

$$I_{ene}(x,y) = \frac{\sum_{i,j} I(x-i, y-j) \times Filter(i,j)}{\sqrt{\sum_{i,j} I(x-i,y-j)^2} \times \sqrt{\sum_{i,j} Filter(i,j)^2}} \quad (17)$$

4.3. Matching of Curvature & Phase Histogram

Now, we define multi-scale feature vector for global similarity:

$$V = \{H_C(\sigma_1), \cdots, H_C(\sigma_n); H_P(\sigma_1), \cdots, H_P(\sigma_n)\} \quad (18)$$

In which, H_C, H_P are the normalized curvature & phase histograms at different scales. In practice, we use 4~5 scales, and all scales are evenly spaced (e.g. $\sigma = 0.8, 0.9, 1.0, 1.1, 1.2$, which can tolerate $\pm 20\%$ scale change), we can get better matching results.

In the end, the normalized distance of two feature vectors V_i, V_j are defined:

$$d_{ij} = \frac{V_i^{(m)} \bullet V_j^{(m)}}{\left\|V_i^{(m)}\right\| \times \left\|V_j^{(m)}\right\|} \quad (19)$$

Two feature vectors are compared using normalized cross-covariance in Eq.19, where $V_i^{(m)} = V_i - mean(V_i)$.

Until now, we can first generate feature vector V_q of query image using Eq.18. We then compare V_q with feature vector V_i of database image using Eq.19. The matching score are sorted and the corresponding images are retrieved from server database and returned to user as the query results.

5. Experiment Results

We describe our simulation process briefly in this section. Fig.4 is example images used in our simulation. There are 36 pictures totally, including 22 pictures of cars for simulation purpose.

Fig 4 Example images for Query

In local similarity retrieval, server database images are processed off-line and their feature vectors are extracted and stored. User select interesting area in query images, then the area is evenly sampled. In every sample point, we obtain feature vector by Gauss filters and construct the corresponding index structure. We obtain corresponding points through point matching, then verify the matching degree through stricter spatial matching. Fig.5 is the result of local similarity retrieval. Rectangle in top-left image is area user selected for query. We retrieve images in database by similarity of the car wheel in the picture. 14 of 22 cars are retrieved, the precision is 63.6%, a satisfying result.

Fig 5 Result of Local similarity（14/22）

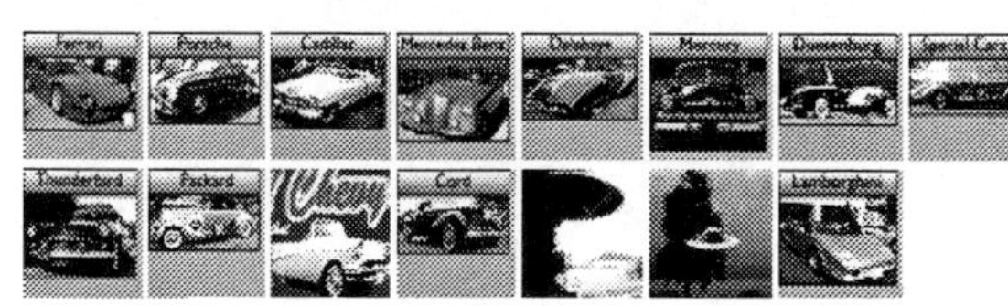

Fig 6 Retrieval results of Ferrari (>20%)

In global similarity retrieval, We obtain the curvature & phase histograms at five different scales using Gauss filtering outputs. We use the matching method in section 4 to retrieve images and the matching degree is marked as similarity percentage (%). Image 'Ferrari' is the template image in this experiment. Fig 6 is the retrieval result. Notice that above 20% similarity percentage, there are two errors. However, we can still retrieve 13 images from the database.

6. Conclusion and Future Work

We discussed feature extraction and representation of images based on multi-scale Gauss filtering. We presented image retrieval by local and global similarity of their visual features. As we see from the simulation, multi-scale Gauss filtering representation of the visual features of images can tolerate length and focus change, image orientation, transformation, etc., to some extent, So it is more efficient for image retrieval. Now, we decide to integrate the local similarity and global similarity method of Gauss derivative filters into a single framework for user's convenience.

7. Acknowledgement

We would like to thank our colleague Dr. Yan Zhang for her constructive comments and refinement on the draft of this paper.

8. References

[1] John R. Smith and Shih-Fu Chang, "*Tools and Techniques for Color Image Retrieval* ", Storage & Retrieval for Image and Video Databases IV. IS&T/SPIE Proc. Vol. 2670,1998.

[2] James L. Crowley, etc. "*A representation for shape based on peaks and ridges in the difference of low-pass transform.*" IEEE Trans. On Pattern Analysis and Machine Intelligence, 6:1984.

[3] Myron Flickner, etc., "*Query By Image and Video Content: The QBIC System*", IEEE Computer Magazine, pp23-30, Sept. 1995.

[4] John R. Smith, Vittorio Castelli etc.,"*Adaptive Storage and Retrieval of Large Compressed Images*", IS&T/SPIE Symp. On Electronic Imaging: Science and Technology-Storage & Retrieval for Image and Video Databases VII. Jan '99.

[5] D. Marr and E. Hildreth. "*Theory of edge detection*". Proc. Royal Society of London (B), B207, 1980.

[6] P. Burt and T.Adlson. "*The laplacian pyramid as a compact image code.*" IEEE Trans. On Communications, 1983.

[7] J. J. Koenderink and A. J. Van Doorn. "*Surface shape and curvature scales.*" Image and Vision Computing, 10(8), 1992.

[8] Tony L., "*Scale-Space Theory in Computer Vision.*" Kluwer Academic Publishers, 1994.

[9] R. Manmatha, "*Gauss Filtered Representation of Images*", CS technical report, University of Massachusetts, 1998.

[10] J. J. Koenderink, "*The structure of images*", Biological Cybernetics, 50:363-396,1984.

[11] A. P. Witkin. "*Scale-space filtering*". In Proc. Intl. Joint Conf. Art. Intell. , Pages 1019-1023,1983.

[12] M. K. Hu, "*Visual Pattern Recognition by moment invariants*", IRE Trans. On Information Theory. IT-8, 1962.

[13] Bernt Schiele and L. Crowley, "*Object Recognition using multidimensional receptive field histograms*", Computer Vision-ECCV'96, Cambridge, U. K, 1996.

[14] L. Florack, "*The Syntactical Structure of Scalar Images*", PhD thesis, Univeristy of Ultrecht, Holland, 1993.

[15] J. J. Koenderink and A. J. Van Doom, "*Surface Shape and Curvature Scales*", Image and Vision Computing, 1992.

Synergetic Collaboration in Networked Virtual Reality System

Yi Chen, Shouyi Zhan
Department of Computer Science,
Beijing Institute of Technology,
Beijing, 100081, P.R. China
chenyi63@hotmail.com, shouyi@public.bta.net.cn

Pengfei Li
Head of BIT Canadian Study Center under
China National Sino-Canadian Study Society,
Beijing, 100081, P.R. China
LiPf@263.net

Abstract

Collaboration of multiple users in a network, which has been a hot topic in computer supported cooperative work (CSCW), is an important problem in networked virtual reality (NVR) systems. In this paper, the authors have covered a series of problems as to the collaboration efficiency in NVR, suggested some solutions for these problems, and as a result present a framework for supporting and developing collaborative networked virtual reality application.

1. Introduction

Though much has been done in the research on synergetic problem with CSCW. Little has been reported on addressing collaborative issues in Networked Virtual Reality (NVR) system. With the ever-increasing spreading of computer network technology, Virtual Reality (VR) system has far exceeded the sphere of single-user and become a powerful social medium. In multi-user NVR system, multiple users that dispersed geographically in different places can shared a common Virtual Environment (VE) thorough computer network. They can interact with shared virtual environment respectively and even engage in collaboration for some common goals on the real-time basis.

NVR system is an actual form of real-time, interactive, 3D application for individual user as well as for social communication. Though much in common, the collaboration issues in NVR are quite different from those in CSCW, which mainly focuses on maintaining the consistency of data and collaborative edit and design. However NVR system focuses on maintaining the consistency of shared virtual world's state and controlling concurrent interactions made by multiple users. For example, when a user grabs an object in a VE and moves it, it can be expected that all the users in this VE would see this object's motion. In order to implement this, there are three basic collaborative problems that must be solved: □communication between all the participants; □the maintenance of shared VE □the control of concurrent or simultaneous operations. However, due to the limitations of existing computer performance and network bandwidth, it is a great challenge to solve these problems. This paper has presented a framework for supporting and developing collaborative NVR system, and suggested some solutions for the synergetic collaboration issues in NVR system.

2. Network architecture and communication model

2.1 Overview

The network architectures adopted by most networked virtual reality systems are of two types: peer-to-peer nature and hierarchical (or client/server) nature. Communication among multiple users can be done in three ways: broadcast, multicast, and unicast. In peer-to-peer architecture, all nodes are peer and there is no central node in the system. Each peer node, which keeps a copy of shared virtual environment data, undertakes the handle of interaction, and is engaged in managing, drawing and displaying 3D scene. When the state of any entity in shared virtual environment is changed, it should be broadcast or multicast to all other nodes in the network. The typical examples of NVR system based on the peer-to-peer architechure are SIMNET [1], which communicated in broadcast way, and NPSNET [2] communicated by means of multicast way and conformed to the Distributed Interactive Simulation (DIS) standard (IEEE1278.1-3) [3]. Peer-to-peer architecture has been proved to be efficient to build NVR system. However, there are some shortages in it. Firstly, with peer-to-peer architecture, each node has to maintain the entities' states in all other nodes. The computing burden is heavy and the implementation of consistency and concurrent control of shared VE are complicated. Secondly, the broadcast of state updates, which forced every node to receive all updates even if it is not interested in them, caused a great deal of unnecessary bandwidth consumption.

In hierarchy or client/server architecture, the server, which is the central node of the system, is in charge of managing and keeping the consistency of virtual environmental state. The clients are, on the other hand, responsible for managing, drawing and displaying the virtual environment. Since all the information has to go through the central host, it has become easier to keep consistency and implement concurrent control of the system. However, this type of system has an obvious weak-point. The central server of the system is much likely to become the bottleneck both in transferring and processing so large amounts of information of all its clients, because all the users of the system have to transfer what they want to their partners through the central server. So, some systems use several servers instead of one server to overcome this drawback. The examples of NVR system with client/server architecture are RING [4] developed by Bell laboratory and BrickNet developed by National University of Singapore [5].

2.2 Client/server architecture with two-tier servers

Aiming to facilitate the communication and management, this paper has presented a NVR system framework on the network architecture, which is called client/server architecture with two-tier servers. The given architecture is composed of a Server Manager (SM), several Application Servers (AS) and multiple clients, as shown in Figure 1. In order to relief the burden of server

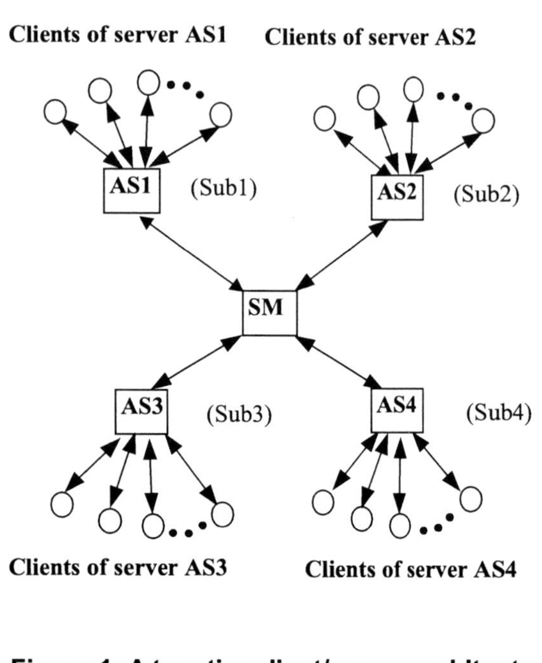

Figure 1. A two-tier client/server architecture

and minimize the unnecessary bandwidth consumption, the whole environment can be divided into several sub-areas, each or several of which can be assigned to an AS for management. Every client can choose a sub-area what it likes based on its own particular interests. All clients are also divided into several groups and every client has its own communication activities in a particular group with no interference from outside at all. And then, all AS are managed by a Server Manager (SM). As with the authors' case, the whole virtual environment is divided into four sub-areas (Sub1, Sub2, Sub3, Sub4), and AS1 is assigned to manage Sub1, AS12 to Sub2, and so on, as showed in Figure 1.

With this type of architecture, the job of Server Manager (SM) is mainly to manage all the ASs in the system, direct a client to the appropriate AS while a client first logs in a sub-area, and migrate a sub-area from an AS to another when load balancing is needed. For managing such business, the SM has to use a configuration file that contains the host address and ports of all AS in the system, and the identification of sub-area managed by each AS. When SM is started up, it reads the above said configuration file and creates a sub-areas allocation table. Then it listens to two ports, one is for accepting connection request from ASs, and the other is for accepting the connection request from clients.

As to the Application Servers (ASs), their duty is to maintain the state of one or several sub-areas and manage all the clients in the same group to keep consistency of their states. When an AS is started up, it first of all has to get itself connected with the SM in accordance with a connect-oriented protocol TCP/IP to inform SM that it is invoked. Then it creates a port for every sub-area, on which it exchanges the state updates with its clients on a connectionless protocol UDP/IP.

As for the Client, what it has to do is nothing but run its own virtual reality application programs, handle interactions, send and receive the state updates, as well as refresh the 3D scene. When a client is started up, the user only needs to input Server Manager's host address, and the identification of sub-area that it wants to enter. When the client has to join an AS, it can send a message to SM, the SM sends back the AS's host address and port number to this client, and directs the client to appropriate AS. After that, the client and AS can communicate directly and exchange state updates each other, bypassing the SM.

From what is shown above, this architecture can have the following advantages: first, the relation among Application Servers is loosely-coupled. Network traffic is mainly done within the sub-area; second, it is easier to implement the consistency and concurrent control; third, the system can be made much more flexible by increasing or decreasing the number of Application Servers according to the requirement of the actual application.

3. The implementation of consistency

Object-Property-Event (OPE) mechanism and programming paradigm, which are widely used in object-oriented systems [6], were used in this framework in order to keep up the consistency of shared VE and simplify programming. In doing so, a given application system is composed of a set of objects that are distinguished from each other by their properties. Property, which is used to enhance the description of object, is identified by a unique name and is of a specific data type. The values of properties decide the state of the object. And the state of object, in turn, changes as the property's value changes. In order to keep the consistency of object's state, property must be shared by the client that wants to change it and the clients that want to receive it. Therefore it is these shared properties that maintain the consistency. When the value of a property changes, an event may occur and then lead to some consequences in the execution of some callback functions. For example, if the position of an object changes, there would be a refreshed view in place of the old presented to the clients.

The so-called event is actually a kind of interactive method between objects that exist in the same computer or different computers. Events in a networked virtual reality system mainly include state update event, interaction event, and network-connection event. Interaction event is mainly caused by user's interaction with virtual environment, such as the change of viewpoint, pressing keyboard or clicking mouse. In most cases, it will result in the update of an object property's value; Update event is caused by the change of property's value, and consequently, results in the submission of the update message and refreshment of shared scene; Network-connection event will result in the connection between Server Manager and Application Server, the connection between Application Server and clients, or the transfer of data between them. When developing an application system, developer can design callback functions on demand and establish the relations between their handlers (pointers to that callback function) and relative events.

The implementation of OPE is showed as Figure 2. In this system, each client is expected to maintain a local-property-list (LPL), which contains information about all local properties. For each property, it includes the property's name, data type, current value, the time and place that the value is set or modified, the name of the object that the given property belongs to, a lock flag, which indicate if the property can be modified in the near future, etc. Such a LPL can be very helpful in reflecting the current states of the shared VE in every client. As soon as the items of LPL of all clients are kept the same, the state of shared VE in all clients can be kept consistent. So, this paper suggests to setting a shared-property-list (SPL) in Application Server, whose structure is the same as that of LPL.

When a user interacts with an object in shared VE at any client-end, an interaction event would occur that would then result in the update of property's value in LPL. This updated situation will then enact an update event, which makes it easy for the client to send its state update message to the Application Server. As the result, the Application Server will be able to modify this property's value in SPL based on update message it receives, so as to be able to send such updated messages to all other clients. In this way, all local-property-list can be kept consistent.

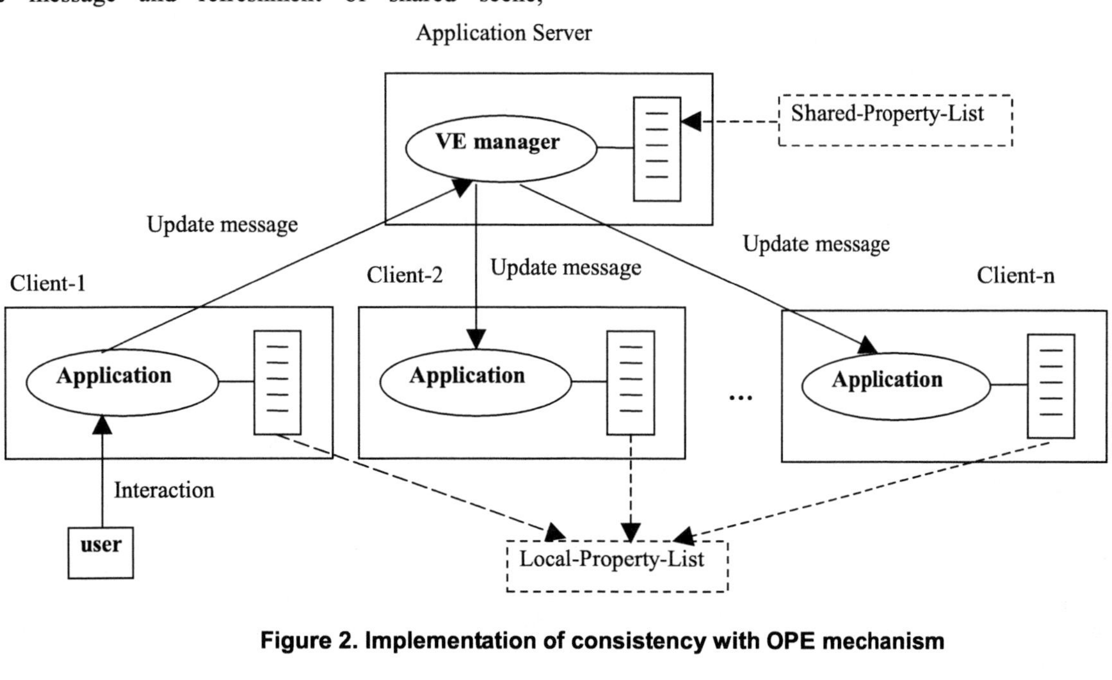

Figure 2. Implementation of consistency with OPE mechanism

4. Concurrent Control

Concurrent control problem need to be solved in NVR system is how to keep the consistency of shared virtual environment when multiple users or multiple applications want to access the same object simultaneously. There are mainly two types of concurrent operations must be controlled in the framework presented by this paper. One is the concurrent access to shared-property-list by multiple threads in Application Server. The other is the concurrent interaction with shared virtual environment by multiple users.

The locking mechanism is basic concurrent control algorithm that is widely used in cooperative (CSCW) system. It can also be used in NVR system to solve concurrent problem. When any thread has intention of modifying a shared property, a lock has to be put on that shared property to prevent other threads' access for this property. After modifying, the lock can be released, see Figure 3. So, only one thread can modify the shared property at a time.

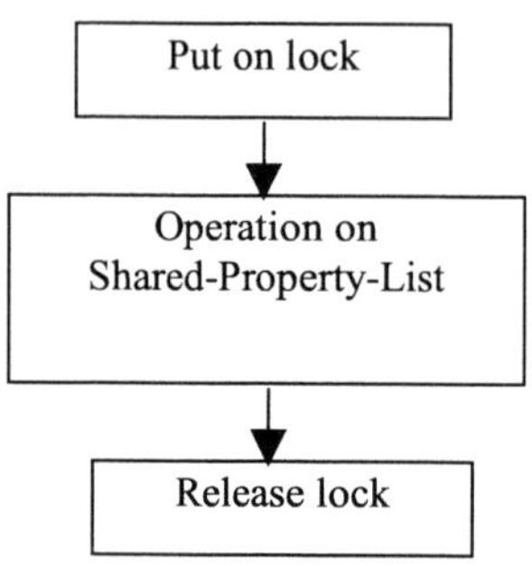

Figure 3. Locking mechanism

Locking mechanism is proved simple but efficient for actual use. However, the method is not good for some very complicated concurrent interactions in NVR system. For example, in a shared virtual environment as shown in Figure 4, if only user 1 lifts the table, the table would turn in clockwise. But, if only user2 lifts the table alone, the table would turn in counterclockwise. However, if both user1 and user2 would lift the same table simultaneously, the table will lift upward. From this example, it can be seen that there may have been varieties of change with the table's state if the interactions are done differently.

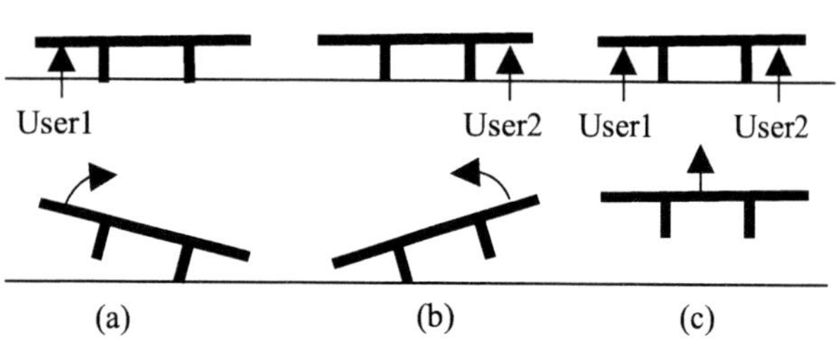

Figure 4. The example of concurrent interactions

The above trouble can be clarified by using the idea of serializing concurrent interaction. The actual procedures suggested by this paper are shown in Figure 5. This paper suggests establishing a shared update-message-queue in Application Server to serialize concurrent operation on shared property list and implement the communication among multiple threads. When multiple users at different client-ends interact with the same object in shared VE at the same time, all of them have to send the updated messages to Application Server. And then, in the given Application Server, there are multiple service threads, each of which exchanges update messages with a client. Thus all the service threads in the case can be expected to modify Shared-Property-List (SPL) at the same time. Instead of directly modifying the exclusive Shared-Property-List, these threads can insert the update messages to the shared update-message-queue particularly based on particular criteria, such as time, priority, or collaborative demand. Then a thread, which called VE manager, retrieves the update messages from the shared update-message-queue and modifies the relative property's value in Shared-Property-List based on this update message. In this mechanism, concurrent interactions can be serialized and handled in order based on criteria that developer defined. In this way, the concurrent interactions problem in Figure 4 (c) can be easily solved.

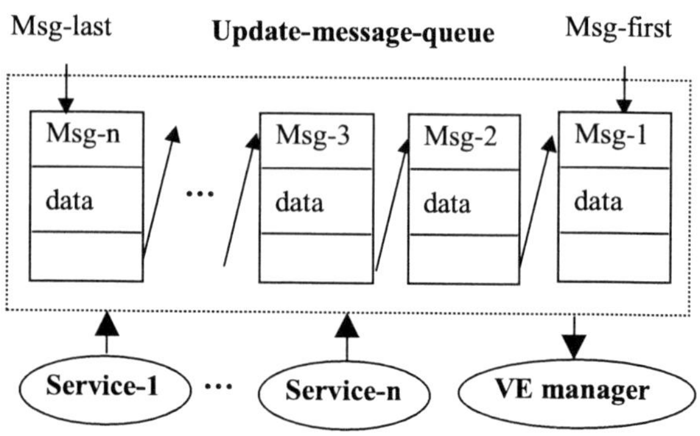

Figure 5. serialization of concurrent interaction

5. Conclusions and future work

In conclusion, the authors want to point out that, with the current state of art for NVR technology, it is still impossible to provide a NVR system fully satisfactory for actual real-time interactive applications. Therefore it is beyond the authors' expectation to take into account all the collaborative problems involved.

In this paper, however, what the authors want to come up with are the two collaborative issues: consistency and concurrent control. In connection with this focus, they also present a framework for supporting, managing and developing NVR system with some practical solutions to the collaborative problems under discussion. The claims for the solutions include:

- Client/server architecture with two-tier servers and grouping clients mechanism is benefit for implementing the consistency of shard virtual environment and avoiding unnecessary bandwidth consumption.
- Object-oriented design method and Object-Property-Event mechanism can implement the consistency and simplify the development of NVR application.
- Locking mechanism can solve the problem of concurrent object access.
- The serialization of concurrent operation can solve the elementary problem of concurrent interactions.

The implementation of these solutions is the part of the development of a network virtual reality toolkit—SVRT[7,8].

6. References

[1] Calvin J, Dickens A, Gaines B. The SIMNET Virtual World Architecture. Proc. of the IEEE VRAIS, Seattle, WA, pp450-455. 1993

[2] Macedonia M. R, Zyda M. Z, Pratt D. R, Barham P. T and Zeswitz S. NPSNET: A Network Software Architecture for Large-Scale Virtual Environment. Presence 3(4), pp.265-287. 1994

[3] IEEE Std 1278.1-3. IEEE Standard for Distributed Interactive Simulation-Application Protocls.

[4] Thomas A.Funkhouser. Network Architectures for Scalable Multi-User Virtual Environments. Proceedings of IEEE VRAIS'96. p222-228. 1996

[5] Singh G, Serra L, Png W. and Ng H. BrickNet: A Software Tookit for Network-Based Virtual Worlds. Presence 3(1), pp. 19-34. 1994

[6] Zhang Guofeng. Object-Oriented programming and C++. Electronic Industry Press. 1997

[7] Su Hongyi, Zhan Shouyi, Chen Yi. Design and Implementation of Simulation-Oriented Virtual Reality Toolkit. Journal of Beijing Institute of Technology. 2001.1

[8] Chen Yi, Zhan Shouyi, Su Hongyi. Representation and Storage of 3D Geometry in Virtual Reality Development Toolkit. Journal of Beijing Institute of Technology. 2001.1

The Development Tool of the Application Cooperation Software

Zhigang Zhang, Zongkai Lin, Yuchai Guo
Cooperative design group, CAD Lab., Institute of Computing Technology, Chinese Academy of Sciences, Beijing, 100080, P. R. China
zzg@baoding.cn.net

Abstract

In the process of cooperative design, we often need to discuss or modify a project generated by traditional CAD software, but there are many limitations in current cooperative design tools, such as White Board Tool and Shared Tool. Considering these limitations, we put forward the concept of Application Cooperation Tool. The tool, a domain-depending one, aims to provide an environment for a group to discuss or modify a project generated by traditional application software. It not only keeps down the original functions of the traditional software but also supports the cooperative work. However, the development of application cooperation tools is a difficult and time-consuming job. During the process of developing application cooperation tools ACPBrush and ACAutoCAD98 we discovered some common features which exist in all kinds of CAD software, such as the analyzing of event-stream and the re-acting of the event. On the basis of those general rules we plan to develop a tool, which can modify traditional CAD software to a cooperation-based design tool. The paper discusses the methods used in it, the structure and components of the tool, and relative problems.

1. Introduction

With researchers' going deep into the technique of CSCW, they developed many general tools to support cooperative work, such as White Board Tool and Application Shared Tool. The purpose of constructing these general tools is to provide a means to intercommunicate between different users and to find and solve the conflict in a group. The more usual case is, however, how to discuss and modify a result generated by traditional CAD software. The existent cooperative tools are not qualified for this job. White Board Tool cannot support a group of members to discuss a project created by traditional CAD software; Application Shared Tool, although it can provide some functions of remote conference to some extent, there are two inborn defects. The first one is that it cannot be used with easy in cooperative environment. Since it can only provide the shared way to support the cooperative work, every member in a group must operate in serial way. Before a member acquires the operation handle he must execute the action of applying, queuing and waiting the release of the handle by other operators. The second defect is that it cannot provide some necessary functions that are essential to cooperative work. For example, it doesn't disassemble the flue of message, so it cannot recur the history operation of any operator. That is Group Undo Function.

To fully support a group of members to discuss and modify the design result in cooperative environment, after analyzing the problem encountered in the development of the Application Shared Tool, We put forward the concept of "Application Cooperative". Application Cooperative tool is a domain tool which aid to support a group to discuss or modify a result generated by another software. It reserves the original function of the software. At the same time I can be used in cooperative way. Under this tenet we developed two instance to express the idea: ACPBrush and AcAutoCAD98. They are built on the base of Windows Paint Brush and AutoCAD98 respectively. The development of this kind of tool is an arduous and time-consuming job. For this reason we developed a prototype of the development tool, through which one can rebuild traditional CAD software into a new cooperation-based tool..

2. Structure of the Development Tool

The structure map of Application Cooperative Tool rebuilt by the development tool is shown as in Figure 1.

The application cooperative tool not only supports members of a group to discuss and modify the design result, but also provides the fundamental functions that are necessary to a CSCW system. The development tool of the application cooperation software realizes the former and the human-human interface development tool (TOGD) actualizes the latter. The solid rectangle in Figure 1 indicates the development tool and the hollow rectangle indicates interface constructed by the TOGD tool and other cooperative functions. The main function of the development tool is to rebuild the edition functions of traditional CAD software to cooperation-oriented functions. On the one hand it disassemble the messages flow into single message, on the other hand it is in charge of the re-action of these messages.

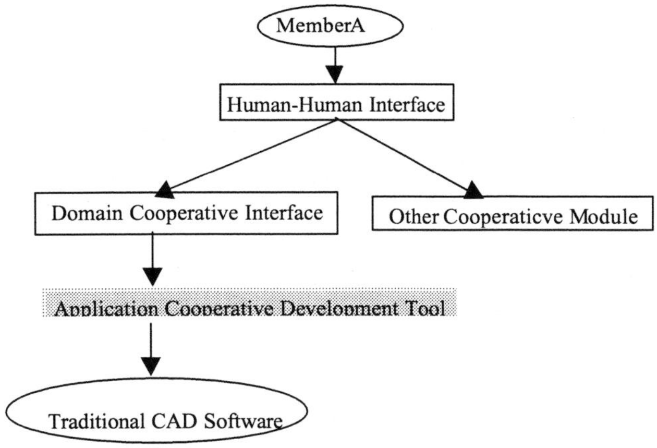

Figure 1. Structure of Application Cooperation Tool.

In the Figure 2 we can know that the main body of the development tool is the message-disassemble-module, which is in charge of acquiring orders of the local user, disassembling the messages flow, assembling these single message and transmitting them to the I/O cooperative component in the human-human interface.

In sum, the development tool must be combined with the human-human interface tool (TOGD) to construct application cooperation software.

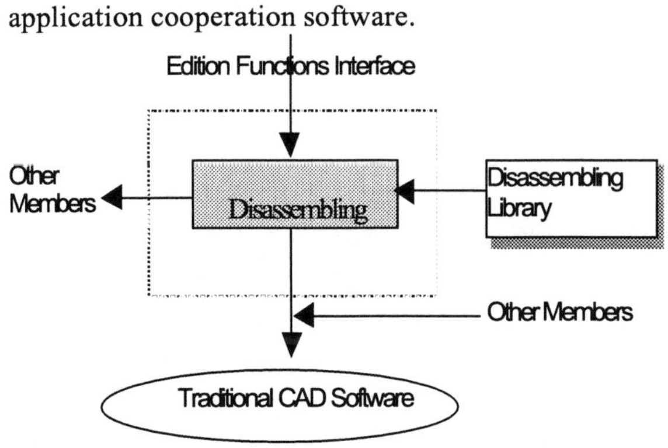

Figure 2. Component of Development Tool.

3. The layout of the development tool of application cooperation software

The development tool of application cooperation software comprises two parts: rebuilding tool (including visual rebuilding tool and manual rebuilding tool) and daemon program. The visual rebuilding tool provides for users with some visual methods to facilitate the progress of collecting information of traditional CAD software. Regarding to some information that cannot be acquired by visual rebuilding tool we design a tool called Manual Rebuilding Tool to act as a supplementation. Another important part of the tool is Daemon Program, which is the kernel of whole system. It is in charge of disassembling the messages flow, assembling these single messages, transmitting them to target user and re-acting them in the remote client.

3.1 The Design of the Disassembling Library

The disassembling library stores the information of the rebuilt traditional CAD software (Figure 3). The information is acquired through the visual rebuilding tool and manual rebuilding tool. During the running period, all the information is the basis of the messages disassembling. The disassembling library comprises two parts: one is the basic information of the CAD software, including the locating data of all kinds of menus, toolbars, and dialog boxes; the other part is order information, which include the initiation information and operation information of orders.

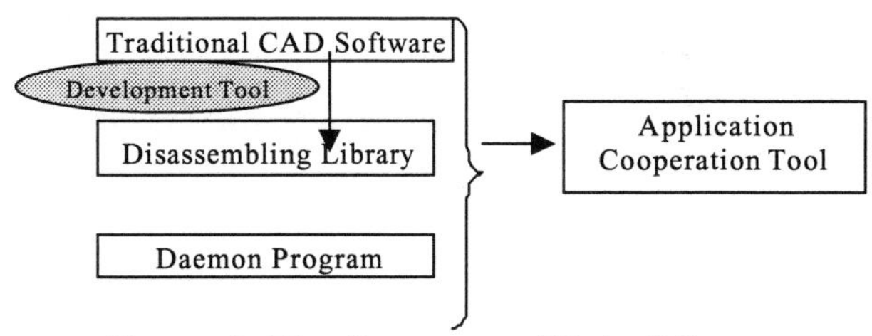

Figure 3. The Progress of Rebuilding a Application Cooperation Tool.

3.2 The Design of The Daemon Program

The daemon program of the application cooperation tool is the running module of the development tool. It disassembles, transmits, and re-acts the messages acquired from the local user. The basis used to disassemble the message flow is the Disassembling Library. The daemon program need be run in every client terminal. The running sketch map is as follow:

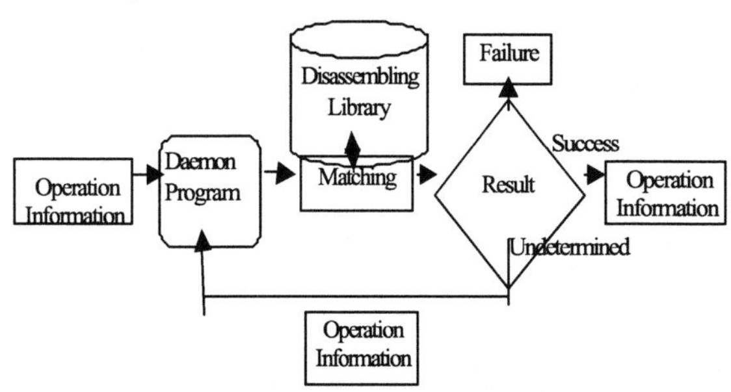

Figure 4. The running progress of the daemon program.

One of the kernel tasks of the daemon program is to disassemble the message flow. The veracity and the efficiency of the disassembling progress is crucial to the whole system. The steps of the disassembling is described below:

Step 1: When the daemon program received the operation messages from the user, it locates the operation ID in the Operation Initiation Table which stores all the initiation information of all the operations and sets it as the value of the Current Tool Variable.

Step 2: The daemon program extracts all the operating information related to the ID to a temporary table called Current Operation Describing Table.

Step 3: According to the succedent operation the daemon program searches the matching item in the Current Operation Describing Table and deletes the un-matching item from the table.

Step 4: If there is only one item in the table and the whole describing character string is matched the disassembling progress is successful; if the table is blank the program fail to disassemble the flow; if there are two or more items remained in the table or the only string isn't matched complete the daemon program continue the disassembling progress at Step3.

3.3 The Describing Method of the Operation Information

The operation information provides dependencies for daemon program to disassemble the message flow. It comprises two parts: one is the initiation information of a operation, the other is the operation method itself. Determinacy must be ensured when describing method is designed. That is every operation is related to only one description. So the veracity of describing method is the guarantee of the correctness of system. Due to the diversity and flexibility of the operation, the design of the describing method the operation is also the difficulty of the whole system. After serious consideration, XML is adopted as the main method to record the operation information.

XML (Extensible Markup Language) is a kind standard and general data describing language, which express the meanings of all kinds of information through TAG. It can be interpreted as the same result by different system through some "self-interpret" text. There are two distinct merits to use XML as the describing method. In the one hand, all kinds of operation information can be recorded accurately in a concise way. In the other hand, all the operation information can be shared in Internet. The developer of CAD software can provide their data in the same time of provide the software itself and the user can get these data expediently.

4. Conclusion

In current time, one must to recast the software respectively to rebuild traditional CAD software to a new application cooperation tool. Aiding to enhance the efficiency and to guarantee the robustness of the software we design and develop a development a prototype system of the tool to fulfill the recasting work. The practice proves that the prototype system can work as it has been designed.

5. References

[1] J.B. Begole, C.A. Struble, et al. Transparent Sharing of Java Applets, A Replicated Approach, In Proceedings of the ACM Symposium on User Interface Software and Technology, Oct. 14-17 1997.

[2] Ahuja, S, Ensor, J, Lucco, S.A. Comparison of Application Sharing Mechanisms In Real-time Desktop Conferencing Systems, in Proc. ACM Conference on Office Information Systems COIS'90, pp. 238-248.

[3] J.B. Begole, C.A. Struble, C.A. Shaffer. Leveraging Java Applets, Toward Collaboration Transparency in Java, IEEE Internet Computing 1(2):57-64, Mar./Apr.1997.

[4] Bly, S. A. Media Spaces, Bring People Together into a Video, Audio, and Computing Environment, Communications of the ACM, Vol.36, No.1, 1994.

[5] Anupam, V.Shastra, Multimedia Collaborative Design Environment, IEEE Multimedia, Summer 1994.

[6] Grudin, J. Groupware and Social Dynamics, Eight Challenges for Developers, Communications of ACM, No.1, 1994.

[7] Kyng,M.Designing for Cooperation, Cooperating in Design, Communications of the ACM, Vol.34, No.12, 1991, pp. 64-73.

The Control and Algorithm of Audio Dynamic Buffer

Chongjia Peng, Yong Tang, Xianji Li, Chengjie Mao
Faculty of Computer Science and Engineering, Guangdong University of Technology
Guangzhou 510090, P.R. China
pengcj@gdut.edu.cn

Abstract

It is a very important work to improve the quality of voice on transferred Internet. In this paper two aspects about voice quality are analyzed, and an algorithm of dynamic setting audio buffer for improving QoS is proposed. Finally, the simulation results of the algorithms and the conclusion are given.

1. Introduction

The QoS parameters can be used to indicate the quality of voice transferred on the Internet [1]:

(1) the lose rate of voice packets.

(2) the delay and deviation of voice packets.

In [2] we put forward a dynamic setting buffer module that can improve the quality of audio. It was described and analyzed with Petri net. In this paper the control method and algorithm of dynamic setting buffer module are discussed, and the simulator program of the algorithm is implemented. The simulation results and the conclusion drawing from the results are shown.

In [2] we put forward a dynamic setting buffer module that can improve the quality of audio. We describe and analyze it with Petri net. This paper discusses the control method and algorithm of dynamic setting buffer module, and implement simulator program and gets a conclusion.

2. Dynamic buffer control

In fact, the dynamic buffer control is a kind of destination quality control. The most time of network communication is unstable. The goal of buffer self-adapting control is to process optimal control at receiving terminal. It is different from the QoS self-adapting control that must be carried out at every router on the path of network.

We use a sending buffer at sending station and receiving buffer at receiving end to smooth the audio [3]. The sending buffer doesn't affect the result. We have sending buffer is to facility to describe and analyze.

Selecting right length of buffers I_0 is the key method of setting buffers. Audio data will be played when the receiving buffer was filled fully. The buffer length I_0 is related to the situation of network, the level of network congestion. The self-adapting control method can adjust the buffer length I_0 for different situation for the optimize effect of playing audio.

The relation of the packet delay jitter n_i and the packet arrival time T_i is follow:

$$n_i = T(i) + c \qquad (1)$$

The follow constant quotient linearity difference equation can describe the self-adapting question of buffer length I_0:

$$y(k)+a_1y(k-1)+\ldots+a_ny(k-n)=b_0+b_1T(k-m-1)+\ldots+b_pT(k-m-p)+e(k) \qquad (2)$$

In the equation, *T(k)* represent the interval from the time control begin $(k=0)$ to the time when the number k sampling packet arrive. $y(k)$ is output representing delay jitter. b_0 is unknown constant. $e(k)$ is module error. The up and down buffer length I_0 is a kind of control use.

In fact, we think that audio packet delay can be acceptable when the delay jitter is set to minimum. This is a kind of strategy that smooth audio through add delay.

The question of buffer length control comes down to follows:

(1) The equation of self-adapting control module. The unknown parameter needs to be estimated online. $a_1, \ldots, a_n$ are constant.

(2) The goal value of destination control: y_r, and the changing range of $y : y_r + \Delta$.

(3) The acceptable risk of jitter: p

In the given conditions, $u(k)$ will be required to:

$$E[y_f] = y_r \text{£¬} E[y_f - y_r]^2 = \min ,$$

According to the following restrictive condition:

$$P\{y_f > y_r + \Delta \mid u(k) \} \le p$$

Here, $u(k)$ is a kind of factor used to control the buffer length I_0. $u(k)$ make the probability of $y_f > y_r + \Delta$ (jitter is more greater) less than the appoint risk p. The output y_f and probability distribution of y_f will determine $u(k)$. We can predict y_f through continuous sampling, so self-adapting control can carry out continuously.

In format (1) and (2), the influential variable which can be used to predict y_f is the interval time between the number $i+1$ and the number i. Thus the key is predicting the time the next packet arrives at time k. We use the algorithm of calculating the time of next packet that Jacobson put forward. Thus we can decide the length of buffer based the result of prediction. At time k, the module will predict the probability of y_f based the module parameter and the estimating time of the next audio packet arriving.

1. If $P\{y_f(k)>y_r+\Delta \mid u(k)\}<p$ (p is small probability), it indicate that the risk of jitter is smaller than the acceptable risk. Thus the length of buffer doesn't increase.

2. If $P\{y_f(k)>y_r+\Delta \mid u(k)\}>p$ (p is small probability), it indicate that the risk of jitter is bigger than the acceptable risk. Thus the length of buffer will increase. There are two ways to increase length of buffer: every time adding one packet length or twice the length of buffer now. Finally, the length of buffer trends to an appropriate value I_0, makes the jitter min.

3. Simulator program and conclusion

We develop a program using C++ 6.0 to simulate the algorithm above. It operates on the Windows98 system. Considering that the sending point enter into Internet through LAN, it obtain the right of sending data immediately when the load of network is very light whether it is Ethernet or Token-ring net. Thus we can think the time of obtaining the right of sending data is a constant. The time is that the simulator program creates a audio packet.

The character of transferring delay of the network we simulator is a probability distribution [4] (*Fig.1*). The program designed simulates the probability distribution as fig.1

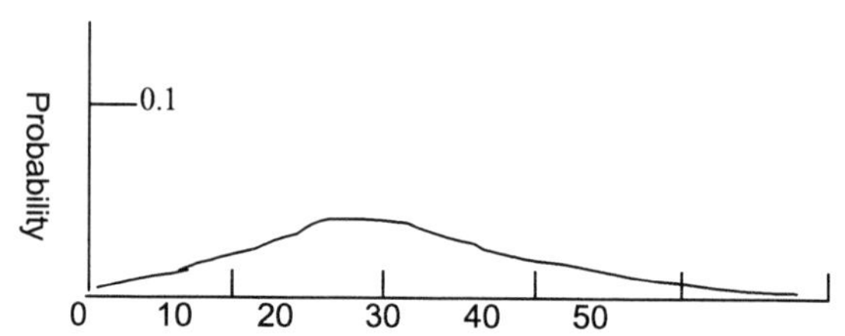

Fig. 1

In simulator program *(Fig. 2)*, the sampling frequency is 11.025KHz, the size of packet sent is 1500 byte, useing 8-bite quantization. The speed of playback is 17ms per frame. The progressbar at the bottom of the Fig2 directly indicate the degree of buffer used.

The meanings other parameters in the interface are as follows:

- Size of original buffer: initialize the size of buffer at beginning.
- Playback value: the playback begins when the number of packets in buffer reaches the value that is called playback value.
- Continuous playback time: the interval from the beginning of playback to the time when there is no data in buffer. The play interval is between two continuous playbacks.
- Size of buffer: represents the size of buffer in opertering when the buffer at receiving end is adjusted dynamically.
- Number of packet: the number of audio packet stored in buffer.

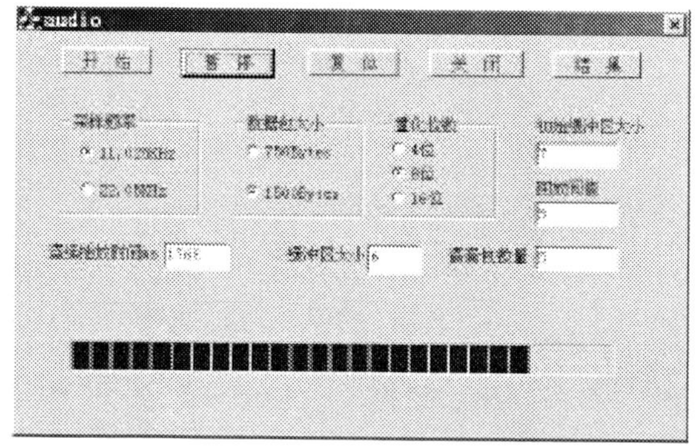

Fig. 2

At the beginning of running the program, those variables can be selected. The program begins to run after the "start" button is pressed. You can observe the results of every variable by pressing the "pause" button. All data will be cleared if the "reset" button was pressed down. It is ready to run the simulator program again. The program interface will be changed if the "result" button was pressed down *(Fig. 3)*. The meaning of the variables in Fig. 3 are as follows:

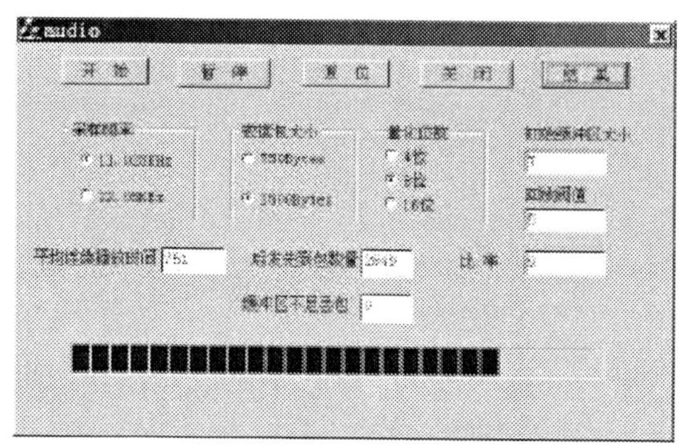

Fig. 3

- Average time of continuous playback: the total time of continuous playback divided by the number of continuous playbacks.
- Number of LSER: the total No. of Packets which are sent later at sending end but received earlier at receiving end. It represents the No. of packets which are out of order after transferred on networks. If the number is greater, the packets out of order will be stored and then played back correctly instead of

discarded simply through UDP protocol.

- Discarded packet for not enough buffers: the audio packets are discarded because the size of buffer is not large enough.
- LSER rate: the definition is (num of LSER)*100/(the total audio packets).

We run the simulator program under the following conditions. Table 1 is the result of average time of continuous playback; the play interval, etc. when the sampling frequency is 11 KHz, the size of data packet is 1500 bytes, using 8-bite quantization, and the speed of playback is 17ms. The results are shown as Table1.

Table 1

Play-back value	The rate of discarded packet (%)	LSER rate (%)	Average time of continuous playback (ms)	Play interval (ms)
6	0	9	1050	102
5	0	9	582	85
4	0.4	9	244	68
3	0	9	121	51
2	0	8	96	34
1	0	9	17	17

If the sampling frequency is 22 KHz, the size of data packet is 1500 bytes, using 8-bite quantization, the speed of playback is 9 ms. Under different playback values the results are shown as Table 2:

Table 2

Play-back value	The rate of discarded packet (%)	LSER rate (%)	Average time of continuous playback (ms)	Play interval (ms)
6	0	71	103	54
5	0	72	80	45
4	0	74	64	36
3	0	70	49	27
2	0	68	25	18
1	0	68	9	9

If the sampling frequency is 11 KHz, the size of data packet is 1500 bytes, using 4-bite quantization, the speed of playback is 34 ms. Under different playback values the results are showed in Table 3.

The simulation results show that the rate of average time of continuous playback to the total time. If the interval between two arriving packets one after another is less than 3ms, We can wait up to 3ms per packet when packets are played back, because people can not detect the break of 3ms. Thus, we increase the rate of average time of continuous playback so as to improve the quality of audio.

We can know the LSER rate is only related to the speed of playback. The speed of playback is bigger and the LSER rate is smaller. Thus, we should make the speed of playback big enough.

Table 3

Play-back value	The rate of discard packet (%)	LSER rate (%)	Average time of continuous playback (ms)	Play interval (ms)
6	0	0	11492	204
5	0	0	7339	170
4	0	0	5813	146
3	0	0	3381	112
2	0	0	1810	68
1	0	0	34	34

4. Conclusion

This paper discusses the control method and algorithm of dynamic setting buffer module, and implements a simulator program. The simulation results show that this method has a good effect on smoothing audio. And we get the general conclusion of the relationship between the size of buffer and the effect on smoothing audio. This method can be also considered to apply to treat video signals.

5. Acknowledgement

This work is supported by the Guangdong Provincial Natural Science Foundation (team project).

6. References

[1] Baohong He, Yecai Guo, Zhongkai Lin, Shouxun Lin, The research of Voice Quality Techniques in the Internet Voice Tools, Proceedings of the Fisrt National CSCW Workshop, 1998.

[2] Yong Tang, Chongjia Peng & et al. Research on quality of audio in desktop conferencing systems, Journal of Zhongshan University, 2000, 39(2): 16-18.

[3] Xianji Li, Yong Tang & et al, GUT: a desktop conferencing system. In: Proceedings of The Fourth International Workshop on CSCW in Design, Sept. 1999, Compiegne,France, 375-378.

[4] Andrew S.Tanenbaum. Computer networks, third edition. Prentice Hall International, Inc, 1998.

T.120/H.32x-based Multimedia Conferencing Design

Minzhe Zhao, Mingshu Li
Institute of Software, Chinese Academy of Sciences
m.zhao@ieee.org

Abstract

Multimedia conferencing technology is becoming increasingly important, as it enables people who are located in different cities (or countries) to share as well as interact with their data, comments, images, and video pictures. This paper introduces the multimedia conferencing standards, such as ITU-T T.120, T.130, H.320, H.323, and H.324, which comprise the core technologies for multimedia conferencing. The paper discusses interoperability and compatibility issues which arise from the deployment of different vendor applications. The paper also presents a proposed multimedia conferencing system with H.32X standard capabilities, which is designed to facilitate interoperability and compatibility between different platforms.

1. Introduction

Multimedia conferencing is an integration of video and data conferencing techniques, with which participants are able to share and exchange their data files, comments, images, and video pictures. Of them, data conferencing plays an important role due to widespread deployment of conferencing tools, such as Microsoft's NetMeeting, Intel's ProShare and DataBeam's FarSite on the desktop, enabling people to work with remote colleagues by sharing documents, images, drawings, and even desktop applications.

However, though, real-life situations are best emulated by sharing data or still images plus audio, in the typical virtual business meeting or classroom-at-distance, participants may communicate by voice, data, and video. Voice is the most common means, but many data files and paper work may also need to be shared, including notes, annotations over a graphical drawing, a spreadsheet, or text documents. Conferencing with video is often required when, for instance, the changing expressions on the participants' faces or their gestures contribute to the discussion [1]. And also, video might be essential when project tradeoffs are discussed, an X-ray image is presented to remote doctors, or a new process start-up/shut-down procedure is shown to a group of operators.

Multimedia conferencing often adopts centralized (or client-server) architecture for implementation, in which the server (or chair) enables a group of participants to share their data in the following scenario: Data packets from the participants arrive at the server at different speeds. The server then sorts, identifies, and, if necessary, replicates the packets for multicast transmission before forwarding them to their destinations. However, the handling of conferencing data packets is always a problem, given a situation where participants often work with different platform desktops. This calls for a standard to which conferencing software in the market can conform, so as to ensure that a uniform procedure is used for the proper functioning of all basic components. On top of these components, different vendors offer different features in server software. The architecture and standards for multimedia conferencing are discussed in the following sections.

2. Trend towards Multimedia Conferencing

Data conferencing technology has been, now, increasingly becoming popular, along with the widespread deployment of conferencing tools, such as Microsoft's NetMeeting, Intel's ProShare and DataBeam's FarSite on the desktop, enabling people to work with remote colleagues by sharing documents, images, freehand drawings, and even desktop applications.

However, though, real-life situations are best emulated by sharing data or still images plus audio, in the typical virtual business meeting or classroom-at-distance, participants may communicate by voice, data, and video. Voice is the most common means, but many data files and paper work may also need to be shared, including notes, annotations over a graphical drawing, a spreadsheet, or text documents. Conferencing with video is often required when, for instance, the changing expressions on the participants' faces or their gestures contribute to the discussion [1]. And also, video might be essential when project tradeoffs are discussed, an X-ray image is presented to remote doctors, or a new process start-up/shut-down procedures is shown to a group of operators.

3. Conferencing Architecture: Centralized versus Distributed

It is suggested that the need for multimedia conferencing will arise when people, who are not physically at the same location, need to work together towards a common goal. Hence, it is evident that any conferencing application operating in such situations will require a well-designed multiuser interface. The key to the successful design of such interfaces is an appropriate supporting architecture, which facilitates various forms of interaction among a group of users [2]. There are a number of architectures being developed for facilitating conferencing applications. Of them, centralized and distributed architectures are two distinctive ones. In some extent, distributed architecture can be fault-tolerant and flexible, as each node creates a replica of the application, making sure that participants can interact freely and actively. However, recently, centralized architecture has been gaining its popularity. Centralised architecture can be viewed as being similar to a "client-server" structure, and is easy to manage and administrate. These advantages have superseded its drawback that any server failure might lead to the collapse of the whole system. The ITU-T endorsed centralized architecture hierarchy in its T.120 family standards on which most commercial multimedia conferencing systems are built.

4. Standards for Conferencing Control and Protocols

The ITU-T T.120 standards address real-time data conferencing (audiographics). It contains a series of communication and application protocols and services that provide support for real-time, multipoint data communications. [http://www.databeam.com/].

5. Standards for Conferencing Communications

To facilitate the interoperability required for heterogeneous networks, and compatibility among various vendor solutions, ITU-T developed H.320 standard series, including H.310, H.321, H.322, H.323 and H.324. The core functions of H.32x series are as the following [http://www.imtc.org/]:

- H.320 – multimedia applications (audio and video) over Narrowband-ISDN.
- H.321 – multimedia applications over Broadband ISDN (B-ISDN), ATM and LAN.
- H.310 – High-resolution multimedia applications over B-ISDN.
- H.322 – multimedia applications over LANs that provide a guaranteed bandwidth.
- H.323 – multimedia applications over LANs that provide a non-guaranteed bandwidth.
- H.324 – multimedia applications over the public switched telephone network and mobile phones.

Especially, the new H.323 and H.324 standards enable networking and applications vendors to support more deployable, manageable and affordable conferencing, and enables multi-vendor interoperability and compatibility. H.324 defines videoconferencing for low bitrate multimedia communication using plain old telephone systems (POTS) lines.

6. A Proposed Conference Setup

A good multimedia conferencing system should be tested over a heterogeneous network to ensure good interoperability among various network components and communication standards. Our proposed multimedia conferencing configuration places emphasis on the various network access media such as Ethernet, ATM, and remote access using analogue modem and ISDN connection (see Figure 1). Two server systems, with different operating systems, will be deployed on the proposed configuration. The UNIX Server with Solaris operating system will house the T.130 server (for multimedia conferencing) while the Windows NT server will house the T.120 server (for data conferencing). The intended servers setup will distribute the client requests among the two servers so as to minimize blockage time.

Ethernet network is the most common local area network in many organizations due to the widespread use, economy and simplicity of this technology. The deployment of multimedia conferencing over Ethernet will enhance collaboration between multiple parties in an organization by using the current existing networks. However, one drawback of conventional Ethernet is that multimedia data must be shared among a number of users, as a result, the network performance is very much limited when many users participate a conference. However, as many organizations have partially or fully migrated to either Gigabit Ethernet or ATM network, enabling us to enjoy the full benefit of multimedia conferencing systems. Here, in this paper, the proposed configuration will be cross-platform, and network independence, demonstrating the usability, flexibility and interoperability of multimedia conferencing systems based on the H.32x standards.

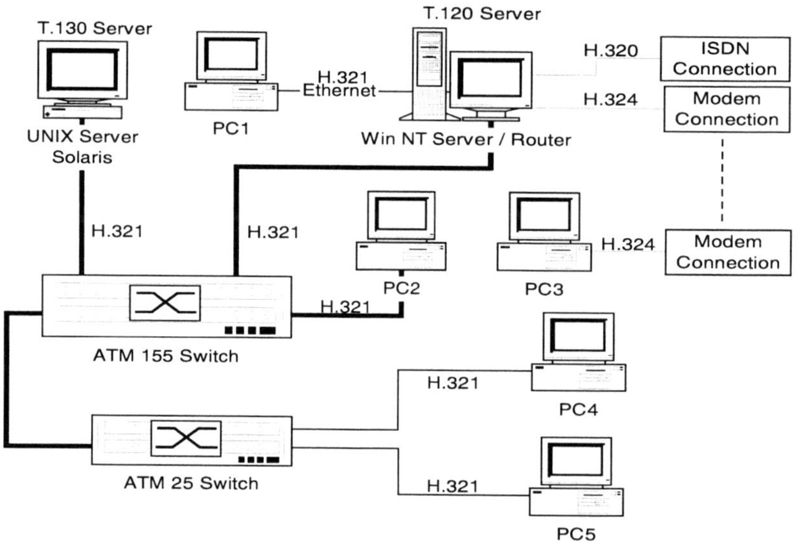

Figure 1. The proposed multimedia conferencing setup.

The proposed configuration will also demonstrate two types of telecommunication connections, namely ISDN connection and analogue modem dialup. Analogue modem connection should be supported, as it is still the most popular remote access method in most countries. Conferencing clients with ISDN connection will use H.320 standard while clients with analogue modem connection will use H.324 standard to link up with the two servers on the network.

The ultimate demonstration on this proposed configuration would be a multimedia conferencing session between a remote access client (PC3), an Ethernet client (PC1) and a client on the 25Mbps ATM network (PC4 or PC5). A successful demonstration will show true network independence and interoperability among the various conferencing standards.

7. Conclusion

Videoconferencing has helped people located in different cities (or countries) communicate more effectively. The early conferencing solutions were based on the ITU-T H.320 standard defining Integrated Switched Digital Network (ISDN) connection-based videoconferencing. Later, videoconferencing was brought to computer desktops, though, it was still dependent upon ISDN and expensive codec devices. The situation was lasted till the mid-nineties, when LAN-based solutions opened up the way to ubiquitous, accessible desktop videoconferencing applications. However, until recently videoconferencing techniques have still remained proprietary, difficult to implement and control, and also expensive. The absence of standards has severely limited widespread deployment due to incompatibilities between different vendor solutions. To resolve this problem, much effort has been made on standardizing the design and implementation of conferencing applications. Our on going multimedia conferencing project is part of this effort, aiming at eliminating incompatibility between different platform-oriented solutions, by exploiting H.32X standards. The construction of the data conferencing part of our proposed multimedia conferencing system has completed, and runs well on both Unix and Windows NT platforms. The implementation of video conferencing part is currently underway.

8. Acknowledgement

The authors wish to thank the Tote Board Singapore for its providing funds for this multimedia conferencing project. Without its financial support, the proposed conferencing setup would not have the opportunity to be implemented. The authors also wish to thank the Department of Electronics and Communication Engineering, Singapore Polytechnic for its support in producing this paper.

9. References

[1] Amitava Dutta-Roy: Virtual Meetings with Desktop Conferencing, IEEE Spectrum, Vol(35):7, pp. 47-58.

[2] Zhao, MZ: The Architecture of a Multimedia System for Use in Industrial Control. *Ph.D Thesis*. University of Wales Swansea, UK.

Research on Adaptive IP QoS Management Framework

Yong Tang [1,2], Chengjie Mao [1], Hai-xiang Ou [1], Xue-liang Yang [2], Xianji Li [1]
[1] *Faculty of Computer Science and Engineering,*
Guangdong University of Technology, Guangzhou 510090
[2] *Department of Computer Science Graduate School,*
University of Science and Technology of China, Beijing 100039
ytang@gdut.edu.cn

Abstract

IP QoS is one of the core technologies of IP network. The mechanisms to achieve IP QoS could be Sub Bandwidth Management(SBM), Multiple Protocols Label Switching(MPLS), IntServ, DiffServ and so on. But some of the above should be integrated for implementing completed IP QoS. At present, one of the most popular ideas is IntServ on the edge of network and DiffServ on the backbone. In this paper, we present a framework for completed end-to-end adaptive IP QoS by integrating DiffServ and IntServ, put forward the negotiation flow-chart of adaptive QoS on the Boundary Router and the QoS control flow-chart on the Interior Router.

1. Introduction

In the recent years, the Internet techniques and applications have developed in an inconceivable speed. The transmitting message on IP network have evolved from single text data to multimedia data, such as graphic, image, real-time audio, and video and so on. However, many difficulties have not been resolved yet, one of those is QoS of multimedia services. Because TCP/IP protocol just provides services of "Best-effort" level, lacking of QoS supporting, it always causes latency jittering, packet loss in real-time multimedia application and affects the qualities of real-time application deeply. It is necessary to research feasible and effective QoS management mechanism over IP network. IP QoS has become one of the core technologies of IP network. We have studied adaptive IP QoS based on IntServ within LAN[1,2], to further the subject, the adaptive IP QoS on DiffServ should continue to be studied. In this paper, we present a framework for completed end-to-end adaptive IP QoS by integrating DiffServ and IntServ of IETF at first, discuss the negotiation flow-chart of adaptive QoS on the Boundary Router and the QoS control flow-chart on the Interior Router in section 3 and 4 respectively. At last, we propose the future work.

2. IP QoS Mechanisms and Techniques

At present, some IP QoS mechanisms and techniques have been proposed; the main techniques include Sub Bandwidth Management (SBM), Multiple Protocols Label Switching (MPLS), Resource reSerVation Protocol (RSVP), Integrated Services Architecture (IntServ), and Differentiated Services Architecture (DiffServ). These techniques and mechanisms do not conflict with each other. We can apply two or more of them to implement more perfect quality of services over IP.

MPLS map the IP QoS to the ATM (or the Frame Relay etc.), but IP QoS mechanisms, for example, IntServ or DiffServ, are needed. IntServ defines three services types: Guaranteed Service, Controlled-load Services and Best-effort, and use RSVP as work protocol. IntServ/RSVP mode could provide QoS guarantee over IP, but its expansibility is not good because of its working bases on per-flow; moreover, the implementation of RSVP depends on the all routers of the routing the packet passed. So that IntServ is applicable only to the LAN or Intranet, not fit to the backbone[3,4,5].

DiffServ solves the problem existing in IntServ and can implement QoS on the backbone. DiffServ provides a framework that QoS can be implemented within the DiffServ Domain. However, DiffServ is proposed by IETF in recently, and is still a draft till now. Different from IntServ, a large quantity of traffic flow in the DiffServ can be aggregated into a small number of Behavior Aggregates (BA). The traffic flows in the same BA are collected and mapped to a PHB (Per Hop Behavior) in Routers and do the same process, and DiffServ provides the QoS guarantee for packets in this way.

These mechanisms and techniques above do not conflict with each other, we can apply two or more of them to implement more perfect quality of services over IP. Such as MPLS, which maps the IP QoS to ATM, and the DiffServ can provide completed end-to-end IP QoS by banding with the RSVP[5,6]; one of the most popular ideas is IntServ on the edge of network and DiffServ on the backbone.

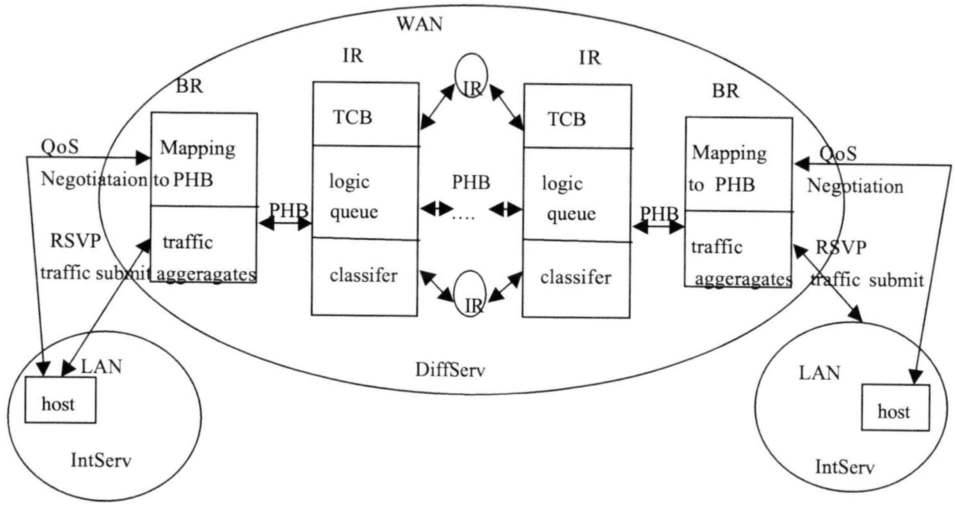

Figure 1. The framework of end-to-end IP QoS Integrating DiffServ and IntServ.

3. Adapted end-to-end IP QoS mechanism based on DiffServ and IntServ

DiffServ can implement QoS on the backbone, but the completed end-to-end IP QoS should integrate DiffServ with a LAN-based QoS mechanism. DiffServ just introduces IP QoS into backbone, the details of QoS control is still implemented within the LAN. We have studied adaptive IP QoS based on IntServ within LAN[1,2], and present a framework for completed end-to-end adaptive IP QoS by integrating DiffServ and IntServ (see Figure 1). In this framework, DiffServ is applied on the backbone to implement the quick exchange while RSVP is applied on the edge of network and admission router to provide end-to-end services. In this way, we can implement the completed end-to-end IP QoS management.

According to the principle that "keeping the simple in the backbone and remaining the complexity in the edge", the adaptive IP QoS mechanism in the backbone should be simpler and more efficient than the one within the LAN. It executes the high efficient queue management mechanism and the queue schedule algorithms (e.g. WF^2Q+ and WRED) in the backbone; and remains the adaptive QoS adjustment (the work of complex and user-oriented) in the edge of networks.

As in Figure 1, before the LAN QoS flow enters the backbone of DiffServ, should pass the Boundary Router to integrate and map LAN QoS to few DiffServ PHB. And in the DiffServ field, implement adaptive IP QoS should be divided into two phases: implement adaptive IP QoS in the Boundary Router node and implement adaptive IP QoS in the Interior Router.

3.1. Adaptive IP QoS on LAN based on IntServ

The traffic that is submitted to DiffServ Boundary Router by users is named micro-flow, which would request QoS dynamic by RSVP in the Boundary Router entrance. So it is necessary to implement adaptive IP QoS based on IntServ.

Applications apply RSVP to request QoS services. The users should negotiate with their service provider about a Service Level Agreement (SLA). During the negotiation service provider must do some detection and disposal, such as user identity certification, available network resource and so on. In literature [8], we have discussed how to implement adaptive IP QoS based on IntServ in the LAN. The QoS priority-processing algorithm is based on multimedia priority session, and the transmission procedure adopts an adaptive QoS control algorithm. The QoS management works on the application level, so the QoS mechanism is separated with low level network protocol.

3.2. Adaptive IP QoS on DiffServ Boundary Router

In the framework, the DiffServ Boundary Router responds to receiving the QoS request message from different applications and implements the QoS integration. In order to implement adaptive QoS, the adaptive network configuration program is necessary in the procedure of mapping LAN QoS to DiffSev PHB. Figure 2 is the flow chart of implementation of adaptive QoS in DiffServ Boundary Router node.

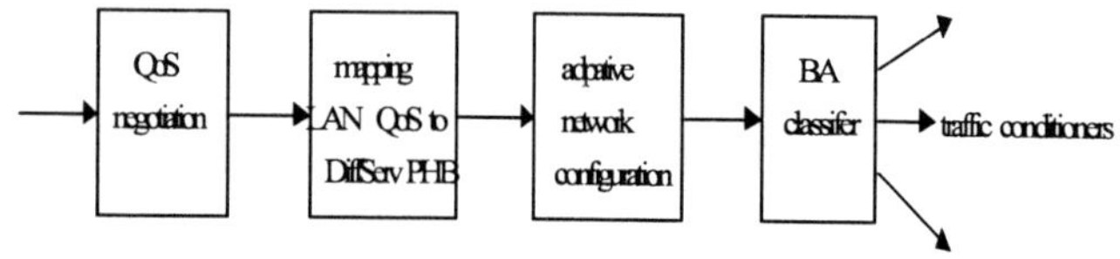

Figure 2. The flow chart of implementation of adaptive QoS on DiffServ Boundary Router.

1) QoS Negotiation

Boundary Router node locates the edge between the customer and the service provider. So it should provide the negotiation between the customer and the service provider, and then reach a service protocol, Service Level Specification (SLS), which defines the general characters and performance about the quality of service. Traffic Conditioning Specification (TCS) is an important subset of SLS, which points out the service parameters for each service level, which include:

- Service mark
- Parameters of service performance, such as the peak value transmission rate, average transmission rate, abandon probability and time delay etc.
- Service range
- The characters profile of traffic provided by service
- The measure for the traffic which is out of the range

SLS can be static as well as dynamic. The static SLS protocol is usually static, which is set by the administrator during the negotiation between the customer and the service provider. However, the SLS should be adjusted automatically by the machine to support adaptive IP QoS management.

2) Map LAN QoS to PHB

The Boundary Router collects QoS flows from different applications and aggregates to several Behavior Aggregates by using Classifier. Each BA is with the same or similar process. The traffic marked with DSCP would be mapped to the corresponding PHB in DiffServ. Different PHB is in relation to different exchange resource and priority schedule, and provides the corresponding QoS management

3) Networks Dynamic Configuration

It should configure network devices after SLS negotiation. The micro-flow request adaptive QoS by RSVP in the Boundary Router based on IntServ. In corresponding, both the network configuring and the mapping of LAN QoS to BA and PHB should also be adjusted adaptively along with the change of the real-time micro-flow in Boundary Router.

3.3. Adaptive IP QoS in DiffServ Interior Router

Figure 3 shows the flow chart of implementation of adaptive QoS in DiffServ Interior Router node.

1) Classifier

The responsibility of Classifier is that decides which logic output-flow the IP packet should enter according to the DSCP in the IP packet head only, but not consider the single micro-flow.

2) TCB (Traffic Conditioner Block)

TCB measures flux of input traffic flow, and compares the quantity with profile. It should adopt certain measure to dispose the traffics, which are out of range, to prevent the traffics from taking more resource than the allocated and to eliminate the influence to other traffic flow. TCB could smooth the burst traffic.TCB includes a shaper, which shape traffic delay until it fit for profile.

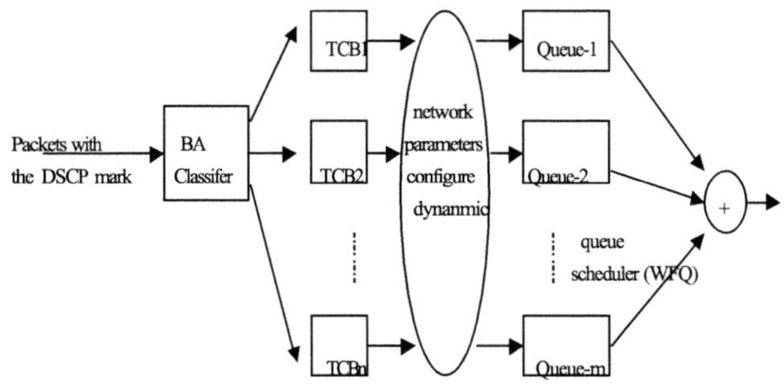

Figure 3. The implement flow chart of adaptive QoS on the Interior Router.

3) Network configuration dynamic mechanism updates the configuration based on the real-time measure. It can feedback the result of measure of current fix quality service traffic to the strategy server, and the bandwidth administrator and the strategy server recomputed the weigh value of different service queue in WFQ node according to the feedback, and then push the newest configuration to the Router. The adaptive algorithm of the mechanism is that measure the transmission rate of traffic and predict the next control time, then make a control strategy to adjust resource configuration and configure the IR node dynamic. The configurations include modify the TCS value of PHB, the value for abandoning packet of RED and the weighted value of WF^2Q+. It can modify the PHB behaviors in real-time by dynamic configuration, so to implement the adaptive IP QoS.

4. Summary

We propose a framework for completed end-to-end adaptive IP QoS by integrating DiffServ and IntServ, put forward the negotiation flow-chart of adaptive QoS on the Boundary Router and the QoS control flow-chart on the Interior Router. However, we now describe only a tentative framework of adaptive IP QoS mechanisms. Our future work is to design a network simulate software based on NS2 (Network Simulator version 2) of UC Berkeley[8] to implement and experiment the IP QoS framework.

5. Acknowledgement

This work is supported by China National Natural Science Foundation under grant No. 69983007 and the Guangdong Provincial Natural Science Foundation (Team project).

6. References

[1] Tang Yong, Yang xueliang, Ou haixiang. An adaptive QoS management mechanism based on Ip network. Chinese Journal of Computer, 2001, 24(1):32-39 (in Chinese)

[2] Tang Yong, Yang xueliang, Ou haixiang. Research and Implementation of QoS Mapping and Adaptive Negotiation Mechanism. Journal of Software , (in Chinese, (accepted))

[3] R.Guerin and V.Peris, Quality-of-Service in Packet Networks Basic Mechanisms and Directions, http://www. seas .upenn.edu /~guerin/publication/qos.ps.gz

[4] Paul Ferguson and Geoff Huston, Quality of Service in the Internet: Fact, Fiction, or Compromise?", 1998, http://www.employees.org/~ferguson/inet_qos.htm

[5]Internet2 Qbone: AtestBed for Differentiated Services Benjamin TEITELBAUM, http://www.isoc.org/ inet99/ proceedings/4f/4f_1.htm

[6] Moh W M, Nguyen B. An optimal QoS-quaranteed multicast routing algorithm with dynamic membership support. In: Proceedings of the IEEE International Conference on Communications, Vancouver, Canada, 1999.

[7] Ou Haixiang, The study of adaptive QoS over IP. Guangdong University of Technology [Master Thesis], 2000 (in Chinese)

[8] UC Berkeley, LBL, USC/ISI, & Xeerox PARC. NS Notes and Documentation, Oct. 20, 1999. http:// www-mash.cs.berkely.edu/ns/nsDoc.ps.gz

Research on Shared Objects Cooperative Manipulating Systems

Wei Li and Hantao Song
Dept. of Computer Science
Beijing Institute of Technology
Beijing, 100081, P.R. China
Email: liwei236@sina.com

Yisong Jiang
Dept. of Computer Science
Shandong University
Jinan, Shandong, 250061, P.R. China

Abstract

Computer Supported Cooperative Work (CSCW) embodies the interaction, distribution and cooperation of the work among people of this information era. Shared Objects Cooperative Manipulating System (SOCMS) is one of the most popular applications of CSCW. We present a framework of SOCMS based on the blackboard model. This framework is composed of one data blackboard and two control blackboards. The data blackboard contains the shared objects and the control blackboards contain the control knowledge. It resolves the concurrency, security of the shared objects and the identification of the roles of cooperators. We also analyze several key techniques of the SOCMS.

1. Introduction

By using computer technique, network and communication techniques, multimedia and interface techniques, Computer Supported Cooperative Work (CSCW) organizes many cooperators who are in different places and at different times, and also organizes their works. CSCW helps independent cooperators to accomplish tasks in distributed computing environments. Among CSCW applications, Shared Objects Cooperative Manipulating System (SOCMS) is particularly attractive. By using it, researchers may cooperatively compose research reports, software engineers may cooperatively write programs, and etc. Many researchers have done a lot of works in this field, and they have developed some prototype systems. But there are still a lot of works to do in many fields, such as the cooperative mechanism, the concurrency control, the cooperative awareness, and etc.

Expert system is the most dynamic branch in Artificial Intelligence (AI).[1] Blackboard model is one important expert system structuring method. By associating the blackboard model with the SOCMS model, we present an effective method to resolve the Shared Objects' coherence, security and concurrency in synchronously cooperative operations.

2. Shared objects' cooperative manipulating techniques

SOCMS definition is: in the shared environment that is composed of multi-user interfaces, the system that is based on computers and allows users to cooperatively manipulate the shared objects is called SOCMS. Keywords of this definition are "the shared environment", "the shared objects", and "cooperatively manipulate". These three elements are SOCMS basic features. By analyzing several existing prototype systems, we conclude SOCMS's main functions as follows:

- Shared workspace: the shared workspace offers users the tools to manipulate the shared objects, such as creating, modifying, deleting and storing. It also copies the shared objects to the private workspace.
- Private workspace: private objects in the private workspace may be independent of the shared objects, and have different structures from the shared objects' structures. This workspace provides the tools to manipulate private objects, such as creating, modifying, deleting and storing. And this workspace also converts private objects to shared objects in the shared windows.
- Seamless linking between the private workspace and the shared workspace makes it easy to convert objects between these two kinds of workspaces.
- Providing effective synchronously or asynchronously cooperative mechanism for cooperative manipulations, and inspecting the cooperation's executing.
- Providing convenient dynamical login in or login out cooperatively manipulating mechanism for users.
- Offering synchronously or asynchronously communicating service for multi-user to exchange information.
- Offering the ability to manage the shared objects' basic versions.
- Offering abilities to verify and check users' IDs.

3. SOCMS based on the blackboard model

Blackboard is an information buffer. It is also the location where every subtask exchanges information.[2] Driven by the main control program, those Knowledge Sources read information on blackboards, and output their generated information to those blackboards. The common blackboard structure has three parts:

1) Knowledge Sources (KSs): the special knowledge of one application field is divided into several independent KSs, and each KS is responsible for one kind of tasks.
2) Blackboard: the problem's answer space, and a database organized in the hierarchy mode. Blackboard is shared by all KSs. Only through the blackboard, can communications and interactions between KSs happen.
3) Control: the control program dynamically selects and activates an appropriate KS, and makes this KS correspond to blackboards' changes in time.

3.1. The double-blackboard structure

In a blackboard structure, if the main control program has many control elements, these elements all have their own duties. And if there is one superior control program at the back of these control elements, and controls their scheduling and applications, we may use the double-blackboard structure: the superior control program's control elements are called Control KSs. And the original KS is called Field KS. We create a new blackboard, Control Blackboard, and write Field KS on this new blackboard. Control KSs access and modify Field KS. The original blackboard is called Data Blackboard. Here, the superior control program drives Control KSs. Control KSs manipulate and drive Field KS, and Field KS manipulates the field knowledge, which is in the sub-area. The double-blackboard structure is shown in Figure 1. This structure is more flexible than the blackboard structure.

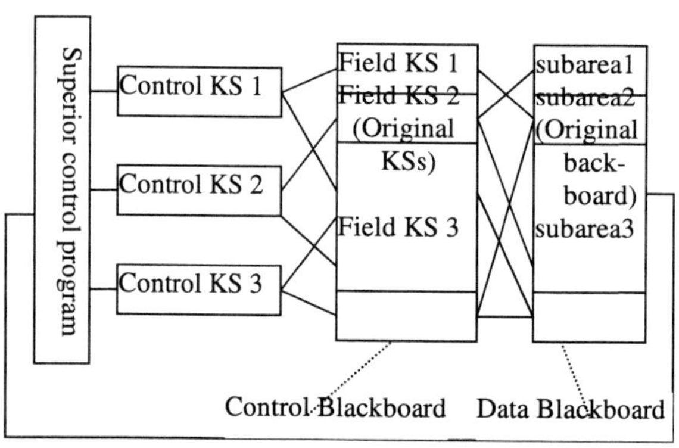

Figure 1. The double-blackboard structure.

3.2. Shared objects cooperative manipulating model based on the blackboard

We believe the shared objects' cooperatively manipulating supports may be accomplished by adding a series of service layers to the shared objects, as shown in Figure 2.

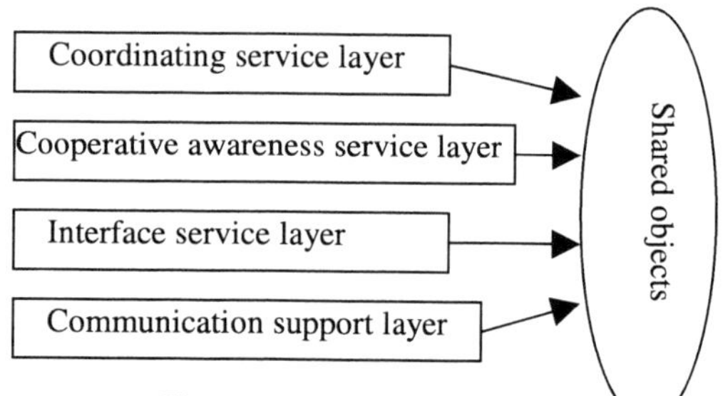

Figure 2. Hierarchy.

- Coordinating service layer provides the following functions: 1. the accessing control checks users' identifications and privileges; 2. the concurrency control guarantees the coherence of the shared objects.
- Cooperative awareness service layer mainly offers the shared objects' cooperative awareness service.
- Interface service layer offer different showing modes of the shared objects for different users.
- Communication support layer provides necessary system communicating means, such as real-time video, audio, and etc.

To implement the above structure, we use the blackboard model. The framework is shown in Figure 3. In this figure, our SOCMS definition is as follows:

SOCMS::= (<Data Blackboard>,
<Control Blackboard>,
<Data Base>,
<Knowledge Base>)
<Data Blackboard>::=({<Shared Object>})
<Control Blackboard>::=(<Blackboard B>,
<Blackboard C>)
<Blackboard B>::=(<Identification Control>,
<Coherence Control>,
<Concurrency Control>)
<Blackboard C>::=(<Awareness Control>,
< Interface Control>)
<Data Base>::=(<DB of Objects>)
<Knowledge Base>::=(<KB of Cooperators>,
<KB of Objects>)

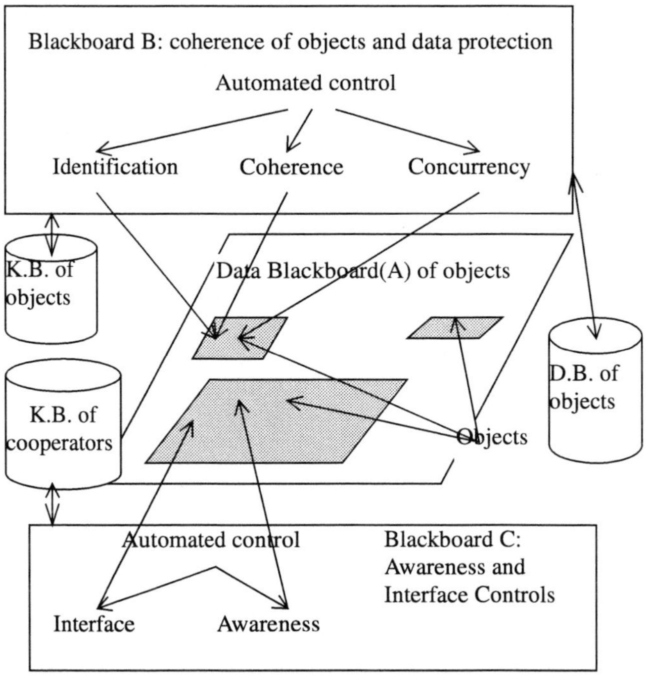

Figure 3. SOCMS based on the blackboard.

Shared objects being manipulated by cooperators and KSs are stored in Data Blackboard, also Blackboard A. Objects are stored in a database, and the database is on the shared objects' data server.

Blackboard B and C are Control Blackboards. Blackboard B has three main KSs: 1. Identification Control is the identification's verifying control. It is used to verify the identification of the cooperator who is going to interact with Blackboard A. Identification Control guarantees the system's security. Before manipulating every object, the Identification Control operates Blackboard A to check whether one cooperator has the object's manipulating privilege. The manipulating privilege information is stored in the Knowledge Base. 2. Coherence Control guarantees shared objects' coherence. After object's every creation, modification, and etc., the Coherence Control checks the resulting objects' coherence. The Coherence Control is associated with a database that stores the shared objects' information. 3. Concurrency Control controls objects' concurrency. It selects an appropriate concurrency algorithm to avoid objects' concurrency on every client side.

Blackboard C contains the Interface Control and the Awareness Control. They offer the shared objects' interface services and cooperative awareness services. They display the actions of one user interacting and manipulating the shared objects on other cooperators' interfaces. Blackboard C manages the message system that is associated with Blackboard A. After the data modification, Blackboard A notifies other cooperators.

Cooperators may manipulate the shared objects on Blackboard A. They may synchronously or asynchronously cooperate with each other.[3] In the 1st case, the Concurrency Control manages cooperators' sessions to avoid their concurrency. In the 2nd case, each cooperator directly accesses the shared objects on Blackboard A (Here, cooperators have no real-time interaction, but the cooperation does exist).

Now we introduce the system's workflow: we create a shared objects' data server to serve as the shared objects' prominent storing place. When a cooperator begins to work, he copies the shared objects and put the copied ones on the local workstation. During the cooperation, he can only manipulate the local copied ones. We use the event handling mechanism to manipulate shared objects. Events are divided into local events and remote events. Control Blackboard's each control program resolves those events. It also controls the concurrency, verifies identifications, controls the awareness, and etc.

3.3. Discussion of several key techniques

3.3.1. Shared objects distribution. In SOCMS, shared objects may be text documents, work schemes, images, multimedia data, and etc.

Shared objects have 2 common distribution modes [4]: the centralized mode and the distributed mode. The centralized mode follows the client/server mode. Shared objects are put in a central place, the server. Cooperators' manipulating requirements are sent to the server, then, the server manipulates the shared objects and broadcasts the results to cooperators.

In our SOCMS model, we use the distributed mode. Blackboards and shared objects' copies are distributed on every workstation. All cooperative workstations have the shared objects' copies. The message of a cooperator's manipulating the shared objects is simultaneously multi-transmitted to every machine. The server manipulates the shared objects. Every workstation manipulates the shared object's copies. This mode greatly reduces unnecessary network transmitting consumption and users' corresponding times.

3.3.2. Designing and selecting concurrency control algorithm. In our model, Blackboard B controls the concurrency problem. In Blackboard B, the Identification Control implements the accessing control, and the Concurrency Control implements the concurrency control.

Prototype system uses the lock mechanism to control the concurrency.[5] There are two kinds of locks in our system: the shared lock and the exclusive lock. When reading shared objects, we use the shared lock. When writing shared objects, we use the exclusive lock. After releasing a lock, the current lock's mode becomes

NO_LOCK. The rule of requiring locks is shown in Table 1.

Table 1. The rule of requiring locks

Current lock mode / Requiring Lock mode	No Lock	Shared Lock	Exclusive Lock
READ	OK	OK	NO
WRITE	OK	NO	NO

Our system maintains a lock information table. This table records the shared information of every cooperator. When a cooperator wants to access a shared object, he firstly checks the lock information table on the server, then decides whether to require the lock or not, and records corresponding information in the lock information table.

3.3.3. Cooperative Awareness. Shared objects' collections constitute cooperators' real shared workspace. Awareness in this level is called the shared objects' cooperative awareness. Research of the CSCW cooperative awareness mainly focuses on this level. [6] In Figure 3's framework, Blackboard C controls the shared objects' cooperative awareness service. We use the event's handling mechanism to implement this function. Blackboard C's procedure of resolving event message may be as follows:

```
{Accepting event message;
CASE event message type:
  Local shared objects manipulating message
    :locally executing shared objects;
    shared objects' concurrency and awareness
    services;
    transmitting message to other users;
  Remote shared objects manipulating message
    :shared objects' concurrency and awareness
    services;
     locally executing shared objects;
  Interface control message:
    locally modifying the interface
ENDCASE
}
```

4. SOCMS implementation

In our lab, we implement this SOCMS. Our SOCMS uses the distributed architecture. The LAN uses Fast Ethernet. The network OS uses Microsoft Windows NT Server4.0 and Microsoft Windows 98, and uses the TCP/IP protocol. Our lab's LAN uses Intel Express 10/100MB Stackable Hub. Through a gateway, the LAN connects with the campus's ATM network. The developing environment of the software is Microsoft Visual C++ 5.0. Every machine is equipped with video and audio cards, microphone, sound box, video camera, and etc. so as to offer abundant tools for coordination among cooperators.

Our system allows multi cooperators to share any common and originally single-user program and interact with other cooperators. Our system also supports the real time audio and video interaction between users. During cooperator's operation, he may feel others' existence and actions.

5. Conclusion

In order to promote the shared objects' cooperative manipulation, we present the SOCMS model based on the blackboard theory, and discuss several associating key techniques. Although it's difficult to implement group level cooperation utterly by computers, we believe with researches' helps, CSCW techniques and products will have brighter futures.

6. References

[1] Shengfeng Tian, and etc., Artificial Intelligence Theory and Application, BIT Publisher, 1993.8

[2] Rujin Lu, Artificial Intelligence, Science Publisher, 1996.9, First Edition.

[3] Xianlin Ma, Zongkai Lin, and etc., "Asynchronous Cooperation's Implementing Method and Application", Small and Mini Computer System, 1997(9), 15-18

[4] X. Wang and G. Zeng, "MCCDS design and implementation of SOCMS based on AGENT", *Computer Engineering and applications*, 1998(9), 49-52

[5] Greeberg S, Marwood D, "Real Time Groupware as Distributed System: Concurrency Control and its Effect on the Interface", Proceedings of CSCW'94, 207-217

[6] Chenhua Feng, Jie Xu, and etc., "Research of Cooperative Awareness Technique in CSCW System", Small and Mini Computer System, 1998(9), 61-67.

HDBIS Supporting E-Collaboration in E-Business

Lizhen Liu, Hantao Song,Yanmei Liu
Beijing Institute of Technology, Beijing, 100081, P.R. China
songhantao@hotmail.com

Abstract

This paper introduces the E-Collaboration in E-Business. It stresses that HDBIS (Heterogeneous Database Integration System) plays an important role for supporting E-Collaboration in E-Business. HDBIS provides a crucial technique for integration of heterogeneous information source in E-Business. The paper further displays the E-Collaboration model supported by HDBIS through application of clothing companies, which include three layers: manufacturers, sellers and customers. The system architecture, functions and characteristics of HDBIS are presented. The implementation and model of HDBIS using COM/DCOM and ASP is also introduced.

1. Introduction

Currently, E-Business has entered into China. It is an existing and developing necessary condition for enterprises to develop E-Business. E-Business = IT + Business [1]. Developing E-Business cannot depart from network environments and information supporting. E-Business always begins with providing information in the net. It is necessary for E-Business entities to connect with other E-Business entities during the activities of E-Business. For E-Business, people pay more attention to the whole process of business activities of E-Business entities (enterprises and consumers) in the network en vironments. E-Business includes the information exchanges among enterprises, consumers, government departments and banks, and so on. The whole process of E-Business relates to information collaboration among many aspects, so it is also named with E-collaboration. E-Collaboration must be considered during the process of establishing complete E-Business strategy. E-Business cannot go on without the support of E-collaboration.

E-Collaboration involves many fields and has deep effects on the development of enterprises. So it is the base for enterprises to reap highly efficiency of operation and high quality of services. Without high efficient E-collaboration, expensive electronic trading systems of enterprise can not make profits [2].

For example, practicing E-Business between costume sellers and costume makers needs the information exchanges and E-collaborations between business entities. At first, costume sellers should provide costume information, to consumers using network environments, such as costume types, styles, materials, makers and prices. Actually, costume makers provide such information to costume sellers. On the other hand, sellers will collect ordering information and transfer it to costume makers. Continually exchanging information and providing are E-collaboration, which is the base of E-Business for costume manufacturing and selling. Costume makers' products information is produced using E-Collaboration inside enterprises by several departments of enterprise, such as designing, planning, material stocking, producing and storing department. Since different makers use different database systems, costume sellers must integrate heterogeneous databases provided by several costume makers. In logic, costume consumers need a friendly interface to get costume information provided by single database. HDBIS *(Heterogeneous Database Integration System)* developed by our research group can support this kind of application, which is the supporting technology in E-Collaboration of E-Business and has been applied in the costume selling and manufacturing system in Wenzhou, China.

2. E-Collaboration supported by HDBIS

2.1. E-Collaboration Supported by HDBIS

The technology of HDBIS belongs to the field of federal databases. It is one of the current development trends of databases. Its many problems are similar to distributed databases. But the member database systems are different kinds of databases and each member database system has the autonomy. There are many kinds of databases used in the applications of E-Business and the HDBIS can get information from heterogeneous information sources. So, It is very useful for E-

Collaboration in E-Business. For example, Figure 1 is an application model based on HDBIS. It has three layers: customers, clothing sellers and manufacturers.

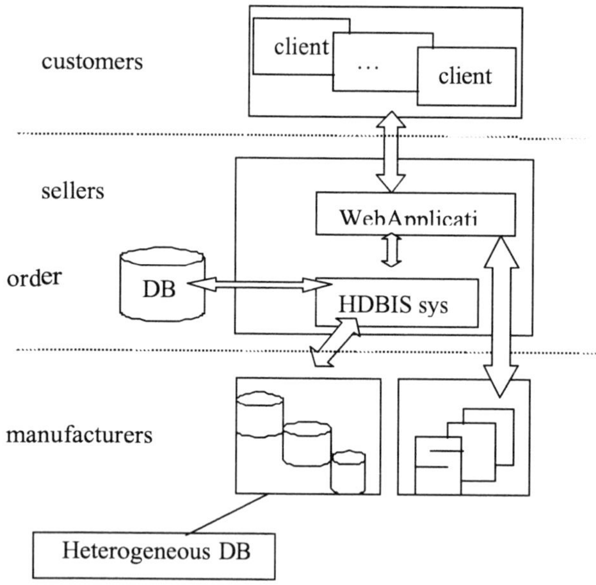

Figure 1. Model of an application based on HDBIS.

HDBIS provides convenient retrieval of clothing information for customers, and receives ordering information from customers.

2.2. Characteristics of E-Collaboration supported by HDBIS

1. It can use the Internet to exchange information efficiently, expand the information exchange scope of enterprise and expedite the process of commerce flow.
2. It can make enterprise to keep in touch with customers and providers with lower costs. It also makes enterprise to face the more global competition environment.
3. It can enhance the business exchange efficiency of enterprise.
4. It helps enterprise to achieve their work more efficiently in a wider scope with more business counterparts.
5. It helps enterprise to distribute their resources more rationally and reduce their circulation cost.
6. It provides a uniform environment to work together according to time, locale, staff and information.
7. It provides collaboration workspace easy to integrate, which can make use the advantages of other Information systems.

3. Architecture, functions and characteristics of HDBIS

3.1. Architecture of HDBIS

HDBIS consists of 9 parts. They are ASP Input Pages Management, Compiler Component, Decompose Component, Parallel Process Component, Query Optimizing Component, GDD (Global Data Dictionary) Management, Result Collection, Error Process, and ASP Result Pages Management [3]. The architecture of HDBIS is showed in Figure 2. The GDD is the description of databases which are integrated in HDBIS. The ordinary database systems (Oracle, Sybase, Informix, MS SQL server, DB2, Foxpro) are included in HDBIS. Users enter GSQL command to access heterogeneous databases. ASP technology allows a browser to communicate with a Web server as follows:

A browser sends a request for an ASP page that contains script language instead of an HTML page. MS Web Server accesses the ASP file, explains the script in it and creates standard HTML, then sends the HTML file back to the browser. Another excellence of Active Server is the code security. What a page visitor sees is the result processed by Web server. So HDBIS uses ASP to collect browser information, to collect and process form data input by users. GSQL commands can be formed by users' selecting menu.

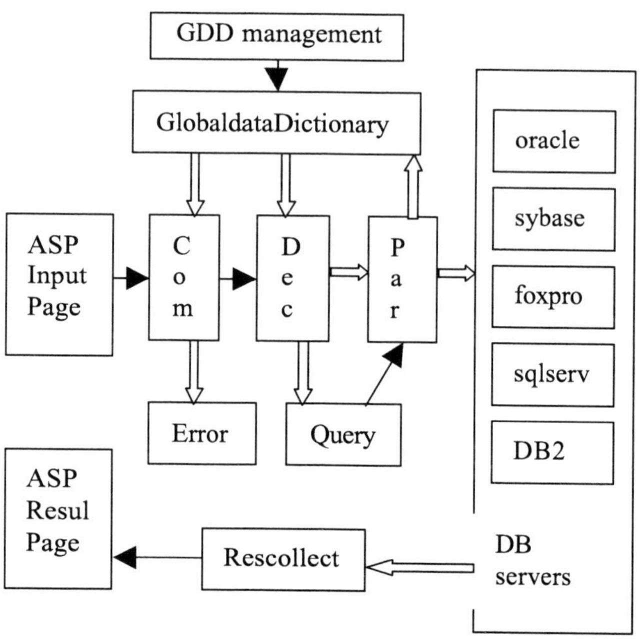

Figure 2. Architecture of HDBIS.

3.2 The functions of HDBIS

HDBIS has the functions [4] as follows:

1. Global Data Model
   It is the foundation for users to access different kinds of databases transparently.
2. Recognizing Data Incompatibility
   H,DBIS can unify the data types and units for different database systems.
3. Global Data Language
   HDBIS uses Global SQL to access heterogeneous databases according to SQL92 standard.
4. Query Decomposition
   The global query with GSQL is decomposed into subquerys, which are in the SQL standard.
5. Query Translation
   Sub query is translated into the form which can be sent to local DBMS systems.
6. Query Optimization
   Query optimization can improve the query speed.
7. Processing Query Result
   HDBIS can collect the results sent by local DBMS and unify them to display the final query result.
8. Security Control
9. Data Conversion

3.3. Characteristics of HDBIS

1. Support of direct and convenient query.
2. Support of distribution.
3. Support of hybrid environment.
4. Support of system dynamics.
5. Support of the sharing of information and knowledge.
6. Support of collaboration.

4. Implementation of HDBIS

4.1. The global data model and local data model

Because different companies use the different data views to support the cooperation in E-Business, HDBIS adopts the Global Layer Data View, Global Concept Data Model, Local Concept Data Model, and Local Inner Data Model showed in Figure 3. The Global Layer Data Model can support the information share in E-Business.

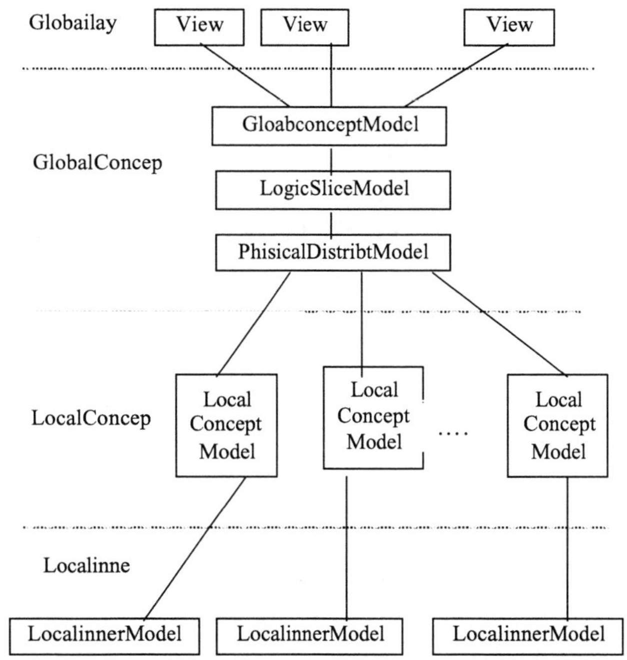

Figure 3. Global Data model and local data model.

4.2. Implementation of HDBIS

HDBIS supporting E-Collaboration in E-Business is a highly complicated application and with the development of IT, new info style will emerge incessantly. So it will make the supporting more simple, flexible and dynamic to develop the application based on thee COM/DCOM standard put forward by Microsoft. ASP introduces ADO (Active Data Objects) to access database background. ADODB embedded in ADO is a COM component of database application. Script language calls ADO components to access database and display results through corresponding interface. ADO is a data access interface of perfect performance based on OLE DB and ODBC, which provides an open data manipulation object model. HDBIS combined HTML, script and ActiveX components into a dynamic and powerful web application. What is especially deserved to refer is that a ActiveX would be programmed in many languages such as VB, VC, Delphi, Power Builder, Java, and so on.

At first, HDBIS sets up GDM (Global Data Model) and maps all heterogeneous databases brought into united use to global member (Yao 1998). This makes users feel that there is only a single global database, therefore it's the base of transparent use of heterogeneous databases. GDM is implemented by GDD (Global Data Dictionary) which describes the distribution of databases on network such as a global table corresponding to local tables, a field in a global table corresponding to fields in local tables, node on which local databases locate and so on. It's the base of global operations. GDD management communicates with databases in order to get the changed information and update the GDD. Users can integrate heterogeneous databases by defining the GDD.

HDBIS provides users with GSQL which is composed of DDL (Data Definition Language) and DQL (Data Query Language) corresponding to global data model. GDDL includes creating, altering and dropping global table. After the definition of GDB, the user can manipulate heterogeneous databases with GSQL according to SQL92.

Pretreatment including accidence and syntax checking is used GDDL and GDML to insure a correct GL sentence processing.

Query decomposition and interpretation will decompose and interpret a correct global query into local database language such as substitution of table name, field name, function name and much more intricate process in hetero generous table join operation. Then local query is sent to local DBMS.

Parallel query processing is an important part of HDBIS. In an application sending out multi-query to multi-DBMS which complete the corresponding local query. We developed a multi-thread component to process multi-query in parallel.

Query optimization is used for heterogeneous database join. We established the effective semi-join operation and the reducible join character processing component to realize the query optimization. So a global query can be finished in 0.3 second.

Query result processing includes data conversion which transforms local query results into global database data form and output process.

As mentioned above, ASP access databases through ADO component based on OLE DB. OLE DB provides interfaces for accessing various types of data sources, such as documents, Excels, E-mails and multimedia. Multimedia is treated as BLOB or OLE object in OLE DB.A user makes the use of store object to access multimedia data as following:

1. Create an accessor processing a binding of store object.
2. Call the method of Irwset::GetData to create a store object corresponding to the multimedia data and return the pointer of the object.
3. Call the method Isequential Stream::Read to read the data. If you want to write, the method should be Isequential Stream::Write.Here multimedia data is treated as an object other than in RDBMS stored in a BLOB field.

5.Conclusion

Combining ActiveX component with ASP is a good way to the integration of heterogeneous databases to support the E-Collaboration in E-Business. HDBIS plays an important role in E-Business. It has applied and got achievement in manufacturing and sales of clothing

6. Reference

[1] Yanmei Liu, Hantao Song. Integration of Heterogeneous Information Sources in Electronic Commerce, Proceedings of Fourth International Workshop on CSCW in Design, Compiegne, France, Sep.29,1999:95-98.

[2] Ning Wing, Implementation of query decomposing and optimization in heterogeneous database system, software journal, china 2000, 1[11]:222-228

[3] Yamei Liu, Heterogeneous Database Integration Technology, Ph.D thesis, BIT 2000.

[4] Lingzhong Yao, Research and Implementation of Union Use of Heterogeneous Database, Ph.D thesis, BIT 1998.

Part III: Collaborative Design

Co-DARFAD—the Collaborative Mechanical Product Design System

Kezheng Huang, Xudong Li, Shukun Cao, Bo Yang, Wei Pan
Province Key CAD Lab., Shandong University, Jinan, P.R.China
cad_center@sina.com

Abstract

Based on a newly developed mechanical product design automation system DARFAD, the D&R (Decomposition & Reconstitution) model based collaborative design concept is proposed first in this paper, and the product development model and collaborative design process model are described. Then the framework of the DARFAD based collaborative design system Co-DARFAD is introduced which is characterized of truly "collaborative design" (not CSCW in design) and formal collaborative design process, lending itself to both loose-coupled and close-coupled collaborative design work in both synchronous and asynchronous modes on Internet. Finally, the conceptual design for mechanical product is shown as an application case.

1. Introduction

With the increasing of global competition, it is important to make the product design process as fast and efficient as possible. Large design problems often require specialized knowledge from many different fields. One way to achieve improvements in efficiency and innovation is through better collaboration of the designers working on a common design task, each with specific core competencies, interacting in the design process, so that unnecessary backtracking and delays are avoided. In an integrated, concurrent design environment, designers interact and reach agreement by sharing common information. Designers bring their own personal viewpoints to the integrated vision of the product design process [1]. This requires us to develop the collaborative design environment and explore the corresponding collaborative design theory so as to facilitate several designers to do creative design work in synchronous and asynchronous ways.

With the development of computer network and Internet, collaborative design is getting more and more attention. However, most efforts are not for truly "collaborative design" but computer tools to assist communication and data sharing in design. The design depends heavily on human designer. Such as the researches on computer supported cooperative design environment and distributed multi-agent design systems.

The DICE [2] system is a centralized multi-user system which pools data and models so that consistency can be maintained. The object-based modeling environment SHARE [3] is a less structured design information system allowing users to exchange engineering data or standard product description files but not providing concurrent system modeling functionality. Agent-based distributed systems provide the most abstract level of system interaction and interoperability, such as agent network and distributed blackboard architecture.

DOME [1] is intended to link heterogeneous design models and tools, assisting designers in evaluating the system performance of design alternatives and making decisions. It maintains generalized communication objects (i.e. modules) and mediates the propagation of explicitly defined messages containing design information. But it does not provide support for alternative generation and design synthesis in conceptual design stage

Actually rapid and wide application of Internet make the design information explode and how to make good use of it efficiently in design is the bottleneck in collaborative design. But in the mechanical product design field, due to complicated design process there are still no effective design theory for automation of product design in true top-down style and with operability, let alone for the collaborative design work on Internet. Further study is needed comprehensively and systematically.

In this paper, based on our own theoretical research and technical development work, Internet and D&R model based collaborative design is discussed, and then the new product and collaborative design process model given, and finally the Co-DARFAD system framework is introduced for conceptual design of mechanical products.

2. Internet and D&R model based collaborative design

The product design process in an interactive single-user CAD system is also the basis for multi-user collaborative design system on Internet. Due to the complexity of design work, interaction between designer and CAD system is vital for successful results with high efficiency, in which designer provides his/her design expertise and computer system will process large amount of data at high speed. The designing process in a CAD system has two problems. One is that each discrete design step has no obvious functional meaning to other designers, and the other is that each meaningful technical design period is relatively longer than needed for effective collaboration asynchronously.

While emphasizing collaborative design on Internet, people have neglected to some extent the development of design theory and artificial intelligence to improve the technical design process in traditional interactive CAD system and to further build new Internet based collaborative design system.

Actually rapid and wide application of Internet makes the design information explode and alternative schemes increase more and more rapidly during design process. Without the effective aid of computer system, human designer will be in greater trouble for best design solutions. The current collaboration design research is based on human decision-making and design scheme evaluation. But in each scheme there are still a large number of problems to be verified. To rely all these synthesis work on human designer is inadequate.

To what extent we should develop artificial intelligence and use them in collaborative design system remains to be addressed. The automatic generation and synthesis of design alternatives are the two of many aspects that can facilitate collaborative design on Internet. On one hand, new collaborative design theory needs to be studied, on the other hand the collaborative design system to reflect these factors need to be developed.

In the following sections, models and frameworks based on Internet and D&R model have been developed according to the collaborative design requirements with the computer system accomplishing some automatic synthesis and collaborative work.

3. Product and design process model

3.1 Collaboration and design work

Analyzing further the roles of and relationships among the participants in collaborative design will help to develop the right collaborating tools. Cuff [4] has observed that participants in collaborative design "did not necessarily participate equally or collaboratively." From their own experience, Sudweeks and Allbritton observed that simply working together or talking about the same subjects does not make the act collaborative [5].

Kvan[6] discussed the nature of collaboration and the important distinction between collaboration and cooperation. Design collaboration requires a higher sense of working together in order to achieve a holistic creative result. It is a far more demanding activity, more difficult to establish and sustain, than completing a project as a team.

Mattessich and Monsey [7] have studied the words more thoroughly. Cooperation is characterized by informal relationships that exist without a commonly defined mission, structure or effort; Coordination is characterized by more formal relationships and understanding of compatible missions. Some planning and division of roles are required, and communication channels are established; Collaboration connotes a more durable and pervasive relationship with full commitment to a common mission.

Most design projects bring teams into a relationship of co-operation or co-ordination, instead of collaboration. While focusing on collaboration, we risk over-emphasizing particular features which may not be needed. For example, synchronicity may be unnecessary for much co-operative work. In place of striving for closer coupling of systems and tasks, we should perhaps pay greater attention to co-operation environment.

For collaborative design study, the basic design problems must be discussed and coped with appropriately. In **creative design** when more ideas and knowledge are required and comprehensive consideration are needed, we typically think of the design process as a continuous close-coupled thinking process in which the participants work closely to realize a design. Here the participants work intensely with one another, observing and understanding each other's intentions. The observer cannot identify a discrete contribution to the design product from one participant or the other.

In **routine design,** there is a relatively well-structured solution or top level scheme for the product to be developed, and the design work can be divided into sub-tasks. Here design is in fact loose-coupled, with each participant contributing what they can in different domains of expertise at moments when they have the knowledge appropriate to the situation. In this situation, we see two or more experts operating in their own domains on a shared problem. The design moves in discrete steps that we can see a step and identify what has happened during that step.

A loose-coupled design process requires a very much different set of tools and conditions than a close-coupled one. The over-riding desire to support collaboration leads to high-bandwidth demands to support close visual linking in order to replicate co-location [8,9,10]. Now that we have over a decade of experimental data, it is becoming apparent that these simulations of co-location do not lead to better work product outcomes [11].

3.2 Generic structural design approach

In the design process of mechanical product, the key stage is the mapping process from functional requirements to product structure, in which a series of design activities performed in sequence.

To facilitate the collaborative design in the structural design process based on D&R Principle[16], we standardized the various design activities so that the mapping process (i.e. Design history) can be easily preserved. Partial categories of design activities and corresponding graphical representations are shown in Fig.1. The fundamental discrete activities at conceptual design stage are a kind of "cell division and growing" process.

Applying D&R principle to product design, a Generic structural design approach (GSDA) was set forth [12] to integrate concurrent design and axiomatic design concepts in a unified and structure-oriented automatic design process for mechanical product. For instance, the concept of Qualitative Closing Space (QCS) was proposed as a theoretical basis for qualitative assembly condition. In the process of decomposition and reconstitution, many alternative schemes occurred. Due to lack of selection knowledge, dynamic tree is used to record multi-scheme data and their relationships. The design process is driven by the assemblability evaluation and proceeds as a growing process of the multi-scheme tree as illustrated in Fig. 2.

The product development model is proposed for design information recording and management during design process, and serves as basic structure for collaborative design. It is a three layered hierarchical structure with Product Series layer at top, Product Schemes layer at middle, and Product Structure layer at bottom as shown in Fig.3.

(1) On top Product Series layer, classification of products is made for better utilization in new design process;

(2) On middle Product Schemes layer, alternative schemes and development history during product conceptual design stage have been generated and recorded as a tree shown in Fig.2 by GSDA method;

(3) On bottom Product Structure layer, product structure in one design scheme is listed for detailed collaborative design.

At conceptual design stage, the goal is to make creative design. Making use of existing product data and searching related product series is important. Generation and selecting of new product schemes is also critical for innovative design. The Product Schemes layer provides a better structure for collaborative design. All participants can contribute to one product scheme generation in some sequence. Since the design time is greatly cut down, collaboration can be made synchronously. And due to the design scheme tree structure, collaboration can also be made asynchronously.

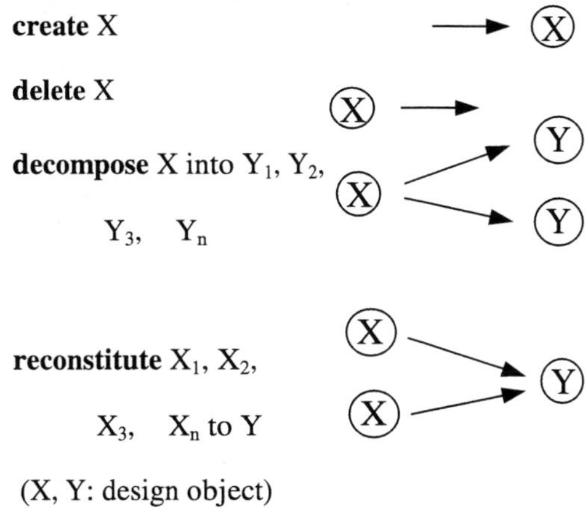

Fig.1 Partial Design Activities Standardized in DARFAD System

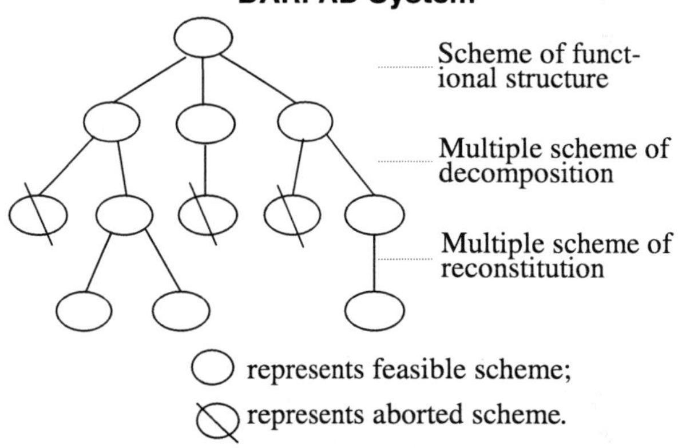

Fig.2 D&R Principle based design process model

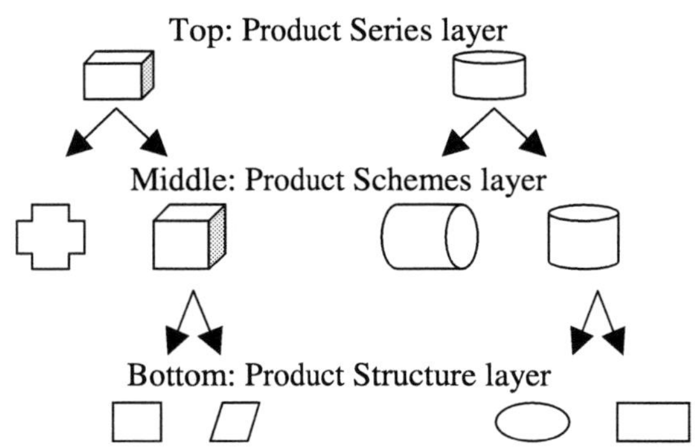

Fig.3 Product development model: three layered hierarchical structure.

At detailed design stage, Product Structure layer provides a well-structured model in which participants can cooperate by selecting appropriate part or sub-assembly to work on.

3.3. Collaborative design process

Gero and McNeill[13] have shown that design can be seen as a series of discrete activities. Collaborators work together for moments, then divide up and go their separate ways[14]. Papers[9,10] suggest that collaborative design consists of parallel expert actions, each of short duration, bracketed by joint activity of negotiation and evaluation. Thus the design activity itself is discrete, individual and parallel, not intimately linked.

The categorization of design collaboration is identified by Maher et al. in their experiment in collaborative design. Mutual collaboration, in which the participants are "busy working with the other"; Exclusive collaboration, in which the participants "work on separate parts of the problem, negotiating occasionally"; Dictator collaboration, where the participants decide who is "in charge" and that person leads the process. Maher et al. note that the "exclusive collaboration" model is the most effective with most productive results. Mutual collaboration led to no result at the end of a busy exchange between the participants, whereas dictator collaboration came to a conclusion as soon as the leader made up her mind.

When collaborating in **asynchronous mode**, one designer can undertake product design via some CAD software system such as AutoCAD, and the corresponding design activities can be recorded by design agent and produce log files. The design agent can form one backup of the log files and other design files such as DWG files. Other designers who join the design in geographically distributed location may register to the ACM , so he has the right to get the backup of log files and other design files. We will also develop a transition interface that can realize the display of design result on web based browser such as IE or Netscape via VRML so that the interested designers can browse the original designer's work without the limitations of hardware and software.

When in **synchronous mode**, several designers can carry out the same design task simultaneously. Here we should mainly solve the two problems: first, how to manage the whole design project; second, how to facilitate communication among designers in the design process.

4. Framework of Co-DARFAD system

The framework of collaborative design system Co-DARFAD built up according to previous concepts is shown in Fig.4. This system allows designers from different fields as well as customers to take part in design activity. The key components of this system include GSDMP-Agent, Design Agent, Agent Cooperation Manager, and Database.

The Agent Cooperation Manager (ACM) monitors and coordinates the activities of the surrounding agents. The GSDMP-agent implemented by extending GSDA to other design requirements is a powerful tool for conceptual design. By the aid of GSDMP-Agent, the designer can create a variety of conceptual structure schemes. The corresponding scheme of conceptual structure is similar to the figure 2. Design agent is mainly used to support the communications with other design agents. Every designer must firstly register to the ACM before taking part in collaboration. It can also convert design scheme to VRML files and store them in common database.

The customer and other participants who are not concerned with product design can directly browse design objects through VRML browser, and summit evaluation information and provide interactive function through Java Applet. All the agents in this system are semi-automated agent that can interact with designers. So when we discuss the interaction among design agents, it is equivalent to the interaction among designers.

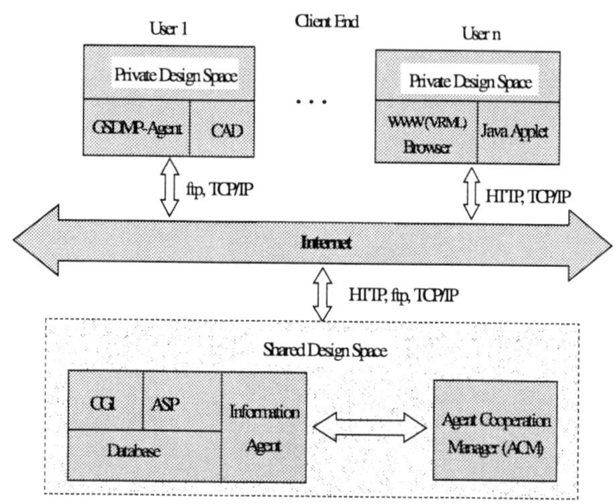

Fig.4 Framework of Co-DARFAD System.

As for the visual representation and communication of product concept, VRML is an ideal tool for data visualizing permitting 3D interaction, platform-independent, low-cost, and of high performance. Functional surface based representation of product concept can be displayed and manipulated by user on WWW browser. The concepts are represented by Functional Surface and it is easy to convert to VRML file, which may be displayed on WWW Browser. The corresponding infrastructure built for the VRML based communication on Internet is shown in Fig.5. Through network communication, design team members can view VRML based visual product concept, discuss problems and exchange design ideas mutually at the same time, so that this mode has intuitive, live, high-interactive characteristics.

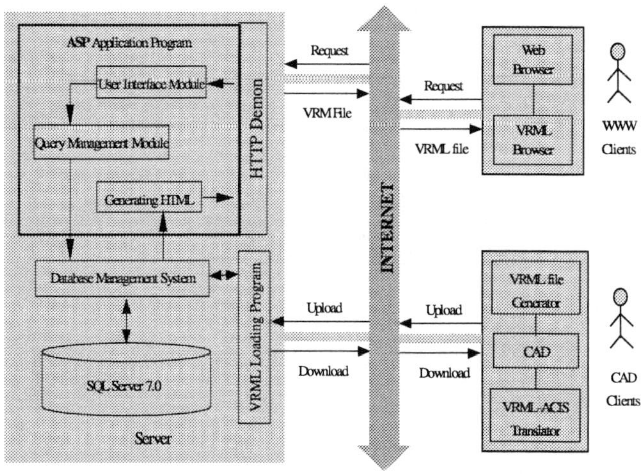

Fig.5 Infrastructure for the VRML based communications among designers.

5. Conceptual design for mechanical product

Product concept is the approximate description of product structure. The generation of product concept should give full play to participants' creativity and collaborative efforts and excavate the potential of computer in supporting conceptual design. This representation scheme should take full advantage of the character of visual thinking in the ideation of mechanical product spatial structure.

Functional Surface based product concept representation scheme [15] was proposed and D&R principle based scheme generating process [12]. Due to the relatively independence of each design activity and growing process, this approach is suited to developing the initial conceptual representation. Participants may independently conceive product concept in his private design space. Thus, this stage allows designers to work individually and is similar to the traditional approach.

Following concept generation, the next stage involves selection of the best concept amongst the concepts generated by the designers. Participants fill in and submit evaluation data by means of WWW browser. The calculation of evaluation result will be executed in due time or when the organizer forces it to execute. This method can synthesize participants' opinions in the determination of optimal product concept. When the optimal product concept is determined, the detailed design stage then follows.

The Internet based GUI of the Co-DARFAD system is shown in Fig.6.

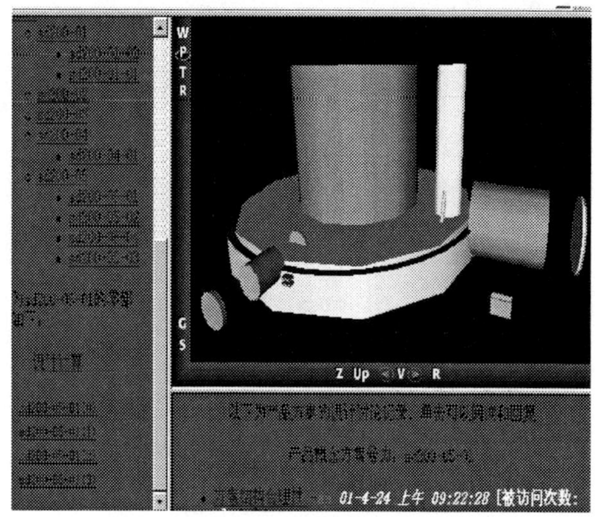

Fig.6 GUI in Co-DARFAD System.

6. Conclusions

Based on the survey of existing research work on collaborative design and on our experience in Co-DARFAD system development, the collaborative design system based on Internet and D&R Model proposed in this paper will greatly improve the effectiveness of collaborating tools. But at the moment further work is needed to make the tool commercial software.

(1) The hierarchical structure of product development model is suited to both loose-coupled cooperative and close-coupled collaborative design work, providing a profound base for practical collaborative design;

(2) The product collaborative design process model based on D&R principle has been proved to be effective during conceptual structural scheme design;

(3) The framework of Co-DARFAD system improves the current collaborative design by providing intuitive evaluation capability on a flexible discrete step-wise design basis.

Further work in automatic evaluation and optimization, and system management etc. is needed to further enhance the current Co-DARFAD system.

7. Acknowledgement

The authors express their thanks to China Hi-Tech Development Program (863) and the Science & Technology Committee of Shandong Province for the financial support.

8. References

[1] N. Senin, N. Borland, David R. Wallace, "Distributed modeling of product design problems in a collaborative design environment", *Technical Report*. CAD Lab., Department of Mechanical Engneering, MIT. 1998.

[2] D. Sriram and R. Logcher, "The MIT DICE Project", *IEEE Computer*, 1993, pp.64-65

[3] G. Toye, M.R. Cutkosky, et al. "SHARE: A Methodology and Environment for Collaborative Product Development", *in Proc. Of 2nd Workshop on Enabling Technologies: Infrastructure for Collaborative Enterprises*, Morgantown, West Virginia, 1993, pp.33-47

[4] D. Cuff, "Architecture: The story of practice", *MIT Press*, Cambridge, MA, 1991

[5] F. Sudweeks, M. Allbritton, "Working together apart: communication and collaboration in a networked group", *in: C.D. Keen, C. Urqhart, J. Lamp(Eds.). Proc. Of the 7th Australasian Conf. Of Information Systems*, Vol.2, ACIS96, Department of Computer Science, University of Tasmania, 1996, pp.701-712

[6] Thomas Kvan, "Collaborative design: what is it?", *Automation in Construction,* 2000(9), pp.409-415

[7] P. W. Mattessich, and B. R. Monsey, "Collaboration: What Makes It Work", *Amherst H. Wilder Foundation*, St. Paul, MN, 1992

[8] S. Harrison, S.A. Bly, S. Anderson, S.L. Minneman, "The media space", in K.E. Finn, A.J. Sellen, S.B. Wilbur(Eds.), *Video-Mediated Communication*, Lawrence Erlbaum Associates, Mahwah, NJ, 1997, pp.273-300

[9] S. Harrison, S.L. Minneman, "Studying collaborative design to build design tools", in: M. Tan, R. The(Eds.), *The Global Design Studio, CAAD Future'95*, Center for Advanced Studies in Architecture, Nat. University of Singapore, Singapore, 1995, pp.687-697

[10] R. Seward, D. Diaper, C. Sanger, "The pot: a purpose - built environment to support group working", in: D. Diaper, C. Sanger(Eds.), *CSCW in Practice: an Introduction an Case Studies*, Springer-Verlag, London, 1993, pp.151-162

[11] S. Whittaker, B. O'Conaill, "The role of vision in face-to-face and mediated communication", in: K.E. Finn, A.J. Sellen, S.B. Wilbur (Eds.), *Video-Mediated Communication*, Lawrence Erlbaum Associates, Mahwah, NJ., 1997, pp.23-49

[12] Huang Kezheng et al. "Generic Structural Design by Assemblability for Mechanical Product", In: *Proc. of 14th Int. Conf. On Computer-aided Production Engineering*, Tokyo, 1998

[13] J.S.Jero, T.McNeill, "An approach to the analysis of design protocols", *Design Study,* 19(1988): 21-61

[14] T. Kvan, A. Vera, R. West, "Expert and situated actions in collaborative design", in: P.Siriruchatapong, Z. Lin, J.-P. Barthes(Eds.), *Proc. of 2nd Intern. Workshop on CSCW in design*, International Academic Publishers, Beijing, 1997, pp.400-450

[15] Huang Kezheng, et al. "Design process model for mechanical product base on D&R principle", in: Sun Guozheng, Ben Qingyuan(Eds.), *Modern Design Science Facing 21st Century*, People's Jiaotong Press, Wuhan, 1997, pp.10-14(in Chinese)

[16] Huang Kezheng, Ai Xing, Zhang Chengrui, "Decomposition and reconstitution principle for complex surfaces and its applications", *Science in China(Series E)*, 1997, 40(1): 89-96

Sharing Learning in a Cooperative Design System

Hong Liu, Xiyu Liu
Department of Computer Science, Shandong Normal University, P. R .China
E-mail: lhsdcn@pub.online.jn.sd.cn

Abstract

Design is a complex problem solving and knowledge refinement process. Learning is a part of this process that can improve computer based design support systems by using the knowledge representing the experience and expertise of designers. Learning from past design examples, and learning new knowledge during the process of design are closely related activities that must be supported by future computer supported design systems.

This paper analyses real design activity and proposes a model of design activity. Then it presents the software architecture of a design agent with learning mechanism. Subsequently, it introduces the knowledge representation in a design system. Finally, the paper presents sharing learning process in a cooperative design system.

1. Introduction

Design is an ill-structured problem in which solution concepts are developed to meet the functional requirements of the design problem. It is characterized by information that is often imprecise, incomplete and inconsistent. Apart from the observation that the generation of the functional requirements and solutions is highly coupled, conceptual design is poorly understood and little support is available.

The experience in design provides a wealth of knowledge of the past, which can be bought to bear on the present. Consequently experience, which consists of knowledge generated from personal exposure to events and artifacts, presents one of the most powerful resources possessed by a designer. Learning is a process that helps to maintain experiential knowledge. It also helps to promote the flexibility of experiences by removing highly specific details and generating more generally applicable knowledge.

Learning provides support for finding a new design solution involves the use of knowledge abstracted from previous designs. Adding learning to a design system is clearly desirable. One of the challenges of developing a design system is to support learning by the use of computational mechanisms that allow for the generation, accumulation and transformation of design knowledge learned from design experts, or design examples. Without learning capability a design system is unable to reflect the designer's growing experience in the field and designer's ability to use knowledge abstracted from past design cases.

There has been increasing acknowledgment that computational design systems can and should include the ability to learn, and there is an increasing amount of research on machine learning in design. Some research work of machine learning in design have been done, such as Explanation-Based Learning, Inductive Learning, Case-Based and Analogical Learning , Genetic Algorithms and Neural Networks etc.

Multi-Agent Design System is concerned with how a group of intelligent agents can cooperate to jointly solve design problems. Recently, work on learning in multi-agent systems has begun to appear [1-2].

The rest of this paper is structured as follows: section 2 analyses the real design activity and propose a model of design activity. Section 3 puts forward the architecture of a design agent with learning mechanism in details. Section 4 describes the design language CDL and knowledge base framework. Section 5 introduces the sharing learning process among designer and multiple agents. The final section summarizes the paper and gives a short outlook for future research work.

2. A model of design activity

It is proposed that a systematic approach to the study of design can be based on analyzing the design activities. In this section, we analyze the real design activity first and make it clear why learning is essential during design. Then, based on the analysis for design activities, we put forward a design activity model with learning mechanism.

2.1 Real design activity

Design problems are widely recognized as being ill-defined problems. It is usually possible to take some steps towards improving the initial definition of the problem, by questioning the client, collecting data, carrying out research, etc. There are also some rational procedures and techniques, which can be applied in helping to solve ill-defined problems. But the designer's traditional approach is to try to move quickly to a potential solution, or set of potential solutions, and to use that as a means of future defining and understanding the problem.

The design protocol data indicates that the specifications are incomplete, inconsistent, or sometimes even infeasible. For this reason, during the design experiments we observed that design processes had different kinds of exception handing operations such as backtracking, addition or relaxation of specifications, and conflict resolution [3]. On one hand, in the early stage of design, designers may spend most their effort in identifying hidden or unclear specifications. In later stages, on the other hand, specification relaxation can be carried out to balance between cost and performance.

A real design process is a process in which the designer has to augment, correct, relax, refine, and compromise the given specification. Since we are designing an artifact that will eventually exist in a physical world, is it even possible to describe every specification before we experiment and test the idea by some physical method (including physical experiments, analysis by hand or by computer, and so forth). We have to accept the reality that specification is never complete, consistent, or feasible before the design is completed.

During the design process, not only the design object but also the specifications are gradually (step by step) refined, improved, relaxed, expanded, or modified and become complete. Specifications become complete only when the design terminates, having arrived at the final design solution. This suggests that CAD systems should accept such a nature of evolutionary specifications.

Adding adaptation to a design system is clearly desirable. Even though learning does not always reach the optimal solution, experience should eventually bring noticeable and worthwhile improvements over initial designs and design processes.

2.2 The basic elements of a design activity

A design activity is an action or cognitive process to achieve a knowledge increment in the state of the design and/or its associated design process in order to achieve a design goal.

Definition 1 a design activity can be defined by a five-tuples (I_d, D_t ,O_d, G_d, T_d).

In which, I_d is the input of design specifications;

D_t is a knowledge-based transformer which transform the input I_s into output O_r;

O_d is the output result;

G_d is the design goal;

T_d is the design activity trigger.

The basic elements of a design activity are as shown in Figure 1.

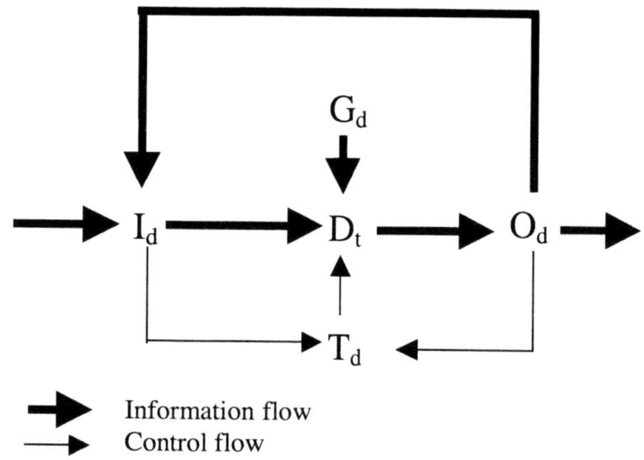

Fig. 1. The basic elements of a design activity

In this figure, it is suggested that I_d is transformed into O_d (such as files, graphics and images), that compares with G_d (such as some constraints) and then feed back together with I_s to produce a design circle. This circle will be continued until G_d is satisfied.

The reasons for driving T_d include mainly:

- Novelty driven: when there is a new design problem or a new design requirement;
- Excellence driven: when the design need to be improved to achieve better quality and reliability;
- Design goal driven: when the design doesn't achieve the desired design goal, D_t is initiated.

The knowledge-based transformer T_d is an operator that derives a piece of design specifications from a given input I_d, design goal G_d and design knowledge from knowledge base and transforms I_d into O_d.

3. Design agent

3.1 The architecture of a design agent with learning mechanism

A design agent is a software assistant that in some way helps users to complete design tasks. It is a designer's colleague and can adapt its own ability via sharing learning among multi-agent and designers.

Definition 2 a design agent with machine learning mechanism can be defined by a eight-tuples (Aid, Input, Communication, Transformer, Leaning, Output, Goal, Trigger).

In which, Aid is the identifier of a design agent;

Iuput is the input interface component. It gets input information (such as design specifications, design constrains) and passes them to design goal and transformer separately. It also gets some information coming from communication component and passes them after translation;

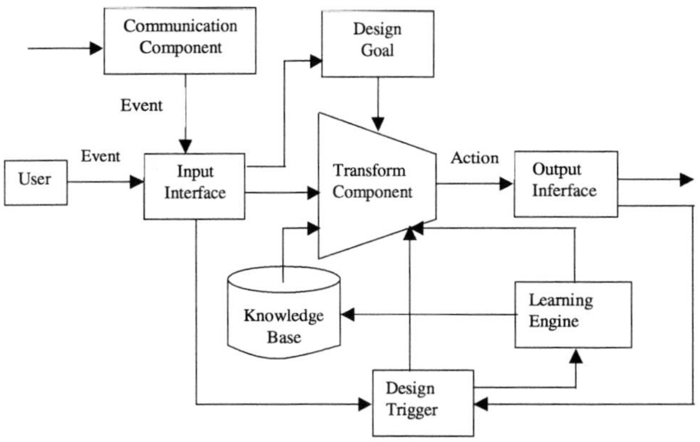

Fig. 2 The Architecture of a design agent

Communication is the communication component. It receives message from other agents or systems and passes them to input interface;

Transformer is a knowledge-based transform component, which transforms the input into output.

Learning consists of Knowledge Base and Learning Engine.

Output is the output interface component;

Goal is the design goal component and consists of some constraints. It gets goal information from other agents or designers. The goals of a design activity determine the type of design activity performed by the design agent.

Trigger is the design activity triggering component. It consists of Event-Condition-Action Rules.

Figure 2 shows the architecture of a design agent.

The agent autonomously performs the action through its transform methods, subject to constraints and checks. Knowledge and design specifications are combined by the transform component to decide on the next action. Finally, learning engine is capable of learning activated by trigger in the agent's world and from events performed by users or communication. This leads to producing new knowledge.

The transformer combines existing knowledge content with new knowledge and apply the transform component and learning engine to determine a desirable action and to extend knowledge base.

3.2 The transformer and learning engine of a design agent

The knowledge-based transformer is implemented by two approaches: research and reuse past knowledge and design by features via alternation with designer.

When transformer is triggered, it performs hill-climbing algorithm to search the reusable information (components or features) first. There are three possible outcomes:

1.The mapping between the new component (or feature) and the known component (or feature) classes is one-to-one.

2.The mapping are one-to-many. This means that the new component (or feature) has common attributes with more than one component (or feature) class. In this case, the designer will have two possibilities: (a) accept one of the proposed classes or (b) create a new class and then activate the learning process that will update the KB.

3.The new component (or feature) does not belong to any known component (or feature) class. A new class should be created and then the learning mode activated to update the KB.

Learning engine is in charge of the KB to be updated and new knowledge to be incorporated into the KB. The refinement learning algorithm is used and which is implemented by some operators such as create, insert, delete, update.

4. Design knowledge representation and storage

4.1Component Description Language CDL

In this section, we introduce a Component Description Language CDL that is used in the multi-agent design system.

EXPRESS has been used to describe data structures and constrains as it is the data modeling language used by STEP (the Standard for the Exchange of Product Model Data, ISO 10303)[4]. This framework allows two aspects of a product to be represented: the level of detail (product, part, assembly, component and feature) and the life cycle stage at each level of detail (specification, definitions and actuals).

The definition of CDL is based on the EXPRESS data specification language ,ISO 10303-11[5] but it is simpler and easier to be learned by the designers. In fact, it is a sub-set of EXPRESS.

The design language CDL (Component Description Language) is a data definition language which facilitates the description of components and products in terms of attributes, relationships and features decomposition. CDL input is in the form of a text stream, which may be stored in a text file or an output pipe from a concurrently executing process such as the graphical user interface. The CDL produces data records to be stored in a component definition file (CDF). This file has a frame structure to simplify description and data retrieval and allows attribute inheritance among produces, components and features.

The hierarchical structure of CDL facilitates the development of new user-defined features or composites of existing library features. CDL allows arithmetic expressions consisting of integer, real and scientific-notational constants, binary and unary operators, exponentiation, sub-expressions, functions and variable names.

4.2 The storage of design knowledge

There are two kinds of design knowledge, long term and short term.

The long term knowledge is organized as files of four types:

- Set of product files consists of a number of past designs described by design product ID, super class ID, sub class ID , product model tree, the components' ID on the model tree and the pointer list to point to component file, and updated date and version.
- Set of component files consist of the component name, class, function, material, roughness, hardness, feature list, tolerance list; relation list, design agent ID and updated date. The data structure is similar with CDL component data frame.
- Set of feature files consists of the feature name, type, location, orientation, parameter list and corresponding datum file names.
- Set of datum files.

The short team knowledge is organized as temporary files of four types:

- New design file consists of the description information of the new design that does not match with any past designs.
- Selected component file consists of components' name and pointer to a set of component files, these components are useful in new design.
- Selected feature file consists of features' name and pointer to a set of feature files, these features are useful in new design.
- Design state file consists of a record of design agent's working state, that is, which work the agent is doing and what step is going along.

5. Sharing learning among designers and agents

5.1. Multi-agent design system

Multi-Agent Design System (MADS) is a computer-aided design system which provides support for a group of design agents and designers to cooperatively solve design problems in distributed environment. Design is a complex knowledge discovery process in which information and knowledge of diverse sources are processed simultaneously by a team of designers involved in the life phases of a product. Complex design generally combines automated software components (Software agents) with human decision-makers (designers), making it imperative to provide support for both human and computational participants in the design process. Software agents provide the necessary support for keeping humans in the loop, and multi-agent system technology offers an appealing framework for combining diverse sources and types of information and reasoning.

The multiple design agents are connected via network and they cooperative work with the help of communication agent, KB maintenance agent and so on [6]. The sketch of a MADS is shown as Figure.3.

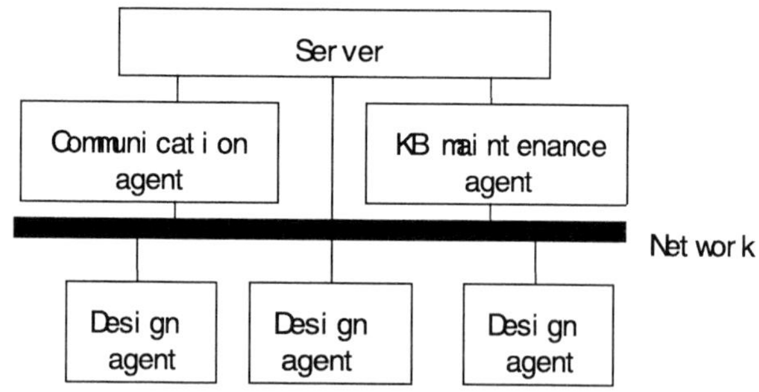

Fig.3 The sketch of a MADS

5.2 Sharing learning

Sharing learning is proposed to empower MADS with more useful learning capabilities that currently available and thereby provide a stronger interaction of learning between human designers and computers in a distributed environment. Sharing learning can get more examples from multiple agents than a single agent. Therefore, it can solve the problem of insufficient examples in single agent learning and accelerate the learning process. The knowledge leant can be shared by multiple agents.

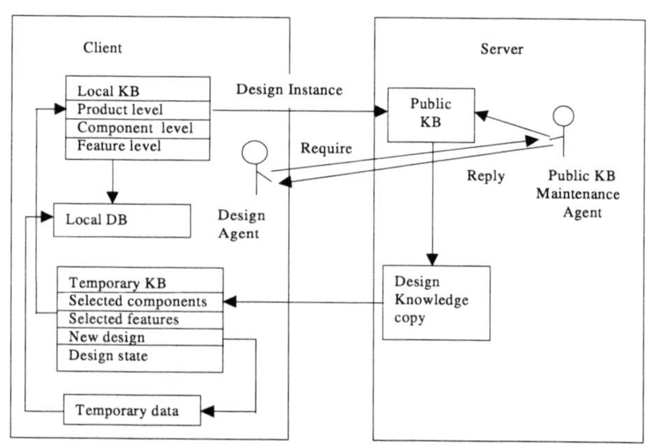

Fig.4 Design knowledge sharing and updating process

Design knowledge is stored on different KBs, the public KB (located on server) and the local KB. Every design agent has a KB of its own. When some design knowledge is needed, the search begins with the local KB. If the local search fails, the requests and replies happen between the design agent and public KB maintenance agent.

If the search is successful, the copy of component design knowledge in public KB is passed to request agent, and is placed on temporary KB (see Figure 4).

Whenever a design modification phase is over, the short term design knowledge will be updated. After all designs are fulfilled and are confirmed by the design engineer, the short term design knowledge will be used to update the long team design knowledge.

When a design agent's long team design knowledge is updated, the corresponding information is passed to public KB maintenance agent by communication. Public KB maintenance agent uses them as instances to classify and update the KB. Classification is implemented by a standard tree-induction algorithm C4.5 [7-8].

5.3 Cooperative design process

During the cooperative design and sharing learning process, the design engineer (the head of the design group) defines design requirements, assigns design tasks to the designers in the design group, directs the design activities, makes judgements about the design drafts guided by design agents, design agents update their design knowledge and improve their design abilities via interaction with designers and other agents.

Step 1. When a product design task is introduced, the design engineer first decomposes it first. After decomposition, the product design task becomes the component design tasks and a product model tree is generated.

This work is done both manually and automatically. The design engineer attempts to find a suitable design product class which matches the current design task. For example, when the design task is a large car, design engineer selects this type from a list (small car, mini-van, large car, jeep and truck). After he/she selects large car, another list will be displayed, which includes general, luxury, sporty, coupe and so on. In the example, the general type is selected. Then, the corresponding product model tree is shown on the screen (See Figure 5).

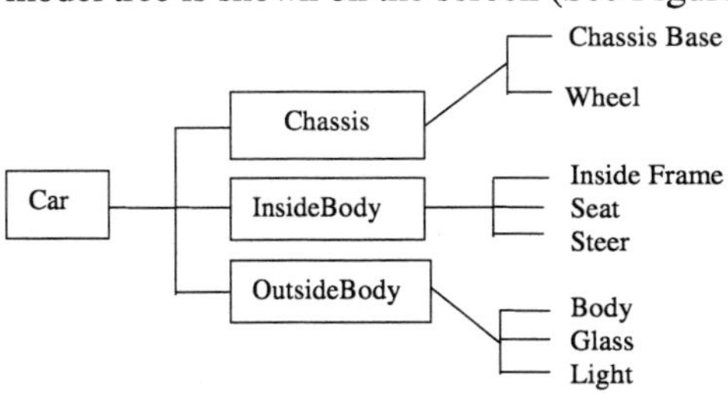

Fig.5 A product model tree

If the design engineer can not find a suitable match for the current design task (for instance, there is no car in the first list), he/she can create a new class and product model tree by selecting **Create** on the menu and following the guide. He/she should answer some questions step by step, such as how many components a product can be divided into and the name of every component. Then the question is how many sub-components are there in each component. The same questions will be repeated. The design engineer can select **Back** to correct any mistakes at every step. He/she can also make modification after the guide process has ended.

Step 2. In Step1, the design task is decomposed and forms a product model tree. The design engineer can add or delete components from the product model tree until it accurately matches the current design task. After the product design attributes and constraints have been input, the component design tasks and corresponding constraints are assigned to several designers via communication.

Step 3. When a designer receives a design task, he/she inputs the design requirement according to the task and constraints. Then, design agent attempts to find a suitable component, which satisfies the requirements. If a component is found, its sketch and corresponding attributes will be shown on the display screen . The designer must then decide whether or not to accept the component. If the component is accepted then the designer will, after possibly having made certain adjustments and alterations, submit it to the design engineer. On the other hand, if the component is not accepted then the designer will create a new component. Whenever the old design is perfected or a new design is finished, the design knowledge of design agent will be updated automatically.

Step 4. The design engineer assembles the components according to their relations and constraints in the design requirements. If conflict among the results from different designer occurs, the design engineer, with the help of agents, takes in charge of negotiation among the related designer and solving the conflict [9]. This initiates the redesign process. This process will repeat until all components are assembled and satisfy the design requirement.

6. Conclusions

In conclusion, design is a complex problem-solution and stepwise refinement process. Learning can improve the abilities of designers and computing systems by utilizing inductive methods that abstract new information from experiential knowledge embedded in the past design examples. This paper puts forward a distributed and shared learning environment within which human designers and computer agents can support each other in a collaborative and learning process during the conceptual development stage of the design process.

There is still much work to be done before the full potential of shared learning can be realized. Future work is the improvement and application of the system in the domain of mechanical engineering design.

This project is supported by NSFC (No. 69975010) and NSF of Shandong Province (N0.Q99G07).

7. References

[1] Tan, M. Multi-agent reinforcement learning: Independent vs. Cooperative agents. Multi Learning: Proc. of the Tenth International Conference, San Mateo, CA: Morgan Kaufmann, 1993.

[2] Schmidhuber, J. (1996). A general method for incremental self-improvement and multi-agent learning in unrestricted environments. Evolutionary Computation: Theory and Applications.

[3] Tomiyama, T. From general design theory to knowledge-intensive engineering. AIEDAM, 1994;8(4):319-333.

[4] Fowler, Julian. STEP for data management, exchange and sharing. Technology Appraisals, Great Britain, 1995.

[5] ISO 10303-11:1994. Industrial automation systems and integration-Product data representation and exchange-Part 11:Description methods.

[6] Liu H., Zeng G. Zh. & Lin Z.K. An agent-based approach to cooperative design. In: Proc. Of Workshop on CSCW in Design'97, Bangkok, International Academic Publishers,1997;191-195.

[7] Quinlan, J.R. Induction of decision trees, Machine Learning, 1(1):81-106.1986.

[8] Quinlan, J.R. C4.5: Programs for machine learning. San Mateo, Calif.: Morgan Kaufmann, 1993.

[9] Liu Hong & Lin Zongkai. A Cooperative Design Approach in MADS. In: Proc.Of. Workshop on CSCW in Design'99:297-303, France,1999.

Tracking of Design Changes for Collaborative Product Development

Helen Xie
Integrated Manufacturing Technologies Institute
National Research Council of Canada
800 Collip Circle, London, ON, Canada N6G 4X8
Helen.Xie@nrc.ca

Abstract

Market demands are moving manufacturing from mass production to mass customization. To quickly respond to customer needs, product development teams have to work together in a collaborative design environment. This paper presents a product data driven approach to tracking of design changes in a collaborative design environment. In this research, design processes are analyzed to identify product data for tracking design changes. The relationships between product data are established using predecessor-successor, master-subordinate, and geometric constraint relationship types. Forward tracking and backward tracking are designed and implemented to identify impacts of a design change. A prototype system has been implemented to verify the effectiveness of the approach. Finally, a milling machine fixture is used as a design example to show how this approach works. The proposed approach provides a fundamental mechanism to promptly notify other designers of design changes in a collaborative design environment.

1. Introduction

Product development teams today are increasingly facing the challenge of meeting the demands for high product variety and quality while these products are manufactured in relatively low quantities [1]. One of the consequences of the demand for mass customization of products has been the necessity for a team of designers to work together in a collaborative design environment. By exploiting each individual designer's expertise, the design team can generate design solutions quickly in response to customer requirements. By analyzing conflicts and constraints earlier from different perspectives, the design team is able to achieve design objectives through optimizing product performance, minimizing manufacturing costs, and ensuring that the product can easily and economically be serviced and maintained [2].

Centric to the collaborative design environment are network-supported computer applications. Recently, advanced information technologies, including Internet related technology and distributed object technology, have opened new possibilities for collaborative design. Several frameworks have been proposed for collaborative design systems. SHARE [3] is an open, heterogeneous, network-oriented environment for concurrent product development. It helps a design team to achieve a shared understanding of its designs and design processes using agent-based tools and services. PEDWorks [4] is a prototype system for the web-based collaborative workspace to integrate design activities in a distributed and heterogeneous computing environment. It uses concept of a process-centric collaborative design for engineering designs.

Collaborative design is the process of designing a product through collective and joint efforts undertaken by a team of designers [5]. In a collaborative design process, customer requirements are first captured and product specifications are determined. The information is shared in a database among members of the design team. A system designer decides the rough structure of the system and several design parameters satisfying the given product specification. Then each team member works on a preliminary design for each assembly and a detailed design for each part in a distributed environment. Even though each team member performs its own activities using different disciplines, the design solution may be tightly coupled. Furthermore, the design solution has to satisfy all the design constraints. If there is any constraint violation, design changes become necessary.

One of the issues in collaborative design is that one must assess the impacts of a design change on other design objects and notify other parties promptly [6]. Due to the lack of tracking of design changes in most collaborative design environments, designers have to manually perform a consistency check for a proposed design change to identify all the impacts of the change. As a result, design consistency and accuracy are not guaranteed and design team productivity is compromised. Hence, an effective approach to tracking of design changes is needed to significantly improve the collaborative design process. Moreover, this approach should be generic so that it can be applied to different products and can be integrated with other application systems.

The focus of this research is to propose an approach to the tracking of design changes in a unified framework [7]. A Web-based collaborative design prototype being proposed allows a team of designers to visualize, manipulate and evaluate the impacts of changes to a design. The remaining sections are organized as follows: Section 2 introduces the approach to the tracking of design changes. Section 3 describes the product information requirements for the tracking of design changes and establishes a change tracking data model. Section 4 introduces a web-based collaborative design prototype and uses an example to demonstrate the efficiency of the approach. Finally, the results of the proposed approach are discussed.

2. Approach to tracking of design changes

A design change is a modification to a product that is necessary to meet the requirements of the product specification, to meet safety standards, or to reduce manufacturing or maintenance costs [8]. From the perspectives of product variation and product improvement, design changes are imperative in collaborative product development processes. Due to the associated relationships between elements in a product, a design change may have an impact on other elements of the product. One of the issues that appears in most of collaborative design systems is the inability to track all the impacts of a design change automatically. Consequently, identifying these impacts usually requires a manual search by designers. This manual process cannot guarantee complete tracking of all design changes. Therefore, automatic tracking of all the impacts of a design change is important in supporting collaborative design environment.

One of the approaches to automatic tracking of design changes is an expert system approach. Design change rules can be stored in a knowledge base so that all the impacts of a design change can be retrieved through an inference engine. However, the rules for automatic tracking of a change are limited to a specific application domain.

Based on product data and the relationships generated during various stages of a product design process, a product data driven approach is proposed to provide a generic mechanism for the tracking of design changes. The product data contains product descriptions for product specification, function decomposition structure, solution principles, layout design, assemblies, and parts. Relationship can be established between two or more elements of the product description, especially between elements having a predecessor-successor relationship, master-subordinate relationship, or geometric constraint relationship to one another. The predecessor-successor relationship represents the logical dependency developed for product data during the design process (Figure 1). The master-subordinate relationship presents assembly-part hierarchical relationship. The geometric constraint relationship defines fit constraints, contact constraints and consistent constraints between parts. The fit constraint exists if there is a tolerance requirement between parts. The contact constraint represents a physical contact between two parts. The consistent constraint exists if two parts hold a dimensional constraint without a physical contact. With these product data relationships, the tracking of design changes can be achieved.

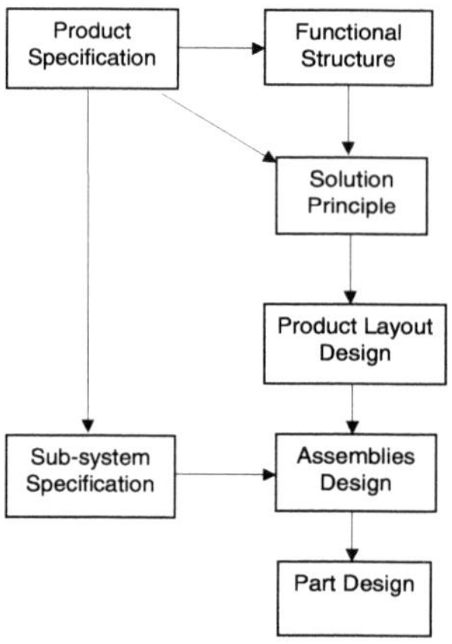

Figure 1. Predecessor-successor relationship.

The data driven approach enables designers to identify the total impacts of a proposed design change on an entire product. It provides a fundamental mechanism to promptly notify other designers regarding design changes in a collaborative design environment. This approach supports forward tracking and backward tracking of a design change. Forward tracking identifies the impact of the change on later design stages if a design change occurs at an earlier stage. On the other hand, backward tracking identifies the impacts of changes on previous stages, if a change occurs at a later stage. These operations can be implemented with database support.

The implementation of the data driven approach to tracking of design changes requires that product data and relationships presented above be retrieved from a product database. However, with most of the existing CAD systems, only the product structure, bills of material, and drawings are recorded in the product database. Product data such as product specifications, function structure, solution principles, and design concepts are usually presented in paper documents. Since the automatic tracking of design change is possible only in a computerized environment, product information in a paper document format has to be converted into a product database. Furthermore, neither the logical dependent relationships between design elements nor the logical intermediate processes are documented in practice. Consequently, final design documents are not linked to initial product requirements without human intervention. Thus, the logical dependent relationships among design elements need to be modeled in the product database. To make all the necessary information available for the tracking of a design change, the data driven approach is implemented by extracting necessary product information from design processes and representing the product information in a data model.

3. Data modeling for tracking of design changes

3.1. Information requirements

Engineering design is the process of devising a system, component, or process to meet desired needs [9]. The design process begins with a product specification and ends up with an artifact. Systematic design methodology [10] suggests a design process through a series of stages including product specification, conceptual design, preliminary layout design, and detailed design. Product specification defines a product's physical and functional characteristics with precise, measurable details. A product concept is a sketch description of the function structure, solution principle, and the form of product design. The function structure is usually expressed as a hierarchical structure in which a general function is decomposed into a group of sub-functions. A solution principle can be represented as an action principle corresponding to a function. After all solution principles are developed for sub-functions, they are combined to form design concepts. The preliminary layout design is the arrangement of a product structure with bill of materials. The detailed design provides complete dimensions, tolerances, and surface property information for assemblies and parts.

3.2. A change tracking model

The product data driven approach to the tracking of design changes is developed using a data modeling methodology. A data model is a set of concepts that can be used to describe the structure of a database [11]. An entity-relationship model, a popular conceptual modeling method, is used to describe product data with such concepts as entities, attributes, and relationships. The entities and attributes describe an abstract object with its properties. The change tracking model includes following entities: *Specification*, *Function*, *Principle*, *DesignObject*, *Assembly*, and *Part*. The Table 1 summarizes the entities and their descriptions for the change tracking model. The associated attributes are shown in Figure 2.

Table 1. Entities and description.

Entity	Description
Specification	Product specification generated from customer requirements
Function	An individual function composed of a function structure
Principle	A specific function implementation
DesignObject	A super class over entity *Assembly* and entity *Part*. It describes common characteristics of assemblies and parts.
Assembly	A subclass of the entity *DesignObject*, It describes the specific characteristic for a assembly
Part	A design object that cannot be further decomposed. A subclass of the entity *DesignObject*. It describes the specific characteristic for a part

The relationships represent a set of associations among entities. They limit the possible combinations of entity instances. Cardinality ratio constraints specify three common combinations for binary relationship types: one-to-one (1:1), one-to-many (1:M), many-to-many (M:N). The relationships in the change tracking model include *Requires, Contains, Previous, Solution, Implement, Belongs, and Constraint.*

Relationship *Requires* is a logical dependency relationship between specifications and functions. According to product design principles, a metric in a specification may demand multiple functions, while a function may be constrained by several metrics. Hence, the relationship *Requires* is of a cardinality ratio of M:N. For instance, clamping force is a metric of a milling fixture specification. It relates several functions including forces accept, forces amplify and forces transmit. On the other hand, fixture alignment to a machine table is a function of a milling fixture. It should satisfy two metrics of the specification, rigid mounting on the machine table and the mounting interface T-slots.

Relationship *Contains* is a master-subordinate relationship of the function entity that represents the function structure. A function may contain several sub-functions, which is represented as a cardinality ratio of 1:M.

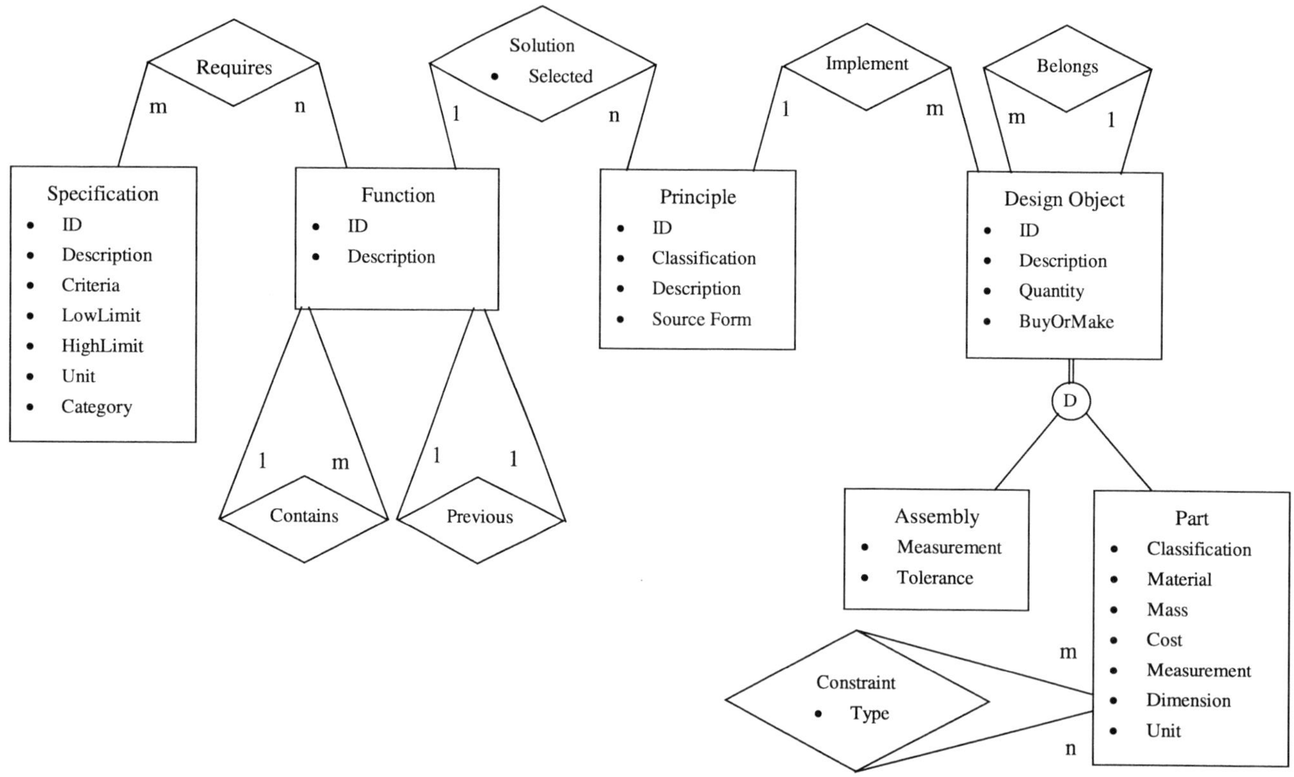

Figure 2. E-R schema diagram for change tracking model.

Relationship *Previous* is a predecessor-successor relationship of the function entity that describes the sequence among the functions in the same level. Since functions are defined in sequence at the same level, a cardinality ratio of 1:1 is used to describe the relationship type *Previous*.

Relationship *Solution* is a logical dependency relationship between *Function* and *Principle.* In conceptual design, several solution principles are usually proposed for a single function. This suggests that the *Solution* relationship type is of cardinality ratio 1:N, where N is on the entity *Principle* side. The proposed principles usually focus on key functions so that part of function instances are related to principle instances. Hence, the participation of *Function* in the *Solution* relationship type is partial. In practice, one principle is usually selected for a specific function. Therefore, an attribute of *selected* is defined in *Solution* relationship to identify the solution principle that is applied to the product design.

Relationship *Implement* is a logical dependent relationship between entity *Principle* and *DesignObject.* Since a principle may be implemented for several design objects, the relationship *Implement* is of a cardinality ratio of 1:M, in which entity *DesignObject* is in the M side.

Relationship *Belong* is a master-subordinate relationship between different design objects. This relationship is defined to represent product architecture. Except for an end product, assemblies or parts in any level are dependent of their upper level. According to this relationship, one is able to know the parts or sub-assemblies that an assembly contains.

Relationship *Constraint* is a geometric constraint relationship between different parts. It describes how a part is associated with other parts through three types of constraints: fitting, contact, and consistency. The constraint type is represented as an attribute of the relationship type. The relationship type *Constraint* is of cardinality ratio M:N because a part may have interfaces with several parts.

Figure 2 shows the E-R schema diagram for all the entities and relationships discussed above. The model identifies a variety of relationships from specification to assembly and part. In this way, a change which occurs in any point of this chain can be tracked forward and backward to find out its total impact.

The data model can be applied in tracking a design change for any product configuration, because it is based on a general design methodology and independent of any specific design context. It can be used to capture the total impact of a design change by backward and forward searching, no matter at what stage a change occurs.

4. Implementation and case studies

4.1. Prototype implementation

A Web-based software prototype called Design Advisor is designed to verify the proposed approach to tracking of design changes. The Design Advisor allows a design team to share product information from a back-end database through a Web server. Java Applet is chosen for the client-side user interface (Figure 3). It provides an enhanced functionality to display a product structure tree at different levels. It also provides designers the ability to manipulate the product database. JDBC API is used to access the database. From the user interface, a designer is able to switch to different design views using the buttons on the toolbars. The Product Specification, Function, Principle, Assembly, and Part views provide designers with the capability to enter, modify, browse, and search product data in a database. These views also help in linking product data for predecessor-successor relationships. The *Requires* view allows the designers to establish many-to-many type of relationships between specification and function. The *Constraint* view allows designers to specify geometric constraints, another many-to-many type, among parts. The *Sketch* view provides designers abilities to review engineering drawings.

4.2. A design case

The design case for a milling machine fixture [12] was studied to obtain its product information from the design process. The design process for the fixture includes product specification, function structure, solution principle, assembly design, and part design. The product specifications are classified as functions, conditions of the technical processes, operation properties, ergonomic properties, appearance, distribution properties, delivery properties, manufacturing properties, and cost. According to the design specifications, functions of the fixture are proposed and decomposed into sub-functions. These sub-functions can be implemented using available solution principles. For each primary sub-function, there are several solution principles available. Only one solution for each primary sub-function is selected for implementation. The layout design can be finalized based on synthesis of these solutions.

The layout design for the fixture is shown in Figure 3. The fixture is designed as three assemblies. The clamp mechanism consists of knob, crank, accessory, vertical plate, grid column, flange, and pin fix. The position device consists of 4-V-slot block, 2-V-slot block, cylinder, pin positioner, and pin chip. The support platform consists of plate and screw. The relationships between the solution principles and parts are identified. A solution principle may require several parts to perform a function. For example, the principle two-column guidance is related to the vertical plate, 4-V-slot block, 2-V-slot block, and grid column. The constraint relationships among parts are also identified in terms of fitting, contact, and consistency.

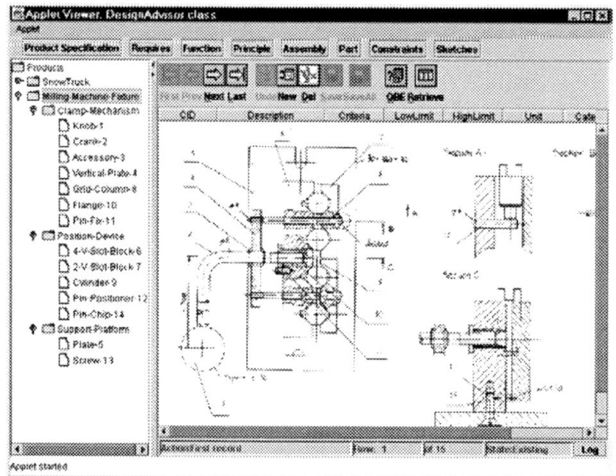

Figure 3. The fixture layout design.

The product information for tracking of design changes involves product specifications, functions, principles and design objects. The predecessor-successor, master-subordinate, geometric constraint relationships among them provide a mechanism to search for all possible impacts of a design change on the product. The tasks for searching for all possible changes are classified into two types based on search directions: forward tracking and backward tracking. The generating sequence of product data is from specification, function, principle, to design object. Forward tracking identifies the impact when a change occurs in upstream. It is applicable for customized product design, variant design and design improvement. Backward tracking identifies the upstream impacts when a change occurs in a part design. It is used for improving manufacturability and maintainability.

For forward tracking, a change appearing in specifications represents the most complex case because the impacts are distributed in functions, principles, and design objects. An example is given to demonstrate how the impacted changes are tracked using the approach.

Suppose a clamping force in the specification has to be changed to meet a customer's requirement, all the impacted items in parts should be located. The query operations are used to select records from individual tables and to combine related records from several tables for a retrieval request. The forward search query can be represented in SQL statement. SELECT section represents related attributes that are to be included in the query result. These attributes consist of ObjectID and RelatedObjectID. FROM section represents the tables involved and their associated relationships. WHERE section specifies the criteria for the query.

The results of this query are shown in Figure 4. The Clamp force is related to function *Force Accept* and *Force Amplify*. The function *Force Accept* is implemented by principle *Handle with X-axis rotation*. The function *Force Amplify* is implemented by principle *Screw/Nut*. The principle *of Handle with X-axis rotation* corresponds to Knob and Crank. The principle of Screw/Nut corresponds to Cylinder and Flange-. The Knob, Crank, Cylinder, and Flange- are constrained by Accessory, Vertical plate, 4-V-slot-Block, and Pin-fix. Hence, If the clamp force is to be changed, Knob, Crank, Accessory, Vertical plate, 4-V-slot-block, Cylinder, Flange-, and Pin-fix should be changed too. By reviewing the description of functions, principles, and design objects, the results are found to be reasonable.

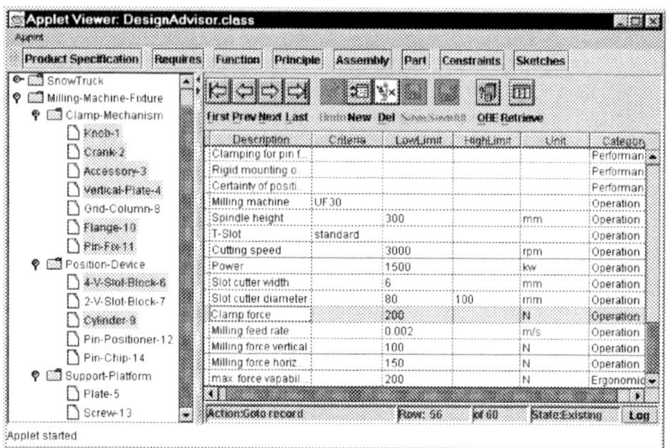

Figure 4. Impact of changes in forward tracking.

Backward tracking is used for searching the impact of changes for parts, principle, function, and specification when a change occurs in a part. For example, the vertical plate needs to be changed to improve manufacturability. Impact of all changes can be searched using a SQL statement. The vertical plate holds a crank and column and is positioned by an accessory. Hence, a change in the vertical plate may result in changes to crank, accessory, and column. These parts implement a solution principle "two column". The principle "two column" performs a function "clamping surface guide". The function corresponds to a metric in specification "clamping of pin for milling". Therefore, changing a vertical plate may result in a change in "clamping of pin for milling".

From the above two examples, it is concluded that the change tracking model can be used effectively for the two types of design change tracking. The query results are found to be reasonable.

5. Conclusions

The analysis of product design processes is important for determining the product data involved in design changes. In this paper, the product data driven approach is proposed to address design change issues in collaborative design environment. The core of the approach is the change tracking model, which represents product data and their relationships in specification, function, principle, design object, assembly, and part. The approach can be used for generic product development because the change tracking model is based on abstract product information and is independent of any specific design context. Although the current model is suitable for tracking changes during the product design process, it can be extended to the product life cycle also.

6. References

[1] U. Roy and S. S. Kodkani, Product modeling within the framework of the World Wide Web, *IIE Transactions*, 31/7, pp.667-677, 1999.

[2] J. Hartley, *Concurrent Engineering*, Cambridge, Mass., Productivity Press, 1992.

[3] G. Toye, M. Cutkosky, L. Leifer, J. Tenenbaum and J. Glicksman, SHARE: A Methodology and Environment for Collaborative Product Development, *Proceedings of 2nd Workshop on Enabling Technologies: Infrastructure for Collaborative Enterprises*, pp. 33-47, 1993.

[4] H. Kim, J. Y. Lee, and S. Han, Process-centric Distributed Collaborative Design based on the Web, *Proceedings of DECT'99: 1999 ASME Computers in Engineering Conference*, pp.1-9, 1999.

[5] W. Shen, Web-based Infrastructure for Collaborative product Design: an Overview, *Proceedings of the fifth International Conference on Computer Supported Cooperative Work in Design*, pp. 239-244, 2000.

[6] L. Wang, W. Shen, H. Xie, J. Neelamkavil, and A. Pardasani, Collaborative Conceptual Design: a State-of-the-art Survey, *Proceedings of the fifth International Conference on Computer Supported Cooperative Work in Design*, pp. 204-209, 2000.

[7] H. Xie, *Effective Configuration Management for Supporting Product Development Process*, Master's Thesis, The Department of Mechanical Engineering. The University of Saskatchewan, 1998.

[8] R. E. Monahan, *Engineering Documentation Control Practices and Procedures*, New York, Marcel Dekker, Inc., 1995.

[9] A. Ertas and J. C. Jones, *The Engineering Design Process*, New York, John Wiley & Sons, 1993.

[10] V. Hubka, *Principles of Engineering Design*, London, Butterworths, 1982.

[11] R. Elmasri and S. B. Navather, *Fundamentals of Database Systems*, New York, The Benjamin/Cummings Publishing Company Inc., 1989.

[12] V. Hubka, M. M. Andreason, and W. E. Eder, *Practical Studies in Systematic Design*, London, Butterworths, pp. 22-37, 1988.

Towards Collaborative Design in an Open Hypermedia Environment

David L. Hicks
Department of Computer Science
Aalborg University Esbjerg
Niels Bohrs Vej 8
6700 Esbjerg, Denmark
Phone: ++45 7912 7632
Fax: ++45 7545 3643
Email: hicks@cs.aue.auc.dk

Uffe Kock Wiil
Department of Computer Science
Aalborg University Esbjerg
Niels Bohrs Vej 8
6700 Esbjerg, Denmark
Phone: ++45 7912 7623
Fax: ++45 7545 3643
Email: ukwiil@cs.aue.auc.dk

Abstract

Open hypermedia systems hold significant potential to facilitate the development of collaborative design applications. In particular, it is the flexible approach utilized by these systems to provide infrastructure services, their ability to accommodate alternative hypermedia domains, and their facilities for information integration across application boundaries that give open hypermedia systems this potential. Current open hypermedia systems, however, have not realized their full potential to support collaborative design. They lack capabilities in one or more of these important areas. Improvements are especially needed in the collaborative infrastructure services provided by open hypermedia systems and in their support for alternative hypermedia domains. A prototype open hypermedia system is currently being extended to address these shortcomings of existing systems.

1. Introduction

The areas of hypermedia and computer supported cooperative work have been closely related since they were first examined by researchers. Indeed, hypermedia functionality was a prominent feature of some of the earliest systems that provided collaborative work capabilities [5, 7]. As research progressed in the hypermedia and computer supported cooperative work areas, each grew into its own research field. Though there have been some notable exceptions, work progressed largely independently in each of the fields, and they began to diverge somewhat as researchers focused on the challenging and difficult fundamental issues inherent in a new research field. Frameworks for computer supported cooperative work have been proposed in which hypermedia capabilities are not provided. Similarly, hypermedia systems have been developed that do not support collaboration. As these fields continue to mature, however, it is important to remember the potential synergism they hold for each other. In particular, the combination of these technologies can be quite effective in supporting the collaborative design process.

This paper examines open hypermedia systems and the potential they have for facilitating computer supported cooperative work, and, in particular, collaborative design. The paper proceeds with an overview of open hypermedia systems in the following section. In Section 3, the specific characteristics of open hypermedia systems are examined that give them potential for supporting collaborative design. This is followed in Section 4 with an examination of how well current open hypermedia systems match up to their potential, and improvements that are needed to enhance their support for collaborative design. Section 5 describes a prototype system and how it is being used to explore the potential of open hypermedia systems for collaborative design. The paper is concluded with a brief summary in Section 6.

2. Open Hypermedia Systems

Open hypermedia systems represent the latest stage in the evolution of hypermedia architectures. They provide extensive computing environments that, increasingly, offer a variety of services in addition to standard hypermedia functionality. This section provides an introduction to open hypermedia systems through an examination of the architectural stages that have preceded their development.

The first systems to provide hypermedia functionality were monolithic in nature. Monolithic hypermedia systems are self-contained in that they explicitly provide all of the functionality that is available to users. Monolithic systems do not support interactions with or make their hypermedia functionality available to external applications. Only the information specifically managed by the system is eligible to participate in hypermedia operations. Several monolithic hypermedia systems have been developed including KMS [1], Notecards [9], and Intermedia [27]. Though some of these systems have been and continue to be used successfully, the lack of interaction with external applications has been identified as a significant drawback of systems built in the monolithic style [13].

To address the problems associated with monolithic hypermedia systems, researchers turned to the client-server architectural style. Client-server based hypermedia systems consist of a hypermedia engine or server that offers functionality that can be integrated into and used within an open set of client applications. The functionality of these systems is made available through a well defined interface enabling a variety of applications to access hypermedia operations. Example client-server based hypermedia systems include HyperBase [18], DGS[20], and the World Wide Web [3]. While this strategy has succeeded in making hypermedia functionality available to external applications, the rules for interacting with the hypermedia servers of these systems are often quite restrictive. For example, some client-server based systems require client applications to adhere to a specific data model in order to use their services. Many of these systems also require that client applications use their specific data storage facilities in order to access their hypermedia functionality. These rules and restrictions make integration of hypermedia functionality difficult since they require significant modification of potential client applications in order to participate in hypermedia operations.

Research on open hypermedia systems started as a result of the drawbacks associated with client-server systems. Like client-server systems, open hypermedia systems also make their functionality available to an open set of external applications. However, the set of potential client applications is larger for open hypermedia systems. This is because the services they provide are designed to be more easily integrated into client applications. Specifically, open hypermedia systems do not require client applications to adhere to a particular data model or to utilize their data storage facilities. Example open hypermedia systems include DHM [6], Microcosm [11], and Chimera [2]. One of the most recent trends in the continuing evolution of open hypermedia systems has been towards the design of component-based systems. These systems adopt a more modular approach to the provision of hypermedia functionality in order to increase the flexibility and usability of the services they provide [26]. Example component based open hypermedia systems include HOSS [14] and HyperDisco[24].

3. Collaborative Design in Open Hypermedia Systems

An important application area to which open hypermedia systems can be applied is to the support of collaborative-based activities, especially collaborative design. There are three primary characteristics of the open hypermedia system environment that make it well suited for supporting collaborative design. Firstly, the open hypermedia system approach to providing infrastructure services is well suited to the provision of facilities to support collaboration. Secondly, these systems are capable of supporting a range of different types of hypermedia functionality, including new types of hypermedia, some of which are especially well suited for the design process. Thirdly, open hypermedia systems support the integration of information across both application and hypermedia domain boundaries. In this section each of these areas is examined in more detail.

3.1 Infrastructure Support

In an influential paper in 1988, Halasz noted several capabilities that were either missing or lacking in existing hypermedia systems [10]. Several of these, including support for locking, long term transactions, event notification, and versioning are important in the development of applications with collaborative requirements [7]. As research on hypermedia architectures has progressed and reached the stage of open hypermedia systems, the lack of support facilities for collaboration has had an increasingly significant influence. Open hypermedia researchers, noting the absence of this functionality, have included it in the collection of desirable infrastructure services for an open hypermedia environment. This section examines open hypermedia systems in more detail, focusing on the way infrastructure services are provided in an open hypermedia environment, and the suitability of the approach for supplying collaboration support services.

Component-based open hypermedia systems represent one of the latest stages in the development of open hypermedia architectures. The conceptual architectural of a component-based system is illustrated in Figure 1. Located at the top layer of the figure are client applications that have been integrated into the environment to utilize the functionality provided by the system. In the middle architectural layer are middleware components that provide hypermedia related services for applications. A storage server resides below the middleware components and supplies basic hypermedia storage functionality.

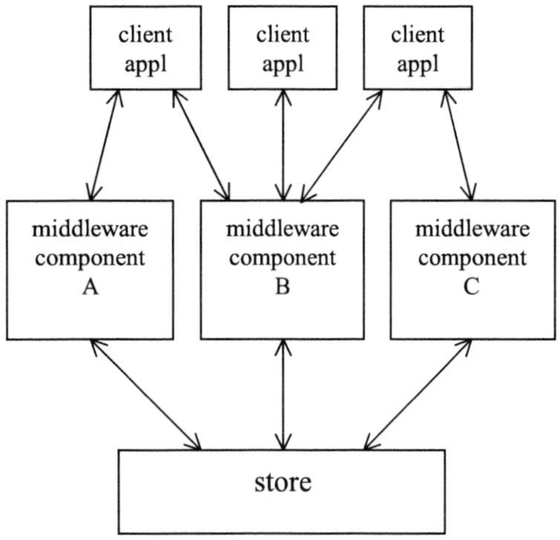

Figure 1. Conceptual architecture of a component based open hypermedia system.

The middleware components of open hypermedia systems can be used to provide functionality beyond conventional hypermedia operations. In addition to standard hypermedia capabilities, such as creating and traversing associative relations or links between information items, a variety of additional services can be provided in this architectural layer. Examples include a middleware component that facilitates the distribution of client applications or a service that enables interoperability between client applications [25]. Note that the functionality provided by middleware components is not intended for direct end user usage, but instead is for use by application developers for building systems.

The characteristics of the open hypermedia environment make it a suitable one for the implementation of a variety of infrastructure services, including those to support collaboration. In particular, the characteristics of middleware components make them an effective way to offer this functionality. As illustrated in Figure 1, middleware services in an open hypermedia environment can be combined and used very flexibly. Client applications can utilize the services of more than one middleware component in various combinations, and middleware components can provide services to multiple client applications. To enhance their accessibility, middleware components have well-defined interfaces making the functionality they offer readily available to be integrated into client applications. This is important for infrastructure related services since it allows for the wide spread use of these commonly needed services. The set of middleware services is also open ended. New middleware components can be added and existing ones can be removed or modified, enabling the services provided in the middle layer to be adjusted when necessary.

Figure 2 illustrates an example component-based open hypermedia environment and how infrastructure services to support collaboration could be implemented within it. The middle architectural layer contains an associative hypermedia service that provides operations for standard hypermedia functionality (creating and traversing links). In addition, the middle layer also contains middleware components that provide collaborative infrastructure capabilities. In this case, the additional middleware components provide a notification service, an access control service, a versioning service, and a transaction service. Since the set of middleware components is open ended, additional ones could be added to the environment illustrated in Figure 2 to extend support for collaboration or provide some alternative infrastructure service. Client application developers can utilize the collaboration infrastructure services in any way necessary. For example, the designer of a software development tool might make use of the facilities of the versioning service to provide a basis for user level asynchronous development activities. Alternatively, notification control and transaction control (as well as additional services) would be useful in the development of an application designed to support synchronous collaborative document authoring.

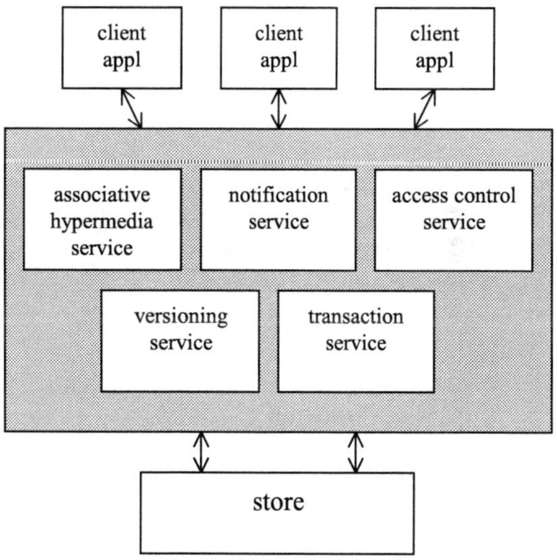

Figure 2. Collaboration support in an open hypermedia environment.

Offering collaboration support facilities through infrastructure services in this way promotes reuse of the functionality since it can be readily incorporated into multiple applications. Doing so in the open hypermedia system environment helps to ensure that the potential set of client applications that can utilize the infrastructure services is as open and large as possible.

3.2 Alternative Hypermedia Domains

The ability to create and traverse links between related pieces of information is often referred to as *associative* hypermedia. This is the conventional form of hypermedia and is by far the most common. The most prominent example of associative hypermedia is the World Wide Web [3]. The ability to link together related information items certainly is an important one for facilitating knowledge work, and can be especially useful in supporting collaborative activities [7]. However, alternative types or domains of hypermedia also exist [15], some of which are well suited for supporting design activities. This section briefly examines two alternative hypermedia domains and their applicability to the design process: spatial hypermedia and argumentation support.

Spatial hypermedia. Spatial hypermedia technology is designed to facilitate information analysis, which is an important subtask in many types of design [16]. Spatial hypermedia applications help users to bring order to an information space. Typically these systems provide a graphical representation of the information space in which the information objects being analyzed correspond to objects on a screen. By arranging and manipulating the object representations visually, users are able to identify similarities between objects, group related objects together, and identify repeated patterns or structures among the objects in the information space. A central component in spatial hypermedia systems is a parser that can interpret the layout of object representations. This spatial parser searches for and helps the user to identify implicit structures that may exist among the information objects [12]. Identifying and classifying these structures enables the system to provide support for emergent structures as the information analysis process proceeds. Examples of spatial hypermedia system implementations include VIKI [12] and CAOS [16].

Argumentation support. The discussion of design decisions and the resolution of the issues that arise is an integral part of the cooperative design process. A specialized method, called Issue Based Information Systems, has been suggested to assist in the design deliberation process. Hypermedia technology has been applied in the development of systems that implement this method and provide online support for discussing and making design decisions [4, 21]. These systems, often called argumentation support systems, typically provide a structured group discussion space in which the issues that arise during the design process can be posed. Individual group members can then take positions on issues and indicate their reasoning through the specification of arguments. This information is maintained as a network of specialized hypermedia nodes and links that help to structure and represent the issues, positions and arguments contributed by group members. In addition to facilitating design discussion, these systems also help to capture and preserve design rationale, which can be valuable information during the later stages of the design and development process. Example argumentation support systems include AAA [17] and gIBIS [4].

Additional hypermedia domains have been suggested to support development in more specialized application areas. Examples include botanical taxonomy and linguistics [15]. Support for each of these hypermedia domains, as well as future ones, can easily coexist in an open hypermedia environment. For example, a straightforward approach would be for each alternative domain to have a corresponding middleware service that provides the abstractions and operations necessary to support the domain. Client application developers would then have a rich development platform from which they could select and build on functionality from whichever hypermedia domain most closely matches user needs.

An important additional advantage for integrating multiple hypermedia domains into open hypermedia systems is that, if necessary, functionality from multiple domains might even be incorporated into the same application. For example, consider an open hypermedia environment that provides support for multiple hypermedia domains. The developer of a collaborative meeting support system working in this environment might decide to integrate associative hypermedia capabilities to provide standard hypermedia linking operations in the application. The application might then be augmented with spatial hypermedia functionality to support the automatic analysis and parsing of meeting documents. In this type of open hypermedia environment, developers can readily explore and experiment with the synergies that might be possible through the combination of functionality from different hypermedia domains.

3.3 Information Integration

The design and development process, like many complex knowledge-based activities, can involve the use of a number of different applications. For example, while a tool designed to support the brainstorming process might be useful in the early stages of design, it is probably not appropriate for the more formal and structured phases that follow. When no single application exists that can support all phases of a process, individual tools must instead be used for each of its component phases. In the case of collaborative design, this situation is exasperated even further. In addition to the multiple phases of design, which each might require the use of different applications, the varied preferences of individual design team members must also be considered. For example, members of a design team preparing notes or other materials for a meeting might each decide to use their own favorite editor or word processor for the task.

Information sharing across the phases of a project is a desirable capability, especially in a collaborative setting. However, this is difficult in an environment in which multiple applications are used. For example, in the collaborative environment described above, differences in storage formats might make it difficult or impossible to incorporate materials produced during the brainstorming phase into formal design documents produced later in the design process. Alternatively, it might be difficult or impossible to integrate the meeting notes produced by design team members who have each used different editors for the task.

Information integration is one of the primary strengths of open hypermedia systems. In the collaborative design setting, information integration can help to bridge the information boundaries that arise from the inherent use of multiple applications. In an open hypermedia environment, structures can be created that span application boundaries enabling integration of information from diverse sources. For example, through the use of associative hypermedia functionality, an engineering diagram produced using a graphical layout design tool might be linked to a document containing the minutes of the design meeting in which it was discussed. Creation of the link is possible even though the underlying information items were produced using different applications. Similar associations or links could be created between the information managed by any of the applications that have been integrated into an open hypermedia environment.

Another direct and important result of the information integration facilities in the open hypermedia environment is the ability to integrate across hypermedia domain boundaries. As described earlier (Section 3.2), certain alternative hypermedia domains have been noted to be well suited for supporting the design process. In an open hypermedia environment, not only can these different domains co-exist, as described earlier, they can also be integrated to essentially share information across domain boundaries. For example, a brainstorming application that is based on a spatial metaphor could be augmented with associative hypermedia capabilities to enable linking to argumentation structures that resulted from an online design discussion.

The powerful integration capabilities of open hypermedia systems enable an information space of isolated knowledge artifacts created by the varied applications involved to be transformed into one in which information can be more easily shared between the different users, applications, and phases involved in the collaborative design process.

4. Discussion

As outlined in the previous section, open hypermedia systems hold significant potential to accommodate the development of applications that support collaborative design. Three primary areas of functionality were described that are especially important for their ability to do so: infrastructure support services, the accommodation of alternative hypermedia domains, and information integration capabilities. Almost none of the current open hypermedia systems live up to their full potential to support collaborative design. They lack capabilities in one or more of these important areas. This section examines each of these areas further with a look at the support offered within current systems and the improvements that are needed to enable open hypermedia systems to support collaborative development more effectively.

4.1 Enhancement of Collaborative Infrastructure Services

Collaboration facilities have been provided within some open hypermedia systems. Examples systems which provide support for collaboration and upon which collaborative applications have been built or integrated into include the Cooperative Hypermedia Server [19] upon which the Dolphin cooperative hypermedia application was built, the DeVise Hypermedia architecture upon which the DeVise cooperative hypermedia application was developed [7], and the HyperDisco open hypermedia environment into which a number of applications have been integrated [23].

Collaboration in the open hypermedia environment, however, is still the subject of research. Increasing the degree and breadth of support is the primary focus. While some early open hypermedia environments did include support for collaboration, the provision of collaborative infrastructure services has not kept pace with more recent open hypermedia architectural developments. Though, as described earlier (Section 3.1), the open hypermedia approach is well suited to the provision of collaborative infrastructure services, few current open hypermedia systems offer sufficient support.

Haake, et. al., note in particular the lack of support for synchronous collaboration modes that exists in open hypermedia systems [8]. They describe the need for and have proposed a comprehensive set of collaborative infrastructure services to support a wide range of cooperative work modes including asynchronous, loosely coupled, and tightly coupled. Support for switching between modes of collaboration is also important. Their approach involves incorporating concepts from cooperative development environments such as session management and active notification services into open hypermedia systems. Regardless of the specific approach taken, the guiding principle in expanding collaboration support should be to open hypermedia systems along the collaborative dimension, thus enabling open hypermedia environments to accommodate an open set of cooperative work modes in much the same way they now accommodate an open set of applications [8].

Once the appropriate functionality necessary to support collaboration has been identified, it must somehow be implemented in the open hypermedia environment and made available to client applications. Strategies for doing so represent another area of continuing research in open hypermedia systems. One approach is to offer services in a very modular way in which the majority of collaboration support is provided by individual middleware components that are positioned in the middle architectural layer of the environment [26]. In this type of arrangement, the middleware services are themselves constructed using the functionality of the lower level hypermedia store as well as peer middleware services (Figure 2). Other approaches have been suggested that concentrate infrastructure services and move them to a lower architectural level. In this arrangement, much of the collaborative infrastructure functionality is provided at the storage level [7]. A detailed discussion of strategies for providing infrastructure services and the tradeoffs between them is beyond the scope of this paper. However, regardless of the specific approach, the goal of a service provision strategy should be to implement collaborative infrastructure services as efficiently as possible, but in a way that provides the flexibility and scalability required to support the variety of potential client applications in the open hypermedia environment.

4.2 Extending Support for Alternative Hypermedia Domains

As described earlier (Section 3.2), alternative hypermedia domains have been suggested that transcend the conventional node/link hypermedia model. Prototype systems, including argumentation support systems and spatial hypermedia systems, have been developed that demonstrate the usefulness of these new hypermedia domains. Although alternative forms of hypermedia hold significant potential for accommodating important tasks including design applications, very few open hypermedia systems currently support alternative domains. The majority of contemporary systems only permit conventional associative hypermedia operations.

Almost all of the prototype systems that have been developed for alternative hypermedia domains share a common architectural disadvantage: they were built in the monolithic style. This limits the effectiveness of these systems in important ways. Since monolithic systems do not interact with other applications, the information these systems contain cannot be easily integrated with information from other sources. For example, it would be impossible to define an associative hypermedia connection between an item in a monolithic spatial hypermedia application and a paragraph or phrase in an external document. Additionally, the alternative hypermedia domain functionality these systems provide, though potentially useful in a range of applications, is only available from within the system itself. It cannot be used by external applications.

The true potential of alternative hypermedia domains will probably not be realized until this functionality becomes available in a more open computing environment, such as an open hypermedia system.

4.3 Augmenting Information Integration Capabilities

Information integration is a category of functionality in which current open hypermedia systems generally excel. By definition, all open hypermedia systems provide at least basic level associative hypermedia operations. These operations enable associative connections or links to be defined between the information items managed by client applications. Information integration is then possible, at least on a basic level, through the creation of hypermedia links between related information items. Linking across application boundaries is permitted so that any related pieces of information can be linked together regardless of their storage format, or the application by which they are managed. In general, the associative hypermedia functionality of current open hypermedia systems provides a good basis for information integration in collaborative design applications.

The introduction of alternative hypermedia domains adds a new and interesting dimension to the information integration process in open hypermedia systems. The integration facilities described above are all based on the associative hypermedia domain. A hypermedia link is somehow created, either manually or automatically, that explicitly associates two pieces of information. Any information items managed by client applications that participate in the associative hypermedia operations of the environment are eligible to be linked together in this way. With the introduction of alternative hypermedia domains, additional forms of information integration might be possible. For example, can spatial hypermedia functionality be used to implicitly link together information items based on their spatial layout or placement? What are the semantics of inter-application spatial links? These questions along with related ones should be explored as alternative domains are integrated to determine their potential for improving the information integration facilities in open hypermedia systems.

5. Construct

Construct is a component-based open hypermedia system under development at Aarhus University and Aalborg University in Denmark. It is based on previous open hypermedia system research efforts that focused on the provision of infrastructure services [14, 24, 6]. Among other things, the Construct system serves as a prototype environment in which the suitability of open hypermedia systems for supporting collaborative design is being investigated. In this section a brief overview of the Construct system is presented which focuses on the architectural strategies it employs to maximize support for cooperative design applications. This is followed with a summary of the current status of development, and a brief look at future work directions.

5.1 System Overview

The Construct conceptual architecture is illustrated in Figure 3. A hypermedia store at the lowest level provides basic hypermedia storage functionality. This includes operations for managing basic hypermedia objects, and defining relations among them. Located above the store, the middle layer of the architecture is structured into two levels: the application level and the infrastructure level. Both of these levels are populated with middleware components. Application level middleware components provide functionality that is primarily useful for and used within client applications. Infrastructure level middleware components provide more basic functionality that, while possibly useful within client applications, is primarily used as infrastructure capabilities to support the construction of application level middleware components. Client applications reside at the top level of the architecture and make use of the middleware components.

As described in Section 3, the three most important capabilities that an open hypermedia system should supply to accommodate cooperative design applications are: collaborative infrastructure facilities, support for alternative hypermedia domains, and information integration facilities. These are accommodated in the Construct environment as follows:

- *Collaborative Infrastructure Facilities.* The Construct approach to the provision of collaborative infrastructure capabilities is to concentrate the most basic collaborative services such as notification control, concurrency control, and access control at the infrastructure level of the middle architectural layer (Figure 3). These operations can then be used in the construction of higher level collaborative functionality for use directly within applications such as a middleware component that supports collaborative session management.
- *Alternative Hypermedia Domains.* Alternative hypermedia domains are supported in Construct at the middle architectural layer. Generally this functionality is provided by application level middleware components that supply the necessary abstractions and operations necessary to support the domain.
- *Information Integration Facilities.* As an open hypermedia system, Construct naturally supplies to client applications the information integration capabilities inherent in an open hypermedia environment. This functionality is provided by an associative hypermedia service located in the application level of the middle architectural layer.

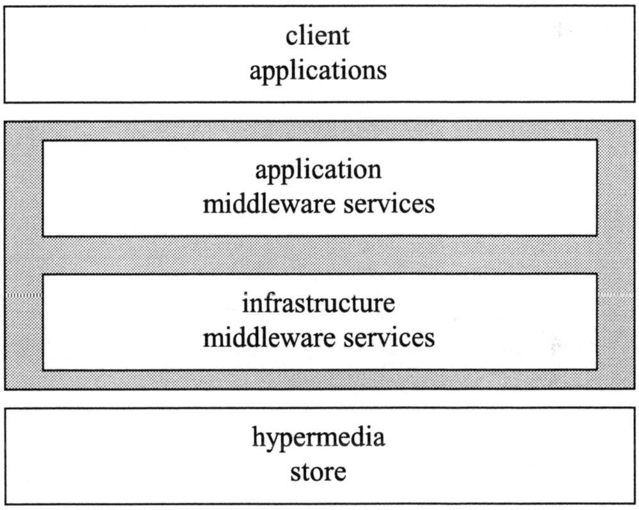

Figure 3. Conceptual Architecture of the Construct Component-Based Open Hypermedia System.

5.2 Current Status

The software components of the Construct framework have been implemented in the Java environment. At the lowest level of the architecture, the hypermedia store of the Construct system currently supports the management of basic hypermedia objects. It also provides the fundamental functionality necessary for the creation of associations between those objects.

A number of middleware components have been implemented in the middle architectural layer. An associative hypermedia service in the application level currently provides functionality for creating and traversing associative hypermedia links. Links can be created between any of the information items managed by client applications in the Construct framework. An application level collaboration service has been implemented to provide abstractions such as session record and session state that are useful in supporting collaborative work sessions. This facilitates the development of collaborative applications, especially those that support multiple modes of collaboration. A spatial hypermedia service has been implemented at the application level that supports information organization according to the spatial paradigm, and the automatic parsing of information spaces. Additional application level services have also been implemented including: a metadata service that provides the ability to define and attach metadata to information items, a taxonomic hypermedia service that provides abstractions to organize information hierarchically, and a data mining service that assists with the knowledge discovery and organization process.

Several client applications have been integrated into or built within the Construct environment to demonstrate its capabilities. The first application integrated into the environment was Emacs, which provides associative hypermedia, metadata, and collaboration services to users. In addition, Microsoft Word, Microsoft Internet Explorer, and the Vital hypermedia based collaborative learning tool [22] have each been integrated into the environment and provide associative hypermedia capabilities to users. Two applications have been built within the Construct framework to utilize the alternative hypermedia domain functionality it provides: a basic taxonomic tool enables users to develop and manipulate classification trees with a hierarchical browser, and the CAOS (Collaborative Authoring using Open Spatial hypermedia) tool [16] which provides spatial hypermedia and associative hypermedia services for users.

5.3 Future Research

One of the primary areas where further research and development is needed on the Construct system concerns its collaborative infrastructure capabilities. Currently only minimal collaboration support is offered in the form of a session management application middleware service. Many fundamental capabilities important in the development of collaborative systems have not yet been implemented. Examples include facilities for notification control, version control, access control, and transaction support. Investigation is needed to determine both the set of services that is needed and the optimal way in which they can be provided in the Construct environment (e.g., at what level of the architecture each service should be offered).

Also important will be to expand support for additional alternative hypermedia domains in the environment. Currently a spatial hypermedia service and a basic taxonomic service have been implemented. Inclusion of additional domains, such as argumentation support, will improve the development platform for design applications, and enable exploration of the use of multiple domain services within an application.

6. Conclusion

The areas of computer supported cooperative work and hypermedia have much to offer one another, as noted by pioneer researchers in these fields. However, the potential synergy between them is often overlooked. Open hypermedia systems offer an opportunity to once again reacquaint these two fields. In particular, the characteristics of the open hypermedia environment – one of the latest developments in hypermedia architectures, give it significant potential to facilitate the development of collaborative design systems – an important computer supported cooperative work application domain.

It is the flexible approach utilized by open hypermedia systems to provide infrastructure services, the ability of these systems to accommodate alternative hypermedia domains, and their facilities for information integration across application boundaries that provide open hypermedia systems with the potential to support collaborative design applications. Current open hypermedia systems, however, have not realized their full potential in this regard. They lack capabilities in one or more of these three important areas. Improvements are especially needed in the collaborative infrastructure services provided by current open hypermedia systems and their support for alternative hypermedia domains.

A prototype open hypermedia system is currently being extended to address the shortcomings identified in existing systems. Several of the software components of the extended prototype are already in place. A middleware component provides a collaborative session management service. Middleware components have also been constructed to support two alternative hypermedia domains: spatial hypermedia and taxonomic hypermedia. As the collaborative infrastructure services of the prototype are extended and additional hypermedia domains such as argumentation support are integrated, a very capable environment for the development of collaborative design applications will be produced.

7. References

[1] Akscyn, R., McCracken D., and Yoder, E. 1988. KMS: A distributed hypermedia system for managing knowledge in organizations. *Communications of the ACM,* 31(7), 820-835.

[2] Anderson, K., Taylor, R., and Whitehead, E. J. 1994. Chimera: Hypertext for heterogeneous software environments. In Proceedings of the 1994 ACM European Conference on Hypertext, (Edinburgh, Scotland, Sep), ACM Press, 94-107.

[3] Berners-Lee, T., Cailliau, R., Luotonen, A., Nielsen, H. F., and Secret, A. 1994. The World-Wide Web. Communications of the ACM, 37 (8), 76-82.

[4] Conklin, J. and Begeman, M. L. 1988. gIBIS: A Hypertext Tool for Exploratory Policy Discussion. In Proceedings of CSCW 88: The Conference on Computer Supported Cooperative Work (Portland, Oregon, Sept) 140-152.

[5] Engelbart, D. C. 1984. Collaboration Support Provisions in Augment. In Proceedings of AFIPS '84, (Los Angeles, CA, Feb), pp. 51-58.

[6] Grønbæk, K., Hem, J. A., Madsen, O. L., and Sloth, L. 1994. Cooperative Hypermedia Systems: A Dexter-based Architecture. CACM, 37, 2, (Feb), 64-74.

[7] Grønbæk, K., and Trigg, R. 1999. From web to workplace – Designing open hypermedia systems. MIT Press.

[8] Haake, J. and Wang, W. 1998. Collaboration Support in Open Hypermedia Environments. In Proceedings of the Fourth Workshop on Open Hypermedia Systems, (Pittsburgh, PA, May).

[9] Halasz, F., Moran, T., and Trigg, R. 1987. NoteCards in a nutshell. In Proceedings of the 1987 ACM CHI Conference, (Toronto, Canada, Apr), ACM Press, 45-52.

[10] Halasz, F. 1988. Reflections on Notecards: Seven issues for the next generation of hypermedia systems. *Communications of the ACM,* 31, 7, (July), 836-852.

[11] Hall, W., Davis, H., and Hutchings, G. 1996. Rethinking hypermedia – The Microcosm approach. Kluwer Academic Publishers.

[12] Marshall, C. and Shipman, F. 1995. Spatial Hypertext: designing for change. CACM 38 (3), Aug, 88-97.

[13] Meyrowitz, N. The missing link: Why we're all doing hypertext wrong. In *The Society of Text:* Hypertext, Hypermedia, and the Social Construction of Information, MIT Press, 1989, pp. 107-114.

[14] Nürnberg, P. J., Leggett, J. J., Schneider, E., and Schnase, J. 1996. HOSS: A new paradigm for computing. In Proceedings of the 1996 ACM Hypertext Conference, (Washington, DC, Mar), ACM Press, 194-202.

[15] Nürnberg, P. J., Leggett, J. J., and Schneider, E. R. 1997. As We Should Have Thought. In Proceedings of Hypertext '97 (Southampton, England, April), 96-101.

[16] Reinert, O., Bucka-Lassen, D., Pedersen, C. A., and Nuernberg, P. J. 1999. CAOS: A Collaborative and Open Spatial Structure Service Component with Incremental Spatial Parsing. In Proceedings of Hypertext '99 (Darmstadt, Germany, Feb), 49-50.

[17] Schuler, W. and Smith, J. 1990. Author's Argumentation Assistant (AAA): A hypertext-based authoring tool for argumentative texts. In Proceedings of ECHT '90, (Versailles, France, Nov) 137-151.

[18] Schütt, H. A., and Streitz, N. A. 1990. Hyperbase: A hypermedia engine based on a relational database management system. In Proceedings of the 1990 European Conference on Hypertext, (Versailles, France, Nov), Cambridge University Press, 95-108.

[19] Schütt, H. and Haake, J. M. 1993. Server support for cooperative hypermedia systems. In Proceedings of the International Hypermedia '93 Conference (Zurich, Switzerland, Mar) Springer, 45-56.

[20] Shackelford, D. E., Smith, J. B., and Smith, F. D. 1993. The architecture and implementation of a distributed hypermedia storage system. In Proceedings of the 1993 ACM Hypertext Conference, (Seattle, WA, Nov), ACM Press, 1-13.

[21] Smolensky, P., Bell, B., Fox, B., King, R., and Lewis, C. 1987. Constraint-based hypertext for argumentation. In Proceedings of HT '87 (Chapel Hill, NC, Nov) 215-245.

[22] Wessner, M., Bech-Wilson, J., and Pfister, H. H. 1998. CLear – a cooperative distributed learning environment. In Proceedings of ED-MEDIA/ED-TELECOM 98, (Freiburg, Germany, Jun), 1876-1877.

[23] Wiil, U. K. and Leggett, J. J. 1996. The Hyperdisco Approach to Open Hypermedia Systems. In Proceedings of ACM Hypertext '96 (Washington D.C., March), 140-148.

[24] Wiil, U. K., and Leggett, J. J. 1997. Workspaces: The HyperDisco approach to Internet distribution. In Proceeding of the 1997 ACM Hypertext Conference, (Southampton, UK, Apr), ACM Press, 13-23.

[25] Wiil, U. K. and Nürnberg, P. J. 1999. Evolving hypermedia middleware services: Lessons and observations. In Proceedings of the 1999 ACM Symposium on Applied Computing, (San Antonio, TX, Feb) ACM Press, 427-436.

[26] Wiil, U. K., and Hicks, D. L.. 2000. HyTech: Development of hypermedia technology for digital libraries supporting scholarly work. In Proceedings of the International Workshop on Cooperative Internet Computing (CIC 2000), (Hong Kong, China, Nov), 64-71.

[27] Yankelovich, N., Haan, B., Meyrowitz, N., and Drucker, S. 1988. Intermedia: The Concept and the Construction of a Seamless Information Environment. IEEE Computer, 21(1), 81-96.

An Overview of Cognitive Factors in Distributed Design

Sherman Lang
National Research Council, Integrated Manufacturing Technologies Institute, 800 Collip Circle, London, Ontario, Canada, N6G-4X8
sherman.lang@nrc.ca

John Dickinson
National Research Council, Integrated Manufacturing Technologies Institute, 800 Collip Circle, London, Ontario, Canada, N6G-4X8
john.dickinson@nrc.ca

Ralph Buchal
Faculty of Engineering Science, University of Western Ontario, London, Ontario, Canada, N6A-5B9
rbuchal@eng.uwo.ca

Abstract

Engineering design is being carried out by distributed design teams in an effort to use expert human resources more efficiently. Efforts to support distributed teams often neglect cognitive factors such as the social functions being fulfilled and limitations in the ability of humans to perceive, contextualise, process and recall information. The support needs of distributed design are reviewed from this cognitive viewpoint using five broad categories: Design Methodology, Collaboration, Teamwork, Knowledge Management, Design Representation.

1. Introduction

The globalisation of manufacturing is based on the trend of using the most efficient resources possible for whatever task needs to be done. In product design, this trend means exploiting the knowledge and expertise of all parties involved, including marketing, engineering, design, management, suppliers, production, etc., in the design team, no matter how these parties are distributed geographically and organisationally.

As such, the support needs of distributed design teams have become an important area of research though the field remains in its infancy. The goal of this paper is to provide an overview of recent research into the underlying cognitive nature of design, from publications in the fields of cognitive science, design and engineering, and review the impact this has had or could potentially have on supporting distributed design. To help group related research work, the support needs of distributed design are classified using a number of broad categories: Design Methodology, Collaboration, Teamwork, Knowledge Management and Design Representation. It is acknowledged that Integration is also highly important to distributed design but a review of research into this area has been left to a future survey, though a brief comment is included in the summary.

The impact of recent cognitive research in each of the five broad categories listed above is reviewed in the following sections, while Section 7 summarises our findings and suggested courses for future research and implementation.

2. Design Methodology

The design activity transforms available information and knowledge to build a mapping from an expressed need to a solution. The transformation has been viewed as an iterative evolution of design from the abstract to concrete. Design problems can vary in type from ill defined to parametric. A broad range of activities may need to be undertaken with the major steps including:

- needs analysis / problem clarification,
- information gathering / research,
- ideation / creative thinking,
- information generation / analysis,
- evaluation, and
- optimisation.

The most commonly prescribed methodology for engineering design can be traced to origins in systems theory, and is often referred to as the systems approach to design. The systems approach can be found detailed in early texts such as by Soulis and Handa [51] and more recently by Pugh [48], Dieter [18], Ullman [64] and Pahl and Beitz [44]. Other authors refer to the methodology as the Analysis-Synthesis-Evaluation (ASE) paradigm. The process granularity may differ but can be seen to have the same general structure, dividing cognitive activities of analysis and synthesis into sequential phases and iteration after an evaluation phase. Braha and Maimon [11] develop a mathematical foundation and frameworks for scientific study of the design process.

Many software tools and systems exist to support individual design tasks and the design process as a whole. The computer support of methods such as design structure matrices, evaluation matrices, QFD and morphological charts allows designers to consider larger numbers of alternatives and makes it possible to perform more design iterations than could be carried out by hand, as in the past. Recent work attempts to automate these software based design tools as well as to transfer them to web-based environments [27]. Web-based implementation has lead to greatly increased accessibility of tools and methodologies, and provides a basis for a more consistent human-computer interface across a range of tools.

The process of design has been recorded and analysed in several protocol studies of individuals and teams. Studies have focused on the cognitive activities of design teams [65, 50, 58] cognitive activities and tasks, [61, 57, 23], the role of experience [25, 1], and the role of personality type [72].

3. Collaboration

Collaboration is an activity where a large task is achieved by a team. Contributions to the work are negotiated and mediated through communications and sharing of knowledge. It is worth noting that the boundary between teamwork and collaboration is not well defined. Successful collaboration requires effectiveness in a number of areas:

- Cognitive synchronisation / reconciliation
- Developing shared meaning
- Developing shared memories
- Negotiation
- Communication of data, knowledge, information
- Planning of activities, tasks, methodologies
- Management of tasks

Cognitive synchronisation is the process of establishing a mutual understanding of issues related to the design. It is recognised as a very important component of collaboration [7, 37] and is significant as a design activity since it takes a large percentage of a designer's time [50, 11]. Current computer assisted cognitive synchronisation is generally limited to scheduled meetings and formalised negotiation methodologies [46] and thus remains an open topic. Theoretically, developing shared meaning requires achieving a mutually accepted and understood lexicology, schema or language in which to communicate, despite differences in backgrounds (education, training, experience, fields, etc.) of the team members. In practice, this needs to be achieved through regular cognitive synchronisation [9, 26, 53] and remains a research issue. Developing shared memories has become more automated and includes the automated generation of the minutes of a meeting or video transcripts of meetings, caching and indexing of email, etc. A formalised approach to this can be found in [10]. Current distributed approaches focus on shared design workspaces [13, 28, 46], which are repositories for shared design forms (e.g. formalised documents like drawings, specs, analysis data, etc.) and do not commonly including videos of meetings or email. Shared design workspaces and forms are discussed in the Teamwork section below.

Negotiation is a vital part of collaboration [6, 13, 32, 40, 46, 59] and computer tools exist that formalise and implement mediation and facilitation techniques to assist in the process. Communication remains a weak link for computer-assisted collaboration due to varying standards, platforms and versions of software between organisations [47] when communicating formal documents (forms). The difficulties are even greater when faced with trying to facilitate productive and natural communication between distributed team members [21, 46]. Room exists for continuing work on facilitating both document and natural communication. In contrast, planning of activities, tasks and methodologies as well as the management of tasks is quite a mature domain and many advanced commercial products already exist to assist in these functions (e.g. Microsoft Project). Prototype tools also exist as integrated parts of collaborative systems [9, 21, 34, 46, 53].

4. Teamwork

The sociological aspect of collaborative design is teamwork. Teamwork is important in maintaining focus and commitment. The development of teams is largely due to organisational factors and decisions, but must be mediated by technology in a distributed environment. Technical aids should be focused on the problems of:

- Ownership and commitment
- Shared design workspaces
- Organisation incentives (team spirit, reputation, co-operation)
- Member assumed roles and responsibilities

Teamwork can make or break a collaborative project and affects all of the design activities, particularly in the selection of design alternatives and resolution of conflicts [17]. Traditionally, team spirit is developed in face-to-face environments with team leaders acting as facilitators encouraging harmonious relations within the group. Communication methods used during the establishment of intra-team relationships are rich and include body-language, physical gestures, and even direct physical contact, predominately centred around an artefact (drawing, diagram, physical model, etc.) representing the topic of discussion. Though modern technology and communication media can bring many of the modes of communication prevalent in face to face meetings to distributed groups, it currently does not bring them together as an integrated, natural, whole [47]. Some consensus seems to exist that technology enhanced communication, at the very least, can improve the usefulness and quality of shared memory and meanings (discussed earlier). To help establish rapport for distributed teams, it is important to support as rich a communication interface as possible, especially at the start of a project. The potential benefits of fostering team spirit include improved effort and innovation by individual "spirited" members, which would in turn encourage others to do the same.

One mechanism that can contribute to team spirit is activity monitors giving constant but unobtrusive feedback of task statuses and team member activity [21] which enhances feelings of contributing to a group effort. To inspire further individual effort, allow members to build and enhance reputations by providing mechanisms to record and rate member contributions and to share these ratings across projects. These mechanisms may also foster improved co-operation, as individual assistance on a task would often be reciprocated later as well as potentially adding to an individual's reputation [28]. Shared design workspaces can also positively contribute to team spirit for distributed groups by further enhancing the feeling that members are contributing to the group effort [9, 53] by providing everyone with the same interface to the project's shared design forms. Shared design workspaces also often becoming another tool for making distributed team communication richer.

Team members often naturally assume different "roles" to assist in the communication process inside and outside team boundaries [59]. These roles include facilitating or mediating intra- and inter-team discussion as well as acting as spokespeople for the team with company or enterprise representatives. Understanding and facilitating these communication roles can result in better team harmony, more support from company officials and better co-operation with external company representatives. Thus, when designing tools to support design groups, consideration should be given to what sort of communication roles will often be performed while or after using the tool in question and efforts made to facilitate those communication roles.

Shared design workspaces are required that can support team collaboration in both virtual and face-to-face situations. In particular, this requires the design of meeting rooms with a shared data display and multiple user interfaces so that team members can interact with the digital design space as well as with each other [55]. For distributed teams, virtual meeting rooms with multiple modes of communication are required. For example: rooms with both a shared electronic whiteboard and video conferencing would give all members access to the whiteboard while others to view both the contents of the whiteboard and how the current speaker is interacting (gestures, facial cues, etc.) with those contents.

5. Knowledge Management

A great deal of knowledge is used and generated in a design. The capture and expression of this knowledge is vital if teams are to be able to use both existing knowledge and generate new knowledge for future activities. There are three basic types of design knowledge that are important:

- design intent,
- design rationale,
- design history.

While design from first principles is always possible and often can result in innovative solutions, the majority of designs are derived from existing designs. One view of design is as a search for an optimal mapping from a problem sub-space to a solution sub-space. Many design problems can be viewed as a search of the solution sub-space near existing designs that are close to the current design problem. A knowledge base of past designs should provide faster design by narrowing the search space and is the basis of case-based design systems. The case-based approach is more important in mechanical designs where there is not a direct mapping from the functional requirements to the system components as in electronic or software design.

It has been shown that even innovative or original design can be supported by a knowledge base of past designs. The Theory of Inventive Problem Solving (also known as TRIZ) uses analogical reasoning and a knowledge base of inventive principles, effects, and known solutions to map a specific problem into an analogous generic problem, generating generic concepts based on a small number of inventive principles, which are then refined to solve the specific problem [62]. It is often found that inventive solutions to the generic problem have already been found in other disciplines, and can be easily adapted to the new problem.

Another important finding is that many design decisions are recorded in documents that do not become part of the final formal design documents, which serve primarily for manufacturing purposes. One very common repository for engineering design information is the design notebook. Designers use the notebook as both a mechanism for recording and transforming information. The design notebook also has the function of juxtaposing knowledge to support creative and analytical tasks. Experienced designers, who make effective use of notebooks, record not only the information and the transformed information, but also the means of transformation and the reasons. In a survey of designers [16] in the UK and the USA, it was found that 50% of design decisions are recorded in hard copy design notebooks. The other major decision repositories were found to be diaries, memos, reports, general notebooks, data/calculation sheets, project files, contracts, design documents and drawings. Kuffner and Ullman [33] propose an intelligent CAD tool that can capture and organise design notebook information automatically. This approach preserves traditionally notebook methodologies but completely different alternatives might be possible leveraging modern technology and data management techniques.

To determine the nature and the importance of information required by designers, the information requests of designers studying blueprints and specifications in order to understand and then modify them to meet altered specifications were recorded and analysed. A large proportion of the information requests was related to the nature or purpose of the design object. Also, the conjectures of the subjects on the design were found to be incorrect about 45% of the time. A detailed study of aerospace designers provides considerable insight into the nature, sources, patterns of usage, and role of information accessed by the designers [38]. It was found that the designers accessed knowledge predominately through colleagues and other people outside their organisation, despite the availability of a large volume of internally available reference and design documents. The author suggests a number of reasons:

- the largely hardcopy format of many of the documents,
- the lack of indexing,
- the ease, cognitively, of accessing personal contacts.

The author speculates that accessing information with people as the point of first contact allows the designer searching for the knowledge to have the person who is the point of contact assess the relevancy of the information sought. The author concludes that the information system needs a browsable structure and the ability to assess the relevancy of the information. Other results include the time spent by designers in acquiring or providing information (24% on average), and the types of queries made by the designers. The information structure of some of the internal reference documents was also analysed and found to contain very little guidance on configuration and concept selection (7% of the documents).

It is clear that traditional paper-based design processes have significant drawbacks, and that effective computer tools do not yet exist to support the entire design lifecycle. Knowledge management is particularly difficult during the period of transition, when some design knowledge is captured on paper and some is digital.

6. Design Representation

Designs are instantiated in a number of different forms. These forms can consist of different representations of artefacts, proto-artefacts, prototypes, process plans, etc. The forms serve as the instantiation of the artefact under design and catalysts for further development and evolution. Representations of the artefact in different domains and at different levels of abstraction and certainty are often needed. Fruchter et al. [20] describe a system for mechatronic design, which provides functional, mechanical and electrical representations, as well as graphic, semantic and symbolic. While the basic types of representation would seem to be the functional, behavioural and structural, no systems exist which support simultaneously the many different types of form representation commonly used in engineering design. Al-Salka et al. [5] provide support for six types: 1) textual, 2) graphical, 3) structural, 4) hierarchical, 5) tabular, and 6) morphological, which serve to capture both the substance of the design artefacts as well as data and knowledge generated in the evolution of the design.

Sketching activity is used as a way of manipulating and developing multiple representations (e.g. functional diagrams, equations, shape, etc.) in the design process. There is evidence that expert designers solve design problems by searching for analogies, and that visual analogies are supported by sketching [12]. The role of sketching in design is studied in [63, 30, 29, 19, 24, 49, 39, 31, 67]. The characteristics of sketching that make it an important process in design are:

- Immediacy and speed of capture of visual representations of ideas and concepts
- Usage of spatial structure to represent relationships
- Evolution of form by facilitating cognitive activities of restructuring and combining.

Despite the recognition of the importance of sketching activity in design, it is not well supported in traditional CAD systems in terms of input capture, parsing, interpretation and storage. While parametric modelling systems allow refinements and constraints to naturally and automatically adjust the design geometry, there is little support for capture and organisation of multi-domain representations and conceptual representations preceding the geometric instantiation. Natural input via pen and pad interface has been recognised as having a number of advantages but has been difficult to implement due to lack of understanding of both the cognitive process that need to be supported as well as the computational requirements of parsing freehand input. To this end, several researchers have developed prototype computer sketching systems like Easel [30], and the Electronic Cocktail Napkin [24] that focus mainly on capture and interpretation of geometry. These systems are faithful to the sketching metaphor, but contain no facility to generate 3D geometry from the sketches. However, CAD systems are beginning to incorporate intelligent sketching capabilities with some of the capabilities described in these and other works, including:

- Automatic constraint recognition
- Automatic conversion of freehand input to regular geometry
- Precise input not required

An important advantage of CAD-based sketches is that they can be used directly to define 3D geometry.

Other authors report experiments with 3D modelling tools that are reported to be superior to sketching in some applications. For example, van Dijk's Fast Shape Modeler (FSD) [66] demonstrated the feasibility of 3D CAD software for conceptual design. Many of FSD's features are found in current surface modelling and solid modelling packages.

The importance of prototyping also reflects the need to be able to create and modify the substance of the design (i.e. form) [43, 56, 42, 8] as part of the creative problem solving process. Prototypes are important in early design to resolve overall form and aesthetic issues. In early design, ease of interaction is a more important factor than precision in specifications, and prototyping is seen as a way of testing assumptions. In the later part of the design, prototypes are used to detect errors in fit, operation and precision, thus support for analytical techniques and detail for production processes become more important. Simulation and visualisation techniques provide realistic virtual prototypes that are particularly important for complex expensive systems where physical prototyping is impractical or prohibitively expensive.

Conventional sketching does not directly support prototyping – the design representation needs to be translated from a sketch to a digital or physical model. Furthermore, until recently CAD systems did not support the design process to allow easy manipulation of imprecise form to be captured and evolved into detailed virtual prototypes. However, the current generation of parametric, feature-based modellers allow designers to create and manipulate approximate geometric representations, which can be refined as the design evolves. During this refinement, rough dimensions are replaced with exact dimensions, and additional geometric features are defined. These systems allow geometry to be defined both through sketches, and by direct manipulation of the solid model. This new approach is fundamentally different from previous modelling methods including wireframe modelling, surface modelling and constructive solid modelling. This shift has only recently been documented in the literature [35, 71].

Collaborative design teams also need to be able to share design forms. Shared designs allow design team members to develop cognitive synchronisation of the state of the artefact and the design process. Digital shared designs are particularly important for distributed design teams where it is impractical to share physical artefacts. Shared design forms can be achieved through the simple importing / exchange of data files or through a use of a shared design workspace [9, 53], and commercial shared design tools and workspaces are beginning to appear (e.g. Alibre Design, www.alibre.com). Shared workspaces have the advantage of avoiding the pitfalls of creating multiple different copies of files and allowing users to always know where the latest design data resides.

The traditional means of communicating design information has been the 2D engineering drawing, including multiview orthogonal projections and pictorial representations. Multiview orthogonal views are difficult to correctly interpret even for experienced engineers because of the lack of spatial cues and the cognitive effort required in constructing a mental image. An interactive 3D display can exploit many known spatial cues, including perspective, light and shadow, texture gradients, depth fading, and motion parallax [70]. Additional cues are available through emerging virtual reality technologies, including stereoscopic displays, immersive environments and haptic interfaces.

Considerable interest currently exists in design in 3D virtual environments. This appears to be a logical extension of 2D sketching to 3D "sculpting". While recent progress has been made, considerable research and development is still required to provide a 3D equivalent of 2D sketching which has similar attributes of low cost and high availability while maintaining and improving ease of use of the interface for trained and inexperienced users alike.

Notwithstanding a 3D equivalent to 2D sketching, it is widely recognised that a computer environment that supports freehand input such as pen based computing environments could have many advantages. A computer environment supporting 2D sketching would allow the integration of the form manipulation (sketching) design activities with downstream modeling, detailed design, analysis, evaluation and optimisation activities. Many of the necessary technologies already exist in devices like pen-based palm computers, digitizing tablets, touch screens, etc., but no platform is currently available that integrates all of these in an effective way. The Tablet PC under development by Microsoft appears to have many of the necessary attributes.

7. Summary

Currently, many of the computational needs of design activities are supported through a large variety of commercially available tools and systems. The integration of the tools and systems is increasing through a variety of research and commercial efforts. However, flexibility to adapt to the range of different design problems and domains is still lacking.

Integration is a highly discussed issue with regards to improving design support systems, let alone collaborative systems. Without integration, data and information needs to be manually transferred between tools adding to the cognitive load, disrupting creative thought processes and leading to the possibility of miss-interpretation or loss of design information. Many of the enabling technologies now exist, and integrated software systems are beginning to appear. However, such systems will continue to have significant limitations until cognitive aspects of design are thoroughly understood and addressed. In particular, research into integration necessarily has to consider work on developing shared meanings in order to encapsulate what information needs to be preserved and communicated throughout the design lifecycle.

There is evidence that designers of technological systems to support design activities do not adequately address the cognitive and human factors of their systems [12]. A survey of the engineering literature provides little evidence of the extensive and highly relevant work being done in psychology and related fields, possibly because these publications do not appear in the engineering databases. Effective design support systems must complement human activities, and must be based on a sound understanding of the human cognitive abilities. Existing knowledge in this key area is thoroughly surveyed by Wickens and Hollands [70].

Tools and systems which support the cognitive and relationship building activities of teams and individuals exist, but are not well integrated with emerging engineering design methodology. Until these tools become easier to use and more widely known, design engineers will continue to rely on traditional methods of collaboration and teamwork.

Knowledge management issues in conceptual design are centred on information gathering, and the capture and use of design knowledge. The capture and representation of design intent, design rationale and design history is required for the purpose of a) capture of design expertise as a corporate asset, b) reuse of design expertise to accelerate future designs, c) facilitating backtracking during complex and ill-defined and ill-structured design problems. Currently hard copy design notebooks are the most commonly used devices to record the design knowledge. Shared electronic design notebooks have been proposed for collaborative design, and a number of efforts have been made at implementing electronic design notebooks. It has been found that designers spend a large portion of their time in information gathering and distribution activities. Even when a large volume of information and knowledge is easily available in electronic form and searchable and indexed, the designer is still faced with the difficult task is of assessing the relevancy of the knowledge in the context of the current design. Methods like TRIZ [62] and Axiomatic Design [60] provide frameworks for this but application of them requires time, effort and knowledge of the method. Further research is required into developing and automating methods for exposing and exploiting relevant knowledge. In addition to developing better software tools, significant gains might also be achieved by studying and improving the human ability to recognise analogies and identify relevant knowledge.

Conceptual design requires different representations of artefacts. Environments that allow designers to work simultaneously with function and form are required to allow capture and structuring of knowledge of different types (semantic, structural, geometric, behavioural, declarative, etc.). Sketching is an important cognitive activity in design because it allows easy capture and manipulation of design forms, and systems are required that support the sketching activity.

The representation and manipulation of shared forms in a shared design workspace is also needed to support distributed collaborative design teams. Design artefacts need to be shared to prevent inconsistency and divergence. A shared design artefact and workspace supports cognitive synchronisation of the state of the design artefact and the design process.

8. References

[1] S. Ahmed, L. Blessing, K. Wallace, "The relationships between data, information and knowledge based on a preliminary study of engineering designers", ASME Design Theory and Methodology, Las Vegas, Nevada, Sept. 12-15, 1999.

[2] M.A. Al-Salka, M.P. Cartmell, S.J. Hardy, "A framework for a generalized computer-based support environment for conceptual engineering design", Journal-of-Engineering-Design, March 1998, Vol.9, No.1, pp.57-88.

[3] P. Badke-Schaub, E. Frankenberger, "Analysis of design projects", Design Studies, Vol. 20, No. 5, Sept. 1999, pp. 465-480.

[4] J. Bardram, "Design for the dynamics of cooperative work activities", Proceedings of the conference on computer-supported cooperative work 1998 (CSCW '98), pp. 89-98.

[5] K. Bechloum, "Intelligent Electronic Mock-Up for Concurrent Design", Expert Systems with Applications, Vol. 12, No. 1, 1997, pp. 21-35.

[6] L.M. Berlin, R. Jeffries, V.L. O'Day, A. Paepcke, C. Wharton, "Where did you put it? Issues in the design and use of a group memory", Human Factors in Computing Systems. (INTERCHI '93), 1993, pp. 23-30.

[7] G. Boy, "The Group Elicitation Method: an introduction". Proceedings of the 9th European Knowledge Acquisition Workshop (EKAW '96), Berlin, Germany, 1996, pp. 290-305.

[8] D. Braha, O. Maimon, "A Mathematical Theory of Design: Foundations, Algorithms and Applications", Kluwer Academic Publishers, Boston, 1998.

[9] H. Casakin, G. Goldschmidt, "Expertise and the use of visual analogy: implications for design education", Design Studies, Vol. 20, No.2, March 1999, pp. 153-175.

[10] M.P. Case, S.C-Y. Lu, "Discourse model for collaborative design", Computer Aided Design, Vol. 28, No. 5, 1996, pp. 333-345.

[11] K.W. Chan, K.H. Lau, W.B. Lee, "Development of a Web-Based 3D Cad System for Automatic Geometry Creation", 5th International Conference on Manufacturing Technology, Hong Kong, Oct. 31 - Nov. 3, 1999.

[12] A. Cooper, The Inmates are Running the Asylum, SAMS, New York, 1999.

[13] A.W. Court, D.G. Ullman, S.J. Culley, "A Comparison Between the Provision of Information to Engineering Designers in the UK and the USA", International Journal of Information Management, Vol. 18, No. 6, December 1998, pp. 409-425.

[14] A.C. Cross, N. Cross, "Observations of teamwork and social processes in design", *Design Studies*, Vol. 16, Iss. 2, April 1995, pp. 143-170.

[15] G.E. Dieter, Engineering Design: A Materials and Processing Approach, 3rd Edition, McGraw-Hill, 2000.

[16] L. Eggli, C-Y. Hsu, B.D. Bruderlin, G. Elber, "Inferring 3D models from freehand sketches and constraints," Computer Aided Design, Vol.29, No.2, Feb. 1997, pp.101-112.

[17] R. Fruchter, K.A. Reiner, G. Toye, L.J. Leifer, "Collaborative mechatronic system design," Concurrent Engineering Research and Applications, Vol. 4, No. 4, Dec. 1996, pp. 401-412.

[18] S.R. Fussell, R.E. Kraut, F.J. Lerch, W.L. Scherlis, M.W. McNally, J.J. Cadiz, "Coordination, overload and team performance: effects of team communication strategies", Proceedings of the Conference on Computer-Supported Cooperative Work 1998 (CSCW '98), pp. 275-284.

[19] A. Giretti, L. Spalazzi, "ASA: a conceptual design-support system", Engineering Applications of Artificial Intelligence, Vol. 10, No. 1, 1997, pp. 99-111.

[20] G. Goldschmidt, "The designer as a team of one", Design Studies, Volume 16, Issue 2, April 1995, pp. 189-209.

[21] M.D. Gross, "The Electronic cocktail napkin - a computational environment for working with design diagrams", Design Studies, Vol. 17, No. 1, Jan 1996, pp. 53-69.

[22] J. Günther, K. Ehrlenspiel, "Comparing designers from practice and designers with systematic design education," Design Studies, Vol. 20, No. 5, September 1999, pp. 439-451.

[23] C.M. Harvey, R.J. Koubek, "Toward a model of distributed engineering collaboration", 23rd International Conference on Computers and Industrial Engineering, Vol. 35, Nos. 1-2, 1998, pp. 173-176.

[24] G.Q. Huang, K.L. Mak, "Web-based collaborative conceptual design", Journal of Engineering Design, Vol.10, No.2; June 1999, pp.183-194.

[25] J. Huang, "Knowledge sharing and innovation in distributed design: implications of internet based media on design collaboration", Design Computing on the Net '99 (DCNET '99), Nov. 30- Dec. 3, http://www.arch.usyd.edu.au/kcdc/journal/vol2/dcnet/sub5.

[26] T-S. Hwang, D.G. Ullman, 'Recognize features from freehand sketches," Proceedings of Computers in Engineering Conference, Sept. 11-14, 1994, Minneapolis, MN, USA., Vol. 1., pp.67-78.

[27] D.L. Jenkins, R.R. Martin, "The Importance of Free-Hand Sketching in Conceptual Design: Automatic Sketch Input," Design Theory and Methodology, Vol. 53, 1993, pp. 115-128.

[28] M. Kavakli, S.A.R Scrivener, L.J. Ball, "Structure in idea sketching behaviour", Design Studies, Vol. 19, No. 4, October 1998, pp. 485-517.

[29] J. Kilker, "Conflict on collaborative design teams: understanding the role of social identities", IEEE Technology and Society Magazine, Fall 1999, Vol. 18, No. 3, pp. 12-21.

[30] T. Kuffner, D.G. Ullman, "The Information Requests of Mechanical Design Engineers" Design Studies, Vol. 12, No. 1, January 1991, pp. 42-50.

[31] B. Kumar, A. Retik, "Towards an integrated engineering design environment", Information Representation and Delivery in Civil and Structural Engineering Design, Int. Conf. on Info. Tech. in Civil and Structural Eng. Design, 14-16 Aug 1996, pp. 127-133.

[32] S.D. Lockhart, C.M. Johnson, "Engineering Design Communication: Conveying Design through Graphics", Prentice Hall, 2000.

[33] P. Luff, C. Heath, "Mobility in Collaboration", CSCW '98 Proceedings, pp. 305-314.

[34] M. Mandviwalla, S.D. Clark-III, K. Sandoe, "Collaborative writing as a process of formalizing group memory", Proceedings of the Twenty-Eighth Hawaii International Conference on System Sciences, 1995, Vol.5, pp. 342-350.

[35] R. Marsh, "The Capture and Utilization of Experience in Engineering Design," Ph.D. Thesis, University of Cambridge, 1997.

[36] A. McGown, G. Green, P.A. Rodgers, "Visible ideas: information patterns of conceptual sketch activity," Design Studies, Vol. 19, No. 4, October 1998, pp. 431-453.

[37] G.N. Mentzas, "Team coordination in decision support projects", European Journal of Operational Research, Vol. 89, 1996, pp. 70-85.

[38] N. Miyake, "Making internal processes external for constructive collaboration", Proceedings from the Second International Conference on Cognitive Technology Humanizing the Information Age, 1997, pp. 119-123.

[39] J. Numata, "Knowledge Amplification: An Information System for Engineering Management," Sony's Innovation in Management Series, Sony Corporation, Vol. 17, March 1996.

[40] S. Ottosson, "Qualified Product Concept Design Needs a Proper Combination of Pencil-aided Design and Model-aided Design Before Product Data Management", Journal of Engineering Design, Vol. 9, No. 2, 1998, pp. 107-119.

[41] G. Pahl, W. Beitz, Engineering design: a systematic approach. Springer, 1996.

[42] H. Park, M.R. Cutkosky, "Framework for modeling dependencies in collaborative engineering processes", Research in Engineering Design, 1999, Vol. 11, pp.84-102.

[43] F. Peña-Mora, K. Hussein, "Proactive meeting management for distributed collaborative design", Advances in Engineering Software, Vol. 29, No. 10, 1998, pp. 839-849.

[44] M. Perry, D. Sanderson, "Coordinating joint design work: the role of communication and artefacts", Design Studies, July 1998, Vol. 19, No. 3, pp.273-288.

[45] S. Pugh, Total Design: Integrated Methods for Successful Product Engineering. Addison-Wesley, 1991.

[46] A.T. Purcell, J.S. Gero, "Drawings and the design process," Design Studies, Vol. 19, No. 4, October 1998, pp. 389-430.

[47] P.N. Robillard, P. d'Astous, F. Détienne, W. Visser, "Measuring Cognitive Activities in Software Engineering," 20th International Conference on Software Engineering, Kyoto, Japan, April 19-25, 1998, pp. 292-300.

[48] P.H. Roe, G.N. Soulis, V.K. Handa, The Discipline of Design, University of Waterloo, 1969.

[49] U. Roy, B. Bharadwaj, S.S. Kodkani, M. Cargian "Product development in a collaborative design environment," Concurrent Engineering Research and Applications, Vol. 5, No. 4, Dec. 1997, pp. 347-365.

[50] M. Saad, M.L. Maher, "Shared understanding in computer-supported collaborative design", Computer Aided Design, 1996, Vol. 28, No.3, pp. 183-192.

[51] N. Sabbaghian, S. Eppinger, E. Murman, "Product development process capture and display using Web-based technologies", Proceedings of the 1998 IEEE International Conference on Systems, Man, and Cybernetics, San Diego, CA, USA, pp.2664-2669.

[52] M. Schrage, No More Teams!: Mastering the Dynamics of Creative Collaboration, Currency Doubleday, 1995.

[53] W. Schweiger, D. Levichev, A. Schön, H. Meerkamm, "Digital Mock-up in the Early Stages of Mechatronic Product Design", Journal of Engineering Design, Vol. 10, No. 3, 1999, pp. 235-246.

[54] J. Singer, T. Lethbridge, "Studying Work Practices to Assist Tool Design in Software Engineering", Proceedings of the International Workshop on Program Comprehension (IWPC '98), Ischia, Italy, June 24-26, 1998, pp. 173-179.

[55] R.P. Smith, A. Leong, "Observational study of design team process: A comparison of student and professional engineers", Journal of Mechanical Design, Transactions of the ASME, Vol. 120, No. 4, Dec 1998, pp. 636-642.

[56] D.H. Sonnenwald, "Communication roles that support collaboration during the design process", Design Studies, Vol. 17, 1996, p. 277-301.

[57] N.P. Suh, The Principles of Design, Oxford University Press, 1990.

[58] M. Suwa, T. Purcell, J. Gero, "Macroscopic analysis of design processes based on a scheme for coding designers' cognitive actions", Design Studies, Vol. 19, 1998, pp. 455-483.

[59] J. Terninko, A. Zusman, B. Zlotin, Systematic Innovation: an Introduction to TRIZ, St. Lucie Press, 1998.

[60] D.G. Ullman, S. Wood, D. Craig, "The Importance of Drawing in the Mechanical Design Process", Computers & Graphics, Vol. 14, No. 2, 1990, pp. 163-274.

[61] D.G. Ullman, The Mechanical Design Process, 2nd edition, McGraw-Hill, 1997

[62] R. Valkenburg, K. Dorst, "The reflective practice of design teams", Design Studies, Vol. 19, 1998, pp. 249-271.

[63] C.G.C. van Dijk, "New insights in computer-aided conceptual design", Design Studies, Vol.16, No.1, January 1995, pp. 62-80.

[64] I.M. Verstijnen, C. van Leeuwen, G. Goldschmidt, R. Hamel, J.M. Hennessey, "Sketching and creative discovery," Design Studies, Vol. 19, No. 4, October 1998, pp. 519-546.

[65] M.B. Waldron, R.L. Brooks, "Analyzing inter and intra group information exchanges in conceptual collaborative design", Design Theory and Methodology – DTM'94, ASME, Vol. 68, pp. 241-247

[66] M.A. Walker, "Testing collaborative strategies by computational simulation: cognitive and task effects", Knowledge-Based Systems, Vol. 8, No. 2-3, 1995, p. 105-116.

[67] C.D. Wickens, J.G. Hollands, Engineering Psychology and Human Performance, Prentice Hall 2000.

[68] E.N. Wiebe, "3-D Constraint-based Modelling: Finding Common Themes", Engineering Design Graphics Journal, Vol.63, No.2, Autumn 1999, pp. 15-31.

[69] D.J. Wilde, "Design Team Roles," ASME Design Theory and Methodology, Las Vegas, Nevada, Sept. 12-15, 1999.

Design and Implementation of an Intelligent Cooperative Design System

Pengcheng Zhang, Renhou Li
System Engineering Institute, Xi'an Jiaotong University, Xi'an, 710049, P.R. China
pc_zhang@263.net, rhli@xjtu.edu.cn

Abstract

Establishing an open intelligent system is one of the goals of the cooperative design system nowadays. In this paper, a system architecture based on the CSCW model with multi-Level and multi-Group is presented. The proposed system is open and permits users to add their special components tallied with the architecture specification to the system. An integration method of the open system architecture based on the agent technology is given. Finally, the implementation of the intelligent management of cooperative task and process is discussed and an example of the application is given in the last section of the paper.

1. Introduction

The concept of Computer Supported Cooperative Work (CSCW) was put forward by Irene Greif (MIT) and Paul Cashman (DEC) in 1984. The main functionality of CSCW is to make people from different places and different fields to solve the common problem utilizing computer technology in any time. Computer Supported Cooperative Design (CSCD) is one of the applications of CSCW. CSCD system provides people who may have different position and play different roles in the work with a platform, in which they can discuss and arrange tasks, and accomplish the specific design task alone or jointly[1].

In the past few years, much research has been done and significant results have been achieved. But up to now, it is found that few commercial CSCD products can fit in with specific cooperative design work. And even the product that fits in with your actual requirement, it still doesn't work well when you want add some own special components. The reason is that the architecture of lots of CSCD system is a close one. In addition, we know that cooperative design includes synchronous cooperative design process and asynchronous cooperative design process. Sometimes, synchronous process plays more important role in the cooperative design. In cooperative process, the change of cooperative information may result in change of cooperative process, which may be more distinct in synchronous process. But current research doesn't pay more attention on it and then the full cooperative information in cooperative design process can't be taken.

In this paper, the system designed by our team is described, its architecture is open and users can add their special components to it according to the specification of this architecture. In addition, in this system, we explain the cooperative information in semantic form and combine with the workflow technology and agent technology to achieve the intelligent management of the cooperative design process. The remainder of this paper is organized as follows. In section II we give the architecture model of the system. In section III, an integration method of the open architecture is presented, including the model of agent and the integration method based on agent. In section IV, the intelligent management of cooperative design is discussed. Finally, we give an example and the conclusion in the last part of this paper.

2. Architecture model of the system

A design task may require many people to join in. In the design process, different people play different roles; one may be a decision-maker, a manager or an executor. One task can be divided into several subtasks by decision-makers or managers, and completing a special subtask may need more than one managers and executors. In a word, a design work is a multi-group cooperative process that may include many states and require many people to take part in.

On the basis of cooperative work model with multi-level and multi-group[2], a three-layer architecture model of a cooperative node is presented as shown in Figure 1.

The architecture model of the system consists of three layers, they are application management layer, object service layer, communication service layer. The detailed description is given as follows:

- Application management layer: There are a set of function that can dynamically change the state and the appearance of the system in this layer. The virtual knowledge base of the top management agent has interface layout knowledge and functional knowledge When interface layout knowledge is updated, the top management agent can access the functions of the layer through executing module of the agent to modify the appearance of the system. On other hand, when the appearance of the system changes, the top management agent can feel the change and update the interface layout knowledge and the functional knowledge.

- Object service layer: The layer consists of lots of service objects, such as multimedia service object, cooperative management object, application sharing service object, whiteboard service object, session management object, file transfer service object, etc. Each of these service objects realizes one or more specific function. Each service object has its own description knowledge. Object can be automatically integrated into the system and can be accessed through the knowledge exchange with the top management agent.
- Communication service layer: This layer provides the system with data communication service that includes H323 data communication and T120 data communication services.

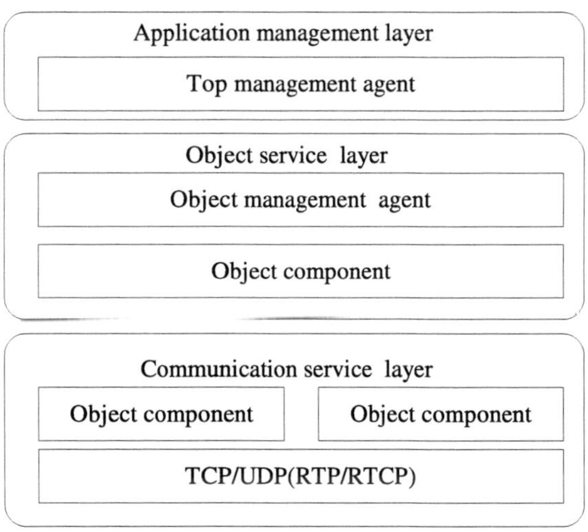

Figure 1. Three-layer architecutre model of the system

The server node has the same architecture model as the client has. In the development, we can combine application management layer with communication service layer as a basic framework, and the service objects are regarded as some relative independent functional agent. These agents can be integrated into the framework through knowledge exchange. So we can express node model as follows:

< Node >::= < Frame > { <Service Object Agent> }
< Frame >::= < Application management layer >
< Communication service layer>

The top management agent of the application management layer can access the functionality of the specific object of the object service layer through knowledge exchange between the top management and the object management agent. Different objects of the object service layer can exchange knowledge through the top management, so they can access the functionality of other object too.

3. Design and implementation of the open Architecture

In this section, we first review the agent technology briefly, then present the architecture based on agent technology.

3.1. Architecture model of the Agent

With the development of the agent theory and technology, agent technology has been introduced into various fields. Agent has features of autonomy, reactivity, collaboration and adaptability, and CSCW focuses on how a group of people who are in different places can utilize computer technology and resources to solve complicated problems that require collaboration in any time. In order to design a flexible CSCW in design (CSCD) we propose a specific agent, which can be regarded a basic component constituting the whole system. The basic architecture of the agent proposed by us is shown in Figure 2. It consists of four modules, they are knowledge base (KB), message interpreter and analyzer (MIA), executor/awareness (EA) and communication module(CM).

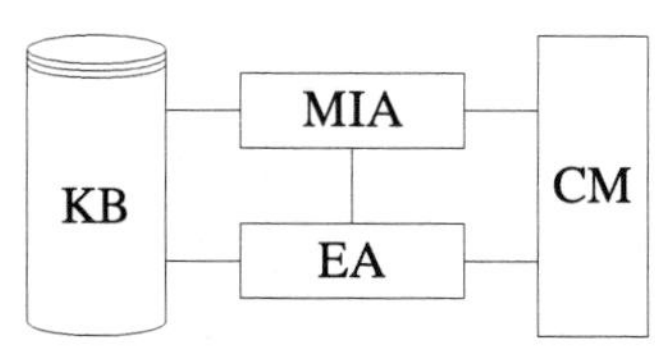

Figure 2. Architecture of the agent

The four modules of the architecture are described as follows.

- Knowledge Base: KB consists of agent state information, rules knowledge required for reasoning, interface layout knowledge and functional knowledge.
- Communication module: The agent can communicate with the outer through this module. At present, the information communication of agent has no formal standard, but KQML(which is developed by ARPA in deal with knowledge sharing.) can be considered as a virtual standard[3]and can be used as a communication language for agents.
- Message interpreter and analyzer: This module processes the message from the outer layout and various awareness components

- Executor/Awareness: This module is a functional part of the agent, different agent may have different executor and awareness functionality

3.2. Implementation of the system architecture based on multi-agent system

On the basis of Figure 1.(architecture model of the system) and Figure 2.(architecture of the agent), the whole system based on multi-agent can be implemented as shown in Figure 3.

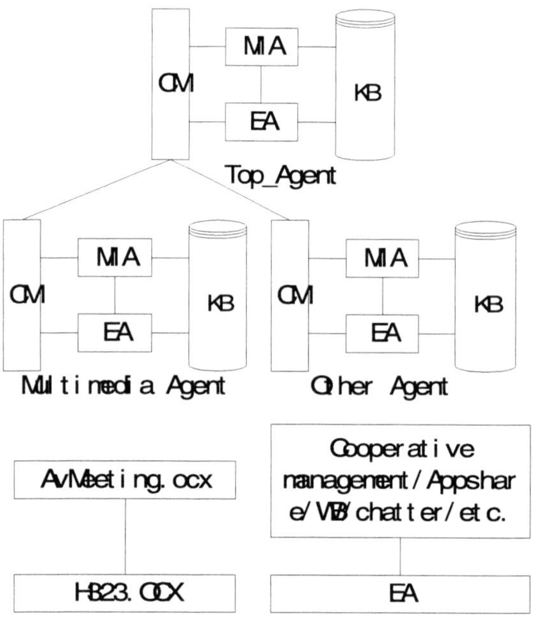

Figure 3. Implementation of the system architecture based multi-agent system

In Figure 3, the architecture of the client node and the server node are given. Where, MCS.OCX and H323.OCX correspond to the communication service layer. MCS.OCX provides the multi-point data communication service complied with protocol T.120. H323.OCX provides the service complied with protocol H323, including call management, session management and audio and video service. The TOP_Agent corresponds to the application management layer that is essential to the system integration and has functionality that contains in the EA. All parts in Figure 3 except the aforementioned one correspond to the object service layer. Here the object corresponds to the agent, and the functionality of the agent corresponds to the component controlled by the agent.

From the architecture model, we know that the client node and the server node have the similar architecture. The difference between them is that they have different components corresponding to the agent. In the client node, the components are the client of the cooperative tools. In the server node, the components are the server of the cooperative tools.

3.3. Features of the System Architecture

The architecture of the system is open and user can add any other components into the system according to user's desires. The components have knowledge describing their own functionality and state. From the description above, we know that the framework of the system presented by us is based on the agent technology and is a multi-agent system. We also know that the agent not only has abundant knowledge but also holds flexible exchange mechanism.

From Figure 3 we can see that the client and server have same structure, with which the functionality of the application management layer and the communication service layer can be realized. The management of the nodes (including clients and the server) and the dynamic integration of the system are accomplished by Top_Agent. Besides, the proposed framework have the ability that can change the layout of the interface according to the knowledge in the KB and can update the knowledge in KB, and can access some functions of the components when the state of the system changes. When a new component is encapsulated into an agent, the agent obtains the knowledge that describes the state and the ability of that component. Therefore through the exchange of the knowledge within the framework, Top_agent gets the state and the ability of the new component. When the knowledge is updated in the KB of the Top_agent, Top_agent will integrate the new component into the framework according to the changed knowledge. For example, there is a function in the functionality set offered by the framework as the following:

Long AddButton(Long interfaceKBID)

A button can be added on the framework according to the knowledge provided by the agent and identified by KBID.

When a new component is added to the framework, the component with the agent templet must be encapsulated. Using the entry function of the agent, the framework can start the agent. Through the message exchange, Top_agent gets the knowledge as follows:

```
KBId1:
(AddButton
 :sender Agent1
 :receiver Top_Agent
 :reply-with AddButton+kbid1
 :language USERDEFINE
 :content "Addbutton(x,y)"
)
```

Where the information is expressed in KQML. The sentences listed above means that when the information exchange is finished, one button will be added in the framework. With the button, a specific function will be accessed.

4. Management of cooperation

Cooperative design is a dynamic process that depends on many factors, so how to manage the information of the state and the environment of the cooperative process are very important[4]. The cooperative management system is presented as shown in Figure 4.

The architecture of the cooperative management mainly consists of three parts:

- Task modeling provides users with the definition method for the topology architecture and the cooperative relationship, such as sequence, concurrent, selection, AND, OR, conditional jump, etc.
- Process definition produces the descriptions of the cooperative process, which can be interpreted by the workflow engine.
- Workflow engine manages the executive process of the cooperative task. The main functions include: interpretation for the process definition, process creation, process activation, process pause, process end, information dispatch, etc.
- The management system has advantages as follows:
- Cooperative task can be modeled with graphical interface and can be expressed in semantic form.
- In the cooperative process, the system can dispatch the information on the basis of the semantic reasoning.
- The system allows users to manage and monitor the cooperative process.

5. Application Example

On the base of the architecture model, we developed the Open Intelligent Cooperative Design(OICD) system. The screen snapshots of the task modeling and synchronized design process using OICD system are shown in Figure 5 , Figure 6. The OICD system has features as follows:

- Open system architecture

  OICD is developed based on integrated framework and multi-agent technology. The application of cooperative node is composed of the application framework and some agents that can realize the goal of user and can be loaded dynamically through knowledge exchange of the agents. User can develop new agent which can be integrated into the system.
- Visual task modeling tool

  OICD provides the visiual task modeling tool. User can expediently establish the transaction modeling. Figure 5 is the screen snapshot of one of the interface of the tool.
- Intelligent cooperative management based fuzzy petri nets and workflow technology

The transaction modeling established by visual task tool can be converted into the workflow definition process expressed by fuzzy petri nets. The workflow server can execute fuzzy reason and schedule the workflow path according to the fuzzy petri nets. Screen snapshot of one of the synchronized design processes is show in Figure 6.

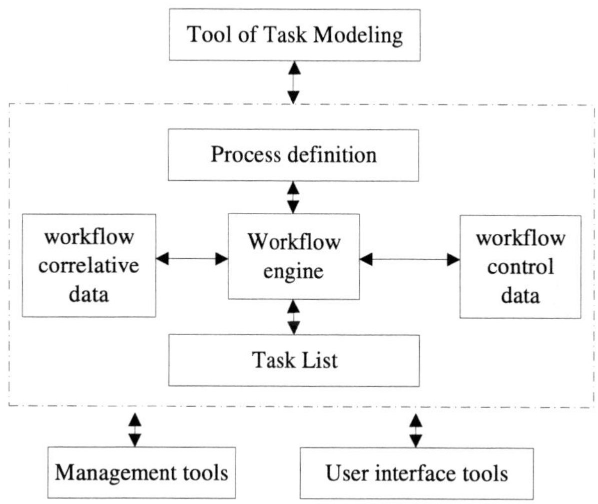

Figure 4. Architecture of the cooperative management component

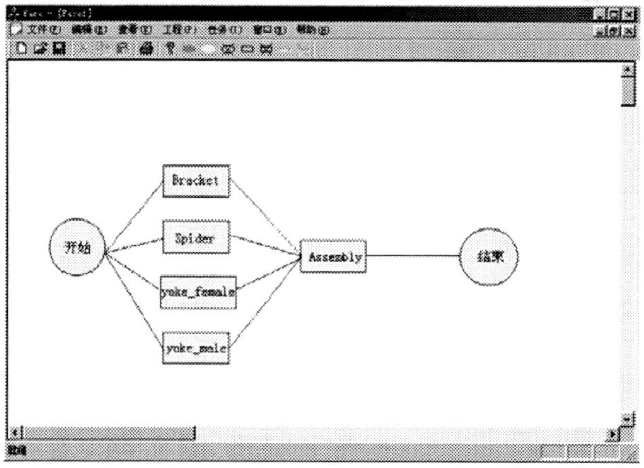

Figure 5. Screen snapshot of the task modeling

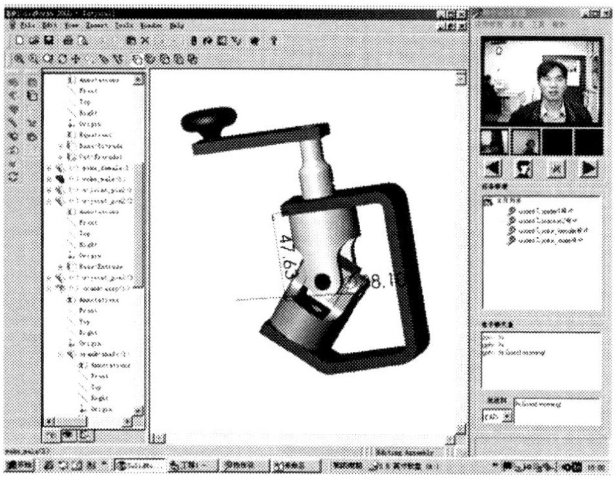

Figure 6. Screen snapshot of the synchronized design process

6. Conclusion

This paper presented the design and implementation of an intelligent cooperative design system. Three-layer model of the system is proposed. Then a model of the agent is given, and on the basis of the model, an integration method of open system is explained. Finally, the cooperative management of the system is described .

7. Acknowledgement

We sincerely acknowledge the support of National Natural Science Foundation of China. We also acknowledge the reviewers of this paper who give us useful revising advices.

8. References

[1] Yu Fu-Jin. "Summarization of the study on computer supported cooperative design". Mechanical & Electrical Engineering Magazine. Vol.17, No.3, 2000:5~7

[2] Zheng Qinghua, Li Renhou, "Study on the model of Multi-Level and Multi-Group Cooperative Work Based on Agent", Proceedings of Fourth International Workshop on CSCW in Design, Sept. 29-October 1,1999 Compiegne France

[3] Tim Finin, Richard Fritzson, etc. "KQML as an agent communication language". The proceedings of the third international conference on information and knowledge management, ACM press, Nov. 1994

[4] Lorraine Sherry, and Karen Madsen Myers. "The Dynamics of Collaborative Design". IEEE Transactions on professional communication. Vol.41, No.2, June 1998 :123~137.

A Visual Part Library – GS-PM

Min Tang, Xinyu Tang, Jinxiang Dong
State Key Lab of CAD&CG, AI Institute, Zhejiang University, Hangzhou, 310027, P.R. China
Email:tang_m@163.com, djx@cs.zju.edu.cn

Abstract

It is necessary to use libraries of parts when using CAD software. But, deficiencies and problems exist in most libraries, so we developed a part manage system – GS-PM. Based on feature, but not geometric information, it not only improves the efficiency but also enables some applications of higher levels, for instance, the parts can be used in different CAD software. This part library can work in Client/Server mode to share information. It can also be applied to establish a customized Visual Part Library (a web part library), which can build parts dynamically according to the user's needs.

1. Introduction

With the widely application of CAD software in so many domains, the need of reusing existent parts as well as the need of establishing custom-built part libraries has become more and more urgent. However, there are some deficiencies in many existent part libraries, for example: they can only be applied on certain CAD software, it is difficult to exchange or extend their contents with other libraries, their efficiency is low, etc. That's why we develop our part manage system: GS–PM. It has been applied on GS-CAD (a feature-based 3D parametrical modeling CAD system developed by ourselves).

Most part libraries are, always based on geometric information , so its implementation is limited. Our systems are based on feature.

Parameters of the parts in given industries are usually different from the parameters used by CAD systems. GS-PM can transform them automatically. Even if the CAD does not support variant design, users need only to input the transform functions when appending the parts.

Furthermore, most part libraries are constructed or used on single computers, so it's hard for users to share information and products of each other. However, GS-PM can work in a Client/Sever mode.

With GS-PM, we can set up a web site to distribute or exchange parts. When the user is browsing the site, GS-PM can send the web browser a VRML file built by the 3D display model. When the user want to get a part, GS-PM will drive corresponding CAD to build the part with parameters provided by the user dynamically and send the result part files to the browser.

2. Feature-based storage

Part libraries often store part files or the geometric information in the database, however, a large amount of these information such as the geometric information of the same feature can actually be shared by most parts. These considerable redundancies not only will devour unnecessary data space but also will surely reduce the efficiency of database operations and confine the capacity of the library.

On the other hand, most CAD software are based on features and provide powerful APIs to employ the feature utilities, so our system stores the feature information in the library and rebuilds the parts with appropriate CAD software when needed.

We decompose the parts into features, and store in the library their types, their location information ,their parameters ,the relations of dependences according to the order in which they are created. When they are needed, we reconstruct the part with such information in the same order.

There is another benefit in this method: we can query parts corresponding to certain feature besides their names and types.

3. Feature extraction

Storage based on features provides more efficiency, but there are other problems: the classifications and the definitions of features in various CAD systems are usually different, and it's almost impossible to define a series of universal features. However because most CAD software support user defined feature, we can set one series of features of one CAD software as default features and realize them as User Defined Features in other CAD systems.

We take the features defined in GS-CAD as the default features in our part manage system. When appending new parts, we use GS-CAD to design desired parts and then we can get the default features directly. We also extract and

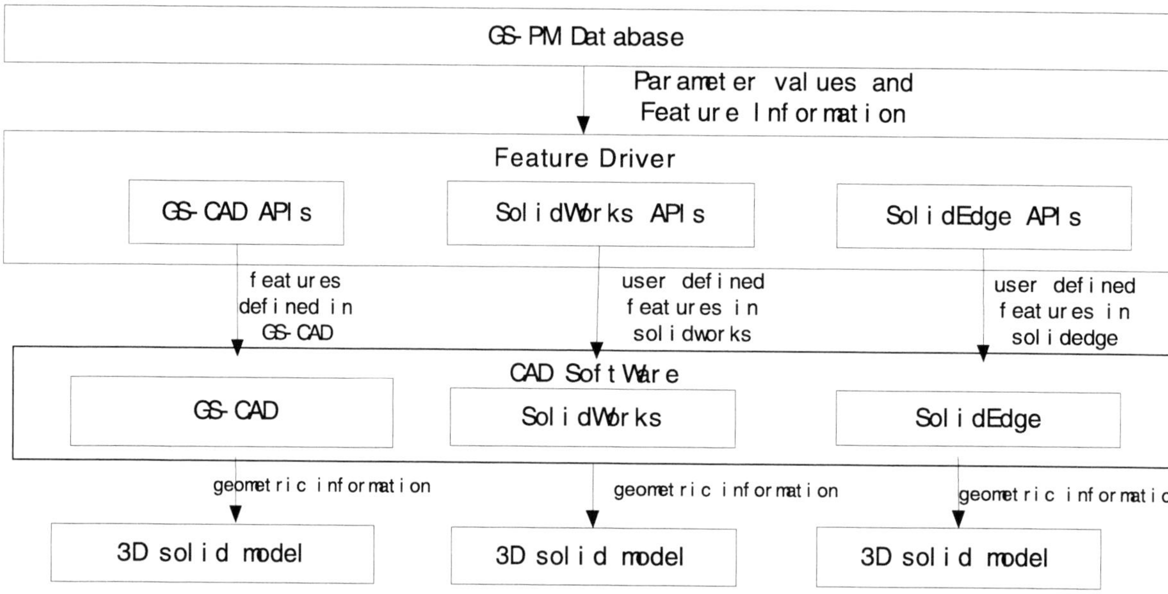

Figure 1 the Structure of GS-PM

store in the database the parameters, locations, directions and types of their features.

Parts designed in other CAD software can be exchanged with standard data exchange files, but the feature information will get lost during this process. It's difficult to recognize a feature automatically and perfectly, so in GS-CAD we provide an utility to enable users to specify features interactively. In this way, we can easily append existent parts designed by other CAD software into GS-PM.

4. Feature rebuild

Part design based on feature is a higher level than merely geometrically rebuilding parts, and it is usually not compatible between CAD systems. However, in GS-PM, features are extracted and stored in the same format, so, we should use an unified interface to drive different CAD systems.

To accommodate various client CAD software, we devised an intermediate level: Feature Driver between GS-PM and client CAD systems. It includes corresponding interface for rendering each feature in various CAD software. The system works as shown in Figure 1.

5. Rebuild geometric model

Feature storage and transformation can thus be finished, but before the geometric solid model can be rebuilt, there is another difficulty: the parameters in given application fields (standard parameters) always differ with the parameters of the features (geometric model parameters).

To meet the specified standard, we should show users only the former parameters in our database and user interface and use the latter to model the parts. What's more, considering that the user may change the value of standard parameters, we should be able to evaluate the model parameters according to the values of standard parameters and the parameter equations of the part. Because the users know only the standard parameters, all these transforming and evaluating operations should be transparent to the users.

It's easy to solve this problem if the client CAD software supports variant design. We can add new constraints to make the geometric parameters consistent to the standard parameters, however in other cases we must transform the parameters by ourselves.

We devise an interpreter to perform this task. When appending the parts into the library, users need also to input the transform formulae and store them in the database. Before rebuilding the part, we utilize the interpreter to interpret the formulae and calculate the values of the geometric parameters before sending them to the Feature-Driver. The details of the data flow from the data in the database of GS-PM to the Feature Driver goes are showed in Figure 2.

6. Part library server

Most part libraries are constructed or used on single computers, so it's difficult for users to share information and products of each other. However, GS-PM can also work in a Client/Sever mode, so it provides a convenient way for users to share their products.

In this mode, GS-PM is divided into two parts: client and sever. There is only one single GS-PM sever in the system while every user has a GS-PM client to drive their local CAD software to reconstruct the parts model. Every GS-PM client can append and fetch new parts in the GS-PM server. This is quite useful for CSCW.

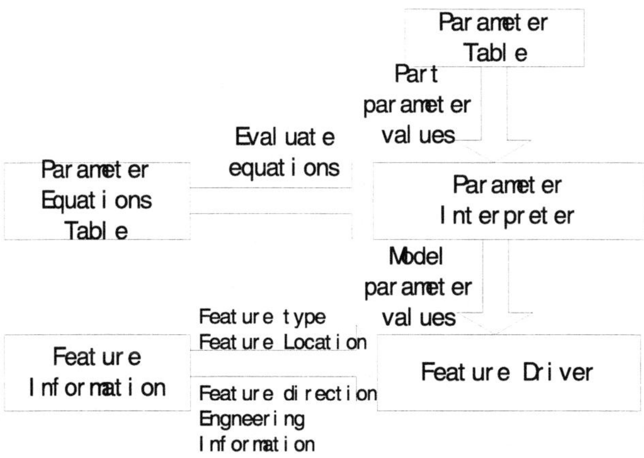

Figure 2 Evaluation of geometric information

7. Visual part library

GS-PM also supports Visual Part Library (VPL), which means a web site which provides part drawings for downloading.

With GS-PM, we can set up a web site to distribute or exchange parts. When the user is browsing the site, What's more, the VPL supported by GS-PM can provide not only the standard part drawing but also

the customized drawing or even the part files created dynamically by GS-PM.

When a user is browsing our VPL, GS-PM can send the web browser a VRML file built by the 3D display model. When the user want to get a part, GS-PM will parametrically drive corresponding CAD to build the customized part with parameters provided by the user and send the result part files to the browser corresponding CAD goes as shown in Figure 3.

8. Conclusion

This paper introduced a Part Library system based on features. After the implementation, we find it extremely useful especially when a group of designers are working in cooperation. It decreases the library size and improves the efficiency. It's also open to users and several CAD systems, what's more, it support information and products sharing very well.

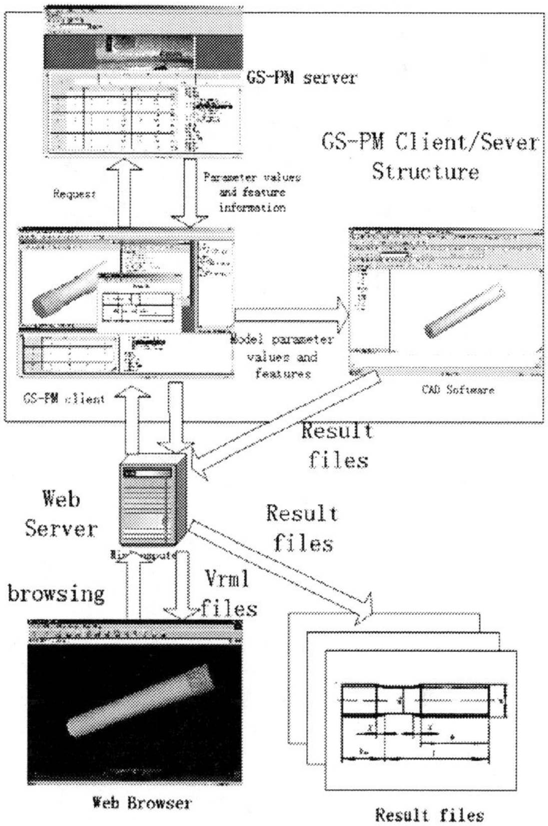

Figure 3 Visual Part Library Structure

9. References

[1] Sun Wanjun ,Visual Part Library, Journal of CAD&CIMS Vol.1, No.1, January, 2000 (in Chinese)

[2] Zhen Fang-Pu, The Research and Implementation of the Library of 3D Parametric Standard Parts, Journal of Computer Aided Design And Computer Graphics, Vol ,11, No.3, May, 1999 (in Chinese)

[3] Li Hailong, Dong JinXiang, Tang Min, He Zhijun, A Schema of developing feature based modeling systems, International Conference on Intelligent Manufacturing, Wuhan, China, SPIE, June, 1995, pp.76-80

[4] Li Hailong, Dong Jinxiang, Tang Mengen, He Zhijun, Feature Based Modeling Systems for Mechanical Product Design, The Fourth International conference on CAD/CG, Wuhan, China, SPIE, October, 1995, pp.508-513

[5] Li Hailong, Han Jianhua, Dong Jinxiang, and Wang Yong, A Feature-Based, Parametric Modeling System for CAD/CAPP/CAM Integrated System, ICIT96, Shanghai, China, IEEE, October, 1996

[6] Li Hailong, Dong Jinxiang, Huang Zhongdong, Feature-Based Parametric Product Modeling System in CIMS Environment, High Technology Letters, Vol.3, No.1, June, 1997, pp.13-16

[7] CV Corporation, Feature-Based Modeling User Guide and Menu Reference,USA,1994

Mechanism Design Staff Room of Northeast Industry Institute, Mechanical Parts Design Manual, Beijing : Metallurgy Industry Press,1974

[8] J. J. Shah. Assessment of features technology. Computer-Aided Design, 23(5):58–66, June 1991.

[9] M. J. Pratt. A hybrid feature-based modeling system. Advanced Geometric Modeling For Engin- eering Applications, International GI-IFIP Symposium '89, pages 177–189, November 8-10 1989.

[10] T. Laakko. Incremental Feature Modeling: Meth- odology for Integrating Features and Solid Models. Ph.D. Dissertation, Technology, Helsinki University of Technology, November 8, 1993

[11] Ranta M. and Mantyla,M., Cut and Paste Based Modeling with Boundary Features, 2nd ACM Solid Modeling '93-5/93/Montreal, Canda

[12] Ghandi, A. and Myklebust, A., A natural language approach to feature based modeling. Advances in Design Automation,Vol.1 of''Computer

[13] von Rimscha, M., Feature modeling and assembly modeling—a unified approach. Advanced Geometric Modeling for Engineering Applications, 1990, pp. 203–213.

[14] Laakko, T. et.al, M. A feature definition language for bridging solids and features. Second Symposium on Solid Modeling and Applications, 1993, pp. 333–341.

[15] Laakko, T. et.al, M., Incremental constraint model- ling in a feature modelling system. Computer Graphics Forum, 1996, 15(3), 367–376.

[16] C. M. Hoffmann, R. Joan-Arinyo, On user- defined features, Computer-Aided Design, 1998, 30(5),321-332

[17] Ranta M. and Mantyla,M., Cut and Paste Based Modeling with Boundary Features, 2nd ACM Solid Modeling '93-5/93/Montreal, Canda

[18] Z. Y. Zeng, et al, Reuse of Model in Design-by- Features Modeling System, CAID&CD'99.

Parametric Modeling with User-defined Features

Min Tang, Ye Wen, Xiaofeng Mi, Jinxiang Dong
State Key Lab of CAD&CG, AI Institute, Zhejiang University, Hangzhou, 310027, P.R. China
tang_m@163.com, djx@cs.zju.edu.cn

Abstract

Feature-based design is the main characteristic of current prevailing parametric solid modeling systems. By means of enabling users to build their own feature libraries, user defined features (UDF) provided as an extensional design method can overcome the shortcoming of inconvenience that exists in the traditional feature-based modeling system. Traditional UDF representation is implemented either by procedural definition, which represents the features implicitly [1], or by an explicit representation or enumerative way [2, 4]. This paper proposes an innovative approach on the representation of UDFs by extending the common procedural UDFs' data structure with the boundary features. Then a unified feature attachment interface is described. Thus, a unified representation of UDFs is introduced.

1. Introduction

Feature-based modeling provides attractive environment to designers. The old feature-based modeling is adequate to create complex shape designs, but they do not necessarily make a good connection between this basic repertoire and application needs for mechanical parts which use complex shape elements that reflect sophisticated functional requirements.

User-defined feature, which is a compound of certain kinds of predefined form features presents a new approach to enable designers to build there own feature library. Besides the benefit to designers of feature model reusing, UDFs themselves can be compounded to become larger functionally meaningful design units.

The construction of predefined feature or reconstruction of user defined feature present a problem before the modeling system designers. A lot of research has focused on the point to give a unified way to assembly any kinds of predefined features or UDFs to a more complex and sophisticate one. There are two major difficulties to resolve the problem: the representation of UDFs and the interface to bring UDFs together.

There are numerous papers reporting on languages to represent features [5, 6]. Laakko and Mäntylä described a language to represent features embedded in Lisp for creating feature and part family classes and support the inclusion of different kinds of application specific information in feature frames [3, 7, 8].

Hoffmann presented a procedural mechanism to generate and deploy user-defined features [9]. The UDFs Hoffmann presented are actually based on volume-based features that are limited by the combi- nation of predefined features. To make full use of the available shapes of the existing product models, a more flexible boundary object [4], boundary feature, is introduced. Ranta M. et.al gave the definition of boundary features, which in fact is an incomplete boundary model.

This paper gives a detailed description of the use of both of the two kinds of feature representation approach. The difference of the underlying implementation of UDFs is transparent to model designers and a unified UDF feature attachment interface is given. We begin with a description of formal UDF definition then we give the representation of the definition in details. Based on the discussion of the representation we then describe how this representation benefit feature-based modeling then a case of such UDF representation is exemplified.

2. Unified representation of user-defined features

Predefined features are predefined by the CAD system. Common predefined features include holed, ribs, slots and chamfers etc. Predefined features, although sufficient for most kinds of applications in principle, this core vocabulary is so 'low-level' that it is lack of necessarily convenience for model designer and should supplemented by customized features defined by the user. Such a UDF, may be specific to an application and therefore complement the core feature vocabulary already provided by the CAD system.

Because a particular set of features (UDF) is used variously in parts that an efficient mechanism is necessary to locate and size the UDFs. Moreover, since different applications have different feature needs, it is natural to archive sets of predefined features in a library. The features so archived may be either standard features or UDFs.

In feature-based modeling systems, feature definitions need to store several types of information. To serve as templates for feature instances, they must specify the attributes to be placed in instances, how the values of the attributes are determined, and the range of permissible attribute values. In particular, feature definitions must specify how the feature instance is constructed on the basis of values of attributes. A geometric model of the part is created by combined the feature instance geometries.

To meet the convenience of both modeling system designers and modeling system users, we implement two kinds of underlying feature representations, say procedural representation of features and editable boundary representation of features.

3. Primitive Feature

If we say that a feature is primitive, we mean that the feature is not any compound of any other features. In our manner of feature representation, primitive features can be divided into two types, one type, which is common, is predefined feature; the other type is boundary feature. The latter is just the one representation of user-defined features, which is described by M. Ranta in his articles [10]. The predefined compound features, such as counter bored holes, is also a kind of primitive feature.
We now give two examples of predefined features and boundary feature respectively.

Figure 1 shows two types of primitive features. (a) of it is an instance of through rectangle slot, which is located by a top face, an entrance face, an entrance point and a vector. While fig.1 (b) shows a compound feature, which consists of two through slot, the axis of which are perpendicular to each other. Such compound feature is defined by user by means of selecting all the faces needed to construct a meaningful feature boundary. In the case of the feature in Figure 1 (b), the user should select all the 8 side faces and the bottom face. Modeling system will search all the external dependence of the selected face and proper constraints will be automatically added in. The algorithm detail will be discussed in later sections.

The use of boundary feature significantly improve the convenience of feature definition while make the modeler more complex. Because of the users' arbitrariness of the selection of the feature, latent invalid case may exist and the model may fail. So the modeler must test the validity of the boundary feature before and after the feature is attached onto existing feature. This work includes constraints checking and non-manifold case detecting.

Further discussion of primitive features, such as location manner, constraint creation etc. will be present in later sections.

4. Procedural representation of UDFs

The procedural feature definition consists of a collection of related procedures, represented in a tree data

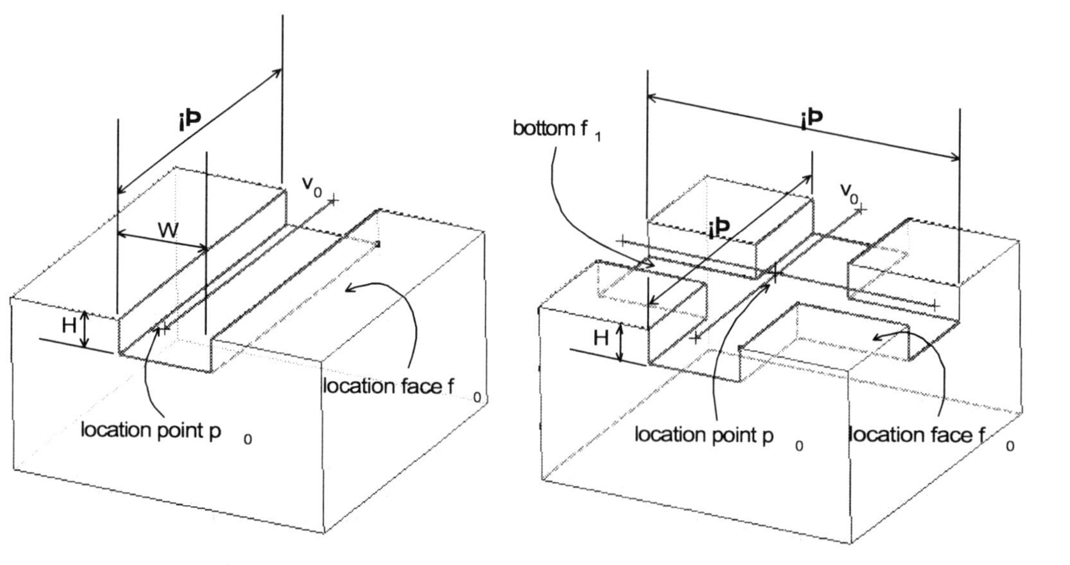

Figure 1. Two types of primitive features. (a) is an instance of through rectangle slot, which is located by a top face, an entrance face, an entrance point and a vector. (b) presents an instance of compound feature, which consists of two through slot, the axises of which are perpendicular to each other. And with a location point at the center.

structure, for the evaluation of dependent feature attributes, relationships, and geometry on the basis of independent attributes given by the user. In order to present a convenient representation of feature definitions, the geometry information is hidden from the representation. It is laborious to define the geometry by this tree, which should present enough information. Concerning the predefined features, the topology information of the predefined features is hidden, the system explains the geometry of the predefined features by these feature types, and the parameters in this representation are explained implicitly.

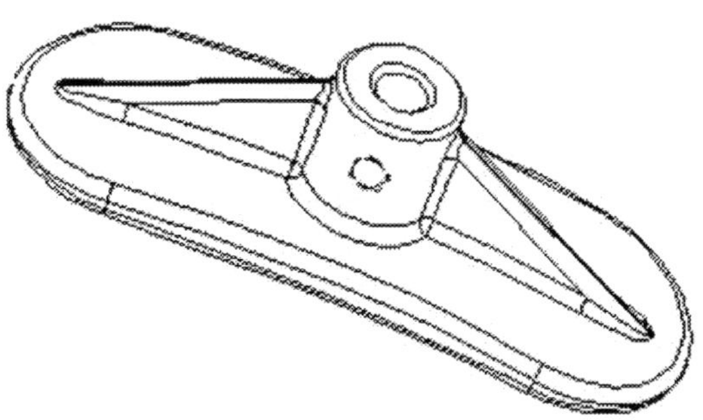

(a) a Rest

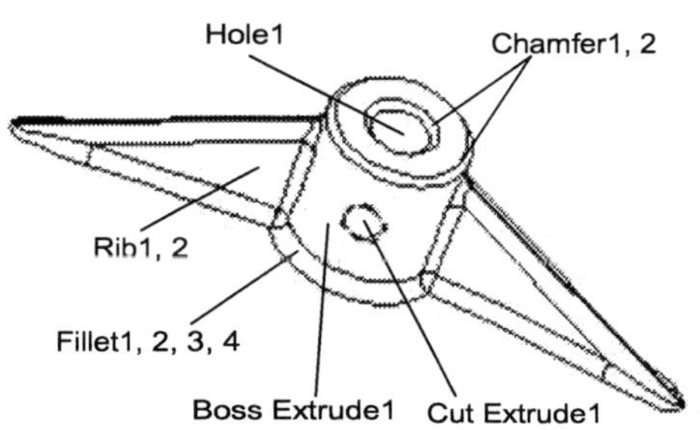

(b) a UDF

Figure 2. (a) is a part and (b) is the extrude portion of it is defined as a UDF, which is re-usable when design other parts.

All the sub-features in a UDF is tree-shaped. That is, there is a root feature that only depends on the location of the UDF and all other features are all depends on the root feature directly or indirectly. Extra datum face, datum line, datum vector or datum point may exist but still be dependent on the root feature directly or indirectly. In fact, all the primitive features in the UDF defined are built procedurally form root to leaves.

Figure 2 (a) shows a rest, the protruding portion of the rest can be extracted to form a UDF which is shown in Figure 2 (b). The primary component of this UDF is the Boss-Extrude1 feature, there are seven component features depending on this feature. Figure 3 illustrates the hierarchical structure graph of the UDF tree. Among the feature nodes, the root's children, including the child features and the geometrical datum elements are all shown. As for the child nodes, only the skeleton of the sub tree is present.

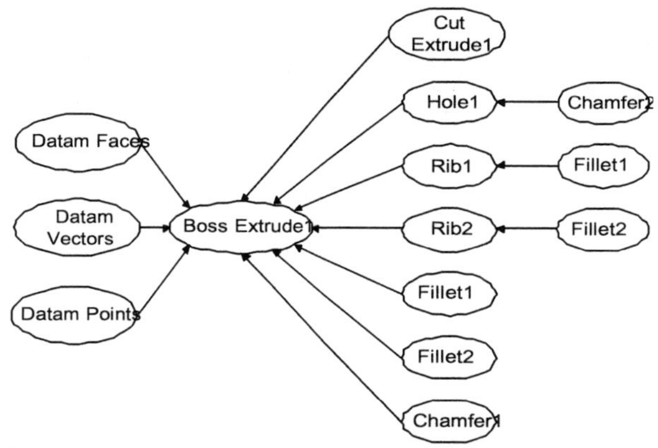

Figure 3. The dependence relationship of the tree node in the UDF in Figure 2 (b). The child features that depend on it is on the right and the datum faces, datum vectors on the left.

5. Unified definition of UDF

Now it's time to give a formal unified definition of UDF. This definition should contain the case of procedural representation and boundary representation. The definition given here will be the base of next section, which will give a detailed description on how to enable users to build their own UDF libraries.

In the view of mechanical engineer, the feature contains not only shape information, but also non-shaped information such as location tolerance, surface quality, etc. As for modeling system, we mostly care the geometry definition of the feature.

Definition: User-defined feature is a collection of the following objects:

1. A root primitive feature. If a local coordination system is given, then the feature is unambiguously located.
2. A set of datum faces, datum lines, datum vectors and datum points each of which directly depends on the root primitive feature or depends on another geometrical datum element and ultimately depends on the root feature.
3. A set of child features of the root features, the locations of which can be evaluated by the datum faces, datum lines, datum vectors or datum points given by 2. The parameters of these features can be either value entities or the datum elements given above. The type of the feature is either a primitive one or a UDF recursively.

The foundation of the user-defined feature is the primitive feature. The representation of UDF definitions is determined by defining every component in sequence. The first component is the primary component. The primary component only makes reference to elements of the existing model. In consequence, the root component must be a generic predefined feature or a user-defined feature, namely boundary feature, the boundaries of which is designated by the user. Subsequent components are defined and placed with respect to the prior component. The representation of the UDF definition has a hierarchical structure which can be described logically as an acyclic directed graph where each node is a component and the edges are direct dependencies between them oriented from the component that makes the reference to the component in which the referenced element is defined. Obviously, the primary component is at the root of the graph, just as fig.3 shows.

6. UDF construction algorithm

There are various methods to construct a UDF including feature recognition or directly feature building environment. In this section, we will give an over view of how to construct a user-defined feature in user interactive manner, which is mostly common used in current modeling systems.

7. Boundary feature construction

As a kind of primitive feature and the augment of the procedural representation of UDF, the construction of boundary features is rather fundamental and essential.

When the user selects some faces and instructs the modeler system to defined a user-defined feature, the following 4 steps should be executed sequentially.

All the features these faces belong to are to be figured out. It is advisable to define a limit of the number of features and the relationship of the entities should also be limited. We specify that the maximum feature number is 3 and the features must have a common parent feature and a common location face, and the location face must be one of the faces of the feature or depends on the feature only. Restrict limitation may do harm to the flexibility of the definition of boundary but anyway enhance the robustness of the feature modeler system.

Mark the property of the newly formed feature as plus feature or minus one. This differ the Boolean operation of the feature when it is attached to existing feature.

Test the validity of the faces. Add proper geometry element to create the feature model. As for the case of fig.1 (b), an extra faces should be added on the top of the cross slot. The selection of auxiliary geometry element set for every predefined feature should be present. As an example, the auxiliary of geometry elements for a blind slot should include the entrance face and the top face whereas the through one only needs the top face of the slot. If the user has not selected enough faces to form a meaningful solid, a message of operation failure will be sent to the user.

After these steps, we should do some work for supporting of parameter-driven feature design. We suggest that all the parameters of the face to be designated by user, with constraints added in. The constraints may be auto detected or added explicitly by means of user interaction. Then all these constraints and the geometry model of the user-define primitive feature is stored into the system's UDF libraries for reuse in other part design session.

8. Procedural feature construction

When the user selects a set of feature to request the system to construct this feature into a UDF, the model system should follow 5 steps below.

Find the entries of all the selected features in the design history tree.

Find out the dependence relationship of all the features, topo-sort the features. If several features is of the same level when they are topo-sorted, then the order in the design history tree is designated. Construct these features into a tree. In this step, left-child-right sibling manner may be applied.

Discard the relationship of the root feature and convert the dependences into a location manner. We implement this by means of local coordination system. For example, if we represent the cross slot presented in Fig. 1 by procedural way, we will select one slot to be the root, and defined a local coordination system, the origin of which is the entrance point of the slot and the extend direction is specified as the vector and the top face as the location face. Thus, when a local coordination is given, we can unambiguously locate the user-defined feature.

Identify the child features of the root and convert the dependence into a location point, a location face and a location vector. Note that these three auxiliary datum elements may be arbitrary except that the point must on the face and the vector must not perpendicular to the face. And the location elements are given by the location parameters of existing location features. Convert the dependence on other features of these datum elements to dependence on the parent feature by means of matrix transformation.

The remaining work to do is the same as the construction of boundary feature representation. Proper parameters are extracted, however, no constraints are to be added for it is the primitive features and feature location manner that determine the shape of the new user-defined feature.

In the course above, extra datum faces, datum vectors or datum points may be created. All these auxiliary elements, together with the transformation matrix, which serve as the location of the child features are all added to their parent entity table.

9. Attachment and validation

As shown in the definition of feature, the interface of user-defined features is very clear, that is, if we supply enough information for the feature location, the feature can be attached to the existing model entity. When we have finished traveling the feature dependence tree, the whole feature is attached on. The order is not unique while the right traveling of the tree is advised for the right child node of the root in a sub-tree is the sibling of the root feature in our representation manner.

After an UDF has been attach to a part or an UDF. The validity of the part or the UDF must be tested. Once a UDF has been defined and before it is stored in the feature catalog. A set of rules of feature definitions should be available to guarantee the validity of the internal attributes of features, including syntactic correctness of constraints, coherence of parameter range definitions, and so on. Some tests should ascertain validity. Route checks include the verification of dimensional, geometric constraints, the absence of self-intersection of profiles, etc.

It is important to check the topological attributes topological attributes to establish the semantic validity of the feature instantiation. It is desirable to re-validate a feature in the design process. After a new feature attachment, the former features that are modified by this new feature must be re-evaluated to ensure its validity.

10. Conclusion

This paper gives a formal definition of user-defined features and detailed discusses the representation of UDFs and introduces an algorithm skeleton of the construction of such features.

We find that this manner of representation of features matches the paradigm of object-orient programming very well. Its clear interface and encapsulated parameters enable the modeler enhancing its work easily and efficiently.

The future work will be focused on how to propose a sophisticate algorithm of parameter extraction and constraint detection.

11. References

[1] J. J. Shah. Assessment of features technology. Computer-Aided Design, 23(5):58–66, June 1991.

[2] M. J. Pratt. A hybrid feature-based modeling system. Advanced Geometric Modeling For Engin- eering Applications, International GI-IFIP Symposium '89, pages 177–189, November 8-10 1989.

[3] T. Laakko. Incremental Feature Modeling: Meth- odology for Integrating Features and Solid Models. Ph.D. Dissertation, Technology, Helsinki University of Technology, November 8, 1993

[4] Ranta M. and Mantyla,M., Cut and Paste Based Modeling with Boundary Features, 2nd ACM Solid Modeling '93-5/93/Montreal, Canda

[5] Ghandi, A. and Myklebust, A., A natural language approach to feature based modeling. Advances in Design Automation,Vol.1 of''Computer

[6] von Rimscha, M., Feature modeling and assembly modeling—a unified approach. Advanced Geometric Modeling for Engineering Applications, 1990, pp. 203–213.

[7] Laakko, T. et.al, M. A feature definition language for bridging solids and features. Second Symposium on Solid Modeling and Applications, 1993, pp. 333–341.

[8] Laakko, T. et.al, M., Incremental constraint model- ling in a feature modelling system. Computer Graphics Forum, 1996, 15(3), 367–376.

[9] C. M. Hoffmann, R. Joan-Arinyo, On user- defined features, Computer-Aided Design, 1998, 30(5),321-332

[10] Ranta M. and Mantyla,M., Cut and Paste Based Modeling with Boundary Features, 2nd ACM Solid Modeling '93-5/93/Montreal, Canda

[11] Z. Y. Zeng, et al, Reuse of Model in Design-by- Features Modeling System, CAID&CD'99.

Parametric Modeling with User-Defined Features

Peng Song, Min Tang and Jinxiang Dong
State Key Lab of CAD&CG, AI Institute, Zhejiang University, Hangzhou, 310027, P.R. China
tang_m@163.com, djx@cs.zju.edu.cn

Abstract

With the demand of distributed designers to cooperatively design a single product, these designers need to simultaneously access a common database, which hold the geometric model and other information of the product. But frequently, their design will lead to invalid result. So it is very important to build a distributed model to coordinate these designers to ensure the consistency of the product. This paper introduces a collaborative model to support this concurrent work. First, the architecture of this system is presented. Then a collaborative design model is described in detail. We analyze two reasons of the conflict and give two collaborative modes to coordinate the design process. These two modes can be used in different environment and can be used in a mixed way. So we can keep the consistency of the common solid model and maintain the maximal freedom of the designers at the same time. Finally, we give some design example to further illustrate our model.

1. Introduction

Due to the internationalization of design process, even in different continents, people who distributed in different regions are involved in this process. And these people are also specialized in different aspects of the product design, such as concept design, detail design, engineer calculation, etc. Even those people who don't participate directly in the product design, such as the managers, manufacturers, and salesmen, have growing demand for accessing the data used in design. But in traditional CAD systems, the way to share these product data is mainly via files, which are often incompatible among different systems. And this process is static and retrieval of the recent data is cumbersome or even not possible.

The development of the distributed technology, especially the emergence of the World Wide Web in recent years, has free the users from the previous constrained systems. The Internet connects those people who distribute throughout the world so they can get the information they need just a click away. The systems can be organized with small modules with different functionality and achieve a flexible, scalable virtual enterprise where people can collaboratively work with much higher efficiency than before. In recent year, there are lots of researches on these distributed design systems.[1,2,3]

But the distributed systems also bring some problems that don't exist in traditional CAD systems. Many people, who have conflictive design intents that will lead to invalid design result, are allowed to work concurrently. So there must be some coordination mechanisms in order to avoid these inconsistent situations. At the same time, these systems must not limit the freedom of the designers too much to take their creativity away.

In this paper, we first introduce a web-based distributed design system that enable lots of users concurrently design a single product. The architecture of the system is outlined. Then we describe the collaborative model used in the system to facilitate the concurrent work and resolve the conflict of design intents among different people.

This paper is organized in the following structure. In the second section, we describe the framework of our web-based design system. In Section 3, the collaborative design model is presented and described in detail. The different aspects of the model are considered and discussed. In Section 4, some examples are presented to demonstrate the process of our coordinating model. At the last part, we give a summary and present the future work.

2. Three-layer collaborative design system

As shown in Figure 1, there are three layers in our collaborative design system: client layer, transaction layer and data layer.

- ***Client Layer***

The client layer is simply a set of Java applet that can be downloaded from the server with only a compatible Web browser and its mainly function is to manage the interaction with the end users. Here we adopt the thin-client distributed model to minimize the function in the client module due to the following two reasons. First we can lessen the size of the applets to reduce the time for download. Second, the computation load, which requires

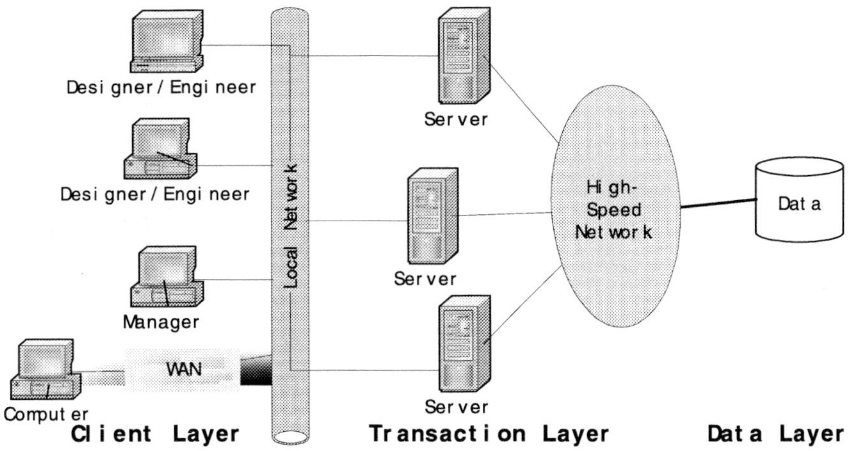

Fig 1 The three-layer architecture of design system

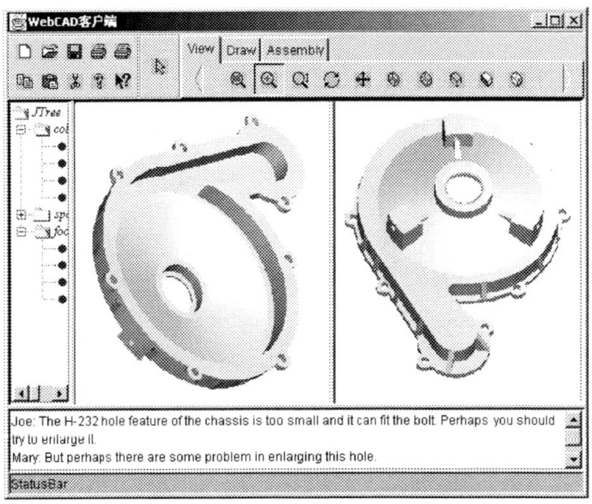

Figure 2 The Client of WebCAD

large mount of computation resource, can be controlled by servers and undertaken by special designed powerful hardware, such as multi-CPU computers.

The information flow between client and transaction layer includes product models and user requests. The product models can be the design shape or other engineer information. Shape model used here is not the same as the model used inside the transaction. It is specially designed for interaction and is compressed before transfer to decrease the load of the network.

Because the end users of the design system, who can be the employees in the enterprise or business partners that access the information through the firewall, are from different departments and perform different works, there need be many applets in the client module to execute different work. It not only allows the designers to create and modify the model, the tools for view or evaluation of the design to non-designer such as manager or salesmen should also be provided.

Figure 2 shows the user interface of client side of WebCAD, which is an experimental collaborative CAD system developed by us to verify the model proposed. This system employs CORBA as the communication framework and uses JAVA3D package as the 3D display engine.

- ***The Transaction Layer***

The transaction layer is the center of the whole design system. It is composed of lots of modules and the relationships among these modules are very complicated. The structure of the transaction layer is shown in Figure 3.

All the sub-modules in the transaction layer are represented in an object-oriented way [4,5] and connected with CORBA ORB. The ORB is responsible for the information communication between different modules. In fact, the communication between server and client are also via ORB to ease the programming of distributed systems on different platform and standardize the information exchange.

The three modules shown in the upper part of Figure 3 communicate directly with the client layer. The Connect Manager is responsible for the creation, renewal and destruction of the multiple connections. After the connection is established, the clients obtain the reference to Request Server from the Connect Server and communicate directly with it. The main function of the Request Server is to get the request from the clients and send the result back to clients. This interaction is organized in form of transaction to ensure the integrity of the core data. And the Coordinate Server's task is to inform the users and coordinate the conflict among these connections when the conflict happens. This approach for coordination will be discussed in the next section detailedly.

Every time the Request Server gets the command from the user, it reorganizes the data structure and passes it to the VPM Server where the request is processed. The VPM

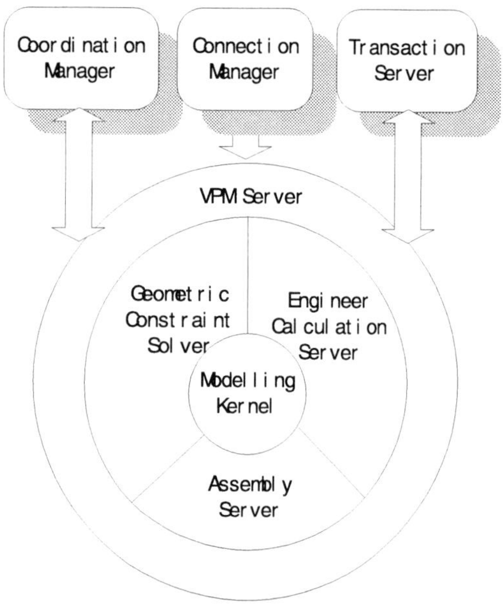

Fig. 3 The structure of the Transaction Layer

[6,7], the virtual product model, is an abstract description of a product proposed by Wen Ye, etc. All the information related to the product, including functions, behaviors and forms that are used in concept design and detail design, are represented efficiently in this model. This model can support the definition of partial model and can be evolved in parallel. The VPM Server manages all the VPM in the enterprise and keeps the validity of the VPM in concurrent environments.

All the modules enclosed in the circle of VPM Server in Figure 2 are support modules that are only called directly by VPM Server. These structures are often the same as those of the corresponding modules in traditional CAD systems and can be transplanted from legend systems. Because of the large amount of computation resources used in these modules, they are often optimized and computed by high-performance computers.

- ***Data Layer***

The data layer is a database where all the information in the design system, such as VPM of the product, the design history and the user security data, etc, is deposited. It is invisible to the clients and can only be accessed by Server Module. The purpose of this indirect access is to maximize the security of the system and facilitate the central control of the Server Module.

3. Two modes for collaborative design

- ***The Design Model***

This design model aims at guaranteeing the validity of the VPM while permitting multi-users concurrently modify a single part with maximal freedom. To implement this aim, we need a mechanism to make the users be involved in the decision-making of the product evolving. So we propose the collaborative design model that is shown in Figure 4. We can see that there are two parts in the model: the central database and the user models for every client. In the central database, a central model and the design histories to every users are deposited.

Distributed Client

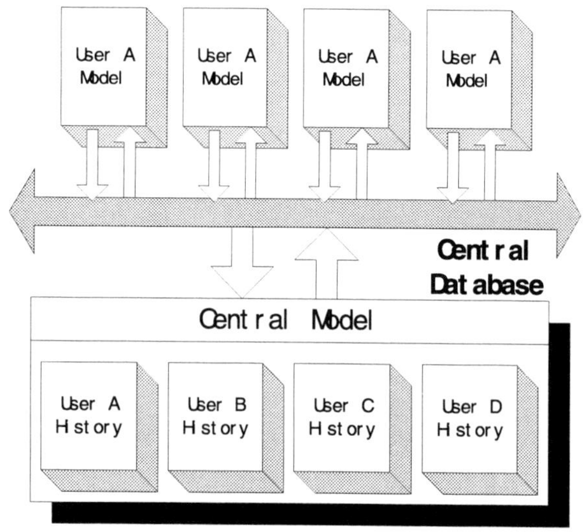

Fig. 4 The collaborative design model

This model supports two modes for collaborative design: the check-in-check-out design mode and direct design mode. In the following, we will first describe the two modes separately. And then discuss various types of conflict between users and compare their advantages and disadvantages.

- ***Two Collaborative Modes***

The check-in-check-out mode means that, each time you want to modify the model, the model must be checked out first. After it is checked out, the server allocates a temporary model for this user as shown in the upper side of Figure 4. The designers modify the model in the same way as in the single-user system. They needn't consider what other user's are working now. After they finish the modification of the model, they check it in and the system automatically update the central model by merging this temporary model in.

Although you may update the central model only when you check it in, the conflict check must be executed every time the temporary models are changed. We do that so frequently because the designers may check their work in after lots of operations and may have to discard it when conflict happens. The users must be informed about the error once the conflict happens.

The direct mode updates the central model every time the user makes any changes on the product. So no matter when the designer updates the model, he will get the latest result, i.e. not only the change made by himself, but also the change made by others since his last update. In this mode, the temporary model for every user can be omitted because the Request Server in server module is responsible for queuing up the incoming requests and ensure the VPM Server process these requests serially.

Obviously in the direct mode, the conflict check must also be executed whenever the updates are carried out.

• *The Reason of the Design Conflict*

Because of the parallelism of the design process, the conflict happens more frequently than single-user design process. There are two reasons of conflict in collaborative design system.

The first one happens when an operation causes the solid of the product invalid. In single-user design system, there are also such kinds of situation. But in distributed system, this invalid operation may be done by other designer, or even by the operations of several designers altogether.

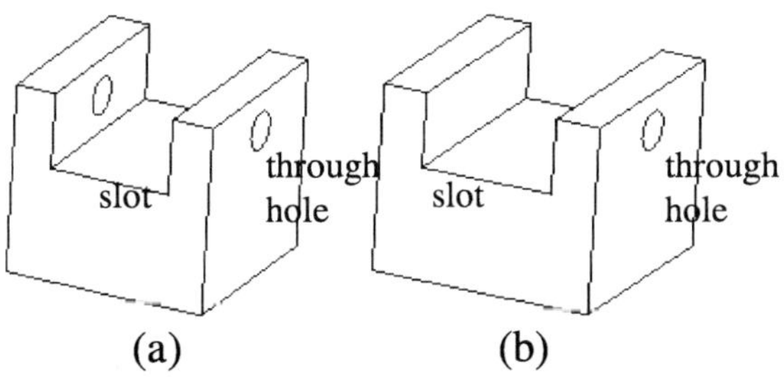

Fig 5. The design results with different operations

The reason for the second conflict is the procedural nature in most of the design systems. Because the design result relates to the order of the operations, the concurrent operations, which are submitted by different designers, may produce unexpected result. For example, in Fig. 5(a), we first drill a through hole on the side face and then make a slot. In Fig. 5(b), we exchange the order of these two features. Apparently, the design results are not the same. To the operations such as deletion of the feature, the exchange of order may even produce devastating result because of the possible loss of reference geometric entity. But in distributed environment, different users may create these two features at the same time. So notification to the designer about the difference is important to keep the design result is the same as the what the designers really want.

The first kind of conflict can be checked by the solid modeler where the operation leading to invalid model will be rejected. And the second conflict type can be checked by examining the dependency between the features. Note that the check of the first conflict must be done in geometric level instead of feature level because the validity of the product is not only determined by the consistency of the features, but also the parameter values of these features.

With the prior analysis and our design model, we can know the conflict check of the model should not only be done between user model and central model, but also between different user models.

When the conflict happens, the design system needs to inform the users related to this conflict about the situation and can provide several options to correct it. The system can provide several communication approaches that are used in different conditions for the designers to negotiate. Fox example, online talk, white board and electrical conference can be used when the designers are all on line. And E-mail can be used to inform the users that are not on line.

• *The Comparison of the Two Modes*

The advantage for the check-in-check-out mode is that you can treat the product in similar manner with the single-user CAD systems. The user needs to consider other designers only when the conflicts happen. It can be used in the environment where the relation of the designed part is not very tight. And the optimization of the speed and space is easy because the user model can be a partial part of the central model where the different users do the detail design on different parts. So the conflict check can also be simplified.

The direct mode is suitable for the situation where the designs of different users are highly coupled. Users can get the latest version of the product and find the other users' modification. Because there are some conflicts that can't be detected by the conflict checker or probably some operations made by others will influence the decision-making of the future design, this mode is valuable for finding this kind of error in time. Another advantage is that the VPM Server needs maintain only one central model and it consumes relatively small memory space. But to this mode, the conflict check is more difficult to implement that the check-in-check-out mode.

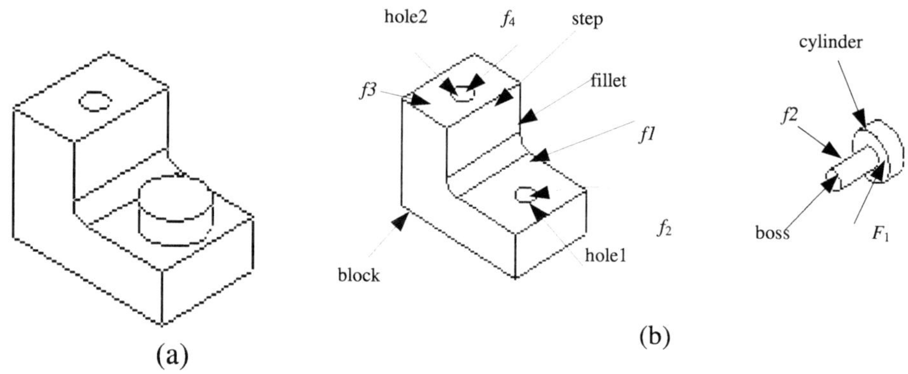

Figure 6 An Example Assembly

Because the previous two modes both have their pros and cons, in practical design, we can use two modes together. For example, the whole design can be divided to several portions and use the check-in-check-out mode. Here every user corresponds to a design group in practice and in these groups, they can cooperate with direct mode.

4. An example for concurrent design

In this section, we present a product model to illustrate the possible design scenario to demonstrate the power of our collaborative design model with both modes.

Figure 6 is an assembly composed of two parts: base and bolt (cited from Reference 7). If different groups of designers independently develop these two parts and the check-in-check-out mode is adopted, the system will construct two temporary models respectively for the two groups. One represents the virtual component of base and another represents the virtual components of bolt. Two groups can only pay attention to their work and not regard to others'. The underlying conflict checker is responsible to examine their relation. When the conflict is tested, it interrupts the design process and reports this error. For example, if one group increases the radius of hole1 on the base and some tolerance constraints are violated, the system will report this error and gives two options: whether roll back the operation or enlarge the boss of the bolt to satisfy these tolerance constraints. If no such things happen, the designers will process the product as normal and not be interrupted.

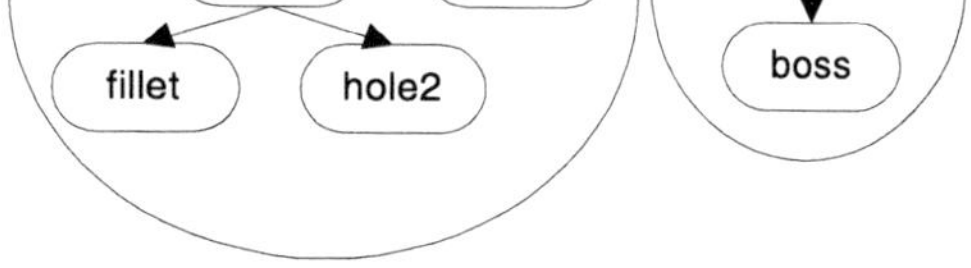

Fig. 7 The Virtual Product Model of the assembly in Fig. 6

If several users in a group jointly design the base part, because of the tight relationship between its features, the users adopt the direct mode. In this situation, only one virtual component is constructed and accessed in sequence. Any invalid modification, which may not violate the explicit constraints or rules and can only be found by the experience of the designers, can be discovered in the shortest time.

As we can see, the combination of the two design modes can find a good compromise between the freedom of design operations and the frequency of design conflicts. Thereby the efficiency of design process can be promoted.

5. The conclusion and future work

In this paper, a collaborative design model is introduced. We construct the system WebCAD based on Web technology. Its architecture and all the modules are described elaborately. And then a collaborative model is presented. The approach for dealing with the conflict and two modes with different advantages and disadvantages are described. The design group can use these modes flexibly to maximize their efficiency and freedom without much conflicts happening.

There are lots of works that need be done in the future as following.

- ✧ The algorithms for the conflict check should be refined to process it more efficiently, especially under the situation where many users concurrently work together.
- ✧ The options for resolving the conflict provided to the designers should be studied to provide more flexible choice.

6. References

[1] M. Rezayat The enterprise-web portal for life-cycle support, Computer-Aided Design Vol. 32 2000 85-86

[2] W.C. Regli., V.A.Cicirello Managing digital libraries for computer-aided design, Computer-Aided Design Vol. 32 2000 119-132

[3] G.Q.Huang, J.Huang, K.L.Mak Agent-based workflow management in collaborative product development on the Internet, Computer-Aided Design Vol. 32 2000 133-144.

[4] Wynne Hsu and Irene M Y Woon, Current Research in the Concept Design of Mechanical Products, Computer-Aided Design, Vol. 30 1998, 377-389

[5] S.R. Gorti, A. Gupta, G.J.Kim, R.D. Sriram and A. Wong, An object-oriented representation for product and design process, Computer-Aided Design Vol. 30, No. 7 1998 489-501.

[6] Li Hailong, Song Peng, Dong Jinxiang. Top-Down Concurrent Assembling Based on Template, Journal of Computer-Aided Design& Computer Graphics, Vol.11□No.5□1999 441□445 (in Chinese)

[7] Wen Ye, Zeng Zhenyu, Tang Min, Dong Jinxiang, An architecture of web-based distributed CAD system, The Second International Conference on CAID&CACD 1999

[8] Tang Min. Research and implementation of Key Technology in Geometric Kernel Based on Sculptured Model Doctorate thesis, Zhejiang University (in Chinese).

[9] C.M.Hoffmann, R. Joan-Arinyo, Distributed maintaince of multiple product views, Computer Aided Design, Vol. 32, 421-431, 2000.

A Framework and Implementation Techniques for Cooperative Architectural Engineering Design Systems

Weiyu Wu, Peng Pan, Hua Li, Wei Yu and Dongmei Liang
Dept. of Civil Engineering, Tsinghua University, Beijing 100084, P.R. China
wuwy@mail.tsinghua.edu.cn

Abstract

In order to realize the Cooperative design for architecture engineering among people in different specialties or in different zones, a computer support cooperative design system based on network environment, which can integrate different professional programs and function modules, was presented. The key implementation techniques, which include solution of data integration based on neutral database, communication mechanism based on standard network protocol, particularly for multimedia system, synchronous and asynchronous control for model data and working rules in cooperative design are also illuminated. 3DMECAD, a prototype cooperative design system was developed by us to prove the validity of our solution. The system has a promising foreground in China.

1. Introduction

Modern architecture design[1] requires tight cooperation and real time communication between people in different specialties and different people in the same specialty to guarantee the data coherence and minimum redundancy. Further more, it also requires many people, who may be in different zones, work together for a complex problem in the design. Traditional computer aided design (CAD) system are mostly isolated and cannot fulfill the requirement. Computer support cooperative design system presented in this paper can be a good solution. In the cooperative design environment, multiple people work together on networks. They can discuss solutions, design cooperatively through multimedia, such as graphics, images, texts, audios and so on. The quantity and efficiency can be highly increased.

Computer supported cooperative design system is a typical application of CSCW (Computer Supported Cooperative work) in the field of architectural engineering. Its design should not only satisfy the principle of common CSCW system, but also has its owner characteristics. Generally, problems on cluster, dynamic, parallel, synchronism and asynchronism should be considered[2]. The occurrence of the system is based on the development of computer networks and multimedia techniques, which includes[3]:

1) High-speed and real time communication network and standard protectoral for multimedia;

2) Real time represent technique for the result of both synchronous and asynchronous events;

3) Distributive database with content-based search function and hypertext technique.

4) Parallel control mechanism for CSCW.

Ups to now, some CSCW systems exist in the world. For example, Fraunhofer graphic institute of German issued distributive CAD system "Distribution Sketchpad" in 1994; Shastra of Purdue University of United State developed a multimedia CSCW environment based on desktop network; Tsinghua University of China also released an integrated service multimedia communication system. Unfortunately, CSCW system specialized for architecture engineering is not adequate and waiting to be developed.

2. Proposed framework and key technique

2.1. Integrated model of cooperative architectural design system[4][5]

Currently, producing 2D drawings independently for each specialty is the most popular method in architecture design in Chinese. That always leads some trouble. For example, if you want to modify the dimension of a window, all the corresponding drawings, such as the plan, front and cross-section have also to be modified. If we use 3D model, we can get the drawing of all views and different drawing for each specialties. 3D model will highly increase the efficiency and quantity of design. It is also the bases of CSCW. Figure 1 shows the integrated model of cooperative architecture design system.

Where Central architecture model is dynamic, designer of all specialties can modify the architecture data through interim processor for the prototype model, thus create and modify the data of each specialty. Different data are put on different layer. Each layer is managed by a specialty. Designers of other specialty can't modify the data, but they can use the data for reference. In this way, the information of data change can be communicated on time. That will do great help to the cooperation in design.

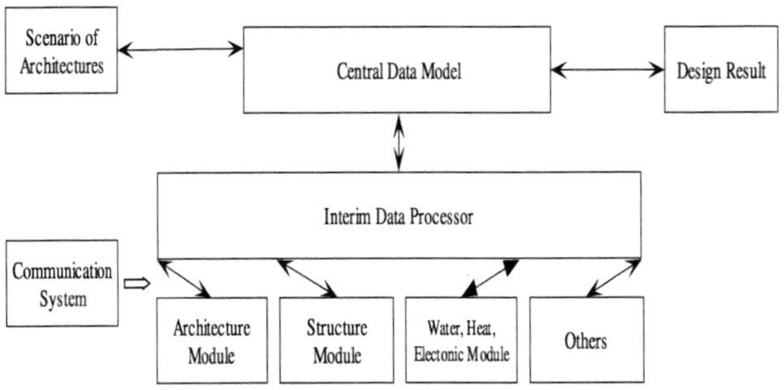

Figure 1. Integrated model of cooperative architecture design system.

2.2 Data integration and sharing model based on neutral database[6]

One of the most important problems in CSCW is data transformation and sharing. On the one hand, data transformation and sharing are necessary when using different software. The output file generated by a program should be able to be recognized by another one.

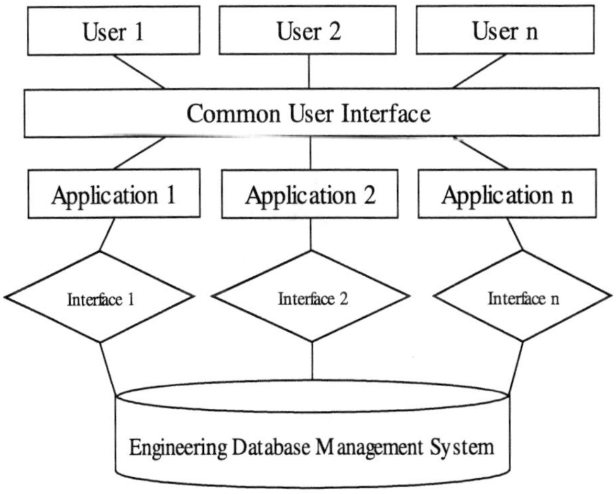

Figure 2. Data integration and sharing model.

On the other hand, data should also be communicated and shared between different specialties. Only in this way, different CAD system and different function modules, such as architecture, structure, water supply, heat supply, manufacture, finite element analysis, information system and so on, can be organized together and controlled by a single control program to guarantee an unblocked data flowing.

For the first situation, we use neutral data as the solution for sharing data between software. Interfaces are constructed for each piece of software. The interfaces are in charge of changing data from special software into neutral format, then saving neutral data into neutral database for other users. It can also take out the data from database and change them into the format legible for special software.

Neutral database can also solve the problem of sharing data between different specialties because all the data for the design are saved in the database after processing by the interfaces. They are transparent to all the specialties.

A common user interface is created for the convenient of communication. Users can modify the data through their own software. But using common interface is preferred because that will be benefit for understanding each other. The data integration and sharing model are illustrated as Figure 2.

In order to acquire real time response and support for unrestricted operation, distributive database are used. There is a copy of the central database on each workstation and all of them are keeping constant with the central database. Users changed the data on their workstations and modified the data in the local database. All the operations were transmitted to central database through cooperative server. Figure 3 shows the distributive database model:

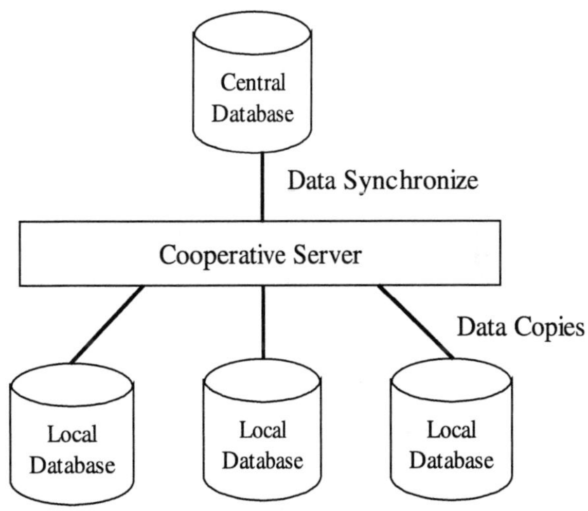

Figure 3. distributive database model.

In the project of architecture design, conflicts often occur between specialties because of influence communication. Data integrated and sharing based on neutral database is a good solution.

2.3 Platforms and communication mechanism of cooperative design system

The platforms and communication mechanism of Cooperative Design System includes three layers. They are browser, web server and database. The system is constructed by series of component works. Through these reusable and plug-in component works, it is easy to design and develop the groupware in CSCW. The structure is very flexible. Java technique and dynamic web pages are also used to realize the difficult mission of communication. Figure 4 illustrates the mechanism of communication.

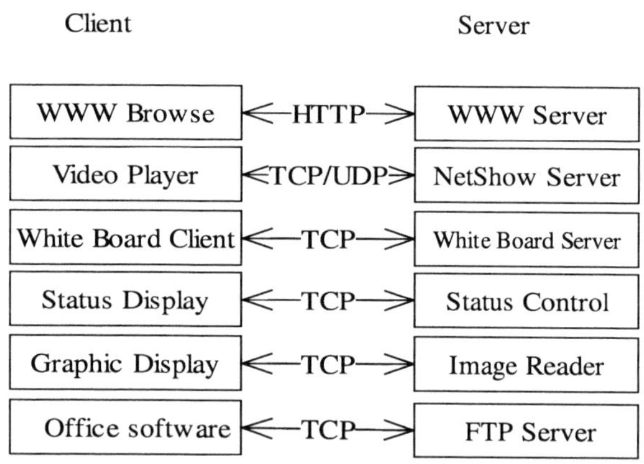

Figure 4. Component works and communication mechanism.

White Board includes an area for drawing. Operation on a client will be transmitted to other clients. All the clients share the same works. The component was developed by Java. The area of white board was divided as shown in Figure 5.

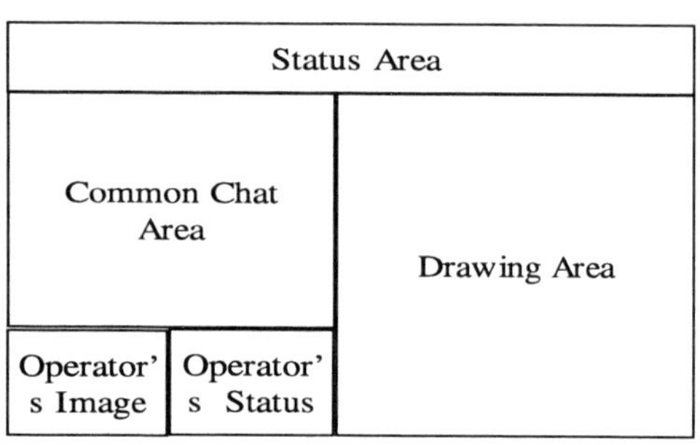

Figure 5. User interface for discuss problems of cooperative design system.

Status area is for displaying the process of the project; common chat area is for exchanging text; operator's Image shows the face of the designer; operator's status area shows the status of all the designers. Drawing area is the place for designers to draw some graphics.

Web paging publication technique was used. Web server can get the input of the users and return the result of the operation to keep the client to get the latest information.

2.4 Parallel control over model data in cooperative architecture design system

Many causes may lead to the requisition of parallel control. As cooperative design occults in different terminals connected with network, applications are also dispersed. While cooperation between applications is accomplished with message exchange, procrastination and contradiction latent in that process will bring on disorder of operations and thus result in parallel conflict (Figure 6).

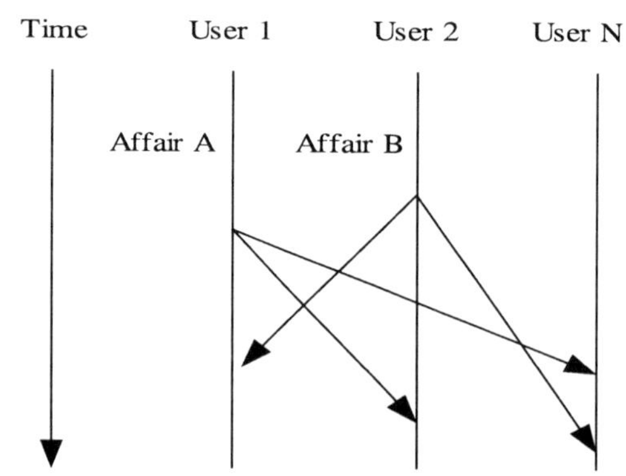

Figure 6. Parallel conflicts.

When a number of users shared a common virtual workspace, two or more users may manipulate one object at the same time. If operations on the object are out of order, errors can even conduce to debacle of the system. For instance, a user may need to move some components that have been removed by another one if operations of users are not concerted. As parallel conflict would dislocate management of cooperative systems, parallel control mechanism must be established to resolve these problems.

Cooperative Architecture Design System successfully set up parallel control layers and divided the whole model into Locked Layer and Network Layer for different types of work. Data on Locked Layer is read-only, and users doing the same type of work shared Network Layer complying parallel control rule.

In our system, users can contain conformity of Network Layer through exchange of operations on object. Operations of different terminals may reach the server in different sequence with occurrence. If operations cannot be interchanged, that difference will make another outcome. Generally, operations should not be interchanged, and we make use of series protocol to reach the ultimate conformity. On account of causality, we designed the scheme on base of status vector. In this way, we can delay some operations selectively and can conform them to orders.

When the system server has received operations out of order, it clears up the prior operations related to the disorder, and newly applies the operations according to the established rule of order.

2.5 Regulations of cooperative engineering design

Cooperative engineering design is far more complex than a discussion face to face in the same room, and we can find many questions unnoticed before. Referring to other CSCW system, we devised a status monitor of design participator, on which movements of participators can be shown. At the same time their operations are recorded in system as files.

Legality of Operation is another problem we have considered. Only one participator is allowed to operate in case that someone assumes the design progress.

In cooperative design system, we need to exchange all types of data with messages. For the convenience of that progress, we must make definitions and explanations of a good deal of messages.

As cooperative designers will unavoidably operate on a common object, we make use of a "Lock" to keep the conformity of shared data. i.e. while the model is being altered by some user, others have only the right to read the same model data.

3. Development and realization of prototype system

According to the needs of Architecture design, we developed an Architecture cooperative design software system (3DMECAD: 3D Modeling Engine-based Architectural CAD System). It is based on structure design, and is contrived to fit the characters and demands of Architecture subentries. And ultimately, our software system aims at purveying cooperation functions of different types of work.

In our software system, we engineered special programming interface which can be used to connect with other Architecture project software. And our system possesses excellent integrating function.

3.1 System scheme

We made a 3D model engine as the kernel of 3DMECAD system. The 3D engine has a CSG/B-rep compatible diagnostic model, which is functioned as the public data source of the whole system. The model and the project (engineering) neutral database connected to it contain all data used in Architecture design and can provide special design with prototype data or figure. The results of special design are returned to the 3D model. The model and the project neutral database keep the data returned back and can purvey other specialty. With the new solution, we condensed special functions such as Architecture design□structure design□structure analysis and construction plot into an integral compatible software system, as in Figure 7 . Thus the system greatly enhanced efficiency of Architecture analysis and design. Also the system can make design more dependable.

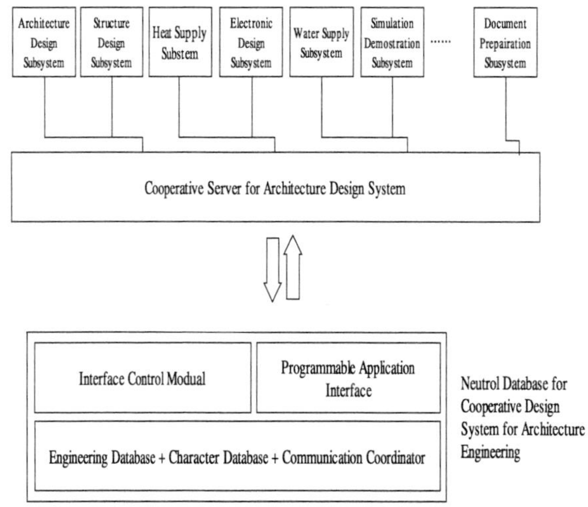

Figure 7. Scheme of 3DMECAD.

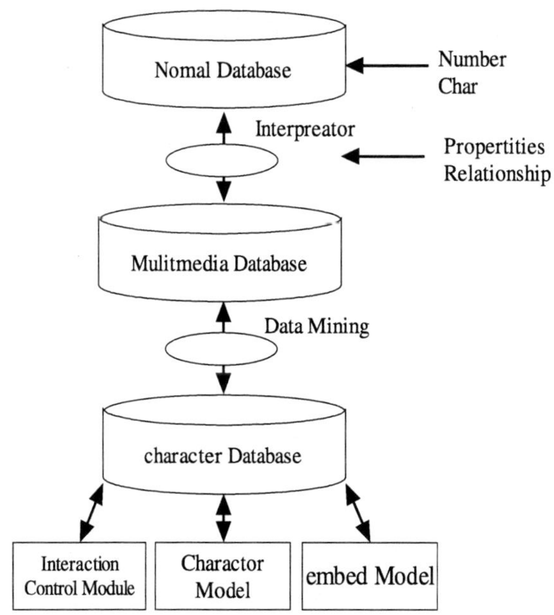

Figure 8. Frame of neutral database.

3.2 System neutral database

Neutral database of the system is a foundation for all types of compatible application. The complex relationship between applications are paralleled on the base of the project database. Our project database is designed on base of frame of a multimedia database. The frame of it is illustrated in Figure 8.

3.3 Architecture CAD cooperative server

Applying VRML, our CAD cooperative server can efficiently exchange design data under SVG standard. Figure 9 shows the working frame of the cooperative server.

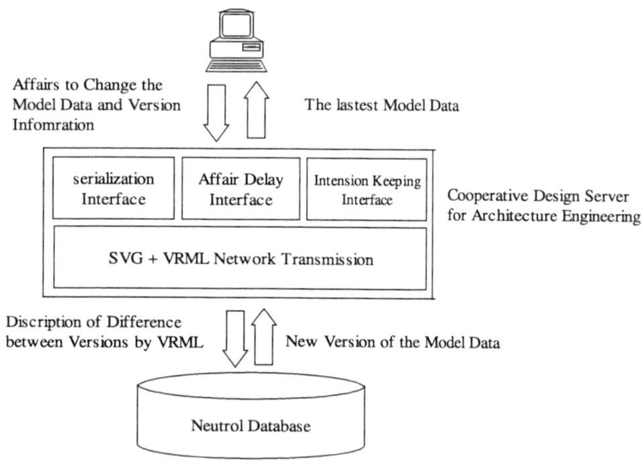

Figure 9. Working frame of cooperative server.

Management of model data editions helps the server to accomplish cooperative function . The detailed functions of the server are:

1) After a subsystem finishes its work on model data labeled with edition number **m**, it sends all the operations done on date edition **m** to its terminal VRML data transportation server. Along with the operations, it sends out demand for renewing data.

2) When receiving the demands from terminals, Architecture CAD cooperative server reorders all these order according to the rules of operation sequence. It delays the operations which are out of precise order and thus it may correct errors caused by disorder operations.

3) On finishing operations with the corrected orders, the server passes the result data to system neutral database.

4) With the result sent to it, system neutral database renews the neutral model data and produces new model data with updated edition number **m+1**. And then the new edition is send back to terminal to meet their needs of cooperation.

During those progresses, Architecture CAD cooperative server kept operations in good order and avoided mistakes which may caused by disorder. Thus it concerted the works of different terminals.

4. Conclusion

CSCW is a new concept, which needs the support from not only high technology but also psychology and sociology. How to make the CSCW system fit the requirement of human beings is a long term and difficult job. Computer supported cooperative design system for architecture engineering is one of the typical applications. Developing a successful system is not easy. 3DMECAD is only a prototype system. Many factors have to be considered in the future, such as the communication between owners and contractors, the security strategy of the system and so on. We can draw some conclusions from the experience of developing 3DMECAD: computer supported cooperative design system for architecture engineering should have a good solution for data transformation and sharing between different software; the parallel control, such as synchronous and asynchronous is the most pivotal job in system; the development of multimedia technique and network is the bases of the system. CSCW technique has made a big progress in recent years. With the development of corresponding technique, such as audio and video transmission through network, database technique, wildly used computer supported design system for architecture engineering will be realized in the future.

5. References

[1] Chenglin Xie, Weiyu WU, Jianqun Hou, "Multimedia Simulation-based Architecture Structural CAD System Model", *Journal of Tsinghua University*, 2000,40(6), pp.12-114, 126.

[2] Hui Wang, Yuzheng Wang, "Multimedia Communication and Long Distance Design and Education", *Theory and Application of Multimedia Aided Engineering*, Chinese Electronic Press, 1998, 8, pp19-46.

[3] Zongkai Ling, "CSCW and Application of Multimedia Technique in CSCW", *Theory and Application of Multimedia Aided Engineering*, Chinese Electronic Press, 1998, pp.90-104.

[4] Francis Pahng, Nicola Senin and David Wallace, "Distribution modeling and evaluation of product design problems", *CAD*, Vol30, No6 1998(6), pp.411-422.

[5] Teresa De Martino, Bianca Falcidieno and Stefan HaBinger, "Design and engineering process integration through a multiple view intermediate modeler in a distributed object-oriented system enviroment", *CAD*, Vol30, No6 1998(6), pp.437-452.

[6] Weiyu Wu, Wei Yu, "Research on Cooperative design system for Architecture Engineering", *Proceedings of the Third Multimedia Aided Engineering Conference*, Shen Zheng, China, 1999, 11.

Volume Modelling in a Feature Based Product Modeling System

Min Tang, Chentao Yu, Jinxiang Dong
State Key Lab of CAD&CG, AI Institute, Zhejiang University, Hangzhou, 310027, P.R. China
tang_m@163.com, djx@cs.zju.edu.cn

Abstract

In this article we propose a hybrid modeling system based on B-rep model and volume model to overcome the inability of representing volume information of modeling object in conversional feature based modeling systems. Using this hybrid representation, we can easily solve the visualization and mass property computation for designing objects with specified volume attribute.

In this article we propose a new method of converting B-rep models into volume models. This converting method is based on marching cube algorithm, and is optimized by octree algorithm. Using this converting method, we can efficiently convert the B-rep models into Volume models. We also discuss the antialiasing technical on those volume model, we analyze different filters used for the voxelization of analytically described objects.

To solve the visualization product problems in modeling systems, such as dynamic simulation, we present the methods of how to add Boolean operations on the hybrid modeling system to solve these problems. Voxelization enables uniform representation of all objects by a single primitive. Thus, instead of dealing with a number of different objects with specific features, dynamic simulation and dynamic interference on volume models only need simple calculations and is more efficient. We show that, using volume modeling, difficult problems in old modeling systems can be easily solved.

The detail of our algorithm is discussed in this article. We have implemented a volume modeling system based on our old feature based modeling system-GSCAD.

1. Introduction

From the initial wire frame modeling to the more efficient surface modeling, then to the fast solid modeling, modeling techniques of CAD have made a rapid progress. With the development of those technologies, such as CIM, VR and visualization, and also with the deeper and deeper implementation of CAD&CAM systems, we feel that those old modeling techniques, though have been developed many years and became matured now, still have their shortcomings. For example, it shows its weakness with these problems such as dynamic simulation and dynamic interference. To overcome those shortcomings of the old modeling techniques, some people proposed a new modeling technique-Voxel based Modeling which would easily solve those problems that might be very difficult to solve with old modeling systems.

In this article, we will introduce to you our hybrid modeling system that is based on B-rep and volume modeling techniques. We'll have a detailed discussion on how to transfer boundary represented models to voxel models, how to solve the antialiasing problem in voxel models. And we'll give out logic algorithms to solve the difficulties in MCAD systems such as dynamic simulation.

2. Model transferring

In B-rep modeling systems, entities are represented by semi-edge data structure, which represent entities with solid, shell, face, loop, semi-edge, edge and vertex data structures. The topological information and the geometric information in B-rep models are saved apart. In voxel-based modeling systems, an entity is represented by a volume buffer, which is formed by volume data sets. So firstly, we must transfer B-rep models into voxel-based models. Using Marching Cube algorithms we can easily do it. The first step of this transferring is to set up a 3D part grid. The boundary of this grid is the same as the solid's of the old B-rep model. After we get the boundary of the 3D part gird, we will divide it to small part of voxels, just very small cubes in this article. Then we must decide which of those voxels are parts of this entity and which are not. Marching Cube algorithm will help us much to solve this problem. The procedure of transferring a B-rep model to voxel mode is listed here as follows:

```
Nf - the number of faces in old B-rep Model;
Begin
        Integer  k;
        For k=1 to Nf Do
    Calculate the boundary box of each face;
    Enlarge the box we get to match the 3D
    Part        Grid;
    For each voxel in this enlarged face-
    boundary box Do
```

Define V1, V2, V3, V4, V5
, V6, V7, V8 as the eight corners of
this voxel;
Calculate whether this voxel is inside
or outside or on the surface of this
face;
Calculate the Normal of this voxel
End For
End For
End All

Now, we've gotten the volume buffer of the model. We also can use the octree algorithm to optimize the algorithm listed.

(1) Using Octree to optimize the transferring

An octree is a tree structure that describes the division of a cubic region of space into smaller cubes. Each cube can be "full" or "empty" or further divided into smaller cubes ("partial"). In the tree structure, a leaf node represents a full or empty cube, and pointers to nodes representing the eight sub-cubes into which it is divided represent a partial cube. Each "full" cube may contain some information about color, density, or some other attribute of the object, but the cube itself is considered to be either fully filled by the object or empty of it. Octree is well known for its dynamic spatial resolution. In fact, we needn't calculate all the voxels in the 3D Part Grid for the entity may posses a very small space of this 3D Part Grid. We can use the octree as a filter that can help us decrease the number of voxels needed to be calculated. But how do we find a particular node in a given octree? We defined the dividing condition, as that-the Volume that the entity interfered with the current voxel is larger than the smallest voxel you defined multiplied by a constant C. And the constant C must be larger than 0 and small than 1.

(2) Advantage of Hybrid Modeling

In our hybrid modeling system, entities are represented by Boundary Represented Models and Voxel-based Models. We use boundary represented modeling techniques, which use solid, curved surface and wire frame to represent integrated entities, to support the designing, rendering and alternating processes of modeling, and use voxel-based modeling techniques, which use uniform small cubes to represent entities, to settle those processes such as dynamic simulation, dynamic calculation, interference detection. The Voxel-based Modeling techniques are made up by two components, one is the representation of voxels we have discussed above, and the other is the Boolean operation that we added to solve those dynamic problems. With these two modeling techniques, we can give out a new MCAD system that may easily solve most of the problems that may occur when designing an entity.

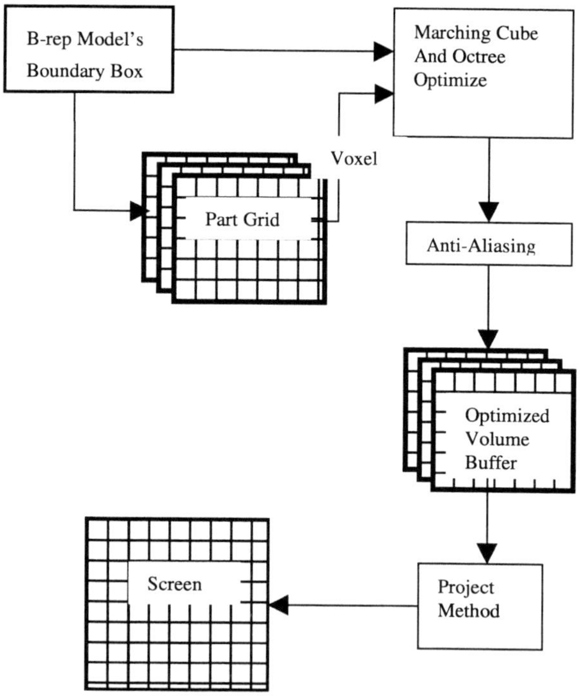

Figure 1. Change B-rep Model to Volume Buffer and Display it.

3. Antialiasing of voxelization

Volume graphics delineates a series of techniques aimed at modeling, manipulation and rendering of objects represented by means of a 3D raster of elementary volume primitives. Volume graphics has several advantages in comparison to the traditional geometric approach. Voxelization decouples object specific issues from rendering and enables uniform representation of all objects by a single primitive. Thus, instead of dealing with a number of different objects with specific features, a rendering process only one primitive object which simplifies its architecture and leads to a more effective implementation. Here we introduce some techniques to solve the antialiasing problem in voxelization.

3.1 26-Adjacent linear interpolation

The easiest algorithm of antialiasing methods may be linear interpolation. Here we introduce the 26-adjacent linear interpolation first. We call two voxels are 26-adjacent if they share a vertex, an edge, or a face. We know every voxel is connected with 26 other voxels adjacent to it. We get the color value of those 26 voxels, and add it with a weight function as below to get the dispersion of current color value of the center voxel:

26-Adjacent Voxels Weighted Color = A1*V1's Color + A2*V2's Color +...+A26*V26's Color.

$$\sum Ai=1. \qquad (1)$$

Then we can get the new color value of the center voxel with this formulation:

Voxel's New Color = C * Voxel's Old Color + (1-C) * 26 Adjacent Voxels' weighted Color

$$0<C<1; \qquad (2)$$

We also can predigest the methods to reduce the calculation as using 18-adjacent or 6-adjacent linear interpolations.

3.2 Other Filtration Techniques

To give out alias-free graphics of a voxel-based modeling system, we can also use the techniques of filtration.

To reduce aliasing, a volume sampling technique has been developed, which estimates the density contribution of the geometric objects to the voxels. The density of a voxel is attenuated by a filter weight function that is proportional to the distance between the center of the voxel and the geometric primitive. To improve performance, pre-computed lookup tables of densities for a predefined set of geometric primitive scan be used to select the density value of each voxel. For each voxel visited by the binary voxelization algorithm, the distance to the predefined primitive issued as an index into a lookup table of densities. For example, we can define a oriented box filtration function in our modeling system as below:

```
if (x, y, z) inside solid then
    f(x, y, z)=1;
else
    f(x, y, z)=0;
```

Other filtrations, such as cone shaped Bartlett filter and Gaussian filter also can be used.

4. Boolean operations on voxel-based modelling

Based on the Boolean Operations of voxels, we can conveniently implement the virtual manufactory and dynamic simulation of current MCAD systems. We will give out a set of Boolean operators and list out their different utilities in MCAD systems.

4.1 Boolean Operators Defined On Voxels

For different utility of CAD systems, different kinds of Boolean operators have defined. For example Milos Sramek and Arie Kaufman Have defined the below operators for a voxel based CSG system (see Table 1).

Table 1: Voxel based CSG operations between objects.(d---density of the voxel, c--- color of the voxel, WA --- weighted average of the colors.)

Operation	Symbol	Density	Color
Intersection	A* B	$d_{A*B} = \min(d_A,d_B)$	$C_{A*B} =$ WA(A,B)
Union	A+B	$d_{A+B} = \max(d_A,d_B)$	$C_{A+B} =$ WA(A,B)
Difference	A – B	$d_{A-B} = \max(0, d_A-d_B)$	$C_{A-B} = c_A$ if $d_{A-B}>0$

To solve the problems in MCAD systems, we defined our Boolean operators and listed its utility in MCAD systems.

Different from the Part-Grid, if we want to use Boolean operation methods in MCAD systems, we must define a Global-Grid. For each entity what will be used to do a Boolean operation must expand its part grid to fit the Global-Grid, that means the corner of those Part Grids must be on one grid point of the Global Grid, and the division precision of the Part Grid must be the same as the Global Grid.

Our defined Boolean operations have shown below (see Table 2):

Table 2: Boolean operations on voxels (A and B are voxels. If Voxel.Inside equals 1, then this voxel is inside the entity or on its surface, if it equals 0, then this voxel is outside the entity.)

Operation	Expression	Value
Intersection	C.Inside = A.Inside * B.Inside	=min(A.Inside, B.Inside)
Difference	C.Inside = A.Inside – B.Inside	=max(0, A.Inside - B.Inside)
Complement	Ā.Inside = I – A.Inside	=max(0, 1-A.Inside)
Union	C.Inside = A.Inside + B.Inside	=max(A.Inside, B.Inside)

Operation	Utility In MCAD System
Intersection	Interference Detecting
Difference	Simulating Incising
Complement	Model Making
Union	Agglutinating

4.2 Dynamic Simulation, Interference Detecting and Molds Making

Now we can use Boolean operators we defined above to easily solve those difficult problems in machinery computer aided design systems, such as dynamic simulation, interference detecting and mold making. For Example, as we want to assemble parts of a object, we must do dynamic interference detecting during the assembling process, and also do static interference detecting after the object is assembled. Using the intersection operator, we can get it. Suppose that we currently want to know whether two parts, part A and part B, are interfered or not. We define that two voxels of two different parts have the same index only when they are the same in the Global Grid we described before. We can now get the result by check each of the voxels in part A. For each of the voxels in part A, we compare it to the same indexed voxel in part B, if voxel B's Inside property is 1, then we get that the two parts are interfered. The two parts are not interfered only when every voxel of part A has no same indexed voxel inside part B. We also can use the Boolean operators to simulate the cutler's working and sculpting. We can transfer the object we want to cut to a voxel model and also transfer the knife-edge to voxel model, then the difference operator and intersection operator can be used on them to simulate the cutting process. Here is the form:

Part A New = Part A Old –
(Knife Edge Current * Part A Old); *(3)*

For the mold making process, we can use the complement operation on the voxels that compound an entity, and then we will get the mold of it.

5. Conclusion

In this article, a hybrid modeling system that can be used to overcome the inability of representing volume information of modeling objects in conventional feature based modeling systems is introduced,. It bases on B-rep and volume models. We introduced the way of transferring the boundary-represented models to voxel-represented models, and also used the octree method to optimize the algorithm. We gave out the way to give out anti-free graphics of voxel-based models. Then we defined a set of Boolean operators on MCAD utilities, and discussed how to use this operator to solve the problems such as dynamic simulation, interference detecting and mold making. Our work is based on the work of GSCAD, which is a feature based parametric modeling system. In our hybrid system, the transferring of any boundary-represented object to voxel-based models can be done, and part of the Boolean operations can be done.

The further work is to implement all the Boolean operators and use them to simulate real world manufacturing. We also will improve our algorithms to reduce the mass calculation in voxel-based modeling and make our system to be efficient.

6. References

[1] D.E. Breen, S. Mauch, and R.T. Whitaker, 3D scan conversion of CSG models into distance volume, In IEEE Symposium on Volume Visualization, pages 7-14, 1998.

[2] R.N. Bracewell, The Fourier Transform and Its Applications, McGraw-Hill Book Company, New York, NY, 2nd edition, 1986.

[3] B. Cabral, N. Cam, and J. Foran, Accelerated volume rendering and topographic reconstruction using texture mapping hardware. In Proceedings 1994 Symposium on Volume Visualization, pages 91-98, Washington, D.C., 1994, ACM SIGGRAPH.

[4] M. Deering, Geometry compression, Computer Graphics, 29:13-20, August, 1995.

AI Supported Computer-Generated Pen-and-Ink Illustration

Yan Gu, Songhua Xu, Min Tang, Jinxiang Dong
CAD&CG State Key Lab
AI Institution of Zhejiang University
P.O.BOX 1549, Hangzhou 310027
guyan@cise.zju.edu.cn

Abstract

In the field of computer graphics there is an increasing demand for non-photorealistic effects. In the paper, we add the idea of pattern recognition guided by theoretical rules into previous traditional NPR system and create non- photorealistic drawings intelligently. The Style Sample Library functions as an expert library and makes it easier to render a picture with non-photorealistic effects even by an amateurish user.

1. Introduction

Recently, a variety of techniques have been proposed to non-photorealistic rendering objects. This work was driven by the impulsion that we wanted to convey necessary information but consuming less storage, so they can be easily reproduced and transmitted.

In this paper, we proposed a method for AI supported pen-and-ink illustration of CAD parts. The system recognized the drawings users created automatically and guided the users towards the patterns predefined in style sample library. Actually, this is an intellectual learning process.

1.1. Related work

Gooch [2] et al. proposed the conception of color illustration, with luminance and changes in hue to indicate surface orientation, reserving extreme lights and darks for edge lines and highlights. Deussen and Strothotte [1] applied depth discontinuities to determine what part of each drawing primitive was to be drawn to constitute the foliage. Winkenbach and Salesin [9] summarized the principles of traditional pen-and-ink illustration and introduced "stroke texture" in their paper. The Premisys Corporation [10] marketed a product called "Squiggle" that adds waviness and irregularities to CAD output as a post-process, lending a hand-drawn appearance to the drawings. In many applications, for example, when we want to describe a complex mechanical hardware, it is time costing to render it including extraneous details. With illustration we can accent on more important parts and achieve effective diagrams, for pen-and-ink illustration, it has well-established conventions. For this reason, we don't duplicate others' work simply.

The main contribution of our work is that we began with important cognitive principles of pen-and-ink illustration and created style sample library and recognized certain patterns based on the mathematical definition of pattern specification. Furthermore, with the guidance of system, we approximated to the predefined patterns.

1.2. Overview

The rest of this paper is organized as follows. Section 2 introduced the idea guiding the design of this intellectualized system and the building blocks of our system. Section 3 explores the traditional illustration principles with AI techniques and gives our synthetic algorithms applied with pattern recognition. Section 4 discusses some of our results and section 5 summarizes our work and looks into the future. Appendix gives details about implementation.

2. Implementation

The most creative investigation of this paper is to integrate AI techniques ingeniously, including specifying style parameters according to the principles of pen-and-ink illustration in pre-process and recognizing patterns with geometry taxonomy.

In pre-process, the system supports three types of stroke textures till now; they are "cross-hatching", "stippling", "glass". We got the idea from [3]. Based on these types of stroke textures, we extract five style parameters. They are L (stroke length), T (stroke thickness), W (waviness), O (stroke orientation), D (stroke distribution).

In our Style Sample Library, we give mathematical definition of these five style parameters to each style of stroke textures. This is an artistic process, not a technical one. No abstract standards rule whether it is non-photorealist or not. We just create this Style Sample Library on our perspective and set it as criterion of latter steps.

The recognized style parameters of the user created drawings compare with the style parameters from Style Sample Library. Recognizing the difference, the system modifies user created drawings under the guidance of predefined modification rules.

The implementation is described in Fig. 1.

3. AI supported computer-generated pen-and-ink illustration

3.1. Style Sample Library

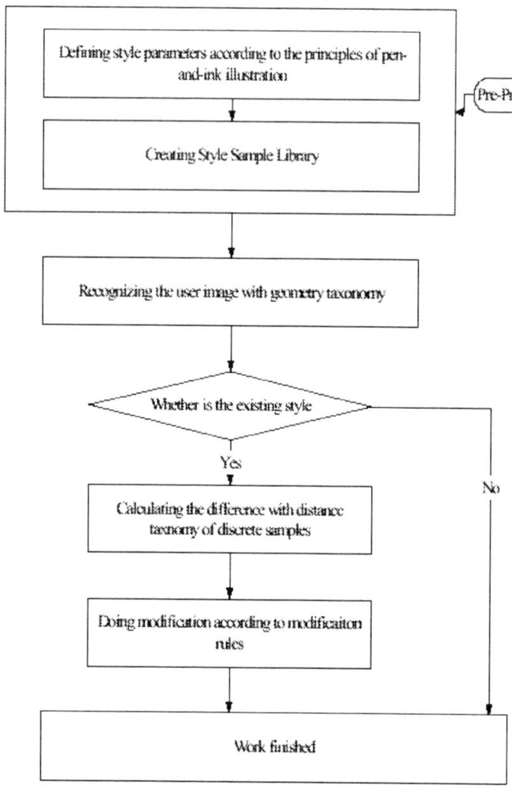

Figure 1. System implementation.

As Winkenbach and Salesin have mentioned in their paper, pen-and-ink illustration is a limited medium. The pen gives off no color or tone, so both color and shading must be suggested by combination of individual strokes. However, pen-and-ink illustration has some advantages such as each individual pen-and-ink stroke can be made expressive by employing small irregularities in its path and pressure, the small difference of stroke distribution can clearly indicate the difference between textures like smooth glass and cross-hatching.

We distill some principles from Winkenbach's paper.

The principles of stroke drawing in our implementation are summarized as below:

1. Even-weight line drawings appear lifeless; the thickness of a line should vary along its length.
2. Wavy lines are a good way to indicate that a drawing is schematic and not yet completely resolved.

The principle of stroke distribution in our implementation is summarized as below:

1. Absence of detail indicates glare.

The principles of outline in our implementation are summarized as below:

1. Crisp straight lines are good for hard objects, while a greater variety of lines quality is better for soft objects.
2. Using indication for drawing outlines is just as important as for drawing tones.

According the above theoretical rules, we extract five style parameters, L, T, W, O and D. And we define these five style parameters for each style sample.

For cross-hatching, we define that:

1. Using indication for drawing outline.
2. For each drawing, $\sum_{i=1}^{n} \frac{L_i}{n} > \alpha$ pixels.
3. Adding noise into stroke path, so W should be perturbed by a small random value.
4. The acute angle of two strokes θ must satisfy the inequality, $\sin\theta \neq 0$
5. The whole recognized area is B, detected area of no stroke is A, then $\frac{A}{B} < \beta$.

For stippling, we define that:

1. Using indication for drawing outline.
2. For each drawing, $\sum_{i=1}^{n} \frac{L_i}{n} < \gamma$ pixels.
3. The acute angle of two strokes θ must satisfy the inequality, $0 < \sin\theta < \lambda$.
4. The whole recognized area is B, detected area of no stroke is A, then $\frac{A}{B} < \beta$.

For glass, we define that:

1. Using outline detection.
2. The whole recognized area is B, detected area of no stroke is A, then $\frac{A}{B} > \beta$.

3.2. Pattern recognition

After transformation of one pattern, the pattern is mapped as a feature vector, also it is a point in feature space. In feature space, a point set belonged to pattern class ω_i can be separated from another point set belonged to another pattern class ω_j in some degree. So using a division function, especially the linear division function independent of conditional distributing density, we can save a lot of calculating time. Out of above consideration, we employ geometry taxonomy in pattern recognition. The implementation is as follows:

We have a 5-dimension feature space, with feature parameters L, T, W, O, D, so we have to define a super-surface to distinguish the patterns.

In the implementation, we change one three-class problem to two two-class problems, and the judging interface of ω_i is adjacent to the judging interface of ω_j.

The judging function is

$g_k(X)=W_k^T X \quad k=1,2,3$ where X is a 3-dimensional feature vector.

The weight vector for judging function $g_i(X)$ is $W_i=(w_{i1},w_{i2},w_{i3})^T$.

The judging surface of ω_1 is

$$g_{12}(X)=g_1(X)-g_2(X)=0$$

$$g_{13}(X)=g_1(X)-g_3(X)=0;$$

The judging surface of ω_2 is

$$g_{21}(X)=g_2(X)-g_1(X)=0$$

The judging surface of is

$$g_{23}(X)=g_2(X)-g_3(X)=0$$

Suppose the unrecognized feature vector is $X=(1,1,1,1,1)^T$. Then substituting X in $g_1(X)$, $g_2(X)$, $g_3(X)$ can get the judgment result.

3.3. Similarity measurement

When the system recognizes the user created drawing, we can judge which pattern it belongs according to the quantitative feature vector, but the vector is a fuzzy area and is difficult to quantify because of distortion and noise. Furthermore, the samples are discrete, which is dependent on the quality of sample, predisposal and feature extraction. So if the system cannot recognize the pattern of user created drawing with geometry taxonomy, the distance d can be used to describe the difference. Considering the discrete samples, the distance $d(X,\omega_i)$ between unrecognized sample and class ω_i can be denoted as the average distance of all samples, that is

$$d^2(X,\omega_i)=\frac{1}{S}\sum_t^s d^2(X,Y_t).$$

And we adopt judgment rule, that is

$d^2(X,\omega_j)<d^2(X,\omega_i)\ \forall i\neq j \Rightarrow X\in\omega_j$

The smallest value of d is the pattern class that X belongs to.

3.4. Modification

According to the most approximate pattern distinguished by similarity measurement, we do some modification on user created drawing. This modification doesn't obliterate the user individuality, but an advanced reformation based on predefined rules as well. As matter as fact, the system provides an intelligent direction to user and helps user to learn the non- photorealistic rendering techniques. Step by step, user masters the styles in Style Sample Library with the help of the experts.

If the most approximate pattern is cross-hatching, then the system will modify the user created drawing with following rules:

1. Adding indication, the indication formula is expressed in appendix A1.
2. Supposing the length of style sample as Ls, the length of user created drawing as Lu, then we adjusted the final stroke length with La, given $La=\frac{Ls+Lu}{2}$.
3. Producing 2-dimension random function and moving each pixel along X, Y axes a random value.
4. Adjusting the acute angle of strokes θ so that θ satisfies the inequality $\theta>\arcsin\lambda$.
5. Supposing the percent of area filled with cross-hatching in style sample stroke is η, β is the percent of area filled with cross-hatching in user created drawing, then the area of $\eta-\beta$ must be filled with cross-hatching.

If the most approximate pattern is stippling, then the system will modify the user created drawing with following rules:

1. Rendering the drawing with stroke length as γ pixels.
2. Adjusting the acute angle of strokes θ so that θ satisfies the inequality $0<\theta<\arcsin\lambda$.
3. Supposing the percent of area filled with stippling in style sample stroke is η, β is the percent of area filled with stippling in user created drawing, then the area of $\eta-\beta$ must be filled with stippling.

If the most approximate pattern is glass, then the system will modify the user created drawing with following rules:

1. Detecting the outline with detail. There are many algorithms published on silhouette detection and extraction.
2. Emphasizing the highlight area. If the percent of highlight area in style sample η is smaller than the percent of highlight area in user created drawing β, then erase $\beta-\eta$ with the procedure Erase_Texture (Polygon3D).

4. Results

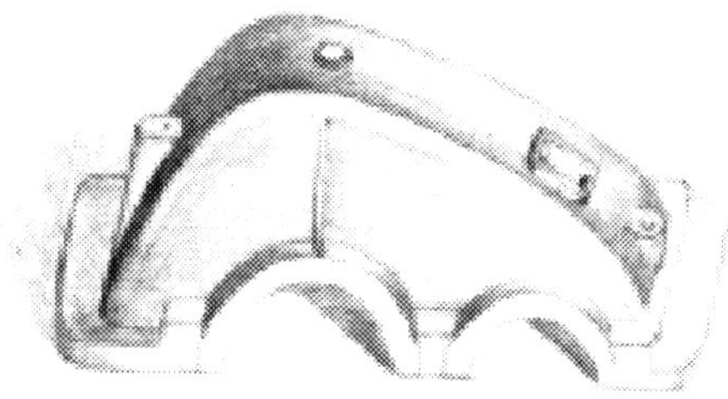

Figure 2. (a) CAD part mapping with cross-hatching texture.

Figure 2. (b) CAD part mapping with cross-hatching texture.

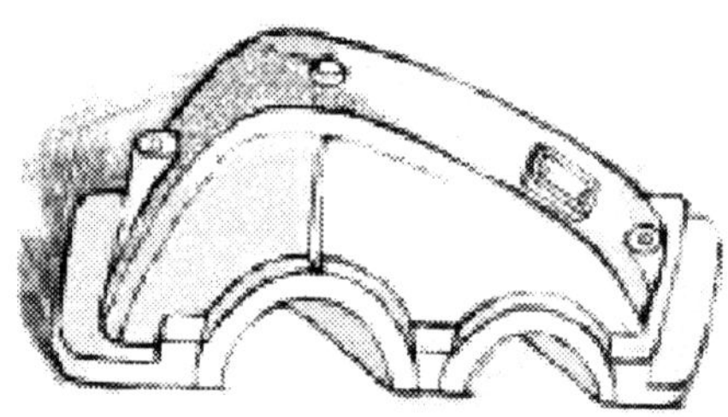

Figure 3. (a) CAD part mapping with stippling texture.

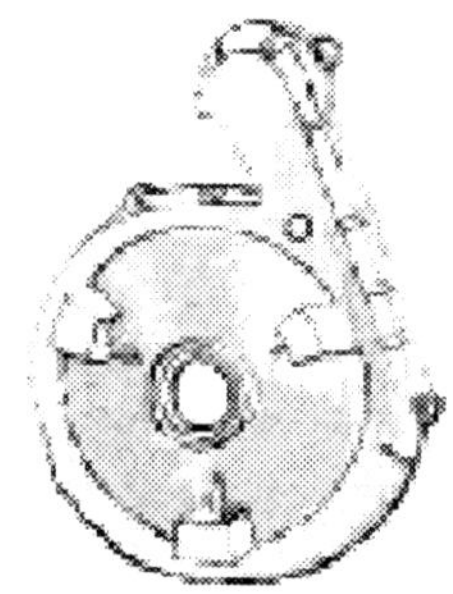

Figure 3. (b) CAD part mapping with stippling texture.

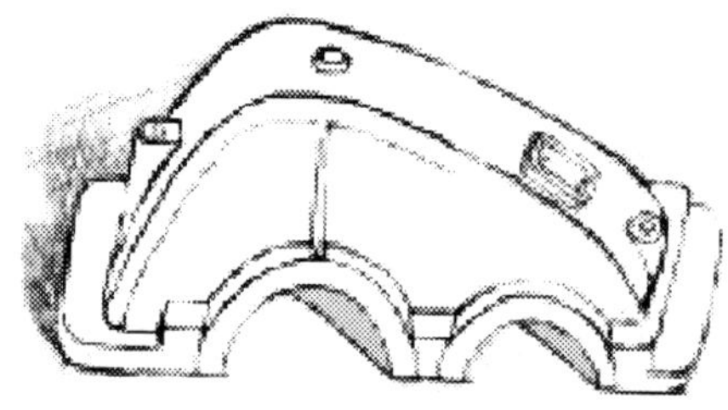

Figure 4. (a) CAD part mapping with glass texture.

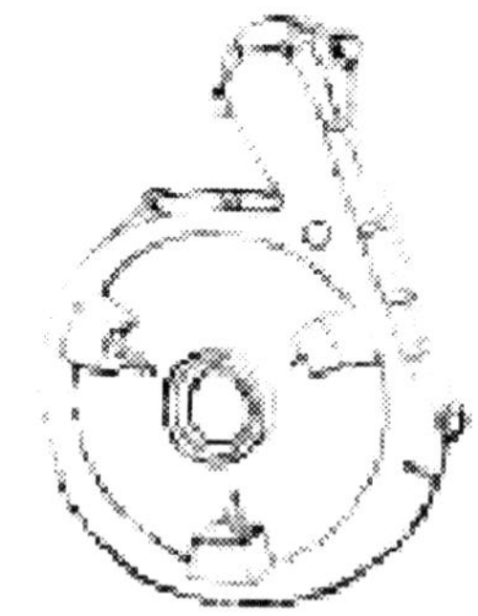

Figure 4. (b) CAD part mapping with glass texture.

5. Future work

This paper doesn't propose any improved algorithms for non-photorealistic rendering techniques except providing an integration of non-photorealistic techniques and AI techniques. The work described in this paper is just an early step in the exploration of pen-and-ink illustration system employed with AI techniques. There are many ways to extend this work, including:

1. Extracting more precise features from the non-photorealistic samples, creating more vivid drawings.
2. Recognizing the patterns with BP neural network, this will be a most challenge brainstorm.
3. Enduing the theoretical rules with self- studying, which means the rules can learn from the human-and-computer interaction dynamically.

6. Acknowledgements

We would like to thank Dai Lu for useful discussion during the early process of this paper and invaluable advice and support.

7. References

[1] Oliver Deussen, Thomas Strothotte, Computer- Generated Pen-and-Ink Illustration of Trees", *SIGGRAPH 2000 Proceedings*.

[2] Amy Gooch, Bruce Gooch, Peter Shirley, and Elaine Cohen,

"A non-photorealistic lighting model for automatic technical illustration", *SIGGRAPH '98 Proceedings.*

[3] Cassidy J. Curtis, Sean E. Anderson, Kurt W. Fleischer, and David H. Salesin, "Computer-generated watercolor", *SIGGRAPH '97 Proceedings.*

[4] M. Salisbury, M. Wong, J. F. Hughes, and D. Salesin, "Orientable textures for image-based pen-and-ink illustration", *SIGGRAPH '97 Proceedings.*

[5] J. Schumann, T. Strothotte, A. Raab, and S. Laser, "Assessing the effect of non-photorealistic images in computer aided design", *ACM Human Factors in Computing Systems, SIGCHI 96: 35–41*, April 1996.

[6] G. Winkenbach and D. Salesin, "Rendering parametric surfaces in pen and ink", *SIGGRAPH '96 Proceedings.*

[7] Adam Finkelstein and David H. Salesin, "Multiresolution curves", *SIGGRAPH '94 Proceedings.*

[8] Michael P. Salisbury, Sean E. Anderson, Ronen Barzel, and David H. Salesin, "Interactive pen-and- ink illustration", *SIGGRAPH '94 Proceedings.*

[9] G. Winkenbach and D. Salesin, "Computer- generated pen-and-ink illustration", *SIGGRAPH '94 Proceedings.*

[10] The Premisys Corporation, Chicago. Squiggle, 1993.

[11] Debra Dooley and Michael F. Cohen, "Automatic illustration of 3D geometric models: Lines", *Computer Graphics, 24(2): 77–82*, March 1990.

[12] Debra Dooley and Michael F. Cohen, "Automatic illustration of 3D geometric models: Surfaces", *proceedings of Visualization'90:307–314*, October 1990.

Appendix

Implementation Detail: Indication

Indication is one of the most notoriously difficult techniques for us to master, but it is also a most important techniques to indicate enough details. It requires putting just enough detail in just the right places, and also fading the detail out into the unornamented parts of the surface in a subtle and unobtrusive way. We referenced the formula of [3], and made some extend of it. A field w (x, y) is generated by the detail segment s at a point (x, y) in texture space according to

$$W(x, y) = (a + b \times dis\tan ce((x, y), s))^{-c}$$

where a , b , and c are non-negative constants that can be used to change the effect of the field; can be educed from the above similarity measurement function, s is the detected detail segment of outline. When several detail segments are present, we define the field at a point (x, y) to be that of the closest segment. So as not to create patterns that are too regular, the field w (x, y) is perturbed by a small random value. We created random value with Gauss Distribution. Textures such as "cross-hatching" and "stippling" evaluate the strength of the field of indication at the outline. The set of strokes for that element is generated only if the indication field is above some preset threshold.

Research on Network Based Conceptual Design

Cuixia Ma, Hongan Wang, Guozhong Dai
Intelligent Engineering Laboratory, Institute of Software, Chinese Academy of Sciences
P.O.Box 8718, Beijing 100080, P.R. China
mcx@imd.cims.edu.cn

Abstract

Conceptual design has played very important role in the design process. Knowledge of all the design requirements and constraints during this early phase of a product's life cycle is usually approximate, imprecise, and unknown. Conceptual Design is now recognized as a highly important process when considering both the concurrent engineering aspects and the commitment of resources necessary to design and develop a product. With the development of the Internet, collaborative design is imperative and effective. The technology of CSCW is adopted widely and necessarily in the domain of the conceptual design. It can fit the nature of the design. It shortens the distance of designers in different places and collects and shares all the information in time. In the conceptual design, collaborative design mainly includes function, principle, layout and initial structure and so on. We developed a system of conceptual design based on the current CAD system, which could realize the seamless link with the later detail design. For the individual design, multi-modal technology, which integrates many interactions such as gesture and speech, supports an advanced interface to assist humans to develop the applications.

1. Introduction

Conceptual design is the initial design phase of the whole design cycle. Decisions made at this stage have significant influence on many aspects such as costs, performance, reliability, and so on[1]. Conceptual design is the kind of creative work which contains designer's inspiration. It is important for the system to capture and record the latest idea during the course of design. It is recognized that computers support tools for designing knowledge representation, database techniques and design layout, which provide an invaluable contribution to the modern needs of conceptual design.

At the same time, the designers' idea is implicit and maybe disappears quickly, so it cannot afford the explicit dimensions and geometry information. However, most current CAD systems require users to give the explicit information, as hinders the natural expression of designers.

It is well known that most of the resources necessary to design and produce a product are committed by decisions taken in the first 10% of the design activity. The performance made at the conceptual design stage will affect the rest process of the product. These demands to reduce the time-to-market of high-quality and innovative products have increased the need for computer support at the conceptual design stage of the engineering design process life cycle.

At a time when there is an increasing concern for improved support to manufacturing industry it is imperative to firmly focus on the continuing research into new applicable design tools for conceptual design.

A team of members of design professionals is usually involved in the development of a product concept. It is usually expressed as a sketch or as a rough three-dimensional model. It is the common way for designers to engage in the process included brainstorming sessions, mind mapping, group discussions, etc. For these activities, the designers usually come together and present their individual designs on a blackboard. Such is the collaborative work. Following the progress and popularization of the Internet, there are exiting new ways for users to make something easy.

The tool provided to designers that is necessary and important should have natural and effective interface. It can support the interaction between human and computer and even between human and human. With the development of HCI technology, the multi-modal technology to a large extent can meet the needs of users.

2. Collaborative Conceptual Design

With the development of collaborative design and CSCW technology, collaborative conceptual design[6] is more important than ever and becomes the trend of conceptual design. Generation of concept ideas and selection of the best concept are two activities involved in any product conceptualization stage of the design process. The requirement of the production is first attainted by the need of the reality. The conceptual design phase begins with the specification of goals of the product. Then all possible solutions that could achieve these goals are suggested, which need all designers to form one team.

Then comments and suggestions are made on the designer's ideas by the team. The goal of the incorporation by the designer is to develop a refined design concept. The procedure repeats for all the members of the design team. Finally, the best concept is selected for utilization in the product modeling stage.

As the traditional and convenient way for designing, it is effective to use with pen and paper. When designers sketch to explore thoughts and find out the solution, freehand drawing has a important role at the early design.

Now that conceptual design is accomplished by one team, the members in it include not only designers but also engineers, managers, users and so on. The work of the team can be divided into two classes, one is the individual work, and the other is collaborative work. The former refers to that each member of the team should accomplish his own idea about the design task. In some separate design periods, the concepts represented as sketches by the designers are not accessible to other members of the design team. Each one of the design team members would come up with their own concepts for the product, which they would represent on the sketch environment provided by this system.

To develop the whole design, the system should support simple and efficient design conditions. For example, it aims to address the development of interfaces. Current existing interactive mode lacks the ability to manipulate objects in three-dimensional space in a direct, inaccurate and intuitive way while it is good enough to explicit two-dimensional operation. The communication bandwidth from computer to human is much more large than that from human to computer. The asymmetry comes from hardware constraints and the mismatch between the manipulation of objects and interactive modality of human. Multi-modal interface is a promising technology to solve this problem. It has been the research emphasis of user interface in the recent years.

The later aspect refers to dealing with the contact one another. Most traditional CAD systems only support the individual job, and the communication among them is carried out by the fact-to-fact talking or correspondence. For that, it is difficult for the team to communicate and discuss frequently and in time. Due to treating the whole design with the separate operation provided inefficient communion, the whole design process lacks the integrated view about all the idea. And owning to the way, the time and cost is so limited that modification cannot get the approving result at most status.

With the network technology fully prepared, multimedia technology has made a great progress. Now it can provide more powerful and convenient human-human communications as well as human-computer ones. To some extent the project has developed an open, network-oriented environment for concurrent engineering, which allows engineers to participate on a distributed team using their own tools and databases. And it also can offer protocols and architectures for enabling cooperative product development and facilitation of knowledge transfer between systems which have their own specialized knowledge bases and reasoning mechanisms[5].

Such comes the collaborative design based on computer technology. Working in this way will overcome some flaw described before and improve the design methods. In the collaborative design process, there are some steps to accomplish it. The very beginning is to disseminate the product requirements to all the team members involved in the design process. They can access existing product concepts and the information gathered could be utilized to generate ideas for a similar mechanism in a product. When finished their own task, the team members start to communicate with each other for a whole design.

Besides, collaborative problems can be solved automatically or partly through the AI technology. So, better design can be expected, repeated design avoided or reduced, design cycle shortened, then better quality of product and lower developing cost attained.

An approach is discussed to collaborative concept generation through negotiations and the best concept selection, with enabling tools and techniques available on and for the Internet. The data organization would reduce the search space involved while the designers are exploring similar designs and concepts documented on the Internet. The webs can also be used to store the common information generated during various design activities of a design team. Existing concepts are combined or modified to improve the overall concept. The collaborative work (Figure1) involves close interaction between the participating design members.

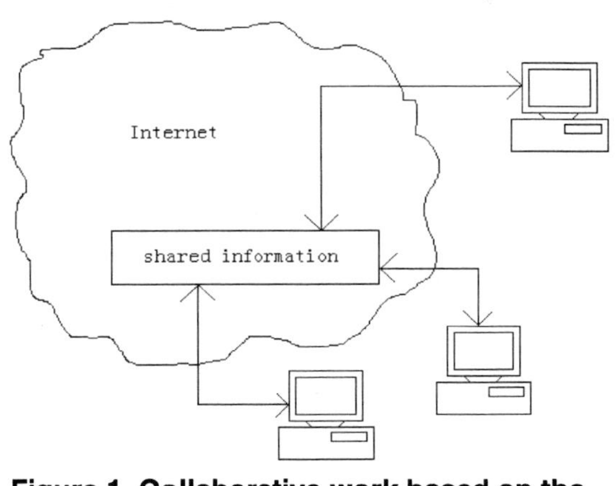

Figure 1. Collaborative work based on the Internet.

The designers distributed in the net can access the drawings through the speech, pen, keyboard and mouse. Each designer should operate the objects of the drawing and notice the behaviors of other designers. The system records the different operations and handles the collision among designers. There are simple collaborative operations in some CAD system. Work related to collaborative concept design in the mechanical design process is sparse.

Much technology and service are offered to support the whole collaborative design based on the Internet. Most current CAD systems cannot support advanced collaborative technology. There is also some simple Internet technology provided by the existing CAD systems. For example, Autocad can support Internet link for its DWF format files. DWF can add the drawings into the Internet with the vector format.

3. Sketch design

In order to pay much attention to design conceptual model, it is imperative to introduce interface with human nature and simpleness, one example of which is pen-based input. Succeeding to traditional design method of pen and paper, this method can get the drawing following designer's intention without defining some explicit dimensions that maybe delay the design thoughts. Pen input technology with tactile feedback has been so advanced to make the design humanized. This way provides an intuitional and swift interactive mode, which focuses on the expression of topologic relation not some material geometry information to sketch 3D view.

Conceptual model is completed through natural sketch, which demands little detail. It tries to design one concept according to designers' thought and express the designers' intention fluently. Freehand sketch is a potent tackle for inspiring design inspiration.

Solid model of traditional CAD systems is set up on geometric representation and corresponding operation, as is one kind of geometric operation in low level without engineer signification. It cannot meet the demands of designers in the high-level applications. Feature-based design supports a flexible interactive mode according to the design thought of high level. Feature includes design intention, which is significant for design maintenance and successive analysis and synthesis.

Gesture is one effective way for designing, which is simple and natural to designers. Gesture [2-3] can be seen as a set of relative strokes, some of which have a stroke and the other of which may have some strokes. To combine feature with gesture gets better design method.

Definition 3.1. feature-gesture operation, the natural, simple and convenient method is that gesture made of strokes can be combined in the process of feature creation and modification.

It supports one fast modeling way of structuring simple geometric body (Figure 2).

The information we get from the conceptual design process based on the pen is imprecise. The strokes reflect the datum with implicit property. So, the internal representation is somewhat fuzzy.

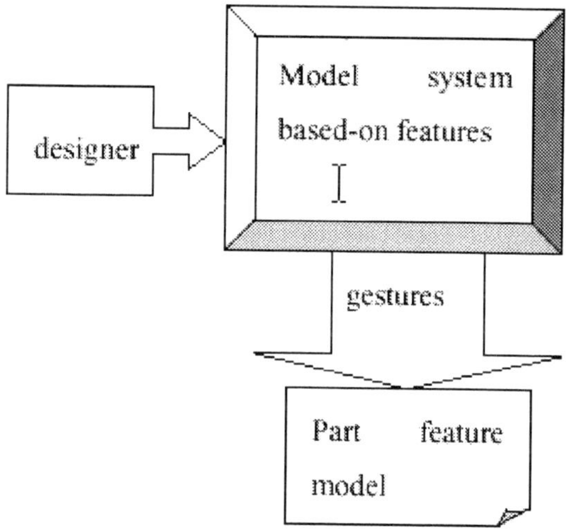

Figure. 2 Architecture of design.

Definition 3.2. Fuzzy model[4] ,represents the fuzzy logical relation in the design process by enduing with the threshold, which is based on the possibility distributing and its relation with fuzzy set.

We define a set of gesture primitives. The set can be extended later following the special application. The prototype system is designed for mechanical engineering design. We just collect some gesture primitives of features in common use. It is convenient to extend the primitive set for some needs.

The features, which can be recognized in 3D input, are:

1. block feature, made up of a parallelogram and an adjoining edge (Figure 3(a))
2. cylinder feature, made up of an ellipse and an adjoining edge (Figure 3(b))
3. sphere feature, made up of a circle and an inner ellipse (Figure 3(c))
4. cone feature, made up of a circle and two intersected lines (Figure 3(d))

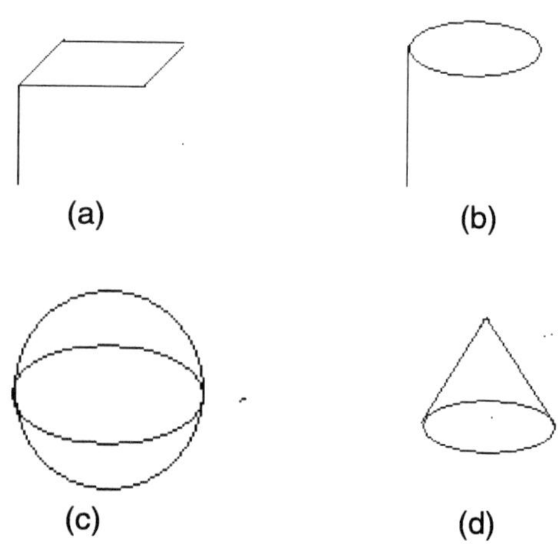

Figure 3. 3D Features.

After the creation of the 3D solid, designers can add new features to them to get new solid features, such as: chamfer, slot or hole and so on.

4. Multi-modal Interface

It is necessary to develop the fast and effective multi-modal interface (MMI) between user and computers and the method of making full of the Internet for the goal of changing the current state of human-compute interaction at all. With new interactive technologies such as voice recognition, gesture et al., MMI allows user to use multiple interactive modalities to interact with the computer system in a parallel and inaccurate way. The goal of MMI is to improve the efficiency and intuition of human-computer interaction. Following the command line interface (CMI) and WIMP (window, icon, menu, pointing) interface, MMI is a new generation user interface.

In our prototype system, we can develop the design efficiency by integrating the pen input and the speech input. We designed the prototype through the secondary development on the base of current CAD system, such as AutoCAD. The part of handling the pen input and the speed input is provided with better encapsulation and integration. So, we can transplant it to other CAD system to boost up its function.

We introduce two interactive agents. One is for pen input; the other is for speech input. The pen agent tracks the input of the pen and recognizes it into graphic element. The speech agent records the user's voce input and translates it into phrase. The modality-synchronizing agent judges the temporal relativity of the pen operation and the speech operation and synchronizes them. The system receives the input of the two interactive agents and integrates them into the drawing tasks.

Speech primitives as well as gesture primitives can also be extended for different application. We just create some base spoken commands for editing the existing features.

We can accomplish a task with enough parameters provided by the pen and the speech. For example, if we want to move an existing solid, we can select it by the pen and speak "move from here to there" as we point the two places by the pen. The three parameters needed for this operation can get successively finish the operation. The rudimental operation can be seen in the Table 1. And can be extended later.

5. Conclusion

This paper discusses an approach to develop and implement an architecture to support geographically dispersed designers, effectively develop and select the product concept, through a collaborative effort, which is mainly used in mechanical design provided the integration of pen and speech based on the current CAD system.

Formal studies on individual and group problem solving have shown that a set of people working alone for a period of time will generate better concepts than the same people working together for the same time period.[5] Collaborative design based on the network platform can increase the efficiency of the whole design. The web is a powerful medium for supporting design teams with more effective, unobtrusive support for storing and sharing information.

Multi-modal technology has developed the interactive mode, which provides more natural and convenient interface. It can express the intention of designers fluently.

Table 1.

Say	Sketch	Result
Select this	Drawing a scope around the solid which you want	You select that solid
Delete this	Drawing a scope around the solid which you want	You delete that solid
Rotate ? angle	Drawing a scope around the solid which you want ,some direction	You rotate that solid with some angle and direction

6. Acknowledgement

This paper is supported by the Chinese National Foundation of Science No.60033020, Chinese High-Tech Plan Project No.863-511-942-012.

7. References

[1] Wynne Hsu,Lrene M. Y. woon, Current research in the conceptual design of mechanical products, ELSEVIER, pp: 377-389

[2] Casper G. C van Dijk, Amnon A.C. Mayer, Sketch input for conceptual surface design, Computers in Industry 34, 1997, pp: 125-137

[3] Hwang,T. And Ullman ,D.,Recognizing features from freehand sketches. Computers in Engineering, ASME, 1994, 1, pp: 67-78

[4] C.L.Philip Chen, Sen Xie , Freehand drawing system using a fuzzy logic concept, Computer-Aided Design, vol.28,No.2, 1996, pp: 77-89

[5] U. Roy, S.S. Kodkani, Collaborative product conceptualization tool using web technology, ELSEVIER, pp: 195-209

[6] Zongkai Lin, Chouqian Sun, Ming tang, Collaborative conceptual design supported by computer, Computer World, 1999, 16(8), pp: 7-9.

Cooperative Product Design Process Modeling

Gang Zhao, Jiati Deng
Institute of Manufacturing Systems (720),
Beijing University of Aeronautics and Astronautics
Beijing, 100083, P.R. China
[zg, jtdeng]@buaa.cims.edu.cn

Abstract

Modeling cooperative product design process is one of key challenges to build up cooperative and intelligent product design information system and environment. Since cooperative design process is characterized with distributed workflow and complicated interaction, and it is constraint oriented involving the recognition, formulation, and satisfaction of constraints. In this paper, the authors first introduce constraint network to model complex relationships in cooperative design process that are both the causes and results of interaction. Based on constraint network, an event-driven constraint-triggered decision-pushed automated design process model is built. Then the authors use Petri net to model interaction behaviors including communication, negotiation, coordination and cooperation. Finally implementation of MAS prototype based on cooperative design process model is described and an application example is provided in brief.

1. Introduction

Product design is an evolving process of problem solving that consists of acquiring requirements, defining the overall goal and task, decomposing the overall task into hierarchical sub-tasks, distributing sub-tasks to designers, solving the sub-tasks, synthesizing the sub-solutions, and finally providing the overall artifact solution, which characterizes with creativity, uncertainty [1]. On the other hand, product design is involved in complicated interaction among multidisciplinary design teams in a distributed, heterogenous and dynamic environment, including communication, cooperation, coordination and negotiation [2]. The goal of this article is to explore these two aspects by cooperative design process modeling.

In order to achieve cooperative design goals, all design units first need to maintain consistent and well-ordered relationships between them. Research on constraint network in AI community has also proposed various theories and algorithms to represent and solve relationships between variables [3]. It is also widely accepted that design is constraint oriented involving the recognition, formulation, and satisfaction of constraints [4]. In this article, the authors thus use constraint network to represent varieties of relationships in cooperative design process. On the other hand, each design unit needs to interact with other design units and adapt its behavior according to dynamic environment. Research on intelligent agent and multi-agent system present promising solution. Researchers have built some models of interaction in multi-agent system mainly classified into two aspects:

- towards implementation: such as GPGP [5];
- towards mental states, including Joint Intention [6], [7] , Shared Planning [8], and STEAM based on Joint Intention and Shared Planning [9].

Some common ideas in these models are: (1) Agent needs to have both a joint intention and a local intention in order to achieve a goal. (2) Agent may share information of other agents by communicating. (3) Building a unified framework to model cooperative process is common attempts for interoperability and reusability. In this article, the authors will use Petri net to model interaction in agent-based cooperative design process that is entwined with a dynamic constraint network.

An established PPOR enterprise model is first reviewed in Section 2, which is the ontological basis of modeling relationships in cooperative design process with constraint network. Section 3 analyzes and builds constraint network in cooperative design process. From a view of workflow, Section 4 attempts to model cooperative design as an event-driven, constraint-triggered and decision-pushed automated process. Section 5 then uses Petri net to model communication, negotiation, coordination and cooperation among agents in a MAS-based cooperative design environment. Section 6 describes the system being implemented and provides an application example to show the effectiveness of the model.

2. An Integrated Multi-View Enterprise Model: PPOR

The Product/ Process/ Organization/ Resource (PPOR) multi_view enterprise model had been proposed [10] to support concurrent design and its management, which is based on object oriented and STEP methodology [11]. Process entities are the core of PPOR model, since they are resigned to responsible organization entities, use resource entities, and operate on product entities. Figure 1 present a simplified and integrated PPOR model with Express_G language that only shows the relationships between different entity models and omits their respective attributes and methods. Although PPOR model has been our basis of building information system supporting design process, it lacks a unified way to sufficiently represents complex relationships inside and between entity models and it doesn't support to model and analyze dynamic interaction and evolving process from requirement to detailed product description.

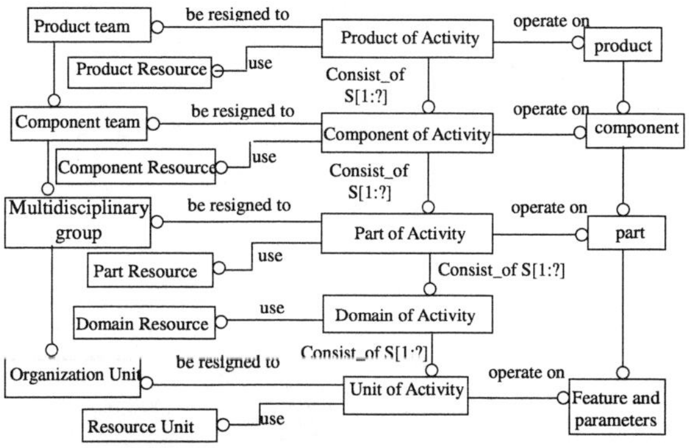

Figure 1. PPOR Enterprise Model.

3. A Unified Relationship Modeling Based on Constraint Network in Cooperative Design Process

3.1 Constraint Network

A constraint network (CN) formally consists of (Dechter, 1992):

(i) $X = \{ X_1 ,..., X_n \}$ is a finite set of variables. According to assigned values, types of these variables can be quantitative such as numerical, symbolic, logic, discrete and vector, or qualitative such as fuzzy interval.

(ii) $D = \{ D_1 ,..., D_n \}$ is an associated set of domains. Each domain is a finite set containing the possible values for the corresponding variable.

(iii) $C = \{C_1 ,..., C_t \}$ is the set of constraints. A constraint C is a relation defined on a subset $\{X_{i_1},..., X_{i_{j(i)}}\}$ of all the variables, that is $D_{i_1} \times \cdots \times D_{i_{j(i)}} \subseteq C$. These constraints can be formulated as equation and inequation (e.g. Cost = (M+P)/W, l/d<100), interval (e.g. x•[15, 50] or d •{20,30,40}), rules (e.g. If(M>10), then C<100), and semantic predicate (e.g. USE(A,B)) etc.

3.2 Constraints in a Cooperative Design Process

Constraints in a cooperative design process can be simply classified into constraints inside each kind of entity model and constraints between different kinds of entity models according to PPOR model.

First, constraints in each kind of entity model include:

- constraints about design product that represent relationships between (sub-)entities defined by product model including design parameters, features, parts, components and product. For example, constraints about dimension, geometry, tolerance, thermal attributes, stress, material of product etc.
- constraints about design process that represent relationships between (sub-)entities defined by process model including unit, domain, part, component, part of activity. For example, time constraints of activity including beginning and ending time, duration etc. Constrains that represent sequence and hierarchical structure of different activities which can be formulated as semantic constraint, such as :
  - Input/Output constraints: activity A's output is one of activity B's inputs, formulated as INPUT(B, A);
  - Parent/Child constraints: activity B is activity A's sub-activity, formulated as PARENT(B, A);
  - Brother constraints: Both activity A and activity B are sub-activities of the same parent activity, formulated as BROTHER(B,C);
- constraints about supporting resources that represent relationships between (sub-)entities defined by resource model. For example, constraints about number of PC, performance of CAD tools etc.
- constraints about organization that represent relationships between (sub-)entities defined by organization model. For example, constraints that organization unit A controls B.

Constraints between kinds of entity models in process oriented design environment may include

- Constraints between design process and product. For example, constraint about which parts are operated by Activity A, or which activities operate on Part A, etc. It can be formulated as OPERATE(Activity, Product).
- Constraints between process and resources. For example, constraint about which resources are

used by an activity, which activities use a resource. It can be formulated as USE(Activity, Resource).

- Constraints between design process and organization, for example, constraint about what is the responsible organization of an activity (or task).
- It can be formulated as RESPONSIBLE (Organization, Activity).

These constraints can together represent complex relationships in a cooperative design process. Meanwhile, most constraints change with the evolution of design process. Therefore constraints in cooperative design process can be formulated as a complicated, hierarchical and dynamic constraint network.

Based on object oriented (OO) and STEP methodology, constraint model with Express_G had be described [12] that is the basis of building constraint satisfaction and management modules.

3.3 Constraint Network in Agent-based Cooperative Design Process

In the context of agent-based cooperative design process, unit of activity can be implemented by software agents [13]. Agents in a design process may interact with each other for many reasons, mainly including:

- An agent is directly or indirectly associated with other agents by constraints about design product;
- An agent is directly or indirectly associated with other agents by some constraints about resources.
- An agent is directly or indirectly associated with other agents by some constraints about organization.
- Mainly due to above constraints with product, resources and organization, an agent may form sequential or structural constraints with other agents, for example Input/Output, Parent/Child or Brother constraints, or more complex semantic constraints, such as "enable", "facilitate", "disable", "hinder" [5].
- An agent can't achieve its goal in isolation or it believes cooperation is more preferred than isolation when evaluating constraints about own capabilities, permitted time, consumed cost etc;

Therefore constraints can be regarded as the main causes of interaction among agents in cooperative design process. Complicated interaction simultaneously results in more complex constraint network. On the other hand, constraint satisfaction problem (CSP) algorithms also can be the basis of agent's internal abilities of problem solving. Constraint network thus can help to build a unified model to represent the relationships among cooperative agents and inside agent.

Figure 2 shows a constraint network in a simple two-agents-based cooperative design process. Some new definitions will be introduced first. The constraint network contained in the range of agent's internal abilities of problem solving is defined as local CN of agent, denoted by CNA. The constraints between CNAs are called cooperative constraints (CC), viz. constraints in shaded region between CNA1 and CNA2 in Figure 2. The variables in CC are called cooperative variables (CV), for example x_{12} and x_{22}; x_{1r1} and x_{2r1} in Figure 2.

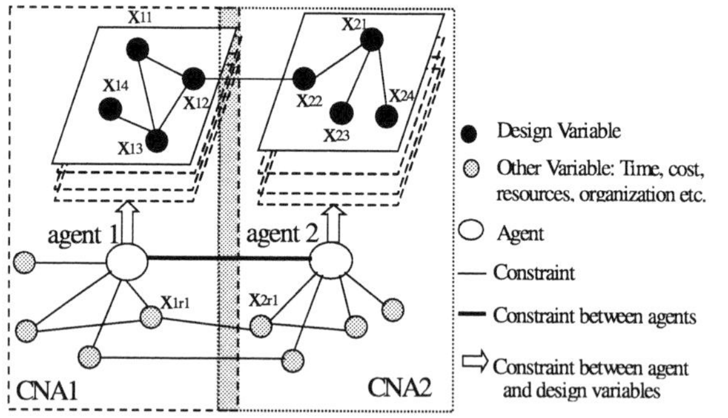

Figure 2. A Constraint Network for a Cooperative Design Process with Two Agents.

Such a model based on constraint network has some explicit characteristics: First, it fit to represent the exact and quantitative relationships in a design process. Semantic constraints between agents also can be quantified in virtue of constraints about design variables or other variables. Secondly, CN can be regarded as an agent's view on changing environments, since it is a dynamic network in which new constraints can be added and old ones are deserted. Thirdly, CN also provides the ability to represent uncertain information and reason about them so as to deal with inherent uncertainty in a design process. Finally, CSP and DCSP have provided a set of algorithm such as backtracking and consistency-enforcing algorithms [3] to propagate the constraints and resolve the conflicts. But it is still necessary to improve semantic supports for relationships among agents in a constraint network. These advantages are found in an application example in Section 6.

4. Event-driven, Constraint-triggered and Decision-pushed Automated Design Process Model

From a perspective of evolution, cooperative design process can be defined as a workflow composed by a set of activities or tasks that are executed in some order. According to WfMC's workflow reference model [14], executing a workflow is generally supported by workflow engine scheduling tasks when they are ready for execution and by assigning them to the proper executing agents, which is based on event-condition-action rules. Although these rules can be fully exploited when predictable and repetitive design processes are executed. Unfortunately, design process with creativity and uncertainty is faced with the need of managing unpredicted and exceptional situations that may arise during their execution [15]. Therefore a more flexible and dynamic automated cooperative process model is needed. Here, based on constraint network, an event-driven constraint-triggered and decision-pushed automated design process model is proposed to meet this need.

Figure 3 give a simplified description of this model. When a state S_{i-1} about object 1 is changed to S_i, an event E_{i-1} about object occurs. Since a constraint exists between state S_{i-1} and state $S_{k,i-1}$ of agent k, in order to continue to meet constraints between state S_i and agent k's state in some matters, agent's decision is to be triggered. This kind of motivation is captured and reasoned by agent k, then a decision is made to generate a new state $S_{k,i}$ to form a new constraint between S_i and $S_{k,i}$, and a new event about agent k also occurs. Such an event about agent k also may continue to cause new event about agent l due to the constraints between them. Since these event-driven processes are involved in many objects and agents in constraint network, an automated design process thus may be formed. However, the process may be chaotic due to complex interaction among large numbers of activities. Reasonable distribution and effective coordination can ensure that design process autonomously converges to its goal.

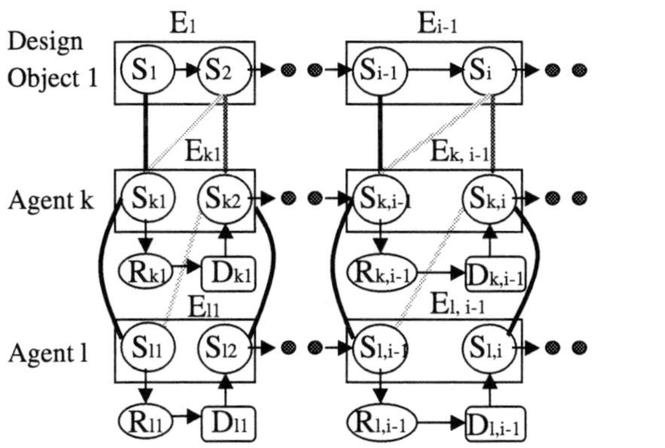

Figure 3. A Simplified Automated Design Process Model.

5. Modeling Interaction in Agent-Based Cooperative Product Design Process

Cooperative design process based on multi-agent system (MAS) is appropriate only when agents integrated in the system are capable of communicating, cooperating, coordinating and negotiating with each other effectively.

5.1 Communication Protocol

Communication is the process of transferring information from one agent to another by means of various media. It forms the basis of the negotiation, coordination and cooperation. In a message-based communication, a communication protocol must be defined that specifies the exact communication process, the format of the messages, communication language and its semantics. KQML is both a message format and a message-handling protocol that provides a general frame for the communication in multi-agent systems [16]. Based on speech-act theory, KQML provides an extensible set of performatives(speech act types), such as *ask-about, tell, reply*, which define the permissible operations that agents may attempt on each other's knowledge, such as belief revision, querying, knowledge base maintenance, actions and services. Every KQML message has the following basic structure:

```
(<Performative>
:content<statement/speechact>
:sender<name>
:receiver<name>
:language<text>
:ontology<text>)
```

5.2 Negotiation Model

Negotiation is a process by which an agent communicates with another agent or a group of agents in order to try and come to an agreement, which underpins attempts to cooperate and coordinate. Research on negotiation can be considered to deal with three broad topics [17]: negotiation protocols, negotiation objects, agent's decision making models.

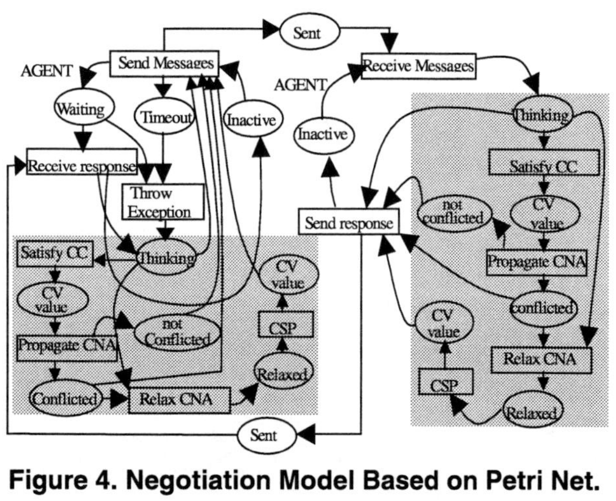

Figure 4. Negotiation Model Based on Petri Net.

Based on constrain network, negotiation among agents is caused by cooperative constraints. Negotiation protocols define exact negotiating process including types of participants (e.g. two or more agents linked by CCs), the negotiation states (e.g. CC satisfied, value of CV sent), the events that cause negotiation states to change(e.g. relaxing constraint). Negotiation objects are cooperative variables between CNAs over which cooperative constraints must be met. Agent's decision-making models describe how the agents deal with local CNA and CC in line with the negotiation protocol in order to achieve their negotiation objectives. Therefore negotiation can be regarded as an iterative process of receiving the information about CVs of the other side, satisfying CCs to get value of own CVs, propagating own CNA, identifying the conflict, relaxing constraints in some matters, solving own CNA, sending the accepted results of own CVs to the other negotiating side. For example, agent 1 is negotiating with agent 2 about respective cooperative variable (Radius r_1 and r_2) because of a cooperative constraint: $r_1^2 = r_2^2 + C^2$ between them. Agent1 has a value to the CV r1 and notify agent2, agent2 receives the result and gets a value of r2 to satisfy CC, and propagates CNA2. If agent2 finds a conflict, then it may relax some constraints in CNA2 to solve the conflict and propose an accepted value of CV r2, and notify agent1. Petri Net (PN) is an appropriate tool to model negotiation behavior. Referred to Cost's model [18], a Petri net model for pair-wise negotiation model is given in Figure 4. The shaded region shows the agent's decision-making models.

5.3 Coordination Model

Coordination is defined as a process by which agents can work together to improve effectiveness. Coordination is based on communication and negotiation, and is the basis of cooperation. Varieties of mechanism have been devised to tackle the coordination problem, and can be divided into three categories [19]:

- Organizational structuring. It can be viewed as a pattern of information and control relationships between individuals.
- Meta-level information exchange. It involves agents sending each other control level information about their current priorities and focus, such as. PGP [20].
- Multi-agent planning. Agents usually form a plan that specifies all their future actions and interactions with respect to achieving a particular objective.

In the context of agent-based cooperative design process, the hierarchical constraints among design activities prescribe constraints among design agents such as Parent/Child, Input/Output, Brothers, which forms the basis of coordination. Meanwhile agents can exchange the information on local CNA by communicating, for example about priorities of constraints, solution of local CNA etc. Figure 5 illustrates a model of coordination using Petri net.

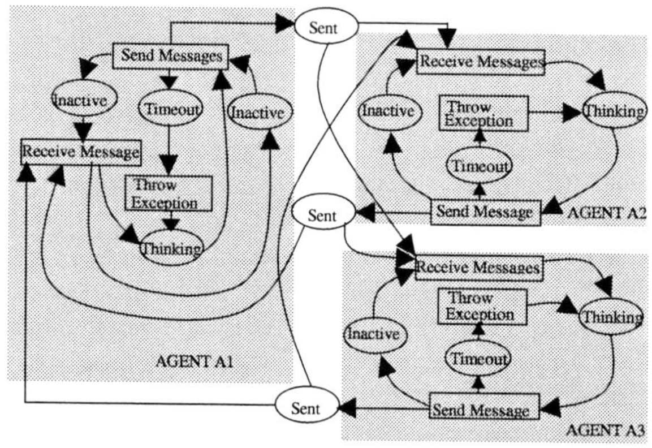

Figure 5. Coordination Model Based on Petri Net.

5.4 Cooperation Model

Cooperation is a process by which agents work together for a common goal. Since cooperation is characterized with agents' autonomous behaviors, agents' mental states such as commitments, intentions are important to build a cooperation model. Wooldridge's cooperative problem solving (CPS) model formalized four stages of CPS based on multi-modal logic, including recognition, team formation, plan formation, and execution [7]. This model is appropriate to be as a specification for future cooperative systems, which attempt to prescribe the way in which a rational, intelligent system should behave. But it is difficult to implement logic-based model. In this paper, a cooperation model is also built based on Petri Net.

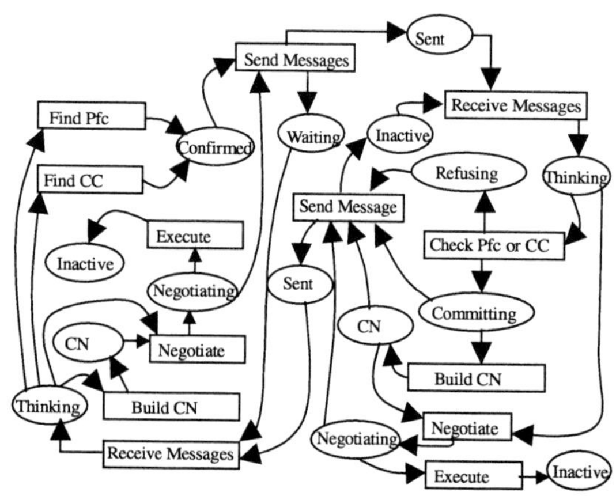

Figure 6. Cooperation Model Based on Petri Net.

In the process of recognition, conditions that characterize the potential for cooperation are found. The potential for cooperation (Pfc *i•*) defined in CPS model makes up of the first cause for cooperation. (Pfc *i•*) thus can be regarded as a potential CC between agent i and another agent or a group of agents. Of courses, the existing CCs among CNAs is the second cause of cooperation. Therefore, identifying potential or existing CCs is task of recognition. Having identified CCs, agent will attempt to solicit assistance from a group of agents that is related with CCs. If such an attempt is successful, an initial constraint network will be built since there will be a group of agents with a nominal commitment to joint action. After such a constraint network is built, a negotiation among cooperative agents is initialed till the agreement is reached. In process of the execution, agreement will be executed by both sides. In this stage, coordination mechanism can help to improve the effectiveness of cooperation.

6. Implementation of Cooperative Design Process and Application Example

A multi-agent system (MAS) is being developed for a cooperative and intelligent design process plan, management and execution: called CLOVER. It is composed by many individual agents (including Design Task Agent (DTA), Process Management Agent (PMA), Product Data Management Agent (PDA), Resource and Tool Management Agent(TA) etc.) who can cooperate to implement design project based on our cooperative design process model, showed in Figure 7. A constraint management system (CMS) is built to manage constraint network among agents in cooperative design process. Various constraint satisfaction modules can be inserted into the module of agents and CMS. These may range from simple constraint monitors to arc consistency to K-consistency and may cover the domains ranging from finite discrete to continuous and both logical and algebraic constraints. Agents and CMS should be able to handle several such modules at the same time and be able to route constraint satisfaction and management requests as appropriate. A Petri net modeling tool is being implemented to aid to model dynamic interaction behavior among agents. How to convert Petri net model into executable program is needed to investigate in the future. In CLOVER environment, authors are attempting to entrust agents with all activities, hence cooperative design process modeling promises to implement a real design automation system. More details can be referred to the authors' another paper [13].

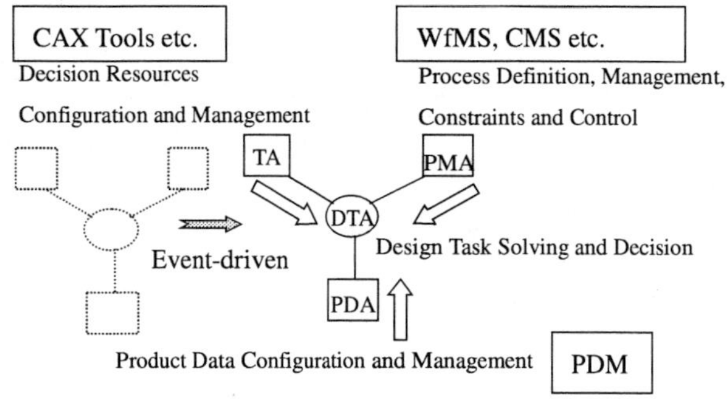

Figure 7. CLOVER Architecture.

Now we will give an application example to show the effectiveness of cooperative design process modeling. Figure 8 show a simplified cooperative design process for blade-disk component of aeroengine. When blade-disk component is selected, design goal and requirements are imported to blade-disk assembly agent. Such an initially formed constraint network will be expanded dynamically with the development of design process. Blade-disk assembly agent first solicits cooperation with blade agents and disk agents. When the requests are accepted, negotiation and coordination among agents are initialized according to expanded constraint network. These dynamic behaviors can be modeled and simulated based on Petri net beforehand in order to control and manage them effectively on the fly.

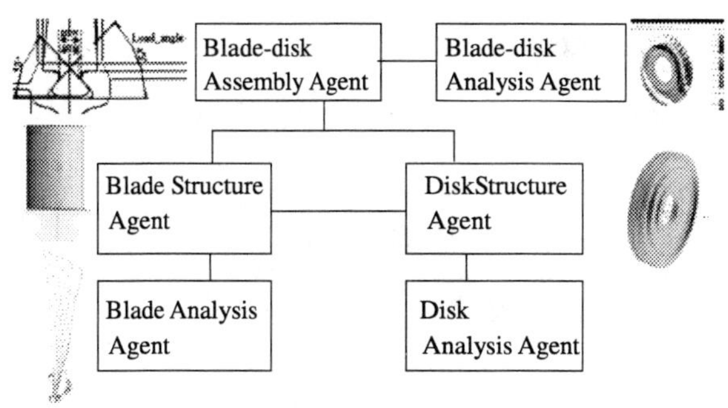

Figure 8. Cooperative Design Process of Blade-Disk Component.

Figure 9 shows part of constraint network in this cooperative design process that is mainly involved in groove feature(gf) of the blade and root feature(rf) of the disk. In this simple example, cooperative constraints exist between cooperative variables such as groove neck width(gnw) and root neck width(rnw), groove load angle(ga) and root load angle (ra), groove width(gwi) and root width(rwi). Values of these cooperative variables will be decided in an interacted way. On the one hand, they should satisfy cooperative constraints (e.g. ga = ra) through negotiation between blade agent and disk agent. On the other hand, they should satisfy constraint network(e.g. ga = f(pa) and gf = f(ga, gwi,gnw)) inside individual agents.

Cooperative design process modeling based on constraint network and Petri net and an integrated application platform are being used in some aeroengine design project. It can ensure to shorten development cycle, improve interoperability among applications, resolve cooperative design conflicts, increase adaptability to dynamic environment and achieve high level of automation [21].

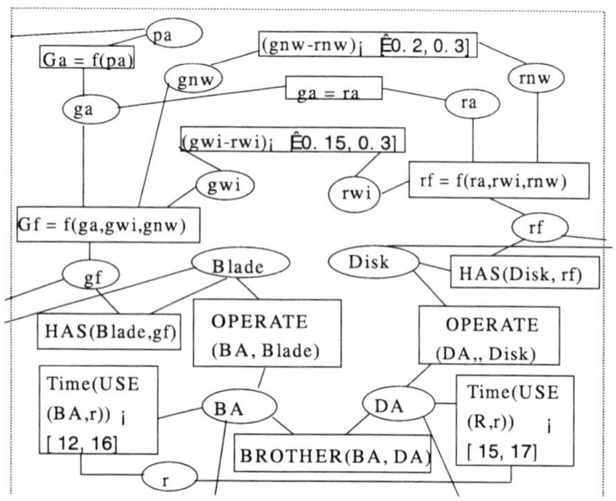

Figure 9. Part of the Constraint Network in a Blade-Disk Design.

7. Conclusion

In this paper, a new modeling framework based on constraint network and Petri net is introduced for a cooperative design process. An event-driven constraint-triggered decision-pushed automated design process model is built for the first time. Furthermore, models of communication, negotiation, coordination, cooperation in a MAS-based cooperative design process are given based on Petri net. These models are basis to a prototype of MAS-based cooperative design environment: CLOVER. Due to complexity, uncertainty and creativity of cooperative design process, more work will be focus on overcoming computing complexity, modeling decision in uncertain environment, controlling and managing conceptual and creative design process, building up self-organizing cooperative design automation system etc.

8. Acknowledgement

Thanks for support of CNSF. Comments from reviewers and discussions with Dr. Weiming Shen at NRC Canada, Dr. Haibo Duan of BUAA are very helpful to this article.

9. Reference

[1] G. Pahl and W. Beitz, Engineering Design: A Systematic Approach, Berlin: Springer Verlag, 1984.

[2] W. Shen, D.H. Norrie and J.P. Barthes, Multi-Agent Systems for Concurrent Intelligent Design and Manufacturing, Taylor & Francis, London, UK, 2000.

[3] R. Dechter, "Constraint Networks", Encyclopedia of Artificial Intelligence, Wiley and Sons, 1992.

[4] L. Gupta, J.F. Chionglo and M.S. Fox, "A Constraint Based Model of Coordination in Current Design Projects", 1996, http://ie.utoronto.ca.

[5] K.S. Decker and V.R. Lesser, "Designing a Family of Coordination Algorithms", in: Proceedings of First International Conference on Multi-Agent Systems' 95, AAAI Press: Menlo Park, CA, 1995.

[6] P.R. Cohen and H.J. Levesque, "Intention is choice with commitment", Artificial Intelligence, 42, 1990, pp.213-261.

[7] M. Wooldridge and N.R. Jennings, "The Cooperative Problem Solving Process", Journal of Logic & Computation, 9(4), 1999, pp.1-33.

[8] B. Grosz, "Collaborating systems", AI maganize, 17(2), 1996.

[9] M. Tambe, "Towards Flexible Teamwork", Journal of Artificial Intelligence Research, 7, 1997, pp.83-124.

[10] X. Jiang, Research on Concurrent Design Process Model and Process Control, Dissertation, 720 IMS, Beijing University of Aeronautics and Astronautics, 1999.

[11] ISO, 10303-1. Product data representation and exchange: Part 1: overview and fundamental principles, 1994.

[12] Y. Huang, Research on Distributed Intelligent Cooperative Product Design System, Dissertation, 720 IMS, Beijing University of Aeronautics and Astronautics, 2000.

[13] G. Zhao, J. Deng and W. Shen, "Interoperability in an Cooperative Intelligent Design Environment", Workgroup on Software Agents and Workflows for Systems Interoperability at The Thirteenth Canadian Conference on Artificial Intelligence, Montréal, May 14-17, 2000.

[14] D. Hollingsworth, "The Workflow Reference Model", Workflow Management Coalition Specification, TC00-1003(Draft 1.0), 1994.

[15] G. Zhao and J. Deng, "Rules in a MAS-based Cooperative Design Environment", in: Proceedings of CSCWD'2000, Hong Kong, 29 November-2 December, 2000.

[16] T. Finin, Y. Labrou, J. Mayfield, "KQML as an agent communication language", Software Agents, MIT Press, 1995.

[17] N.R. Jennings, S. Parsons, C. Sierra, P. Faratin, "Automated Negotiation", http://www.ecs.soton.ac.uk/~nrj, 2000.

[18] R.S. Cost, Y. Chen, T. Finin, Y. Labrou, Y. Peng, "Modeling Agent Conversations with Colored Petri Nets", Autonomous Agents'99, Seatle, Washington, 1999.

[19] N.R. Jennings, Cooperation In Industrial Multi-Agent Systems, World Scientific Publishing Co. Pte. Ltd, 1994.

[20] E.H. Durfee, Coordination of Distributed Problem Solver, Kluwer Academic Publishers, 1988.

[21] H. Duan, Research on Integrated System of Aeroengine Rotor Structure Design in Distributed Collaborative Environment, Dissertation, Beijing University of Aeronautics and Astronautics, 2001.

Supporting Negotiation in Concurrent Design Teams

Robin Barker
Bartec Systems,
Barnsley, UK
RBarker@bartec-systems.com

Leigh Holloway
University of Sheffield,
Environmental Business Network,
5 Palmerston Road,
Sheffield S10 2TE, UK
l.p.holloway@sheffield.ac.uk

Anthony Meehan
Sheffield Hallam University,
Pond Street,
Sheffield S1 1WB, UK
a.s.meehan@shu.ac.uk

Abstract

This short paper reports on that part of a design support system, which assists each individual in a concurrent design team to negotiate with their colleagues during the materials selection stage of design. It identifies a difference in the way some concurrent design team members conduct negotiation compared to one of the standard economic models. It seeks to build upon this to suggest ways in which designers may be supported in identifying alternative materials, the choice of which would improve the design from their own perspective, and which are more likely to be acceptable to partners in negotiation.

1. Introduction

We are engaged in the development of computer-based systems that support concurrent design teams. Typically, such a team has members who maintain a variety of perspectives on the evolving design artifact. These perspectives range from the aesthetic, through design for manufacture, use, maintenance, reuse, recycling and disposal. Our current focus is supporting the design team in selecting materials which optimise the product life cycle from an ecological perspective – so called eco-design.

Negotiation is one of the fundamental social processes that characterise the behaviour of the concurrent design team in arriving at a design decision [1]. This short paper first looks at the extent to which traditional models of negotiation adequately characterise the negotiation behaviour we see in concurrent design teams. It then describes a project to develop a designer-centred system to support human design teams. The paper outlines a negotiation protocol that can be used to support individual members of the team during negotiation with their colleagues.

2. Models of Negotiation

Research in computer-based negotiation has drawn extensively from game theory and decision science. However, the extent to which these economic approaches characterise human-oriented negotiation is very much an open question [2]. Even advocates of these approaches acknowledge that applied theorists have not yet developed models of bargaining that adequately capture complex bargaining situations [3].

As valuable as the foundational theory is, much of it is of limited applicability because of strong assumptions or constraints. Assumptions commonly include shared representation of issues and information, truthfulness, rationality, adherence to agreements, stationarity, non-stochastic environments, independence of issues and absence of linkages in a sequence of bargaining situations. Constraints include bilateral as opposed to multilateral protocols, sequential turn taking, and prohibition of coalition formation. Much is poorly understood, e.g. reputation building, commitment tactics, adding value without directly amending the offer, intention inference.

3. Computational Support for Negotiation

In terms of computational support for human interaction during negotiation, it does not significantly misrepresent the current situation to assert that there is a largely bimodal distribution of approaches. There are systems that seek to facilitate efficient and effective human negotiation, leaving members of a (distributed) team firmly in control of their associated communication agents [4]. Other systems implement more fully autonomous, agent-based negotiation e.g. in the context of e-commerce or communications systems [5, 6, 7, 8, 9, 10].

4. Designer-centred Support for Negotiation

In seeking to design negotiation support systems, it is important to avoid some of the problems that have prevented the adoption of other design support systems. Many systems do not correspond to the motivation and cognition of the individuals who are expected to use them [11]. Computational techniques may focus on what machines can do and not on how people actually reason and behave [12].

In seeking to realise flexible computational approaches which support negotiation we have been led to exclude many of the economic models which, whilst valuable in many other contexts, do not allow us to support negotiation dynamics as we have witnessed them in design teams. We have also sought to avoid imposing a particular negotiation protocol from the outset. However, we have identified some approaches that suggest a way forward. We have built upon these to develop a prototype system, which supports negotiation in the context of materials selection for product design.

In previous work with product designers we have developed a model to guide the development of design support. This involved using a technique to re-represent the designers' linguistic representation of their knowledge, behaviour and their organisational context in a semi-formalism based upon the CommonKADS model set [13]. The methodology is described more fully elsewhere [14]. The outcomes were a representation of how individual members of the design team viewed materials as design solutions [15] and an understanding of the tactics used by designers in negotiating materials selection [16, 17].

5. Designers' Viewpoints

Designers' viewpoints appear to be partially shared at the level of syntax in so far as they share labels for some of the material attributes [15]. Table 1 records some of the attributes of materials considered from three different concurrent design perspectives. In some cases, it is possible to have considerable confidence that designers share semantic representations in that the interpretation or meaning ascribed to the attribute is genuinely shared, e.g. 'specific heat'. Air Pollution Index (API) and Water Pollution Index (WPI) are examples of concepts that are deliberately constructed in order to facilitate shared meaning [15, 18]. However, the semantic interpretation of some shared labels may be far from agreed between designers, e.g. toughness, suitability. Even if attribute meanings are shared, the values attributed to them may differ (e.g. when comparing 'high' and 'low' for the concept 'durability').

Table 1. Concurrent designers represent materials in an attribute space. For many attributes, designers share labels (if not meanings) and this facilitates dialogue and argumentation in negotiation. Many attributes are not shared: some belonging to just one designer's perspective on the materials under consideration.

Material Property (HDPE)	**Materials Analyst**	**Mechan. Analyst**	**Disposal Analyst**
API/WPI	✔	✔	✔
Applications	✔	✔	
Chemical Resistance	✔	✔	
CO/CO2	✔		✔
Corrosion Resistance	✔	✔	
Density	✔	✔	
Durability		✔	
Dust	✔		✔
Hardness	✔	✔	
Incineration Waste			✔
Lubricants		✔	
NO/NO2			✔
Permeability	✔	✔	
Recovered Energy			✔
SO/SO2			✔
Specific Heat	✔	✔	
Strength	✔	✔	
Temperature Range		✔	
Thermal Conductivity	✔	✔	
Toughness	✔	✔	
Transparency	✔	✔	

Designers who perceive they share common concepts are able to conduct a dialogue in which proposals and counter proposals are made, with supporting arguments. A dialogue (in the context of the design of a drink container) illustrates this:

Designer 1. "*Can we consider HDPE instead of PET: I think we can live with the loss in strength and we will achieve a significant reduction in WPI.*"

Designer 2. "*I'm happy to look at that but I think the permeability of the HDPE will be too great.*"

Designer 1. "*Not if we reduce the gas pressure, or perhaps the shelf life.*"

The materials mentioned (HDPE, PET) are the highest level of shared concept (without shared terminology at this level, there can be no negotiation). The attributes of these materials that the designers appear to share include 'strength', 'WPI' and 'permeability'. The values these take in the dialogue are qualitative and have an implied order e.g. 'too great'.

When designers do not share syntactic or semantic representations they tend (or perhaps learn) not to feature these in dialogues. These concepts may be important to the individual concerned, and a high priority may be attached to optimising them in a negotiation, but, they do not naturally form the basis of argumentation. This implies that a computational agent representing a specific designer should distinguish between the (apparently) shared representation and a private representation of any artifact.

6. Co-operative Negotiation by Designers

Drawing upon a template suggested by [10], we have examined negotiation between members of design teams and identify features of the situation which inform the choice of a negotiation model.

6.1 Co-operation

Co-operative design does not always imply co-operative behaviour by team members. Designers often pursue their own objectives within a project; not least because they have different perspectives on the current design solution.

6.2 Information sharing

The model cannot assume information sharing. If designers have different representations of an evolving artifact, then, of necessity, some information is non-sharable.

6.3 Proposal Evaluation

In evaluating a proposal, the agents will, in general, rely upon a private evaluation function. Analysis suggests that, in respect of design proposals and counter-proposals, designers use the concepts of *similarity* between proposals and *value* of proposals. The fact that designers do not necessarily, or even normally, share the same private representation of a design solution also implies that the above measures may usefully be applied over a designer's *private* representation or over some subset of attributes that they believe are shared.

6.4 Resource Constraints

Resource constraints do not always feature as attributes of the design. They feature in reasoning about proposed changes to a design; e.g. change may imply additional costs or impact on other projects. Resource costs can be incorporated into the concept of *utility*, which is a function of the value attributed to a proposal by an agent and the cost to that agent of agreeing to the proposal.

6.5 Tactics

Negotiation can be viewed as search for agreement. The tactics of the negotiators determines the character of the negotiation. In observing design negotiation, it is possible to discern a spectrum of tactics. At one end of this spectrum is the classic monotonic concession protocol [2, 9]. In this situation, negotiators start from their goal position and make concessions from that ideal seeking a compromise (often Pareto optimal) that all can accept. At the other end of the spectrum is an approach in which is very different and seems to characterise co-operativity. In contrast to the monotonic concession protocol, the negotiator does not always have an ideal in mind but formulates a 'local' goal (better than the current solution) and attempts to draw others towards agreement of that goal. We call this strategy *enticement*. The strategy seems to involve searching for a counter-offer that is similar to the current proposal but which has added-value, both for the proposer, and, where possible, for others in the team. This continues until other team members cannot be drawn any further towards the designers actual goal.

The above suggests it is important not to be proscriptive in respect of tactics allowed to the users of any negotiation support system. Finally, it implies that there is scope for supporting a designer in modelling their peers by inferring which attributes of a design solution (private and shared) are important to others in the team and using this to select counter-offers perceived as similar. The next part of this paper describes how each of these components of a negotiation support tool can be realised.

7. A Negotiation Support Tool

Our negotiation support tool for material selection is specified as follows. We support designers with personalised computer-based agents each representing an individual designer's viewpoint on a design artifact. Agents share the same architecture. They represent domain objects (materials in this case) in different spaces (see description above and Table 1). [15] and [18] document the materials related knowledge of the system and the sources of the data. Agents share the same competencies derived from the analysis above: they are able to assess similarity, and value and use these to contribute suggestions to the user about counter-proposals that may be made in negotiation. In doing this, they attempt to infer which attributes of materials are of significance to others in the team and which are not. In this way, an agent can attempt to identify whether or not their objectives (in respect to any one dimension of the solution) are coincident with the intentions of another agent. This helps to identify possible win-win proposals.

7.1 Similarity

From the perspective of any *one* member of the design team, the materials under consideration share a set of attributes or properties. The values attributed to each property will vary between individual materials. Given two materials (m, n) from a set of materials, M, under consideration, the similarity of these materials over a set of properties, P, associated with the materials may be given by:

$$Sim(m, n) = \Sigma_{p \in P} w_p Sim_p(m, n) \qquad (1)$$

where w_p is a normalised weight associated with a property, p, by the designer ($\Sigma_{p \in P} w_p = 1$); and Sim_p is the similarity function for property, p [19].

The values associated with some properties vary continuously (subject to natural limits or constraints). For other properties, discrete values only may be found. The values of a property may be explicitly ordered but sometimes an order is determined by the preferences of the designer (see dialogue above). In these circumstances, a fuzzy measure of similarity is appropriate.

Following [20], a similarity function can be defined as:

$$Sim_p(m, n) = Inf_{1 \le i \le m} \, (h_i(m) \leftrightarrow h_i(n)) \qquad (2)$$

where Inf is the infimum and $h_i(m) \leftrightarrow h_i(n)$ is a fuzzy equivalence relation induced by the ith criteria or comparison function, h_i: examples include:

$$h(m) \leftrightarrow h(n) = 1 - | h(m) - h(n) | \qquad (3)$$

and

$$h(m) \leftrightarrow h(n) = min(h(m)/h(n), h(n)/h(m)) \qquad (4)$$

We emphasise that there are many alternatives to the above similarity function. It is our intention to explore the extent to which the behaviour of the system is sensitive to the similarity funtion that is used.

7.2 Value

The concept of value is an attempt to condense the desires and aspirations of the designer into a single discriminant term. Given some material, m, with a set of properties, P, a measure of the value that a designer attaches to the material, in some context, is suggested by [19]:

$$Score(m) = \Sigma_{p \in P} w_p Score_p(m(p)) \qquad (5)$$

again, w_p is a normalised weight ($\Sigma_{p \in P} w_p = 1$) associated with a property, p, by the designer: these weights are not, in general, the same as those used in the similarity function. $Score_p$ is the score assigned to a material property, p, arising from its value, $m(p)$. For each property, a scoring function mapping to [0,1] is chosen.

Both *Sim* and *Score* produce values in the range [0,1] as this in evaluating proposals and counter proposals.

7.3 Utility

Utility is related to the value of a proposal but makes explicit any cost associated with its acceptance [9]. A suitable expression is:

$$U(m,n) = (Score(m) - Score(n)) - c(m,n) \qquad (6)$$

where $(Score(m) - Score(n))$ is the difference in value between choices m and n and $c(m,n)$ is the cost associated with making the change, if agreed. In general, explicit representations of these costs are not available, mainly because they tend to be non-stationary. Thus, our system does not support users in assessing *utility*.

7.4 Determining Weights

It is necessary to determine various weights for the above expressions. It is possible to use machine learning to determine such weights although they too may be non-stationary. An approach which gives more control to the user is to derive the weights from user preferences at the time of use.

Each member of the team may rank materials using a graphical interface in which sub-trees and leaves of a class/instance hierarchy of materials related concepts can be ordered. They can also rank the properties associated with materials although these seem less prone to change with each design context). These rankings enable the various weights in the similarity and value functions to be obtained.

The easiest way to express preferences is by ranking. If normalised weights are required, they can also be obtained directly from the ranking. If a simple ordering does not adequately express the preference weightings, then relative 'distances' between ranked objects can be used.

7.5 Peer Modelling

Over a sequence of offers made by another team member, the values taken by any one attribute constitutes a time series. If a design agent can learn to predict changes in attribute values resulting from a future offer from a peer (usually the next offer), then that attribute can be inferred to be of some (indirect) significance to that peer. In turn, this information can be used to inform which counter proposals, from a set of similar or equally valued proposals, may be more attractive to the other parties in the negotiation and hence more likely to be accepted.

One approach to measuring prediction success is based upon correctly predicting directional changes in the value of an attribute. Considering, for illustration purposes, the values taken by a continuously variable attribute, over a series of n offers from a peer, the statistic:

$$d = \frac{1}{n}\sum_{1}^{n} a_i$$

$$a_i = \begin{cases} 1 if \left(x_{t+1} - x_t\right)\left(x'_{t+1} - x_t\right) > 0 \\ 0 \qquad otherwise \end{cases} \tag{7}$$

where x is the actual value of an attribute and x' the predicted value) measures success compared to a guess based upon a coin toss [21]. If d>0.5 then the agent can infer that the party making the offers is attempting to take the attribute in the direction predicted. As d approaches unity, one is able to predict the value a peer seeks in an attribute with near 100% success.

If certain attributes are thus inferred to be of significance to another party in the negotiation, then an assessment may be made as to whether both parties are seeking similar outcomes on that attribute or whether their aims are contradictory. It may be that the attribute is of value to one party, and not another. This offers scope for counter-offers to add value for one without loss to the other - a form of conduct which features in co-operative negotiation.

In summary, given a list of materials currently proposed for the design, the system offers similar alternatives to individual designers based upon both estimated peer and private representations. The system uses expressions (1), (2) and (3) to determine similarities. It does this for each agent and gives two similarity values, one over all (private) properties, the other using what are perceived by system to be the significant attributes of the peer making the offer. Expression (5) is used to evaluate similar materials. Expression (7) maintains a model of peer intentions in respect of individual attributes of materials and can be used to wieight the extent to which any one attribute contributes to a werithed similarity measure.

The human user remains responsible for the communication of any proposal he or she wishes to make, and for the supporting argumentation.

8. Discussion

The system described above is intended as a step towards a materials selection support tool for concurrent designers. Negotiation, is a key sub-task in the co-operative design process: designers evaluate proposals from their peers; consider alternatives based upon their preferences and choose a counter proposal in the event of non-acceptance. This paper offers a partial model for this process and suggests computational means to support designers, which are sensitive to the designers changing preferences and criteria. The system described is intended for use by individual members of the design team but could be used by smaller teams, even individual designers, interested in accessing other design perspectives.

Future work on this project has a number of different objectives, but at this stage a significant empirical trial of the system is needed to verify (or otherwise) the extent to which it succeeds in supporting the negotiation dynamics that are observed in design teams.

We are current exploring the extent to which it is important to manage the interaction of many of the factors involved. Standard theories assume, for example, that the values of attributes are independent. But, this is overly restrictive. The question is, does the relaxation on this restriction significantly inhibit the quality of the support offered by the tool as specified.

Currently, initiative in negotiation lies very much with the human user. More flexible/adjustable design-agent autonomy would seem to offer increased scope for reproducing the dynamics of collaborative design and perhaps increased cost effectiveness. It might also increase the likelihood of adoption of the design support system.

9. References

[1] Barker, R., Tranter, I. and Meehan, A., 1998. Towards a knowledge-level model of concurrent design. In IEA-AIE'98, *Proceedings of 11th International Conference on Engineering Applications of AI and Expert Systems*, Mira, J, del Pobil, A.P., Moonis Ali (eds), Lecture Notes in Artificial Intelligence 1415; Springer-Verlag, pp57-67, ISBN 3-540-64582-9.

[2] Huns, M.N. and Stephens, L.M., 1999. Multiagent Systems and Societies of Agents in *Multiagent Systems: A Modern Approach to Distributed Artificial Intelligence*. Weiss, G. (ed.), MIT Press. 79-120

[3] Muthoo, A., 1999. Bargaining Theory with Applications; Cambridge University Press, UK, 334

[4] Wong, S.T.C., 1994. Preference-based decision making for co-operative knowledge-based systems. *ACM Trans on Information Systems*, 12:4, 407-435.

[5] Sycara, K., Lewis, C., 1991. Modelling group decision making and negotiation in concurrent product design. *International Journal of Systems Automation* 1:217-238.

[6] Ephrati, E. and Rosenschien J.S., 1996. Deriving concensus in multiagent systems. *Artificial Intelligence* 87:21-74

[7] Kraus, S. Sycara, K., Evenchick, A., 1998. Reaching agreement through argumentation: a logical model and implementation. *Artificial Intelligence* 104:1-69.

[8] Zlotkin, G., Rosenschein, J.S., 1996. Mechanisms for automated negotiation in state oriented domains. *Journal of Artificial Intelligence Research* 5:163-238.

[9] Zlotkin, G., Rosenschein, J.S., 1996. Mechanism design for automated negotiation and its application to task oriented domains. *Artificial Intelligence* 86:195-244.

[10] Faratin, P., Sierra, C., Jennings, N.R., 1997. Negotiation Decision Functions for Autonomous Agents. *Int. J. of Robotics and Autonomous Systems* 24 (3-4) 159-182.

[11] Landauer, T.K., 1995. The trouble with computers. Academic Press, Cambridge Ma.

[12] Smithers, T. 1996. On knowledge level theories of the design process in Artificial Intelligence in Design. Gero, J.S., Sudweeks, F. (eds.) Kluwer Academic Publishers. 561-579.

[13] de Hoog, R., Martil, R., Wielinga, B. et al. 1992. The CommonKADS Model Set, ESPRIT Project P5248: Document Id. KADS-II/WP I-II/RR/UvA/018/4.0, University of Amsterdam.

[14] Barker, R., Meehan, A., Tranter, I., 1999 A Knowledge-level Model for Concurrent Design. *Applied Intelligence* 10:113-122, ISSN: 0924-669X.

[15] Holloway, L., 1997 A methodology and support tool for environmentally conscious design and manufacture. PhD Thesis, Sheffield Hallam University, Sheffield S1 1WB, UK.

[16] Barker, R. and Meehan, A., 1999 Supporting Concurrent Design Teams with Adjustably Autonomous Agents, AAAI Spring Symposium 1999, Stanford, Ca., USA.

[17] Barker, R., Holloway, L., Mardell, J. and Meehan, A., 2000 Supporting Knowledge-based processes using flexible intelligent agents. AAAI Spring Symposium 2000, Stanford, Ca., USA.

[18] Holloway, L., 1998. Materials selection for optimal environmental impact in mechanical design. *Materials & Design* 19; 133-143; Elsevier Science Ltd.

[19] Sierra, C., Faratin, P. and Jennings, N.R., 1997. Deliberative Automated Negotiators Using Fuzzy Similarities. Proc EUSFLAT-ESTYLF Joint Conference on Fuzzy Logic, Palma de Mallorca, Spain, 155-158.

[20] Valverde, L., 1985 On the structure of F-indistinguishability. *Fuzzy Sets and Systems* 17:313-328 (cited in Sierra, et al 1999.)

[21] Refenes, A.N. 1995. Neural Networks in Investment Management: Testing Strategies and Performance Metrics. in Zenios, S. (ed) Qualitative methods, supercomputers and AI in Finance, Stanley Thornes.

A Case Study on Distributed, Collaborative Design: Investigating Communication and Information Flow

Steven P. MacGregor, Avril I. Thomson, Neal P. Juster
Design, Manufacture & Engineering Management, University of Strathclyde, Glasgow, UK, G1 1XJ
steven@cad.strath.ac.uk, avril@cad.strath.ac.uk, n.p.juster@dmem.strath.ac.uk

Abstract

A current research project at the University of Strathclyde is introduced which aims to better understand the role of distributed engineering design in industry and address present problems. An element of the first major industrial case study of the project, completed within a multinational in the oil and gas industry, is then described and shown to address the main research questions. This multinational will be described in the text as company A. Sampling and analysis of distributed activity over time is served by the completion of daily diaries and direct observation. This detail design phase of a distributed design project finds that most collaborative work consists of simple information exchange supplementary to the main design activity. Furthermore, concentrated periods of collaborative design are found to follow these information exchange 'cycles'. 62% of distributed activity is found to be with the main design partner and asynchronous communication tools are the most popular. Distributed problems are also listed, including misinterpretation and finding the right person for information.

1. Introduction

The use of globally distributed engineering design teams continues to increase as companies aim to boost profits and decrease lead times by effectively leveraging knowledge and communication from dispersed locations. However, the benefits offered by distributed design are often marginalized by the problems that are inherent in the process [1]. The following problems have been communicated from recent pilot studies [2]:

- Knowledge bases or Domains (KB) are "confused" and "fragmented";
- KB 'islands' exist internally within organizations and externally within projects;
- There is a lack of common understanding between departments and problems with other project partners;
- An additional issue regards awareness. In order to effectively re-use knowledge from past experiences people have to know that it exists.

Currently, for the purposes of this research the manifestation of these problems have been termed *inconsistencies.* This seems to be the most appropriate term for factors that cause disruption to the smooth running and understanding of the design process. They arise, and are most abundant when knowledge and communication are not utilized correctly. These beliefs are re-enforced by the current industrial climate worldwide. Continual technological advancement is set against a backdrop of international mergers, take-overs and partnerships, resulting in a distinct lack of consistency on an organizational and design project level. There is a definite need for measures that will ensure a satisfactory level of continuity across all levels and particularly within distributed working scenarios. These problems have been re-enforced by the findings in the literature. The two main study domains of the project, Computer Supported Co-operative Work (CSCW) and knowledge management, detail the present shortcomings in working at a distance and maximizing the use of knowledge as a company asset.

A set of industrial case studies is required in order to focus the many issues in the field and to give adequate focus to problem areas to be addressed. The overall aim of these case studies is to investigate the distributed design activity within a large multinational that designs complex technical products. Particular attention is to be placed on the role of knowledge in supporting distributed communication and any supporting knowledge or information systems that the companies use. The first case study, discussed here, was carried out within Company A, a company which supplies pressure control and drilling equipment for the oil and gas industry. It is primarily a descriptive case study which aims to provide a snapshot of a part of the distributed design process.

2. Background and motivation - KITE

KITE (Knowledge Integration and Transfer for Engineering design) is a three-year research project based in the fields of Computer Supported Co-operative Work

(CSCW) and knowledge management. The main aim of the project is to safeguard against inconsistencies in the distributed design process by the effective integration and transfer of knowledge. This will be achieved through investigating a number of issues and encapsulating these findings in a Knowledge Integration and Transfer environment for Engineering design (KITE). It is envisaged that continuous consistency will be a key requirement for real progress to be made in future distributed engineering design projects. This will be addressed through investigating the following objectives:

- appropriate knowledge that can be used to enhance the design process;
- the most efficient and useful methods of integrating and transferring knowledge;
- types, levels and location of inconsistencies within the design process;
- the needs of various distributed design teams and required modes of communication.
- define the requirements of KITE and produce a specification and prototype;

These objectives form the core of the research questions. Although these questions were not asked directly during the case they were at the center of all data collection method development as they maintain a strong link between the overall aims of the research and the case study findings.

3. Case study methodology

3.1 Rationale

Company A are worldwide market leaders in the production of pressure control and drilling equipment for the oil and gas industry. The main case study site is in Aberdeen, Scotland. Aberdeen is the head office of the Eastern region of Company A, a region including the UK, Europe, Africa and Russia. The two other main sites are in Houston, USA and Singapore. The company represented an ideal opportunity for the first case of the KITE project for a variety of reasons. Firstly, the company is widely distributed. With regional headquarters in Aberdeen, Houston and Singapore many issues within the CSCW field will be present, including those of distance, time-zones, cultures and the resultant knowledge, information and data which exists within three widely dispersed regional headquarters. In addition to these three regional headquarters Company A has engineering and manufacturing centers, and sales/service offices in 59 locations and 24 countries around the world. Secondly, in designing and engineering pressure control and deep sea drilling equipment for the oil and gas industry, the work that engineers do is necessarily complex. This technical complexity adds another challenging dimension to the company's distribution and necessitates the optimal use of knowledge. Additionally, Company A is a market leader worldwide and already has in place world-class systems and people. This places extra onus on the project and case study in terms of finding problems. It is believed that if a suitable problem domain can be identified in this advanced domain, then it should be applicable on a general scale.

3.2 Overall methodology

A core team of ten engineers were involved in the case study while informal discussions took place with other employees. The core team was requested to take part in semi-structured interviews, complete questionnaires and a daily diary of five short questions for a period of thirty days. These data collection methods were supplemented by other informal discussions and direct on-site observation. All research subjects were either involved in the design process or some level of distributed engineering work. The case study lasted for a period of ten weeks. Record sheets were completed for significant events, interactions or studies. The types of records and their codes are shown below:

A - Informal staff meetings or interactions (e.g. discussion at lunch, in corridor)
B - Data collection methods
a Interviews, b Questionnaires, c Diaries
C - Documentation analysis notes
D - Significant events (e.g. presentations)
E - Observation notes
F - Impressions and interpretation notes

A significant effort was made to record all relevant data, to create a sizeable and searchable case study database and separate actual events and facts from interpretation and opinion. These initiatives are in line with recommendations from the appropriate literature [3]. In total 21 interviews and 19 incidental interactions took place. This included the core research group of 10 plus other employees at managerial level. This totaled over 130 record sheets and nearly 12 hours of interaction. 24 completed questionnaires and 174 days worth of activity recorded through daily diaries also resulted from the case. This equated to a response rate of 87% for the diaries. Interviews helped with investigation of the main case study issues, while questionnaires profiled current activity and perceptions. Diaries were used to sample distributed projects over time and were preferred to direct observation, although observation augmented the data. Two main iterations were employed during the case. The results of the diaries only, will be discussed in this paper.

3.3 Data collection methodology

The use of daily diaries afforded the opportunity to analyze the engineering design activity over a period of time without having to utilize observation techniques, although observation augmented the findings in the diaries. The diary consisted of five short questions. An e-mail message, comprising the questions, was sent from the research base each evening, allowing the engineers to complete the diary each morning, on opening their mail client (Lotus Notes). This was the best available method, decided in conjunction with the research group. As such the response rate for the diaries (30 days data for ten engineers) was 87%. Each of the engineers completed the diaries with varying levels of detail. The analysis proffered in this paper comes with the aid of one engineer who completed to a relatively high level of detail. The diary questions, together with a typical response are shown below:

1. Did you take part in any distributed activity (yesterday)? If YES, what was it?

Further attempts to gain information about subsea connectors, eventually being passed on to the correct person, still awaiting details. Received Pro/E simplified model of Company B's template, onto which our tree fits. In exchange, sent them drawings of our tree with dimensions in inches and mm.

2. Did you encounter any problems from this distributed activity? If YES, what were they?

Getting hold of the correct person to get the required information. Company B's Pro/E template model was created in mm, which did not convert very well to inches. The geometry converted, but all reference entities stayed at the original size.

3. Did you take part in any activity which involved the use or exchange of data/information/knowledge? If YES, what was it?

Pro/E model and Engineering bid drawings.

4. Did you encounter any problems with this data/information/knowledge activity? If YES, what were they?

As above.

5. Was there any particular piece of knowledge that proved useful yesterday? Details please

The template model was very useful.

The 30 days (6 weeks) of data available for this scenario concerned distributed collaboration between the engineer on site (Aberdeen, Scotland) and a sister company in Norway. The team in Scotland was part of the Subsea production equipment group and the data focused on a graduate engineer with 4 years experience. The teams were working together on a major bid for an upcoming Invitation to Tender for a major oil and gas company based in France. Both teams were working on the design of separate products, which were required to fit together, resulting in an *interface* design project. The work carried out could also be equated to the detail design phase in light of the level of complexity. The data collection period coincided with the start of the project and continued through to an advanced stage. If the contract were to be awarded the design activity preached was to follow specification – concept – embodiment - detail – manufacture. However, investigation showed that engineers just "responded to pressing requirements" working backwards from a contract hand over date. Specifications were rarely tackled thoroughly due to substantial re-use from previous projects. Initial emphasis for the project presented in this paper centered on the interactions which took place over the six weeks. Essentially, this information sharing supported the collaborative design activity between Scotland and Norway. Interactions were identified as being one of five types:

Information Out (IO): Information was sent to the collaborating party in support of ongoing work;

Information In (II): Information was received from the collaborating party;

Request Out (RO): A request for information was sent to a collaborator. In most cases this information was required to progress the design activity;

Request In (RI): A request for information was received, and;

Information Exchange (IE): A congested activity where numerous pieces of information were sent and received over a short period of time.

Each of the above identifiers was specialized by the tool adopted. For example, IOe would signify sending something out by e-mail while IIt would signify receiving some information by telephone. Distributed activity identified in the diaries took the form of either simple information exchange (for example, with a supplier) or collaborative design, as supported by information exchange (for example, a level of discussion and trade-off in agreeing an aspect of a design).

4. Case study findings

As would be expected the highest level of activity took place with the main collaborating partner in Norway (see figure 1). This accounted for 62.5% of all activity. Next highest was collaboration with suppliers at 20%, followed by other parts of the Company A group at 10%. Sub-contractors and customers at 5% and 2.5% make up the rest of the activity. This variety of collaboration calls for measures to be adaptive to certain situations. Furthermore, activity with the Norwegian design team

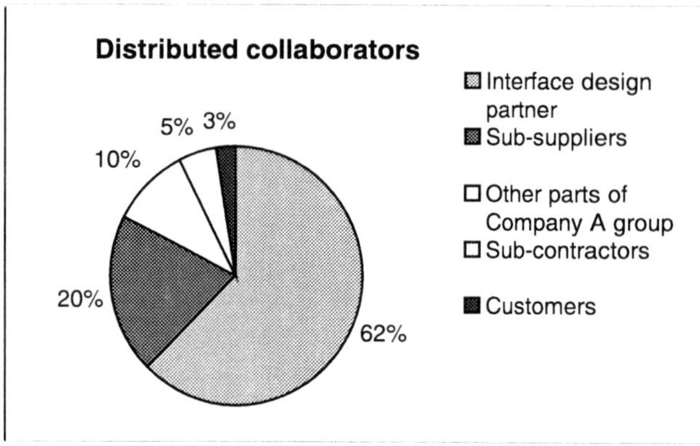

Figure 1. Distributed collaborators

was by far the most prevalent in all groups with the notable exception of *Requests Out (RO)* which consisted of 70% interaction with suppliers. Distributed problems throughout the sample period will be listed below. Examining these types of interactions in more detail (see figure 2), asynchronous communication tools are found to be significantly more prevalent than synchronous tools. This concurs with other research in the field [4]

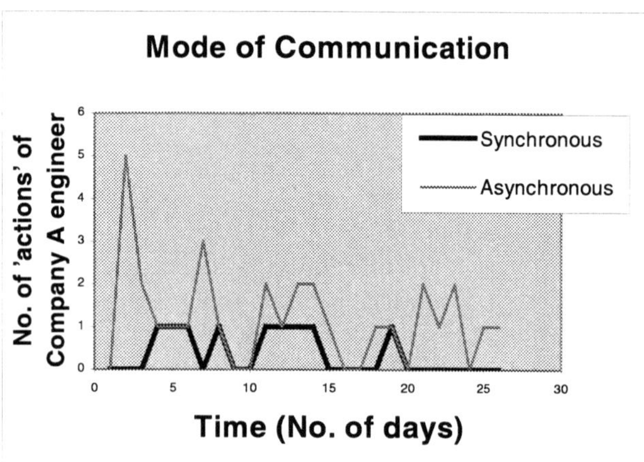

Figure 2. Mode of communication

and could be explained by the slight time difference and the difficulty inherent in the availability of dispersed colleagues. More significantly, it may be that the bulk of the subject of interaction is only *supplementary* to the ongoing design activity with only periods of core designing done distributively, especially in a synchronous mode. These cycles can be examined in more detail by charting the occurrence of each of the types of interactions (see figure 3). A high level of activity in all in all information types is seen at the start of the project. *Requests Out* shows a high level of requests for information being sent out over the first 11 days. *Information Out* shows a consistent occurrence of

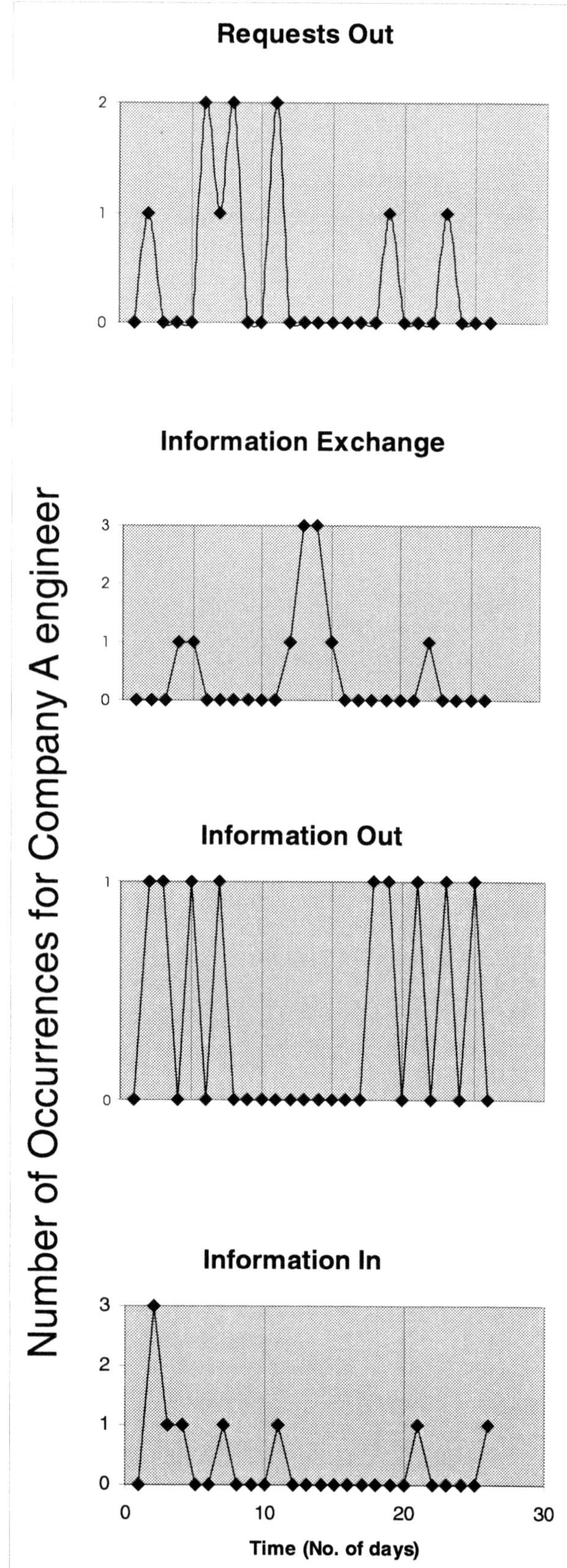

Figure 3. Information cycles

information being sent out over the first 7 days while the amount of information being received is consistently high over the first 11 days and especially on day 2, as shown by *Information In*. This activity followed on from a meeting which called for a series of actions. Most of *Requests Out* occurred during the first two weeks and were sent to sub-suppliers. The rest of the information categories involved, on the most part, collaboration with the main design partner in Norway. *Requests Out*, *Information Out* and *Information In* all have their highest level of activity at the beginning and the end of the sample period. The middle phase is taken up by a focused period of *Information Exchange (IE)*. This indicates periods of simple information exchange followed by collaborative design with respect to the information shared, and then on to the next cycle. We propose that the results for the time period after 40 days would include another period of concentrated IE. Distributed activity is also *continual.* On very few days was there no distributed activity at all. It is important to note that the data collection period coincided with the beginning of the project and ended at about 65% completion. The most popular type of interactions week by week were:

- Week 1 – Information In;
- Week 2 – Requests Out;
- Week 3 – Information Exchange;
- Week 4 – Information Exchange;
- Week 5 – Information Out;
- Week 6 – Information Out.

Requests In (RI) only occurred once throughout the thirty day period on day 12. Further studies are required to detect how each of these activities *combined* to make up the distributed design activity. Further investigation would also estimate the type of information interaction in the rest of the project, from week 7 onwards. A series of problems were also communicated during this sample activity.

Week 1 - Design indecision, use of unfamiliar abbreviations and acronyms, not knowing whom to contact for information, unachievable design detail.

Week 2 - Misunderstanding during call, no reply/response (2 instances), getting hold of correct person for information, geometry conversions from metric to imperial.

Week 3 - Ill informed design changes, poor distribution of information, misinterpretation of e-mails.

Week 5 - Describing and interpreting, poor distribution of information, ambiguities.

Week 6 - Disagreement, no reply/response, colleague with important information absent.

Obviously, some problems are human based and will occur during any collaborative or work activity, regardless of distribution. However others such as the problems with abbreviations and ambiguities may be solved through the aims of the KITE project, by making best use of the design knowledge abundant in the work activity. Others such as disagreement with the design are compounded with distribution, making it harder to come to an agreement. Taking the framework proposed by Gutwin and Greenberg [5] some of these problems may be associated with the *taskwork* and others, the *teamwork* or the work of working together.

5. Summary

5.1 The findings

The findings so far from the case study increase the understanding of distributed engineering design within the modern day workplace, giving an overall perspective of the role of collaborative design and distribution. Time series analysis is an important facet of any case study investigation. Daily diaries were the main method in this case as opposed to direct observation. Valuable results for level of work with different partners came from analyzing the diaries:

- 62.5% of distributed activity with main design partner;
- 20% with sub-suppliers
- 10% with other parts of the ABB group;
- 5% with sub-contractors, and;
- 2.5% with customers.

Adaptability of any implemented systems seems to be a must. Cycles of information exchange and collaborative design are also apparent from figure 3. Significantly at this stage, for this type of *interface* design the bulk of distributed activity is supplementary or in support of the design, rather than collaborative design itself, which only occurs in concentrated periods. Furthermore, the identification and analysis of this type of activity may be reflective of much collaboration in the modern marketplace where companies concentrate on their own specialisation which is interdependent on other companies and functions. Problems in the sample project are also interesting. Some are solely attributable to conflicts inevitable in teamwork. Others may also be attributable to a poor use of design knowledge while most, although in some cases not caused by distribution, are certainly made worse by distance and are harder to solve. The design studio concept as practiced by Maher et al. [6] is one possibility of addressing some of these problems, primarily within education, while Regli et al. have attempted to leverage specialized design knowledge for distributed work [7]. Other papers *compare* the educational and industrial arena [8].

5.2 Project progression

It is important to view the case study in the correct contextual light and relate the findings to the overall aims of the research project. It is crucial to address any perceived level of tenuousness between the case study findings and the project. Firstly, the findings, in general, have confirmed the results of the pilot studies undertaken during the initial stages of the research project, while obviously adding significantly to the problem domain. The following points have been confirmed:

- KB 'islands' exist internally within organisations and externally within projects – *issues of this nature were apparent* in *the diaries where particular problems occurred when trying to source information from outwith the company. Internal problems were not as frequent*;
- There is a lack of common understanding between departments and problems with other project partners – *misunderstandings, ambiguities and misinterpretation occurred from time to time in the diaries. This happened on a distributed level only. Collocated problems were rare*;

Secondly, the case study findings have begun to address the overall aims of the project. The research questions (reflective of the aims and objectives of the project) have been addressed in the following areas:

- What are the needs of various distributed design teams and their required modes of communication? – *the diaries have shown a live project showing that asynchronous communication is more prevalent and the information profile charts have addressed the needs of engineers at different stages of collaboration;*
- What are types, levels and location of inconsistencies within the design process? – *a series of problems relating to distribution and knowledge were communicated. Problems of ambiguity and misinterpretation communicated through the diaries may, in turn lead to inconsistencies if they are not continually checked and confirmed;*
- What is appropriate knowledge that can be used to enhance the design process? – *an understanding has been advanced through the diary questions and times of greatest progress during distributed activity;*
- What are the most efficient and useful methods of integrating and transferring knowledge? – *understanding afforded through the case will indicate various options suitability or otherwise in the future.*

6. Further Work

The findings to date address the overall aims and objectives of the project. However, further material from the case will augment the present results. This will come specifically from interviews, questionnaires and an additional in-depth scenario from the daily diaries completed from a different perspective. Furthermore, the medium term will witness the completion of a further case that will enable a different perspective on industry and enable the required cross case analysis. One element of focus will include the identification of types of inconsistencies within the distributed design process. Generally, the case studies will help to produce a *description* of distributed design activity within industry today. Finally, the medium to long term will include some form of prescriptive solution (KITE) in response to this descriptive phase.

7. Acknowledgements

This research project is supported by the Engineering and Physical Sciences Research Council (EPSRC), a UK government funded research body. Research studentship number 99309292.

8. References

[1] J. Huang (1999), "Knowledge sharing and innovation in distributed design: implications of internet-based media on design collaboration", *International Journal of Design Computing: Special Issue on Design Computing on the Net (DCNet'99)*, Sydney.

[2] S.P. MacGregor (2000), "KITE – Knowledge Integration and Transfer for Engineering design: a PhD proposal", *CAD Centre technical report*, DMEM, University of Strathclyde, Glasgow, UK.

[3] R. K. Yin (1994), *Case study research: design and methods*, 2nd ed. (Applied social research methods series; v.5), Sage publications, London.

[4] A.I.Thomson, W.J.Ion, (1997) "Integration of design specialists through shared workspaces", *UK Design Council Final Report.*

[5] C. Gutwin, S. Greenberg (2000), "The mechanics of collaboration: developing low cost usability evaluation methods for shared workspaces", *IEEE 9th International Workshop on Enabling Technologies: Infrastructure for Collaborative Enterprises*. June 14-16, NIST, Gaithersburg, MD, USA.

[6] M.L. Maher, S. Simoff, A. Cicognani,, (2000), *Understanding Virtual Design Studios*, Springer-Verlag, London.

[7] W. C. Regli, V. Zaychik, T. Hewett, J. Sevy (2000), "Issues in building and evaluating networked engineering environments", *The fourth IFIP working group 5.2 workshop on knowledge intensive CAD KIC-4*, May 22-24, Parma, Italy.

[8] S. P. MacGregor, A. I Thomson, W. J. Ion, "CSCW in industry and education - transferring knowledge in engineering design" in proceedings of *Concurrent Engineering, CE2000*, Lyon, France, July 2000.

POPIM: Pragmatic Online Project Information Management for Collaborative Product Development

GQ Huang, XB Feng, KL Mak,
Department of Industrial and Manufacturing Systems Engineering,
The University of Hong Kong, Pokfulam Road, Hong Kong,
gqhuang@hku.hk

Abstract

POPIM (Pragmatic Online Project Information Management) is a prototype web-based platform for managing collaborative product development projects within an extended enterprise environment. A suite of facilities are provided for geographically dispersed project team members to communicate, share, and collaborate on a project in a common workspace where they enjoy online access to the most up to date project information and maintain a high-level data consistency and accumulate experience and knowledgebase. In addition to standard project management functionality such as defining work structure breakdowns, determining work schedules, teaming up with specialists, and allocating resources, POPIM incorporates workflow management (including dependency management), and deliverable management (document management if documents are considered as one kind of deliverables). Individual members have their personalized accounts according to their skills and roles/responsibilities in a project. A project team and its members may maintain their own journals/records. More application-specific functions such as product design review and engineering change management can be implicitly performed through online document forms.

1. Introduction

Collaborative Product Development (CPD) has been an area for intensive research for two decades. Among the many success factors are the teamwork, better communication, better project management, better information sharing and consistency.

One direction has been to develop methodical techniques or tools that reinforce these success factors. Quality Function Deployment (QFD), Failure Mode and Effect Analysis (FMEA), and Design for X (DFX) are just three examples. The other direction is to computerize these methods or techniques to enable better application of the methods or techniques in practice. As a result, majority of these methods are computerized.

Paradoxically, significant inconsistency exists between these two directions. Let us consider the example of evaluating concepts in early product definition. On the one hand, methods such as QFD, FMEA, DFX, etc have the tenet of nurturing teamwork as well as requiring teamwork for their successful applications. On the other hand, however, their computerized systems have mostly been standalone. Standalone systems defeat their original tenet of teamwork because they do not allow simultaneous access from multiple users from geographically dispersed locations.

This research has been conceived based on the above observation. A web-based framework, called POPIM (Pragmatic Online Project Information Management), is under development to promote and facilitate the management of collaborative product development projects in the extended enterprise environment. The focus of this paper is not on how to develop web applications for formal methods and techniques to overcome the limitations of standalone systems. Instead, the emphasis is placed on how such web applications and their uses and results can be incorporated dynamically into both strategic and tactical management of product development projects.

This paper summarises the results of the initial investigation. Section 2 briefly reviews the related literature along three dimensions of project management, product data management and workflow management. Section 3 presents the overview of the POPIM methodology and framework. Section 4 demonstrates the POPIM facilities through an illustrative case study. Section 5 briefly summarises implementation issues. Some of the important insights learnt from the research are highlighted in the final section.

2. Literature Review

There have been enormous efforts in developing computerized systems for supporting the management of collaborative product development projects [2]. A large number of research and commercial systems have now appeared [3-7]. In the context of this research, three categories of systems are most relevant. They are computer systems for project management, workflow management, and product data management.

Early computerised project management systems were mostly standalone while most of the commercial systems on the present market have adopted the client-server architecture. They have overcome many of the problems and difficulties faced with paper-based approach. However, the geographical barrier, especially at the international level, has not yet been adequately addressed. This leads many to investigate into web-enabled or web-based systems that are capable of resolving the time and space barriers. Some of the notable systems are as follows:

- SquareWorks is a web-based suite of tools for IT project development enables all team members to contribute more efficiently to planning projects, tracking defects, and resolving time-critical problems (http://www.squareworks.com/).
- Account4 is yet another web-based project management system for an enterprise to initiate, receive, evaluate, process, and assign current and forecasted project and activity requests.
- OnProject.com is a service provider where organizations and individuals can create their projects and maintain all the relevant information at the web site at a rental cost.

In addition to computerised project management systems, most PDM (Product Data Management) systems offer functionality related to project management, together with workflow management, and document (hard copy, electronic copy or online form) management. Product Data Management (PDM) is also widely known as Document/Drawing Management (DDM), Engineering Data Management (EDM), Enterprise Data Management (EDM), Product Information Management (PIM), and so on. More recently, the term Knowledge Management is used to cover wider scope.

The PDM technology is no longer new. It enjoyed rapid developments in the 1990s. There have appeared hundreds of research and commercial PDM systems in various forms for various industrial sectors. The functionality provided by these systems and the technology used for implementing these systems are similar to large extent. During this period of expansion, some companies have implemented PDM systems to accelerate product development cycles, improve time-to-market, improve quality and reduce costs. However, the majority of these implementations were in pilot tests which only recognized the great potentials but siginificantly limited the scope of functionality. Very few reports exist on companies that have fully implemented PDM systems.

However, previous investigations into various what can be called first-generation PDM systems [10] have identified a number of factors that limit the widespread use of the PDM technology in industries. Although the usefulness of PDM systems has been confirmed, their useability has been a major problem. The level of complexity of the comprehensive PDM functionality and system administration have almost eliminated the useability. Although PDM systems are supposed to enhance communications and information access between geographically dispersed locations to facilitate teamwork, the traditional client-server architecture used for implementing these systems fails to maximize such potentials. Systems such as SmarTeam are primarily not web based although some facilities are provided for web access. On the other hand, systems such as onProject.Com [5] are 100% web-based but their functionality and business model are limited.

After the period of expansion in the 1990s, the PDM developments seem to enter a perod of consolidation at the turn of this century. One of the most significant developments is the use of the web technology in implementing PDM systems. However, recent evidence in this direction is that only web-enabled approach has been demonstrated while 100% web-based approach has not yet been widely used by leading software developers.

Based on the above analysis, the following aspects must be adequately addressed before the technology can take grass-root in industrial practice:

- The balance of usability and usefulness must be achieved. A sophisticated system with poor usability will not be adopted without substantial investment in training and start-ups.
- The integration of facilities for project management, workflow management, data/knowledge management, and user skill management is essential. Product, process, and resource data are just treated as different types of information in the system.
- The system must be very easy to maintain and administer without specialist skills or knowledge.
- The system must be wide accessible from different locations and at different times by users with varying roles in the projects.

3. The *POPIM* Framework

This proposed project is mainly concerned with the development and implementation of a 100% web-based framework for managing product (both manufactured and service product) development projects among geographically dispersed team members. The main aim of the framework is to integrate functionality usually provided in separate systems such as project management, PDM (Product Data Management), and workflow management. Emphasis is placed upon the pragmatism and usability of the system while the functionality is not compromised.

The project management methodology underlying the POPIM system follows what is generally available in the literature or practice [9] although some customization has been done. However, the discussion in this paper focuses on the web-based implementation of the method and its application in a product development project.

3.1. POPIM System Architecture

The prototype system called POPIM has been developed after our initial efforts. Figure 1 shows an overview of the POPIM system. The figure shows a typical 3-tiered model of the web-based application:

The right-hand part of Figure1 is the data services tier. It provides a device – independent interface to data storage devices way be physically distributed throughout the network. Repositories of all run data and configure data include data of works, users, resources, documents, and so on.

The middle part of Figure 1 is the business services tier. It is divided to two parts to implement business process. One is based on IIS (Internet Information Server) that is responsible for receiving input from the presentation tier, interacting with the data services to perform the business operations that the application was designed to automate, and sending the processed results to the presentation tier. Some components such as Work, Gantt chart, Teaming, Resourcing, Setting up deliverables, Online forum, are built on the web server. The other is a DCOM server that is responsible for providing a remote object named Workflow for application client and controlling the information communication between various tiers.

The left-hand part of Figure 1 is the client tier that runs on the web browser with which the users connect to POPIM system. The client side of Workflow is embedded to web page.

All of these three tiers may be physically distributed among different machines at different locations.

3.2. Case Study

To simplify the discussion, a case study of using POPIM is presented to support collaborative product development. The case is taken from a textbook (Roozenburg and Eekels, 1998). A project was based on an assignment for the sales department of a major producer of shoe-care products, leather paint, and insoles. For keeping the market leader position, the company decided to develop a new product. A new shoe polish in combination with a new packaging is considered. The whole process of the project to design and develop a shoe polish packaging was summarized in the textbook. The project took almost 10 weeks to implement the plan. It consisted of 5 sub-works: Product planning; Clarification; Conceptual design; Embodiment design; Detail design.

This case study is borrowed here to demonstrate how the POPIM system can be used for defining and executing the project. The project team members are the primary users of the system. The project manager is a super-user who has the overall access authority while the other member users have the access authorities appropriate to their individual roles.

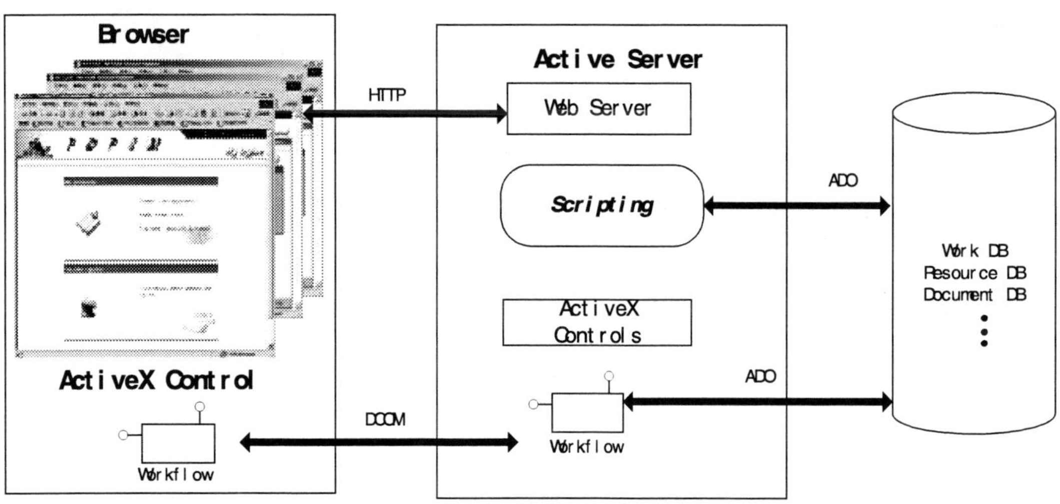

Figure 1. POPIM architecture.

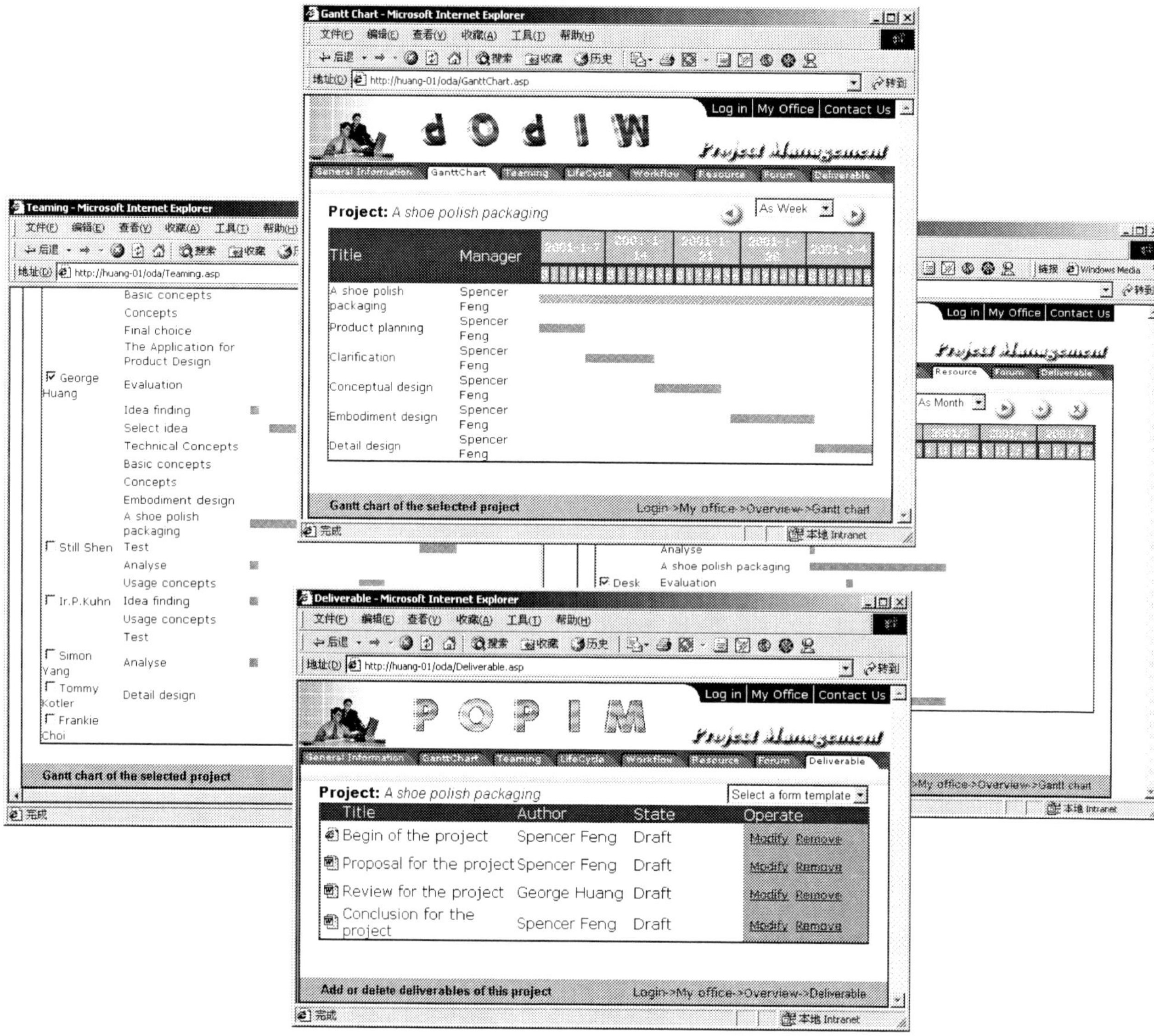

Figure 2. Defining the project.

3.3. Procedure of Defining a Project

The POPIM system itself provides some facilities for managing the lifecycles of important objects including the projects and their works. For example, the lifecycles of a project include "proposal", "definition", "execution", and "completion". The following discussions will focus on the definition stage only, involving defining the work structure breakdown, determining the workflow (dependency), identifying the personnel, allocating resources, specifying deliverables, and finally presenting the project plan in the Gantt chart form.

3.3.1. Starting a New Project and Activities

A new project must be created in the POPIM at the beginning. This is a straightforward task, just the same as adding a new work to an existing project with the Project Explorer. Information needed at this stage is very general and more detailed to be determined later.

A new Product Development project usually includes a series of activities or work packages. In POPIM, projects, activities, and work packages share the same "work" data structure. Therefore, their definition is exactly the same.

The project definition is represented as a tree-like work breakdown structure. The user may choose one item to proceed into the next step of project definition.

3.3.2. Scheduling

Once the work breakdown structure of a project has been defined, it is the time to determine the start and due dates for the project and its works.

In the POPIM "work" data model, these dates are treated as properties of the data object, just like other descriptions. Therefore, the process of scheduling the start an end dates of the project works is an extension of the work definition. This time-related information is displayed in the Gantt chart.

3.3.3. Teaming up

Organizing a project team is an dynamic process. The membership of the project team may change from time to time depending on the nature and progress of the project works. Initially, a team must be established anyway. The following considerations must be taken:

- The requirement of the work
- The expertise of the candidate
- The availability of the candidate during the specified period.

POPIM provides some facilities for the project manager to carry out the above tasks. Firstly, the work requirements are matched onto the skills of the personnel in the database. A list of the candidates is produced for further consideration. The work requirements are defined in the template of the specific work while the human skills are described in their profiles.

Next, the candidate list is further reduced by considering the personnel availability. The project manager is able to look at the loading profiles of all the candidates and choose one eventually from the list.

3.3.4. Resourcing

The human resourcing aspect has been dealt with during the process of teaming up for the project. This step of resourcing is mainly concerned with allocating material and equipment resources to project works. Resources of a company are shared.

The process of allocating equipment resources to project works is basically the same as allocating human resources. Firstly, the equipment capabilities stored in the equipment profiles must be compared against the work requirements defined in the equipment profiles. Next, the loading or capacity of the matched equipment must be considered.

3.3.5. Setting up Deliverables

Deliverables are visible outcomes from project works. If all the deliverables are delivered satisfactorily, then the project work is considered completed successfully. Although most commercial project management computer systems do not support deliverable-based management, POPIM places heavy weight on this aspect.

When defining a project, POPIM allows the user to define deliverables for any items in the project work breakdown structure. The definition of work deliverables includes the name, the description, the author, and the format of a deliverable. Three kinds of format are provided in pop-down list. They are online form that a URL needed to be specified, Word document and Adobe Acrobat document. The last two kind of deliverable will be upload to database after the user submits it.

Project deliverables are often not documents. Instead, physical product prototypes, test rigs, etc. are some of the common forms of deliverables. However, POPIM only captures their descriptions in the form of document.

3.3.6. Putting Together in Gantt Chart

Gantt charts are commonly used to present project plans. Most of the project work definition information can be displayed concisely and systematically in this chart. Figure 2 shows one snapshot of the Gantt chart prepared in the case study. This chart can be used in the following steps of project execution and monitoring.

3.3.7. Online Forum

POPIM provides online Forum facilities as a multi-purpose project loggers/journals. Any comments or notes can be recorded in the POPIM database regarding a project or its works. Such notes can be private to the current user only, shared by the project team, or made public to all the users. While the private property is selected, only the user himself can read the article. Thus the Forum could be a personal logger. When the team sharing property is selected, only team members can read the article. Thus the Forum could be a team logger. When the public property is selected, all users can read the article.

4. Concluding Discussions

This paper has discussed how a web-based framework such as the prototype POPIM system can be used to manage collaborative product development projects within an extended enterprise environment. The framework provides a common workspace for geographically dispersed project team members to communicate, share, and collaborate on a project through online access to the most up to date project information so that a high-level data consistency can be maintained, and experience and insights can be accumulated to form the knowledgebase. In addition to standard project management functionality such as defining work structure breakdowns, determining work schedules, teaming up with specialists, and allocating resources, POPIM incorporates workflow management (including dependency management), and deliverable management (document management if documents are considered as one kind of deliverables). Individual members have their personalized accounts according to their skills and roles/responsibilities in a project. A project team and its members may maintain their own journals/records. More application-specific functions such as product design review and engineering change management can be implicitly performed through online document forms.

Let us now return to the issue of gap between the methods and their computerized systems. The POPIM framework resolves ths issue by first of all incorporating the methods into web applications, and then relate these web applications to deliverables in project management. These deliverables are outcomes and results of the web applications, usually related to product, resource and process information usually managed in Product Data Management (PDM) systems.

The resulting framework has its balanced features, mostly reflected by the three keywords "Pragmatic", "Collaborative Design Management", and "World Wide Web". By "Pragmatic" it is meant that the methodology and system must be affordable and usable by the targeted users and for the intended functionality. By "Collaborative Design Management" it is meant that majority of the useful and usable functionality such as Project Management, Workflow Management, and Document (Project Deliverables) management provided in the traditional PDM systems have been integrated into one framework. By "World Wide Web" it is meant that this POPIM fully relies on the innovative use of the Internet and web technology. Our emphasis is on 100% web-based and low-cost, in contrast with other expensive and web-enabled systems on the market.

5. Acknowledgements

The authors are grateful to the Hong Kong Government Research Grant Council and the Hong Kong University Committee on Research and Conference Grants for the financial supports. The authors are also grateful to other researchers in the team for their discussions, especially Mr. Jun SHI for his contribution of the workflow management module.

6. References

[1] Corbett, P., Atkinson, J., Hinduja, S. (2000) A computer-based design decision logger, ImechE 2000, 13-22.

[2] Davey, K. (2000) Networking for project success, IEE, 45-48.

[3] Draeger, E. (1998) Project management with SAP R/3, published by Addison-Wesley.

[4] http://www.Account4.com/

[5] http://www.OnProject.com/

[6] http://www.squareworks.com/

[7] Microsoft: http://msdn.microsoft.com/workshop/author/dhtml/dhtmlovw.asp.

[8] Roozenburg, N. F. M., Eekels, J. (1995) Product design: fundamentals and methods, published by John Wiley & Sons, 345-360.

[9] Rosenau, M. D. (1998) Successful project management, published by John Wiley & Sons, Third Edition.

[10] Yee, W. Y. (1998) Development of a web-based system for engineering change management, M.Phil. Thesis, Department of Industrial and Manufacturing Systems Engineering, The University of Hong Kong.

Web-based Product Design Review: Implementation Perspective

GQ Huang and KL Mak
Department of Industrial and Manufacturing Systems Engineering,
The University of Hong Kong, Pokfulam Road, Hong Kong, PR China
gqhuang@hku.hk

JY Shen and JQ Yan
CIMS Institute, Shanghai Jiao Tong University, Shanghai, PR China

Abstract

Product design review is one of the typical scenarios of collaborative product development. We developed a web-based prototype framework for supporting the collaborative tele–product design review on the Internet. It provides a suite of tools to establishing and management the new product design review process. This paper discusses the issues related to the development and implementation of web application in prototype implication. The typical 3-tier architecture is explained to show how the CyberReview components work together to achieve the intended functionality. The VRML EAI and Java Applet-Servlet pair technology was included to support the implication.

1. Introduction

Product design review is one of the typical scenarios of collaborative product development [4, 8, 11]. A team is tasked with the design and development of a new product. The team consists of members from multiple disciplines. Some members are lead users (key customers), some are core (key) suppliers, and others may come from various functions and units of the organisation. The members are typically geographically dispersed.

Traditionally, the design review is conducted by circulating the documents related to the product design to be reviewed from one member to another [2]. After all members have investigated into the subject, a meeting is then arranged to resolve the different opinions. This process is very inefficient, especially when some external members such as key customers and suppliers are involved from other regions.

On the one hand, design review is a mandatory requirement by ISO9001 [7, 10]. On the other hand, Information Technology (IT), especially Internet/web Technology, has provided opportunities for effective design review in the extended enterprise [3].

Despite the fact that Information Technology (IT) has been employed to facilitate this process, the progress is far from satisfactory. This observation has been verified from our contacts with industrialists involved in both IT and manufacturing sectors. The roles of word processors and email messages are limited. Instead, a new paradigm is needed to completely reengineer this vital process of design review.

The aim of this research is to develop an overall methodology for enabling more efficient and effective design review system in new product development process, and to demonstrate the framework through a prototype web-based platform on the Internet/intranets using the web technology.

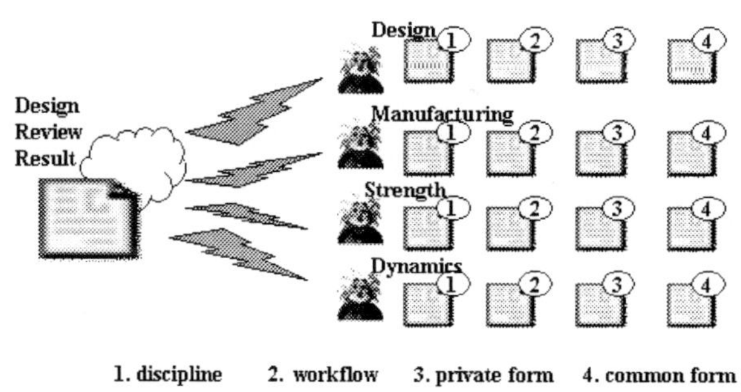

Figure 1. Web based product design review.

2. CyberReview Framework and Components

After the initial round of investigation, a prototype system called CyberReview has been developed. CyberReview is a web site where design documents are submitted and subsequently reviewed, and the resulting review documents are archived. A rich set of facilities are provided to facilitate the entire design review process. The system includes the following main components:

- *CyberReview Explorer.* This is the main component to which all the other components and facilities are attached.
- *Project Explorer.* This lists all the CDR projects.
- *Review Coordinator.* This component provides a set of facilities for the project manager to plan and manage the activities and resources involved in the design review process, in particular, for establishing the review committee, defining design documents, and preparing review documents.
- *Design Explorer.* This component is provided for the product development and design team to upload the product design onto the CyberReview database for future access. The review team uses the similar facilities to download design documents for review.
- *Comments Explorer.* This is an electronic discussion forum for review team members to submit their comments freely concerning the design documents submitted by the design team.
- *Review Explorer.* This collects all the online and offline forms and documents that are prepared during the review process.
- *Meeting Explorer.* This is a sophisticated module providing a variety of facilities to support holding review meetings for both the chairperson (project manager) and the team members before, during and after the meeting.
- *BOM Explorer.* This is a module associated with the design explorer. BOM is treated as a special type of design document required for design review when it becomes available. VRML files, comments and reviews may be directly related to BOM items.
- *VRML Whiteboard.* This is a whiteboard based on the VRML display of product features. This module would be necessary during the review process.
- *CyberReview Utilities.* In addition to the above main modules, WeBid provides a set of general utility facilities such as user registration, searching for necessary information, etc.

The review procedure using the CyberReview system in this case study generally includes the following main activities:

- With the help of the Review Coordinator, the project manager establishes a review team or committee and prepares review documents (pro-forma and procedure).
- The design team uses the Design Explorer to upload the design in the VRML format onto the database.
- Individual members in the review committee use the Comments Explorer to carry out their reviews by submitting comments and suggestions to the CyberReview database. This is generally asynchronised.
- With the help of the Meeting Explorer, a review meeting is called upon to resolve the comments from individual members.

3. System Design Considerations

Numerous factors must be addressed during the implementation of CyberReview facilities. Some of them are:

The industrial practices in design review vary widely from one company to another. The web-based collaborative product design review framework must have sufficient flexibility to incorporate and reflect good practices across the industrial spectrum.

Design review naturally involves 3D geometries of the product features in addition to other textual descriptions. It has been decided to employ VRML format in CyberReview for this purpose.

Comments made during the design review process are usually specific to the perspectives or positions of the 3D product drawings from which the comments have been made.

Design review is usually interactive and collaborative. Discussions and arguments are usually conducted based on topics or threads. The electronic forum is adopted in CyberReview for this purpose.

Design review can be either asynchronized or synchronised or both. Synchronization poises a great challenge in implementation.

Majority of the CyberReview facilities can be implemented based on various standard web and Internet constructs in relatively straightforward ways. The following considerations have been taken when implementing CyberReview:

- The standard 3-tiered architecture of web application has been used for building CyberReview framework.
- The CyberReview application servers and database have been deployed in the MS Windows 2000 environment.
- HTML and ASP (Active Server Pages) have been used as the basic language for implementing the application servers and clients, supported by Java and VB scripts when necessary.

- Java Applets and Servlets have been extensively used in the implementation of application servers and clients.
- A relational database is used. The connection to application servers is through ODBC and ADO (ActiveX Data Object). No direct connection is allowed from application clients to the database.
- CosmoPlayer is used for the VRML browser although other alternatives have also been experimented.

However, some complication has been encountered when implementing the synchronization mechanisms for the VRML Whiteboard and Meeting Explorer. The synchronization of the Meeting Explorer has been achieved by periodically refreshing the web pages. The time can be set at 1-10 seconds.

4. System Implementation

4.1 System structure realization

WWW-based Java solution is one of very important styles of the Internet computing model. With the help of Java programming, here, client-side computing can be implemented with the combinations of Applets and HTML files under the control of browsers and server-side computing with ones of Servlet, URL calls, or extended HTML files under the control of Java-enabled web server. It will also be available to integrate a background database, to develop drivers for networked devices, to connect with client-side local standalone applications, etc. So, it is the very important technology solution for the Design Review in External Enterprises.

Our system used a 3-tier database application that used Java Servlet and the Java Database Connection (JDBC). The application is partitioned into three tiers: user interface layer, the web server layer and the data store layer.

- First Tier: The first tier is a web browser, which serves as our universal client. In the first phase of the application, an HTML front-end was used for user-input and displaying the review comments, which submitted by the participator and stored in the database.
- Second Tier: The second tier of the application is implemented with a Web server capable of executing Java Servlet. The Java Servlet harnesses the power of JDBC to access the database to store/retrieve information as needed. A dynamic HTML page is generated by the Servlet based on the database results.
- Third Tier: The third tier is composed of our back-end database server. The database server stores the information that is used by the application. Thanks to the JDBC API, the Servlet can access the database in a portable fashion by using the SQL call-level interface.

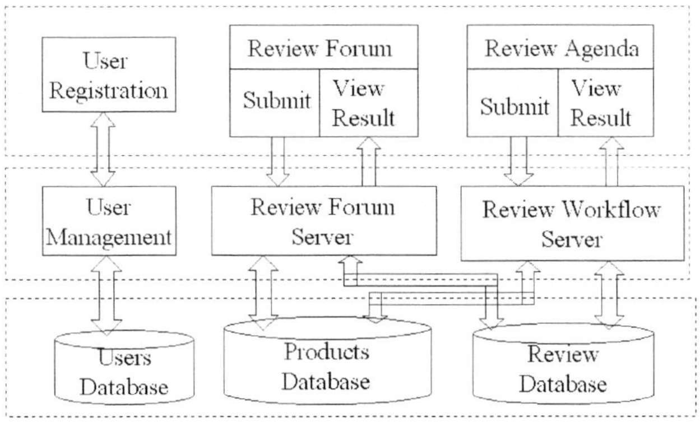

Figure 2. The framework for collaborative design review.

4.2 EAI Realization

4.2.1 Applet initializes VRML EAI

EAI (External Authoring Interface) is an API for connecting a VRML plug-in and external program [1, 5, 6, 9]. EAI allows for an external program to control VRML object in VRML plug-in, in our system, the review workspace is a VRML file, and the review Applet is an external program. The use of VRML EAI requires that several class be initialized when the connected Applets. Therefore the execution of collaboration Applet begins with the initialization. First of all, A browser class in initialized, which makes is possible for Applet to access the VRML plug-in used, then node calls in the VRML file, such as transform and touchsenser nodes are obtained from the browser class. There by, the VRML nodes attribute values can be referenced or change in the Applet while controls the VRML object. Change of node's attribute is called an event, There are two event: EventIn and Event Out. EventIn is an event that change and attribute from an external program. EventOut sends out a change attribute to an external program. Both events are derived form the node class. With the class initialization, is enable the transmit user input on the Applet to control VRML object to VRML plug-in.

4.2.2 Sending events

Once an instance of EventIn class is obtained, an event can be sent to it. EventIn is abstract class so the cast to appropriate EventIn subclass is needed. This subclass contains a method for sending events of given type. If a VRML scene contains the folowing node:

```
DEF SCALER Transform { ... }
```

the scale can be set to values (1.0, 2.5, 1.0) from Applet like this:

```
Node scaler = browser.getNode("SCALER");
EventInSFVec3f scale =
    (EventInSFVec3f) scaler.getEventIn("set_scale");
float sc[3] = { 1.0, 2.5, 1.0 };
scale.setValue(sc);
```

4.2.3 Reading events

Once an instance of EventOut is obtained, a value of event or exposedField can be read. Since EventOut is abstract class it needs a cast to appropriate subclass which contains method for getting events of given type. Current value of scale field can be read like this:

```
EventOutSFVec3f scale =
    (EventOutSFVec3f)
scaler.getEventOut("scale_changed");
float current_sc[3] = scale.getValue();
```

4.3 Applet-Servlet Communication

In our application, all the user's registration information, the position information got form the VRML EvenOut and EventIn message was encapsulated in a Java object. For example, in the meeting room, when a participator changes the position of the VRML prototype, the new position message was sent to the Servlet. Also, it is our desire for the Servlet to send the Applet an updated message list as a vector of Msg objects. This will allow the Applet to quickly and easily display the changing of the VRML. And the Servlet could add a choice position message to the database. Figure 10 displays the object interaction between the Servlet and the Applet.

4.3.1 Sending object from an Applet to a Servlet

The Applet sends a position object to the Servlet when a participator change the VRML in the COSMO player. In order to send a Java object from the Applet to a Servlet. We should make a Java object to be serializable, its class must implement the java.io.Serializable interface which allows an object to be flattened and saved as a binary file.

The Applet is actually sending a POST method to the Servlet. This client-side code fragment opens a URL connection to the Servlet URL. We inform the Servlet connection that we are sending output data over the connection and receiving input. Methods are also called such that the connection will not use cached versions of the URL. An important call in this code fragment is setRequestProperty(...). This method sets the content-type in the HTTP request header to the MIME-type application/octet-stream. The application/octet-stream MIME-type allows us to send binary data. In our case, the binary data is our serialized Student object.

4.3.2 Sending object from a Servlet to an Applet

The Servlet can return an updated list of changed positions. The updated positions list is returned as a vector of position objects. When the Servlet returns the vector of position objects, there is no need to iterate through the vector and serialize each position object individually. The Servlet can simply serialize the entire vector in one step, since the class java.util.Vector also implements the java.io.Serializable interface. And the Servlet is now capable of storing the position message into the database.

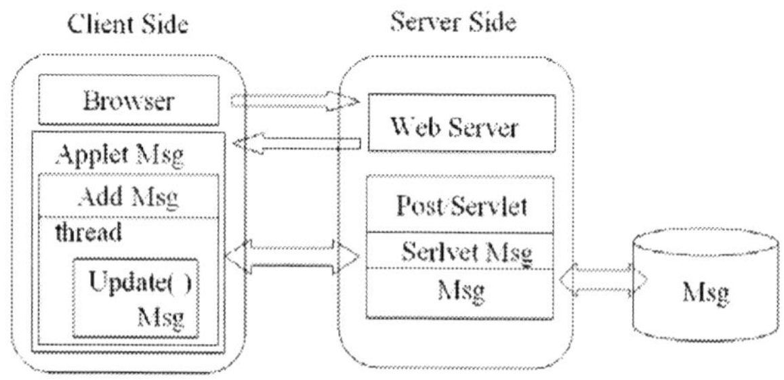

Figure 3 Applet-Servlet communication.

5. Case Studies

This section presents two typical scenarios of collaborative design review.

5.1 Scenario 1: Review Forum

The first typical scenario of collaborative design review is that team members are free to make their comments on a product design in asynchronous mode. Typical steps are as follows:

1. Participators access the URL of the product design review Web site individually.
2. The VRML file of the new product downloads to the different participators' browser. The participator chooses the part form the BOM of the product
3. The review list of this part was displayed, which include the author and the date. The user clicks on the hyper-link of the topic.
4. The contents of the review topic will display. The user can check the VRML by click on the "View VRML" hyperlink on this part.
5. The user can submit response to this topic by click on the "response to the topic" hyperlink.
6. In the response interface the user can define the view point of the part VRML file, input the

response topic and the content, then click on "submit". The review result will store in the review database.

7. The user can also create a new topic on the part by click on the "new topic" hyperlink.

In the new topic interface, the user can define the view point of the part VRML file, input the response topic and the content, then click on "submit". The review result will store in the review database also.

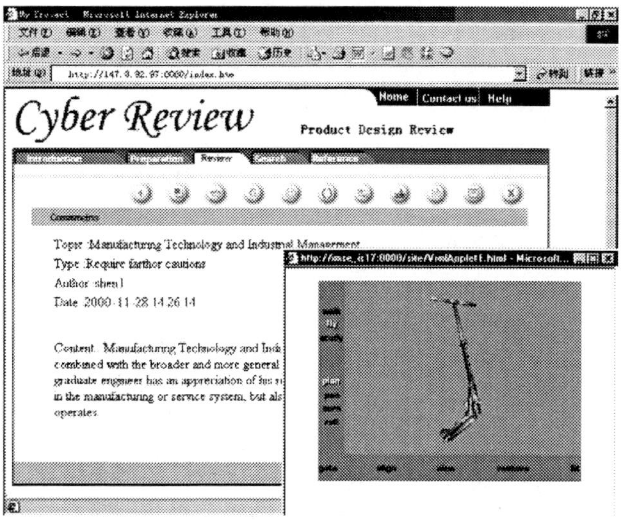

Figure 4. Review forum display.

5.2 Scenario 2: Review Agenda

The second typical scenario of collaborative design review is that team members hold a meeting to make their comments on a product design in the synchronous mode. Typical steps are as follows:

1. The chair enters the system to prepare the relative works of the review, which including select a product, and establish review policies. Then send the message to the relative participators.
2. After get the message, the predicators access the website, and enter into the review system.
3. The relative participators submit the review comments according the agenda.
4. When they meet the problem during the review, they can discuss in the meeting room. Usually, The chair control the view of the product's VRML making it display on all the participator at the same view in the same time. All the participators can view the same content and submit they idea on the topic.
5. At this time, the participator can also apply to create a new view and a new topic. This applying was sent to the chair first. The chair decided if this topic was submitted to discuss. If the administer agree on the new topic, he will control all the user's browser display the same view of the VRML file. And they can discuss on the topic now.
6. At final, all the participators fill in one form in the browser to make a conclusion of this review.

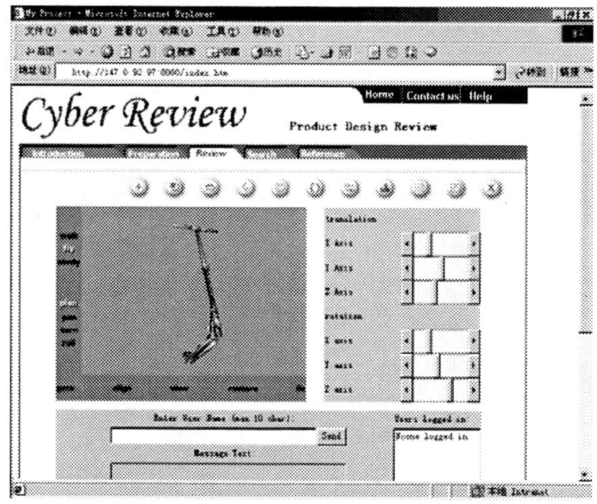

Figure 5. Meeting explorer.

6. Concluding discussion

The Web based collaborative review system is a prototype system whose ultimate goal is providing a platform for the product design review in the external enterprises. According to the design review procedure, the system allow the manager define the review agenda, and organize the review process easily. A VRML based synchronize meeting system was provided to support the product design review process. This PDR system is mainly distinguished by allowing it to run in the Web environment. This is important because the participators undertaking the product design review works can collaborate in a distributed and real-time environment without any additional client program.

The major contribution of this research is that it provides a framework of the design review on the Web. The framework may be used in such wide application areas as mechanical product design review, building construction design review, electronic components design review and software package design review. The usability of such a system would greatly increase as the network bandwidth improves.

Our further researches mainly focus on two aspects: the first is trying to integrate the review knowledge to the review system, making the review can be completed more convenience. The second is trying to integrated the agent technology into the review system, to improve the system agility and intelligent.

7. Acknowledgments

Authors are most grateful to the Hong Kong Research Grant Council and Hong Kong University Committee on Research and Conference Grants for financial supports that made this research possible.

8. References

[1] Ames, A.L., Nadeau, D.R., Moreland, J.L. (1996), *VRML 2.0 SourceBook*, Wiley, John & Sons, Incorporated.

[2] Dhillon, B.S. (1996), *Engineering Design: A Modern Approach*, Richard D. Irwin, a Time Mirror Higher Education Group, Inc. Company.

[3] Hardwich, M., Bolton, R. (1997), "The Industry Virtual Enterprise," *Communications of the ACM*, Vol.40, No.9, 59-60.

[4] Hartley, J.R. (1998), *Concurrent Engineering: Shortening Lead Times, Raising Quality, and Lowering Cost*, Productivity Press.

[5] Hartman, J. (1996), *The VRML 2.0 Handbook: Building Moving World on the Web*, Addison Wesley Longman, Inc.

[6] Hunter, J., Crawford W. (1999), *Java Servlet Programming*, O'Reilly & Associates, Incorporated.

[7] Hutchins, D.C. (1992), *Achieve total quality*, Cambridge: Director Books.

[8] Ichida (1989), *Deizain Rebyu Jireishu*, Productivity Press, Portland, Oregon.

[9] Lea, R., Matsuda, K., Miyashita K. (1996), *Java for 3D and VRML Worlds*, New Riders Publishing.

[10] Schoonmaker, S.J. (1996*), ISO 9001 for Engineering and Designers*, The McGraw-Hill Companies, Inc.

[11] Voigt, E.C. (1996), *Product design review, a method for error-free product development*, Productivity Press, Portland, Oregon.

Web-based Concurrent Design Using a Distributed System and Database Modeling Approach

D. Xue* and Y. Xu
Department of Mechanical and Manufacturing Engineering
University of Calgary
Calgary, Alberta, Canada T2N 1N4

Abstract

In this research, systems, libraries, and databases for modeling different product development life-cycle aspects are distributed at different locations that are linked through the web. Product modeling libraries and databases at different places, described by class features and instance features, are associated by defining their relations. Product modeling systems at different locations for manipulating different product development life-cycle activities are integrated by using the system components at one location as the components at other locations.

1. Introduction

Concurrent design is an approach to incorporate the considerations in down-stream product development life-cycle aspects, such as manufacturing, maintenance, etc., into the early design stage for reducing product development lead time and improving product overall life-cycle quality [1,2]. Because the product development life-cycle activities usually take place at different locations, modeling of these life-cycle activities at different locations and integration of these activities into the same environment using Internet technologies and distributed computing methods are required [3].

Many web-based concurrent design systems have been developed in recent years [3]. Huang et al. studied the web techniques that can be used for developing design-for-X systems [4]. Rezayat discussed the methods to enhance the life-cycle support of products by the Enterprise-Web portal in terms of information and resource sharing and management [5]. Kim et al. developed a system called WebCAD to model product geometry using web browsers and associate CAD models with manufacturing models [6].

Despite of the progress, many problems have to be solved for developing the Internet/Web based concurrent design systems. These problems include:

(1) Although distributed object modeling techniques allow implementation of a system using remote programs as the components [7], relations among these distributed programs are pre-defined. Therefore an open architecture programming environment that allows the programs developed at different locations to be linked in a dynamic manner has to be introduced.

(2) Although some distributed databases have been developed, relations among the databases at different locations are also pre-defined. Therefore a more flexible distributed product database modeling method that allows the relations among the databases at different locations to be associated dynamically has to be developed.

In our previous research, a distributed database and knowledge base modeling system has been developed for concurrent design [8]. In this system, product modeling libraries and databases are modeled by class features and instance features using the schemes introduced in [9,10]. The class features and instance features are preserved at different locations called Internet nodes. When an accessibility relation between two Internet nodes is defined, the class features and instance features at the remote node can then be used at the local node automatically.

The research presented in this paper is a further development of this concurrent design approach by modeling these distributed systems and databases using web technologies, so that these distributed systems and databases can be accessed anywhere in the world.

2. A Web-based Distributed System and Database Modeling Approach

Architecture of the web-based distributed system and database modeling approach is shown in Fig. 1. In this architecture, different product development systems and databases are modeled at different locations represented as Internet nodes such as Design, Manufacturing, and so on. Each Internet node is defined by its address and port number. A node address can be described either by an Internet Protocol (IP) address, such as 136.159.105.85,

* To whom all correspondence should be addressed.
Tel: (403) 220-4168, E-mail: xue@enme.ucalgary.ca

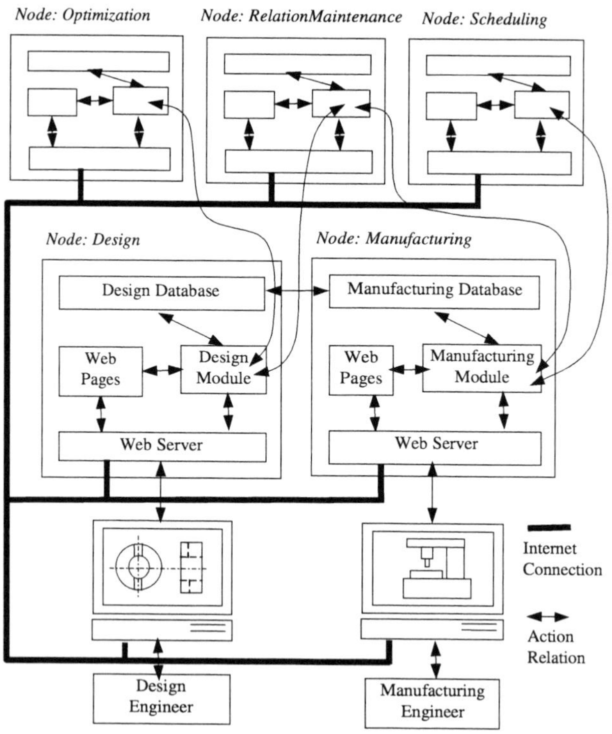

Figure 1. System architecture

or by a structured name supported by Domain Name Server (DNS), such as enme-z.enme.ucalgary.ca. A port number is a 16-bit digit, such as 9876. End users access these nodes through web-browsers.

Each node is modeled by the following components: (1) a web server to link this node with other nodes and users, (2) a number of web pages used as interfaces to access the product modeling module, (3) a product modeling module to get input from users through web-browsers, create/modify product database, and send the output back to users through web-browsers, and (4) a product database modeled by instance features.

This system is different from other web-based concurrent design systems in the following aspects: (1) A class feature at one node can be used as the super-class feature for defining a sub-class feature at another node. (2) A class feature at one node can be selected for creating an instance feature at another node. (3) Functions of a module can be executed from other locations, therefore a module can be implemented using other modules at different locations as the components. (4) Relations among instance features at different locations can be defined.

The system was implemented using HTML, Java, and Perl on MS Internet Explorer 5.0. TGS 3Space Assistant VRML Browser 3.0 and AutoCAD 2000 are used for displaying the design geometry. A snapshot of the implemented system is shown in Fig. 2. In the web-based collaborative concurrent design, usually different product

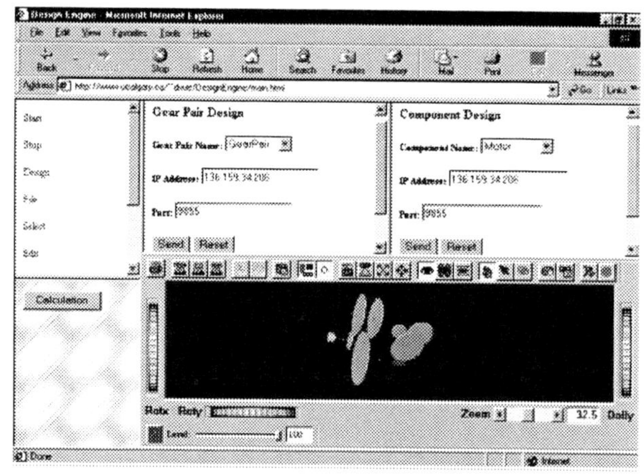

Figure 2. A snapshot of the implemented system

life-cycle modeling modules at different locations are used by different users for modeling different life-cycle databases at these locations. These distributed systems and databases are integrated into the same environment through the web. By evaluating the created design using down-stream life-cycle aspects, the optimal design considering these life-cycle aspects can then be identified.

3. Modeling of Distributed Class Features

3.1. Components of a Class Feature

A class feature is defined by element-features, attributes, qualitative relations among features, and quantitative relations among attributes, as shown in Fig. 3. Element-features are described by variables and their feature types, representing the features used to compose the feature currently being defined. An attribute takes the form of attribute[Feature]. Qualitative relations among features are described by predicates in the form of $(p, x_1, x_2, ..., x_n)$, which are organized in groups. Quantitative relations among attributes are described by functions, each having several input attributes and one output attribute.

A class feature is defined as a sub-class of an existing super-class feature. All descriptions of a super-class feature are inherited by its sub-class features automatically. When an instance feature is generated using a class feature as the template, all the descriptions in that class feature and its super-class features should be added to the database automatically.

3.2. Relations among Distributed Class Features

In this research, a class feature at a remote node can also be used as a super-class feature for defining a sub-class feature at the local node. Fig. 4 shows 6 class features defined in 2 nodes. A class feature at a remote node is described using the format of

<node name>%<class feature name>

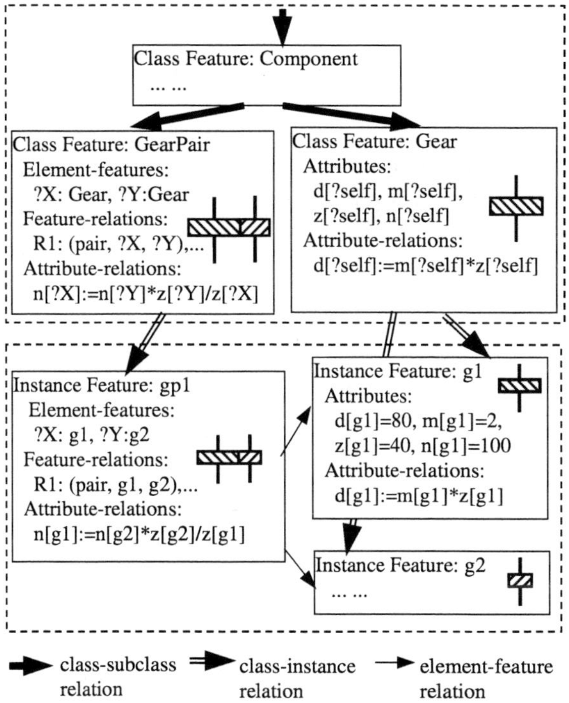

Figure 3. Class and instance features

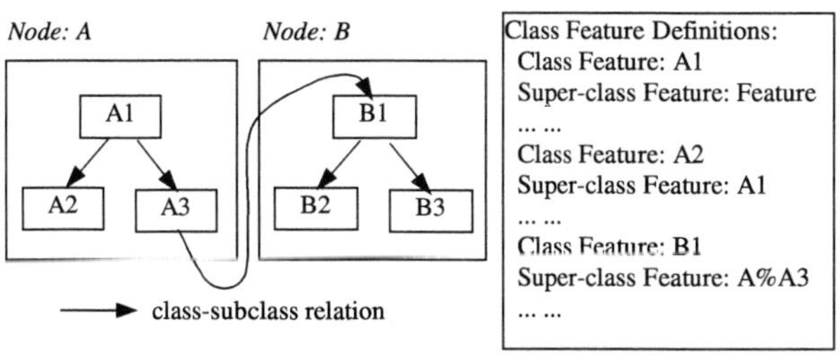

Figure 4. Distributed class features

at the local node.

3.3. Modeling of Class Feature Geometry

Two geometric modeling schemes, Virtual Reality Modeling Language (VRML) and AutoCAD 2000, have been employed for modeling product geometry.

In VRML modeling, a class feature is associated with a VRML coordinate system, described by 3 location attributes, x, y, and z, and 3 orientation attributes, rx, ry, and rz. Feature geometry is defined in this local coordinate system. Description of geometry follows the VRML scheme, except that features and attributes can also be used in these VRML descriptions. Since a feature can be composed of element-features, the VRML coordinate system of an element-feature, related to the VRML coordinate system of the feature being defined, can therefore be described by the six location and orientation attributes. Fig. 5 shows a mechanism modeled by 7 instance features. Each feature is associated with a VRML

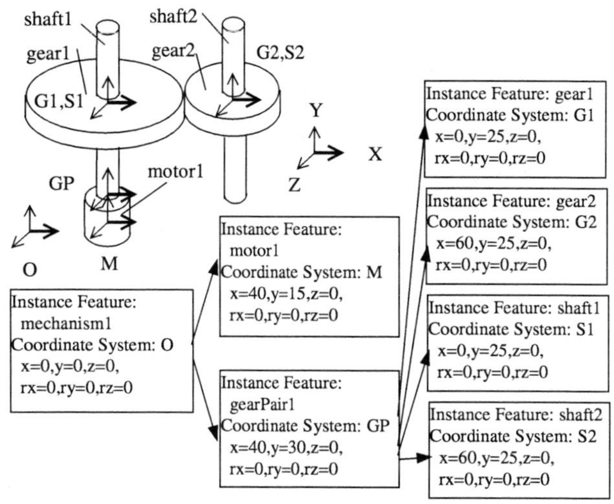

Figure 5. Coordinate systems of instance features

```
Class Feature: Mechanism
Element-features:
  ?m1: Motor, ?gp1: GearPair
Attribute-relations:
  x[?gp1]:=x[?m1], z[?gp1]:=z[?m1], y[?gp1]:=y[?m1]+height[?m1]/2
```

```
Class Feature: Motor
Attributes:
  height[?self], radius[?self]
Geometry:
  VRML Cylinder { height height[?self] radius radius[?self] }
```

Figure 6. VRML descriptions in class features

coordinate system that is defined using the relations with the VRML coordinate system of its super-instance feature. Fig. 6 gives definitions of 2 class features for modeling this mechanism. The VRML geometric descriptions created in instance features will be introduced in Section 4.4.

Modeling of AutoCAD geometry in class features is conducted using a similar method.

4. Modeling of a Product Database at One Location Using Distributed Product Libraries

In this section, modeling of a product database at one location using class features at other locations is discussed. The user interface web page for selecting class features and creating instance features, and the relations of this web page with the system and the class feature libraries at different locations are shown in Fig. 7. A snapshot of this web page displayed in a web-browser during a design process has been given in Fig. 2. This web page is composed of 5 frames: two control panels, two class feature selection panels, and one geometry display panel.

4.1. Selection of Class Features

Each node is associate with a group of web pages explaining the functions of the class features defined at this

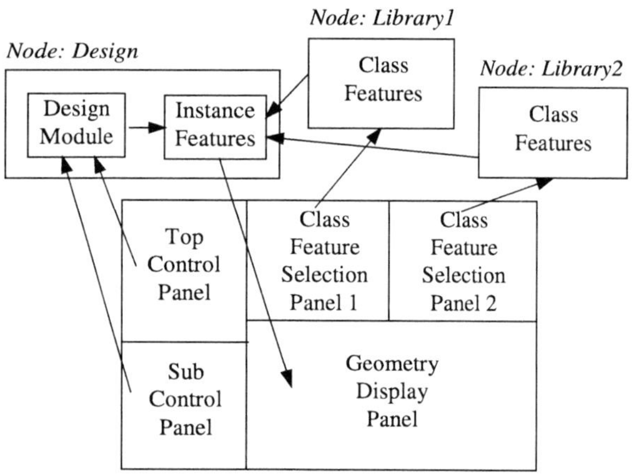

Figure 7. A web interface for selecting class features and creating instance features

node. A user can select a class feature from a node and ask this node to send the class feature definitions to a product modeling node for creating an instance feature. By linking the relevant descriptions of class features distributed at different locations using HTML hyper-link functions, the available class feature libraries in the world can be identified through web links and web-based search.

4.2. Creation of Product Descriptions

A product database is modeled by instance features. Each instance feature can be composed of other instance features. The root instance feature in the database is a built-in instance feature called System. When a class feature is selected for creating an instance feature, this newly created instance feature is then described as an element-feature of the System instance feature.

4.3. Maintenance of Attribute Relations

The quantitative relations among attributes in the generated instance features form a network called *attribute relation network* [10]. An attribute relation network is composed of two types of nodes: attribute nodes and function nodes. Each attribute node is associated with an attribute value. A function node is linked with one or several input attribute nodes and one output attribute node. When an attribute value is changed, the functions that use this attribute as the input node are then activated to update this change to the output attribute nodes. This mechanism is also called *attribute dependency relation maintenance mechanism.*

4.4. Modeling of Product Geometry

The geometric descriptions in instance features are created using the geometric descriptions in their class features as the templates. For VRML-based geometric modeling, an instance feature is associated with a VRML coordinate system. The relation between the local VRML coordinate system and its super-instance feature's VRML coordinate system is described by six built-in attributes: 3 location attributes, x, y, and z, and 3 orientation attributes, rx, ry, and rz. Values of these six attributes are obtained during the design process. The instance features and their VRML geometric descriptions for modeling the mechanism shown in Fig. 5 are given in Table 1.

Table 1. VRML descriptions in instance features

Instance Feature	Attributes	VRML Description

mechanism1
Class Feature:
Mechanism
Element-features:
?m1: motor1
?gp1: gearPair1
	x=0, y=0,
	z=0, rx=0,
	ry=0, rz=0
		TransformSeparator {
		<Insert VRML description for motor1>
		< Insert VRML description for gearPair1>
		}

motor1
Class Feature:
Motor
	height=30,
	radius=10,
	x=40, y=15,
	z=0, rx=0,
	ry=0, rz=0
		TransformSeparator {
		Translation { translation x[motor1] y[motor1] z[motor1]}
		Cylinder { height height[motor1] radius radius[motor1] }
		}

... ...	... ...	... ...

The VRML descriptions in an instance feature are created from the definitions in its class feature using the following format:

```
TransformSeparator {
 Translation { translation <x> <y> <z> }
 Rotation {
    rotation 1.0 0.0 0.0 <rx>
    rotation 0.0 1.0 0.0 <ry>
    rotation 0.0 0.0 1.0 <rz>
 }
 <Copy VRML descriptions in the class feature>
 <Insert VRML descriptions in the element-features>
}
```

Modeling of product geometry using AutoCAD format is conducted in the same manner.

5. Modeling of Distributed Systems and Databases

5.1. Modeling of Distributed Systems

In this research, a module at one Internet node is a piece of Java program, including classes and instances, for accessing product database. At one node, a number of Java classes and instances are allowed to be used as components at other nodes, i.e., data members and functions of these classes and instances can be accessed by programs in the remote nodes. These classes and instances are called open classes and instances.

Class methods and instance methods of open classes and open instances at one node can be executed from another node through the client-server type communication between these two nodes. Execution of a function is described in a form of

<return>=<node>%<object>.<function(arg_1,...,arg_n)>

Where <node> is the node name, <object> is either an open class name or an open instance name, <function(arg_1,...,arg_n)> is the function call with a number of argument values, <return> is the return value variable. Arguments and return value are described by Java primitives (e.g., int, float, etc.), instance features, and attributes. For instance, a class function call is described as:

Design%Gear.diameter(2,z[gear1])

5.2. Modeling of Distributed Databases

In this research, an instance feature at a remote node is described as

<node>%<instance feature>

at the local node. An instance feature at a remote node can be used as an element-feature to define an instance feature. When an instance feature at a remote node is used as an element-feature at the local node, a proxy instance feature is actually created at the local node to communicate with the true instance feature at the remote node. When the components of the remote instance feature need to be accessed (e.g., to obtain attribute values, to change attribute values, etc.), messages are then sent from the proxy instance feature to the true instance feature through the client-server communication channel.

The attributes of instance features distributed at different locations are associated by their relations. Maintenance of the attribute dependency relations within one Internet node is carried out using the attribute relation network [10]. When attribute values of remote instance features are changed, the attribute dependency relation maintenance mechanisms in these remote Internet nodes are then activated to propagate the changes using the attribute relations defined in these nodes. This process is carried out continuously until all the relevant attributes in these nodes are updated.

Modeling of distributed product geometry using the distributed databases is conducted using a similar method as the one introduced in Section 4.4. When an instance feature is composed of element-features modeled at different locations, first the six location and orientation attributes of each instance feature are calculated based upon the relations defined in instance features using the distributed attribute relation maintenance mechanism. When creation of geometry for a remote instance feature is required, a message is then sent to the true instance feature through the proxy instance feature using the client-server communication mechanism.

6. A Case Study Example

In this section, the power sieving mechanism design example given in [8] is used and modified to illustrate the effectiveness of the introduced method. In this design example, the customer requirement is decomposed into two tasks: (1) to design the sieve device, and (2) to design the power driven device. The objective of this work is to identify the optimal design with the minimum production cost from all feasible design candidates.

Five modules in five nodes are used for modeling this concurrent design problem, as shown in Fig. 8. The design module is responsible for modeling design candidates using instance features. The manufacturing module is used for modeling manufacturing processes to produce the created designs.

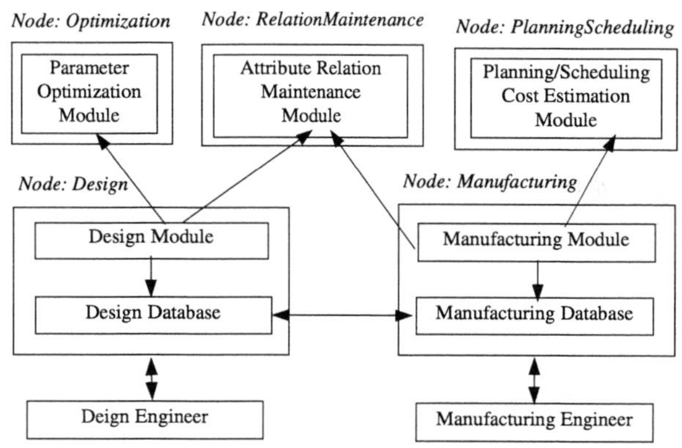

Figure 8. Five modules for the concurrent design

Two alternative power driven mechanisms, a pulley-belt drive mechanism and a gear-pair mechanism, and one sieving mechanism are identified from the class feature libraries for satisfying the two design functions. Two design candidates are created using the class features distributed at different locations and modeled using instance features: (1) candidate 1 using the pulley-belt drive mechanism and the sieve mechanism, and (2) candidate 2 using the gear-pair mechanism and the sieve mechanism. The databases for modeling these two design candidates are shown in Fig. 9.

The design database and manufacturing database are associated by attribute relations. When design attribute values are changed, the production processes and cost for this design are updated automatically using the attribute relations. Identification of the optimal attribute values is conducted using optimization approach. The parameter optimization problem for the design candidate 1 shown in Fig. 9 considering 3 design parameters, diameter d of pulley2, length l of sieve1, and width w of sieve1, is formulated as follows:

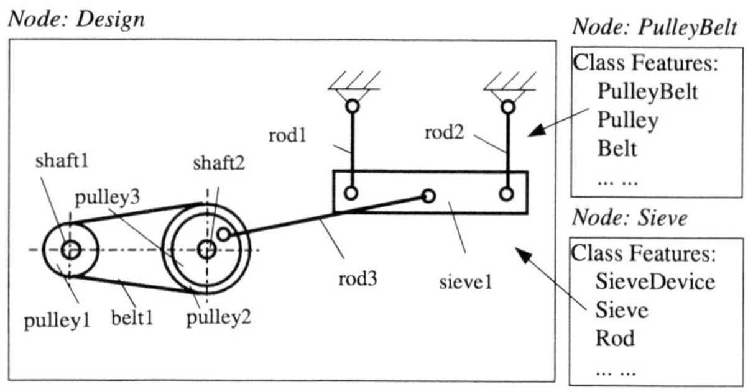

(a) Design candidate 1

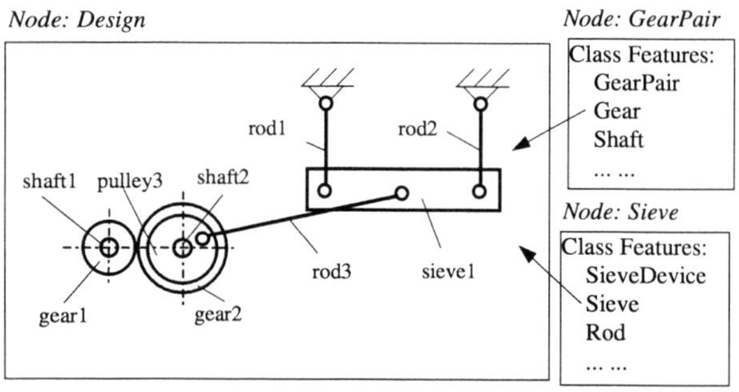

(b) Design candidate 2

Figure 9. Modeling of two design candidates

Min $F(\mathbf{X}) = cost[Manufacturing\%process1]$

$\mathbf{X} = \{d[Design\%pulley2],\ l[Design\%sieve1],\ w[Design\%sieve1]\}$

Subject to:

$l[Design\%sieve1] - 1500 \leq 0$

$w[Design\%sieve1] - 1000 \leq 0$

$400 - w[Design\%sieve1] \leq 0$

$w[Design\%sieve1]*3/2 - l[Design\%sieve1] \leq 0$

$1000 - capacity[Design\%sieve1] \leq 0$

The optimal parameter values and the minimum cost are identified as:

$\mathbf{X}^* = \{164.785,\ 990.789,\ 443.888\}$

$cost[Manufacturing\%process1] = 994.776$

The optimal attribute values for design candidate 2 can be achieved using the similar method. Since the production cost for the gear-pair mechanism is higher than the cost for the pulley-belt mechanism, the design candidate 1 is considered better than the design candidate 2 based upon the evaluation from the manufacturing aspect.

7. Conclusions

The characteristics of this web-based concurrent design method using the distributed system and database modeling approach are summarized as follows.

(1) By modeling generic product libraries and special product data using class features and instance features, both the geometric descriptions and non-geometric descriptions can be modeled in this system. This product modeling function allows different product life-cycle aspects, including design, manufacturing, service, etc., to be modeled in the product database.

(2) Since the program at one location can be implemented using programs at other locations as the components, a dynamic open architecture distributed system framework has been introduced. This framework allows the different product life-cycle modeling systems to be integrated into the same environment.

(3) By defining relations among instance features and their attributes distributed at different locations, the product life-cycle modeling databases at different places are integrated into the same environment. This distributed database modeling function allows the different life-cycle aspects of the same product to be modeled at different locations and the optimal design to be identified considering these relevant life-cycle aspects.

8. Acknowledgements

This research is supported by the Nature Sciences and Engineering Research Council (NSERC) of Canada through Individual Research Grant OGP 0185788. The authors would also like to thank F. Zhang for his valuable literature review on web-based concurrent design.

9. References

[1] Kusiak, A. (ed.), *Concurrent Engineering: Automation, Tools, and Techniques*, John Wiley & Sons, 1993.

[2] Prasad, B., *Concurrent Engineering Fundamentals*: Volume I, Prentice Hall, 1996.

[3] Huhns, M. N. and M. P. Singh, *Readings in Agents*, Morgan Kaufmann Publishers, 1998.

[4] G. Q. Huang, S. W. Lee, and K. L. Mak, "Web-Based Product and Process Data Modeling in Concurrent Design for X," *Robotics and Computer-Integrated Manufacturing*, Vol. 15, 1999, pp. 53-63.

[5] M. Rezayat, "The Enterprise-Web Portal for Life-Cycle Support," *Computer-Aided Design*, Vol. 32, 2000, pp. 85-96.

[6] J. H. Kim, F. C. Wang, C. H. Sequin, and P. K. Wright, "Design for Machining over the Internet," *Proceedings of the 1999 ASME Design Engineering Technical Conferences*, Las Vegas, Nevada, 1999.

[7] McCarty, B. and L. Cassady-Dorion, *Java Distributed Objects*, SAMS, 1999.

[8] F. Zhang, and D. Xue, "Distributed Database and Knowledge Base Modeling for Concurrent Design," *Proceedings of the 2000 ASME Design Engineering Technical Conferences*, Baltimore, Maryland, 2000.

[9] D. Xue and Z. Dong, "Feature Modeling Incorporating Tolerance and Production Process for Concurrent Design," *Concurrent Engineering: Research & Applications*, Vol. 1, No. 2, 1993, pp. 107-116.

[10] D. Xue, S. Yadav, and D. H. Norrie, "Knowledge Base and Database Representation for Intelligent Concurrent Design," *Computer-Aided Design*, Vol. 31, No. 2, 1999, pp. 131-145.

An Optimal Concurrent Design Model Using Distributed Product Development Life-cycle Databases

F. Zhang and D. Xue*

Department of Mechanical and Manufacturing Engineering
University of Calgary
Calgary, Alberta, Canada T2N 1N4

Abstract

An optimal concurrent design approach is introduced in this research using a previously developed distributed product development life-cycle database modeling method. In this approach, product realization process alternatives and relevant activities are modeled at different locations that are connected through the Internet. Genetic Programming (GP) and Particle Swarm Optimization (PSO) are employed for identifying the optimal product realization process alternative and the optimal parameter values of the feasible alternatives, respectively.

1. Introduction

Although many computer-aided concurrent design systems have been developed by integrating different product development life-cycle aspects, such as design, manufacturing, etc., into the same environment, these systems are primarily used on single computers [1]. Since product development life-cycle activities usually take place at different locations, integration of these distributed activities must be conducted.

The recent advances in the technologies of Internet, distributed modeling and computing, and multi-agent systems provide new approaches to associate the distributed product life-cycle modeling systems into integrated environment [2]. For instance, the SHARE project allows designers to gather, organize, and communicate design information over computer networks to establish shared understanding of the design [3]. Sriram and Logcher developed a computer-based design system that provides a shared workspace where multiple designers work in different engineering disciplines [4]. Pahng et al. proposed a framework for modeling and evaluating product designs in a computer network environment [5].

In these distributed systems, relations among modules at different locations are pre-defined. Since in product development, alternative processes at different locations are considered to identify the best one, a concurrent design system, which can dynamically model and change relations among the modules at different locations, has to be developed.

In our previous research, a distributed database and knowledge base modeling system was developed [6]. In this system, databases and knowledge bases at different locations are connected through the Internet. When a relation between two Internet nodes is established, the database and knowledge base at the remote node can then be used at the local node automatically. The research presented in this paper is the development of an optimal concurrent design system based upon the previously developed distributed database and knowledge base modeling method.

2. A Distributed Product Development Life-cycle Database Modeling Method

2.1. A Feature-based Product Modeling Scheme

The product life-cycle aspect models are built using primitives called features [7,8]. Features are described at two different levels, class level and instance level, corresponding to generic product libraries and specific product data respectively, as shown in Fig. 1. Instance features are created using class features as their templates.

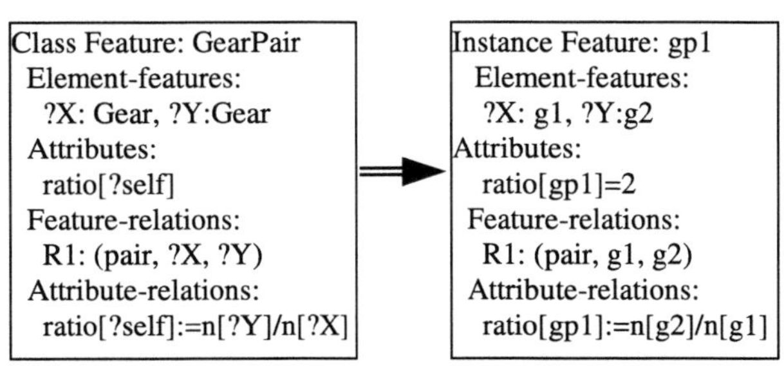

Figure 1. Class and instance features

A class feature is defined by element-features, attributes, qualitative relations among features, and quantitative relations among attributes. A class feature is

* To whom all correspondence should be addressed.
Tel: (403) 220-4168, E-mail: xue@enme.ucalgary.ca

defined as a sub-class of an existing super-class feature. When an instance feature is generated using a class feature as the template, all the descriptions in that class feature and its super-class features should be added to the database automatically.

2.2 Distributed Feature-based Product Life-cycle Modeling

In the distributed feature-based product life-cycle modeling system [6], the different product life-cycle databases are preserved at different locations called Internet nodes, as shown in Fig. 2. The accessibility relations of the databases among these nodes are specified during the concurrent design process. When node A is defined to be able to access node B, the database in node B can then be used in node A automatically.

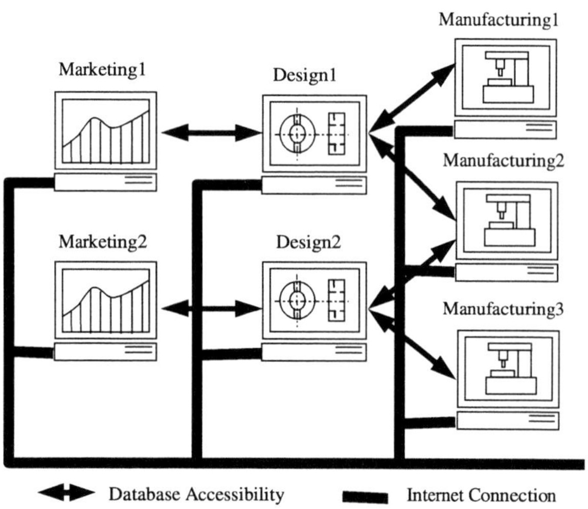

Figure 2. Distributed database modeling

When a node A is defined to be able to access node B, the class features and instance features in node B are then considered as virtual class features and virtual instance features in node A automatically, as shown in Fig. 3. A virtual class feature at a remote node can be used for creating a true instance feature at the local node. A virtual instance feature at a remote node is treated the same as a true instance feature at the local node in product modeling. By defining relations among true instance features and virtual instance features, the distributed databases are then integrated in the same environment.

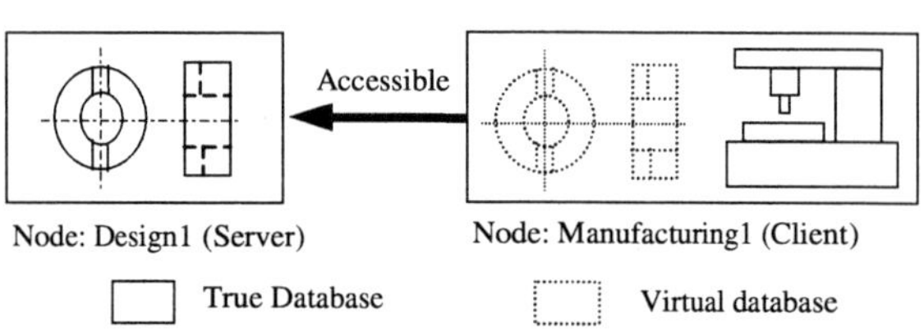

Figure 3. Accessibility relation between two nodes

2.3 Maintenance of Relations among Product Descriptions at Different Locations

Maintenance of the attribute dependency relations within one Internet node is carried out using the attribute relation network [8]. In this network, when an input attribute of a function is changed, the output attribute of this function is then updated automatically. When attribute values of virtual instance features are changed, the attribute dependency relation maintenance mechanisms in these remote Internet nodes are then activated to propagate the changes using the attribute relations defined in these nodes [6]. An example of attribute relations defined in 4 Internet nodes is shown in Fig. 4.

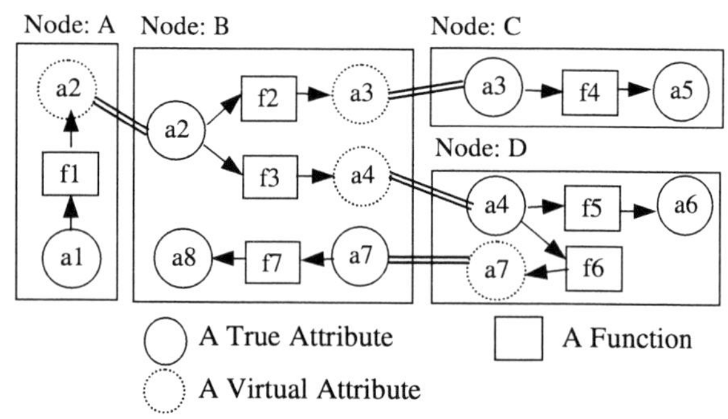

Figure 4. Maintenance of the database relations

3. A Distributed Optimal Concurrent Design Model

3.1. Modeling of Product Realization Process Alternatives and Their Parameters

The alternative realization processes of a product are described by an AND/OR graph, as shown in Fig. 5. In this graph, the different product development life-cycle activities are modeled at different locations represented as Internet nodes. When a product development activity defined at one node requires several subsequent activities that are defined in other nodes, these nodes are then associated using an AND relation. When an activity defined at one node requires only one of these subsequent activities, these nodes are then associated using an OR relation.

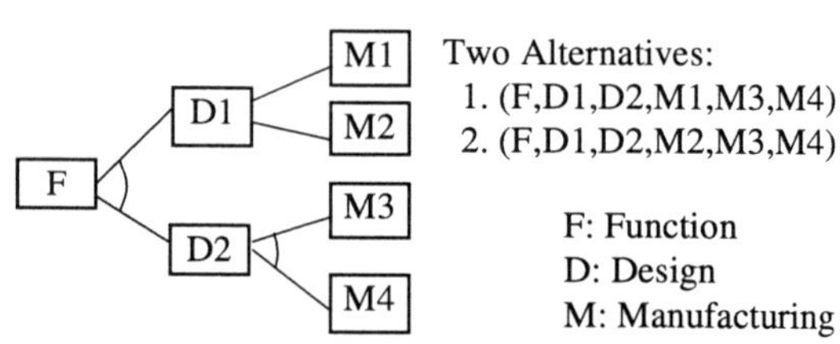

Figure 5. An AND/OR graph for modeling product realization process alternatives

The alternative product realization processes are identified from the AND/OR graph. Each alternative is defined by a number of Internet nodes, in which instance features are used for modeling the product development life-cycle databases. When an Internet node is selected for modeling a product realization process alternative, all the subsequent AND nodes of this node should also be selected for modeling this alternative. When this Internet node has subsequent nodes with an OR relation, only one of these nodes should be selected.

3.2. Identification of the Optimal Product Realization Process Alternative and Its Parameter Values

The optimization is conducted at two different levels: (1) alternative optimization level and (2) attribute optimization level.

A feasible alternative is described by a number of Internet nodes, which are modeled by instance features and their attributes. Since the qualitative descriptions of instance features remain the same for a design alternative, a feasible alternative, $\boldsymbol{P_i}$, can therefore be described by a collection of attributes:

$$\mathrm{P_i} = \{X_{i1}, X_{i2}, \cdots, X_{in_i}\} \quad (1)$$

The optimal attribute values considering one design alternative are obtained using constrained optimization approach:

$$\underset{w.r.t.\ X_{i1},X_{i2},\cdots,X_{in_i}}{Min} \quad f_i(X_{i1}, X_{i2}, \cdots, X_{in_i})$$

subject to:

$$h_{ij}(X_{i1}, X_{i2}, \cdots, X_{in_i}) = 0, \; j = 1,2,\cdots,k_i$$

$$g_{ij}(X_{i1}, X_{i2}, \cdots, X_{in_i}) \le 0, \; j = k_i+1, k_i+2, \cdots, m_i \quad (2)$$

The objective function, $f_i(X_{i1}, X_{i2}, \ldots, X_{in_i})$, is an evaluation measure of the design from a certain product life-cycle perspective, such as manufacturing.

The optimal objective function evaluation measure is described as $f_i(\boldsymbol{P_i^*})$. The optimal alternative is identified from all possible alternatives using

$$\underset{w.r.t.\ \mathbf{P_i^*}}{Min} f_i(\mathbf{P_i^*}) \quad (3)$$

where $\boldsymbol{P_i^*}$ is iterated among the feasible alternatives with the optimal attribute values.

4. Identification of the Optimal Design Alternative Using Genetic Programming

Genetic programming is an extension of genetic algorithm [9]. Compared with genetic algorithm in which a solution is modeled by a fixed-length binary string, a solution in genetic programming can be described by a data structure such as a tree. During the evolution process, elements and lengths of the solutions are changed dynamically. Problem solving using genetic programming is conducted through the following steps:

1. Create members in the initial generation randomly. The number of members in one generation is pre-defined.
2. Evaluate each individual member using the predefined objective function called fitness.
3. Create a new generation using the 3 operations: reproduction, crossover, and mutation.
4. If the pre-defined maximum number of generations is reached or a stop criterion is satisfied, the best individual encountered in the evolution process is selected as the solution. Otherwise go to Step 2.

The AND/OR graph used in the examples to explain the genetic programming method is shown in Fig. 6.

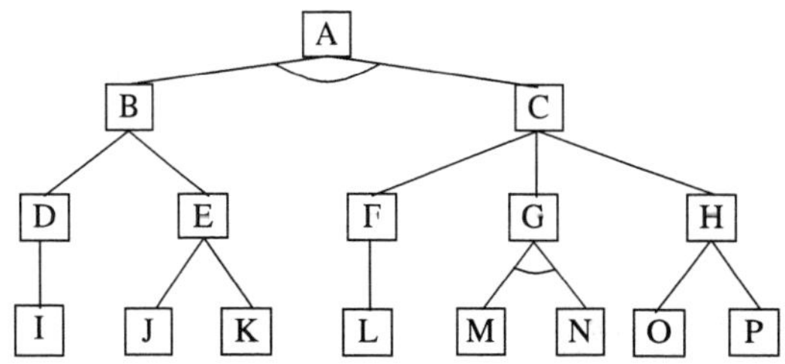

Figure 6. An AND/OR graph for alternative optimization

4.1. Creation of the Initial Generation

The individual members, representing product development alternatives, of the initial generation are randomly created from the AND/OR graph. An alternative is created through the following steps:

1. Identify the root node. Add the root node to an empty list.
2. Pick up a node from the list. This node should be neither a leaf node nor an expanded node. Identify its sub-nodes.
3. If these sub-nodes have an AND relation, add all of them to the list. If these sub-nodes have an OR relation, select one of these nodes randomly and add it to the list.
4. Check whether all the nodes in the list have been expanded. If no unexpanded node can be found, terminate this individual creation process, otherwise go to Step 2.

This process is continued until all the individuals in one generation are created. Creation of a random alternative from the AND/OR graph given in Fig. 6 is shown in Fig. 7.

4.2. Reproduction

In reproduction, the individuals with better fitness measures have better chance to be duplicated in the next generation. The number that each individual should be

Step	Nodes in List	Picked Node	All Sub-Nodes	Relation	Selected Sub-Nodes
1	A	A	B,C	AND	B,C
2	A,B,C	B	D,E	OR	D
3	A,B,C,D	C	F,G,H	OR	G
4	A,B,C,D,G	D	I		I
5	A,B,C,D,G,I	G	M,N	AND	M,N
6	A,B,C,D,G,I,M,N				

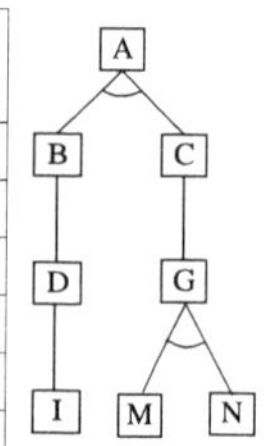

Figure 7. Creation of a random alternative

duplicated in the next generation is calculated based upon the individual's normalized fitness n_i using

$$N_i = \text{int}[m \cdot n_i], \quad i = 1, 2, ..., m \tag{4}$$

where m is the number of individual members in one generation. The function *int[]* converts a real number to its closest integer. The reproduction operation process is illustrated in Fig. 8. The fitness of each individual a_i is calculated from cost C_i using

$$a_i = \frac{1}{1 + C_i}, \quad i = 1, 2, ..., m \tag{5}$$

The normalized fitness n_i is calculated using

$$n_i = \frac{a_i}{\sum_{i=1}^{m} a_i}, \quad i = 1, 2, ..., m \tag{6}$$

No.	Individual	Cost: C_i	Fitness: n_i	Copies
1	Ind1	300	0.18	1
2	Ind2	600	0.09	0
3	Ind3	100	0.54	2
4	Ind4	280	0.19	1

Average Cost: 320

(a) A generation before reproduction

No.	Individual	Cost: C_i
1	Ind1	300
2	Ind3	100
3	Ind3	100
4	Ind4	280

Average Cost: 195

(b) A generation after reproduction

Ind1: (A,B,C,E,G,K,M,N), Ind2: (A,B,C,E,G,J,M,N),
Ind3: (A,B,C,D,H,I,O), Ind4: (A,B,C,E,F,J,L)

Figure 8. Reproduction operation to alternatives

4.3. Crossover

In crossover, two parental individual members are selected from the current generation. A crossover point is selected randomly from each parent. The sub-tree rooted at the selected crossover point is then removed from one parent and replaced with the sub-tree from the other parent. By swapping the two sub-trees, two new individuals are created in the next generation.

For concurrent design problems considered in this research, the individual members of a generation must be valid product development alternatives. Therefore the crossover point, the location at which the crossover operation is conducted, must satisfy the following conditions:

(a) The node at the selected location should be neither a root node nor a leaf node in the AND/OR graph.
(b) The node at the selected location should have an OR relation with other nodes. Another node, that has an OR relation with the selected node, can be found in the other alternative.

If the selected crossover location does not satisfy these conditions, the location is moved one step forward or backward to a new location. The changing direction is determined randomly. If the node at the new location still does not satisfy the crossover conditions, the location is continuously moved in the determined direction until a location that satisfies these conditions is found. If the location has reached to the top (or the bottom) of the node list, the location is changed to the bottom (or the top) of the node list to continue this process. The process of crossover is illustrated in Fig. 9.

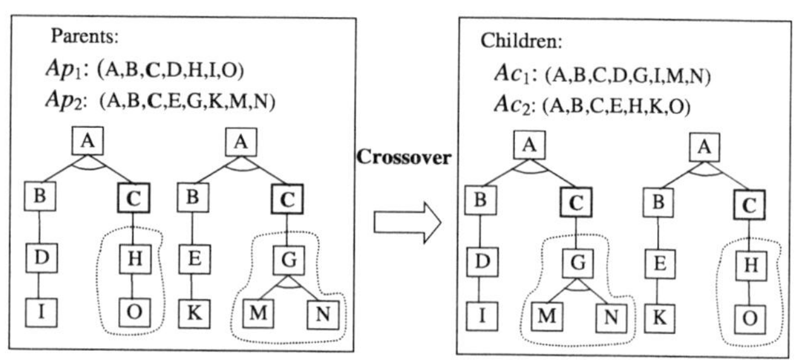

Figure 9: Crossover operation to two alternatives

4.4. Mutation

In mutation, a single individual is randomly selected from the population. The mutation point is also chosen randomly. The sub-tree rooted at that point is replaced with a new sub-tree, which is randomly created.

To ensure the sub-tree rooted at the selected node can be replaced with a different sub-tree in mutation operation, the selected node has to have an OR relation with other nodes. When the selected location doesn't satisfy the mutation condition, the location is then moved forward or backward until a valid location is found. When a mutation location is identified, all the nodes on the sub-tree rooted at the selected node should be deleted. A new sub-tree is created randomly using the selected node as its root node. This mutation operation process is illustrated in Fig. 10.

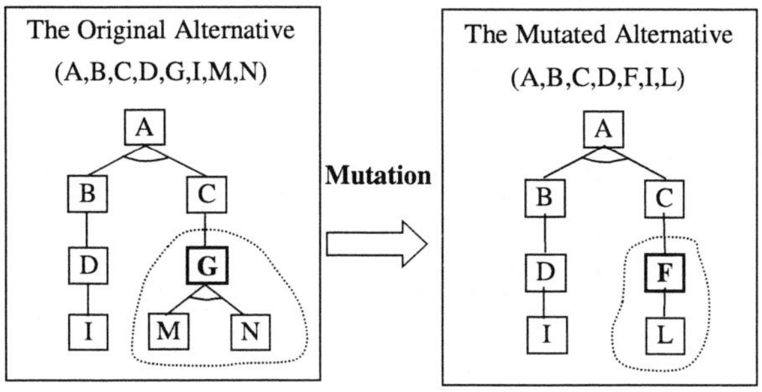

Figure 10. Mutation operation to the alternative

5. Identification of the Optimal Design Parameter Values Using PSO

In this research a population-based optimization method, Particle Swarm Optimization (PSO) [10], is employed for design parameter optimization. PSO method simulates social behavior of organisms, such as bird-flocking and fish-schooling. The idea is that when a bird in a flock tries to find food it uses not only its own knowledge and experience but also its neighbors' (other birds') experiences. PSO can provides high optimization quality, while being simple to be implemented.

In PSO, the position of *i*-th particle is represented as

$$X_i = (x_{i1}, x_{i2}, \ldots, x_{id}), \quad i = 1, 2, \ldots, N \qquad (7)$$

where N is the number of particles in the space and d is the dimension of the space. The best previous position of X_i is recorded and represented as

$$P_i = (p_{i1}, p_{i2}, \ldots, p_{id}), \quad i = 1, 2, \ldots, N \qquad (8)$$

X_g and P_g are used to represent the best particle (the one with the best position) and the best previous position of the best particle respectively. Then we have:

$$X_g = (x_{g1}, x_{g2}, \ldots, x_{gd}) \qquad (9)$$

$$P_g = (p_{g1}, p_{g2}, \ldots, p_{gd}) \qquad (10)$$

During flying, the position change (velocity) for X_i is represented as

$$V_i = (v_{i1}, v_{i2}, \ldots, v_{id}) \qquad (11)$$

The next position of X_i is then:

$$X_i' = X_i + V_i \qquad (12)$$

The velocity V_i is calculated by the following equation:

$$V_i = wV_{ip} + c_1P_{r1}(P_i - X_i) + c_2P_{r2}(P_g - X_i) \qquad (13)$$

where V_{ip} is the previous velocity and w is a weighting number. A greater w value results in strong global search ability and a smaller w value leads to more local search. Coefficients c_1 and c_2 are positive constants. P_{r1} and P_{r2} are two random numbers between 0 and 1.

The distributed attribute relation maintenance mechanism is used for updating attribute value changes during parameter optimization process.

6. System Implementation

The distributed optimal concurrent design system was implemented using Smalltalk, an object oriented programming language [11]. Smalltalk was used due to its excellent user interface environment and large class libraries. A snapshot of the implemented system is shown in Fig. 11.

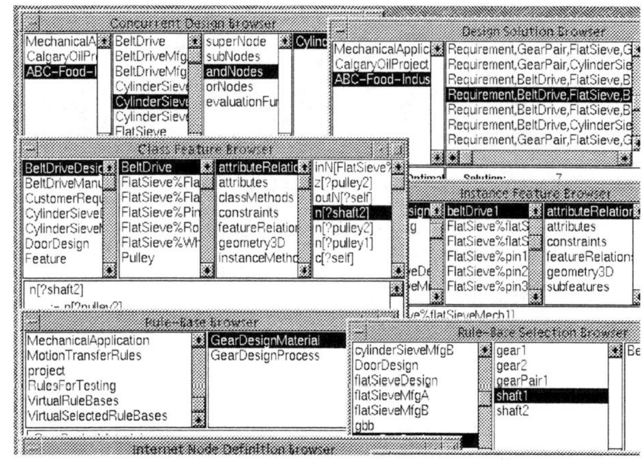

Figure 11. A snapshot of the implemented system

7. A Case Study Example

In this section, the power sieving mechanism design example given in [6] is modified to show how the distributed optimal concurrent design method is used. This design problem is composed of two tasks: (1) the design of a power transmission mechanism, and (2) the design of a sieving mechanism. Two alternative power transmission mechanisms, a belt drive mechanism and a gear pair mechanism, and two alternative sieving mechanisms, a flat sieving mechanism and a cylinder sieving mechanism, are considered in this example. These mechanisms are modeled at different Internet nodes as shown in Fig. 12.

The objective of this design is to identify the design alternative that requires the minimum manufacturing cost, while satisfying the design requirements. To reach this objective, the manufacturing aspects for the design models shown in Fig. 12 must be considered. The relevant manufacturing processes related to the alternative mechanisms are modeled by Internet nodes, as shown in Fig. 13. The design requirements are modeled in an Internet node called Requirement. The node R1 and node R2 in this graph are not real Internet nodes. They are used to form the AND/OR relations. These nodes are called pseudo nodes in this research.

From the AND/OR graph, different design alternatives can be created using genetic programming for identifying the optimal one. These alternatives are evaluated through

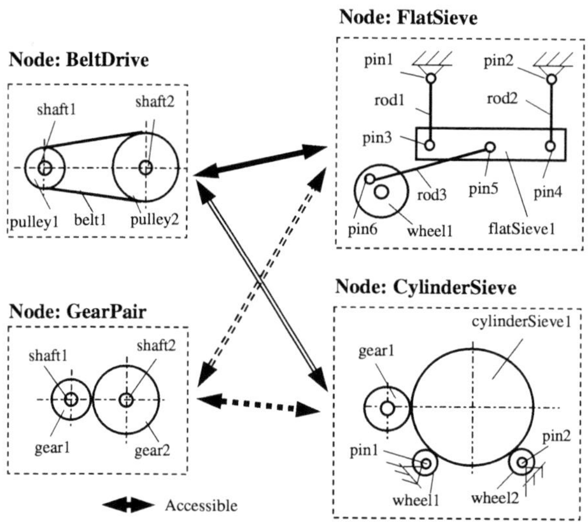

Figure 12. Possible design alternatives

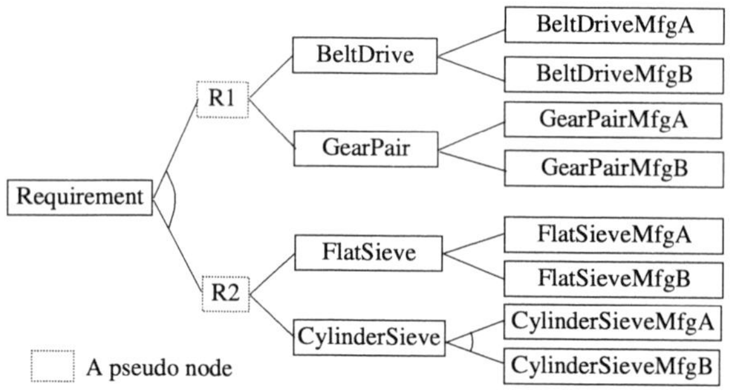

Figure 13. The AND/OR graph for the example

parameter optimization using Particle Swarm Optimization method.

The parameter optimization problem for the alternative (Requirement, BeltDrive, FlatSieve, BeltDriveMfgA, FlatSieveMfgA), considering 3 design parameters, diameter d of pulley2 at node BeltDrive, length l of flatSieve1 at node FlatSieve, and width w of flatSieve1 at node FlatSieve, is formulated as follows:

Min *F(X) = cost[Requirement%customerRequirement1]*
  X *= {d[BeltDrive%pulley2], l[FlatSieve%flatSieve1], w[FlatSieve%flatSieve1]}*
Subject to:
l[FlatSieve%flatSieve1]-1500[0
w[FlatSieve%flatSieve1]-1000[0
400-w[FlatSieve%flatSieve1][0
*w[FlatSieve%flatSieve1]*3/2- l[FlatSieve%flatSieve1][0*
1000-capacity[Requirement%customerRequirement1][0

The optimal parameter values and the minimum cost are identified as:

X = {164.785, 990.789, 443.888}*
cost[Requirement%customerRequirement1] = 994.776

The optimal alternative is identified using Genetic Programming. The alternative (Requirement, BeltDrive, FlatSieve, BeltDriveMfgA, FlatSieveMfgA) is identified as the optimal alternative for developing the power-sieving system considering manufacturing cost.

8. Conclusions

The advantages of this approach are summarized as follows. (1) By modeling the relations among the different product development life-cycle databases at different locations, all the possible product realization process alternatives are integrated into the same environment. (2) Genetic Programming (GP) is effective for modeling the alternative product realization processes and identifying the optimal alternative. (3) Particle Swarm Optimization (PSO) is an effective optimization method with high quality, while being simple to be implemented.

9. Acknowledgements

This research is supported by the Nature Sciences and Engineering Research Council (NSERC) of Canada through Individual Research Grant OGP 0185788.

10. References

[1] Kusiak, A. (ed.), *Concurrent Engineering: Automation, Tools, and Techniques*, John Wiley & Sons, 1993.
[2] Huhns, M. N. and M. P. Singh, *Readings in Agents*, Morgan Kaufmann Publishers, 1998.
[3] T. G. Cutkosky, M. R. Tenenbaum, and J. Glicksman, "SHARE: A Methodology and Environment for Collaborative Product Development," *Proceedings of IEEE Infrastructure for Collaborative Enterprise,* Morgantown, 1993, pp. 33-41.
[4] D. Sriram and R. Logcher, "The MIT DICE Project," *IEEE Computer,* Vol. 26, No. 1, 1993, pp. 64-65.
[5] F. Pahng, N. Senin, and D. Wallace, "Distribution Modeling and Evaluation of Product Design Problems," *Computer-Aided Design,* Vol. 30, No. 6, 1998, pp. 411-423.
[6] F. Zhang and D. Xue, "Distributed Database and Knowledge Base Modeling for Concurrent Design," *Proceedings of the 2000 ASME Design Engineering Technical Conferences*, Baltimore, Maryland, 2000.
[7] D. Xue and Z. Dong, "Feature Modeling Incorporating Tolerance and Production Process for Concurrent Design," *Concurrent Engineering: Research & Applications*, Vol. 1, No. 2, 1993, pp. 107-116.
[8] D. Xue, S. Yadav, and D. H. Norrie, "Knowledge Base and Database Representation for Intelligent Concurrent Design," *Computer-Aided Design*, Vol. 31, No. 2, 1999, pp. 131-145.
[9] Koza, J. R., *Genetic Programming: On the Programming of Computers by Means of Natural Selection*, MIT Press, 1992.
[10] J. Kennedy and R. Eberhart, "Particle Swarm Optimization," *Proceedings of IEEE International Conference on Neural Networks*, Perth, Australia, 1995.
[11] Goldberg, A. and D. Robson, *Smalltalk-80: The Language and Its Implementation*, Addison-Wesley, 1983.

A Data-Centric Distributed Framework for MDO Management

Bin Chen, DaoJun Liu, Babak Mahdavi, Qun Zhou
Djamel Bouhemhem, Amadou Ndiaye, François Guibault
Benoît Ozell, Dominique Pelletier, Jean-Yves Trépanier
CERCA (Centre de Recherche en Calcul Appliqué) *
firstname.lastname@cerca.umontreal.ca

Abstract

This paper presents the overall architecture of the VADOR application framework. The purpose of the VADOR framework is to enable the seamless integration of commercial and in-house analysis applications in a heterogeneous, distributed computing environment, and to allow the deployment of automatic design optimization algorithms based on the framework. A multi-tiered client-server architecture has been devised, which comprises a client GUI for interactive data definition and execution launching, separate data and execution servers, and autonomous remotely executable application wrappers.

1. Introduction

Computational-based engineering design-and-analysis now constitutes one of the main building blocks within engineering disciplines. In mechanical engineering, and particularly in the aerospace industry, computational fluid dynamics (CFD) and computational structural dynamics (CSD) are now part of the daily activities of engineers. Analysis and design specialists often rely on additional "in house" programs taylored to the specific requirements of their tasks. The computational-based design environment then becomes highly complex and targetted to each application. Even within a single organisation, each department has its own set of computational tools and programs which have little in common with those from other departments.

As a result of this complexity, the application of multidisciplinary analysis and optimisation (MDO) practices [6] faces a number of significant challenges, including collaboration and data sharing. In this context, there is a need for a software infrastructure which will facilitate data sharing. This is the subject of the VADOR (Virtual Airplane Design Optimisation frameRk) project[3]. The specific objectives of the VADOR project are:

1. To develop a state-of-the-art software framework capable of supporting an MDO paradigm in a collaborative design environment.
2. To implement, within the framework, data management capabilities to closely follow the design data used and shared by the design team.

2. Component-Based Framework

The object-oriented methodology is used in the development of the framework. The implementation is done using the JAVA computer language. Our choice is motivated primarily by its portability, its internet capabilities, as well as its high-level language capabilities and tools. The object model and global system architechture will be described in this paper.

An effective framework enabling MDO [5] should provide users with a flexible and configurable data model, and also, more than adequately satisfy the evolving requirements of engineers using computational-based analysis-and-design programs. Such a system must provide capabilities for the automation and integration of various processes used by engineers. It must also support and promote collaboration and data sharing.

In the VADOR framework, collaboration and data-sharing are enabled through the usage of the concept of Components and through data standardization. Two kinds of components [3] are defined: *DataComponents* and *StrategyComponents*. *DataComponents* encapsulate the design-and-analysis data while *StrategyComponents* encapsulate the design-and-analysis workflow. In order to provide data management capabilities, the components have a set of attributes (such as owner, access permission, history, comments, and status). In order to promote standardization, user defined data types are introduced, which are then used for validation and documentation.

DataComponents and *StrategyComponents* are hierarchically arranged and they may contain other components. The leaves of this composition are called the atomic

* www.cerca.umontreal.ca

components. The atomic *DataComponents* encapsulate (exactly) one data file, while the atomic *StrategyComponents* encapsulate one executable or script. *DataComponents* and *StrategyComponents* each have a *type*, which serves as a mechanism for data standardization and validation. These characteristics provide the flexibility required to represent a variety of design-and-analysis processes starting from the execution of a simple utility program to a complex process with different stages, involving several distributed programs.

The formal association between a *DataComponent* and a *StrategyComponent* (which is able to fill an encapsulated data file) provides a link between the behaviours and the services they provide. In addition it provides a tool for documenting the methodologies in use.

2.1. *DataComponent*

DataComponents are objects that encapsulate design-and-analysis data which are usually contained in data files. They comprise an appropriate set of attributes required for data management. A *DataComponent* can be atomic (i.e. only encapsulate one data file) or composite (i.e. encapsulate other composite or atomic *DataComponents*). A *DataComponent* requires a *StrategyComponent* to fill its encapsulated data file. Note that the data files are only known by their URL.

2.2. *StrategyComponent*

StrategyComponents encapsulate the design-and-analysis methodologies or processes. They represent the basic methods and the data flow required to transform data in a given process. *StrategyComponents* are constructed using the basic elements of a procedural language including: the instantiation, test, and while constructs. *StrategyComponents* can be sequential or parallel and an *IfStrategy* or a *WhileStrategy* enables the description of complex conditional processes.

StrategyComponents allow one to define methods used to create *DataComponents*. A *StrategyComponent* has a *type*, which indicates the type of *DataComponent* that the strategy can create. *StrategyComponents* also enable the data flow standardization and documentation. The Atomic *StrategyComponents* encapsulate programs which can create the datafiles encapsulated in the *DataComponents*. The programs are usually executable legacy programs to be executed on a specific machine on the network. Note that the execution time required to run these programs can vary from a few milli-seconds to many days, depending on the specific engineering analysis to perform.

3. VADOR Architecture

Figure 1 illustrates the VADOR architecture which is composed of: a VADOR GUI, a VADOR Librarian, a VADOR Executive, a Database Management System (DBMS) and a set of CPU Servers. Communication between servers is socket-based, and the concept of object serialization is used to transfer components across the system.

The VADOR Architecture is based on *peer-to-peer and Client-Server* concepts. These concepts are very useful for the design of distributed architecture, because they spare the user from having to be aware of the complexity and details of the system. Users interact with the whole system through a GUI. Process integration and automation in this heterogeneous network is achieved through the use of Java RMI [4] in servers for *StrategyComponent* execution.

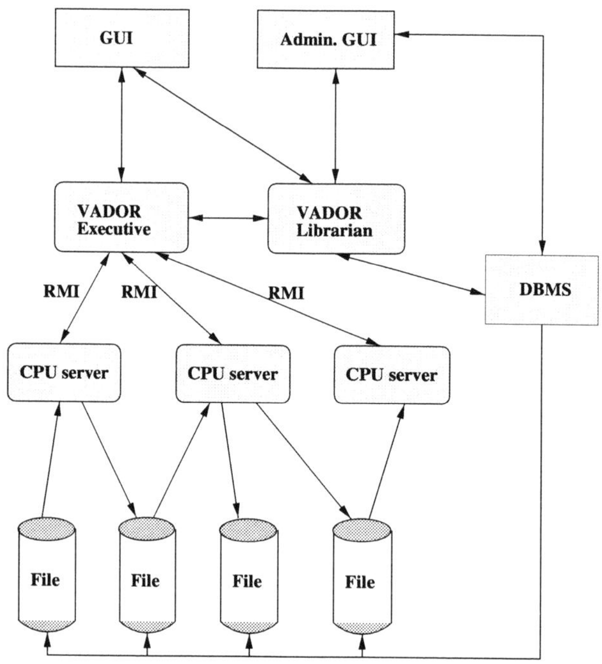

Figure 1. VADOR architecture

3.1. VADOR GUI

Most users interact with the VADOR framework through its GUI. The VADOR GUI (see fig 2) is a Java program which runs on the user's machine. The GUI is used as an interface between engineers and the VADOR Executive server, and Librarian server. Because of Java's portability, all users will run the same program, independently of their platform.

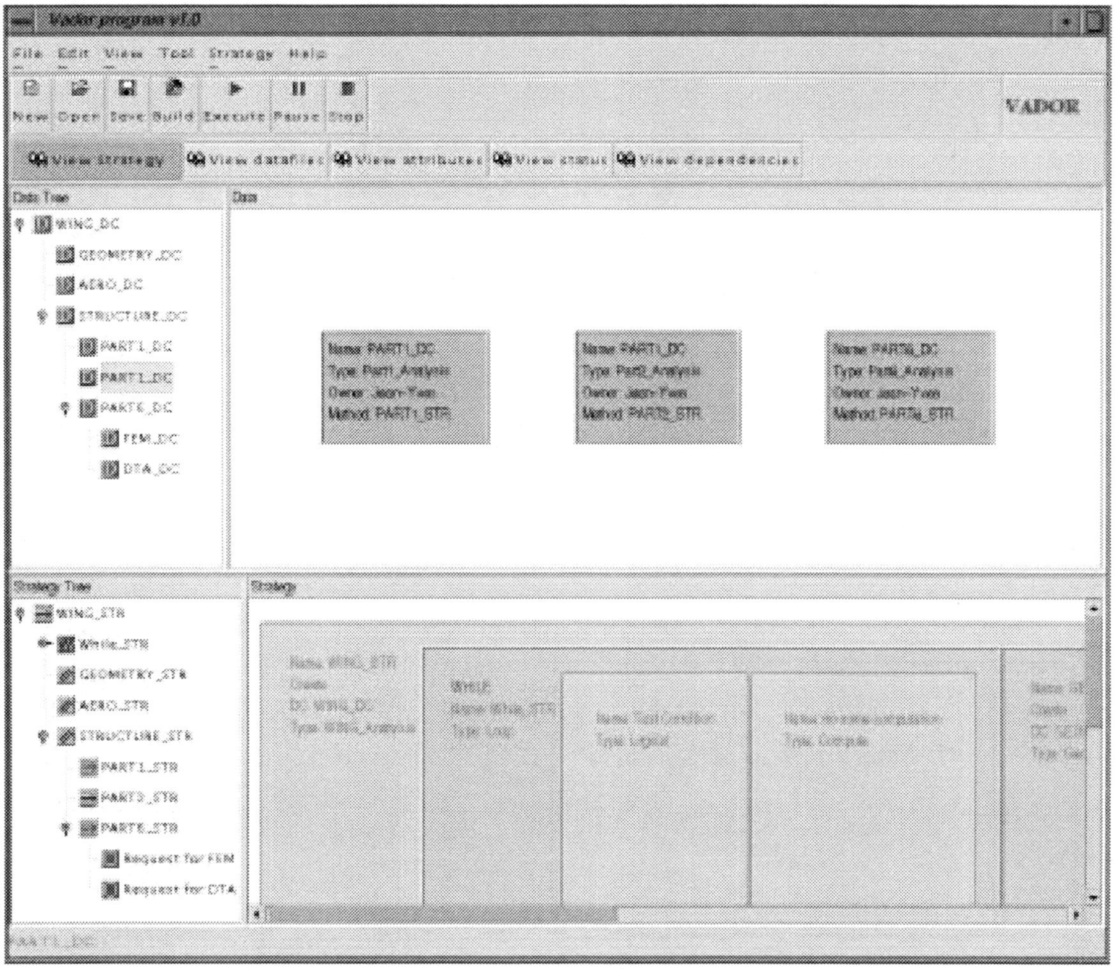

Figure 2. A view of VADOR GUI

The upper half of the main window provides a graphical representation of *DataComponents*. A tree structure is used to display the composition of a hierarchic *DataComponent*. Individual rectangles identify individual *DataComponents*, enabling selection and queries to be made on their contents. Utility programs can be invoked from the menu in order to edit or visualise the contents of the encapsulated data files.

The lower half of the main window provides a graphical representation of the *StrategyComponents*. A tree structure is used to display the composition of a hierarchic *StrategyComponent*. The complete *StrategyComponent* is represented by recursively including Components within each other.

The GUI also provides the user with tools to contruct new *DataComponents type* and to define new *StrategyComponents*.

3.2. VADOR Librarian

The VADOR Librarian decouples the VADOR API from the underlying Database server and FileServers capabilities. The VADOR Librarian is a Java server program handling Components. The Librarian provides capabilities including an interface with a suitable database to store the information about each Component. It is important to emphasize that the present architectural design supports the separation of basic data (potentially a rather large amount), usually contained in files, from descriptive information. Typically, only the descriptive information, or metadata, will be stored in the database. The large datasets will be stored on FileServers and will usually reside where they have been created.

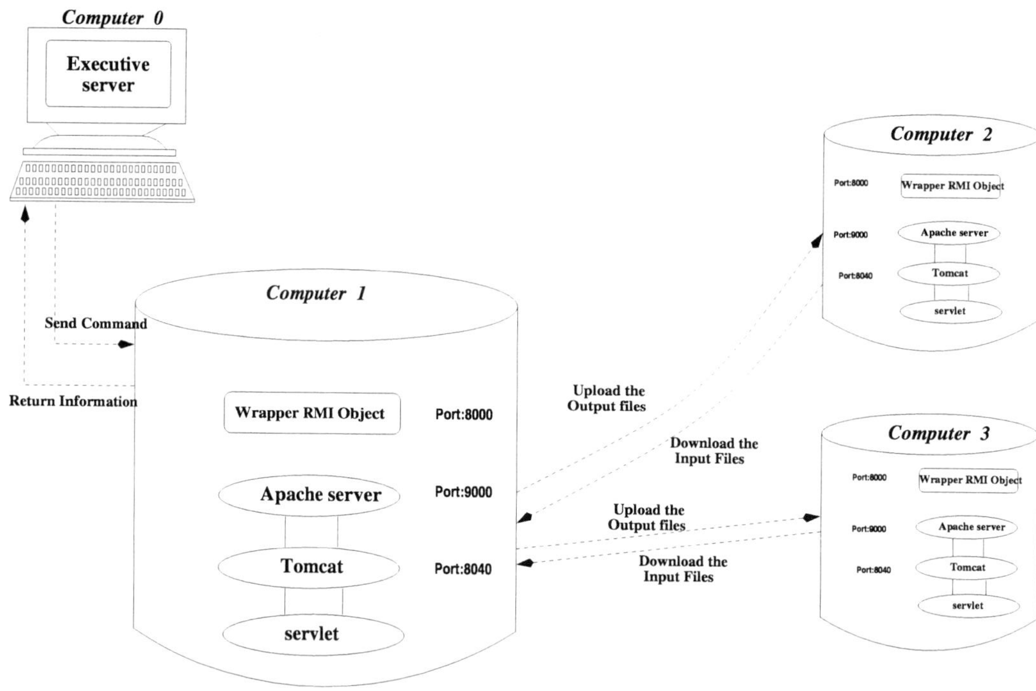

Figure 3. Global architecture of wrappers

3.3. VADOR Executive

The VADOR Executive is a Java server program that manages the execution of *StrategyComponents* to create *DataComponents*. It answers the needs of process automation in a heterogeneous distributed environment. The Executives are multi-threaded (i.e. are capable of handling multiple tasks) and can exist in any number. The Executive interacts with the Librarian to retrieve *DataComponents* to be created and to update the database contents. Interaction of the VADOR Executive with the whole system can be described as follows:

- It receives simple requests from the GUI to create *DataComponents*;
- It communicates with the Librarian to retrieve *DataComponents*;
- It communicates with the CPU Servers on different hosts through Java/RMI or CORBA communication mechanisms to run analysis codes;
- When execution of the *StrategyComponent* is completed, it notifies the Librarian to update the status of the components in the database, and eventually the user, via the GUI or e-mail.

3.4. CPU Servers

The CPU Servers can be installed on any suitable machine. These servers will typically run analysis codes summoned by the Executive Server, using Java/RMI or CORBA services, to fill or create the data files encapsulated in the *DataComponents*. A CPU Server is a Java program that includes a method to execute a program or a script with a list of I/O files. It also contains a *file transferer* that can move input files for the execution and movement of the output to required locations. The file transferer uses standard web servers for downloading and uploading files. The global architecture of the wrappers is illustrated on Figure

3.5. VADOR Database

In the VADOR framework, a Database Management System (DBMS) is used to store, maintain and provide access information for Components. The DBMS used in the framework is the MySQL DBMS [8] which stores component information using a relational model - the most popular data model on which DBMSs are built. The

To access the database, *Java DataBase Connectivity*(JDBC) [9] is used. In our architecture, the Java application, namely the Librarian, requires a JDBC driver that

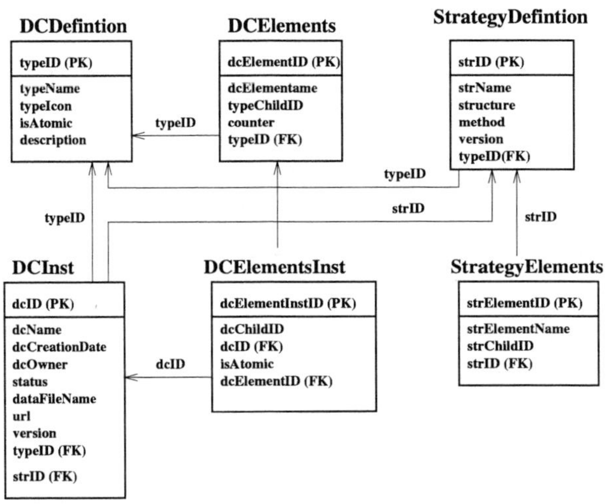

Figure 4. A simplified view of the VADOR relational database model

List of tables in VADOR database
DataComponent **definition**
DataComponent **instances**
DataComponent **Elements**
DataComponent **Elements instances**
DataComponent **Comments**
DataComponent **History**
DataComponent **Access Permission**
StrategyComponent **definition**
StrategyComponent **Elements**
Input *DataComponent* definition
External InputDC Instances
StrategyComponent **method Host**
StrategyComponent **Access Permission**
UsersList
Utilities

Table 1. Non-exhaustive list of tables

can communicate with the particular database management system being accessed (MySQL server in our case). A user's SQL statements are delivered to the database, and the results of those statements are sent back to the user. The database can be located on another machine to which the Librarian is connected via a network.

4. Examples

To illustrate the object model used in the VADOR framework, a simple example will be used from a Computational-Fluid-Dynamics (CFD) application. Fig 5 illustrates a flow chart for this example. Starting from a *geometry*, a first program is used to define some parameters for mesh generation. This results in a file containing the *mesh parameters*. The *geometry* and the *mesh parameters* are then fed to a mesh generator, which will produce a file containing the *mesh*. A different GUI reads in the *mesh* file and allows for imposing boundary conditions for CFD. It produces a file containing *solver parameters*. The CFD solver reads in the *mesh* and the *solver parameters* files and produces a *CFD Solution* file.

In the context of our data model, five atomic *DataComponents* TYPES will be defined: *GEOMETRY*, *CFDMESHPARAM*, *CFDMESH*, *CFDSOLVERPARAM*, *CFDSOLUTION*. The composite *DataComponent*, *CFDANALYSIS*, will contain four atomic *DataComponents* of *type*: *CFDMESHPARAM*, *CFDMESH*, *CFDSOLVERPARAM*, *CFDSOLUTION*.

The design strategy is also represented in Fig 5. It is a sequential *StrategyComponent*, with no conditional branching, calling in sequence four programs which are labeled: *Mesher GUI*, *Mesher*, *Solver GUI* and *CFD Solver*. In the context of our data model, four atomic *StrategyComponent* will be defined: *MESHGUI*, *MESHER*, *SOLVERGUI*, *CFDSOLVER*. The composite *StrategyComponent*, *CFDANALYSIS*, will contain the four atomic *StrategyComponents* in sequential order. The *CFDANALYSIS StrategyComponent* requires as input a *DataComponent* of *type GEOMETRY* and produces a *DataComponent* of *type CFDANALYSIS*.

5. Conclusion

The object model and framework architecture for an MDO framework have been described in the present paper. The framework has been designed to suits the needs of a large engineering department with hundred of engineers working concurrently on an engineering design project. The framework is highly flexible and configurable and is expected to be adaptable to the needs of every engineer. The framework enforces standardization of the data and methods, which will result in a self-documenting system. The attributes tagged to the data and strategies will provide an easy access to all critical information concerning the design.

References

[1] I. Kroo. "Computation-based design – a white paper". http://aero.stanford.edu/ComputationalDesign.html, 1996.

[2] S. Allwright. "Multidisciplinary design, analysis and optimisation of aerospace vehicles - the MDO project". In *Royal Aeronautical Society MDO conference*, 1998.

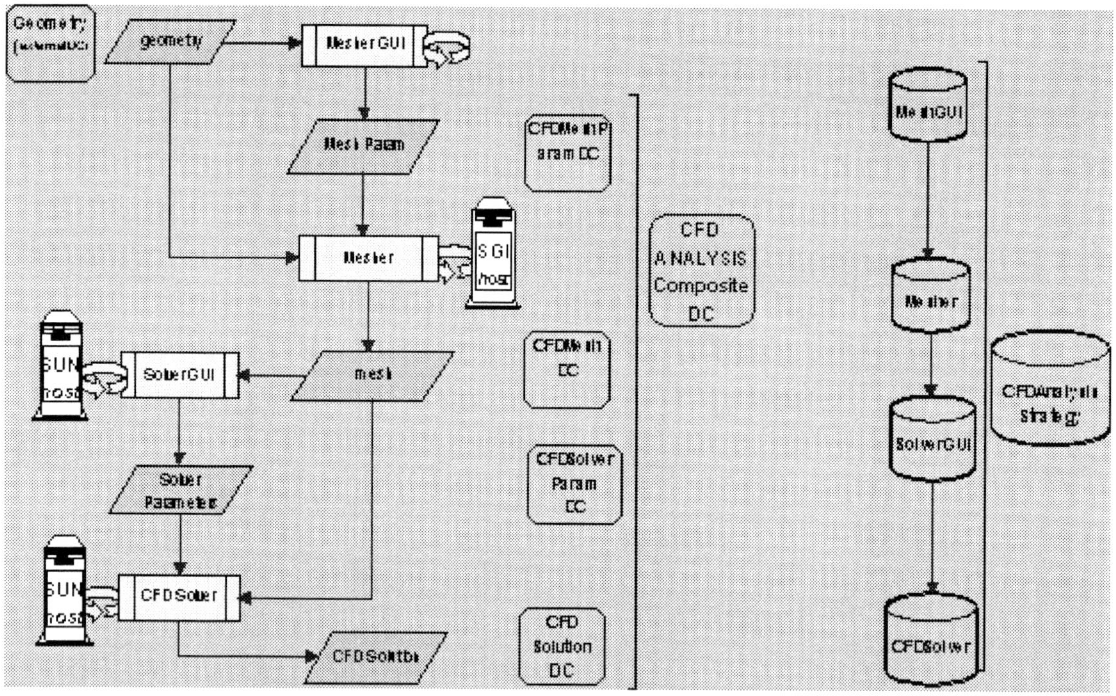

Figure 5. An example of CFD Analysis composite *DataComponent* with the *StrategyComponent* defined to create this CFD Analysis

[3] A. Ndiaye, J.-Y. Trépanier, F. Guibault, B. Ozell, and B. Mahdavi. "Database requirements for an mdo software framework". In *CFD2K, 8e conférence annuelle de la Société canadienne de CFD*, Montréal(QC), Juin 2000.

[4] A. Alzubbi, A. Ndiaye, B. Mahdavi, F. Guibault, B. Ozell, and J.-Y. Trépanier. "On the use of JAVA and RMI in the development of a computer framework for MDO" In *8th AIAA/USAF/NASA/ISSMO Symposium on Multidisciplinary Analysis and Optimization*, Long Beach(CA), September 2000.

[5] 0.A. Salas and J.C. Townsend. "Framework requirements for MDO application development". In *AIAA-98-4740*, 1998.

[6] J. Sobieszczansk-Sobieski. "Multidisciplinary design optimization MDO methods: their synergy with computer technology in the design process". *The Aeronautical Journal*, 1999.

[7] R.P. Townsend, J.C. Weston and T.M Eidson. "A programming environment for distributed complex computing. An overview of the framework for interdisciplinary design optimization (fido) project". *Technical report, NASA TM 109058*, 1993.

[8] P. Du Bois. "MySQL". *New Riders Publishers*, 1999.

[9] G. Reese "JDBC & JAVA, Database Programming, Java 1.1". *O' Reilly*, 1997.

Part IV: Agents and Multi-Agent Systems

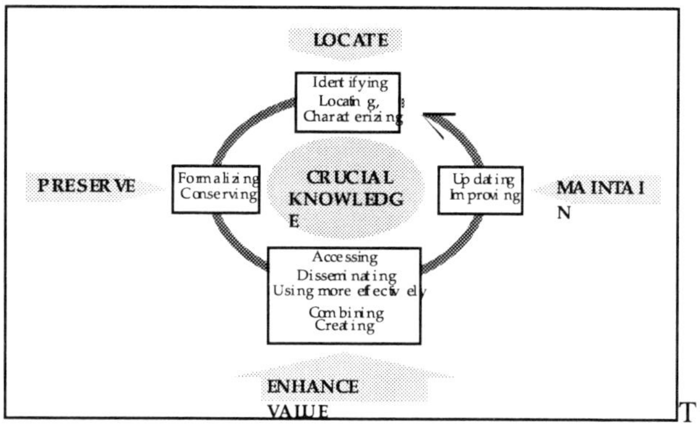

Figure 1. Knowledge capitalization cycle.

Two technologies are good candidates for developing such systems, namely groupware or agent systems. Both were shown to be able to support the collaborative process. We first undertook a project for testing this approach, developing a system based on a groupware approach. The result will be shortly described in Section 3. However, from the experience we had gained in developing multi-agent projects, we found the result not entirely satisfactory. Although another system could be developed around cognitive agents and personal assistants using the approach discussed in Section 4, we feel that it would be better to combine the groupware and the agent approaches to offer agent-supported portal systems with interesting characteristics.

3. A Groupware Portal for Complex R&D Projects

We first developed a project portal based on Lotus Notes™ to be tested in the automotive industry. The objective of giving a portal to each member of a project was twofold: (i) to reduce the time needed to access relevant information; and (ii) to collect knowledge and to organize the information exchanges.

The main requirements from the user's side were the following:

-the system must cover as much of the R&D activity as possible,
-the system must save time in an obvious manner
-the system must be reliable and persistent
-the system must fit into the existing organization

Regarding the content, De Azevedo showed that everyday's knowledge is as important if not more than technical knowledge and, when poorly organized, an important source of wasted time [5]. Thus, in addition to technical knowledge, the system must contain all the administrative and organizational knowledge deemed necessary for "smoothing" the user's work.

Regarding access to knowledge, several spaces were defined with different levels of confidentiality:

- a personal space, private to the user
- a project space, shared by all members of the project
- a division's space, shared by all the projects with the same objectives

3.1. Portal Features

The main features of the portal are the following:

- access to email
- possibility to create documents (email, personal, project, public)
- task management with notification (e.g., milestones, subtasks)
- directories (personal, project, general)
- calendars (personal, project)
- subscriptions (to documents, web searches)
- favorite references
- workflow
- discussions

The use of the resulting project portal is quite simple and emphasizes easy access to information (i.e., with as few clicks as possible). Figure 2 shows the personal page for a user involved in several projects (Home page).

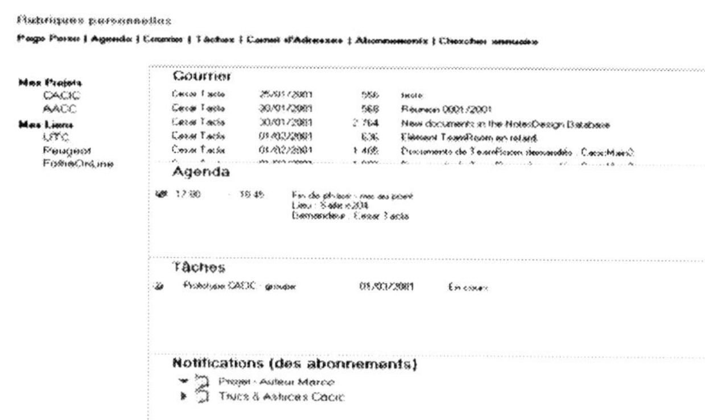

Figure 2. User's portal.

Figure 3 shows access to team and personal documents that can be displayed by date, category, keyword, project event, group, etc.

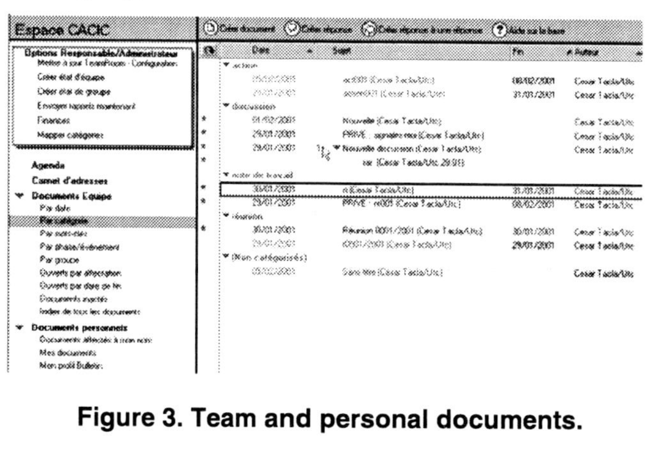

Figure 3. Team and personal documents.

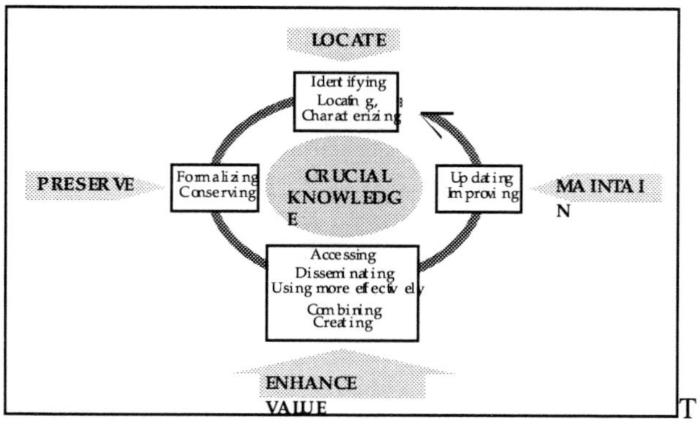

Figure 1. Knowledge capitalization cycle.

Two technologies are good candidates for developing such systems, namely groupware or agent systems. Both were shown to be able to support the collaborative process. We first undertook a project for testing this approach, developing a system based on a groupware approach. The result will be shortly described in Section 3. However, from the experience we had gained in developing multi-agent projects, we found the result not entirely satisfactory. Although another system could be developed around cognitive agents and personal assistants using the approach discussed in Section 4, we feel that it would be better to combine the groupware and the agent approaches to offer agent-supported portal systems with interesting characteristics.

3. A Groupware Portal for Complex R&D Projects

We first developed a project portal based on Lotus Notes™ to be tested in the automotive industry. The objective of giving a portal to each member of a project was twofold: (i) to reduce the time needed to access relevant information; and (ii) to collect knowledge and to organize the information exchanges.

The main requirements from the user's side were the following:

- the system must cover as much of the R&D activity as possible,
- the system must save time in an obvious manner
- the system must be reliable and persistent
- the system must fit into the existing organization

Regarding the content, De Azevedo showed that everyday's knowledge is as important if not more than technical knowledge and, when poorly organized, an important source of wasted time [5]. Thus, in addition to technical knowledge, the system must contain all the administrative and organizational knowledge deemed necessary for "smoothing" the user's work.

Regarding access to knowledge, several spaces were defined with different levels of confidentiality:

- a personal space, private to the user
- a project space, shared by all members of the project
- a division's space, shared by all the projects with the same objectives

3.1. Portal Features

The main features of the portal are the following:

- access to email
- possibility to create documents (email, personal, project, public)
- task management with notification (e.g., milestones, subtasks)
- directories (personal, project, general)
- calendars (personal, project)
- subscriptions (to documents, web searches)
- favorite references
- workflow
- discussions

The use of the resulting project portal is quite simple and emphasizes easy access to information (i.e., with as few clicks as possible). Figure 2 shows the personal page for a user involved in several projects (Home page).

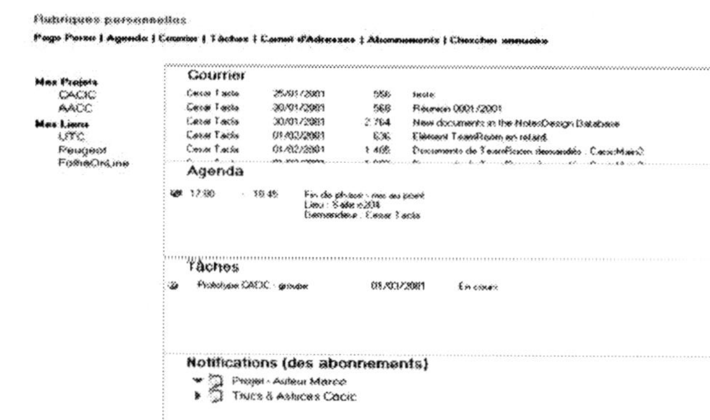

Figure 2. User's portal.

Figure 3 shows access to team and personal documents that can be displayed by date, category, keyword, project event, group, etc.

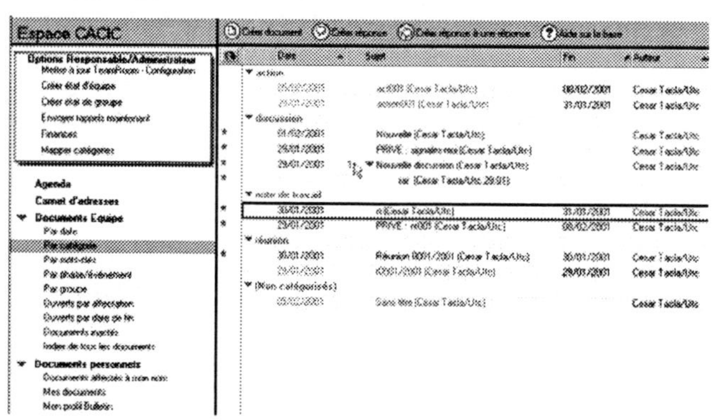

Figure 3. Team and personal documents.

Agent-Supported Portals and Knowledge Management in Complex R&D Projects

Jean-Paul. A. Barthès, Cesar Tacla
CNRS UMR 6599 HEUDIASYC
Université de Technologie de Compiègne, France
{barthes, cesar.tacla}@utc.fr

Abstract

This paper is concerned with organizing Knowledge Management in complex R&D projects where time is the prime factor. We argue that specific portals developed using groupware technology and products should be augmented by agents in order to increase the overall system reactivity and to achieve the global objective, namely to save time. We describe a portal we have developed using a groupware approach, we give the structure of a system of cognitive agents, and discuss the possibility of bringing the two technologies together.

1. Introduction

Our paper analyzes the research being done in the context of a French national project for the automotive industry. The main problem to be solved is to organize knowledge management in complex high priority urgent R&D projects, so that the gained experience can be used by the following complex high priority urgent R&D projects. A similar but more general concern was expressed in the MOKA European project [1]. In order to offer possible solutions, one has to take into account the reality of the situation, and in particular organizational constraints, rather than devising elegant but impractical systems. Since we are dealing mostly with information, two technologies are useable: *groupware* for document management and cooperative work, and *multi-agent systems* for flexibility and dynamics. The point of view taken in the paper is that of the application.

The paper is organized in four parts: Section 2 introduces the context of the work. Section 3 describes the features of a portal we developed for complex R&D projects. Section 4 discusses the multi-agent architecture that we have developed recently. Section 5 discusses the possibility of putting the two technologies to work together, mentioning some problems that remain to be solved.

2. Complex R&D Projects

Complex R&D projects belong to routine design although they cannot be equated with catalog design [2]. Indeed, designing a new vehicle consists of reusing a number of past designs and old parts, but also assembling new parts developed for the project. The projects we consider lead to the development of prototypes that will eventually be built in a volume production. The main criterion in such projects is time. Everything is expressed in terms of time including costs (man-year, man-month, man-hour). The projects of interest have a 2-5 year time-to-market delay. While working on the prototype, there is very little time allocated for producing detailed documentation. Information ends up spread in a multitude of documents on various supports. Most lessons learnt on the project are buried in the heads of the project team members. Then, since R&D teams change from a project to the next, a great deal of experience can be lost. Thus, it is important to organize and manage knowledge developed in R&D projects.

A classical approach to developing knowledge management systems is to follow a knowledge capitalizing cycle as advocated by many authors. Figure 1 shows such a cycle taken from [3]. The process consists of locating crucial knowledge, formalizing it using some information technology so as to be able to distribute it, and to maintain it. In R&D projects, there is simply no time to do that. However, since R&D projects end up producing a great number of documents and require the collaboration of many people, a good way to capture information without turning project engineers into clerical staff, is to offer them a good system supporting collaborative work. If they like it, they will use it, and if they use it, then part of the information can be organized without requiring too much effort. A similar idea has been developed in the CAIRO environment [4], where information is collected by software agents from the messages exchanged during project meetings.

Part IV: Agents and Multi-Agent Systems

3.2. Portal Architecture

The portal was designed around a number of different Lotus Notes™ databases containing various types of documents. Four bases were defined for the user (personal, mail, address, subscriptions), two more were defined for the project (project space, project directory), and one for public information. Since all bases contain documents related to the project activity it is important to organize them in a consistent manner. To this purpose, a *domain ontology* was defined to structure the space of concepts expressed by category names or keywords. The ontology is administered by the project manager and constitutes the backbone of the system. All pro-active behavior is implemented as Lotus™ agents, which in fact are demons attached to particular databases.

It is worth noticing that the project portal structure can be viewed as a set of services, each form of service being attached to a particular database. Thus, we are not far from a multi-agent structure, which is one of the reasons why we want to compare the two approaches.

4. Multi-Agent Systems

Multi-Agent Systems offer a very flexible architecture for developing complex applications. The idea is to specialize autonomous agents to specific services and to let agents interact (negotiate) in order to solve problems. No wonder that many projects were developed in the domain of design. An overview of such projects, analyzed from the point of view of negotiation, can be found in [6]. A broader and more general review is detailed in [7].

In our laboratory we have been developing agent systems for a long time, in particular in the domain of complex design. Monceyron (1992) developed a blackboard design environment [8], EXPORT, applied to harbor design, and more recently Shen developed a generic multi-agent platform, DIDE, for mechanical engineering design [9]. This led to the development of the OMAS platform (Open Multi-Agent System) [10]. The main features of our approach to multi-agent architecture are summarized in the two following sections.

4.1. Agent and System Structure

Our approach consists of organizing agents tightly linked within groups we call *coteries*. Each agent inside a coterie receives all messages, allowing it to update and maintain its representation of the world. Inside a coterie there are two types of agents:

- service agents, that provide a particular type of service corresponding to specific skills (one of the services is the interface to other coteries or to foreign platforms),
- personal assistants, in charge of interfacing humans to the system (their particular skills are devoted to understanding their master and presenting the information intelligently and in a timely manner).

A view of this architecture is given on Figure 4.

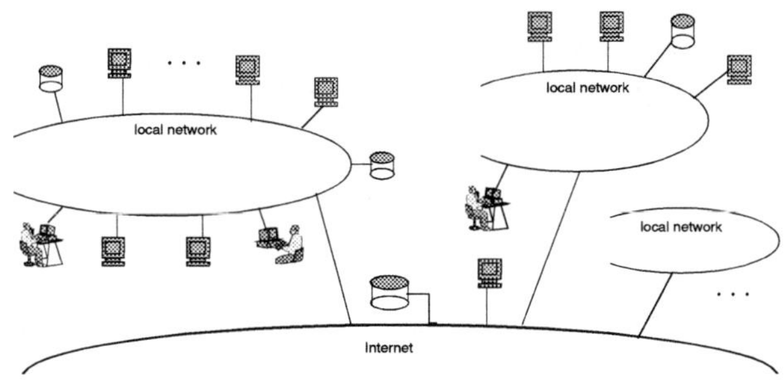

Figure 4. MAS (DIDE) Architecture for design .

The agent internal architecture is given on Figure 5. It can be seen that an assistant agent is a service agent augmented with a user interface (gray area).

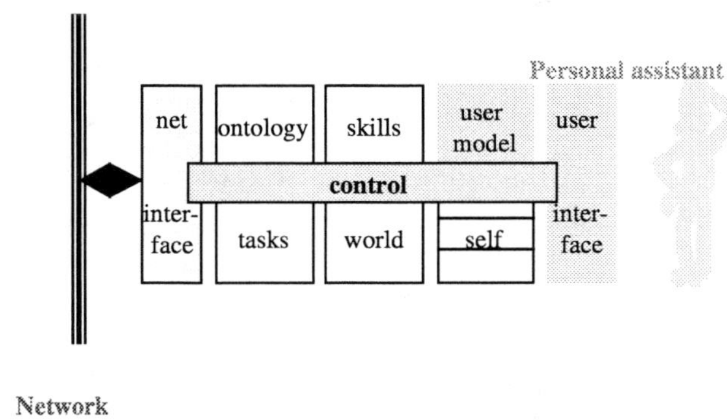

Figure 5. Internal agent structure.

It would be too long to detail here the various components of a typical OMAS agent, but for our purposes the important ones are:

- skills implementing a particular service,
- ontology (dealing with an agent ontology and domain ontologies),
- self, containing the memory, goals and skill description,
- user model for assistant agents.

Communication within a coterie occurs asynchronously in broadcast mode (messages can be directed, e.g., point to point, but communication mode is broadcast). Representations are built dynamically as the result of the analysis of the different messages. Ontologies play a significant role to this purpose.

4.2. Organizing Knowledge Management with Agents.

It is quite straightforward to imagine that a multi-agent organization is an ideal structure to support knowledge management, since each typical service required by the system can be implemented as a service agent and each user can be assigned an assistant agent. A multi-agent system can display an intelligent behavior, since agents are not limited in complexity. For example, using techniques inherited from artificial intelligence, agents can maintain a representation of what is happening and offer help spontaneously (pro-active behavior). To do so, they can use past information, e.g., through case-based reasoning techniques. In addition, since agents are developed independently, a multi-agent system can evolve easily by simple addition of new agents.

5. Mixing the Two Approaches

Although our practical experience is yet limited, we already see some problems with each of the two approaches for managing knowledge. We also see the possibility to overcome some of the problems by combining both approaches, as discussed in this section.

5.1. Advantages and Limits of the Portal Approach

A project portal based on a commercial groupware platform has a number of advantages:

- it is proven technology, thus providing a reliable and persistent environment,
- it can be easily integrated with the rest of the organization information systems,
- it can be well fitted to the organization structure,
- it allows excellent levels of security,
- it is easily accepted

On the other hand its major drawbacks are the following:

- the replication mechanism allowing a specific user to have a copy of the information on her machine can lead to large transfers of data,
- the pro-activity one can install with standard demons is limited,
- extensions, i.e., addition of new features may require a significant reorganizing and reprogramming effort.

The first restriction can be lifted with a careful organization of the different databases containing documents. The others are more difficult to resolve. Indeed, extensive pro-activity calls for artificial intelligence techniques and symbolic processing, which is difficult to program within the framework of commercial products. Extensions do pose problems when one needs to introduce new features, leading to significant reprogramming.

5.2. Advantages and Limits of the Agent Approach

Agent systems and, in particular OMAS agents, do not suffer from the same limitations. Each is totally independent from the others and can be added at any time. Activity is controlled by exchanging messages, each concerned service answering when needed. Therefore, the main advantages are:

- flexibility,
- possibility to evolve by simple addition or removal of service agents,
- possibility to define pro-active "intelligent" services running in the background easily,
- possibility to interface to legacy software through encapsulation,
- possibility to run several options for the same service in parallel using contract-net protocol.

If the flexibility of a system of intelligent cognitive agents needs not to be proven any longer, the actual use of MAS is not an obvious solution for the following reasons:

- MAS are difficult to organize or to control due to their fuzzy contours, i.e., the system is evolving continuously,
- most experience with MAS systems was gained in academic projects, that do not always reflect industrial constraints,
- an agent system is by nature "open", which may lead to security problems.

In addition, in the knowledge management perspective, a simple request may lead to uncontrollable activity through automatic propagation of tasks and subtasks. Coordination is thus necessary. However, as noted by Whitfield and coworkers [6]: "What is apparent [...] is that there is no clear way of managing coordination, and despite the many attempts to generalize it, such that it may be applied to any domain, it is debatable whether the rules obtained and used within one domain may be of use within another."

5.3. Combining a Portal Approach with Cognitive Agents

A possible solution we are currently investigating is to keep a groupware approach and to introduce some of the flexibility that can be found in multi-agent systems. The approach that we have chosen consists of

- viewing portal features as services, leading to independent "intelligent" agents,
- augmenting the autonomy of groupware elements (databases) by encapsulating them into units having

richer interfaces,

- adding external agents, in particular personal assistants to interface users to the system.

Indeed, groupware environments tend to operate as a centralized client/server architecture. Central services need to be distributed as independent processes capable of intelligent pro-active behavior. Some of the services, in particular those handling large amount of documents can be regarded as legacy applications and encapsulated by defining a precise exchange protocol defined in terms of semantically meaningful actions, a somewhat higher level intelligent version of an API. In some cases, it is even possible to adopt free-style communication [11]. The system can be extended by including progressively more complex agents or agents providing redundant services. While retaining some of the original features of the groupware approach, namely handling volume and providing security, the system could be made progressively more like one of the agent coterie discussed in Section 3. The resulting combination of artificial agents and human actors would then embody the concept of *ba* presented by Nonaka and Konno (1998) [12].

The system we are developing is sketched on Figure 6 showing a combination of Lotus™ bases together with a number of agents. The Lotus™ environment is used to record and to display information, the multi-agent structure is intended to provide background services to the user, e.g., organizing and retrieving information, helping the project member to prepare reports by providing references dynamically, searching other sources of information, etc.

While doing that, we hope to achieve a better cooperation by allowing independent pieces of software to run according to sophisticated models of their human master. We also want to introduce in the groupware approach a better service by developing more sophisticated ontologies than the ones we are using today. Indeed, our ontologies are used to organize and retrieve information, but also to infer what type of services is requested by the user at the personal assistant level.

The major difficulty for building the new system is that our agents are developed with tools from artificial intelligence that do not fit well into commercial groupware environments. We are designing a loose coupling through message passing a possible solution to this problem. However, when this is resolved, another problem consists in redistributing system features to two quasi-independent environments.

6. Summary and Conclusion

In this paper, we discussed the problem of collecting and organizing knowledge during complex high priority R&D projects. We first developed a project portal to be given to the project members. The portal system is currently being deployed in several projects: some are industrial, some more administrative, some are research projects. In the coming months we hope to obtain a good feedback from a wide usage of this approach. Already, early results indicate some rigidity and lack of flexibility, compared with our experience with cognitive multi-agent systems. However, multi-agent systems alone do not offer enough robustness and security for industrial applications. Consequently, we argued that a combined approach mixing groupware and agents would present more interesting features, providing flexibility and overall more intelligent behavior. We are currently developing such an approach.

7. Acknowledgments

This work is funded in part by a grant from CNRS in the French PROSPER national research program.

Cesar Tacla has financial support from CAPES and CEFET-PR, two Brazilian governmental entities for research and education.

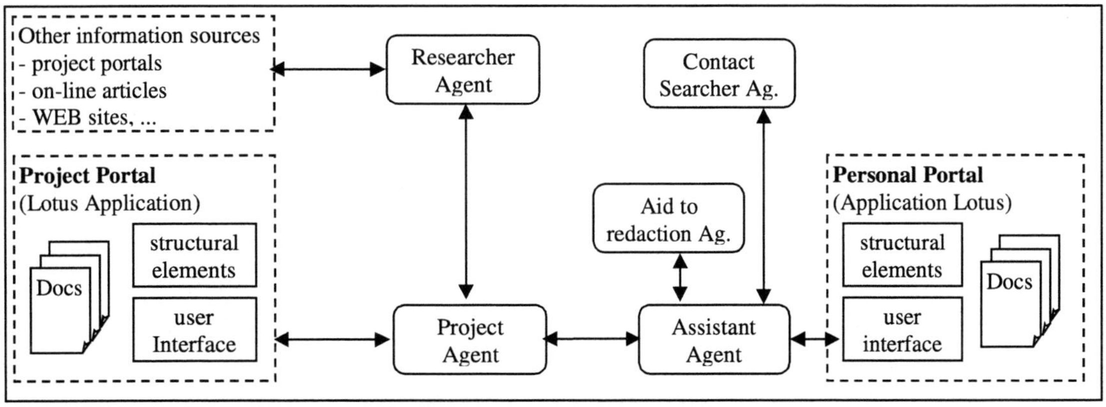

Figure 6 - Mixed architecture.

8. References

[1] K Oldham, S Kneebone, M Callot, A Murton, and R Brimble, in N. Mårtensson, R. Mackay and S. Björgvinsson. "MOKA - A Methodology and tools Oriented to Knowledge-based engineering Applications', Changing the Ways We Work", Advances in Design and Manufacturing, Volume 8, Proceedings of the Conference on Integration in Manufacturing, Göteborg, Sweden, IOS Press, Amsterdam, October, 1998, pp 198-207.

[2] S Carlson-Skalak., M.D., White Y. Teng. "Using an Evolutionary Algorithm to Catalog Design". *Research in Engineering Design*. 1998, 10:63-83.

[3] M Grundstein. and J-P.A. Barthès "An Approach to Enterprise Knowledge Capitalization. in Knowledge Management". Enterprise, Network and Learning. Advances in Knowledge Management. Vol 2. J. Schreinemakers, J-P. Barthès (Eds). Ergon Verlag. Germany. 1999, pp 151-159.

[4] F Peña-Mora, K Hussein, S Vadhavkar, K Benjamin.. "CAIRO: a concurrent engineering meeting environment for virtual design teams." *Artificial Intelligence in Engineering*, 2000, **14**:203-219.

[5] H. De Azevedo. "Contribution à la modélisation des connaissances à l'aide des systèmes multi-agents." Thèse de doctorat de l'Université de Technologie de Compiègne. Spécialité Contrôle des Systèmes (in French), 1997.

[6] R.I. Whitfield, G Coates., A.H.B Duffy, B Hills.. "Coordination Approaches and Systems – part I: A Strategic Perspective." *Research in Engineering Design*. 2000, 1(**12**):48-60.

[7] W. Shen, D. H. Norrie, and J-P. A. Barthès. *Multi-Agent System for Concurrent Intelligent Design and Manufacturing*. Taylor and Francis. 2001.

[8] E. Monceyron and J-P.A. Barthès. "Architecture for ICAD Systems: an Example from Harbor Design." *Revue Sciences et Techniques de la Conception*, 1992, vol **1**:49-68.

[9] W. Shen and J-P. A. Barthès. "An Experimental Multi-Agent Environment for Engineering Design." *International Journal of Cooperative Information Systems*. 1996 Vol 2&3(**5**):131-151. World Scientific Publishing Company.

[10] E.E. Scalabrin, L. Vandenberghe, H. de Azevedo, J-P.A. Barthès.. "A Generic Model of Cognitive Agent to Develop Open Systems." In Advances in Artificial Intelligence. Series Lecture Notes in Artificial Intelligence. D.L. Borges, C.A.A. Kaestner (Eds). Springer. 1996, pp 61-70.

[11] M. Ramos. "Structuration et évolution conceptuelles d'un agent assistant personnel dans les domaines techniques." Thèse de doctorat de l'Université de Technologie de Compiègne. Décembre. (in French). 2000.

[12] I. Nonaka and N. Konno. "The Concept of "Ba": Building a Foundation for Knowledge Creation." *Califormia Management Review*. 1998. 3(**40**):40-54.

Web Agents for Spatial Mining on Air Pollution Meteorology

Vincent Ng, Stephen C F Chan, Sandra Au
Department of Computing, Hong Kong polytechnic University, Hung Hom, Hong Kong
cstyng@comp.polyu.edu.hk

Abstract

This paper describes an agent framework to support spatial data mining of air pollution data. A web-based solution, APMF (Air Pollution Mining Framework), is developed to investigate the effect of meteorological and air pollutant elements on air pollution. It makes use of three types of agents: Collect agents, Coordinate agents and Query/Mining agents. Query/Mining agents interact with the user, receive user queries and mining request, and deliver the results. Coordinate agents help to prepare the data resources for the querying tasks. Collect agents provide access to a heterogeneous collection of Web sources. In order to standardize data exchanges and query/mining process descriptions, an XML-based language, DQML, is developed. A prototype of the APMF is implemented with Java servlet and JavaLite.

1. Introduction

The heavy air pollution in Hong Kong has greatly affected public health, especially respiratory problems. In 1996, the Department of Community & Family Medicine of the Chinese University of Hong Kong conducted a survey on the effect of increasing air pollution on additional hospital admissions [5]. The result showed that the number of patients suffering from respiratory illnesses, heart diseases and asthma in hospitals increased when the air pollution got worse.

Since most of the air pollution meteorological data is available on the Web, agent technologies and spatial data mining techniques can be applied to support public queries and spatial data mining. Agent technologies can be used to communicate and coordinate between different agents; whereas spatial data mining is aimed to discover spatial and non-spatial relationships about the Hong Kong air pollution situations. In our work, Hong Kong is divided into 18 districts according to the Planning Department of HKSAR. We use our agents to collect the air pollutants concentration and other meteorological data through different web sites online. The results after mining the spatial data can help decision making, planning and forecasting. For example, tracing the good or bad air quality to their causes can assist the making of strategies to control the air pollution. Town planners can plan to develop residential areas in districts with good air quality. Environmental engineers can concentrate on improving the districts with bad air quality, and the public can estimate or forecast the air quality and meteorological condition by providing some parameters, and so on.

Section 2 describes the current situation of air pollution in Hong Kong, and provides a brief overview of the air pollution meteorology and air pollutants. Section 3 gives an overview of the literature of spatial data mining and agent technology that can be applied in this project. Section 4 discusses the design of the agent framework. Section 5 presents on our current prototype and Section 6 concludes our contributions..

2. Air Pollution In Hong Kong

Hong Kong has been experiencing months of persistently high levels of outdoor air pollution recorded. The air pollution index (API) has frequently been hovering around 100. Under the present system, people are warned to avoid outdoor activities and physical exertion when the API hits more than 200. "*It is accepted that air pollution in Hong Kong caused 2,000 premature deaths a year and an array of chronic respiratory illnesses*" [12]. As air pollution affects every one of us, it can cause health problems and, maybe, death. It is important for us to find out the factors affecting air pollution and to control the air quality.

The factors that are related to air pollution are wind, rainfall, temperature, fog radiation, air pressure and atmospheric stability. According to Environment Protection Department (EPD), the Air Pollution Index (API) ranges from 0 to 500 and is divided into 5 bands according to the potential effects on health. An API number is particularly important since it corresponds to the short-term Air Quality Objective values (AQOs) (i.e. 1-hour and 24-hour limits). An API higher than 100 would mean that one or more pollutants poses immediate health effects to some susceptible members of our community. Figure 1 lists the potential health effects and API ranges are as follows:

Air Quality (AQ) Status	API	Health Implications
AQ significantly worse than both short-term and long-term AQOs.	201 to 500	People with existing heart or respiratory illnesses will experience significant aggravation of their symptoms and there will be also widespread symptoms in the healthy population. These include eye irritation, wheezing, coughing, phlegm and sore throat.
AQ worse than both short-term and long-term AQOs.	101 to 200	People with existing heart or respiratory illnesses will notice mild aggravation of their health conditions. Generally healthy individuals may also notice some discomfort.
AQ within the short-term AQOs but worse than the long-term.	51 to 100	Very few people, if any, may notice immediate health effects. Long-term effects may, however, be observed if you are exposed to such levels for a long time.
AQ within all AQOs.	26 to 50	None expected for the general population.
AQ well within all AQOs.	0 to 25	None expected.

Figure 1. The potential health effects and API ranges.

3. Background Review

3.1 Agents

Agents fundamentally provide a mediation of access to any number and types of services. They are often used to retrieve, transform and process information. A large number of agent infrastructures, such as OAA [2], JATLITE [6], RETSINA [4] and Infosleuth [13], have been developed. For many of these systems, agent communication languages are designed specifically to facilitate communication between two or more agents. One of them is the KQML (Knowledge Query and Manipulation Language) [1,7,14]. It is a language and a protocol for exchanging information and knowledge. The KQML can be viewed as consisting of three layers: the content, message and communication layers. The content layer specifies the actual content of the message. The message layer provides the performative and specifies the protocol for delivering the message that subsumes the content. The communication layer encodes low-level communication parameters.

Amongst different types of agents, there are the collaborative agents which emphasize autonomy and cooperation (with other agents) so as to perform tasks for their owners [7]. In order to have a set of collaborative agents work together, they need a common language. We therefore propose DQML which is a XML dialect to send and receive information. It is similar to KQML/FIPA ACL-like format, except encoded in XML.

3.2 Spatial Data Mining

Spatial data mining methods allow extension of traditional spatial analysis methods by putting emphasis on efficiency and knowledge discoveries. One can discover spatial association rules in the form X→Y (c%), where X and Y are sets of spatial or non-spatial association rule and c% is the confidence level [8]. Another general approach is to find spatial clusters, such as the CLARANS (Clustering Large Applications based upon RANdomized Search) algorithm developed by Ng and Han [11]. In 1996, Knorr and Ng presented a method which enables fast finding of the features that are close to clusters [10]. They used multiple level approximations in the algorithm CRH (where C is for encompassing circle, R for isothetic rectangle, and H for convex hull) to gradually reduce the candidate features.

In our work, we are interested to apply spatial generalization-based method [9]. The generalization-based knowledge discovery assumes the existence of background knowledge in the form of the concept hierarchies. Concept hierarchies can be explicitly given by the experts, or in some cases, can be generated automatically by data analysis. With the ascension of concept hierarchy, information becomes more and more general, but still remains consistent with the lower concept levels.

4. An agent-based Air Pollution Mining Framework

Information available from the Web is unorganised, multi-modal, and distributed on server sites all over the world. Furthermore, the availability, type and reliability of information services are constantly changing and being updated. Therefore, information is becoming increasingly more difficult to collect, filter, evaluate, and use in problem solving. As a result, the problem of locating information sources, accessing, filtering, and integrating information in support of spatial data mining as well as coordinating information retrieval has become a very critical task.

Our work aims at designing an agent-based client-server framework in order to support spatial data mining of air pollution data. A web-based solution, APMF (Air

Pollution Mining Framework), will be provided to investigate the effect of meteorological and air pollutant elements on air pollution using spatial data mining technologies. It would make use of three types of agents: interface agents, collaborative agents and information agents. Interface agents interact with a user, receive user specifications and deliver the results, i.e., the query agents and mining agents. Collaborative agents help a user to perform tasks by formulating problem solving plans and carrying out these plans through querying and exchanging information with other software agents, i.e., the coordinate agents. Information agents provide intelligent access to a heterogeneous collection of information sources, i.e., the collect agents. These agents are designed to support querying and spatial data mining of air pollution data (as shown in Figure 2).

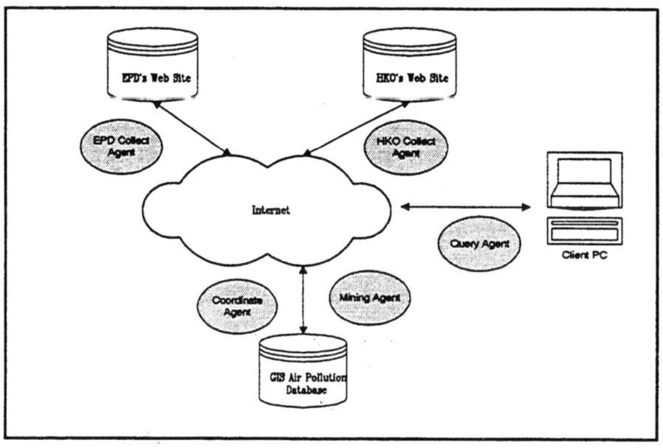

Figure 2 Air Pollution Mining Framework.

The APMF is divided into 2 layers. The front layer accepts a user query request and then displays a detailed report on the meteorological and air pollutant data. The results can either be mined results or ad-hoc queries. The back layer is to collect meteorological data, transform it and prepare it for the front layer. The interactions of the various agents in the system are:

- **Scheduled events**
  - o The Collect agents access the Internet resources to collect the meteorological and air pollutant data regularly. The Collect agents send raw data or HTML files to the Coordinate agent.
  - o The Coordinate agent transforms the raw data from the Collect agents and stores it the DQML format.
  - o The Coordinate agent passes the structured meteorological and air pollutant information to the Mining agent.
  - o The Mining agent carries out spatial data mining on the data.
  - o The MiningaAgent passes the results as DQML files to the GIS databases.
- **On-demand events**
  - o A user inputs a request to the Query agent.
  - o The Query agent formats the query into a DQML message and get the corresponding results from the database.
  - o If the Query agent cannot find the wanted results from the GIS database, it transfers the request to the Mining agent. Then the Mining agent collaborates with the Coordinate agent in extracting, transforming and analyzing the data. Finally the results of the Mining agent will be returned to the Query agent for display.

In the next section, we first describe DQML (Data Mining and Query Markup Language) and then each type of agents in turn.

4.1 DQML

DQML is developed to model the general data querying and data mining process by defining corresponding XML entities and elements. There are 2 main types of DQML documents. One type is the request document that is used either to initiate the data mining process or to query the data placed in a set of heterogeneous databases. The second type of documents contains the query or the mining results. In the DQML data model, an important class of objects is Dataset. Each Dataset element can be first mapped to some external data sources before the data is passed and manipulated by data mining processes. A Dataset element is generally composed of 2 parts. The first part is mandatory and concerns with the data schema definitions. The second part is optional and contains the actual data set.

In our design, besides the QueryProcess, two other types of processes are defined in DQML, TransformationProcess and DataMiningProcess. The primary function of QueryProcess is to retrieve the data from different interfaces and to convert them into a Dataset element. Every interface defines the data source location, connection method, data fields required and any selection criteria of the data. In the current version of DQML, three types of query interfaces are defined. SQLDBInterface describes the connection and database location of any SQL database systems. XMLInterface describes the location of XML location and RawHTMLInterface that handles HTML data sources. Frequently, raw data cannot be used before filtering and transformation. These functions are provided by the second process type, TransformationProcess. It is used to split, merge and manipulate the data schema of input Datasets and filter out any unwanted records from the resultant Dataset(s). The third type of process defined in DQML is MiningProcess. Its primary function is to describe the procedures of passing the input datasets to a data mining agent.

4.2 Collect Agents

Both meteorological and air pollutant data can be obtained from the web sites of Hong Kong Observatory (HKO) and Environmental Protection Deportment (EPD) of the government of HKSAR. Meteorological data, such as daily temperature are available through the Web from the Hong Kong Observatory (HKO). There are ten stations of HKO: King's Park, Lau Fau Shan, Wong Chuk Hang, Shatin, Tuen Mun, Ta Kwu Ling, Lau Fau Shan, Sai Kung, Cheung Chau, Chek Lap Kok and Tseung Kwan O. The entire series of daily values are to be captured for all stations. For these data, they are frequently updated at different times according to different data sources. Further complication is the data does not exist in the same format, some in image format while some in HTML format. Figures 3 and 4 list the data information available on the Web and their corresponding data formats.

	Update Frequency	**Data format**	**Localized or Territorial**
Wind	Daily	HTML	Territorial
Rainfall	Hourly	Image	Divided into 11 districts
Temperature	Hourly	HTML	Divided into 18 districts
Air pressure	Daily	HTML	Territorial
Humidity	Hourly	HTML	Territorial

Figure 3 Data from HKO's web site.

	Update Frequency	**Data format**	**Localized or Territorial**
Nitrogen dioxide (NO_2)	Hourly	HTML	Divided into 16 districts
Sulphur dioxide (SO_2)	Hourly	HTML	Divided into 16 districts
Ozone (O_3)	Hourly	HTML	Divided into 16 districts
Carbon Monoxide (CO)	Hourly	HTML	Divided into 16 districts
Respirable suspended particulates (RSP)	Hourly	HTML	Divided into 16 districts
API	Hourly	HTML	Divided into 16 districts

Figure 4 Data from EPD's web site.

Collect agents are designed to collect meteorological and air pollutant data from the web sites of HKO and EPD. They are used to plan the retrieval of data, actual extraction and archival in a GIS database.

4.3 Coordinate Agents

The major function of the Coordinate agent is to convert the raw data to a usable data format for querying and mining agents. In many cases, the Coordinate agent needs to clean the unnecessary information in the HTML. Moreover, raw data comes in at different frequencies. Hence, the agent is also necessary to re-configure and coordinate the data in order to get a consistent data format. It is done by using the DQML (Data mining and Query Markup Language). Figure 5 shows part of the DTD of DQML.

4.4 Spatial Mining Agent

We choose the generalized-based algorithm of the classification-based mining method as our spatial mining algorithm because it can easily represent the meteorological and air pollutant information when level information is involved. For instance, a concept hierarchy of three levels is defined for air pollution. The leaf level contains different API ranges; and the intermediate level has values of "severe", "very high", "high", "medium" and "low", while the top level has two values, "acceptable" or "serious".

The main usage of the Mining agent is to carry out the spatial data mining process. Each mining agent upon request will send a DQML message to the Coordinate agent. In return, the agent receives the dataset back from the Coordinate agent, and then performs the corresponding spatial data mining. When the mining process is done, the results are shown to users or can be stored in the database. An example of the mining requested done via DQML is shown in Figure 6.

```
<!ELEMENT DataField (Extension*, (Interval*|Value*)) >
    <!ATTLIST DataField
          name          %FIELD-NAME;    #REQUIRED
          displayName   CDATA           #IMPLIED
          optype        (categorical |
                         ordinal |  continuous|
                          spatial) #REQUIRED
          isCycle       %BOOLEAN;       #IMPLIED>

<!-- DTD for Interval -->
<!ELEMENT Interval EMPTY >
    <!ATTLIST Interval
           closure   (openClosed | openOpen   |
                    closedOpen |closedClosed )   #REQUIRED
                    leftMargin  %NUMBER;         #IMPLIED
                    rightMargin %NUMBER;     #IMPLIED >

<!-- DTD for Value -->
<!ELEMENT Value (Extension*) >
     <!ATTLIST Value
           value           CDATA    #REQUIRED
           displayvalue    CDATA    #IMPLIED
           property        (valid | invalid |
                            missing ) "valid" >
```

Figure 5. Data fields definition of DQML.

```
<DQML>
   <Header copyright="DL" description="Process Status">
   <Application name="DQML Air Pollution Spatial Mining System"
              version="1.0"/>
   </Header>

   <DataMiningProcessStatus dataMiningRequestID="DMR00001"
                            numberOfProcesses="2">
      <ProcessStatus>
         <ProcessID>DMP000001</ProcessID>
               <DataSet ID=EP0001 NAME=FIRST>
                     <Datafield>TEMP</Datafield>
                     <Datafield>DISTRICT</Datafield>
                     <Datarecord>1-1000</Datarecord>
               </DataSet>
               <DataSet ID=EP0002 NAME=SECOND>
                     <Datafield>HUMIDITY</Datafield>
                     <Datafield>DISTRICT</Datafield>
                     <Datarecord>1000-1999</Datarecord>
               </DataSet>
            <ProcessType>association</ProcessType>
         <Status>submit</Status>
         <SubmissionTime>2000-01-26 08:12 GMT
+08:00</SubmissionTime>
         <EstimatedRunningTime>500</EstimatedRunningTime>
         <ProcessLocation>www.iphk.com:2000</ProcessLocation>
         <ProcessMessage>Air Pollution Mining</ProcessMessage>
      </ProcessStatus>
   </DataMiningProcessStatus>
</DQML>
```

Figure 6. A mining request represented in a DQML document.

4.5 Query Agents

A Query agent allows users to input their ad-hoc request or to retrieve data mining results. The Query agent passes the request messages in the DQML format to the server side to search for answers. If the answers can be found in the GIS database, the Query Agent will retrieve and send the results to the users. If the results cannot be found in the GIS database, the Query agent will trigger the Coordinate agent to get the data. Then, after the Coordinate agent returns the data, the Query agent performs the query and displays the results on the web browsers. Figure 7 shows an example of how the Query agent issues a query to the Coordinate agent.

It is also possible that a Query agent asks for a mining result. Similar to the above procedure, if the result is not available, the Mining agent will put the Coordinate agent to work and the result will be passed back to the Query agent.

```
<SQLInterface>
            <DataSet fieldName = "NO2", fieldName = "NT", DName = "source1"
/>
            <DataSet fieldName = "O3", fieldName = "NT",    DName =
"source2" />
            <DBJoin JoinID = "Join1", JoinType = "Equi">
              <DBJoinField fieldName = "NT", tableName="source1",
tableName=="source2" />
            </DBJoin>
            <QUERY SELECT>
                        <SelectedField fieldName = "NO2" />
                        <SelectedField fieldName = "O3" />
            </QUERY>
</SQLInterface>
```

Figure 7. An SQL example in DQML.

5. Implementation

There are several tools necessary to implement APMF, such as compilers, data conversion tools, agent tools, GIS software, data mining tools, and spatial data mining tools. Figure 8 summarises the tools and software chosen in our work.

	Tools
Client/Server support	Java Servlet
Web Server	vqServer
Agent Tools	JATLite
GIS	ArcView GIS
Data Mining Software	Intellix Analyser and Java

Figure 8. Summary of the tools and software chosen.

We implement the APMF in Java. In order to incorporate and utilize pre-existing agent tool, namely, the JATLite [6], KQML will be used as a language for inter-agent communication. Java servlet is used to support client-server interactive web accesses. A Query agent will process a user's query through Java servlet. The agent displays simple HTML forms with combo-boxes and radio buttons as the user interface. The query results are presented in HTML documents generated by the servlets. Figure 9 show the answer to an example query of asking ". Which district has the highest or lowest API on April 24, 2000?"

The Mining agent is implemented with Intellix Analyser and Java. Intellix Analyser is a data mining and modeling tool [3]. It is capable of finding patterns in unstructured data and to produce models for prediction or classification. For example, it can predict which customers are likely to leave the portfolio from direct marketing campaigns. In Intellix Analyser, the unlearned error probabilities mean that the software can guess the result of API with the mistakes probability. In the figure, the lowest unlearned error probability of API range is from (30.5 to 40.0). The unlearned error probability is 17.3% which means that Intellix Analyser will guess the result of API in mistakes by the probability of 17.3%.

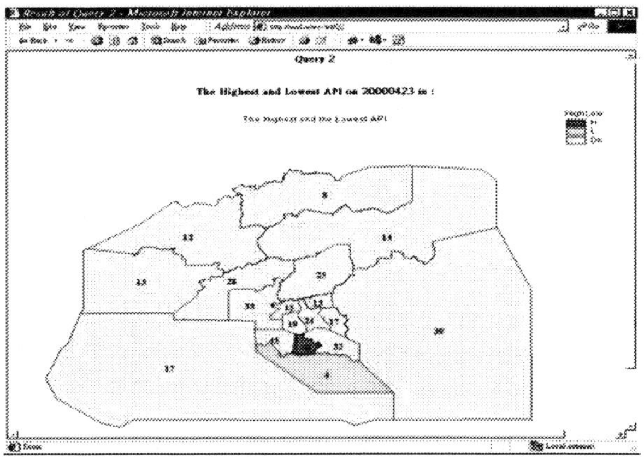

Figure 9. Query result.

For spatial data mining, we implement the generalized-based algorithm in Java. It is to discover the association patterns from the air pollution data. In the current version of our prototype, the spatial mining part is still under-development.

6. Conclusion

Although the current system cannot control the air pollution, it allows the public to access the information anytime to increase the public awareness of the issue. In our system, the APMF (Air Pollution Mining Framework) is adopted to allow web-access for users to investigate the effect of meteorological and air pollutant elements on air pollution. In order to support data manipulation and process definitions, we have developed the DQML (Data mining and Query Markup Language). It is used as a means to give specifications for agents to perform querying, data extraction and transformation. The framework includes Collect agents, Coordinate agents and Query/Mining agents. A prototype of the APMF is implemented with Java servlet and JavaLite. We are planning to support more spatial mining algorithms in our next version.

7. References

[1] D.N. Chorafas, "Agent technology handbook", New York, N.Y. : McGraw-Hill, c1998.

[2] PR Cohen, A Cheyer, M Wang and SC Baeg. An Open Agent Architecture. AAAI Spring Symposium, pp 1-8, March 1998.

[3] M. Copenhagen, "Data Mining using Intellix Analyser Research Version 3.0.6", Intellix A/S, 1999.

[4] K Decker, K Sycara and M Williamson. Matchmaking and Brokering. Proc. of the Second International Conference on Multi-Agent Systems (ICMAS-96), 1996.

[5] Economic Forum, "Environment in Hong Kong", Hong Kong Trade Department Council, February 2000.

[6] JATLite. http//java.standford.edu

[7] N. Jennings, M.J. Wooldridge, "Agent technology: foundations, applications, and markets", Berlin; New York: Springer, c1998.

[8] K.Koperski, J.Han, and J.Adhikary. "Mining Knowledge in Geographical Data", In Communications of ACM. 1998.

[9] K.Koperski, J.Adhikary, and J.Han "Knowledge Discovery in Spatial Databases Progress and Challenges", In Proceedings of Workshop on Research Issues on Data Mining and Knowledge Discovery, Montreal, QB, June 1996.

[10] E.M. Knorr, R.T. Ng, "Finding Aggregate Proximity Relationships and Commonalities in Spatial Data Mining", IEEE, 1996.

[11] R Ng and J Han. Efficient and Effective Clustering Methods for Spatial Data Mining. In Proc. of the 20th VLDB Conference, Santigo, Chile, 1994.

[12] "Clean up our air!", South China Morning Post, 21 June 1998.

[13] M Nodine, B Perry and A Unruh. Experience with the InfoSleuth Agent Architecture. Proc. of AAAI-98 Workshop on Software Tools for Development Agents, 1998.

[14] M. Wooldridge, N.R. Jennings, "Intelligent Agents: Theory and Practice", Knowledge Engineering Review, October 1994.

Implementation of Multi-Agent System Based on CORBA and COM

Xinming Tan
Department of Computer Science
Wuhan University of Technology
Wuhan 430063 P. R. China
tanxming@public.wh.hb.cn

Shaomei Wang
Port Machinery CAD/CAE Center
Wuhan University of Technology
Wuhan 430063 P. R. China
sunwang@public.wh.hb.cn

Abstract

This paper discusses implementing multi-agent systems with component software techniques. Studies on how to implement autonomy, adaptability, interaction, and interoperability of agents and multi-agent systems based on CORBA and COM are carried out and presented respectively. It elaborates that those main features of agents and multi-agent systems could be obtained naturally by applying component software techniques. It indicates that CORBA based or COM based development approach is a practical way to develop large scale and complex multi-agent systems running under distributed heterogeneous environments.

1. Introduction

Agents are widely used in Artificial Intelligence (AI) for they can exhibit intelligent behavior involving perception, reasoning, learning, communicating, and acting in complex environments. Nils J. Nilsson considered AI systems as agents and discussed various agents in [1]. Faratin et al considered agents are autonomous or semiautonomous entities that can perform tasks in complex and dynamically changing environments [2].

In this paper, we are interested in those agents that could be implemented in computer systems called software agents. By the term Software Agent (SA), we mean a computer program that simulates an agent or acts like an agent when it runs. And we also call a Multi-Agent System (MAS) consists of a group of software agents a Software Multi-Agent System (SMAS). The most interesting features of software agents and of software multi-agent systems are autonomy, adaptability, interaction, and interoperability.

There are some pioneer researches about software agents and their implementations in computer systems, such as [3] and [4], but with many limitations and far from practical application [5]. D. A. Moreira and L. T. Walczowski developed specific purpose software agents to solve practical problems [6]. Nowadays, component software techniques come into practice. A great number of large-scale distributed heterogeneous applications has been developed by using component software techniques. We adopt component software approach to deal with the implementation of SMASs.

Component software approach has two main standards CORBA [7] and COM [8] at present days. We study on implementing SMASs in these two frameworks.

2. Software multi-agent system

Software agents in a SMAS can run in one computer system or be distributed in heterogeneous computer systems and are of following main characteristics:

□ Autonomy

A SA's computing capabilities, such as planning, learning, and reasoning, are determined by the SA itself.

□ Adaptability

A SA can percept its environment and the changing of its environment, anticipate other SAs' capabilities, and adapt its behavior according to the changing of its environment and the other SAs' behavior.

□ Interaction

A SA can affect the other SAs' behavior in its environment. It can ask the other SAs to do actions it wants by send messages to them.

□ Interoperability

By communicating with each other, SAs in a SMAS can do their own jobs and cooperate to achieve a common goal.

Therefore, a SMAS can be considered as a distributed heterogeneous application with special conditions. And a SMAS runs in one computer under one operating system platform is the simplest case.

3. Implementation policies

Component software techniques, which evolved from traditional object-oriented programming by improving Object Model and Object Reference Model, makes component objects could be implemented and used in different programming languages, under various operating systems, and across computer networks so that found a solid foundation for distributed heterogeneous application development.

The improved Object Model defines a component object (or simply object) as a uniquely identified, encapsulated entity that can be accessed only through its strictly defined interfaces. And how the object is implemented and where it runs is unknown to its client whose only way to get the target object's servers is sending the object requests. The improved Object Reference Model specifies how component objects interact with each other via various brokers.

The following features of component software techniques suggest they are good candidates for implementing SMASs:

- Components are building blocks in developing software.
- Adopting generalized client/server computing model in which an object may easily change its role as client or server. Clients send their requests to servers to ask for services while servers can send back results to clients if needed.
- Objects' behavior is strictly defined by their supported interfaces using Interface Define Language (IDL). The use and implementation of objects are separated in the way that users use objects according to the interface specifications while implementers implement objects in any possible means to satisfy the interface specifications.
- Clients' invocations do not act on objects directly, but send these invocation requests to some brokers. By such kind of mechanism, clients' invocations on objects achieve following advantages:
  - Location transparency: need not to know in which processes the target objects are or weather the target objects are on the same machine or not.
  - Server transparency: need not to know which servers have implemented which objects.
  - Programming language independency: do not care servers are coded in which programming languages.
  - Implementation independency: need not to know how objects do their work so that can use different ways to implement objects.
  - Architecture independency: need not to know what kinds of CPU architectures servers are using.
  - Operating system independency: need not to know under which operating systems servers are running.
  - Protocol independency: need not to know what communicating protocols are using to send messages.
  - Transmission independency: need not to know what kinds of networks are using.
- Objects manage their life cycles by themselves.
- Providing interface query mechanism so that clients can query the objects supported interfaces and functions in these interfaces.

Our general implementing policy is mapping software agents to component objects and putting those relative component objects together to construct a given SMAS.

Crossing programming languages, operating systems, and networks interoperability of component software is achieved by applying standards. There are two such kinds of standards that have come into practice. One is CORBA promulgated by Object Management Group (OMG), the other is COM released by Microsoft Corporation and Digital Equipment Corporation. The rest of the paper will discuss the implementation of SMASs with the two standards respectively.

4. Implementing software multi-agent system based COM

Implement SAs as COM objects (Technically, should be COM components but we use the term COM objects just for presentation simplicity and avoiding tedious details.). For a given SA, doing this involves several steps:

- Classify its functionality into method groups and then define one interface for each group or several groups until all the groups have been put into interfaces. The interfaces can be divided into two categories: one is called common or standard interfaces usually defined by COM itself, the other may be called special interfaces which represent object peculiar functionality, often only owned by itself, and should be defined by developers. After these special interfaces have been well defined, promulgate them in every possible means to let other developers known.
- Develop a COM object that implements those interfaces. In this paper, we also call this COM object a SA object.
- Register the SA object and its interfaces in the Windows System Registry on those computers that the SA object supposes to run.

How can these SA objects be built in and exhibit their autonomy, adaptability, interaction, and interoperability?

□ Autonomy

A given SA object uses the CreateInstance method in its interface IClassFactory to perform actual creation. Uses reference counting managing its life cycle, counting performed by AddRef and Release methods in its interface IUnknown. When counts to zero, the SA object will be destroyed. The other functions are all encapsulated inside the SA object so clients know nothing about its implementation.

□ Adaptability

A given SA object can be perceived through its QueryInterface method in its interface IUnknown. One given SA object can know the functionality and running states of the other SA object with interface querying and distinguishing the return values of invoked methods, and then decide which actions to take.

□ Interaction

One SA object acts on the other SA object by sending invocation requests which are of the same form as local procedure calls of most programming languages, but the real progressions are quite different. Firstly, the calling SA object initializes the COM Library if it has not been initialized. Secondly, the called SA object creates instance and the caller get callee's one interface. Thirdly, the caller invokes methods in that interface. Then the caller may ask other interfaces of the callee and commit more method invocations. After finishing with one interface, the caller reduces callee's reference count. Finally, the caller un-initializes the COM Library.

□ Interoperability

With interacting, one SA object can ask other SA objects to perform actions or get services from other SA objects and this can be down cross processes, computers, and networks by the nature of COM. Therefore, these SA objects may cooperate to achieve a common goal.

For a special application purpose, implementing and deploying every SA objects over platforms supporting COM, then getting the wanted SMAS.

5. Implementing software multi-agent system based CORBA

Implement SAs as CORBA objects. This work changes greatly as using different development environments. For a given SA, implementing it as CORBA object involves following main steps:

- Specify your IDL definitions containing interfaces.
- Generate the client stub and the server skeleton from your IDL definitions.
- Implement servants that actually implement the interfaces that are the functions the SA supposing to have.
- Develop a server program that initializes ORB (Object Request Broker) and POA (Portable Object Adaptor), creates servants, and starts monitoring requests.

We also need to discuss how can these SA objects be built in and exhibit their autonomy, adaptability, interaction, and interoperability.

□ Autonomy

CORBA proposed life cycle service is much more complicated than the reference counting mechanism in COM, but it also suggests using factory model to create objects and defines method copy, move, and remove in interface LifyCycleObject to manage object life cycle. These methods should be finally implemented in servants just like other functions of a SA object so they are all encapsulated. Although you can apply life cycle policies in POA, clients have no means to know the implementing details for they are all in server programs.

□ Adaptability

A SA object can dynamically query interface repository and obtain return values of invoked methods to know the functionality and running states of the other SA objects and accordingly take its actions.

□ Interaction

One SA object acts on the other SA object by sending invocation requests to ORB, then ORB works with POAs to find correct servants to perform actual operations. By the nature of CORBA, this kind of client invocations is location and server transparent, programming language, implementation, architecture, operating system, communicating protocol, and transmission independent.

□ Interoperability

Developed with CORBA, SA objects can obtain even better interoperability. Because, at present stage, COM builds its basic functionality in operating system, this can be considered as a specific implementation of ORB.

6. Conclusion

Component software techniques found a solid foundation for developing distributed heterogeneous multi-agent systems. Implementing software multi-agent systems based on COM or CORBA, those main features of agents are obtained naturally, so do the whole multi-agent systems. It could be expected that many large scale, complex, practical using multi-agent systems will be built by applying CORBA or COM in very near future.

Component objects fit agents' autonomy quite well, but we also noticed the fact that component software techniques do us little help in promoting the individual agent's intelligent capabilities, such as learning and reasoning, except for borrowing other agents' capabilities.

Interface querying and returning invoked method values lets agents be aware of the other agents' functionalities and running states, which makes building powerful adaptabilities possible.

The client invocation model adopting in component software techniques provides agents interacting with each other under distributed heterogeneous environments with unified and easily be used way.

Interoperability under distributed heterogeneous environments provided by component software techniques is the most powerful interoperability available nowadays. The multi-agent systems developed by applying component software techniques will naturally inherit the best interoperability.

7. References

[1] N.J. Nilsson, *Artificial Intelligence: A New Synthesis*, Morgan Kaufmann Publishers, Inc., 1998.

[2] Faratin, Peyman, Sierra, Carles, and N.R. Jennings, "Negotiation Decision Functions for Autonomous Agents", *International Journal of Robotics and Autonomous Agents*, 24(3-4), 1998, pp. 159-182.

[3] Y. Shoham, "Agent-oriented programming", *Artificial Intelligence*, 60(1), 1993, pp. 51-92.

[4] R.G. Michael and P.K. Steven, "Software Agents", *Communications of the ACM*, 37(7), 1994, pp. 48-53.

[5] C. Petrie, "Agent Based Engineering, the Web, and Intelligence", *IEEE Expert*, 11(6), 1996, pp. 24-29.

[6] D.A. Moreira and L.T. Walczowski, "Using Software Agents to Generate VLSI Layouts", *IEEE Expert*, 12(6), 1997, pp. 26-32.

[7] Object Management Group, *The Common Object Request Broker: Architecture and Specification*, Revision 2.3, 1998.

[8] Microsoft Corporation and Digital Equipment Corporation, *The Component Object Model Specification*, 1995.

Exception Handling and Structure Recasting in Distributed Process-Oriented Information Systems

Hong Xu, Guang Shu, Diji Zhou
Research Institute of Logistics Science
Beijing, 100071, P.R. China
shu_guang@netease.com, jerryyi@sina.com

Qingzhong Lu
Beijing University of Aeronautics and Astronautics,
Beijing, 100083, P.R. China

Abstract

The era where business rules are buried in Cobol code is coming to an end, and users themselves may seek to dynamically change their business rules. Customers require systems that more easily adapt to changing business needs, meet their unique requirements, and scale to large and small installations. On the other hand, the same technique is adequate for the slightly different purpose of producing a whole line of software products. Of course, a line of products may be obtained by variously instantiating an abstract model, and by adapting a given initial system to various requirements that appear simultaneously instead of evolving in time. Moreover, the diversification of a successful product may also be seen as a form of re-engineering. This paper analyzes the so-called exceptions and adaptive changes in modern enterprise-wide information systems and advocates a generic mechanism for handling them.

1. Introduction

The desire for an enterprise-wide system integration and business process automation has been the driving force of many research initiatives and new technological advances, including application integration and interoperation technologies [11] [12] and workflow technology [4] [9]. Even though business process engineering and automation require more than just a technical solution, workflow technology still offers a promising computational paradigm for effectively constructing process-centric, distributed and cooperative systems that involve people, software tools, application data and other information resources.

Workflow technology is characterized by extracting embedded process information from application programs and manipulating it separately in the form of workflow processes. Thus, a major concern of workflow technology is how to model business process logic properly and how to provide a suitable software support for its execution. As an emerging new kind of system software, workflow management systems (WfMS) provide an operational environment for workflow modeling as well as coordinated execution, monitoring and administration of workflow processes [1]. Workflow processes were once commonly distinguished and classified as production workflow, administrative workflow, and ad-hoe workflow [4]. As a consequence, WfMS products are often targeted at a special kind of workflow processes. Now, it is generally realized that different process types often coexist in enterprise-wide process management [8].

We believe that successful WfMSes should be able to support business processes at a knowledge and strategic level so that the application of situational knowledge and adaptive changes can be reasonably considered. In fact, upon the recognition of the above-stated problems, the newer generation of workflow systems, under the buzzword - adaptive workflow or adaptive workflow management [2] [13] [7], strives to make workflow systems more adaptive and flexible. HOON project [5] [6] is a research effort, aiming at establishing a conceptual and technological framework for adaptive workflow management. Based on the research results and experiences of HOON, this paper puts emphasis on a generic handling of workflow exceptions and adaptive changes. In this paper, we discuss workflow process modeling and analyze the so-called exceptions in workflow management, present some ideas that help synthesizing definable exceptional situations, and advocate a generic mechanism for the handling of ad hoc exceptions and adaptive changes. Then, we further investigate these issues together with the main features of HOON. Finally, we conclude with a discussion of our future work.

2. Processes, Process Models and Exception in Process Management

When applying workflow technology, business processes are implemented with workflow processes that are composed of a set of causally and temporally related workflow tasks. To describe and enforce workflow processes, executable workflow models are built with a workflow modeling language. A workflow modeling language may be pragmatically designed or based on a formal formalism like temporal logic, Petri nets, state and activity charts, or process algebra. The essence of workflow modeling includes: specifying business tasks and their interdependencies like control and data flow; specifying work context of tasks; and specifying application of workflow resources to individual tasks, such as deployment of application programs, data usage and assignment of roles. Task specifications define a set of logic units of work, their types and interfaces. Task scheduling reflects task inter-dependencies and execution sequences. Workflow resources included 1) software programs or workflow model fragments, which implement tasks, 2) persons who are in charge of executing tasks, and 3) data that are required for performing tasks. Resource application to a task specifies specific resources or resource types to be applied for performing the task. Depending on workflow modeling languages, the specification of resources is included in workflow models in different ways. While some approaches ac to specify all related aspects in one compact language at a single abstraction level [3], the majority of workflow modeling approaches builds workflow models with different abstraction layers. Some workflow models have a specification part and a task specification part (like in METEOR [10] [14], for example), covering the scheduling of tasks and task specifications respectively. Still others use an extra third layer for describing resources and their allocation in executing tasks [6]. Most workflow models are "executable" in the sense that they can be directly interpreted. Execution of a workflow model results in a workflow process, also called a workflow instance.

Exceptions of a workflow instance are abnormal events that may occur during the execution of the workflow instance and may cause the corresponding workflow process to deviate from the expected, predefined plan. Exceptions can be desirable because they provide a means to reflect environmental changes and other events that need immediate attention and reaction. Exceptions in workflow management can be classified as:

- Errors (information errors, operation errors, system errors, etc.) that result in incorrect outcome or failure in task and workflow execution.
- Violation of defined constraints, such as time out alarm, resource conflicts, etc.
- External events that require immediate ad-hoc adjustment or run-time refinement of workflow tasks in response to continuous process improvements of a long-running workflow process.
- Uncertainty in workflow execution. Occasionally, certain tasks can only be completely and definitely specified at run-time according to the concrete work context. The same applies for interdependencies as well as resources required for task execution. Some process elements can be parameterized, but it is often impractical or even impossible to enumerate all possible cases for a class of applications.
- Application-specific events that may require customization according to certain specific cases that are not reflected in the current workflow model. Such kind of customization will in turn require a reconfiguration of the workflow model and even the underlying software system.

Exception handling deals with detecting the events that raise exceptions, responding to them and providing solutions. How much workflow exceptional situations are considered during workflow modeling and how they are handled at run-time are very critical to the flexibility of a Whole workflow system. A majority of contemporary workflow management systems tries to handle exceptions at task level and by a certain aromatic means, often ignoring work contexts and human decisions. However, instead of being passively controlled, certain users in a workflow system should be treated as process owners. In reality, a user has frequently to adjust and situate his/her activities depending on specific circumstances under the influences of external events and his/her personal work style.

In the following section, we discuss an exception handling strategy that relies on a generic adaptation mechanism and situated user participation. Besides the development of more suitable modeling formalisms, tool support is an absolute necessity. Although infrastructure issues are not treated in depth in this paper, they should not be neglected. Recent advances in, for example, distributed computing infrastructure, software architecture, agent technology, reuse methodologies based on middle-ware and component, as well as methodologies for version and configuration control can all contribute to a more flexible and configurable workflow infrastructure.

3. Excepion Handling and Workfloew Adaption in HOON

Suitable workflow models and a flexible system infrastructure are very essential for adaptive workflow management and exception handling. To develop flexible workflow models that are open for changes, we need to develop a more flexible workflow modeling language. Methodologies are also needed to realize various changes and guarantee correctness. The HOON project has established a technological framework that facilitates a synthesized treatment of workflow adaptation and exception handling.

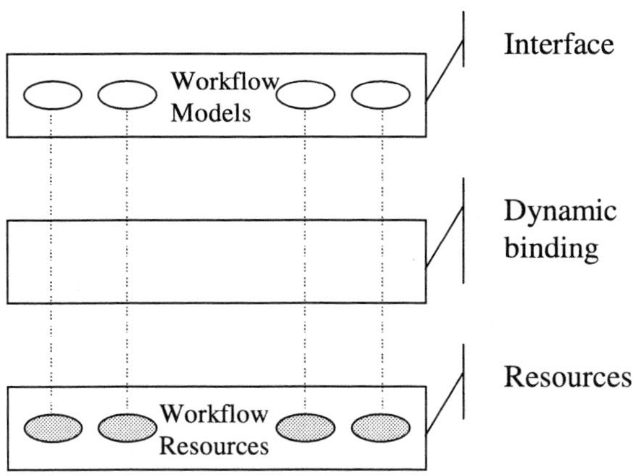

Fig. 1

Figure 1 illustrates the referential system architecture of HOON. A workflow system at run-time is viewed as a dynamic system configuration driven by a set of workflow models. There is a built-in open interface for each task in a workflow model. Workflow framework provides an architectural and operational basis for the dynamic configuration and interoperation of the underlying workflow resources. Workflow resources are wrapped as distributed objects. Workflow models play a central role hereby. The following issues are well supported in HOON: tiered workflow system architecture, separation of a task specification and its corresponding task implementation that is treated as a kind of workflow resources; encapsulation and dynamic binding of workflow resources. In the subsequent paragraphs, the main features of the HOON-based approach are introduced, with a focus on its adaptation and exception handling mechanism.

3.1 Flexible Composition and Dynamic Hierarchy of Workflow Models

Business processes are often hierarchically organized. Most workflow languages support process hierarchy by allowing a sub-process to be represented as a single task at a higher hierarchical level. However, the hierarchy tends to be totally fixed. In reality, While a business task can be accomplished by a workflow process in one situation, it may be accomplished by an atomic application program or another workflow process in another situation. Generally, there can exist multiple choices for accomplishing a business task, each of which is suitable for or applicable to a specific subset of all possible situations. Flexibility is significantly reduced if a business task is associated with only one solution in a fixed manner. In such a context, the first thing we have done is to make a business task more abstract by separating its specification from its implementation, and by allowing for a dynamic binding between a business task and its various potential implementations.

3.2 Systematic Management and Dynamic Binding of Workflow Resources

Resource management (resource collection, allocation, assignment, recollection, etc.) forms an extra internal process. From the resource management point of view, workflow models should also be a plan of resource allocation and utilization. Tasks are executed by binding and applying various related workflow resources. A task in HOON declares a unit of work, which can be performed by a software component, a database transaction or by a workflow process at a lower level. Through the separation of an extra resource management layer, dynamic binding between workflow resources and workflow tasks is made feasible and also safe from resource conflicts. Though most modeling languages define the application of resources to tasks in a fixed way, we advocate that resources be encapsulated in an object-oriented way, systematically managed and dynamically bound to tasks. With suitable tool support, certain changes of workflow resources can be made independently of workflow models.

3.3 Local Decision Making and User Involvement

Exception handling and adaptive changes, especially ad-hoc changes, have often to be strongly associated with user involvement. Users are either process owners Who can make structural changes to processes or process participants who can only make local changes regarding their tasks. Ownership that identifies a task owner and embodies change rights is associated with each business task. At run-time, the ownership is dynamically resolved by concrete users who become run-time task owners. Based on their ownership, users can initiate changes for modifying the deployment of all potential resources for performing their tasks, including application programs or workflow models (see Figure 2). Structural changes may affect a number of participants. Only process owners can initiate structural changes to workflow models accessible to them, either making new versions or changing running instances of the models. Normal users are not encouraged to make structural changes to the workflow model in which they are only process participants. Since a process owner understands the logic of the workflow model as a whole, he/she knows the implication of changes in contrast to a participant who is unaware of the consequences of a change. As a participant of a workflow process, a user delegates the needs of structural changes to the process owner instead of changing the flow structure by him/herself As such, there is a dynamically evolving responsibility hierarchy for handling exceptions and changes. On the other side, a process owner can delegate ownership to a user if appropriate, and subsequently the user can apply structural changes. Note that a participant is the owner of all the resources associated with his/her task. In particular, if the task is to be implemented by a workflow process, the user becomes the owner of the workflow process and can make structural changes in the corresponding workflow model to create a new version. To sum up, suppose a user is assigned a business task that is to be realized by a workflow process in which multiple users are involved. Only he/she, not one of the participating members, then has the right to make new versions of the corresponding workflow model by making structural changes.

3.4 Exception Handling and Adaptive workflow Management

As mentioned earlier, business tasks in HOON are made more abstract in the sense that their execution semantics can only be completely determined at runtime. Task structure and exception handling mechanism in HOON are sketched in Figure 2. In HOON, the handling of workflow exceptions is realized through a generic adaptation mechanism. In other words, exception handling and workflow adaptation are considered and treated in a unified way. Note that the occurrence of exceptions should be clearly distinguished from exception handling. The later is much more specific to languages and methodologies. Also, the detection and handling of an exception may be at different sites. We are convinced that mechanisms for exception handling and adaptive workflow can be tightly combined into a generic adaptation handler as shown in figure 2. Through the generic resource binding and adaptation handling mechanism embedded in each task, a user can try different variants, dynamically acquire new resources, report problems, issue commands to retry, postpone, or abort tasks, and seek help from the process owner or a system manager. Mechanisms for adaptive workflow management can help converting certain definable exceptions into normal cases. For example, if certain exceptions happen too often, then it may mean that the corresponding workflow model has flaws and needs to be modified to incorporate these cases into the normal flow. The responsibility hierarchy is used to report problems to a higher position if one is unable to handle it. For example, if an exception deals with more than one task and therefore a run-time task management component is unable to resolve it, the task management component can report it to a corresponding run-time workflow management component that might have been initiated by another task management component at a higher level. Exceptions that nobody is the definite target for its reception will be delegated to the system manager.

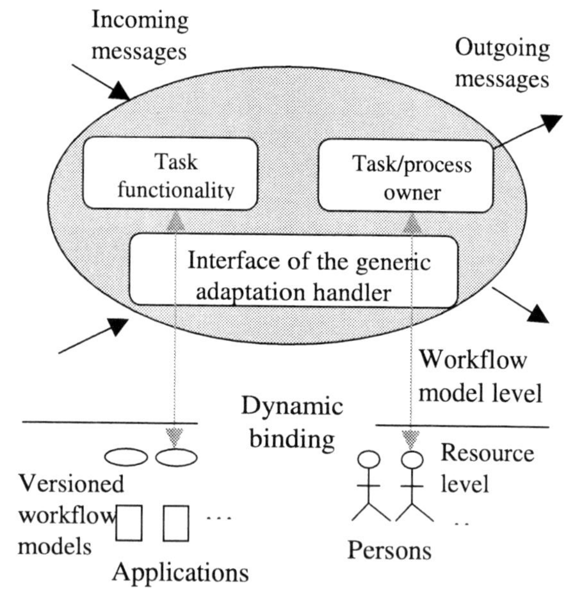

Fig. 2

4. Conclusion

So far, the successful rate of workflow management systems is still low, and many of them are limited to certain specific problem scenarios. The paper identified exception handling and adaptive changes as one of the major hurdles to their successfulness, and discussed some useful ideas for a potential improvement. In fact, solution for an enterprise-wide business process engineering and application integration is multidisciplinary in nature. We believe that multi-paradigm approaches, such as WCCS (Work Coordination and Collaboration Systems) [7], can bring new light to workflow management. HOON should go one step further toward an integration of coordination, collaboration, information management, and infrastructure technologies. Meanwhile, our intermediate enhancement of HOON is to introduce agent technology to re-build the generic exception handling mechanism and to further investigate the correctness issues associated with adaptive changes.

5. References

[1] Bussler, C.; "Specifying Enterprise Processes with Workflow Modeling Languages", In: Concurrent Engineering Research and Applications. Special Issue; Application of Enterprise Modeling Languages for Concurrent Engineering. Vol. 4, No. 3, September 1996

[2] Ellis C. A., Keddara K., Rozenberg G. "Dynamic Change within Workflow Systems", Proceedings of the Conference on

Organizational Computing Systems, pp. 10-21, 1995

[3] Forst A., Kuhn E., Bukhres O. "General Purpose Workflow Languages", Distributed and Parallel Databases, 3(2), pp. 187-218, April 1995

[4] Georgakopoulos D., Hornick M., Sheth A. "An Overview of Workflow Management; From Process Modeling to Workflow Automation infrastructure, Distributed and Parallel Databases, vol.3, pp.l 19-152,1995

[5] Han Y., Himrnighoefer J., Schaaf T., Wikarski D. "Management of Workflow Resources to Support Run-take Adaptability and System Evolution", Proceedings of PAKM'96, Switzerland, Oct. 1996

[6] Han Y. "HOON -- A Formalism Supporting Adaptive Workflow", UGA Report UGA-CS-TR-97005, 1997

[7] Han Y., Sheth A. "On adaptive Workflow Modeling", Proceedings of 4th international Conference on information systems, Analysis and Synthesis, Florida, July 1998

[8] Han Y., Sheth A., Bussler C. "A Taxonomy of Adaptive Workflow Management", Proceedings of CSCW'98 (Also to appear in the CSCW Journal), Seattle, Nov. 1998

[9] Jablonski S., Bussler C. "Workflow Management: Modeling concepts, Architecture and Implementation", International Thomson Computer Press, 1996

[10] Krishnakumar N., Sheth A. "Managing Heterogeneous Multi-system Tasks to Support Enterprise-wide Operations", Distributed and Parallel Databases, 3(2), April 1995, Pp. 155-186

[11] Open Application Group, www.oag.org

[12] Object Management Group, www.omg.org

[13] Pareschi R., Michelis G., Sarin S. (organizers) "Workshop on Adaptive Workflows" in Proceedings PAKM'96, Basel, Switzerland, 1996

[14] Sheth A., Kochut K. J., Miller J., Worah D., Das S., Lin C., Palaniswami D., Larch J., Shevchenko i. "Supporting State-wide Immunization Tracking using Multi-Paradigm Workflow Technology", Technical Report TR-96-001, LSDIS Lab, Dept.of CS, Unly. of GA, February 1996.

Research on Multi-Agent for General Group Decision Support System

Chaozhen Guo, Wu Dong, Jia Wu
Department of Computer Science and Technology, Fuzhou University, Fuzhou, Fujian,360002, P.R. China
gczwfm@pub5.fz.fj.cn

Abstract

The research work presented in this paper is based on the concrete background of the Group Decision Support System (GDSS), modeling and abstracting the process of the GDSS. The paper presents a design of General GDSS based on multi-agent approaches. In this design, we have abstracted a suit of procedures which can hold the communication between General GDSS system's coordinator (server) and GDSS system's decision-maker (customer). It overcomes the limitations of traditional systems, and can be used for the group decision in many fields. This paper focuses on introducing the design that many components were giving by the form of "COM", and the idea of "agent" was embodied in the design of these components. It offers some universal components for GDSS system developers, including: GDSS communication component, GDSS site monitor agent component, cryptogram agent component, preference agent component, alternation agent component, and cooperation agent component, etc. A prototype has been implemented.

1. Introduction

GDSS is a rising filed. It's an information system obtained by the decision-makers working as a team and going through a series of decision process. At present time, most researches on GDSS are aiming at a particular using background, so these software can not be widely used and spread, it makes the development more complex and less effect. So from the specific need, we have done some specific and model study on GDSS and present a suit of plan on multi-agent for general GDSS .We also have designed series components which is exist as COM and the idea of "agent" is embodied in it.

Classified according to the function, agents can be divided into intelligent and unintelligent, that is cognitive agent and reactive agent. The main difference between them is cognitive agent has inference ability, it can speculate from its own activities and other agent's activities. Reactive agent can do different thing just depend on the environment. We adopt some ideas of reactive agent, and has designed a series of general components that can be used by the GDSS developers. Main components are: GDSS communication component, GDSS site monitor agent component, cryptogram agent component, preference agent component, alternation agent component and cooperation agent component etc.

Combined with the theory of GDSS, we can divide the construction of GDSS into two parts. First, make sure a certain model that can fit for the communication in the group and also can make a decision that based on the group. After comparing all models carefully, we selected a typical model for GDSS that called Delphi model, which has been proved to be very successful. We have done some improvement on the Delphi. First, aiming at the characteristics of unstructured and multi-target decision, we introduced multi-target decision targets into decision process. These multi-targets have different weights so it can effect the decision result differently. On the process of decision making, every decider would give different marks on different targets. On the last stage of assessing plans, it has provided many prefer algorithm for choice, not as the traditional weighted averaged algorithm. One of the focal point of the second job of building model is a suit of intact communication mechanisms that adopted by group decisions. It can realize the information communicate in-group. The using state of GDSS mainly depends on the rationality and integrality of the communication mechanisms. Formally GDSS always made for a particular target, so developed GDSS system that can only used for a certain purpose. It wasted a lot of manpower and material. So it's very necessary to develop a suit of general GDSS. And it also needs a suit of general communication mechanism. After researching on some typical GDSS using facts, we can abstract a suit of communication procedure that can support group decision. This procedure breaks through the former limitation, it can be used on different jobs and areas' group decision. General GDSS communication protocol is base on TCP protocol, and it's a reliable transfers self-defining protocol between the server of GDSS and customer of GDSS. This protocol has adopted the ideas of TCP/IP protocol. It can be divided into two parts, one is encapsulation of transceiver's information, and the other is acceptor's explanation for information. We adopt relatively separate design module for information, and have integrated some agent software design ideas in it. So that it makes the whole communications part relatively independent and make it convenient for the information's transplant.

2. General structure of based on multi-agent

From the logic viewpoint, GDSS is mainly made of coordinator and decision-maker. Coordinator is very important in the whole decision making course. Its main duties are:

1) Collect every decision-maker's idea, question, advice during the whole decision making course.
2) Arrange, classify and store all the information, so as to facilitate the decision-making.
3) Coordinate the decision-makers so as to make the decision course more successful.
4) Issue information of all the decision-makers and the decision result.
5) Set associative parameters for whole decision environment.
6) Estimate all decision plans according to prefer algorithm, then work out the decision result.

Decision-makers participate the whole decision making course as individuals who provide all information is from decision-makers. Their main duties are:

1) Monitor decision network and attend meetings in time.
2) Bring forward plans, questions and advice.
3) Communicate with coordinator and other decision-makers.
4) Mark every decision plan.

Carry out information processing and show it to decision individual. Correspond to the duties of decision-makers and coordinators, using the agent idea. We draw the logical diagrams as shown in Figure 1 to Figure 3.

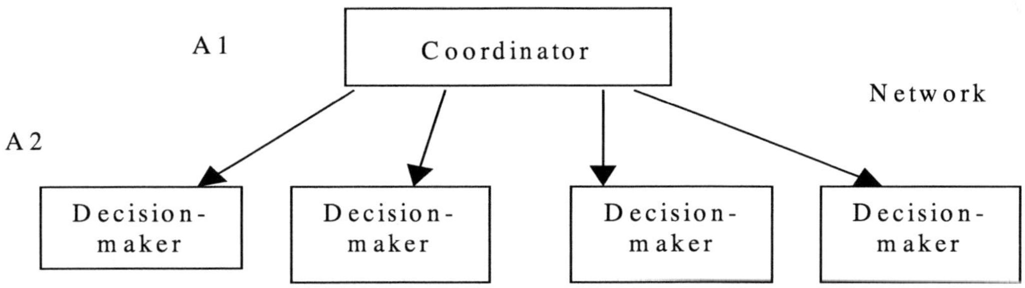

Figure 1. The first structure of coordinator and decision makers.

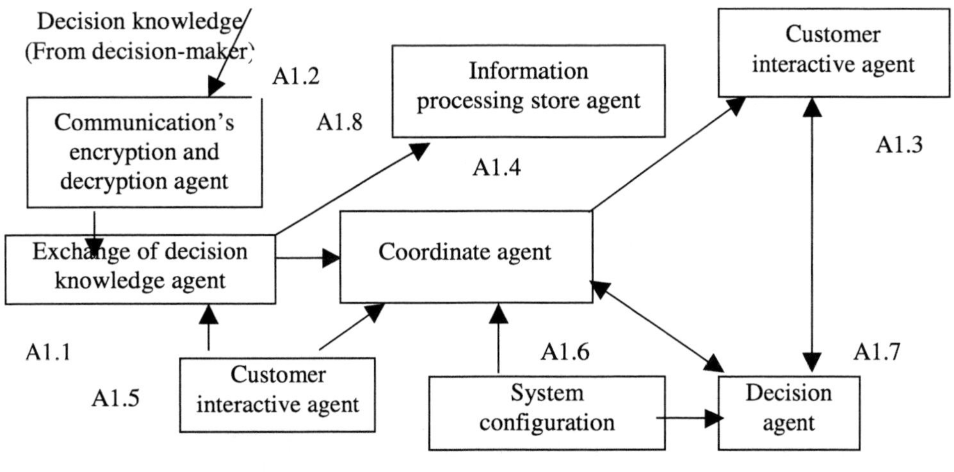

Figure 2. The related structure of every component in coordinator (Outspread for A1).

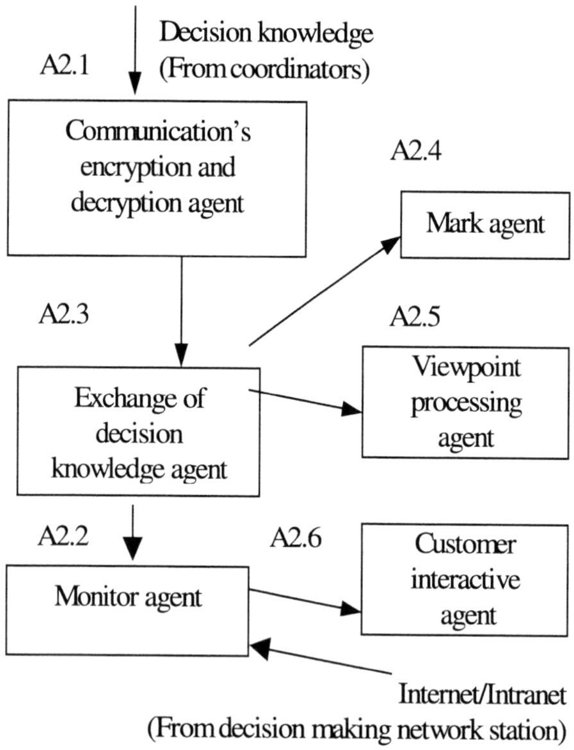

Figure 3. The related structure of every component in decision-makers (Outspread for A2).

From Figure 1 to Figure 3:

Communication's encryption and decryption agent: responsible for exchange information's encryption and decryption in decision making course.

Exchange of decision knowledge agent: responsible for exchanging information between server and customer or among customers.

Coordinate agent: responsible for coordinating decision-makers.

Information processing agent: processing the information from communication agent, then classify and put it into database or files.

Decision agent: using some prefer technology to make decision according to the information from exchange information agent (marks for different plan) and some relate parameters (such as decision-makers weights, index weights).

Customer agent: provide customers with decision information, let customers know progress of decision making course.

Control agent: dispatching all agents and submodules , control the progress of decision making (when to show viewpoint, when to discuss freely, beginning and ending of every decision making course).

3. Design on main components of general GDSS

After research on GDSS, we combine COM with Agent, modularize all components during the group decision-making course. We use both component technology and agent technology in system design. The main components are: GDSS communication component, network station monitor agent component, information encryption agent component, preference agent component, alternation agent component, and cooperation agent component, etc.

3.1 Network station monitor agent component

This component is based on idea of agent. It is mainly used for gaining information from long-distance resources, especially from Web network stations. The initial information obtained from the networks is processed into discriminable data format. Then we let information that gained and the information that you interested in do pattern matching, to find some information that is needed. Then tell customer the monitor result. It's a scheduled cyclic course.

This component's main use in GDSS is that all decision-makers can monitor network station, when they find something they interested in, this component will tell the decision-makers, so they can process the information. One typical use of the components all decision-makers monitor the network station's "The announcement of decision-making meetings". If their invited information is included in it, this component will inform decision-makers whether they should attend the meeting. By using this component, we avoid the traditional way that we have to inform the decision-makers one by one. The announcer only needs to send announcement information to the fixed place in network station.

The principle of this component is: first use application layer communication protocol

File Transfer Protocol(FTP) which is based on TCP/IP, to regularly download certain file's content from fixed network station., then put the content into buffer register and process it .Then delete the HTML text's id information and some useless information, and adopt some certain unification algorithm to match key elements and something that customers are interested in. If it find something fit for the elements, it will inform customer, or after a certain fixed period, download other FTP, and the process is repeated.

3.2 Information's encryption and decryption agent

This component's main use is information's encryption and decryption. To some certain data and secret key, first encrypt them by using normal DES arithmetic, then encrypt result again by using the encryption arithmetic which self-designed. Output the finally result. Decryption course is to decrypt the encrypted information by using its own decryption arithmetic, then decrypt the data by using normal DES decryption arithmetic, to get source information.

This component's main use in GDSS is protecting information during the whole decision-making course. At the beginning of sending information, it encrypts information by using encryption arithmetic, and decrypts the information in the receiver, in order to assure the security of information in transfer courses.

The normal input of DES arithmetic is eight bytes, but Chinese characters and other characters exist in information transfer course. If the input information is not processed effectively, the DES input may include half of a Chinese character. It may cause a lot of trouble in decryption course. So before encrypt a certain information, we need to do some special processing. Likewise by inserting some characters into proper places, in order to avoid half of character. Just the same, at the decryption course, we need to process the information again, in order to delete the special characters that inserted at encryption course.

3.3 Preference arithmetic agent component

This component's main use is that it can provide much preference arithmetic for "multi-index for group decision", and help the customers elect some better plans from all plans. Now this component provides two kinds of typical preference arithmetic. One emphasizes on the individual and group eccentricity, and the other is based on the fuzzy theory of group decision.

(1) Multi-index decision arithmetic that is based on plan's eccentricity

Multi-index decision problem is an important problem in area of decision. The analyze that is based on effective function is a direction in it. But it needs every decision individual can provide its own effective function expressions, which is very difficult. So to some certain problem we can avoid using function expressions. Combined with some common multi-index decision arithmetic and the universality of GDSS, by using a kind of interactive method and the improvement of a typical decision arithmetic, we adopt this idea of arithmetic:

Set standby plans set as $D=\{x^1, \ldots, x^s\}$, and $x^j \square R^n(j=1, \ldots, s;s\square 2)$ is the j th plan. Mark decision group as $G=\{DM_1, \ldots, DM_l\}$ and DMr($r=1, \ldots, l;l\square 3$) is the r th decision individual. Set DMr's estimate index vector in G is

$$f^r(x)=(f_1^{\,r}(x), \ldots, f_m^{\,r}(x))^T, \qquad r=1, \ldots, l$$

And $f^r_k:D\text{->}R1(k=l, \ldots, m;m\square 2)$ is DMr's the k th estimate index mark for plan.

By using plan's conception of good distance number and bad distance number, and make group's preference degree for plan by using variance. Then make the effective method of doing group sort to a discrete group multi-index decision-making problem(GVP).

By using individual preference information that is G's decision individual DMr($r=1, \ldots, l;l\square 3$) gain the plans $x^j(j=1, \ldots, s)$ according to multi-index $f^r_k(k=1, \ldots, m)$, we can ensure the preference degree for the plans.

(2) The preference arithmetic which is based on perfect point displacement theory:

Perfect point displace theory is considered to be one of the most important theories, it's based on conformance theory. Zadeh said that: the plan which is near the perfect point have more preference than the plan that is far from the perfect point, and it's a rational behavior for people electing the thing which is near the perfect point. Now let's show the point displacement module and some relate theories, by using fuzzy set theory.

There are two ways that can ensure the weights. One is every decision 's weight has already been fixed before making decisions. The other is to collect the mark that decision –makers give during the decision making course, then get every decision's weight by using average arithmetic. And the first one is more common, so in this paper we also adopt this method.

On the design of group decision preference arithmetic, to every decision index's fetch area, we make it normal as $X=\{x, 0<x<=100, x\square n$ (n is natural number)$\}$ in order to mark and statistic more conveniently. At the same time to every index, the first kinds of it's fuzzy set's subject function, we set $d^{ki}=x^{ki}/x^{*i}$. When ensure the subject function, we elect the Hamming distance when p=1, and Euclidean distance when p=2.

So to the final result of the final plan, we can find the smallest mark is the nearest to the perfect plan, and it's the best result, the rest of plans' taxis can be found out by the result of counting. In this way we can have a taxis about plan's quality.

These two kinds of preference technology have their own characteristics, they can provide the customers with the best consultation in the finally decision by taxis of every plan's preference.

The excellence of providing multi-arithmetic in the term of component is that we only need to improve the intrinsic component but needn't to change the system's source programs. And we only need to insert the new preference arithmetic into the components and change some information of the number of arithmetic and the method of explain. So we can buildup the functions of system, and it can show the superiority of the software technology that is based on component.

3.4 Coordinate agent component

This component's main use is to controls the decision-makers behavior by some coordination methods in order to control the decision-making course and let whole decision-making course develop well.

Considering the common use of the system, this component's main use is do the effective verification on the ground of plans that decision-makers raised. As the presenter always presents the ground of plans when presenting the plans, in order to improve the plan's reliability. But every plan's ground has its own dependency relation and consistency. So not every ground is feasible and reliable. By using coordination agent component, we can estimate the ground's validity and reliability, to provide the information for the decision-making course, and make it as one of the objective ground for setting the reliability and weights. Just like preference arithmetic, this component is good for system's upgrade and strengthen.

3.5 Customer interactive agent component

This component's use is that it provides a good and visual man-machine interactive interface. The realization of this component is based on multimedia agent technology that is provided by Microsoft Company, using COM, then insert some special functions which is fit for group decision-making. This agent is responsible for showing the customers the information in term of pictures and characters. It provides beautiful multimedia customer interfaces.

4. Conclusion

This paper is about the study on the problem in decision-making course. We abstract the decision-making course and design common components and common communication rules for the decision-making course. We also adopt two kinds of special preference arithmetic as the grounds of decision-making. We provide a series of COM components for GDSS developers by combining COM and agent technology. These common components are used in forecasting GDSS and better plan election GDSS. The practice has proved that these components are feasible, but it still needs further research and discussion on multi-agent coordinate management.

5. References

[1] Chaozhen Guo, Yandong Kang, Research and Implementation of Communication Components for Group Decision Support System, Journal of Software, Beijing, 2000,11(3): 421-425.

[2] Wu Zhaohui, Mo zhenrao, Pan Yunhe, Group Decision Support System(GDSS): module and sort, Computer Science, 1997

[3] Hu Yuda, The preference arithmetic for GDSS,Theory and Practice of Systematic Project, 1996.3

[4] hen Min, Wang Jin, Shun Hongkun, "Distributed Self adaptive Agent System Moduel", 1998

[5] Lu Kaichen, Computer Cryptography: the safety in communication, Tsinghua University Press, 1990

[6] Wang Xinzhong, LinYi,Yu Lei, Research on Calculation Method of Combined Evaluation, Theory and Practice of Systematic Project, 1999.11

[7] Li Hongchen, Yang Guangxia, Shi Meilin, Orgnization and Communication of Agent in DICSE, 1999.

A Strong Migration Method of Mobile Agents Based on Java

Hong Wang
Institute of Computing Technology, Chinese Academy of Sciences, Beijing 100080, P.R. China
sunzfc@jn-public.sd.cninfo.net

Guangzhou Zeng
Dept. of Computer Science Shandong University Jinan, 250061, P.R. China
gzzeng@sdu.edu.cn

Shouxun Lin
Institute of Computing Technology, Chinese Academy of Sciences, Beijing 100080, P.R. China
sxlin@ict.ac.cn

Abstract

In many CSCW systems, mobile agents are used to do some important work. However, mobile agents add a notable amount of complexity to all common problems of distributed systems, and cause a couple of new ones, such as location management, frequent disconnection and migration, etc. In this article, we focus on the technique for strong migration of mobile agents. Capturing and reestablishing the agent's state is the most important problem in strong migration of mobile agents. In this paper we describe a way to save and restore the state of a running Java-based agent. We achieve this on the language level, without modifying the Java Virtual Machine, by instructing the Java source programs with a preprocessor which is used in a mobile agent scenario to offer strong migration for mobile agents based on Java. The automatically inserted code saves the runtime information when the program requests state saving and reestablishes the program's runtime state on restart.

1. Introduction

Recently research on mobile agents has attracted much interest. Mobile agents facilitate the support of heterogeneous distributed systems, the reduction of network traffic and a smaller overall delay by encapsulating protocol steps into an agent. Today applications for mobile agents have existed in many fields such as information retrieval[1], management of large heterogeneous networks[2] and mobile computing[3][4]. However, applications are still evolving. In many CSCW systems, mobile agents are used to do part work[5]. However mobile agents add a notable amount of complexity to all common problems of distributed systems and cause a couple of new ones such as location management, frequent disconnection and migration, etc. In this article we focus on the technique for strong migration of mobile agents.

Mobile agents are actually programs that can move from one host to another. They can initiate their own transfer by executing a special instruction in their code. The agent's state can be divided into two parts: the runtime state and the date state. The runtime state refers to the control information for the agent running, consisting of the program counter, the stack, etc. The data state refers to the data information for the agent running, consisting of the agent basic attribute, the local variables, etc.

The agent migration can be divided into two kinds according to whether it transfers the runtime state or not: Strong Migration and Weak Migration[6]. Strong Migration also called Transparent Migration, means: When the agent requests migration, both the runtime state and the data state have to be saved and shipped to the destination together with the agent program code. At the target destination the moved agent can be restarted in exactly the same state and at the same code position as it was before migration. Weak Migration also called non-Transparent Migration, refers to: When the agent requests migration, only the data state is saved and transferred to the destination together with the agent code. At the target destination the moved agent can not be restarted at the same code position as it was before migration but from the program entrance. Both mechanisms require the capturing of the state information and the reestablishment of the saved state during restart.

Capturing and reestablishing the state of a running program is a well-known issue in different areas of computer science[7]. For example, it is used in distributed operating systems to provide load balance functionality. In such a scenario, the state of a program in execution (i.e., the process state) is captured and sent to some other hosts with low load. The receiving host creates a local process that has exactly the same state as the process whose state was captured. State capturing can also be used to provide fault tolerance or persistence[8] in a distributed system. The state of programs or processes is captured at regular intervals and is written to a stable secondary storage. When the system restarts after a crash or a regular system shutdown, the saved information is used to reestablish the processes and the processes can continue to work. Our application scenario is the migration of mobile agents from one host to another. This differs from the process migration for 'traditional' purposes (e.g., load balancing) in the sense that a migration is initiated by an agent itself but not by an external control instance.

One problem in capturing the state of the program is

that the required information is located in different places: the program variables are accessible from within the program itself (i.e., on the language level), but in contrast to this, all the runtime information is located in lower hierarchy levels. The state capturing mechanism has to collect all that information from different places.

Java is an object-oriented programming language. It is very suitable for agent programming. Firstly, it can directly support agent programming by its attribute of platform independence, security, static type test, distinct type transformation, automatic storage administration, etc. Secondly, Java provides a lot of flexible extended mechanisms such as Object Serialization, Threads, Reflection, etc. Although Java does not support state capturing, process migration now, we can realize these functions by using the flexible extended mechanisms in Java[8].

This paper is structured as follows: Section 2 introduces three approaches for realizing Java-based agent migration. Section 3 describes a method of state capturing and reestablishment of Java-based agents, including the method of state capturing, the method of state reestablishment and the method in the scenario of multi-threads. Section 4 discusses the limitation and the cost of this method and provides concluding remarks.

2. Migration methods of mobile agents based on Java

There are mainly three ways to address the problem of capturing/restoring the state of mobile agents based on Java.

In the first approach, which is called explicit management, the programmer has to explicitly manage backups in his agents. Managing a backup consists in storing in a memory area the data managed on the stack on which the agent depends. In Java, this memory area is a Java object that belongs to the state of the agent. When the agent state is restored, this backup object is explicitly used by the agent code in order to restart the agent at the point it was interrupted. For instance, in applications which implement weak migration of mobile agents, the programmer usually has to manage his own program counter[9].

The two other approaches in which a transparent mechanism is provided are called implicit management. The mechanism is independent from the agent code and is able to capture the agent state (data state and runtime state). These two approaches differ by their implementations: The first approach is implemented by extending the JVM in order to make threads' state accessible from agents based on Java. This extension provides a facility for extracting the thread state and storing it in a Java object which can be later sent to another machine. This extension also provides a facility for building a new thread initialized with a previously captured state. Currently, there are some systems provide the required state capturing of agents based on Java by modifying the Java Virtual Machine[10][11]. The second approach is implemented by pre-processing the source code of the agent in order to insert statements that back up the thread's state in a backup object. The main motivation of this approach is not to modify the JVM. When an agent requires a snapshot of the thread state, it just has to use the backup object produced by the code inserted by the pre-processor in the agent code. The restoration is achieved by re-executing a different version of the agent code (produced by the pre-processor) which rebuilds the stack and initialized the local variables with the values stored in the back up object. This is the basic idea of our mechanism.

3. Capturing and reestablishing state in Java

As Java is an object oriented programming language, the state of each Java program comprises the state of all the objects that exist at the time when the capturing takes place, the method call stack resulting from the method invocations during the program execution, and the program counter. Java is an interpreted language that requires an interpreter (the Java Virtual Machine, or VM for short) to execute Java programs. In fact, the method call stack and program counter information is located in the VM, It would be sufficient to have access to the information inside the VM to capture the method call stack and program counter information. And the state of all the objects that exist at the time when the capturing takes place can be obtained by using Java object serialization. So we found that it is indeed possible to capture the state of a Java program at the language level.

3.1. Capturing the state

Java object serialization offers an easy way to dump the state of all Java objects that exist in the Java-based agent. This state consists of the values of all variables (i.e., class and instance variables) of each object, which represent each object's internal state, and the information about the type of each object. By using object serialization, a large part of the information (all language level information) required to reestablish the agent's state can be captured. What is missing, however, is information located in the VM: the method call stack with the values of each method's local variables, and the current value of the program counter. So, we developed a preprocessor. By using it, we can instrument the user's Java code by adding code that does the actual state capturing, and reestablishes the state on restart at the target machine. We do this instrumentation by parsing the original program code using a Java based parser generated with the JavaCC-tool[12] from a Java 1.1 grammar. The preprocessor uses and modifies the parse tree from which the new code is generated.

Since additional code introduces time and space penalties, we only instrument the code where it is necessary and make sure that the additional code is executed only when necessary (i.e., when state capturing occurs). Now, a special method that is responsible for saving all local variables of methods and the value of the program counter is described as follows:

protected void saveState() throws Migration
{save every local variables; throw mig;}

Consider the program in Figure 1 which defines local variables for the method mymethod of class Myprogram and uses saveState() to save the value of new variables.

Since all the current state is located in the method call stack of the VM, we have to be able to traverse that stack and execute the state saving code which the preprocessor inserted in each method on the stack. In addition, no further code of the program must be executed after the state saving process is initiated. To conform to both requirements, we use the Java error mechanism. Similar to Java exceptions, errors can be thrown and successively caught. When thrown, the normal flow of execution stops immediately. An error can be caught by a **catch** clause of a **try** statement. If not caught, the error is propagated up the method call stack. This is automatically done by the exception/error handling mechanism of the Java VM. We make use of this behavior to traverse the method call stack and save all local variables of each method currently on the stack. The approach is:

(1) The method that initiates the process of saving state throws an error.
(2) The preprocessor inserts an encapsulating **try-catch** statement for each method that might initiate the state saving in order to save the local variables of this method.
(3) After executing the code related to saving the local variables in **try-catch** statement, the error is re-thrown. Thus each method in the stack in turn catches the error leading to the execution of the variable saving code in this method.

Thus the code of class Myprogram shown above is transformed to the code depicted in Figure 2.

Using an error instead of an exception to realize state saving has the advantage that errors do not have to be declared in the method's signature.

To save all local variable values we use a special **save** object that is inserted by the preprocessor in the top level class of the Java-based agent. In addition to that, all methods that might be part of the state saving process are

```
Class Myprogram{ //definition of variables
        ...
        public void mymethod(int I, real j, MyObject m)
        {int k; //k can be given any value anywhere in this program. //
        Hashtable h;
        ...
        saveState(); ... //save the value of I, j, k, m, h //
        if (k==5)
       { Vector x = new Vector(); ... saveState(); ...}
        // save the value of x //
        ...
        int v=10; //v is also a local variable in this method. //
       ...
        saveState(); //save the value of v //
      ...}
}
```

Figure 1. Program using state saving.

```
class Myprogram { // variables are saved by normal serialization
        ...
        public void mymethod ( int I, real j, MyObject m )
          { int k;
          Hashtable h;
          ...
          try { saveState() ;}
          catch ( Migration mig){ save(h); save(k); save(m); save(j); save(I);
                                  throw mig;} ...
          if (k==5) {Vector x = new Vector(); ...
                  try{saveState();}
                  catch ( Migration mig ) {save(x); save(h);
                 save(k); save(m); save(j);
                 save(I); throw mig; }...
                  } ...
          int v=10;
          ...
          try{ saveState(); }
          catch (Migration mig ){ save(v); save(h); save(k); save(m);
                               save(j); save(I);
                               throw mig; }
... }
```

Figure 2. Transformed code of class Myprogram.

provided to this special object. By this way, the method can be called as a local method. Because of this, all relevant method signatures have to be instrumented. Unfortunately, this leads to problems in the inheritance tree when such methods are parts of an interface that the class has to implement. Our solution to this problem is to generate a new interface incorporating the instrumented method signatures.

After saving and deconstructing the stack, all state information is held in the special **save** object. Since this object is part of the top-level program class, its value can be saved by the normal object serialization mechanism. Depending on the purpose of the state saving mechanism, the serialization information can be written to a file (in the case of check-pointing) or to a network socket (in the case of state transfer, as in mobile agent applications).

3.2. Reestablishing the state

Capturing the state of a running Java-based agent is only half the way. It must also be possible to construct a program state from the saved state information that is equivalent to the state of the agent from which the state was saved.

From the program's point of view, the flow of control should be continued directly after the statement that initiated the state saving process. This task requires rebuilding the program's object graph and its objects states, rebuilding the method call stack, and reestablishing the values of the local variables of each method on the rebuilt stack. Most of the program's state can be automatically reconstructed from the serialization information provided by the normal de-serialization process Java offers. This process results in an object graph which exhibits the same connectivity and object state properties as the object graph that represented the program at serialization time. What is missing is the method call stack that is not automatically rebuilt.

Rebuilding the method call stack. Since the **save** object (which keeps the relevant information) is part of the program's object graph, we can make use of that information to fill all the local method variables with the correct values once we recreated the method call stack. To do so, we just call again all relevant methods in the order they have been on the stack when the state capturing took place. To prevent re-execution of already executed method code, we have to skip all the code parts of each method which have been executed before the state capturing took place. So we introduce an artificial program counter. The artificial program counter indicates for each modified method which statements are already executed and therefore have to be skipped when the method call stack is rebuilt. It is not necessary to modify the artificial program counter after every instruction, since successive statements that do not initiate state saving can be treated as a compound one. Therefore the artificial program counter is updated only before and after such a compound statement region. Each compound statement region is guarded by an **if**-statement that checks whether the artificial program counter indicates that this compound statement region has to be entered or skipped.

Setting the values of local variables. The method call stack has been rebuilt as described above. Now we have to set all local method variables to the correct values (i.e., the values that we saved in the **save** object). To achieve this we insert declaration code for each variable that sets the correct value. For a variable there are two possibilities to get its value: the original initial value as provided by the programmer in the case of a normal agent start, and the value stored in the **save** object in the case of agent restart. To satisfy the Java compiler, all variables are initialized to a default value. The actual value assignment is done in an if statement. Figure 3 show the transformation.

3.3.Threads

In Java it is not possible to transfer the state of running threads by the means of object serialization. Since each Java program is executed as a thread by the Java VM, the converter is able to save the state of a single thread. To save the state of all program threads, we simply use a new **save** object for each thread, which stores the method stack information of the associated thread. On restart, all threads that existed at the time of state saving are newly created and read their runtime information from their **save** object. Saving the runtime information of each thread is simple, but there are other problems that require attention: Since threads run concurrently, one cannot predict at what time a thread that requests state saving will initiate state saving and in which state all other threads will be at this moment. From the point of view of all other threads, state saving could occur at every instruction. But it is impossible for us to be prepared to save the thread's state after each instruction, since it is rather inefficient.

So, we provide the programmer with a pair of a new

```
real I; int j = 7; Integer x = new Integer(5);          is transformed into:
    // init I
    real I = 0.0; if ( restart) I = so.restore (i );
    //init j
    int j = 0; if (restart) j = so.restore(j); else j = 7;
    //init x
    Integer x = null; if (restart) x = (Integer)so.restore(x); else x = new Integer(5)
```

Figure 3. Transformation for local variables.

```
setflag(): public synchronized void setflag()
   { flag[I]=1; resume();}//flag is an array of integer, Initiated to 0.//

allowgo(): public syuchronized void allowgo()
     { for (int I=1; I<=N; I++)
              {if (flag[I]=0 suspend(); }} //N is the number of current running thread.//
```

Figure 4. Synchronization methods of threads.

methods called setflag() and allowgo(). Any thread should execute allowgo() before it requests state saving. The method allowgo() checks if another thread requested state saving. If not, it returns immediately, otherwise it blocks the current thread until all running threads called the synchronization method setflag(). By this way, state saving occurs only if all running threads have called the method setflag(). Actually, this can be seen as barrier synchronization of all running threads. We describe these methods as follows:

4. Summary

We provide a mechanism by which it is possible to collect and reestablish the state of a Java-based agent. In the application area of mobile agents this allows strong migration. By this way we allow a programmer to program code that assumes transparent migration for a system that only provides non-transparent migration.

It is the most important that strong migration greatly lessens the programmer's burden. Then the question is the cost (in terms of run time penalties and additional code). Inserting code introduces time and space overheads, such as the time penalty at compile time and the time penalty at run time that is needed to collect the program state after the state saving process is initiated and reestablish the agent state before normal agent execution continues. All overheads depend on how often in the code the state saving method is called, how many other methods call a method that initiates state saving, and how many local variables have to be saved. Since we instrument only those methods that could be on the stack while saving the state, the instrumentation overhead is as small as possible. It is up to the programmer to decide, whether he or she is willing to pay the cost of the mechanism in order to program conveniently.

In order to realize the agent strong migration, we have to consider other two questions. First is environment. When moving an agent to another environment, this environment will usually differ from the former one. However, changing the environment between saving and restarting introduces additional difficulties when providing strong migration[1]. For example, it is only the local system layer that can realize opening a file, disconnecting temporarily from the file, and reconnecting to the open file some time later. It is required that the system should accord with some criterion such as CORBA or tag variables that carry local references. Secondly, the preprocessor requires that all methods that might initiate state saving are instrumented. This causes a problem when using standard program libraries. Normally, standard program libraries come without source code. In most cases this is not a serious problem, since these calls do not initiate state saving by themselves. But if the library call results in a callback to a program's method that can initiate state saving, the callback method could raise a flag that indicates to initiate state saving to another method.

The essential difference between our mechanism of capturing and reestablishing the state of a Java-based agent and other mechanism is that our mechanism realizes these functions at the language level and not at some lower hierarchy level (i.e., from outside the program). Our mechanism supports agent's strong migration even in the presence of multiple program threads. We think the price of our method is reasonable. Our work in future is eliminating some limitations in the preprocessor and using the reflection classes Java offers to do some parts of the transformation at runtime.

5. Acknowledgement

Our research is supported by the Chinese natural science fund (69975010) and Shandong province natural science fund (Y98G07103).

6. References

[1] D. B. Lange and M. Oshima. Programming Mobile Agents in Java-With the Java Aglet API, 1997.

[2] M. Baldi, S. Gai, and G. Picco. Exploiting Code Mobility in Decentralized and Flexible Network Management. Lecture Notes in Computer Science, 1219:13-26, 1997.

[3] D. Chess, B. Grosof, C. Harrison, D. Levine, C. Parris, and G. Tsudik. Itinerant Agents for Mobile Computing. IEEE Personal Communication, 2(5): 34-49, October 1995.

[4] Robert S. Gray, David Kotz, Saurab Nog, Daniela Rus, and George Oybenko. Mobile agents for mobile computing. Technical Report TR96-285, Dartmouth College, Computer Science, May 1996.

[5] Koch M. The collaborative Multi-user editor project Iris. TUM Info XI, Technical Report, 1995.

[6] Baumann J., Hohl F., Rothermel K., Straber M., Mole-Concepts of a Mobile Agent System, WWW Journal, Special issue on Applications and Techniques of Web Agents,1998

[7] Smith J.M., A survey of Process Migration Mechanisms, Operating System Review, Volume 22, Number 3, July 1988, pp 28-40

[8] Mira da silva M., Mobility and Persistence, Chapter in

Mobile Object Systems. LNCS 1222, Springer-Verlag, 1997, pp 157-175

[9] Baumann J., Hohl F., "Mole-Concepts of a Mobile Agent System", http://mole.informatik.uni-stuttgart.de/

[10] Peine H., Stolpmann T., The Architecture of the Ara Platform for Mobile Agents, Proc. of MA'97, Springer-Verlag, Berlin, April 7-8, LNCS 1219, pp 50-61

[11] Ranganathan M., Acharya A., Sharma S., Saltz J., Network-aware Mobile Programs, Proceedings of Usenix'97, Anaheim, CA, 1997

[12] Java Compiler Compiler, http:// www.suntest.com/JavaCC/

CoDesign - A Collaborative Pattern Design System Based on Agent

Bo Jiang, Chun Chen, Jiajun Bu
Dept. of Computer Science and Engineering, Zhejiang University,
Hangzhou, Zhejiang, 310027, P. R. China
bjiang@cs.zju.edu.cn, chenc@cs.zju.edu.cn, bjj@cs.zju.edu.cn

Abstract

For effective collaborative working between the parties in a pattern design project team, it is essential that a highly-efficient and feasible cooperative platform is available. This paper presents an agent-based infrastructure for the automated collaborative design of textile industrial pattern. Cooperative awareness among distributed participants is one of the most important issues associated with collaborative pattern design. Several techniques based on the management of intensity of awareness are proposed. In addition, the Pattern Knowledge Library enables intelligent design on the WEB greatly enhances the function of CSCD. The prototype is intended to serve as a useful cooperative system for designers, and should allow faster, better, and more economic, collaborative design of patterns.

1. Introduction

Due to the meta-components' regularity and tasks' complexity, pattern design often requires collaborative working between members of a project team in several fields, such as textile printing and weaving industries. A typical project involves a wide range of disparate professionals – clients, pattern designers, project manager, technical specialists, etc. – working together for a relative short period on the design. In many cases, the participants are geographically distributed making the need for effective cooperative information. With the rapid development of CSCW technologies and WEB-type environments, the effective and high quality collaborative design tasks will be carried out in a seamless way.

Increasingly, the benefits of concurrent and collaborative working with the design environment are now being recognized and much work has been down on collaborative building design [1][2], part design, and etc. However, little research has been down on pattern design while collaborative design in the textile industry is in great need. Globalization of the economy and the resulting competitive industrial environment demand that the textile industry develop new products with shorter lead times and better overall performance. As pattern design is often considered to be one of the most critical stages of the production process, designing more creative patterns in least time ensures the enterprise to survive. Therefore, the pattern design system with better cooperative awareness functions that facilitate collaborative design on the net and with certain intelligent design tools to generate multiple patterns automatically will be a good assistant to the textile enterprises. To construct such a system, we adopt the technique of software agents[3][4], which are autonomous entities capable of acting independently and cooperatively with other agents to achieve pattern design objectives.

The rest of this paper is structured as follows. Section 2 introduces the meta-model of workspace of CoDesign. Section 3 presents the infrastructure of the prototype based on agent. Section 4 proposes a number of techniques related to cooperative awareness and Section 5 discusses the Pattern Knowledge Library associated with intelligent collaborative pattern design. Finally, Section 6 concludes the paper.

2. Meta-model of workspace

The notion of workspace of CoDesign system provides its users with environment for collaborative design on the computer network. The main elements defining workspace in our meta-model are roles, actions, artifacts, and cooperation messages. They are described below:

- Roles refer to the organizational entities involved in the workspace and are assigned to participants in a workspace. Multi-role and right control of members are supported in our system. Project manager, task manager, common members (including designer and end-user) and etc. is assigned corresponding rights.
- Actions are various activities performed and process employed by each participant according to his/her role in a project. The action performed by participants may be individual or collaborative in nature.
- Artifacts are any objects consumed or produced in an action. The artifacts defined in this system can be presented in a hierarchical structure (See Figure 1).
- In Figure 1, P is project, D is document, L is layer, G is drawgroup, and O is drawobj. The definition is shown as follows [5]:

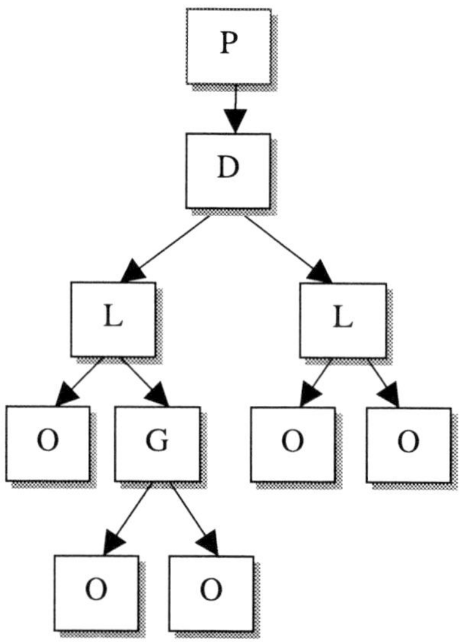

Figure 1. An example of hierarchical structure of artifacts.

project ::= attribute | < document | ...> | operation
document ::= attribute | < layer | ...> | operation
layer ::= attribute | < drawgroup | drawobj | ...> | operation
drawgroup ::= attribute | < drawgroup | drawobj | ... > | operation
drawobj ::= attribute | operation
attribute ::= data | < access_user | ...>

The project is composed of several pattern files. To increase the parallelism, the pattern is divided into several transparent layers. The graph, image or text objects rendered by members are contained in the layer. The overwhelming advantage of the structure is that it can support various granularity of concurrency.

- Cooperation messages are short strings exchanged between two peers to signal the occurrence of diverse events.

3. The multi-agent system architecture

As shown in Figure 2, the software architecture of CoDesign consists of a number of agents. An agent is a software component that has a role to play in the functioning of the system. We describe below each agent by its general functionality in the system, the services it provides and the data and knowledge it owns:

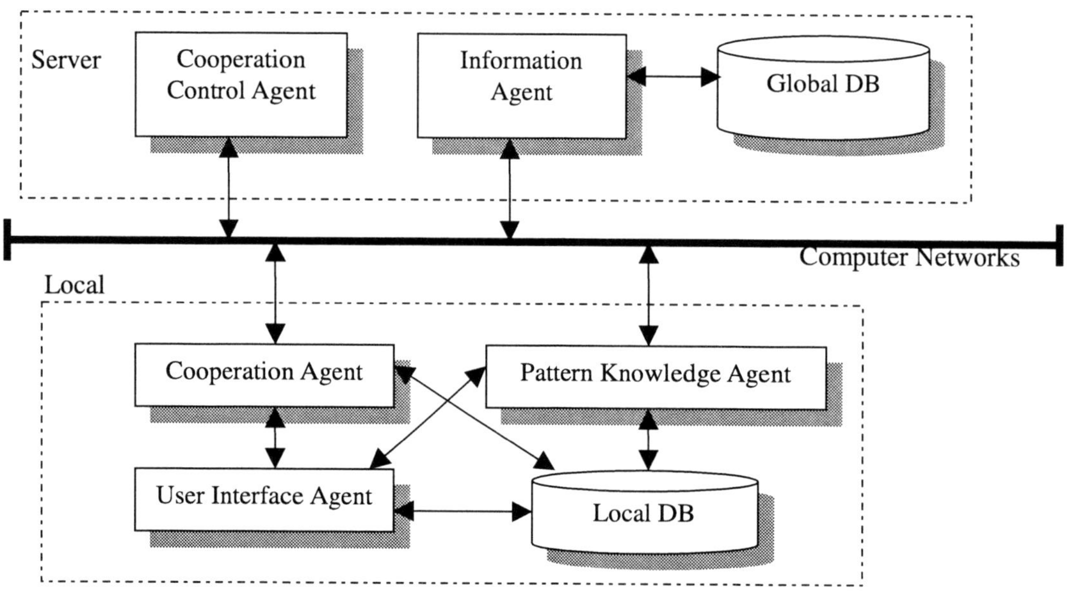

Figure 2. The architecture of CoDesign.

- Cooperation Control Agent (CCA): it comprises session manager and floor control manager. Session manager determines the manner in which the users join together into a collaborative session, i.e. who is allowed to join and leave a cooperative activity (scratch discussion, task assignment, pattern design, output evaluation) as well as how he joins (as the chair or just a participant). Floor control allows users of collaborative applications to utilize and share resources without access conflicts. It mainly works as a concurrent control mechanism. CoDesign supports several floor policies: Chair-decision, FIFS, and Turn-Taking.
- Cooperation Agent (CA): it consists of the following modules:
  1. Cooperative message streams handler: it

communicates with user interface agent to provides multi-user interface in WYSIWIS (What You See Is What I See) mode while in synchronically cooperative design. Messages formed in the predefined format will be optimized before UIA is notified to refresh the screen.

2. Cooperative awareness management: it supplies several means to support awareness. For example, action management, pattern navigation and telepointer. Details are given in Section 4.
3. Task Manager: it decides whether a task should start on the base of the state of the design workflow. Notify the user if it's time for a task to begin. Task manager also monitors the progress of tasks assigned to the user to ensure the task to be fulfilled in time.

- User Interface Agent (UIA): it interacts directly with the end-user and takes the responsibility to receive users' input and trigger the corresponding functions of CA to complete the design. Apart from react to the mouse click and key stroke, it also gain certain self-learning capability (such as remembering peers that frequently connect to) by "looking over the shoulder" of the user and detecting patterns and regularities in the user's behavior.
- Global Information Agent (GIA): it includes global information access services, as shown in Figure 2. CoDesign provides information access services by combining external database management information in order to avoid quick expansion of the database and improve the performance of transaction of image object. Global information database stores information related to user and right, artifact, project and task, application tool, pattern knowledge, and etc. The content of each database is described as follows:
  1. User database: stores personal information, its roles in each project and corresponding right.
  2. Artifact database: stores information of pattern document, layer, drawgroup, drawobj and their access requirements, such as access control, concurrent control, and etc.
  3. Project and task database: stores models of cooperative project and the executing state of each model.
  4. Pattern knowledge database: stores elements, samples, layout, and constraint knowledge related to intelligent pattern design.

  The data in local database maybe a subset of the global database or belongs to a particular user.
- Pattern Knowledge Agent (PKA): it plays an important role of developing intelligent design. Section 5 describes it in detail.

4. Cooperative Awareness

CoDesign supports both asynchronous and synchronous cooperative work modes. In the stage of synchronous design (for example, scratch discussion and pattern design phase), it is most important to supply sufficient ways of cooperative awareness. Or else, the cooperative design will lie in a chaotic state and the resultant lower efficiencies, more errors and more communication cost are unavoidable. In this paper, we propose some techniques accomplished by CA and UIA described below to support group awareness.

- Telepointer: Each peer is assigned a telepointer indicated in a different color to display the ongoing design activity of other users. Some telepointers can be turned off easily if the user is not interested in them or the workspace is cluttered.
- Pattern navigation: It provides the user with overview of the entire workspace context while he is concentrating on a certain part of the drawing canvas. Besides, it also supports awareness of the other user's current activities and locations.
- Collaborative action control: In order to support the facilities of cooperation (Redo / Undo) on the shared drawobj and action overview, collaborative actions are well organized and fine controlled. By developing collaborative action management, a participant editing a shared drawobj can undo the actions originally made by his partners if he is assigned the corresponding rights.

It is not new to simply introduce telepointer and pattern navigation into a collaborative design system. Some applications realize these two functions to enhance the ability of cooperation [6]. It is obvious that a large number of collaborative messages brought about by these functions will be transmitted and handled, which may result in low efficiency in the cooperative design. Hence, we propose above three techniques based on the intensity of awareness management.

The intensity of awareness supported by agents mainly represents the degree that participants affect each other at a time. Figure 3 shows the mutual relationships among the intensity of awareness and other miscellaneous modules. The intensity between two participants varies inversely with the distance between the drawobjs controlled by the corresponding peers. While the value of intensity predefined or set by users, the actions of "check in" and "check out" of users, the clicks on a certain part of the pattern navigation captured by the agent, the telepointers which selected by users, and etc. will also have some influence on the intensity. The intensities of awareness are stored in the local databases. The core unit – agent taking the responsibility of intensity variation management will automatically alter the knowledge database. Like the algorithm of progress schedule – multilevel queue scheduling in OS, we classify the received cooperative awareness messages into two queues according to the intensity. The queues owning higher priority will be handled first. Thus, the intensity management enables users to collaborate more efficiently and productively.

These three techniques based on the intensity of awareness will overcome the shortcoming of previous applications. Telepointers can be selected automatically by the agent with the variation of intensity. The drawing-areas and the navigation diagram will be refreshed according to the priority of the queues. Not all users but those with high intensity in a certain period of time will be stored to realize efficient collaborative action control. The feedback of telepointers and pattern navigation will affect the intensity of awareness through agents. The techniques that based on agent provide an effective method for collaborative awareness.

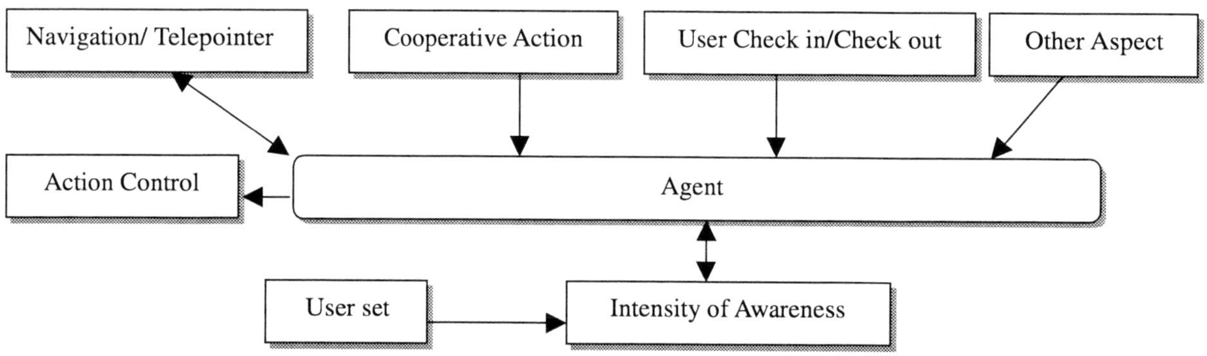

Figure 3. The relationship between agent and other modules.

5. Pattern Knowledge Library (PKL)

5.1 The content of PKL

Pattern Knowledge Library concretely describes the knowledge that assists users to design multiple creative patterns quickly, including resource library, layout knowledge library and constraint knowledge library.

Resource library includes element library and sample library. The former stores the elements of pattern that can not be segmented and can be used as an object in a pattern. The library stores about 5000 elements that are classified. The latter maintains the samples that can be utilized as models of new patterns. So it will be one of the most convenient ways to generated new patterns just by simple modification of the samples.

Layout knowledge library stores the strategies of layout (such as random, regular and symmetry) most frequently adopted. The position, size, angle of each element for the random strategy, the tracks of layout for regular strategy and the axis of the symmetry are recorded in the library. The knowledge will be helpful to generate multiple patterns with identical elements.

Constraint knowledge library records the information of colors (the color set that can be used to a certain pattern style or the base color) and describes the rules about how elements can be coupled with each other.

5.2 A Single User's Agent

To make full use of a single PKL, agent is related to the library. The responsibilities taken by Pattern Knowledge Agent are described below:

- Information search: The agent retrieves information from the library.
- Insertion of the resources: Abundant pattern element resources like graphics element CD published regularly will be installed and the well designed elements or assembled elements can be stored in the library.
- Knowledge capture: The novel knowledge of layout or rule of a new pattern can be captured by the agent.
- Intelligent design: The agent will provide users with a sufficient number of patterns based on the knowledge automatically.

It is obviously that intelligent design based on agent should integrate the technique of human machine interaction to acquire novel patterns. A pattern will be generated after the layout, the elements and the base color are selected. It can be changed into a large number of other patterns in minutes.

The information stored in PKL is considered to be the experience that owned by the agent. An agent starts out with limited experience. But with time goes, it can learn a lot from the user and become more intelligent to assist the user.

5.3 A Framework for Collaboration

The framework of multi-library is presented in Figure 4. As shown in Figure 4, Global Library maintains the basic knowledge that is public to the users in a certain area (e.g. LAN), whereas Local Library, which belongs to a particular user, may have more novel knowledge. Instead of each pattern knowledge agent re-learning what other agents have already learned and work independently, agents can simply ask for help in such cases. This gives each agent access to a potential vast body of experience that already exists. Over time each agent builds up a trust relationship with each of its peers analogous to the way we consult different experts for help in particular domains and learn from each other. With the assistance of multi-agent, a quick creative pattern design will become realized.

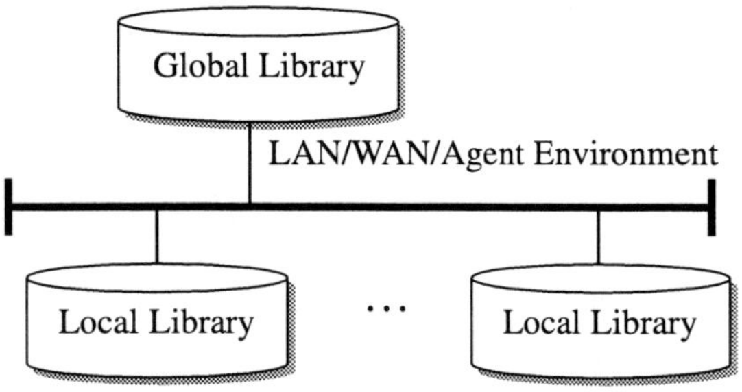

Figure 4. The framework of multi-library.

6. Conclusion

With the current proliferation of agent technology, it seems natural to adopt multi-agent techniques to bolster cooperative design activities. In this paper, we present an agent-based software system to support collaborative pattern design. The framework of the prototype system is introduced and the agents are described in detail. Furthermore, the paper discusses the critical parts of the system. The technique of cooperative awareness that mainly based on the management of intensity of awareness leads to more efficient collaborative design. The intelligent design relying on the PKL and the design knowledge shared on the web can bring out more novel pattern quickly. The framework of the system based on local network has been established. The techniques of cooperative awareness and intelligent design (now restrict on a single workstation) have been realized. Further research work continues to develop and implement CoDesign system in the wide area network environment.

7. References

[1] W. Dai, R. Drogemuller, "Collaborative Framework for Building Design", Proceedings of 1999 IEEE International Conference on Systems, Man, and Cybernetics, Tokyo, Japan, 1999, Vol. 4, pp. 23 - 26.

[2] B. Dave, "Teamwork Constructs in Architectural Design", Proceedings of 1998 Australasian Computer Human Interaction Conference, Adelaide, South Australia, Nov. 1998, pp. 45 - 52.

[3] N. Jennings, M. Wooldridge, "Software Agents", IEE Review, 18 Jan. 1996, Vol. 42, No. 1, pp. 17 - 20.

[4] D.T. Ndumu, H.S. Nwana, "Research and development challenges for agent-based systems", IEE Proceedings Software Engineering, 1997, Vol: 144, No.1, pp. 2 - 10.

[5] Yi Zeng, Duanqing Xu, Chun Chen, "Research on Internet based Collaborative Pattern Design System", in Chinese, Proceeding of 1999 National Workshop on Computer-Aided Industrial Design and Conceptual Design, 1999, pp. 213 - 318.

[6] Sun Wei, Lee Bu Sung, Yeo Chai Kiat, "JMS: A Flexible Collaborative Environment", Internet Workshop, 1999. IWS 99, Suita, Osaka, Japan, 18-20 Feb. 1999, pp. 195 - 202.

Of Networks, Interactions and Agents: An Approach for Social Network Analysis*

P. Temdee and L. Korba
Institute for Information Technology, National Research Council of Canada
Punnarumol.Temdee@nrc.ca, Larry.Korba@nrc.ca

Abstract

In the modern office environment, workers use computers, networks, and various software applications to allow them to work together. Measuring and understanding how effectively people work together in the performance of tasks can be used to assess the appropriateness of the tools they are using for collaborative work, or the effectiveness of their work environment in general. In this paper, we describe an approach for determining how people interact together in the performance of computer mediated tasks. We also provide the background for this work and some caveats concerning its application.

1. Introduction

Computer-supported cooperative work (CSCW) refers to the cooperative work carried out by a group of individuals with computer and network support [1]. It has particular application in the situation [2] where people work together in dynamically formed groups to accomplish a particular task. A work plan is quickly established. The plan is then divided into many subtasks and integrated together. The means for coordination and self-regulation are established.

There are four modes of the CSCW operation [1,3]: synchronous, distributed synchronous, asynchronous and distributed asynchronous. Group members can work together in any combination of these modes, but their work must still be coordinated effectively. Also, computer networks form the most common means for participant interaction. Attention has been paid to distributed approaches for CSCW [4] for some time. These approaches offer some potential advantages in scalability, reliability, extensibility, maintainability and flexibility of the resulting systems.

Computer networks form an environment that is generally open and distributed [5]. In order to make it easier to design, to implement and to maintain distributed systems, agent technology has received growing interest. Agent technology has also matured significantly in the last few years [5,6,7,8,9]. Agents have the ability to interact with their execution environment and to act asynchronously and autonomously. In addition, agents provide the advantage of problem partitioning and scalability. The number of applications using agent technology is wide spread and includes applications in domains such as security and information retrieval [9].

Social network analysis [10] provides a very powerful tool for understanding how people work together. In the recent past, there was much research demonstrating the use of social network analysis for different purposes [10,11,12,13,14,15,16]. Many of these approaches examine how work behaviors change when participants are supported by different kinds of media [11,12,13,15,16]. In some cases, there was an attempt to understand how such media affect the relationships between people [13,14,16].

Social network analysts seek to describe networks of relationships as fully as possible. They characterize prominent patterns associated with relationships set up within such networks. Many kinds of tracing modules have been proposed to extract information about the relationships facilitated by the networks. Statistical data processing tools have been used to discover interaction patterns. Finally, analysts have used these patterns to interpret and to uncover what effects the relationships among people in small groups have upon the organization as a whole. In our work we are developing automated approaches for discovering and analyzing social networks that are developed by way of computer and computer network mediated interactions.

Figure 1 illustrates an example of a view of interactions based upon an analysis of computer network interactions between individuals. In the diagram, all of members of the group are displayed as circles and arranged in relation to the relative frequency of their interaction. People who communicated more often with each other are placed closer together. The connecting lines indicate communication direction (arrows) and communication frequency (length of the lines). For

* NRC pub number: NRC 44198

example in the Figure 1, there are eight people, who are A, B, C, D, E, F, G and H, working together in the same group. Two kinds of relationships are shown: one way interactions like those between A and C, and C and H, as well as two-way interactions such as A and B, B and C, and C and G, etc. There may be a number of inferences taken from this example. For instance, the interaction between A and C may indicate the ordering by A of a task to be completed by C rather than an exchange of information. The interaction between A and B indicates an information exchange. Relationship information, as available from an organizational chart may be used to augment the analysis by way of an interpretation of the nature of participant interactions. The information used in the social network analysis can be gathered either manually or automatically.

Collecting information manually involves both the observation and recording of activities, or the use of

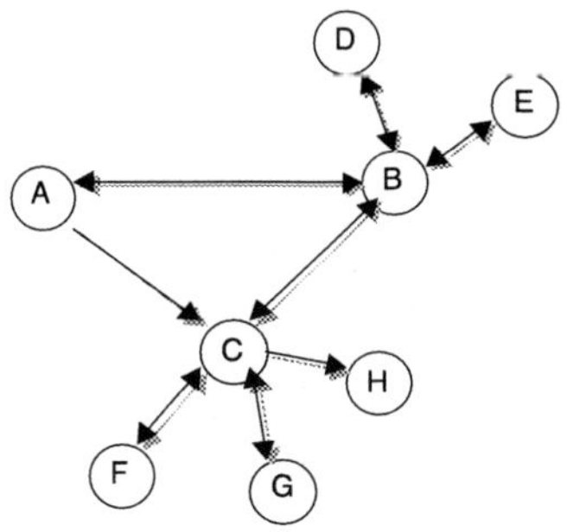

Figure 1. A typical diagram indicating an analysis of a social network based on network interactions.

questionnaires, interviews and diaries. Even though a large quantity of information can be obtained, its reliability is often questioned. Incorrect reporting, either intentionally or not, often occurs when participants record their activities. For instance, the obtained information may be biased because not all interactions are remembered equally well.

The automatic information gathering approach is often termed as an "electronic data gathering method". Normally, this approach obtains an overwhelming amount of data. While the data has high reliability and high accuracy, determining appropriate association and correlation between multiple collaborators is difficult. Effective data processing tools are required.

For CSCW, the information gathered is in two broad categories: network information and local information. Network information has been used widely in monitoring tasks [17,18,19]. This information can also be used in social network analysis [12]. Message headers, traffic information, etc. have been used for this purpose.

The other source of interaction information is local information [20,21,22]. Typically this information includes computer input and output activities generated by the participant while performing his or her work. Some examples of local activities include: file accesses, keyboard and mouse activity, etc.

2. Methodology

In this paper we present an approach for automatically monitoring and analyzing individual activities in computer network mediated interactions for thc purpose of inferring social or organizational interactions and

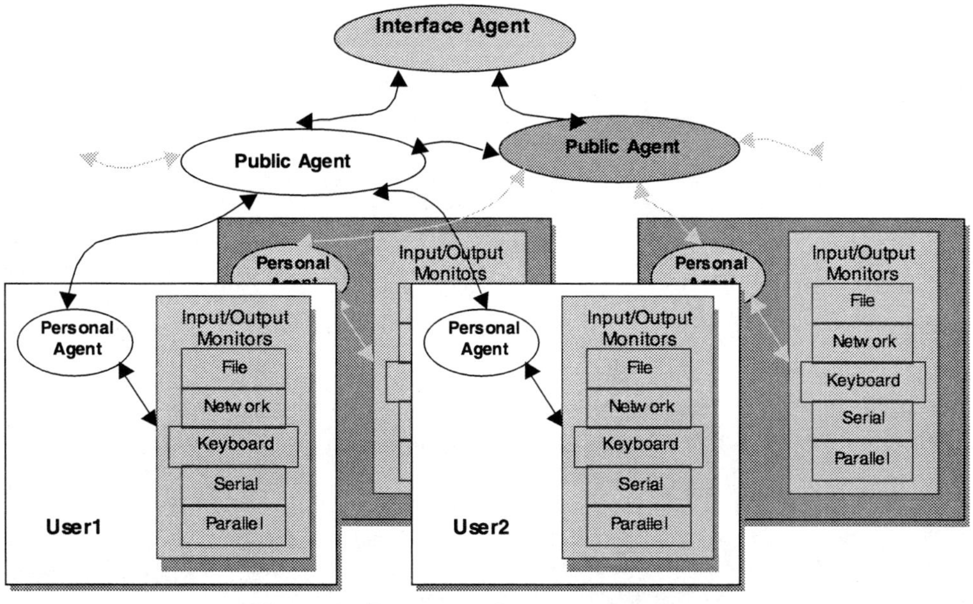

Figure 2. The general system architecture

relationships. As illustrated in Figure 2, the architecture is distributed and hierarchical. Located on every computer where a user might participate, is a personal agent. The personal agent gathers local information regarding participant activity. The personal agent may monitor a variety of different activities including file system, network, keyboard, mouse, serial or parallel ports. In our current work we only monitor file activity. In a hierarchical manner, operations performed by and information gathered by personal agents are coordinated by and aggregated by public agents. Public agents are arranged to handle zones of personal agents based upon the organizational structure as well as network requirements. Above all of the Public agents is the Interface Agent. The Interface Agent displays the mainly graphical results of the social network analysis. It also offers a user interface to manage the functions of the Public and Personal Agents.

Individual behavior patterns are determined by analyzing the information gathered while users perform either local or network operations. This arrangement makes it possible to monitor activity not only for social network analysis, but also of discovering security threats to the networks or on interconnected computer platforms.

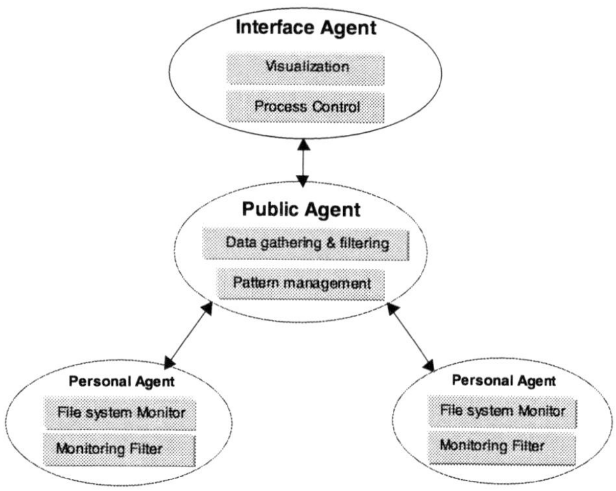

Figure 3. Contents of personal agent, public and interface agent for our prototype.

The public agent determines interaction patterns. It correlates the individual activities monitored by the personal agents. A neural network is used to identify particular kinds of interactions between the users.

Every public agent can communicate with all of its peers in search of correlation among activities of all participants. The personal agent has the function of acquiring input and output computer activities of each user and performing some filtering on the information based on feedback from the public agent. The public agent determines relationship patterns. It does this by examining the correlation between the individual behavior patterns among multiple personal agents. The public agent handles a variable number of personal agents depending upon the physical network configuration and the organization of the company. Public agents communicate with their peers to determine all of the personal agents involved in an interaction pattern. All of the public agents are connected to the interface agent. The contents of personal, public and interface agents are shown in Figure 3.

3. Example Analysis and Results

As shown in Figure 3, from a practical point of view, the prototype personal agent contains a means for monitoring file system activity and for filtering its monitoring. Activities related to file operations at each computer are tracked and time-stamped using Filemon [23]. File operations vary depending upon the applications being used and documents being operated upon by the user. The personal agent filters the information from Filemon depending on a few different criteria. These include:

1) The type of file.
2) The location of the file in the file system.
3) A list of candidate files that are active on other computers. The public agent feeds a list of files that appear on other personal agents to provide a distributed correlation function.
4) File life cycle information. Life cycle information includes the times when files are created, renamed, or saved with a new size. One of the objectives here is to be able to track files as they are modified in the editing process.

The pattern recognition modules use neural network and the decision modules based on the straightforward candidate relationship rules to evaluate the proposed approach. More complicated approaches will be attempted if this approach does not achieve the desired accuracy and reliability.

A neural network provides the recognition of file operations. This approach offers a way of dealing with the overwhelming amount of information obtained from file system tracing mechanism. The identification is based upon the predefined file operation patterns. There are three kinds of file operations: Edit, Read and NOP. The Edit operation is a concatenation of the following operations: the file is opened, edited and saved with the same or a different file name. After editing, files often have a new file sizes. The Read operation corresponds to a file being opened and then closed without alteration. NOP means that there were no operations during a particular time frame.

In the recognition module, a back propagation neural network [24] incorporating one hidden layer, is used. The network first uses a training input data set to produce its own output and then compares this with the target output. Learning occurs when the weights are adjusted in order to reduce the error between the actual output and the target output. A recognition module consists of two phases: training phase and recall phase.

In the training phase, sequences of file operations are used as an input to neural network. Before the neural network can identify any file operation in the recall phase, it must be trained with meaningful and reasonable number of stimulus patterns representing the range of operations. Operations are first labeled to the numerical data before being fed to the input layer of the neural network. Also the target outputs are set to the output layer. The target training output is shown in Table 1. An example of the training process for the Edit operation is shown in Figure 4. Once trained, the neural network uses its generalization ability to correctly identify each file operation in the recall phase.

Table 1. Target training output.

Target Output	Operation
1 0 0	Edit
0 1 0	Read
0 0 1	NOP

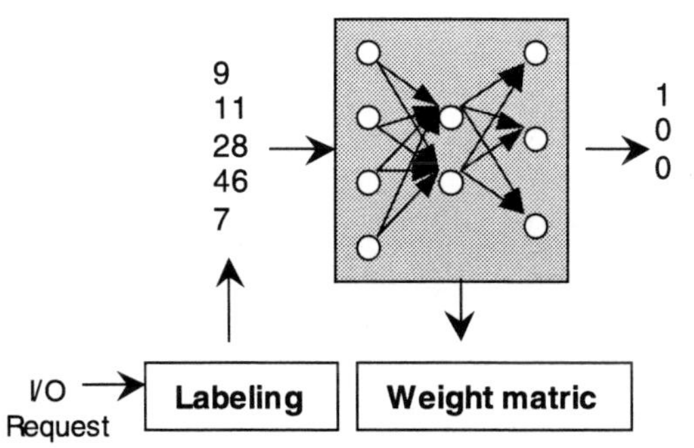

Figure 4. Traning process for Edit operation.

File operations are examined continuously. Examination is based on the assumption that the response operation occurs in the following tracing duration. The candidate relationship rules between two people are shown in a simplified form in Table 2. For the purpose of this explanation, the candidate rules have been simplified. The degree of change in the documents and the sequence and timing of document related activities are also used to characterize social network relationships.

In the example illustrated in Table 2, two people are considered to be sharing in the development of a document if and only if: they edit files that have a file name, time of use and file size correlation.

Table 2. Candidate relationship rules.

Operation of A	Operation of B	Candidate relationship between A and B
Edit file	Edit file	Share file together
Edit file	Read file	Share file together
Edit file	NOP	No connection
Read file	Edit file	Share file together
Read file	Read file	Share file together
Read file	NOP	No connection
NOP	Edit file	No connection
NOP	Read file	No connection
NOP	NOP	No connection

The file name offers an obvious correlation factor. Unfortunately this is not a reliable approach.

During ad-hoc editing and document development, either participant may change the file name to indicate a new version of the document for instance. The personal agent establishes connections between file names that might have been changed as a result of the editing process.

The time of an operation on a file is another parameter used for activity correlation. The personal agent monitors edit sessions, establishing start and finish edit times. These times are associated with root file names and are used to characterize an edit relationship.

The personal agent also monitors changes in file sizes as a parameter for correlation. The public agent uses information about file size, operation timing and name changes to track activities related to document development.

3.1. An interaction Scenario

This is an example of two people working together asynchronously in the development of a document. The two collaborators are A and B. The document upon which they are working has the file name X. Operations they perform on file X during eight different sessions are shown in Figure 5. Each session has a start and end time. Operations performed on X form a life cycle for the file. This simple example is illustrated in Figure 6. File X is first edited and then saved under a new file name, X1, by A. File X1 is then read by B. During this life cycle, the file name is changed three times, starting with X1 and ending with file name X4. Table 3 illustrates the summary of operations performed during the relationship established between A and B in the development of file X as shown in Figure 5.

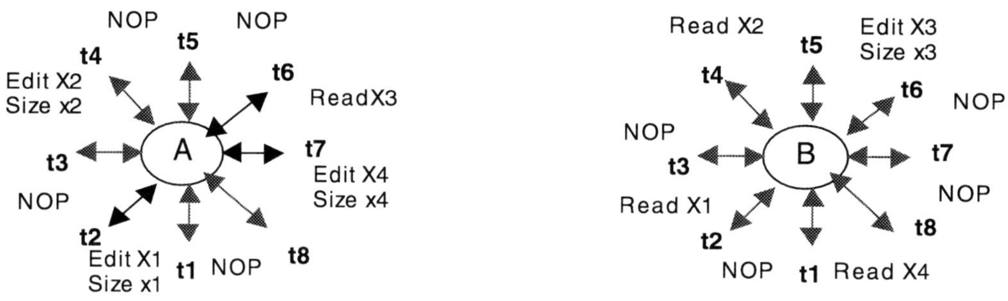

Figure 5. Operations in different time, which two collaborators perform with file X.

Table 3. Interaction table.

Time	Operations of User A	Operations of User B	Operations and relationship displayed at monitor
1	EditX1	NOP	A is editing file X1 B is reading file X1 A&B are sharing file X
2	NOP	Read X1	
3	EditX2	NOP	A is editing file X2 B is reading file X2 A&B are sharing file X
4	NOP	Read X2	
5	NOP	EditX3	B is editing file X3 A is reading file X3 A&B are sharing file X
6	Read X3	NOP	
7	EditX4	NOP	A is editing file X4 B is reading file X4 A&B are sharing file X
8	NOP	Read X4	

From Table 3, during the first time period, A is editing file X1 while B is not directly involved with file X. During the second time interval, A is otherwise occupied while B is reading file X1. The relationship rules (Table 2) determine the participation relationships. The system builds relationships as the work on document X develops. In this example, there is a time lag before B is recognized as a collaborator in the development of X. This is simply due to the asynchronous operations performed by A and B on the file.

4. Discussion and future work

Understanding how people work together in the performance of tasks can be used to assess how well collaborative tools perform, or the effectiveness of their work environment among other things. In this paper, we present an approach for automatically determining how people interact together for computer mediated tasks.

We use an agent-based approach. In our system, three kinds of agents interact with each other to analyze participant interactions and to present the results. The

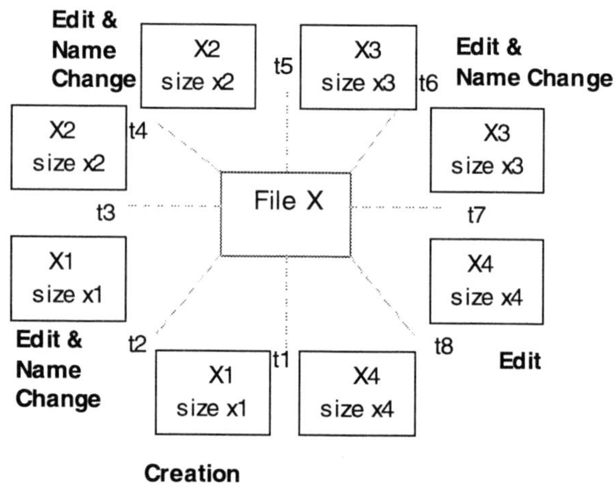

Figure 6. Life cycle of file X.

personal agent traces file system activity on every monitored workstation. The public agent gathers the individual behavior pattern, filters inconsequential patterns and produces interaction patterns between multiple public agents. The interface agent provides a single monitoring and control point for the social network patterns discovered by the system.

We also show an example life cycle of a file during the process of collaborating on a document. While the example is simplified by examining the interactions between only two people, we will be extending the example to multiple participants where there is an overlap of time response between the different participants. Complications abound when considering the fact that multiple participants may operate on the same file on different computers at the same time. We are currently developing algorithms and implementing solutions to deal with a wide variety of interaction scenarios.

A challenge in the development of this system is the preservation of some level of individual and group privacy, while at the same time gathering information useful in the analysis of social networks in CSCW. While social network analysis may be used for a variety of beneficial reasons, the information gathered would be considered by many to be quite private. It may be analyzed to reveal interactions or individual activities that may be used by management to flag activities (or inactivity), prompting further investigations, and leading to dismissals. For instance, patterns indicating unexpected "collaborations" that do not correlate with organizational activities or project structures may flag tasks related to on-the-job moonlighting. As well, the level of interaction and contribution within groups become readily apparent when subject to social network analysis. Rather than determining the underlying causes for interactions indicated in the analysis, an overzealous manager may unceremoniously root out someone flagging the person as a "weak performer".

Ideally a privacy solution would offer a flexible way for organizations and individuals to control how sensitive information is shared inside or outside of an organization. We are currently developing approaches for privacy management in distributed systems that may be applied to social network analysis. Some of this work will be based upon our collaborative work in the Privacy Incorporated Software Agent (PISA) project [25]. PISA is a European Union Fifth Framework project involving the development of privacy enhancement technologies for multi-agent based systems for e-commerce applications.

5. References

[1] B. Bruegge and A. Houghton, "Computer-Supported Cooperative Work: Discussion Summary", Software Engineering: Education and Practice International Conference, 1996, pp. 516-519.

[2] K.L. Mills, "Introduction to the Electronic Symposium on Computer-Supported Cooperative Work", ACM Computing Surveys, Vol. 31, No. 2, June, 1999, pp.105-115.

[3] J.D. Palmer and N.A. Fields, "Computer-supported cooperative work", Computer, May, 1994, pp. 15-17.

[4] T. Rodden, "A Technical Framework for CSCW", IEE Colloquium on CSCW: Some Fundamental Issues, 1991, pp. 7/1-7/5.

[5] F. Hattori, T. Ohguro, M. Yokoo, S. Matsubara and S. Yoshida, "Socialware: Multi agent systems for Supporting Network Communities", Communications of the ACM, Vol. 42., No. 3., March, 1999, pp. 55-61.

[6] D.B. Lange, "Mobile Objects and Mobile Agents: The Future of Distributed Computing?", Proceedings of The European Conference on Object-Oriented Programming '98, 1998.

[7] D.M. Hilbert and D.F. Redmiles, "Agents for Collecting Application Usage Data Over the Internet", Proceedings of the second international conference on Autonomous agents, 1998, pp. 149-156.

[8] T.S. Perraju, "Agents and Autonomous Distributed systems", The Fourth International Symposium on Autonomous Decentralized Systems, 1999, pp. 264-266.

[9] N. Pissinou, R. Benton, B. Bhargyavati and S. Kurkovsky, "A Roadmap to the Utilization of Intelligent Information Agents: Are Intelligent agents the link between the Database and Artificial Intelligence Communities", Proceedings of Knowledge and Data Engineering Exchange Workshop, 1997, pp. 196-205.

[10] B. Wellman, "For a social network analysis of computer networks: a Sociological Perspective on Collaborative Work and Virtual Community", Proceedings of the 1996 conference on ACM SIGCPR/SIGMIS conference, 1996, pp. 1-11.

[11] L. Garton, C. Haythornthwaite and B. Wellman, "Studying On-Line Social Network", Journal of Computer Mediated Communication, Vol. 3, Issue1, 1997, pp. 75-105.

[12] J.D. Eveland and T.K. Bikson, "Work Group Structures and Computer Support: A Field Experiment", ACM Transactions on Office Information systems, Vol. 6, No. 4, October, 1988, pp. 354-379.

[13] C. Haythornthwaite and B. Wellman, "Work, Friendship, and Media Use for Information Exchange in a Networked Organization", Journal of the American Society for Information Science, 1998, pp. 1101-1114.

[14] B. Wellman, J. Salaff, D. dimitrova, L. Garton, M. Gilia and C. Haythornthwaite, "Computer Networks as Social Networks: Collaborative Work, Telework ,and Virtual Community", http://soc.annualreviews.org/cgi/content/full/22/1/213.

[15] S. Qureshi, "Supporting Electronic Group Processes: A SOCIAL PERSPECTIVE", Proceedings of the 1995 ACM/SIGCPR conference on Supporting teams, groups, and learning inside and outside the IS function reinventing IS, 1995, pp. 24-34.

[16] C. Haythornthwaite, "Collaborative work Networks among Distributed Learners", Proceedings of the 32nd Annual Hawaii International Conference on Systems Sciences, 1999, pp. 1-16.

[17] B. Barr, S. Yoo and T. Cheatham, "Network Monitoring System Design", Proceedings of the twenty ninth SIGCSE technical symposium on Computer science education, 1998, pp. 102-106.

[18] R. Talpade, G. Kim and S. Khurana, "NOMAD: Traffic-based Network Monitoring Framework for Anomaly Detection", IEEE International Symposium on Computers and Communications, 1999, pp. 442-451.

[19] M. Abrams and S. Williams, "Complementing Surveying and Demographics with automated Network Monitoring", http://vtopus.cs.vt.edu/~chitra/docs/96w3j/96w 3j.html.

[20] M.G. Baker, J.H. Hartman, M.D. Kupfer, K.W. Shirriff and J.K. Ousterhout, "Measurements of a distributed File system", Proceedings of the thirteenth ACM symposium on Operating systems principles, 1991, pp. 198-212.

[21] K.K. Ramakrishnan, P. Biswas and R. Karedla, "Analysis of File I/O Traces in Commercial Computing Environments", Proceedings of the 1992 ACM SIGMETRICS and PERFORMANCE '92 international conference on Measurement and modeling of computer systems, 1992, pp. 78-90.

[22] W. Vogels, "File system usage in Windows NT 4.0", 17th ACM Symposium on Operating systems Principles (SOSP'99), 1999, pp. 93-109.

[23]Filemon File monitoring utility from Sysinternals, http://www.sysinternals.com

[24] A.K. Jain, J. Mao and K.M. Mohiuddin, Artificial Neural Networks: A Tutorial, Computer, March, 1996, pp.31-44.

[25] Privacy Incorporated Software Agent (PISA) Project, http://www.tno.nl/instit/fel/pisa.

A Dynamic Web Agent for Verifying the Security and Integrity of a Web Site's Contents

Soroush Sedaghat
Division of Planning & Quality Information Management and Analysis
University of Western Sydney
New South Wales, Australia
S.Sedaghat@uws.edu.au

Josef Pieprzyk
Department of Computing
Division of Information and Communication Sciences
Macquarie University
New South Wales, Australia
Josef@ics.mq.edu.au

Ehsan Vossough
School of Computing and Information Technology
University of Western Sydney
New South Wales, Australia
E.Vossough@uws.edu.au

Abstract

To harness safe operation of Web-based systems in Web environments, we propose a Server-based SHA-1 Page-digest Algorithm (SSPA) to verify the integrity of Web contents before the server issues an HTTP-Response to a user request. In addition to standard security measures, our Java implementation of SSPA, which is called Dynamic Security Surveillance Agent (DSSA), provides further security in terms of content integrity to Web-based systems. Its function is to prevent display of altered Web contents, through malicious acts of attackers and intruders, on client machines. This is to protect the reputation of organisations from cyber attacks and to ensure safe operation of Web systems by dynamically monitoring the integrity of sites content on demand. We will discuss our findings in terms of applicability and practicality of the proposed system. We also discuss its time metrics specifically in relation to its computational overhead at the Web server as well as the overall latency from the clients' point of view, using different Internet access methods. SSPA, DSSA, experimental results and related work are discussed.

1. Introduction

Continuous attack on Web servers and alteration of Web contents by intruders and hackers, who exploit system weaknesses, are prime concerns of the organisations, owners of the sites, as well as users who access them. Groups such as AusCERT [1] and CERT [6] monitor and report on Internet security breaches and so do not require further elaboration. To strengthen the security of Web systems, we have developed and implemented a server side Web content verifying agent, which automatically and dynamically intercepts and processes the user's request before the Web server responds to the client. In this scenario the organisations running Web servers are the primary parties interested in the integrity of their site's contents.

This system is different from the currently used systems where clients use technologies such as digital signature or encryption keys, as part of client's trusted software, to perform the verification of requested pages. In the latter case, an example such as Micropayment system [23][24], relies on its clients to verify the integrity of the electronically transmitted material. Therefore, if a Web page had been maliciously altered, the user would still see the modified content or would notice it after the event when the authenticity checks were performed. At this stage the damage is already done. Our algorithm running at the server prevents this by blocking material that does not match its true fingerprint (page-digest). By introducing the verification agent, our aim is to further secure Web systems in Cooperative Internet Computing and other Web-based environments, where still clients can verify digitally signed Web pages via other methods for their confidence to conduct online transactions. Conformance to the proposed algorithm does not assure complete security. Web owners should ensure that an overall system security and associated implementation provide an acceptable level of security suitable to the function and security policy of their site.

Although application of fingerprinting techniques, such as hash algorithms, are not new, different circumstances with Web permits adoption of known strategies not only to provide solutions to situations as described above, but also to evaluate their performance outside the conventional distributed computing systems.

This paper is structured as follows. Sections 2 and 3 discuss types of Web content integrity and their verification methods. Related work is described in section 4. The SSPA algorithm and its implementation, DSSA, are proposed in sections 5 and 6. Experimental and simulation environments are discussed in sections 7 and 8 followed by interpretation of experimental results in section 9. Effect of Web proxies on SSPA is explained in Section 10 and finally we conclude the paper in section 11.

2. Types of Web Content Integrity

Web content integrity has been perceived differently by different groups of users and Web developers.

We classify them into three categories:

- *Integrity in terms of Web page development*: this refers to checking the integrity and workability of links, existence of image and other files used in the Web page [15] as well as the syntactic validity, correct use of the HTML language, associated conventions and Web authoring techniques [11][32].
- *Integrity in terms of data validation and security*: this refers to the collection of secure valid data over the Internet using Web forms. We have fully discussed this in our previous publications [25][26]. Or verification of Web contents based on semantic contents of the site [10].
- *Integrity in terms of content authenticity:* This is to ensure that no tampering is done to the page content since it was created. This is related to our work in this paper. We will explore it further in this section [8][21][33].

Verifying the integrity of Web contents is not confined to HTML source code and accompanied texts. But also all of the utilised components such as image files, sound and video files, Cascading Style Sheets (CSS), frames, and other tags, which must remain free from unauthorised alteration. Overall content should remain secure before transmission to Web users. The proposed algorithm looks at the embedded components and every detail in the page. It automatically detects any alteration in the requested page.

3. Verification Methods

The following methods (the first three for use on the client side and the last one, the proposed method, for use at the server side) can be used for integrity verification:

- Each page is authenticated (signed) using a digital signature as a separate document. This is computationally demanding. One can sign two messages per second or verify 200 signatures per second [23]. This is not suitable for the type of work we are proposing at the server side.
- A sequence of pages are traversed and signed as a single authenticated path. This is more efficient than the previous method but verification is done at the end of the path. This is time consuming and inappropriate for our purpose.
- The site has a tree structure of pages. Then it is possible to check integrity in a more flexible and natural way, which requires application of digital signature.
- Just use the hash functions on-demand to the full content of a web page including all embedded objects. If the calculated hash value is the same as the previously calculated and stored value then the page is authentic and Web server would respond to clients. Otherwise it assumes corrective measures about the breaches in page security.

4. Related Work

An MD5 [22] message digest for Web pages was developed for the BIBLINK [3] project to investigate the bi-directional flow of information between publishers and UK National Bibliographic Agencies (NBAs), where embedding resources metadata into Web pages, using the HTML META element, was found to be appropriate for the requirements of the project [21]. This was done by placing the checksum of the resource for which metadata is described, as one of the metadata elements, in the Web page to determine if the resource was modified.

The checksum algorithm in the BIBLINK project is limited to extraction of specific sections of the Web page like <META> elements of the Web page (where metadata and checksum are stored) and embedded objects such as resource images. It ignores the body of the document that may include tabs, white spaces, etc. It then applies MD5 hash function to each of the above sections and calculates their message digest. A message-digest (hash value), h, is a fixed size numeric value representing a message of variable size p [23]. Finally, it rehashes the previously hashed message digests for each sections of the document in the order of appearance on the page. If the resultant hash value equals the stored hash value embedded in the page, we conclude that the resource, for which metadata was defined, was not modified since its creation.

Storing the message-digest in the page may assist with the verification task, but there would be some degree of doubt on how embedding the hash value in the same page would not alter its original hash value. Verification may be possible if there are mechanisms to verify the overall hash value of the page, or, the stored hash value is excluded from the parts to be hashed, which in this case it would be difficult to rely on its integrity.

Hash functions together with digital signatures are also used in Micropayment systems [33] and Commerce and Conventional Electronic Payment Systems.

In Micropayment Systems, such as MicroMint, the efficiency of hash functions is exploited to allow the parties involved in transaction to quickly verify the legitimacy of electronic coin [23][24]. These systems rely on the client's trusted software and application of digital signature to verify the integrity of the displayed contents. Normally the required security values, against which authenticity checks and system initialisations are made, are embedded in the same Web page.

These systems operate based on chain of hash values. The user authenticates a complete chain to the vendor with a single public-key signature, and then successively reveals each hash value in the chain to the vendor to make micropayments [23].

There is also work on using PGP [34] to sign Web pages [2][16]. Different signing methods are suggested in this area [9]. The problem with these methods is that they are only capable of signing the HTML Source code but they cannot sign the referenced components in the page. For example a page may pass signing test if nothing in the page source has changed, but since it does not sign the referenced components such as images, an illegally modified image remains undetected. In our suggested algorithm we have addressed this shortcoming. More elaborate versions of the Electronic signature are already developed by the introduction of E-SIGN legislation in US [14]. Examples are: PKI-based digital signing of Web Pages [5] or e-lock [8].

E-lock supports signing texts and images only. Hence other embedded objects within Web pages are left unprotected [27]. This solution has been mainly provided for business users and consumers using the Internet for secure Internet-based banking and insurance, e-commerce, and customer account information access.

Browsers such as Netscape Communicator® v4.04 and above introduced the java script crypto.signtext( ) method, when embedded in a HTML Page, requests the user to enter his private key to digitally sign a text string as persistent proof of a transaction [28].

There has been developmental work on verifying the content of Web pages in terms of analysing the semantic contents of the site [10]. This is a knowledge-based verification of Web pages where Web developers exploit *semantic markup* in Web pages (which is hard to find in the current Web) to formulate rules and constraints that must hold on the information in a site. An inference engine subsequently uses these rules to categorise Web pages in ontology of pages, while the constraints are used to define categories of pages, which contain errors. This is mainly a Web maintenance tool and does not provide automatic detection and blocking of altered content by unauthorised users.

IBM has developed a document protection solution called Cryptolope (Cryptographic Envelope) [12]. However, if page contents were modified at the server, Cryptolope would assume that the page was generated at the source and delivers the invalidated page to the user.

There are also various intrusion detection tools and firewalls, which aim to protect web systems if properly configured. However, they lack the ability to check Web contents of organizations that implement these technologies. Moreover, they are prone to software and hardware exploitations common to any other electronic systems.

In summary, work done to date does not guarantee any altered page not to be displayed on the clients' machine. Currently, users who utilize Web specific systems conduct their own integrity checks on downloaded pages. Although these mechanisms would provide some degree of confidence and allows clients decide on their intention to conduct on-line activities, it does not prevent side effects from unintended material sent to the clients.

5. Server-based SHA-1 Page-digest Algorithm (SSPA)

SSPA is an algorithm to enable automatic detection of alterations to Web contents before delivery to site users. We have used a hash function in our algorithm in preference to digital signatures because it is 1000 times faster than RSA signature generation and it is 100 times faster than RSA signature verification [23].

A hash function H, takes a variable-size Web page and its embedded objects, p, and returns a fixed-size string, called hash value h, such that $h = H(p)$. h is used for integrity checks.

We have chosen NIST's SHA-1 hash function [29], because of its ability to create a strong hash value (a binary digest of 160-bit) against 128-bit long numeric string computed by MD5 [22]. This has been the strongest hash function at the time of writing this paper. Better hash functions such as SHA-256, SHA-384 and SHA-512 may replace SHA-1 in the near future [30]. Other reasons for choosing SHA-1 have been availability of its source code and its Java implementation methods [17].

SSPA algorithm works as follows:

- o Initialise the hash value for the hash function
- o Detect arrival of page URL from the client's site. (We discuss its implementation later).
- o Search for HTML page for this URL on the host server (local organizations owned server).
- o When file found, read the HTML source code including accompanied text on character-by-character basis.
- o Apply the hash function to calculate the hash for each character and update the hash value.
- o While reading the Page detect the tag character "<". Process the contents between the tags "<...>" including white spaces.
- o Parse the content for keywords such as APPLET, EMBED, OBJECT, IMG, IMGE, LINK rel = "stylesheet", A HREF, a href, etc.
- o Look for secondary tag identifiers such as SRC followed by IMG, APPLET, etc. We do not process contents of external sites in this page as referenced by this URL. If user decides to select another URL that resides on another server, the server hosting that page processes contents of the page. We only hash the URL link string itself. However, images or applets that reside on an external site are processed.
- o Read the associated file on a character-by-character basis. Apply the hash function as before and update the hash value.
- o Apply the same hashing method to the remaining page contents and embedded URL links, image files, multimedia files, etc, in the page.
- o Produce the final hash value in hexadecimal format. Compare this hash number with the previously calculated number stored in a secure database on another local server. Storing hash values in a different server provides better security and also its communication channel with the Web server can be secured using Secure Socket Layer [31].
- o If an external image or applet is not available, the server compares the final hash number with another previously stored number for the same page without the image or applet.
- o If result of comparison is positive, set the server's flag to send the requested page to user.
- o If it failed send an email to alert site's owner of the situation. Also transmit a message to client showing unavailability of the site for the time being.

6. SSPA Implementation (DSSA)

The above algorithm is implemented in Java and runs dynamically on our test Web server (The operation of this

applet at the server is shown in `Figure 1`. This applet is named Dynamic Security Surveillance Agent (DSSA) and performs the following:

- o Dynamically listens to port 80 or 8080, which is normally used for Web-based communication between Web servers and Web browsers to intercept client's request for a URL specifying a Web resource.
- o Calculates page-digest as specified by the algorithm.
- o And finally takes appropriate action based on the resultant hash value.

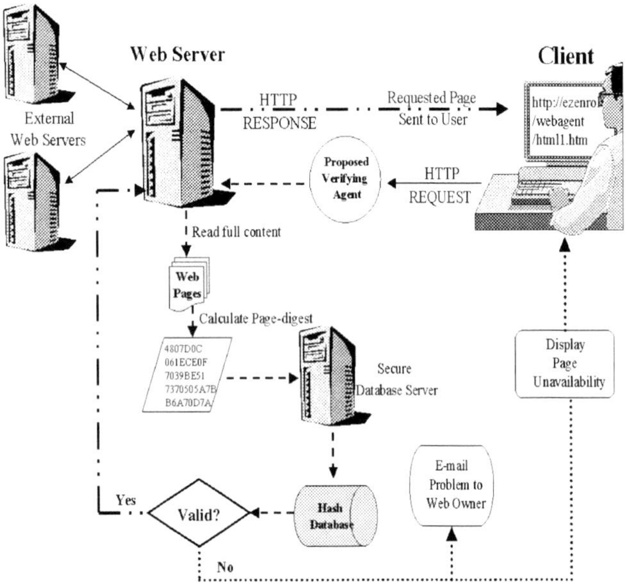

Figure 1. Operation Flow of the proposed Security Surveillance Agent (Web Page Verifier)

7. Experimental Environment

Software and hardware set-up: For the experimental work, we used a PCDirect 690®,AX100 with CPU speed of 300 MHZ as our test Web server on NT® v4.0 platform, Microsoft Internet Information Server® (IIS) v4.0, Microsoft Internet Explorer® and a 56kb/s NetComm® internal Modem as a dial-in system. Our network ring operates on 16 Mbits/sec IBM Token Ring over TCP/IP.

Web Server and Web Page Configuration: In order to ensure that during the course of experiment all requested Web pages are directly requested from the test Web server, we set up the HTTP Headers in IIS® Server to avoid caching Web pages [18] in our university SQUID® Web Proxy. In conjunction with the IIS set-up, we have added the required META instructions (PRAGMA, CACHE-CONTROL and EXPIRES) in the HEADER section of the HTML Pages to control caching [19]. This is to analyse the time metrics in absence of Web proxies in actual Internet packet switching medium. We will discuss the effect of proxy on our experiment and suggest ways to utilise the advantages of Web caching through proxies.

Experimental Web-pages: we constructed an experimental Web site and created several different test Web-pages as follows:

- o Plain *text* HTML pages of various sizes (0.5- 35 KB).
- o Plain text files as above, each containing a replaceable image of various sizes (image size 1 KB to 21 KB).
- o Plain text files as above, each referencing an internal link to either a plain text HTML file or to a Page containing an image.
- o HTML files containing a number of internal and external links as well as different images.
- o HTML Pages containing multimedia, frames, embedded objects, applets, action URLs and other common tags in Web systems.
- o HTML pages containing JavaScript code, executables, cookies, and other programmable objects.

The reason for building different pages of various contents has been to mimic the practical situations as well as exhaustively test the workability of the proposed algorithm. This also permits to analyse Web server's overhead (hash calculations) with respect to time metrics. The average size of a Web page has been reported to be 10K in size, having 23 internal links (Median = 4), 6 external links (Median = 1) and 14 image files (Median = 1) [20]. It may appear rather hard to believe that the average size of a Web page is 10K, especially for corporate Web sites and Web-based educational training sites. However we have constructed pages with 3-4 times the reported average size to test for bigger pages as well. We also established a mathematical equation to be used for prediction of hashing time as well as latency measurement for any Web page sizes with our settings. These are discussed in the following sections.

8. Simulation Environment

To study the performance of the hashing algorithm we implemented a simulated Web agent called *listener* and used Internet Explorer browser as *transmitter*. Also, we set the DSSA so that another instance of the listener can simulate a browser. Therefore it can be used as a transmitter for testing purposes.

Listener is an intermediate agent between the IIS Web server and the Browser (transmitter). It intercepts the incoming URLs from the browser and processes the requests as described before. Figure 2 shows the relationship between the transmitter and the Web verifying agent.

9. Experimental Results, Time Metrics

In this section we describe the result of our measurements in terms of the required time for the Web server to apply the SSPA algorithm by DSSA. Presented results are in the absence of the web proxy and non-utilization of page caching in order not to distort the results. Because of the variability of network bandwidth and also because of the effect of clogged

network on time metrics, we have performed extensive experiments on the real network to obtain our data. In order to measure the time more accurately, DSSA processed each URL interception exhaustively for 1000 times. Then the actual response time is averaged for one pass. We have measured the time for processing the requested pages with and without application of SSPA.

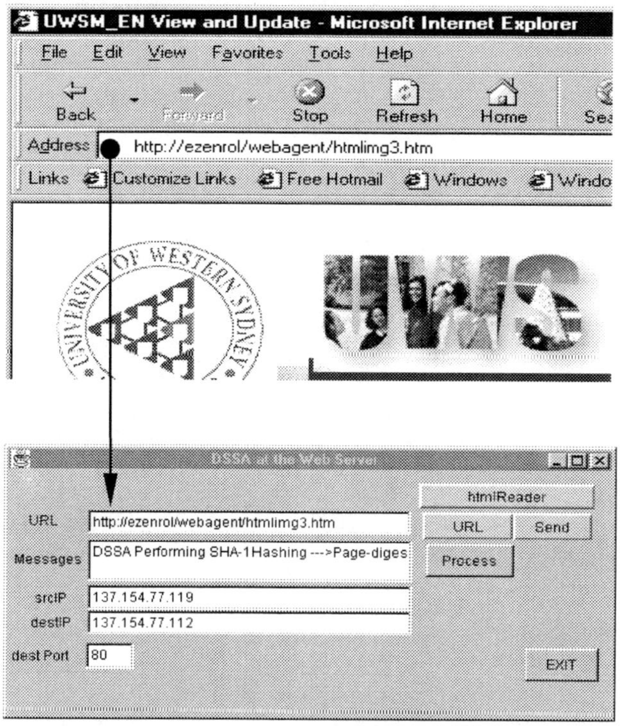

Figure 2. Simulated Web Verifying Agent dynamically listens to user HTTP request from a browser.

Time metrics obtained for the following environments:

- o Processing time for the Page at the Web server before transmission via HTTP Response to the client.
- o Overall time (presence and absence of DSSA processing time plus network latency for each case from client to server over intranet). This is processing time when directly accessing the network - using Intranet (no proxy caching).
- o Processing time when accessing the test site by Dial-in Modem (presence and absence of DSSA plus network).
- o Deducing the effect of Proxy.

Graph 1 shows the time measurements at the Web server only. We have found that for HTML pages that are less than 5000 bytes in size, there are no significant differences between presence and absence of DSSA application. However, there have been significant differences (LSD < 0.05) when the size of Web page exceeds 0.5 KB. We have also found that there is a linear relationship between the processing time and the file size.

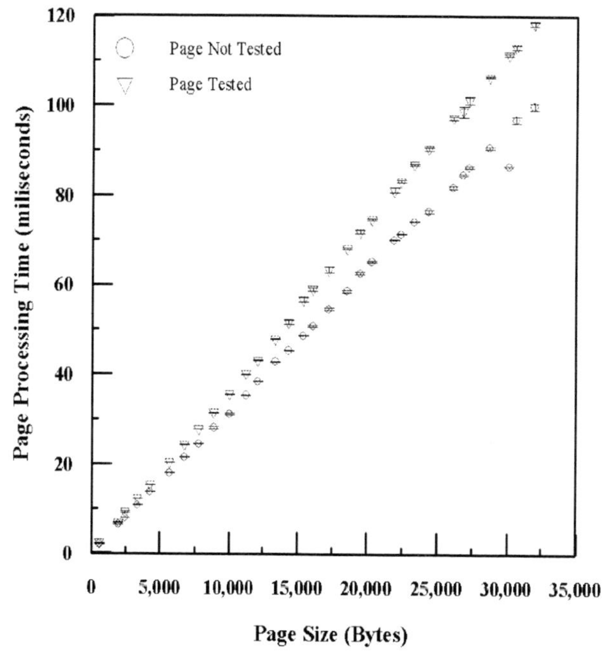

Graph 1 - Web Page Processing at the Web Server

Since the results are mainly dependent on the CPU speed of the Web server and the overall IIS configuration and system set-up before IIS dispatches Web Page to the client, it is difficult to find a generalized equation representing time vs HTML file size that applies to every machine. However, for our particular set-up, in Graph 2, we have shown relevant equations for estimating the processing time T for different HTML file sizes F (F_h for DSSA and F_{nh} for no-DSSA):

$F_h = 0.00369\ T$ [R-Squared = 0.99] (equ. I)

$F_{nh} = 0.00314T$ [R-Squared = 0.99] (equ. II)

Graph 3, shows the SSPA execution time for calculating page-digest for different file sizes at the Web server. We have also obtained the mathematical relationship between hashing time and file size variables ($F = 0.00055\ T$, R- Squared=0.970). The results obtained by this equation exactly confirm the difference of the results using F_h and F_{nh}, equ. I and equ. II, above.

When processing at the server is completed and authenticity of the computed hash is successfully determined, then Web server starts responding to user's request.

Graph 4 shows the cumulative time (DSSA processing time plus time spent on Internet transmission) for different size Web pages. The same graph also shows file transmission in absence of DSSA overhead. The results show three distinct clusters of data: the first cluster is associated with files less than 10 KB in size where they are either text pages or pages with simple images and links.

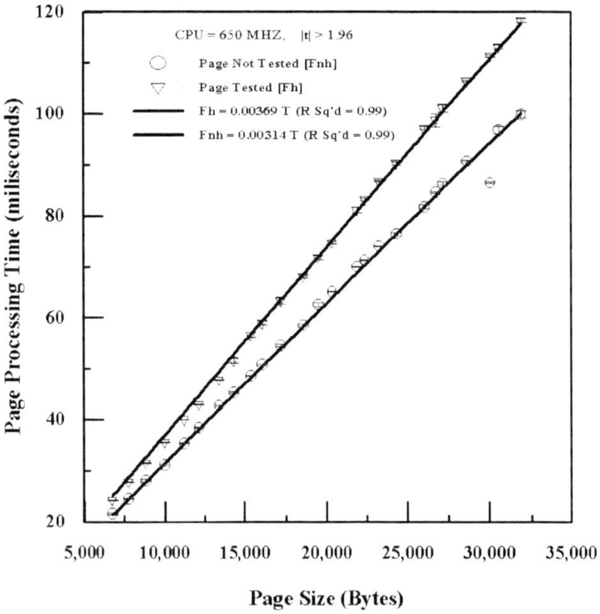

Graph 2 - Relationship between page size and time at the Web server

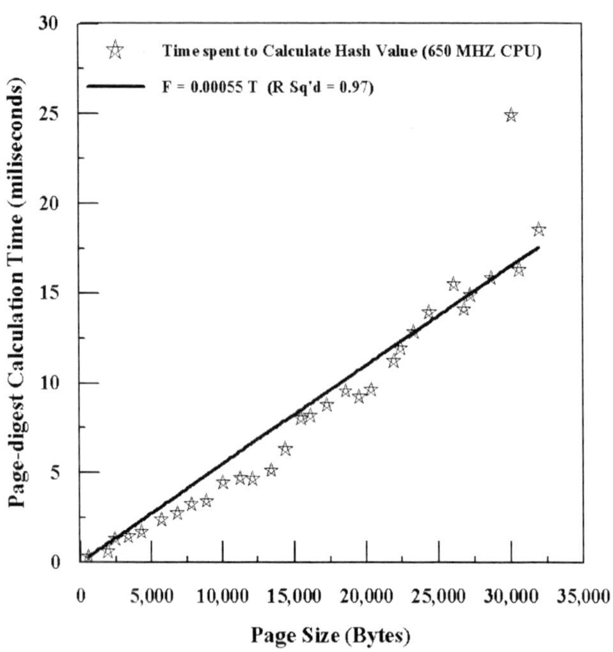

Graph 3 - Time to calculate Page-digest by DSSA at the Web Server

Here, almost a linear relationship between files and the latency (time before a page is displayed on user's machine) is observed and there are no significant differences between the absence and presence of the DSSA.

The measured latency, when compared with the actual processing time done at the Web server (Graph 2), is significantly higher.

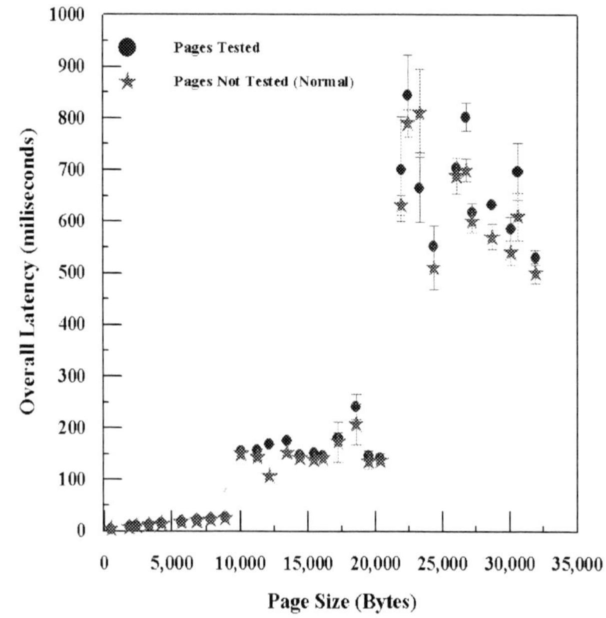

Graph 4 - Effect of Internet on overall Page processing (Server's Page test + transmission to the client)

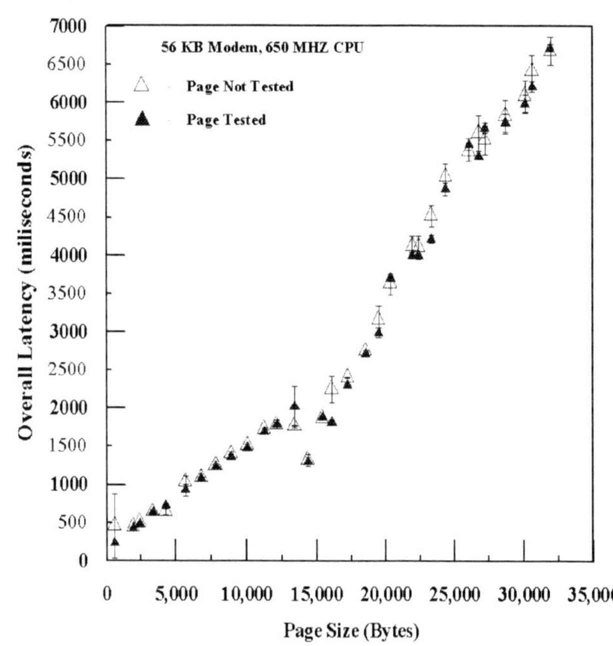

Graph 5 - Effect of Dial-in Modem on Overall Latency

For a 33 KB page size, server's overhead is around 118 to 120 milliseconds, while for the same file it took an average of 8 to 9 times higher before it reached the user's browser. This is mainly due to effect of interconnected packet switching network of computers. The second cluster with a jump in measured time refers to a moderately sized group of Web pages having medium size objects such as images, as commonly used Web components. This is because of the bandwidth problem to transfer more complex data across the Internet. This

cluster shows a moderate tooth saw effect when file sizes of greater than 10 KB and less than 25 KB are analysed.

The third cluster is further confirming the effect of Page size with more complicated data (objects where higher transmission time is required for their transmission. e.g. large compressed image files, executable applets etc).

In this group (file sizes greater than 25 KB and up to 35 KB) there is a pronounced tooth saw effect on time to send the files to the client. However, for a 35 KB Web page it was found to be not more than 1 second latency, taking into account parameters contributing to the measurement as described in the experimental set-up section of this paper.

The graph also shows that there is no significant difference between the applications of SSPA or normal transfer of the files by our test Web server. We have concluded, in this situation, that the main processing task happens very rapidly in a very small amount of time at the Web server, but its effect (although significantly different at the Web server) would diminishes when a Web page is on its way to clients over the Internet.

The latency over the Internet is much more noticeable when user tries to access a Web site, running with or without DSSA, using a modem. This is shown in Graph 5.

Graph 5 shows a high jump in latency when this is compared with situations where a client uses the Internet in Intranet environment. In this situation, because of excessive time delays due to bandwidth problem or clogged network, the clustering effect, which was shown up in Graph 4, is almost smoothened to a more linear and less fluctuating pattern. However, the latency is much higher, in order of 8 to 9 times, when compared with Intranet operation. Hence comparing measured latency with the small SSPA calculation overhead at the Web server makes this overhead negligible.

10. Role of Proxies

Web proxies normally try to cache Web pages to respond to HTTP requests from the clients on behalf of the Web server. Proxies assist to reduce latency by 26% [13], depending on the technology including Internet access of the client. This is in comparison with the direct response from the Web server to user requests. Also when this is combined by other techniques such as prefetching of pages at the client side latency is reduced by 81% in a homogenous environment and 63% in a heterogenous environment [7]. Despite this benefit, centralised Web page caches may be tempting targets for hackers if not protected securely [4].

Conventionally Web page signing has been a solution to protect them and to detect modifications. DSSA can also be used at proxies to prevent alteration of cached files and to speed up subsequent operations. But if users refresh the page on their browsers, a new copy of the page is fetched from the Web server and integrity test is performed at the Web server.

11. Conclusion

We have suggested a Server-based SHA-1 Page-digest Algorithm (SSPA) for verifying the integrity of Web pages from malicious attack and described its implemented results. We have proved that Web server's CPU overhead due to application of SSPA is negligible. This is in relation to the effect of Internet and also dial-in modems on DSSA enabled Web servers. Moreover extra time spent to perform SSPA is well spent protecting organization's reputations from damage done by a hacker. Depending on experimental parameters and CPU speed, our findings on time metrics provide a means to adopt this technique to Web sites with different configurations.

We recommend incorporation of such algorithm at low-level assembly language in Web server software or generally incorporating it into Server and Proxy software to fully speed-up its operation. This system is applicable to almost all Web-based and computer supported cooperative systems over the Web. It protects the interest of Web server providers as well as clients. This together with other available integrity verifications at the client's site would provide confidence to site users where sensitive on-line transactions are concerned.

References

[1] AUSCERT. (2001). [Internet]. Available from: < http://www.auscert.org.au/>. [Accessed: Feb 2001].

[2] Bell, E. (1997). "*PGP Signed Web-pages*". [Internet]. Available from: < http://members.aol.com/EJNBell/pgp-www.html> [Accessed: Nov. 2000].

[3] The British Library. (1996). "*BIBLINK Project*". [Intrnet]. UKOLN. Available from <http://hosted.ukoln.ac.uk/biblink/ > [Accessed: Feb 2001].

[4] Brooks, V. (1998). "*Web Caching Helps Intranet Performance*". [Internet]. CMP net. Media Inc. Available from: < http://www.lantimes.com/98/98aug/808c022b.html > [Accessed: Feb, 2001].

[5] Celo Communications. (2001). "*Celo Enable PKI-Based Digital Signing of Web Pages*". [Internet]. Available from: < http://www.celocom.com/web/celo/d.asp?p=667> [Accessed: Feb. 2001]

[6] CERT. (2001). "*Incidents, Vulnerability and Reporting*". [Internet]. CERT Coordination Centre. Available from: < http://www.cert.org/> [Accessed: Feb 2000].

[7] Eden, A.N., Joh, B.W. and Mudge, T. (2000). "*Web Latency Reduction Via Client-Side Prefetching*".In Proceedings of the 2000 IEEE International Symposium on Performance Analysis of Systems and Software. IEEE. Pages 193-200.

[8] E-Lock Technologies. (2001). "*Digital Signature and Electronic Signature Software for e-commerce security, E-Sign Compliant*". [Internet]. Available from: < http://www.elock.com/ > [Accessed: Feb. 2001].

[9] Goldberg, J. (1996). "*Methods of signing Web pages using PGP*". [Internet]. Available from: < http://www.cranfield.ac.uk/docs/email/pgp/html/signed_html.html> [Accessed: Nov. 2000].

[10] Harmelen, F.V. and Meer, J. v.d. (1999). "*WebMaster: Knowledge-based Verification of Web-pages*". In Proceedings of Practical Applications of Knowledge Management, PAKeM'99. The Practical Applications Company, London. Pages 147-166.

[11] Hower, R. (2001). "*Web Site Test Tools and Site Management Tools*".[Internet]. Available from: http://www.softwareqatest.com/qatweb1.html > [Accessed: Feb, 2001].

[12] IBM (1997). "*Cryptolope*". [Internet]. Available from: < http://www-4.ibm.com/software/security/cryptolope/ > [Accessed July 2000].

[13] Kroeger, T.M, Long, D. and Mogul,J.C. (1997). "*Exploring the Bounds of Web Latency Reduction from Caching and Prefetching*". [Internet]. University of California at Santa Cruz. USA. Available from: < http://www.cse.ucsc.edu/~tmk/publications/ideal/> [Accessed: Feb, 2001].

[14] Kunze, C. (2000). "*The Electronic Signatures in Global and National Commerce Act, S.761 (E-SIGN)*". [Internet]. Uniform Electronic Transactions Act (UETA). Available from: <http://www.uetaonline.com/hapcong.html> [Accessed: Nov. 2000].

[15] Maslov, I. (1999). "*Web Integrity Tool*". [Internet]. Available from:< http://www.cs.columbia.edu/~hgs/teaching/ais/1998/projects/wlc/report.html > [Accessed: Feb, 2000].

[16] McKinnon, C. (1996). "*Signing Web Pages with PGP*". [Internet]. Available from: http://www.wolfenet.com/~aderhaag/signedpages.html [Accessed: Nov. 2000].

[17] McManis,C. (2000). "*SHA-1.JAVA. A class to implement the SHA1 message digest algorithm*". [Internet]. Available from: < http://www.mcmanis.com/~cmcmanis/java/ > [Accessed: April, 2001].

[18] Microsoft Corporation. (2001). "*Controlling the Caching of Web Pages with IIS 4.0*". [Internet]. Available from: < http://support.microsoft.com/support/kb/articles/Q189/4/09.ASP " [Accessed: Feb, 2000].

[19] Microsoft Corporation. (2001). "*Howto: Prevent Caching in Internet Explorer*". [Internet]. Available from: http://support.microsoft.com/support/kb/articles/Q234/0/67.ASP [Accessed: Feb 2000].

[20] Murray, B. and Moore, A. (2000). "*Sizing the Internet, A White Paper*". [Internet]. Cyveillance artificial-intelligence search-and-analysis development company. Available from: < http://www.cyveillance.com> [Accessed: Feb. 2000].

[21] Peacock, I. And Powell, A. (1998). "*BIBLINK.Checksum – an MD5 message digest for Web pages.*" [Internet]. Ariadne. Available from:< http://www.ariadne.ac.uk/issue17/biblink > [Accessed: 15 Feb 2001].

[22] Rivest, R. (1994). "*The MD5 Message Digest Algorithm*". Internet RFC 1321. [Internet] . Available from: < http://www.landfield.com/rfcs/rfc1321.html > [Accessed: 1 Feb 2000].

[23] Rivest, R. and Shamir, A. (1997). "*PayWord and MicroMint--Two Simple Micropayment Schemes*". In Proceedings of 1996 International Workshop on Security Protocols, (ed. Mark Lomas), (Springer, 1997), Lecture Notes in Computer Science No. 1189, pages 69-87.

[24] RSA Security. (2000). "*What are micropayments?*". [Internet]. RSA Security. Available from:< http://www.rsasecurity.com/rsalabs/faq/4-2-5.html. [Accessed Feb 2001].

[25] Sedaghat,S. and Pieprzyk, J. (2000), "*Secure On-line Data Acquisition Systems in a Web-based Environment.*", In Proceedings of the International Workshop on Cooperative Internet Computing, CIC 2000. Hong Kong Polytechnic University. Hong Kong. Pages 37-42.

[26] Sedaghat,S. (2001), "*Designing Electronic Forms in Web Applications: Integration of Form Components*", The Second International Conference on Mobile Data Management. 8-10 January 2001. Hong Kong. Pages 275-276. [Also (ed. Lee, T.L. et. al. (2001). "Mobile Data Management". Springer Verlag. Vol. 1987. Pages 245-246.]

[27] Shah, N. (nimish@elock.com). (23 March 2001). RE >> AlphaTrust. E-mail to Soroush Sedaghat (S.Sedaghat@uws.edu.au).

[28] Sun Microsystems, Inc. (2001). "*Signed Form Example*". [Internet]. IPlanet. Available from: < http://developer.iplanet.com/tech/security/formsign/stex.html > [Accessed: Feb. 2001]

[29] US National Institute of Standards and Technology (NIST). (1995)."*Secure Hash Standard, SHA-1*". [Internet]. US Federal Information Processing Standards (FIPS) Publication 180-1. Available from: < http://www.itl.nist.gov/fipspubs/fip180-1.htm > [Accessed: Feb. 2000].

[30] US National Institute of Standards and Technology (NIST). (2001). "*Secure Hash Standards – New Hash Algorithms (SHA-256, SHA-384 and SHA-512)*". [Internet]. Available from: < http://csrc.nist.gov/cryptval/shs.html> [Accessed: Feb 2001].

[31] VeriSign (2001). "*SSL Server Certificate*". [Internet]. Available from:< http://www.verisign.com > [Accessed: Feb 2001].

[32] Vertical Sky. (2001). "*Vertical Sky Content Manager*". [Internet]. Available from: < http://www.verticalsky.com/solutions/content.htm > [Accessed: Feb. 2001]

[33] W3 Consortium. (2000). "*Micropayment Systems*". [Internet]. Available from: < http://www.w3.org/TR/WD-Micropayment-Markup/ > [Accessed February 19, 2001].

[34] Zimmerman, B. (2000). "*PGP User Guide, Volume I: Essential Topics*". [Internet]. Available from: < http://pgp.rasip.fer.hr/pgpdoc1/pgpd1.html > [Accessed: Feb 2001].

A Framework for Intelligent Agents within Effective Concurrent Design

Kuo-Ming Chao
School of Mathematical & Information Sciences, Coventry University, UK
k.chao@coventry.ac.uk

Anne James
School of Mathematical & Information Sciences, Coventry University, UK
a.james@coventry.ac.uk

Peter Norman
Engineering Design Centre University of Newcastle-upon-Tyne, UK
peter.norman@newcastle.ac.uk

Abstract

One feature of concurrent engineering is the notion of distributed design, and the ability to communicate changes to all participating disciplines. However, successful engineering design is constrained by the complexities of design activity. The differences in system architectures or methodologies, the diverse knowledge and information generated by these distributed disciplines, all compounded by the problems of co-ordinating the communication process, will reduce the effectiveness of distributed design. Current thinking indicates that multiple agent systems (MAS) could alleviate the complex and large engineering design problems. A drawback of these approaches is the failure to recognise the significance of agent autonomy and proactiveness. In this paper, it is suggested that agent attributes such as, proactiveness and autonomy are ideally suited to resolving the problems. They provide a more flexible and dynamic approach to multiple disciplines situated within distributed design environments, which can reduce redundant design activities, and thereby reduce co-ordination management and design versions. It is proposed that an agent-based framework incorporating CORBA technology should be used to maximise the proactive nature of autonomous design agents.

1. Introduction

The successful design of large and complex made-to-order (MTO) products such as ships, offshore oil platforms and aeroplanes, require the collaboration of multi-disciplinary design teams. These design teams use specialised computer systems to aid their design process. Unfortunately, such computer systems may use different data representations of the product model. Such systems may utilise different design software packages. These packages may be written in dissimilar languages, for instance C++, Java or other languages, and be installed on different hardware systems. Consequently, any collaborative communication or co-ordination between such diverse and different models, languages and system architectures may prove difficult.

Within the last decade, a number of design tools have been developed at the Engineering Design Centre, University of Newcastle. The following briefly describes these systems.

- Process Flow Diagram (PFD) System [1], responsible for the design of a gas-condensate separation process. The system generated a PFD frame and rule model of the process.
- Electrical System [2], computes the power demand and power generation dimensions. The system can also select equipment from the supplier's catalogues. This selection is based on the PFD specifications.
- Plant Operation System [2], produces diagnostic rules for functioning plant. These rules are based on the output of the PFD system.
- Associativity Data Generation (ADG) [3], calculates the strength of the relationship between any two pieces of equipment based on connectivity, function, and cost. The equipment and connectivity data are carried from the catalogue and the PFD system, to the ADG system via a knowledge sharing and reuse tool [4]. The ADG system also generates the input for the Spatial Layout Design (SLD) system.
- Spatial Layout Design (SLD) System [5], responsible for producing an optimised graphical layout in two dimensions.
- 3D Parametric CAD System [6], improved layout system which generates 3D Parametric CAD information. The Parametric CAD system can forward any design changes to the Cost Estimating system.
- Cost Estimating System [2], allows for a revision of costs following a design change.

Some of these tools are implemented in Unix and Windows using C++ and Java. Some of them use commercial CAD, PFD packages, or expert system tools to implement their methodologies. Consequently no source code (apart from the Application Programming Interfaces) will be accessible. The prime aim of the research was the provision of a single collaborative environment that allows for the dynamic interaction of such tools. Therefore a means for providing that collaborative interaction was needed. Initially, the authors proposed a CORBA-based framework as a wrapper to facilitate communication between design tools. Within the CORBA-based framework an Interface Definition Language (IDL) defines the system interfaces.

Each system has a client and a server program, so it can trigger other systems and be invoked by yet others. The association between these systems is identified as a sequential flow. The relationship between different data models is monitored by a knock-on effect mechanism [7]. The mechanism traces the information flow between models, and is based on the interdependent relationships between objects in the models (i.e. equipment items). When a change to information held within one system is made, changes to the related information held within other systems are indicated by this knock-on effect mechanism.

2. Analysis of a Scenario

The following scenario taken from a case study of an offshore petrochemical plant design is used to assess the effectiveness of the originally proposed framework.

A pump needs to be replaced by a bigger one in order to produce more pressure. The change from the PFD is passed to the power system for re-calculating and it produces a new specification of power generators in order to accommodate this change. The simulated PFD diagnosis system also takes this change to analyse the potential faults. The ADG system takes the result from both systems to produce the new associativity data. The SLD starts to simulate the optimised layout according to the new data from the ADG system. A new optimised 3D layout and cost were produced to reflect this change.

The process described above is considered to be a concurrent engineering approach. The participating systems in the framework can initiate changes and then trigger other systems to act accordingly. However the following limitations of using the CORBA-based framework were identified in this exercise.

The systems were acting passively. The new pump, for example, could be expensive and its replacement may result in an excessive budget. The framework could not identify this problem until the last process. A system that can observe the changes and proactively provide the solution is required in order to reduce the unnecessary tasks.

- The systems were not autonomous. They could not control their own behaviour. Each system did what it was told by others. A system was unable to refuse invocation. The SLD system, for example, may generate a new layout and the designer may not have a chance to look at it. The ADG system with new data may trigger the system again and overwrite it. A level of autonomy is required.
- This architecture requires that users or other systems have a good knowledge of the functions that other systems provide. Precise function names and the required arguments need to be specified by the requested system before the functions can be invoked. The APIs supported in commercial packages include functions of grain size and those of a complex nature. The other system required, is one to control the others in the correct sequence, in order to produce the result. A level of abstraction is required.

An intelligent agent that includes properties i.e. proactiveness and autonomy as well as reactiveness provides a promising solution to overcome the above limitations. Intelligent agents with the most sophisticated capabilities are known as *proactive* agents. Proactive agents have the ability to assess their surroundings and decide what action to take. An example of this type of agent can be seen in engineering design environments where the agent monitors process flow design, prices of equipment items needed, and estimates the cost in order to avoid exceeding item costs [8]. These agents can gather the related information and exercise reasoning skills that will then allow the agent to determine a specific course of action. Thus, the autonomous agents perform tasks without the need for constant human interaction or other agent interaction. Because this agent controls its behaviour within its domain, it can choose whether or not it will approve requests made by other agents [8]. A reactive agent can respond to the requests made by other agents. An agent is a level of abstraction that represents the application and enables it to interact with its collaborative environment by reacting to the requests from other agents in order to solve common problems. Reactiveness without autonomy would imply obedience and the same problem as that mentioned above, where a system blindly follows instructions from another system. Reactiveness with autonomy means that an agent can react to a request in a number of ways, one of which may be to obey the request and another to reject it.

Observing is an important mechanism to achieve the agent's proactiveness. The agent being observed is called the observable agent, and the agent observing is called the observer agent. Every agent can act in both roles at same time according to need, so it can perceive other agents' behaviour and pass its states to other agents. The observer agent specifies the information in which it is interested within the observable agents. A filter mechanism is used to reduce the unnecessary information to be passed. The observer agent sets the conditions in the filter mechanism. For example, the cost agent is only interested in a list of expensive equipment items in the PFD, because upgrading any of equipment items in the list can be over the budget. It tells the PFD agent that these items need to be observed so given any change to these items the cost agent will be informed. As a result, the observer agent can proactively provide the advice to the observable agents about possible solutions.

The mechanism resides in the observer agent's mental model and the ORB layer in the observable agent. So, the observer agent can observe the states of the applications in the observable agent. The other advantage of using ORB to implement observing function instead of threads is to reduce the communication traffic. The ORB is client/server architecture. In this case, the server, the observable agent (sever side) pushes the data out to the observer agent (client side) when any change has been made in the observable agent. So there is no need for the observer agent to monitor the observable agent constantly. The IDL module shown below provides a flavour of the type of interface required for the observing function.

```
module ORBObserver
{ interface Observer
  {  void update(in any observables, in any message);};
interface Observable
  {  void notifyObserver(in any data);
         void addObservers(in Observer objRef);
         void deleteObservers(in Observer objRef);
         long countObservers();
         boolean hasChanged();      };                };
```

3. The Proposed Framework for Cooperation

The system (see Figure 1) includes three parts, communication, mental model and application. The communication mechanism is a message passing mechanism built upon an Object Request Broker (ORB). It transports the agent's message in the syntax of Agent Communication Language (ACL) to the recipient and the recipient parses the message. The mental model interprets the content of the message, reasons with it and asks the underlying application to perform the task. The underlying application returns the result to the mental model. The mental model generates the appropriate reply and forwards it to the request agent. Even though the message passing is used for the communication, and the mental model reasons with the message content, the remote method invocation is inevitably used to invoke the functions in the application. The interface between the application and the mental model uses the ORB.

The mental model is responsible for accessing the functions in the application through the ORB. This also applies to multiple agents' conversation.

The agents communicate with each other through a protocol called the Agent Communication Language (ACL). Since the agents are intended to perform some action by virtue of being sent [9], then the ACL messages are actions or Communicative Acts (CAs). CA's, are a special class of actions, that are modelled on speech act theory [10].

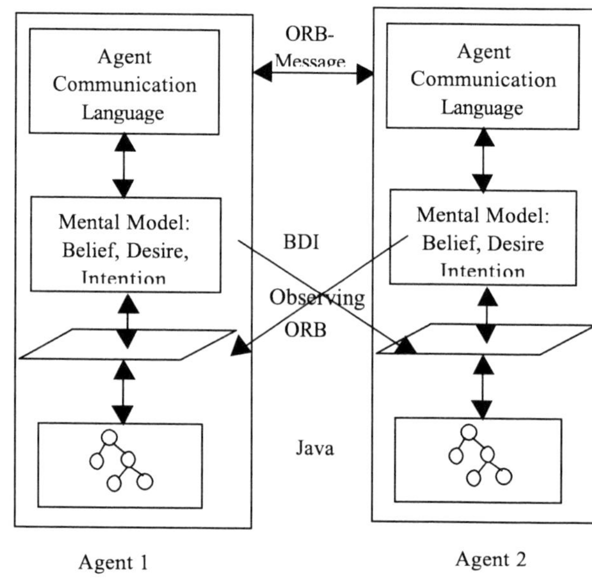

Figure 1. The Proposed Three-level Framework.

3.1. The Mental Model

The BDI implementation parses the incoming message from ACL and reasons with it. The BDI then invokes the server methods accordingly as a client side. BDI is a reasoning mechanism that interprets the information, motivational, and deliberative states of the agents. "These mental attributes determine the system's behaviour and are critical for achieving adequate or optimal performance when deliberation is subject to resource bounds" [11]. The following formulas describe the behaviour of the Belief-Desire-Intention (BDI) methodology.

$B' = B(P, B)$ where P is the agent's current perception

$D' = D(B, D)$

$I' = I(B,D,I)$

Belief is about facts or information that the agent receives from its internal states or environment status when currently participating in some process. Desire is a list of goals that the agent can or tries to achieve. Intention is a plan for the agent to achieve its goal. These attributes with three functions (*B*, *D* and *I*) allow the agent to behave dynamically, according to some change in their environment. A new belief (B') is derived from the old belief (B) and the agent's current perception (P) of the environment and its internal state. The new desire (D') is derived from the new belief and the old desire (D). The new Intention (I') comes from belief, desire and old intention (I).

3.2. An Example

The applications, the multiple disciplinary design tools including their domain knowledge, are implemented in various forms such as CAD, PFD packages, expert system shells and programming languages (i.e. C++, Java). These tools provide different levels of APIs for the developers to control the tools. So the APIs provided by the tools are wrapped by the ORB (Object Request Broker) in order to communicate with BDI. BDI invokes the ORB interfaces to activate the functions in the applications and obtain the results. This approach is similar to the previous Common Object Request Broker Architecture (CORBA)-based solution but here the applications do not directly invoke other applications. Interaction is carried out through agents. The course of the application actions is controlled by its BDI. Moreover, the knowledge and reasoning mechanism in the mental model can assess and determine whether the agent should fulfil the requests. Consequently, this mechanism allows the agents to have a certain degree of autonomy. Let us consider another example from the petrochemical plant design case study with three scenarios.

Scenario 1: The process engineer is required to change the operating temperature drop across the Low Temperature Exchanger and so adjusts the flowsheet parameters accordingly. The new equipment is selected and the change is passed to the other design agents. The layout and electrical systems report that this does not conflict with their existing models. They accept these changes and update their models accordingly. However, the cost engineer decides that this exchanger configuration would be cheaper if replaced by a set of two exchangers.

Scenario 2: The process engineer responds to this suggestion, accepting it and changes his model accordingly. The electrical system also accepts this change. However, the extra space required for the two exchangers configuration cannot be accommodated in the existing area. In model terms, constraints in the layout design have now been violated and the layout engineer cannot accept this change.

Scenario 3: In response to this conflict the electrical system produces a new design for the platforms power supply. This alternative uses a generator set up with similar power output but which is smaller. However this generator configuration is more expensive. The cost engineer states this violates a global project constraint.

A number of design processes have been carried out. We use these scenarios to examine the proposed system effectiveness. In Scenario 1, there is no improvement compared with the CORBA-base solution. The Scenario 2 has little improvement. Before the electrical system starts to run the program, the layout system has also already issued the infeasibility. So the electrical system rejects the request from the PFD system. The diagnosis system, however, has already carried out the simulation. In Scenario 3, the cost agent observed the changes in the electrical system, its mental model determined that the new design has violated the global constraint and autonomously issued it to the electrical system through ACL at early stage before the new design propagates to other agents. In this scenario, the numbers of design processes were reduced with the proposed systems. The layout, 3D layout, Associativity Data systems were not invoked, after the cost system identified the problem.

In the Scenarios 2 and 3, the proposed system produces better results by reducing the number of design activities by one and three respectively over the CORBA-base system. It is difficult to represent the improvements in time scale, because the layout system uses random seeds to generate optimised layouts, so the system process time varies. These scenarios, however, demonstrate how the use of middle layer i.e. mental model in the federation creates the proactiveness and autonomy of the underlying systems. The representation of beliefs, desires and intentions allows conflicts to be identified by participating systems before expensive work had been undertaken by the system.

4. Discussions

There are a number of research groups using agent technologies to tackle the problems in concurrent engineering. However none of them really explores or emphasis the potentials of intelligent agents' mental attributes such as proactiveness and autonomy to alleviate redundant design tasks.

PACT [12] is one of pioneers using agent technology to achieve collaborative engineering design in a distributed environment. The agent in the PACT encapsulates program modules and invokes them remotely as network services when needed. The aim of the experiment demonstrates its framework enabling the knowledge reuse and sharing among design agents. Olsen [13] and Khedro [14] use the same architecture to demonstrate its applications in car design and building engineering environment. The facilitator, a task-independent mechanism, co-ordinates the runtime activities of the individual software agents. The agent communication language allows the agents to communicate in the same expressive language. An ontology is used to define common terms in the participating agents in order to alleviate the misunderstanding. Edmonds et al [15] proposed a multiagent group support framework to provide agents with context-sensitive dialogue and to enable group interaction with a remote geographic information system.

Agent-based blackboards [16], like federation systems, use grouping to manage agent interactions. Each local group of agents shares a data repository that is provided specifically for the efficient storage and retrieval of active shared data. Along with design data, tactical control knowledge can be represented in the shared repository, enabling reasoning about how to proceed with the design process to share the same status and priority as reasoning about the design itself. Within an agent group, a control shell (analogous to the federation's facilitator) notifies appropriate agents of relevant events [16].

Mori et al [17] proposed reactive design agents that internally maintain states and rule-based knowledge and they do not yet believe that autonomous agents are practical for most engineering applications. A new initiative from DARPA [18]. proposed a control strategy to enable the autonomous agents to form super applications at run time. This results from the recognition that that the autonomy and partial knowledge make agents useful for addressing the problems of complex engineering design environment. The power of the autonomous agent, however, can cause problems such as misinterpreting the request, making poor decisions, interacting with other agents in destructive ways [18]. Thus, it requires a control strategy to eliminate the above problems.

The A-Design theory [19] is founded on the notion that engineering design occurs in an ever-changing environment and therefore computer tools developed to aid in the design process should be adaptive to these changes. The use of agent is to perceive their environment (design states) through sensors (i.e. functional inputs) and act upon their environment through effectors (i.e. modifications to the design states). The agents in A-Design do not fully conform to the definition of agency since its agents do not display specific behaviours of autonomy, mobility, or sociability [19].

Shen [20] carried out a comprehensive literature survey in agent-based systems for intelligent manufacturing that provides a good classification of existing approaches. The survey, however, did not highlight the potential of proactiveness in the agent that can fundamentally reduce faults that occur in the process of manufacturing.

The agents used to tackle large and complex concurrent engineering problems can be classified into software and intelligent agents. They focus on knowledge sharing and reuse, and design coordination. However, our view about intelligent agents does not only achieve their goals, but they can also minimise the redundant design activities. Reducing design activities, benefits other areas in design. It reduces the interaction of systems and minimises the number of design versions [21]. Before the agent formally publishes its design output, the feasibility of solutions is assessed. It reduces the number of design versions and the required space for design archives. A fundamental aspect of research in the area of Distributed Artificial Intelligent is coordination [22]. The need for coordination exists because tasks are completed in a distributed environment [2]. An oversight in this research area is the failure to identify and develop the idea of agent-oriented systems acting proactively and autonomously.

We suggest that the proposed system can significantly alleviate the demand of coordination. It could be applied to other fields. However, a number of changes in the framework would be required for new application areas. Domain knowledge in the applications and the mental models represented by the agents would need to be changed according to the characteristics of the new application areas. The IDL used to link the mental model and the applications would also need to change to represent the required new functions.

The tools, FIPA-OS [9] and JAM agents [23], used in the experiment are prototype systems, so the reliability is not sufficient for real applications, but they are able to demonstrate the effectiveness of this research. The monitored attributes (such as the cost and functions of equipment items) in the agents and the workflow are determined according to the human expert's arbitrary experiences in the current design environment. These attributes may not always be critical through the design life cycle due to the environment changes even though this issue did not occur in this case study. A further investigation needs to be carried out in order to identify dynamically essential attributes and devise an appropriate workflow for such. The two characteristics of proactiveness and autonomy in agents have been shown to have significant impact on effectiveness and efficiency of engineering design. However, more attributes having positive impacts on reducing design activities need to be explored.

5. Summary and Conclusions

One requirement of concurrent engineering is the ability to communicate the effects of a design change (within a distributed environment) to all concerned disciplines. However, the successful communication of such a change faces many hurdles. For example, different design tools may be based on different system architectures or methodologies, such tools will often be located at different sites, and each tool may have a different data model of the design process. Consequently, the difficulties in understanding the diverse knowledge and information generated from these multiple design disciplines may reduce the effectiveness of a distributed design environment. Initially, the authors proposed a CORBA-based framework to address these problems, however as the work progressed, the limitations of such a framework became self-evident. As a result of these limitations an agent-based framework (incorporating CORBA technologies) was developed. Agent attributes (for example, proactive, reactive and autonomous properties) are ideally suited to this particular problem. They provide a more flexible and dynamic means by which multiple disciplines situated within distributed design environments can successfully design large and complex made-to-order (MTO) products. It is the intention of this paper to demonstrate how intelligent agent technologies can be used in a distributed design environment to design MTO products.

6. References

[1] King, B. *Automatic Extraction of Knowledge from Design Data*, Ph.D. Thesis, University of Sunderland, 1995

[2] Guenov, M., Chao, K-M, Florida-James, B., Smith, N., Hills, B., & Buxton, I., "Tracing the Effects of Design Changes Across Distributed Design Agents", *The Second World Conference on Integrated Design and Process Technology*, 1-4 Dec. Houston, Texas, USA, 1996.

[3] Chao, K-M, Guenov, M., Hills, B., Smith, P., Buxton, I., & Tsai, C-F, "An Expert System to Generate Associativity Data for Layout Design", *Journal of AI in Engineering*, 11(2) pp. 191-196, 1997.

[4] Chao, K-M, Smith, P, Hills, W, Florida-James, B, Norman, P, "Knowledge Sharing and Reuse for Engineering Design Integration", *Journal of Expert System with Applications*, 14, 1998, p 399-408. ISSN 0957-4174.

[5] Smith, N., Hill, W., & Cleland, G., "A Layout Design System for Complex Made-To-Order Products", *Journal of Engineering Design*, 7, 1996.

[6] Chao, K-M, Florida-James, B, P. Norman, P. Smith, B. Hills "Use of Virtual Reality and Agents in Engineering Design" published in *Conference Proceedings of EXEPERSYS-98*, Virginia, USA, 16-17 Nov. IITT-International, p 183-188

[7] Guenov, M. "Modelling Design Change Propagation in an Integrated Design Environment", *Journal of Computer Modelling and Simulation in Engineering*, 1, 1996, pp. 353-367.

[8] Greenstein F. *Electronic Commerce: Security Risk Management and Control*, McGraw-Hill, 2000

[9] FIPA, *Agent Communication Language Specifications 97*, http://www.fipa.org, 1997

[10] Searle JR, *Speech Acts*, Cambridge University Press, 1969

[11] Rao, S A, & Georgeff. M. P., "BDI Agents: From Theory to Practice", *Conference Proceedings of 1st international conference on multiple agent system*, 1995, pp. 312-319

[12] Cutkosky M., Engelmore R.S., Fikes, R.F., Genesereth, M.R., Gruber, T.R., Mark, W.S., Tenenbaum, W.S., Weber, J.C., "PACT: an Experiment in Integrating Concurrent Engineering System", *IEEE Computer*,. 26(1), 1993, pp28-37.

[13] Olsen, G.R., Cukosky, M., Tenenbaum, J.M., Gruber, T.R, "Collaborative Engineering Based on Knowledge Sharing Agreements", *Journal of Courrent Engineering: Research and Applications*, 3(2), 1995 pp 145-159.

[14] Khedro, T; Genesereth, M.R., "The Federation Architecture for Interoperable Agent-based Concurrent Engineering Systems", *Journal of Concurrent Engineering: Research and Applications*, 2(2), 1994, pp125-131

[15] Edmonds, E.A., Candy, L., Jones, R., Soufi, B., "Support for Collaborative Design: Agents and Emergence", *Communication of ACM*, 37(7), 1994, pp 41-47.

[16] Lander, S.E., "Issues on Multiagent Design Systems", *IEEE Expert*, 12, 1997, pp 18-26

[17] Mori, T., Cutkosky, M. "Agent-Based Collaborative Design of Parts in Assembly", *Proceeding of ASME design engineering technical conference,* Sept 13-16, 1998, DETC98/CIS-5697.

[18] Dyer, D.E, "Multiagent Systems and DARPA", *Communication of ACM*, 42(3), pp 53, 1999.

[19] Campbell, M. I., Cagan, J., Kotovsky, K., "A-Design: An Agent-Based Approach to Conceptual Design in a Dynamic Environment", *Journal of Research in Engineering Design*, 11(3), 1999 pp 192-192.

[20] Shen, WeiMing, Douglas H. Norrie, "Agent-based Systems for Intelligent manufacturing: A State-of-the-Art Survey", *International journal of Knowledge and Information Systems*, 1(2) pp 129-156

[21] B Florida-James, N. Rossiter, K-M Chao "An Agent System for Collaborative Version Control in Engineering", *Special Issue- Journal of integrated manufacturing System (The International Journal of Manufacturing Technology management)*, 11(4), 2000 pp. 258-266, MCB University Press.

[22] Coates G, Ritchey I, Duffy AHB, Hills W, Whitfield RI. "Integrated engineering environments for large complex products", *Concurrent Engineering-Research & Applications*, 8(3), 2000, pp.171-82. Publisher: Technomic Publishing, USA

[23] Huber, M. J., *JAM Agents in a Nutshell*, http://members.home.net/marcush/IRS, 1999.

Research on Integrated Model Based on XML, CORBA and Agent

Hong Guo, Rui Chen
Department of Computer Science and Technology, Fuzhou University, Fuzhou, Fujian, P. R. China 350002
ghlzh@pub5.fz.fj.cn

Li-ming Hu
Fujian Suncom Intelligent System Institute, Fuzhou, Fujian, P. R. China 350002
suncom@pub1.fz.fj.cn

Abstract

In the past two decades, perspective on system integration (SI) has changed greatly. Several disciplines have done much work on SI, and advance new techniques unceasingly to resolve problems that arose from distribution, heterogeneity and autonomy. Studied several new techniques widely used in SI, such as extended markup language (XML), CORBA and Agent. We propose an integrated model based on these techniques to improve performance and apply it in intelligent building system integration.

1. Introduction

In recent twenty years, changing perspective on SI usually goes with new breed of techniques and applications. In the mid to late 1980s, one used to understand programs and extraction essential functionality from them in order to develop a new two-tier client-sever program, which is more robust and maintainable. But soon defects of this kind of architecture emerged. In the early 1990s, three tiers replaced the two-tier architecture to make programs more flexibility, scalability and improve system performance. However it made programs more complexity. Later, with the maturity and acceptance of object-oriented languages, people began to use wrapper to hiding legacy internals while providing a well-defined interface view to the outside world. And now, middleware and framework are broadly used in SI. SI has changed its focus from legacy migration to legacy integration and has come into its own.

Target of SI is to build a new application that is adaptable to business and technology changes while retaining legacy applications and technology as reasonably as possible. A variety of information technology (IT) disciplines are studying problems and solutions for SI on different respects, but they all focuses on three issues: autonomy, heterogeneity and distribution. Then obviously SI is highly interdisplinary.

Studied pros and cons of several popular integration techniques, we, using CSCW ideas, propose an integration model, which fuses XML, CORBA and Agent, and apply the model in SI of intelligent building. In second part of this paper, we will analyze these techniques and specify the reason why we choose them. Architecture and constituents of the model will be described in the third part. Then in the forth part we will discuss communication and coordination mechanism of agents in the model.

2. Analysis of techniques used in SI

Normally, an information system can be structured in three layers: the business architecture, the application architecture and the technology architecture [1]. Accordingly, the solution for SI can also be achieved in three layers. The top layer is Inter-organizational processes that aims at business reengineering; Enterprise application integration integrates independent enterprise resource planning (ERP) or workflow management system (WFMS). At this layer, the concrete syntax and the semantics of standardized messages is needed; Middleware integration addresses the syntactical level and aims at building componentized systems with state-of-the-art infrastructures. Several techniques, including distributed object, XML, mediator and agent, are wildly employed in the later layers in SI.

Distributed object, such as CORBA and DCOM, integrates distributed computing and object-oriented design, plays an important role in SI. Each object in such environment interoperates with others using the same mechanism provided by distributed object computing (DOC). In SI, wrapper acts a crucial role by encapsulating the legacy applications [1][2][3]. When used solely, however, such approach has the same shortcoming as two-tire client-server architecture. Besides, this solution is code-centric and easy to arise semantics issues because of building system based on interfaces.

To address semantics issues, XML that essentially free data from its dependence on software infrastructures becomes popular and has become the leading data format definition language in e-commerce [4][8]. This approach is data-centric and lets applications communicate regardless of the programming model and thus reduces heterogeneity on homonyms and synonyms. Although

XML now is wildly used in data exchange between different systems, it cannot be used solely and must marry other techniques for system controlling.

In distributed artificial intelligence (DAI), one can integrate heterogeneous resources with mediator and agent that managed by logical rules. Mediator and agent bear the responsibility for resolving conflicts and avoiding redundant tasks in order to improve autonomy and intelligence of each legacy system [7][8]. But such solution is platform dependant and highly complex when implementation.

When we think which technique should be chosen in our model, we select all the above to integrate their merits and avoid their respective defects. CORBA, which is selected to address syntactical issues, is applied to enable an agent to communicate with other agents in heterogeneous environments. There are three reasons for choosing XML instead of wrappers as common language to represent data, request and responses for three reasons. First is because XML lends itself to automatic generation of wrappers through its transportability between heterogeneous information systems in a neutral and system-amenable manner. The second reason is many XML parsers, tools, and libraries are readily available. The last reason is that XML structure data would not change over time, so these data become more stable and enduring. To deal with semantics issues in communication between agents, Agent Communication Language (ACL) is needed [2]. We use XML as inter language and Knowledge Query and Manipulation Language (KQML) as outer language.

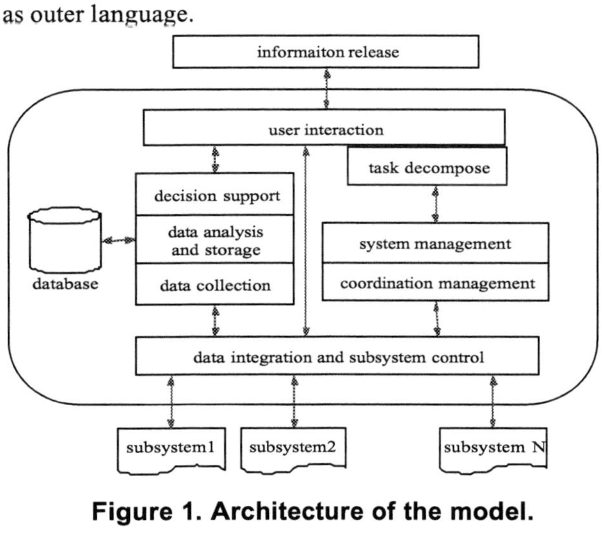

Figure 1. Architecture of the model.

3. Structure and constituents design of the model

3.1. Target and function of the model

Facing so many legacy systems, we need an open framework which can not only make them smoothly work together but also achieve new tasks which can not be resolve by any single one. By integrating legacy systems, which are seen as subsystems in the model, time and money will be saved, early investment will be revived, and life of these legacy systems will last, thus it will strengthen stability of the new system. Architecture of model is shown in figure 1. The model has mainly three functions. The first is data integration and subsystem control, which means collecting data from subsystems and sending message to control them; The second is comprehensive data analysis and decision support, which cataloging, analyzing and storing the data have collected, and make decisions according to these information. The last is user interaction, which interacts with users such as result displaying, decomposing user request and then sending them to relevant agents.

3.2. Construction of the model

Our model is composed by several kinds of agent each has different tasks: Core Agent supervises all the agents in the model; Local Agent, which administrates a fixed legacy system, acts as a mediator among the subsystem and other agents; Integration Agent, plays an important role in the model, subscribes and collects data from Local Agent, analyzes them and makes decision according to them, and is in charge of storage of these decisions and data. We will briefly describe structure and function of these agents in following.

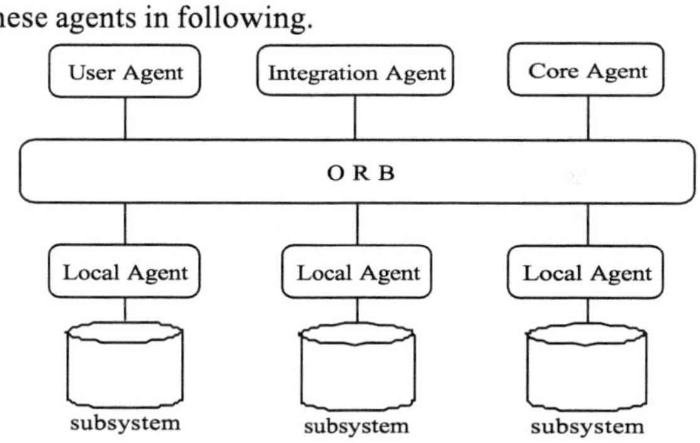

Figure 2. Constituents of the model

Before describing these agents separately, we should discuss four common parts of them: communication unit (CU), planning unit (PU), administration unit (AU), and executive unit (EU). CU is in charge of communication with other agents and performs translation information between KQML and internal representation in agent. With a knowledge base, PU decomposes and plans the task that it should finish, and then submit them to CU or EU. AU is in charge of administration of the agent, analyzing messages from other unit in the agent, controlling and coordinating jobs of units. EU performs tasks requested. To support concurrent operation, a task pool is needed. But tasks should be executed according to its importance.

3.2.1. Core Agent. Core Agent, which is core constituent in the model, supervises all agent in the model, and coordinates activities of them. It maintains, for instance, an active Agent list to record status and services of an active agent. Any agent should register on Core Agent when initiating. When receiving requests for services from other agent, Core Agent responses the handle of an relevant agent which can provide that kind of service. It also coordinates overall agents activities, makes plan for cooperation and then send commander to corresponding agent. Thus, Core Agent acts as an mediator among agents in the model. Figure 3 illustrates structure of Core Agent.

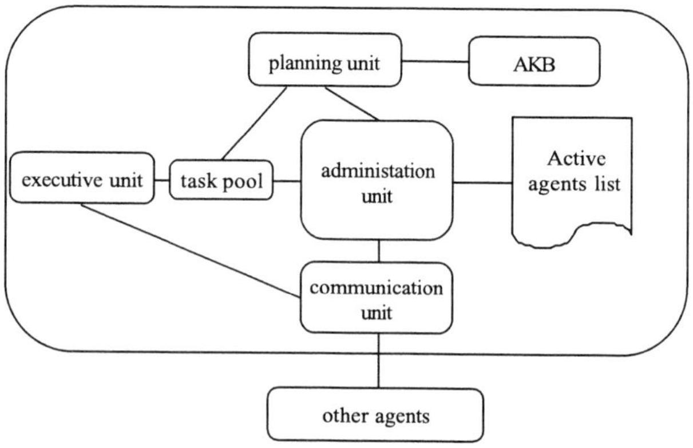

Figure 3. Structure of Core Agent.

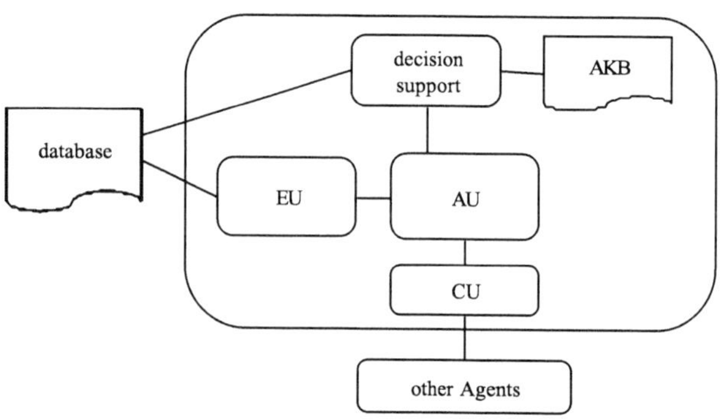

Figure 4. Structure of Integration Agent.

3.2.2. Integration Agent. Structure of Integration Agent is shown in figure 4. The main function of Integration Agent is to subscribe and collect data which is used for making decisions, and to store these history data and decisions requested by other agent such as Core Agent and User Agent. Integration Agent sends data they demand. Thus, Integration Agent is crucial because it performs data analysis and decision support in the model.

3.2.3. Local Agent. Local Agent can be seen as a plug-in system in legacy system and a mediator in the model. It administrates the status and activities of the legacy system by interface which legacy system exposes. Structure of Local Agent is shown in figure 5. For example, Local Agent collects real-time data from the subsystem, then translates and stores them in XML format. Local Agent also responses other agents' request, sends these update data at fixed intervals or any time on demands, or controls subsystem at other agent's will.

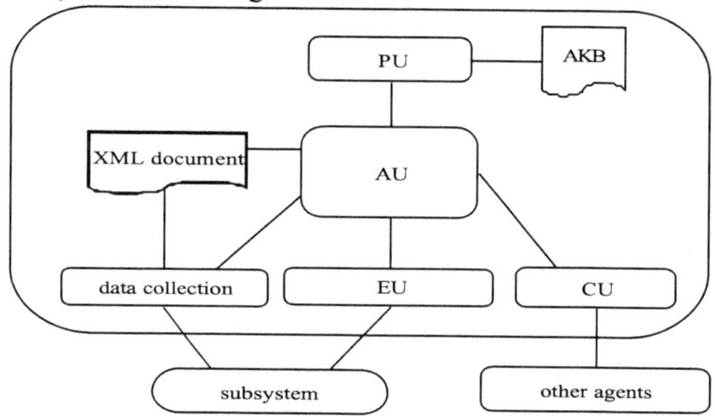

Figure 5. Structure of Local Agent.

3.2.4. User Agent. User Agent takes charge of interaction with users, including receiving demands from users and display data user request. After receiving user's request, User Agent decomposes them into several smaller tasks, sends them to relevant agent such as Core Agent, and synthesizes results of these tasks to show them to users. In order to speed up responses, User Agent subscribes real-time data from Local Agent and stores short-term history data by itself. User Agent, thus, is a mediator between model and users.

4. Communication and coordination design

4.1. Basic communication mechanism between agents

As brokers in distributed coordination models for client-server computing [5][7], which can free clients from having to maintain information on where and how to obtain particular services, there are three mechanisms to perform coordination of agents: forwarding broker, handle-driven broker and hybrid broker. In first approach, Core Agent acts as a forwarding broker, illustrated in figure 6a, relays others agent's request to the relevant server agent, retrieves the response, and relays this reply back to the agent requesting. The approach provides one locus of control for all requests and response, so it is easy for fault tolerance and recoverability. But excessive message traffic and request logging can degrade overall performance. In second approach, Core Agent acts as a handle-driven broker, illustrated in figure 6b, returns a service agent's "handle" back to the request agent, which contains all the information required to interact with the server agent for the given service; the request agent uses this handle to issue its request directly to the server agent, which then replies back to that client. This approach reduces the potential for bottlenecks and client agent can store handles for later reuse, which can reduce broker traffic if client agents request the same services repeatedly. However, handle-driven broker shift the burden of detecting and managing faults to the clients. Hybrid broker approach exploit the benefits of both models by offering the flexibility to selectively adjust control and performance trade-offs for individual client agents.

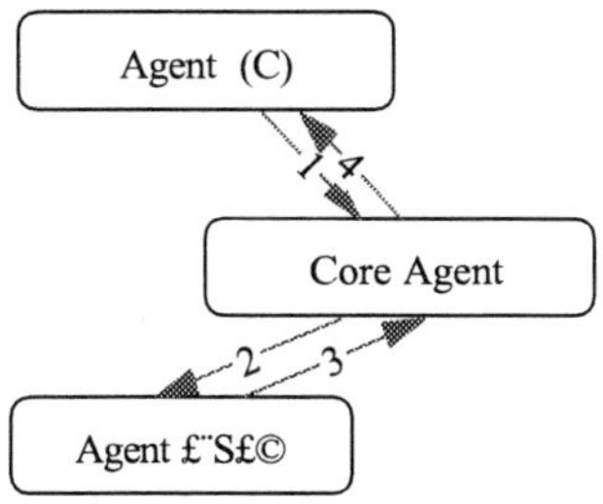

Figure 6a. Forwarding broker.

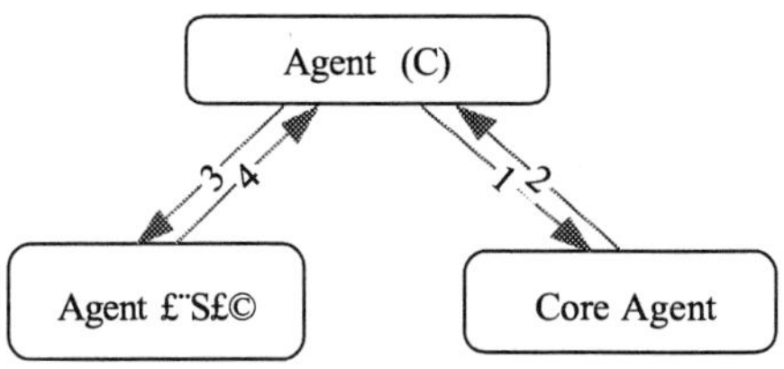

Figure 6b. Handle-driven broker.

We employ hybrid broker model as a control mechanism that mediates interactions between client agents needing services and server agents capable of providing them. All applications that belong to a distributed system register the services they support as well as their locations and client interfaces with the broker. As to data transmitting which is heavy-traffic and do not need global knowledge, we use handle-driven broker approach; for interoperation of agents, we employ forwarding broker approach and use Core Agent, which has global knowledge, as a mediator of these agents. Thus by hybrid broker, we avoid shortcoming arise from using single approach of forwarding broker and handle-driven broker.

4.2. Communication language

Although XML defines metadata of each heterogeneous subsystem in the model and avoid different modeling of the same real-world concepts such as homonyms and synonyms, it cannot represent actions and is far away from enough to perform the work well. A communication layer is needed to build above XML structure data to describe the action type that agent request such as subscribe, tell and reply. So KQML is chosen to fulfill the task. When a agent in the model received request in KQML, CU of the agent will translate contents into XML format and submits them to AU. In general, contents of the request represent the status the agent should be in. Then the agent controls the subsystem according its current status. Some message used in the model is following.

perform (request contents in XML);
subscribe (request contents in XML);
tell (request contents in XML);
ask-for (request contents in XML);
reply (request contents in XML);

For example, in our project on system integration of intelligent building, system is composed by several independent sub-systems such as Communication Automation (CA) subsystem, Office Automation (OA) subsystem and Building Automation (BA) subsystem. In those earlier days, coordination between these subsystems should be performed manually. So they cannot build up an efficient integrality. In our system, when alarms occurs in a subsystem, Local Agent will immediately inform Integration Agent, and inform Core Agent if it is a major alarm. The message that Local Agent sending to Core Agent has the following format:

tell (<alarm> alarm contents < \alarm > < site > alarm locale < \site >);

Local Agent will repeatedly send this message until it receives Core Agent's reply. After answering Local Agent, Core Agent analyzes alarm contents, produce individual tasks by PU of the agent, and then send them to relevant agents. The message that Core Agent sending to a Local Agent has the following format:

Perform (< site > alarm locale < \site > <device-type> device type need changing its status <\device-type> <status> status the device should be in <\status>);

The Local Agent will handle the request immediately, perform an operation according to current status of the subsystem, and then inform Core Agent.

5. Conclusion

The model has been applied in SI of intelligent building. The result is satisfactory. Comparing with model proposed in [5], the new model can also shield heterogeneity comes from operation system (OS) and hardware and so suits to distributed environment. By introducing XML and agent technique, our new model would fuse merits of them, prolong the life of legacy systems, and easily adapt to change of business rules. Different agent performs coordination of subsystems and data integration, thus it enhances reliability and scalability of the system that is build on the new model.

6. Acknowledgments

The work described in this paper was supported by Science Council and Education Council Grant of Fujian province, P.R.China: 2000H065, JA00146

Reference:

[1] Wilhelm Hasselbring Information System Integration. Communications of the ACM. Jun. 2000/Vol.43, No.6 P33-38

[2] Michael R. Genesereth and Sieven P. Ketchpel. Software Agents Communications of the ACM. Jul. 1994/Vol.37, No.7 P48-53

[3]Gho Hong, Guo Chao-zhen, Chen Rui and Liu Zhi-hui. Research on heterogeneous resource integrated model used in intelligent building system. Proceedings of The Fifth International Conference on Computer Supported Cooperative Work in Design (CSCWD2000). Nov.29 to Dec.1,2000 Hong Kong,330-335.

[4] Frank P.Coyle. Legacy Integration--Changing Perspectives. IEEE Software. Mar./Apr. 2000 37-41

[5] Richard M. Adler. Distributed Coordination Models for Client/Server Computing. IEEE Computer, Apr. 1995 pp14-22

[6] Narinder Singh. Unifying Heterogeneous Information Models. Communications of the ACM. May. 1998/Vol.41, No.5 P37-44

[7] Paolo Ciancarini el. Coordination Multiagent Application on the WWW: A Reference Architecture. IEEE Transaction on Software Engineering. May 1998 Vol.24 No.5 362-375

[8] Nabil R. Adam, VijayaLakshmi Atluri, and Igg Adiwijaya. SI in Digital Libraries. Communications of the ACM. Jun. 2000/Vol.43, No.6 P64-72.

Architectural Properties of CSCWD System Based on Multi C^2D^2-Agent

Xiaosu Zhan, Zongkai Lin, Guang Xu, Yuchai Guo
CAD Lab, Institute of Computing Technology,
Chinese Academy of Sciences
P.O. Box 2704, Beijing 100080, P.R. China
xszhan@public.bta.net.cn

Mingye Liu, Meilong Ye
Department of Computer and Engineering
Beijing Institute of Technology,
Beijing, 100081, P.R. China

Abstract

In order to accommodate the new change brought by CSCD application systems, a new CSCWD system integration platform, based on Multi C^2D^2-Agent, is required. In this paper, a method in the form of MAS-based model for designing CSCWD system and C^2D^2-Agent concepts are proposed. The architectural property of this type of CSCWD system is studied.

We show the importance to construct C^2D^2 agents and explore an architecture model for multi-agent based CSCWD system. We illustrate the structure of C^2D^2 agents, especially autonomous interface agents, and discuss some design principles for such agents. In the CSCWD system, autonomous interface agents can expediently support synchronous applications.

1. Introduction

CSCW system shows that computer application has got a new breakthrough, from supporting essentially individual work to supporting group work, especially the cooperative work the people of different specialties participate in, and discover and solve conflicts [2]. Recent developments in CSCW and MAS promote the research on computer supported cooperative design, i.e. CSCWD (CSCWD in Design) system. CSCWD system is an important application of CSCW system

CSCWD system aims at improving the effectiveness and usability of the way in which people work with computers and communication in design process [7]. Agents in CSCWD system are one of the most important factors [1][4][5][6]. They can provide synchronous and asynchronous communication functions, group collaboration paradigms and strategies, concurrent shared resource assignment, system management, task process transaction, etc.

Based on the MAS theory, we have introduced a MAS system to establish framework of CSCWD environment. This paper studies the multi-agents from the viewpoint of building CSCWD systems, including the software architecture style of multi C^2D^2-agent system, the vocabulary elements and descriptions of C^2D^2-agent. The meaning of software architecture, the pattern or the characteristics of software, is the term of some kind of system repeatable and shared organizing pattern, for example Client/Server (C/S), Browser/ Server (B/S), etc. It is important to study the software architecture style from constructing CSCWD system methods.

Some issues associated with multi C^2D^2-agent system are investigated in this paper including architectural model, description model, running control mechanism, environment factor representation, multi C^2D^2-agent infrastructure and communication.

Architectural properties is one the most important elements when building C^2D^2-agent systems and MAS frameworks.

2. A Panoramic Overview of the Different Agent Types

In this section, we overview the different types of agents. Based on our typology, agents can be summarized into two types: cognitive and reactive.

```
Agent =
{
cognitive agent;
reactive agent
hybrid agent;
};
cognitive agent=
{
smart agent;
mobile agent;
collaborative agent;
human agent
};
hybrid agent=
{
heterogeneous agent systems;
};
```

reactive agent=
{
information agent;
interface agent
};

3. The Architecture and running control mechanism of Multi C^2D^2-Agent System

The architecture we adopted is a mixed one of client/server, browser/server, and MAS (Multi-Agent System). The client/server is adapted to local network for cooperative work for small scale. As browser/server uses multi-tiers, multi-levels server structure, its processing ability raises in a wide range. It is more suitable to meet the needs for cooperative work in design of large scale on Intranet and the Internet.

Dynamic characteristics of the CSCW system are another cooperative feature. Multi-agents structure and meet the needs of dynamic change and open ability. The personal and various programs people use for performing cooperative tasks are divided into some groups and basic units and allocated to different computers. These personal groups and software tools may constitute independent autonomous agents that can be activated when certain conditions are satisfied, then the agents can perform the specified tasks. Therefore, this architecture adapts to the demands of dynamic change quite well.

We do research on reactive agents. This type of agent consists of two parts: basic unit and management unit. To constitute agents of this kind, groups and software tools are named basic unit (BU), and each of them are covered with a layer of interface, called management unit (MU). The management unit is composed of four parts: **C**ontrol module and **C**ommunication module (C^2), **DRM** (Descriptions of Running Mode) module and **DWE** (Descriptions of Working Environment) module (D^2). So we call this type of agent C^2D^2 agent (described below).

C^2D^2 agent=
{
BU(basic unit);
MU(management unit);
}
BU=
{
human group;
software tools;
...
}
MU=
{
Control module;
Communicating module;
DRM(Descriptions of Running Mode) module;
DWE (Descriptions of Working Environment) module;
...
}
DRM
{
active agents list;
centralized agents list;
decentralized agents list;
status of server;
...
}
DWE
{
personal list;
group member list;
task list;
...
}

In C^2D^2 agents, the control module is responsible for the running mode, i.e. to analyze the message from other agents and information from D^2, to decide whether the agent will be running and to activate or trigger the agent when the D^2 are satisfied and it is capable of initiating and reacting to events. Communicating and I/O module, whose aims are to receive the other agents' message, is used for communication and I/O processing interface to the related agents directly. DRM and DWE manage current cooperative working environmental state, running environmental state, and other control information. Therefore, the C^2D^2 agent structure we used is a dispersible mechanism

3.1. Multi C^2D^2 Agent based Models

Multi agent-based models structure a CSCW system as a collection of specialized combinatorial units called C^2D^2 agent. C^2D^2 agent models stress a highly parallel modular organization and distribute the state of current tasks among the agents. Modularity, parallelism and distribution are convenient mechanisms for supporting the design of coordination manipulation, for implementing physically distributed applications and for handling multi-threaded procedures. In order to apply the separation at every level of abstraction and refinement, C^2D^2 agent systems distribute the functional separation of concerns among distinct cooperating agents.

3.2. A Hybrid Multi C^2D^2-agent Based Architecture Model

Figure 1 gives a hybrid Multi C^2D^2-agent based Architecture Model for CSCW System on client/server and browser/server structure. The agents on server side use system software and management system to support synchronous and asynchronous cooperation.

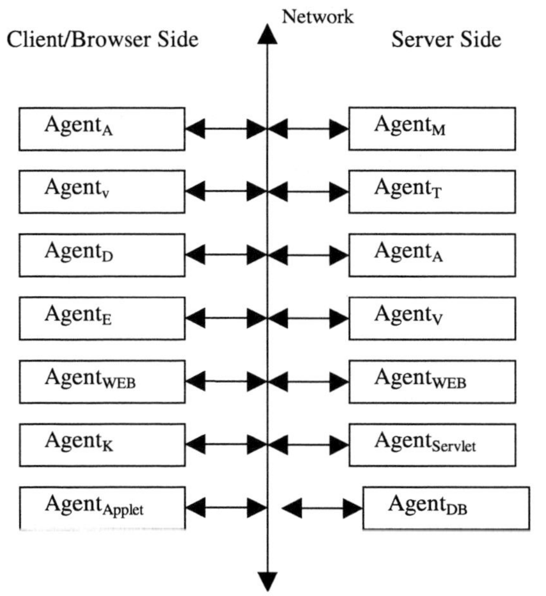

Figure 1 An architecture model for CSCW system.

Initializer has two primary functions: one function is capable of compiling the descriptions of cooperation to encapsulate MU (management unit) at every basic unit and another function is used to generate Agent$_M$, Agent$_T$, Agent$_A$ and to activate.

Agent$_M$. Agent$_A$ is an administration agent. It can mediate the whole running state of the system. Agent$_M$ manages the working flow of the cooperative task and informs every agent the working status of the other agent. Agent$_T$ monitors the working status of the agent group.

3.3. Description of C^2D^2 agent

In this architecture, an agent is defined as a specialized object whose state is viewed as consisting of attributes, capabilities, and running modes.

<agent> ::= AGENT(<agent_name>)
attributes
capabilities
running mode
END_AGENT
<attributes> ::= agent_type
site
related agent
<capabilities> ::=<action><mntlcond>
<action> ::=DO <private sequences> <condition>
| INFORM <agent><fact><time>
| REQUEST <agent><action><time>
|UNREQUEST<agent><action><time>
| STOP <agent><action><time>
<fact> ::= (<predicate><arg><time>)
| <variable>
<mntlcond> ::=< mntlcond > OR <mntlpattern>
| < mntlcond > AND <mntlpattern>
| <mntlpattern>
running mode ::= <excute>
| <wait>
| <stop>
| <end>
...

This structure of agent provides the basic components to construct agents and communication primitives such as INFORM, REQUEST, UNREQUEST, and STOP to support agents to communicate with each other. Each agent is controlled by itself.

4. HHII C^2D^2 agents

Direct manipulation graphical interfaces display visual representations of physical or conceptual objects, and allow the user to issue commands that change the state of the objects. In a direct-manipulation interface, the changes to the display states are more or less one-to-one with the commands explicitly invoked by the user.

We define HHII agents to be a program that can also affect the objects in a direct manipulation interface, but without explicit instruction from the user. The HHII agents reads input that the user presents to the interface, and it can make changes to the objects the user sees on the screen, though not necessarily one-to-one with user actions. The agent may observe many user inputs, over a long period of time before deciding to take a single action, or a single user input may launch a series of actions on the part of the agent, again, possibly over an extended period of time.

HHII agents could be considered to be a "robot" whose sensors and effectors is the input and output capabilities of the interface, and for that reason are sometimes also referred to as "softbots" [3].

The best-known examples of HHII agents are electric conferencing systems and context-sensitive help systems. In such systems, the user may operate the interface with complete disregard for the agent. However, if called upon, the agent may also display suggestions, or perform direct-manipulation actions on objects in the displayed interface, based on input implicitly collected from the user. HHII agents may critique the user's behaviors, or augment the user's direct-manipulation actions with extra computed information that the user may find helpful.

HHII agents are becoming more and more attractive due to the increasing complexity of user interfaces and the tasks to which they are applied. If we insist on maintaining the one-to-one correspondence between interface actions and capabilities of the user interface, the hallmark of direct-manipulation interfaces, we will soon reach the point where no more functionality can be added to our systems. HHII agents provide a way out of this dilemma.

5. Autonomous C^2D^2 agents

An autonomous agent (AA) is an agent program that operates in parallel with the user (Figure 2). Autonomy says that the agent is, conceptually at least, always running. The agent may discover a condition that might interest the user and independently decide to notify him or her. The agent may remain active based on previous input long after the user has issued other commands or has even turned the computer off.

Why autonomous agents? An assistant may not be of much practical help if he or she needs very explicit instruction all the time and constant supervision while carrying out actions. Assistants can be time-savers when they are allowed to act independently and concurrently. Allowing an interface agent to run off-line and in parallel with the user directing attention to other activity enables the user to truly delegate tasks to the agent.

By contrast, a traditional command-line or menu-driven interface is conversational. The user performs input, enters it, the system accepts the input, computes some action, displays the result, and waits for the next input. This has the property that the system is doing nothing while the user is preparing the input.

And the user is doing nothing in the interface while the system is running. It is clear that an agent may exhibit each of these characteristics independently. An interface agent may observe user interface actions and make changes to user interface objects displayed on the screen without having the capability to continue running in parallel with the user. The electric conferencing systems, context-sensitive help systems, intelligent tutoring systems, and monitoring systems fall into this category. They may have no need to independently notify users of external conditions or run while the computer is off.

An autonomous agent may have the capability to run autonomously without needing to interact through graphical interface operations. For example, many programs send members in groups e-mail when there is an update to Web pages you have marked as interesting; these agents are autonomous, but operate outside the user interface. Java is one of "agent languages" in the sense that they are suited to programming asynchronous, distributed activity. Though these agents may act autonomously off-line, the agent does not operate the interface itself.

In order for an agent to be considered an autonomous agent, it follows that there must be some part of the interface that the agent must operate in an autonomous fashion. The user must be able to directly observe autonomous actions of the agent and the agent must be able to observe actions taken autonomously by the user in the interface.

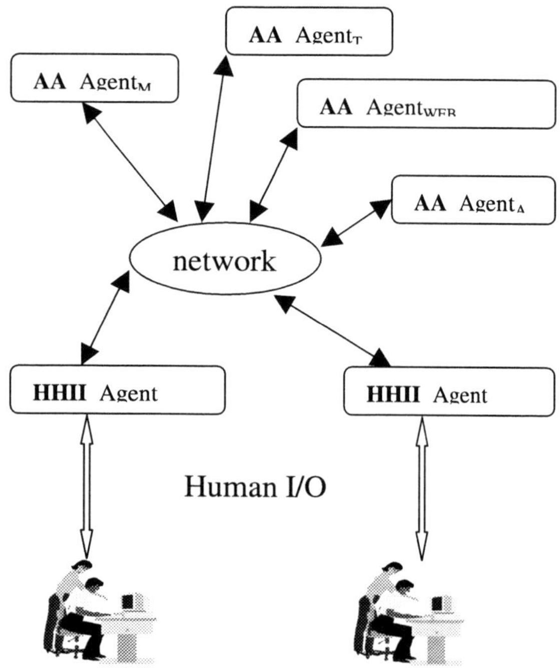

Figure 2. AA in multi C^2D^2-agents based system.

6. Agent$_{WEB}$

Agent$_{WEB}$ presents its results to users continuously, using a kind of IE-like interface style. It can automatically sample information sources in the search for something of interest to the viewer.

Agent$_{WEB}$'s usual interface consists of three IE-like windows. The right-hand window is reserved for human browsing activity. The human may browse in this window in a normal manner, and can completely ignore any other agents' activity. The two windows on the left side are under control of the agent. The top window displays search candidates, those pages that Agent$_{WEB}$ is considering to recommend to the user. The bottom left window displays those pages that Agent$_{WEB}$ actually decides to recommend to the user, passing Agent$_{WEB}$'s tests for user interest. Agent$_{WEB}$ notices which window is being controlled by human and only being used which are not being actively used by humans.

Accepting a suggestion from Agent$_{Web}$ simply consists of switching to a window that contains a page recommended by Agent$_{Web}$, and continuing to browse.

$Agent_{Web}$ is not limited to this 3-window style. In practice, a less obtrusive way of running $Agent_{Web}$ is just to use two windows, a human window and one $Agent_{Web}$ window. Displaying the search candidates is not strictly necessary, but is often useful for demonstration and debugging purposes. Displaying more than two agent windows are also possible, for example, where the two windows could each be working off different profiles.

7. Conclusion

The research of C^2D^2 agent and architecture, which is corresponding to them, is still in its initial stage. We have solved and tested parts of the key technology by designing a prototype system. This paper discusses some problem of structure of C^2D^2 agents, and we debate running control mechanism and decentralized control model of CSCW system based on MAS.

The ability of autonomous agents to operate the same interface operated by human, and the ability of a software agent to act independently of, and concurrently with, human will become increasingly important characteristics of human-computer interaction. Agents will observe what human do when they interact with interfaces by using HHII agents, and provide assistance by manipulating the interface themselves Increasingly, applications will be designed to be operated both by HHII agents and corresponding agents synchronously.

This paper has illustrated some of the principles behind autonomous agents with a description of $Agent_{WEB}$, an agent for assisting automatically human's browsing the Web. It reports its results by continuously browsing Web pages in an IE-like window. It is autonomous, in that it browses concurrently with other autonomous agents, searches and analyzes the information it has searched. Autonomous agents for interface like $Agent_{WEB}$ will be the next step in the evolution of interfaces for CSCW system.

8. Acknowledgments

This research is supported by the National High Technology Development Program of China under Grant No. 863-511-946-011 and No. 863-511-942-016.

Heartfelt thanks are due to Professor Lin Shuoxun , Professor Wu Quanyuan and Professor Di Qiguang , who nurtured this project from the beginning and patiently sustained it to a successful end. I am grateful to Ph.D. candidate YU Lei, Ph.D. candidate Li Feng and Ph.D. candidate Li Qiujing for their contributions for this paper and for many stimulating discussions. Zhang Chendong, Su Zhong, Zhang Shaohua and Deng Guilong provided helpful feedback on an earlier draft of this paper. Significant support for this effort was provided by Management Group of CAD Lab, Institute of Computing Technology, the Chinese Academy of Sciences. Important support for this paper was provided by my Cooperative Design Group colleague in AFCC: Lai Jianrong and Li Shiwei.

This research was supported in part by project 1999-2000 of AFCC , PLA of China

9. References

[1] Coutaz, J., Bérard, F. and Crowley, J, Coordination of Perceptual Processes for Computer Mediated Communication. In *Proceedings Second International Conference on Automatic Face and Gesture Recognition,* IEEE Computer Society Press Pub. Oct. 1996.

[2] Ellis, C. and Wainer, J. A. , Conceptual Model of Groupware. In: Furuta, R., Neuwirth, C. Eds. *Proceedings CSCW'94,* ACM Conference on Computer Supported Cooperative Work,1994, 79-88.

[3] Henry Lieberman, From Human-Computer Interface to Human-Computer Collaboration, *CHI Research Symposium, 1995.*

[4] Nigay, L. and Coutaz, J. Building User Interfaces: Organizing Software Agents. In: *Proceedings ESPRIT'91 Conference,* 1991, 707-719.

[5] Patterson, J.F. A taxonomy of Architectures for Synchronous Groupware Applications, Workshop on Software Architectures for Cooperative Systems. *CSCW'94, ACM Conference on Computer Supported Cooperative Work.*,1994.

[6] Shaw, M. and Garlan, G. Software Architecture, *Perspectives on an Emerging Discipline,* Prentice Hall, 1995.

[7] Lin Zongkai. The Appearance of CSCD Making CAD Technology to Leap onto a New Step. In: Zongkai Lin et al eds. *Proceedings of Second International Workshop on CSCW in Design.* Bangkok, Thailand: International Academic Publishers, 1997. 162-167.

Agent-Based Cooperative NC Conceptual Design

Zhi Gang Xu
Design Technology & Research Center,
School of Design,
The Hong Kong Polytechnic University,
Hung Hom, Kowloon, Hong Kong
sdzhgxu@polyu.edu.hk

Ke Zheng Huang, Wei Min Liu
CAD Center
College of Mechanical Engineering,
Shandong University,
Jinan, Shandong, P.R.China
k_z_huang@yahoo.com, taolwm@sdu.edu.cn

Abstract

Creativity is a core issue of product design and development, so as is the machine tool that is implemented to make them. This paper presents a NC conceptual configuration model based on agent. Modularization is the fundamental of NC conceptual layout design. It takes each module as an autonomous entity participating in design activities, and the product design process is described as the organized cooperative work of agents. The Cooperative Muti-agent NC Configuration (CMNC) system will greatly increase the designing efficacy as well as creativity by offering a large number of design options and a rudimentary evaluation mechanism. The prototype system, which is under further development, has showed great potential of creativity.

1. Introduction

Creativity is the most competitive issue in product design and development, so does the machine tool that is implemented to make them. To date, the significant progress has been made in manufacturing technology, and a large number of FMS have been implemented arround the world. How ever with the ever increasing trend of individulization of products, agile manufacturing, rapid product design and develoopment etc have attracted more and more attention. Quick and creative design of NC (Numeracal Controlling Lathe etc) system is becoming even more significant. Conceptual design which affect 70 % (or more) the whole cost, but only accounts for 10 % of the budget, is under the way of in depth investigation.

Modularization is an effective way of NC system conceptual design and development[1], which is a simple way to express the structure of the system, also has enough flexibility in developing new structures. To incease the design efficacy, expert system is introduced to the mechanical design systems. While traditional expert systems, which are mostly sequential single-purpose systems definately lead to low efficiency. Agent and multi-media techniques make it possible to cooperate several designers to do the same designing task on the basis of Computer Supported Cooperated Work (CSCW)[2]. This article proposed a new approach of agent oriented Creative NC Configuration Design system.

The rest of this paper is organized as the followings: section 2 is a brief introuction of NC system modularization, section 3 describes the architecture and the function of agent. Section 4 proposes the agent oriented NC system design model. In section 5, 6 one design example is exploited to demonstrate the design process.

2. NC Modularization

2.1. Module Definitions

Three typical modules are introduced to express the NC conceptual layout. They are: translation module, rotation module and base module (as Fig.1 (a)). Their definitions are as follows:

Definition 1: Base module, support the translation or rotation modules which can not move or rotate, denoted by B, where H_B denotes the horizontal module, V_B denotes vertical module.

Definition 2: Translation module, supported by base module or rotation module which can move along one corrodinate direction x,y,z, denoted by H_X, H_Y and H_V accordingly.

Definition 3: Rotation module, supported by base module or translation module which can rotate around one corrodinate axis: x,y,z, denoted by R_X, R_Y, R_Z accordingly, while R_H, R_V means module rotating on the horizontal or the vertical base module.

Deduction: One rotation module can not be connected with another rotation module.

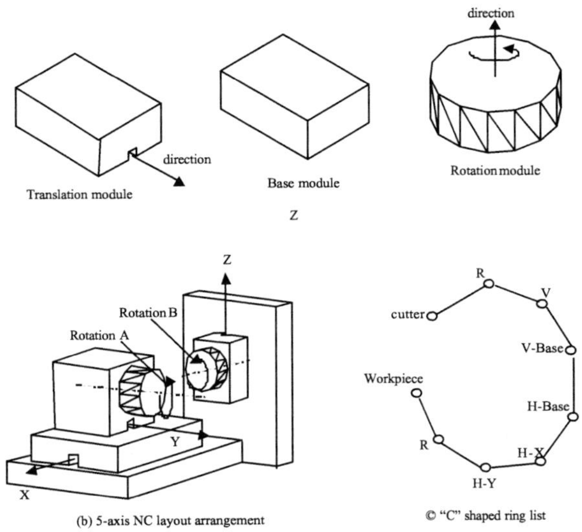

Fig.1 Module definition and 5-Axis NC conceptual layout.

2.2. Module Constraints

From the above definitions, the following Module Constraints can be deduced:

$$\begin{cases} R_i \cap R_j = 0; \\ B_i \cap B_j = 1; \quad R_i \cap H_j = 1; \\ R_i \cap B_j = 1; \end{cases} \qquad 0:$$

illigle :1 :ligile

According to the design specifications of lathe, the following design constraints can be got:

1) if the workpiece is heavy, then the workpiece can not move vertically;
2) if the workpiece is in high precision, then the workpiece is fixed or has only one translation movement;
3) if the workpiece is heavy and in high precision, then the workpiece jig must be adjacent with the base; ....

2.3 Module Display

The NC conceptual layout model can be expressed precisely by arranging different modules in an explicit way. Fig 1.b is an example of five-axis NC layout model with x,y,z translation movement and A, B rotation. Fig1.c is the "C" shape ring list model. It is a close ring from cutter to workpiece.

3. Function and Structure of the Agent

An agent contains five modules (Fig.2). The function and performance of the modules are as follows: *Acquaintance module* is the representation of other agents and of the local domain level system respectively. *Cooperate module* is responsible for managing the agent's social activities, the need being detected by the *situation assessment* module. *Control module* is the interface to the domain level system. *Situation assessment* module decides which activities should be performed locally and which should be delegated to other agent, which requests for cooperation should be honored, how requests should be realized, what action should be taken as a result of freshly arriving information, etc. It issues instructions to, and receives feedback from the other modules. The remaining components provide support function. The *self-module* is a repository for all the data which the underlying domain level system has generated or which has been received as a result of interaction with others. The *communication manager* sends and receives messages to/from *other* agents in the community.

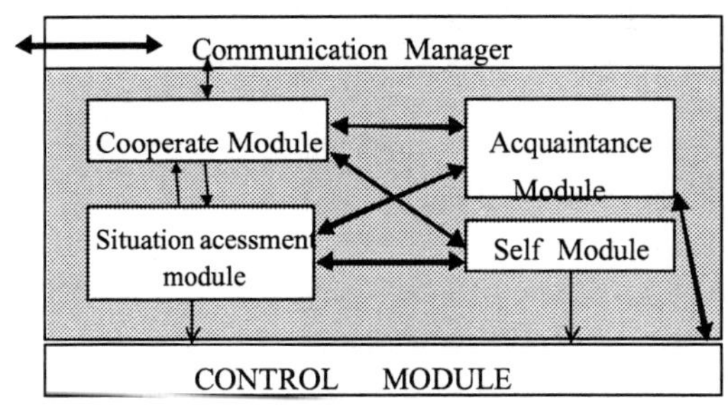

Fig.2 Agent Architecture.

Definition 4: Agent is described as a sextuple:

(Aid, Communication, Acquaintance, Cooperate, Assessment, Control).

<Agent>::=<Aid><Communication><Acquatence > <Cooperate ><Assessment ><Control>

<Aid>::=<Agent name>

<Communication>::={<Communication primitives> (<Communication content>)}

<Communication primitives>::=<Command>| <Request>| <Accept>|<Reject>|<Inform>|<Cancel>

<Communication content>::=<Sender> <Receiver> <Time><Information flow>

<Acquaintance >::={<Aids>(<Agent State>)}

<Agent State>::= <Active>|<Inactive>|<Participant>| <Wait>

<Cooperate>::= <Detect>|<Legil>|<Illegile>

<Assessment>::=<High>|<Media>|<Low>|<Reject>|<Inform>|< Cancel>

<Control>::={<Control command>}

4. Management of the Design Process Based on Agent

The engineering design is viewed as a distributed system consisting of various autonomous entities completing the special task through cooperation [3]. There are two kinds of basic elements in the system: entities and activities. The entity includes the CAD systems (dealing mainly with the graphic issues), users and various agents (information handling). Activity is the process of the entity performing the design process. In the designing process there are also two categories of information: Foreground and Background [4], Table 1 is the list of information and the corresponding activities and participants.

Definition 5: The agent oriented cooperative design system DS is a quintuple {A,Tr,Ev,OR}, among which, A is the set of agents, Tr is the design Transaction, E is the set of events, OR is the organization (including CAD system , various users etc) .

Event is the state change of the system triggered by users, CAD system and agents taking part in the design process, cooperation between agents and the communication with the outside world. In this conceptual model, the product design processes is an activity set of a group of agents and activities undertaking by the cooperative work of agents with the interference of users.

5. Illustration of the Designing Process

Definition 6. Event is a sextuple (EID, ESTATE, EPRE, EAG, EAC, EPOST), where EID is the name of events, ESTATE is the present system state, EPRE is the precondition triggering the event. EAG is the set of activated agent, EAC is the action taking by the agent, EPOST is the state caused by this event. Fig.3 give the event description.

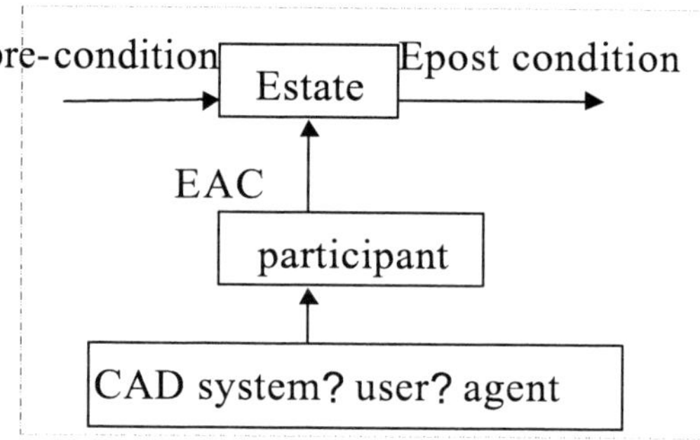

Fig.3 Event Description.

If the design of a complex NC miller is wanted, which can process turbine blade, the cutter should move simultaneiously in five axis, say x, y, z, A, B, i.e the cutter has 5 DOF (Degree Of Freedom). This is the design precondition. If the part is light, from the designing constraints in chapter 2.2 , the movement of the part has no limitations.

Then the EPRO has two hues: 1. x,y,z,A,B five DOF of cutter, 2. the part property is light.

Definition 7 : Transaction is a sequence of events, one transaction is an acceptable solution which can satisfy the module constrains and the designing constraints.

T::<EID1><EID2>......<EIDn>
EID::<EPRO ><ESTATE><EPOST>
<ESTATE>::<EAG><EAC>
<EAG> ::<agent state>: 1: allowed; 0: forbidden
<EAC>::<user state><CAD system state>: 1: interactable; 0: uninteractable.

NC milling machine has a "C" list structure which take the workpiece as the starting point and the cutter as the ending point [5] or vice versa. So the design activity process with workpiece, it first deal with the module which hold the workpiece.

Table 1.

Information	Activities	Participants
Design requirements	user input, file input	CAD system, user
Design intents	Deduction, input	user, agent, CAD system
Design methods	Search, inquire	user, agent, CAD system
Design standards	Input, display	user, agent, CAD system
Standard parts	data base inquire	user, agent, CAD system
Design history	Record, compare, modify	agent, CAD system
Engineering analysis records	Calculate, state change	agent, CAD system
Design modification records	Record, compare, modify, decide	agent, CAD system
Design process controlling	begin, end, interrupt	user, CAD system
Massage	Receive, send, accept, refuse	agent, user, CAD system
Participants state	Active, freeze	agent, user, CAD system

For event1 : EID1:

- EPRE: x,y,z,A,B five cutter DOF & (is light part_property)
- the agent state EAG:

H_B	H_V	H_X	H_Y	H_Z	R_V	R_X	R_Y
1	1	1	1	1	1	1	1

- user: 0 waiting.
- CAD system: 0 waiting.

At this point all the modules can be triggered, if R_Y is triggered which can realize the "y" translation. Then the condition will be changed.

- EPOST: cutter lost one DOF (Degree of Freedom)

So for the event2: EID2

- EPRE: x,z,A,B four movement &(is light part property)
- agent state EAG: from the module constrains , R_Y can not connect with H_B,H_V,R_V,R_Z, so the EAG is like this:

H_B	H_V	H_X	H_Y	H_Z	R_V	R_X
0	0	1	1	1	0	0

- user: 0 waiting
- CAD system state: 0 waiting

  At this time, if the H_Z is triggered , then:
- EPOST: cutter lost one DOF (Degree of Freedom)

....

After six events , the DOF of cutter reached 0, one design transaction has been finished. The result of one transaction is viewed as one acceptable solution. Various possible solutions will be recorded and compared. Among each transaction, user will interact with CAD system, compare and talk about the existing solutions with other designers. Fig.1 (b) is one transaction, which denotes the 5 axis NC layout, though it may not be the optimal solution.

6. Example

Fig.4 gives the tree structure multi-solution paradigm based on agent. In Fig.4, "Base-H" isthe input or beginning. From this point, the whole system will work under the monitor of modular constraints. In the design process, lower modules derived upper modules, and it is divergent. More often than not, the result may go beyond the least conserved traditional conventions, some may seem to be ridiculous, but that may means creativity. By introducing design constraints, the solutions may be converged greatly. With more constraints being introduced, the result may be more and more realistic.

For example, the thickened curve followed by R-Y->T-Y->T-X->H-Base->V-Base->T-Z->R-X in Fig.4 is one feasible NC conceptual structure in Fig.1 (c).

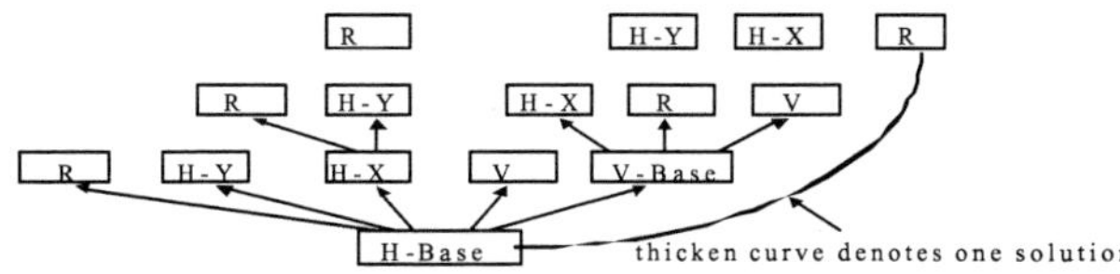

Fig .4 Multi-Solution in NC layout design

7. Conclusion and Acknowlegement

The NC structure conceptual design model based on agent described in this paper is now under intense development. The future work will focuse on the solution qualification mechanism, now it depends mainly on the designers to compare the various conceptual structures, only after the computer tell which is feasible and which is better, can we realize some real form of creativity.

This project is supported by the Shandong Provincial Excellent Young Scientist Award.

8. Reference

[1] HU Weigang Development and Practice of Machine Tool Modularization and the Intelligent Supporting System, Doctoral Thesis, Huazhong Institute of Technology, WuHan, 1994, 6.

[2] Xu Zhigang, Huang Kezheng, Ai Xing, Cooperative Problem Solving in Mechanical Multi-Agent Systems, International Pro of Workshop on CSCW in Design, 1996, 6, Beijing, P.R.China: 251-258.

[3] Zeng Guangzhou, Sun Boqi. A method for software requirement engineering based on system simulation and its supporting environment. *Chinese Journal of Computer*, 20(3):280-283(1997).

[4] H.Suzuki,et al Modeling Information in Design Background for Product development Support, *Annals of the CIRP* , vol 45/1/1996: 141-144

[5] Yotaro Hatamura, et al, Actual Conceptual Process for an Intelligent Machine Center, *Annals of the CIRP,* vol 44/1/1995:123-128.

Cooperative Work Mechanism in UUHDB System Based on MAS

Lei Ma, Mingye Liu, Hantao Song
Beijing Institute of Technology, Beijing, 100081, P.R. China
leima2001@sina.com

Abstract

In this paper, a United Use of Heterogeneous Database System based on Multi-Agent structure is introduced. We expatiate the means and strategies which can enhance the self-adaptive learning ability of single Agent and Multi-Agent. Cooperative work mechanism of Multi-Agent in UUHDB is presented. We also research the means of cooperative learning in Multi-Agent structure.

1. Introduction

The concept of CSCW (Computer Supported Cooperative Work) was brought forward by MIT's Iren Grief and DEC's Paul Cashman in 1984. It describes a way that peoples in a community work cooperatively to reach a goal with the computers' assistance. CSCW mostly describes how the people in a colony cooperate to accomplish a task in a computer supported environment. The colony should be a or several units, projects or groups. The people in the colony must be execute the same task. CSCW provides a shared working environment and a friendly flexible human-machine interface. As we know, Database System is the foundation for CSCW to save and manage information. It is quite important for CSCW.

With the development of Database System, all kinds of Database Systems have great differences in model and structure. For this reason, users require to use all these kinds Database Systems as a single system in logic. Heterogeneous Database System, also can be named Federal Database System, which is a relaxed system composed of several heterogeneous Database Systems. These systems should be heterogeneous and in different sections. These systems must be integrated by some formula. And the Heterogeneous Database System can provide section-ignored global data serving. The Heterogeneous Database System can solve the share and management of distributed heterogeneous information in CSCW.

2. United Use of Heterogeneous Database System based on Multi-Agent Structure

UUHDB system is a method used to resolve the heterogeneity and distribution problems of database systems. It pays more attention to resolving the heterogeneity of the supporting environment of the information management system.

We design a UUHDB system based on Multi-Agent Structure. In this system, single agent is designed according to the universal structure. The system's structure is shown in Figure 1.

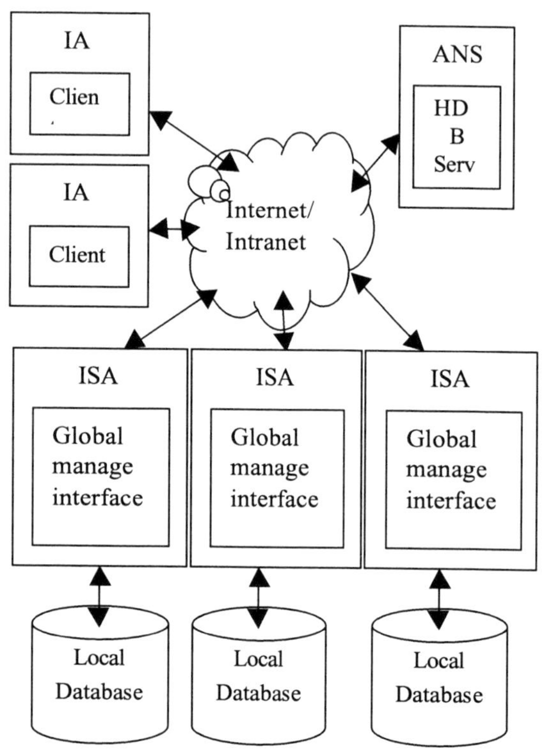

Figure 1. The structure of UUHDB system.

This system's structure combine the MAS structure with Client/Server structure. In this way, we can reserve the aptitude, agility and opening characters of MAS system, and use Client/Server structure's advantage of managing and controlling[3].

We design three kinds of agents in this system. They are ANS, IA and ISA. ANS (Agent Name Server) provides registering serving and messages transmitting serving for incoming agents; IA (Interface Agent) provides interface for users; ISA (Information Source Agent) provides heterogeneous information serving.

Agents communicate with each other through the network based on TCP/IP protocol. IA communicates with other agents through ANS. ISA can communicate with other ISAs directly.

In UUHDB this service is provided by ANS. To some degree, the ANS is the soul of UUHDB. It is responsible for corresponding each agent's working, this correspondence is implemented not through sending command directly to agents in UUHDB, but through command-parsing, task-programming and plan-forming. The formed task plan will be transferred to IA which will cooperate with other agents according to the task plan and complete the task that user brings forward step by step.

ANS is the cooperating and managing center of Multi-Agent System, and provides the global managing model for heterogeneous database system. ISA is the interface which manages the communication between heterogeneous database system and composing database systems. ISA can also deal with cooperative work with other ISAs. IA mainly deals with the interfacing work with users.

- ANS (Agent Name Server): To register the incoming agents and services provided by agents, provide the map service from agent name to IP address, be in charge of decomposing and scheduling task according to the user's requirements transported by IA.
- IA (Interface Agent): To manage receiving the user's requirements and dealing with them simply, monitoring the user task's executing.
- ISA(Information Source Agent): To provide local data-transform requirements coming from other agents. It's the foundation of the heterogeneous database system.

In the prior research, we have completed the framework of the heterogeneous database system based on Multi-Agent System. But there are several deficiencies in this system. For example, the system could not adapt the variable environment perfectly. Also the excellence of MAS structure hasn't be perfectly realized in this system. So we continue the research mainly in the self-adaptive learning ability and cooperative work in this system.

3. The Self-Adaptive Learning Strategy of System

MAS emphasizes the cooperation and coordination of independent and relative units in distributed environment. At this view, MAS is very similar to the demands of CSCW. So the UUHDB system based on MAS is very significant to solve the share of distributed information in CSCW.

To improve the cooperation of MAS, two aspects of work should be done. One is to reinforce the self-adaptive and learning ability of single agent. MAS is the integration of the action, ability and cooperation of single agents. The self-adaptive and learning ability of single agent directly affects the cooperation and consistency of the whole system. So reinforcing the self-adaptive and learning ability of single agent is very important to improving the cooperation of MAS. The other is to design the cooperative means and layout standards of MAS. Only by resolving this problem perfectly, we can improve the unitary adaptive learning ability of MAS, and improve the efficiency of cooperation.

In the factual using heterogeneous database system, there exist various changes around the system, such as the join or disjoin of the comprised database system, the changes of the users, and the changes of the user's requirements. The agents in this system must adapt all of the changes rapidly. We adopt two steps to realize this target:

- By the self-adaptive ability of the agents, the agents can program and adjust their activities themselves. Finally agents adopt the correct active model which fit into the changed environment.
- By the cooperative work with other agents, agent can learn and use reference about the countermeasure from other agents in this system. Even more the agent can assign some part of the work to other agents.

The first is named centralized learning, because in this way, agents mostly depend on their ability to adapt and learn. The second is named distributed cooperative learning, because in this way, agents mostly depend on the ability of cooperative learning in MAS structure.

3.1 The Figurc of Knowledge Library

The knowledge library is the foundation of the agent's learning ability. The knowledge library consists of several parts:

- Background Knowledge Library: It saves the common knowledge about the outer objective environment. The common knowledge includes basic information about the known congeneric agents, and the characteristic parameters about the objective outer environment.
- Field Knowledge Library: It saves the special knowledge about the field. In this system, field knowledge means the correlative knowledge about the comprised database system.
- Ability Knowledge Library: It saves the knowledge about the agent's ability.

The learning of agents includes two modes: increased and updated. In increased mode, the knowledge library merely expands itself while learning new knowledge. During other time, the knowledge library is stable. In updated mode, the knowledge library can update itself by two means. One is using the time mark, this system deems that the newer knowledge is more effective, and deletes older knowledge. And the other is adopting delayed management. Agent learns new knowledge while it's idle. While the agent is working, its' knowledge library is relative stable.

3.2 The Centralized Learning of Single Agent

1) The Express of State Space and Action Space

In our heterogeneous system, according the objective environment around the system, the state space is comprised by characteristic parameters sequence, which includes environment characteristic parameter and user characteristic parameter. The state space can be defined as follows:

S={E, U}

E represents the environment state characteristic parameter, U represents the user characteristic parameter.

E={C, S_c, S_a}

C represents the characteristic identifiers of the comprised database system. S_c represents the current state identifiers of C, and S_a represents the current state identifiers of agent.

U={I, M}

I represents identifiers of users. M represents the identifiers of user's active model.

According to the actions frequently executed by the agents in this system, we define the action space:

A={a, s}

In this definition, s? S, a represents the agent's action in the current state.

2) The Use of Reinforcement Learning

We adopt the reinforcement learning method which gradually be applied to deal with the self-adaptive ability of single agent. The reinforcement learning is developed from self-adaptive control theory. The basic principle is: If one active strategy of the agent lead to positive reward from the environment, then the agent will enforce the trend of adopting this strategy. The basic model of reinforcement learning is shown in figure 2[1]:

We can express the process of agent's action as followed: According to the agent's learning strategy, the agent selects the correct action to execute. Thereby it can complete the task, at the same time, it also can affect the environment. Also it can be rewarded or inspired.

Generally, the process is non-remembered model. But in order to improve the efficiency of this system, we adopt remembered model. According to the environment's state information, we define the state space. According to the agent's action, we define the action space. When the agent finished a new type of task, we save the relevant characteristic state parameters into the state space, also we save the action and strategy the agent executed into the action space. When the environment have new changes, at first the agent collects the new state information of the environment. Then it contrasts the new state information with state space. If there exists historical information records which match the similarity degree, then the agent directly acquires the relevant historical action records to execute. Otherwise, there have two ways to choice: One is to adopt the cooperative work strategy, the agent can acquire the strategy from other agents; And the other is to adopt the reinforcement learning, the agent can decide the strategy itself.

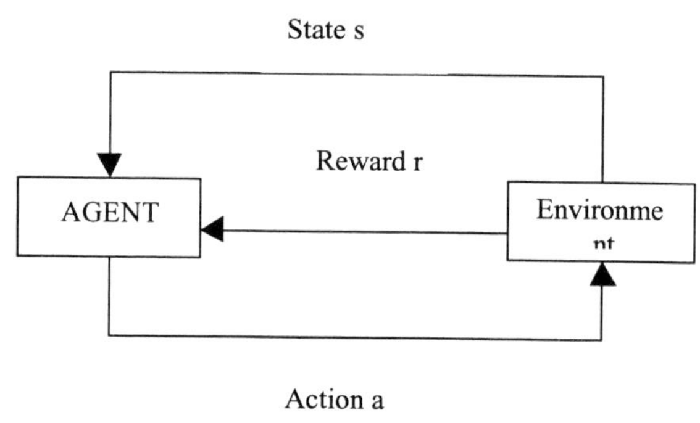

Figure 2. The model of reinforcement learning.

We define the environment state space as S, the action space as A. Then we can define the following formulas:

T(s,a,s') represents the probability of state transition. It means the probability from state $s \in S$ translating to state $s' \in S$ by executing action $a \in A$.

r(s,a) represents the rewards executing action $a \in A$ at state $s \in S$.

The object of agent is to find the action which can maximize the future rewards during state transformation. According to the definition of translating probability and reward function, we can adopt the reinforcement learning to solve the maximum reward problem. We adopt a typical reinforcement learning arithmetic in this system: Q-learning arithmetic.

Q-learning arithmetic solves the reflection of state/action couple to Q value. There exists a relevant Q value at every state or action. The definition of Q value is the accurate summation of the future rewards[2]. We have the following formula:

$$Q(s,a) = r(s,a) + \gamma \sum_{s \in S} T(s,a,s') \max_{a' \in A} Q(s',a')$$

If r and T are known, we can acquire the final Q value. We replace the reward by the Q value. The selection of the max Q value means adopting optimum strategy. In the practice, usually r and T are unknown. So the agent must refine the Q value from the experiential knowledge acquired from the interaction with the environment.

According to the relevant Q value, the agent can decide the optimum action fitting to the changing environment. We also define an action filtration process which can decrease unnecessary exploration and improve the efficiency of learning.

4. The Distributed Cooperative Work of Multi-Agent

In the traditional machine learning field, more attention is paid to the learning of single agent. The cooperation and learning of MAS is always being ignored. Because this system is based on MAS structure, we also must solve the problem of MAS's cooperative work and learning.

During the execution of a task, the task should be divided into several sub-tasks. Because of the dependence in sub-tasks and absence of information, agents must cooperate with each other in this system. The cooperation makes system based on MAS more efficient than those based on single agent, and fulfil the requirements of CSCW.

In the interaction of agents, agent must think of its benefit firstly. Any of the knowledge acquired from other agents must be paid back. So in practice, agent doesn't interact with other agents in any condition. We introduce the conception of reliability quantity C[agent1,agent2]. It represents the reliability that agent1 request and acquire the available information or help from agent2. If the value of C[agent1,agent2] is higher, more available information or help can agent1 get from agent2. On the contrary, less available information or help can agent1 get. The cooperative work in this system is shown at figure 3:

The first step: When the agent need to acquire information or help from other agents, firstly it consults its reliability quantity list, selects the agents whose value of C more than the referenced value N, then send the request to these agents. The request includes the characteristic state parameters or the characteristic user parameters. The target agent receives the request, according to the parameters in the request it matches the similarity in its state space. If there exists relevant action and strategy in its action space, the target agent sends the action and strategy back with the matched similarity value to the request agent. The request agent compares all of the response, then selects the response which has the highest similarity value, adopts the action and strategy in this response to execute. At the same time, increases the relevant C of this response agent.

The second step: After selecting the satisfied response, the request agent may analyse some parts of the task relevant with the response agent. Through the request/response communication action, the request agent arranges with the response agent. The response agent will help the request agent to finish some parts of the task.

If there aren't satisfactory result after received all of the response, the agent has to depend on the learning ability of itself.

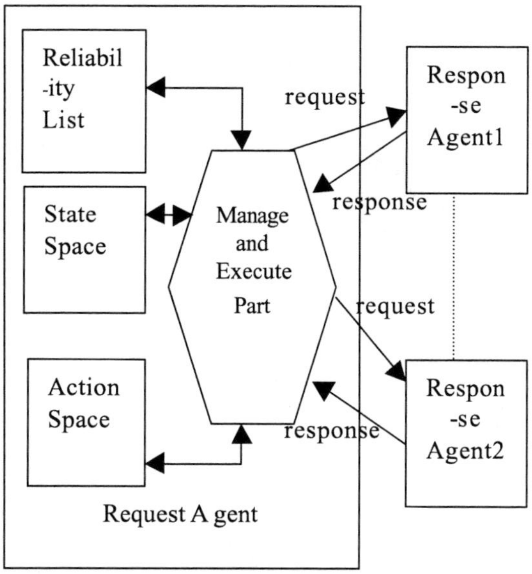

Figure 3. The model of cooperative work.

5. Conclusion

This article presents the structure of UUHDB system based on MAS, discusses how to improve the self-adaptive, learning and cooperative work ability of agent in the system. By using the MAS structure which can provide perfect ability of cooperation, the UUHDB system provides an effective information support to the application of CSCW.

6. References

[1] Gao Yang, Zhou Zhi-Hua, "Research on Markov Game-Based Multi-Agent Reinforcement Learning Model and Algorithms", Journal of Computer Research and Development, 2000 Vol.37,No.3,P.257-263.

[2] Cai Qing-Sheng, Zhang Bo, "An Agent Team Based Reinforcement Learning Model and Its Application", Journal of Computer Research and Development, 2000.9 Vol.37,No.9.

[3] Xing yanhui,, Song hantao, "A United Use of Heterogeneous DataBase Based on Cooperation", Proceedings of second International Workshop on CSCW in design. 2000 Hongkong

[4] Xing Yanhui, Liu Yanmei, Song Hantao, "Communication of multi-agents used in UUHDB system" Proceedings of international workshop on CSCW in design. Jul17-25 1998, Tokyo Japan.

[5] Xing Yanhui Guo guisuo Song Hantao, "A Union Used of Heterogeneous Database Based on Multi-agents System", Proceedings of second International Workshop on CSCW in design. Nov26-28 1997, bangkok thailand.

[6] Jennings N R ,Sycara K,Wooldridge M J. "A roadmap of agent research and development". Journal of Autonomous Agents and Multi-Agent Systems, 1998, 1(1):275-306.

The Modeling of Cooperative Environment in Digital Video Recorder Monitoring Control System

Dan Sun, Hantao Song
Department of Computer Science and Engineering, Beijing Institute of Technology
Beijing, 100081, P.R. China
Songhantao@hotmail.com

Abstract

Digital video recorder monitoring control system (DVR-MCS) consists of digital hard-disk video recorders, which integrates the most advanced digital video technology, real-time digital video compression technology and multiplexer software. The cooperative monitoring control function in DVR-MCS is required which enables operators to configure, monitor or control the digital hard-disk video recorders. This paper discusses the modeling of multi-agent based cooperative network environments in DVR-MCS and related technologies.

1. Introduction

With the development of digital video technology, real-time video compression technology and software engineering, Digital Video Recorder Monitoring Control System (DVR-MCS for short) will grow up in the future few years. The goal of DVR-MCS is to replace the traditional simulative video recorder system which couldn't implement some monitoring control functions, for instance, video triggered alarm, alarm triggered recording and sophisticated playback search functions etc. Now the development of DVR-MCS technology focuses on designing digital video recorder host with hard disk supported. The cooperative function of current DVR-MCS is so weak that is not suitable for large video recorder monitoring control system [1].

Figure 1 shows the hardware structure of digital video recorder host.

Modern DVR-MCS is required to implement the following functions:

- configure and control any devices in the DVR-MCS;
- monitor any channels in the DVR-MCS;
- operate transparently by users;
- be independent of OS;
- be uniform in system structure.

As described above, DVR-MCS in a whole is in a distributed, dynamic interdependent network environment. To setup effective network structure of DVR-MCS, the modeling of cooperative environment is a paramount concern. The conception of Computer Supported Cooperative Work (CSCW for short) is suitable for modeling DVR-MCS.

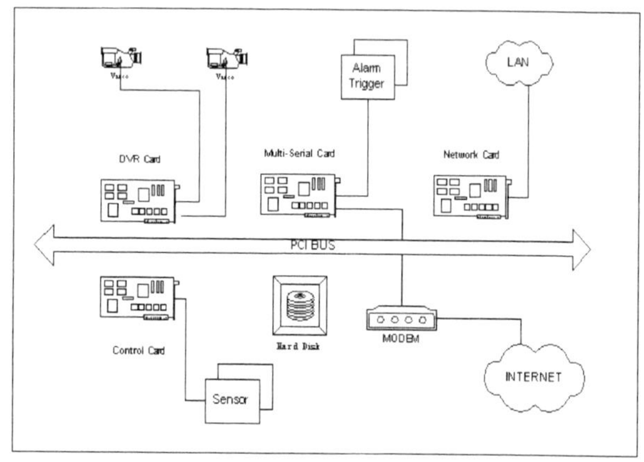

Figure 1. Digital Video Recorder Host.

The present work is organized into sections as follows: in section 2,we give some conceptions about CSCW and Multi-agent system; in section 3, we detail the modeling of multi-agent based cooperative environment in DVR-MCS and section 4 presents the communication protocol between agents. In section 5, we outline the conclusion.

2. CSCW and Multi-Agent System

Computer Supported Cooperative Work (CSCW) is a new research area focusing on the use of Information Technology to facilitate productive activities that are carried out by groups or ensembles of people. Its focus is on both the nature of cooperative work forms and current work practices and on the way information technology can change, augment and support cooperative work patterns.

CSCW is inherently interdisciplinary, with major influences from Software Engineering on the one hand

and Work Organization and Sociology on the other. It also overlaps with fields such as Management Information Systems (MIS), Systems Development, Ergonomics and Human-Computer Interaction (HCI), and Organizational Behaviour.

There are two dimensions that make up the CSCW domain: time and place. The goal of CSCW is to discover ways of using computer technology to further enhance the group work process through support in the time and place dimensions. The following notions and technologies help to satisfy CSCW goals:

- WYSIWIS
- Integration across different tasks
- Personal Secretary
- Time Management
- Multimedia
- End User Program

The CSCW system structure divides into two patterns: concentrated and distributed. Usually concentrated pattern uses client/server structure or whiteboard, and that distributed pattern uses multi-agent architecture. In view of the goal of CSCW, multi-agent architecture is more adapted to build CSCW system.

The kernel of multi-agent architecture is intelligent agent[2]. Intelligent agent is a system that can perform a task with a certain amount of "intelligence" due to some specialized skill (learning systems) or knowledge (knowledge bases). Basically, intelligent agents don't just blindly do what they are told. They can be set up to evaluate choices and help the user do things. The multi-agent system is an aggregation of many agents with various functions, and their combined actions allow a greater range of action. The selection of agents allows the multi-agent system (MAS for short) to pick the right one for the job. But for most uses people don't really need to know which agent they are using.

The advantage of MAS is as follows[3]:

- hide complexity of tasks
- perform tasks on the user's behalf
- can train or teach users
- help different users collaborate
- monitor events and processes

As above described, multi-agent architecture is fully suitable for the modeling of cooperative environment in DVR-MCS.

3. Cooperative Environment Model of DVR-MCS

The main task of DVR-MCS is to collect video or audio data, to capture various condition-triggered or video exception-triggered signals, and then to store video or audio data according to condition defined, and furthermore, to control corresponding devices in order to respond to alarm signals. In view of users, DVR-MCS is the uniform whole architecture. The base of DVR-MCS is Virtual Monitor Control Operation Platform (VMCOP for short) we named.

Figure 2 shows the abstract structure of DVR-MCS. The information of IS (Information Source) may come from video preview/compression cards or audio collection cards which belongs to various DVR hosts. The signals of SS (Signal Source) may come from triggers or sensors. As to TSA (Target Storing Area) includes all hard disk space in DVR-MCS and TOC (Target Output Controller) includes device controller, trigger or video display.

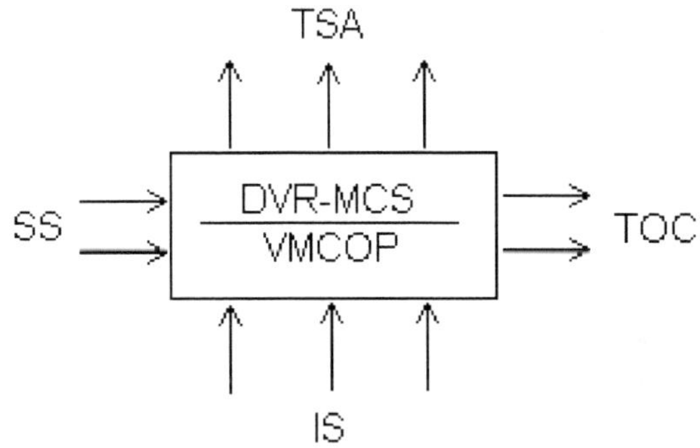

Figure 2. Abstract Structure of DVR-MCS.

According to VMCOP, the cooperative environment model of DVR-MCS in theory is divided into three layers: instruction defined layer (IDL), instruction parsed layer (IPL) and instruction executed layer (IEL).

- IDL: where SS,IS,TOC and TSA are uniformly defined and catalogued; moreover, feedback is acceptable so that system expands or corrects instruction defined by self-study and self-define.
- IPL: where instruction is parsed, interpreted as instruction code.
- IEL: where instruction code is dispatched and executed.

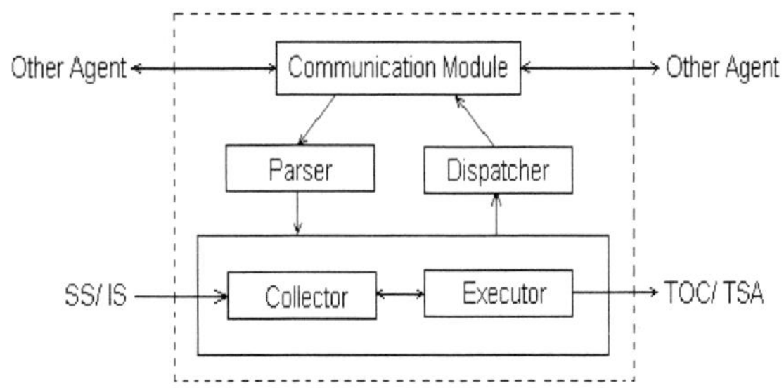

Figure 3 Basic Agent

To implement multi-agent system, agents in DVR-MCS are divided into four types as follows:

- Basic Agent (BA): BA provides basic services of DVR-MCS. All layers in the cooperative environment model are involved by BA (Figure 3).

- Register Agent (RA): RA is kernel of DVR-MCS which not only takes charge cooperative work between agents, but also acts as Agent Name Servers (ANSs). RA accepts the register requests which come from other agents, at the same time, maintains an uniform share resource and restricted resource list.
- Search Agent (SA): SA takes charge searching resource and agents.
- Cache Agent (CA): CA provides cache service in some special conditions.

Figure 4 shows Multi-Agent based DVR-MCS Model.

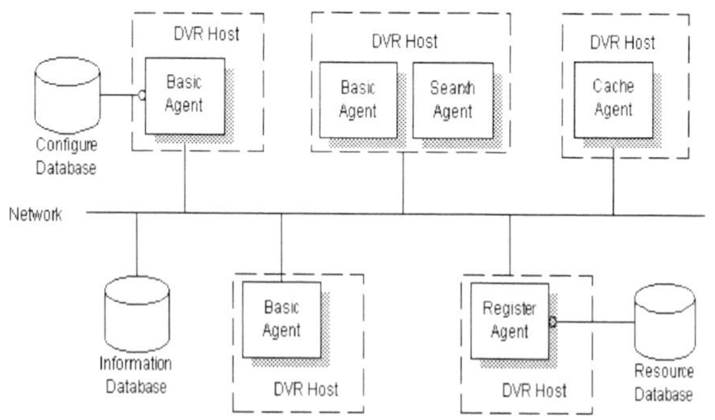

Figure 4. Multi-Agent based DVR-MCS Model.

4. Agent Communication Mechanism

The footstone of CSCW and MAS is communication. So a hale communication mechanism is very important to DVR-MCS. Agent Communication Language (ACL for short) decides the strength of agent communication mechanism. Currently KQML and FIPA ACL are at large applied. KQML and FIPA ACL are no less different. Based on analysis of DVR-MCS, KQML is selected for ACL in DVR-MCS Model.

The performative in DVR-MCS model is divided into four types: query-style, command-style, discuss-style and feedback-style.

The implementation of performative bases on UDP of TCP/IP set. UDP is simple and efficient, but reliability of UDP is not good. So UDP must be changed to add reliability functions so that is applied to transmit video data stream.

Another important problem which DVR-MCS should consider is security. Open-network multi-agent systems face security threats from malicious agents. These agents may try to unregister their competitors from Agent Name Servers (ANSs), eavesdrop on supposedly private communications, and spoof other agents. To avoid hidden trouble in security field, system demands that agents must be certified by an agent certification authority when they want to join a system. And then system binds each agent to a unique Agent ID [4].

5. Conclusion

This paper introduces a digital video recorder monitoring control system and the modeling process of multi-agent based cooperative environment in DVR-MCS. CSCW and MAS technology are suitable for DVR-MCS.

6. References

[1] R C Gonzalez, R E Woods. Digital Image Processing[M]. Addison Wesley, 3rd. Ed., 1992.

[2] K Sycara, et al. Distributed Intelligent Agents [EB/OL]. http://www.cs.cmu.edu/~softagents/.

[3] M Wooldridge, N R Jennings. Intelligent Agents: Theory and Practice [J]. The Knowledge Engineering Review, 1995, 10(2):115-152.

[4] Wong, H. C., and Sycara, K. Adding Security and Trust to Multi-Agent Systems. In Proceedings of Autonomous Agents '99 (Workshop on Deception, Fraud and Trust in Agent Societies). May 1999, Seattle, Washington, pp. 149-161.

Part V: Databases

Update Algorithms for the Sketch Data Model

Michael Johnson
Macquarie University, Australia
mike@ics.mq.edu.au

Robert Rosebrugh
MountAllisonUniversity, Canada
rrosebru@mta.ca

Abstract

The authors have developed a new approach to database interoperability using the sketch data model. *That technique has now been used in a number of applications, but an important question remains: What are the algorithms that support updates in the sketch data model? The question has significant content since the sketch data model uses* EA-sketches *to specify data structures, and these include constraint and other information not normally supported by relational database management systems. In this paper we answer the question by using the framework of EA sketches to give a detailed mathematical treatment of database updates, providing a formal definition of insert update together with an algorithm which provably achieves updates. The algorithm is new as it is the first to treat data and constraints on an equal categorical footing. We also note that further exactness properties (limits and colimits) can aid specification, and we provide algorithms for updates of EA sketched databases with finite limits. These are the first update algorithms for such databases. The sketch data model is being used in industry for designing interoperations for computer supported cooperative work and computer assisted software engineering (CASE) tools are under development. The paper is predominantly theoretical, and provides an important link needed for CASE tool development.*

1. Introduction

For a number of years the authors, together with others, have been developing the theoretical foundations for a new semantic data modelling technique — the *Sketch Data Model* (SDM). Meanwhile, Dampney and Johnson have been applying the insights gained from the theoretical work to data modelling problems in consultancies with major Australian Companies [8], [28], [10], [7].

At CSCWD2000 the authors presented a new technique for supporting database interoperability based on the sketch data model [20]. The theoretical underpinnings of that technique, which depend on the solution of the view update problem, have since been further developed [22] and [21], and the technique has been used successfully in industry [9]. Furthermore, the technique was extended to *half-duplex interoperations* once the need for that extension became apparent in industrial applications [19], and these half-duplex interoperations are being used now to support interoperations on semi-structured databases used for design.

However, there remains a fundamental question which has not, until now, been addressed. While it is possible to use the relational data model to implement database interoperations which have been designed using the sketch data model, what are the update algorithms which would be the basis of a sketch data model database management system? This paper provides a first answer to that question, by developing a formal treatment of updates along with algorithms that can be proved to be compliant with that treatment.

The sketch data model (defined formally below) can be viewed as an extension of the widely used Entity-Relationship (ER) approach [5] to information modelling. In addition to the graphical advantages of the ER approach, the SDM incorporates many of the insights of categorical universal algebra [3] (chapter 4), [24], allowing a much more detailed specification of constraints, and a commensurate improvement in the quality of the specification and design process. Thus, the principal successes of the SDM approach have been in the early design stage. If the design is intended to be implemented, the model is usually just viewed as an ER model (forgetting the detailed constraint specifications) and then it is implemented as a relational database in the usual way.

The problem addressed by this paper is to develop the theory to bridge the gap between the SDM based design phase and implementations, while taking into account the full set of constraint specifications.

We need not concern ourselves with the details of the data model used for the implementation, which might be the common relational model [6], or even more appropriately the functional model [13] (Section 5.4), since for constraints the critical issue is the update process. Thus we seek (and in Section 4 provide) a formal definition of elementary update in the SDM framework, and algorithms which can be proven to achieve such updates (Section 5).

Before proceeding we briefly review the early results in the previous theoretical work on the SDM, and the work of some other authors who have also proposed sketches as a basis for the study of semantic data models.

Dampney, Johnson and Monro [11] were the first to note that the categorical universal algebra approach implied that the classifying category for a data specification included objects representing all of the query results for the corresponding database. This important fact gives the SDM a categorical closure akin to relational completeness: Any of a class of categorical operations (which includes the standard relational operations) can be applied to objects in the classifying category and the results will always be objects of the classifying category. Dampney et al also recognised the importance of update algorithms, but their approach to updating was clumsy and incomplete. Dampney and Johnson [18] demonstrated the importance of commutative diagrams in constraint specification (taking advantage of the query objects), and gave a range of examples indicating the breadth of the constraints that can be dealt with. Johnson [16] began the development of the logic for the query objects and [17] provides a detailed development of the categorical logic. Johnson, Rosebrugh and Wood [23] provide a theoretical treatment of updates in terms of the span construction, and prove the important technical result that a span of SDM models in a lextensive category $\mathcal{C}$ is an ordinary SDM model in $\mathbf{spn}\,\mathcal{C}$.

Meanwhile, Piessens [26, 27] used sketches as part of his notion of 'data specification'. Motivated by the problem of view integration he has obtained results on the algorithmic determination of equivalences of model categories for certain classes of data specifications. Diskin and Cadish [14], [15] describe enhancements to entity-relationship models also using sketches. Their perspective differs from ours in using 'diagram operations' on sketches to model queries.

The remainder of this paper is arranged as follows. In Section 2 we review the category theoretic view we take of ER models and of semantic data modelling in general. In Section 3 we present the definition and develop the elementary properties of EA sketches — the basis of the SDM. Section 4 introduces the theory of insert updates (delete updates can be analysed in a dual manner). That section, and Section 5, form the core of the paper as they develop the mathematical theory of an update in the SDM framework, and demonstrate the connection between that theory and the algorithms presented in Section 5.

2. Background

This section contains a brief review of our categorical view of ER models. For a fuller description see [18] or [20].

An *entity* is a set about which a database owner has information. For example a university might include STUDENT, UNIT, LECTURER, ... as entities in its information model. Entities have certain *attributes* or properties — for example students have a name, address, degree program and so on. Among the attributes of an entity there may be a specified *key attribute* — a student identification number would be an example. A *relationship* among entities is a (many-many) relation — for example students are enrolled in units. Typically there are many students in any one unit, and also any one student is typically enrolled in several units.

An ER model is a set of entities, together with their attributes, and any relationships among them. Notice that the model records the structure of the information, but not yet any instances — no data. In essence such a model is a structured datatype specification, or in mathematical terms a specification for a variety of (flat, multi-sorted) algebras.

ER models are generally presented graphically, but with a range of notations and extensions. We will use a specific directed multigraph (hereafter such graphs will simply be called "graphs") which can be obtained from each ER model as follows. Given an ER model, the nodes of its graph G will be the disjoint union of the entities, their attributes, the relationships and a special node denoted 1. The (directed) edges of G will consist of an edge from each entity to each of its attributes; an edge from each relation to each of the entities that it relates; and for each attribute A with domain having n elements (that is there are n different values for that attribute), n distinct edges $1 \longrightarrow A$.

Notice that in this graph entities and relationships are treated similarly. Henceforward we drop the distinction between entities and relationships. This point of view is also espoused by other writers on database theory, for example C. J. Date, [12]. The name EA sketch emphasises the technical importance of the distinction between entities and attributes, first noted in [18].

3. EA Sketches

In this section we introduce EA sketches, the fundamental notion of the sketch data model. To establish our notation we begin by briefly describing the language of sketches and their models. For fuller treatment we refer the reader to [1] or [2].

A *cone* $C = (C_b, C_v)$ in a directed graph $G = (N, E)$ consists of a graph I and a graph morphism $C_b : I \longrightarrow G$ (the *base* of C), a node C_v of G (the *vertex* of C) and, for each node i in I, an edge $e_i : C_v \longrightarrow C_b i$. *Cocones* are dual (that is we reverse all the edges of G which occur in the definition, so the new definition is the same except that the last phrase requires edges $e_i : C_b i \longrightarrow C_v$). The edges e_i in a cone (respectively cocone) are called *projections* (respectively *injections*).

Definition 1 *A sketch* $\mathbb{E}$ *is a 4-tuple* $\mathbb{E} = (G, \mathbf{D}, \mathcal{L}, \mathcal{C})$ *where* G *is a directed graph,* $\mathbf{D}$ *is a set of pairs of paths in* G *with common source and target (the commutative diagrams) and* $\mathcal{L}$ *(respectively* $\mathcal{C}$*) is a set of cones (respectively cocones) in* G.

Definition 2 *A* model M *of a sketch* $\mathbb{E}$ *in a category* $\mathbf{S}$ *is a coherent assignment of nodes and edges of* G *to objects and arrows of* $\mathbf{S}$ *so that the images of pairs of paths in* $\mathbf{D}$ *have equal composites in* $\mathbf{S}$ *and cones (respectively cocones) in* $\mathcal{L}$ *(respectively in* $\mathcal{C}$*) have images which are limit cones (respectively colimit cocones).*

For a sketch $\mathbb{E}$ we denote the category generated by G subject to the relations $\mathbf{D}$ by $\widetilde{\mathbb{E}}$. Using the evident inclusion $G \longrightarrow \widetilde{\mathbb{E}}$ we will sometimes refer to nodes of G as objects, edges of G as arrows and (co)cones of $\mathbb{E}$ as (co)cones in $\widetilde{\mathbb{E}}$.

A model M of $\mathbb{E}$ in $\mathbf{S}$ extends to a functor $\widetilde{M} : \widetilde{\mathbb{E}} \longrightarrow \mathbf{S}$ which takes every (co)cone in $\mathcal{L}$ ($\mathcal{C}$) to a (co)limit (co)cone. If M and M' are models a *homomorphism* $\phi : M \longrightarrow M'$ is simply a natural transformation from $\widetilde{M}$ to $\widetilde{M'}$. Models and homomorphisms determine a category of models of $\mathbb{E}$ in $\mathbf{S}$ denoted by $\mathrm{Mod}(\mathbb{E}, \mathbf{S})$, a full subcategory of the functor category $[\widetilde{\mathbb{E}}, \mathbf{S}]$.

We speak of (limit-class, colimit-class) sketches when $\mathcal{L}$ and $\mathcal{C}$ are required to contain (co)cones only from the specified (co)limit-classes. For example, (finite-product, $\emptyset$) sketches correspond to (multi-sorted) algebraic theories.

Definition 3 *An* EA sketch $\mathbb{E} = (G, \mathbf{D}, \mathcal{L}, \mathcal{C})$ *is a (finite limit, finite coproduct) sketch such that*

- *There is a specified cone with empty base in* $\mathcal{L}$. *Its vertex will be called* 1. *Arrows with domain* 1 *are called* elements.
- *Nodes which are vertices of coproduct (discrete) cocones whose injections are elements are called* attributes. *An attribute is not the domain of an arrow.*
- *The underlying graph of* $\mathbb{E}$ *is acyclic and finite.*

A node which is neither an attribute, nor 1, *is called an* entity. *An EA sketch is* keyed *if each entity* E *has a specified monic arrow* $k_E : E \rightarrowtail A_E$ *to some attribute* A_E. *An entity is called a* base entity *if it is not the vertex of any cone nor in the base of any cocone of* $\mathbb{E}$. *A* finite limit EA sketch *is an EA sketch with empty* $\mathcal{C}$.

Remark 4 i) Recall that in general an arrow m is *monic* if $mf = mg$ implies $f = g$, and that in **Set** such an arrow is an injective function. We say that an arrow m in an EA sketch is monic when its image in every model is monic. This can be assured in a number of ways. Choosing one for definiteness, we call m monic in $\mathbb{E}$, and denote it $m : X \rightarrowtail Y$, when there is a cone in $\mathcal{L}$ whose base has image $X \rightarrowtail Y \leftarrowtail X$ where both arrows are m, and whose two projections onto X are equal.
ii) The base entities correspond to the *strong entities* in ERA models, while an entity with a monic arrow to a base entity or which is in the base of a cocone with vertex a base entity corresponds to a *weak entity*. In practice we expect that when the EA sketch is keyed, the key for a weak entity will be the composite of the key for its base entity and the inclusion.
iii) Notice that every ER model yields an EA sketch: Let G be the corresponding graph as described in Section 2, let $\mathbf{D}$ be empty and let $\mathcal{L}$ contain only the empty cone with vertex 1. Let $\mathcal{C}$ be the set of discrete cocones of elements of each attribute. If we wish to ensure that the relations can only be modelled as true relations, add for each relation a product cone with base the discrete diagram containing the entities that it relates, and a monic arrow from the relation node into the vertex of the cone. If the ER model is extended to include indications of key attributes add cones to $\mathcal{L}$ to ensure that the corresponding arrows to those attributes are monic.
iv) Now the extra power of the sketch data model is clear: $\mathbf{D}$ can be used to record constraints, and $\mathcal{L}$ and $\mathcal{C}$ can be used to calculate query results from other objects, and these query results can in turn be used to add constraints, etc. Furthermore, all this takes place with a firm and well-understood mathematical foundation developed for universal algebra.

The appropriate receiving categories for models of EA sketches (and hence for their databases' states) should have finite limits and finite coproducts. The coproducts will be assumed to be disjoint and universal. Such categories are called *lextensive* and have already been studied [4]. They are examples of *distributive categories*, and they have many applications in theoretical computer science [29]. For the record,

Definition 5 *A category* $\mathbf{S}$ *is* lextensive, *if it has finite limits and finite coproducts which are disjoint and universal.*

For the balance of this article, $\mathbf{S}$ will be a lextensive category. We note that the definition of lextensive in [4] differs from that above, though the definition there is shown to be equivalent to Definition 5.

Remark 6 Notice that the definition of EA-sketch does not rule out the possibility of an *inconsistent sketch*, i. e. a sketch with no non-trivial models. For example, the sketch might include attributes A_2 and A_3 with 2 and 3 elements respectively and then specify that their sum is 1 which implies that the sketch has no models in $\mathbf{S}$ unless $\mathbf{S}$ is degenerate, that is $0 \cong 1$ in $\mathbf{S}$.

Definition 7 *A* database state *D in $\mathbf{S}$ for an EA sketch $\mathbb{E}$ is a model of $\mathbb{E}$ in $\mathbf{S}$. The* category of database states of $\mathbb{E}$ in $\mathbf{S}$ *is the category of models* $\text{Mod}(\mathbb{E}, \mathbf{S})$ *of $\mathbb{E}$ in $\mathbf{S}$. For brevity we sometimes denote it by* $\text{Mod}(\mathbb{E})$ *when $\mathbf{S}$ is the category of finite sets,* $\mathbf{Set}_0$

4. Insert updates

Under what circumstances can we say, given two database states for the same EA sketch, that one is an *update* of the other? The question is important, and requires a mathematical answer, if we are to prove that update algorithms have properties corresponding to consistency and completeness.

Of course, in most databases, any state can be obtained as an update of any other state: We can for example delete all the data from the first state, and then insert all the data for the second. So, we restrict our attention here to *elementary updates* — those that can be obtained via a single instance insertion or deletion.

As is common practice, we will not deal theoretically with updates which modify existing values. Instead, such modifications are expected to be obtained by (elementary) deletions and insertions. Furthermore, a formalism for elementary deletions can be obtained from our treatment of elementary insertions. Thus, we focus our attention on elementary insertions.

Finally, a few words about *cascades.* There remains some controversy in the database community about whether elementary updates should cascade. (In the following it is convenient to speak about entities and relations even though relations are themselves entities in our framework — whenever $X \longrightarrow Y$ we can say that X is a relation involving the entity Y.) Suppose we attempt to delete an instance of an entity, which is related to another instance of another entity. Some systems will prohibit this operation, requiring the relation instance to be deleted first, while other systems will cascade the delete, deleting the relation instance (and any relation instances that it takes part in etc.) automatically. Similarly insertions may cascade (an insertion of a relation instance may automatically lead to the insertion of a new entity instance to be related by the relation instance, etc.) or the system may prohibit cascades and require that all instances to be related are inserted before the relation instance can be inserted.

Now to the formal treatment of insertions. We will insert a new instance x in a database state D at a given entity E.

Let $\mathbb{E}$ be an EA sketch, and E be an object of $\mathbb{E}$. We define $\mathbb{E}[x_E]$ to be $\mathbb{E}$ augmented with a new arrow $x : 1 \longrightarrow E$ (and no new commutative diagrams, cones or cocones). Write J for the evident inclusion $\widetilde{\mathbb{E}} \longrightarrow \widetilde{\mathbb{E}[x_E]}$.

Write $\text{Mod}(\mathbb{E})$ for the category of database states. Thus objects of $\text{Mod}(\mathbb{E})$ are certain functors $D : \widetilde{\mathbb{E}} \longrightarrow \mathbf{Set}_0$. Write, abusing notation, $U = J^*$ for the functor $[\widetilde{\mathbb{E}[x_E]}, \mathbf{Set}_0] \longrightarrow [\widetilde{\mathbb{E}}, \mathbf{Set}_0]$ given by composition with J, and for its restriction $U : \text{Mod}(\mathbb{E}[x_E]) \longrightarrow \text{Mod}(\mathbb{E})$. Now J^* has a left adjoint computed as the left Kan extension of the functor along the inclusion $J : \widetilde{\mathbb{E}} \longrightarrow \widetilde{\mathbb{E}[x_E]}$ and in the case of finite limit EA sketches, the adjoint restricts to the category of database states. Call the left adjoint L and let η be the unit of the adjunction.

Definition 8 *Suppose D is an $\mathbb{E}$ database state, E an object of $\mathbb{E}$, and D' an $\mathbb{E}[x_E]$ database state. Then $m : D \rightarrowtail UD'$ is an* insert update *of D at E if there exists an epi $\phi : LD \longrightarrow D'$ such that $(U\phi)\eta_D = m$.*

Remark 9 i) Intuitively, LD should be thought of as the universal database state containing D and a new constant x of type E. The definition requires D' to be "squeezed" between D and LD, coherently with the way D is included in both D' and LD.
ii) Our definition permits, but does not mandate, cascades. The Kan extension, LD, cascades maximally. Requiring D' to be an epimorphic image of LD means that what in the Kan extension is a cascaded insert, may in D' be a relation instance assignment to an already extant entity instance or attribute value.
iii) In fact, LD will not in general be a database state if the sketch includes coproduct cocones, and it will certainly not be a database state if the sketch is keyed. This is easy to see by using the Kan extension formula [25] (page 236) to calculate the value of $LD(A_E)$, where A_E is the key attribute for E. The "maximal cascade" mentioned above corresponds to $LD(A_E) = D(A_E) + 1$, and so LD can't be a database state. Nevertheless, D' is a database state, and ϕ indicates how cascaded inserts which have disrupted coproducts in LD are replaced by assignments to extant instances or values in D'.
iv) Elementary properties of adjunctions ensure that if ϕ, as in the definition, exists, then it is unique.

5. The algorithms

We provide pseudocode algorithms for the two most important cases of insert updates, linear and finite-limit EA sketches. In each case we remark briefly on some of the finer points of the algorithms, and in the linear case we establish in detail the connection between the theory (Section 4) and the algorithms.

In the following sections we write $A + B$ for the disjoint union of A and B when A and B are sets. We remind the reader that we compose functions from the right, thus $pqa = p(q(a))$ means evaluate q at a, and then apply p to the result.

5.1 The linear case

A sketch is called *linear* when its sets of cones and cocones are both empty. Of course this can never occur with EA sketches which are required to have a specified cone (Definition 3). The corresponding notion for an EA sketch is

Definition 10 *An EA sketch $\mathbb{E} = (G, \mathbf{D}, \mathcal{L}, \mathcal{C})$ is called* linear *when $\mathcal{C}$ is empty and $\mathcal{L}$ contains just a single cone (which must then have an empty base).*

We provide an insert update algorithm (Figure 1) for the linear case. The algorithm forms the basis for algorithms which deal with further exactness properties (limits and colimits) and is useful for abstract data specifications which typically record no attributes. Because it is so fundamental, we will, after a brief discussion of some of its facets, establish in detail some of its properties, and its relationship with the theory presented in Section 4

The algorithm deals with a database state `D`, and constructs a new database state `D'`. It thus requires storage for sets, which appear in the algorithm as, for example, `D'(E)`, and for functions, which are presumed to be stored as their graphs (ordered pairs of (domain, codomain) values). As an example of a function, the line `D'(x_E)1 = t` says that the pair $(1, t)$ should be added to the graph of the function corresponding to x_E in the database (partial) state `D'`.

The algorithm begins by adding the element `t` and then prompts the user for all the images of the new `t`. The user's proposed images are checked to ensure that commutative diagrams are preserved (and this needs to be done carefully, because at this point many of the functions are only partial functions). The user is also given an abort option. This is necessary since it is possible to begin an insert which can never be completed without violating commutative diagram constraints.

We now explore some of the mathematical properties of the algorithm.

Proposition 11 *If the procedure* `INS` *returns with* `ins = TRUE`*, then* `D'` *is a database state for $\mathbb{E}[x_E]$.*

Proof. Suppose `INS` returns with `ins = TRUE`.

To show that D' is a database state for $\mathbb{E}[x_E]$ we check explicitly that D' is a graph morphism into the underlying graph of **Set** and that the required diagrams commute.

Firstly notice $D'(1) = D(1) = 1$, since E being an entity is not 1 (Definition 3). Note the abuse of notation: As is customary, we write 1 for the vertex of the empty cone, for a set with one element, and as the name of that element.

Next notice from statement 3 of `INS` that $t = D'(x_E)1$ (and by acyclicity, this is never modified). Thus $D'(x_E)$ is a total function.

The node part of the graph morphism is straightforward: For all E', $D'(E')$ is a set, defined by D unless $E' = E$ whence $D'(E') = D(E) + \{t\}$.

The arrow part of the graph morphism follows since for all $a : E'' \longrightarrow E'$, $D'(a)$ is a total function: If E'' is not E and a is not x_E, then $D'(a) = D(a)$. If $a = x_E$, $D'(a)$ is the total function defined above. The only remaining case is $a : E \longrightarrow E'$. For such a, the required mapping is defined by $D(a)$ on $D(E)$ by the first statement of `INS`, so it remains to see that $D(a)t$ is defined. Since `ins` returns `TRUE`, but `ins` is set `FALSE` at the beginning of `INTER_ASSIGN`, either there are no a's to check in the loop in `INS`, so we are done, or $D'(a)t$ must have been assigned in `EXTEND_FUNCTION`. Indeed, `INTER_ASSIGN` had to return `ins = TRUE` on every loop iteration (otherwise we exit with `ins FALSE`, contrary to assumption). For `INTER_ASSIGN` to return `ins = TRUE`, `EXTEND_FUNCTION` returned `ins = TRUE` (since `ins` is initialized to `FALSE` in `INTER_ASSIGN`). `EXTEND_FUNCTION` only assigns `ins = TRUE` after assignment of $D'(a)t$.

Finally, all commutative diagrams in $\mathbb{E}[x_E]$ are respected by D': Any commutative diagram not involving E is unchanged. If a commutative diagram involving E does not start from E, then in the diagram E is in the codomain of at least one arrow b not equal to x_E, but for each such b, t is not in the image of $D'(b)$ since $D'(b) = D(b)$, so the diagram still commutes. So suppose we have a commutative diagram starting at E, say $(qb, q'b')$, with q, q' (possibly empty) paths. Suppose the diagram fails to commute in D'. This can only be because $D'(qb)t$ is not equal to $D'(q'b')t$. Since `INS` returned `ins = TRUE`, `INTER_ASSIGN` must have returned `ins = TRUE` on each iteration of the main loop of `INS`. Thus `EXTEND_FUNCTION` must have set `ins = TRUE` (for each call to `INTER_ASSIGN`). Hence, `CHCD` evaluated to `TRUE` when called with b and when called with b' (each as second parameter). But then if $b' = b$, the first `if` in `CHCD` found $D'(qb)t$ not equal to $D'(q'b)t$, and so set `CHCD` to `FALSE`, contradicting the assumption that `INS` returned `TRUE`. On the other hand, if b is not equal to b', without loss of generality, suppose the call to `CHCD` with b was the latter, then $D'(b')t$ was already defined when this call occurred, so the second `if` in `CHCD` found $D'(qb)t$ not equal to $D'(q'b')t$, and so set `CHCD` to `FALSE`, contradicting the assumption that `INS` returned `TRUE`. ∎

For the next proposition, we will need an explicit description of LD.

Lemma 12 *For E' not equal to E*

$$LD(E') = D(E') + \{\beta : E \longrightarrow E'\}$$

```
/* INSERT ALGORITHM FOR LINEAR EA SKETCHES

Algorithm INS(t, D, E, ins, D')
/*  pre: D a database state, E an object of D, and t not in D(E).
/* post: if ins is FALSE, D is unchanged, and D' is garbage;
/*       if ins is TRUE t is in D'(E), and D' is a database state for
/*       the extended sketch E[x].

D' = D                    /* D' will be the new database state
D'(E) = D'(E) + { t }     /* This is disjoint union
D'(x_E)1 = t              /* Abuse notation writing 1 for the element of 1
ins = TRUE
for each a : E --> E'
/* loop completion means D'(a)t is validly defined for each arrow a ie insert successful
   INTER_ASSIGN(D', a, t, ins)
   /* Let user choose D'(a)t 's value. Does not cascade (allow new element creation)
   if not ins
        exit   /* If INTER_ASSIGN is aborted we exit with ins FALSE

/* end of INS

Algorithm INTER_ASSIGN(D', a, t, ins)
/* Interactively asks user for a valid value of D'(a)t

ins = FALSE
repeat
   Output('type value of D'(a)t in D'(E') or CTRL-Q to quit')
   Input(t')
   if t' not in D'(E')
      output( 't' not in D'(E'); try again')
      break
   if t' = CTRL-Q
      exit    /* Needed since it's possible to attempt an insert
              /* which would never satisfy CD requirements.
   EXTEND_FUNCTION(D', a, t, t', ins)
until ins = TRUE                  /* so user must have succeeded
/* end of INTER_ASSIGN

Algorithm EXTEND_FUNCTION(D',a, t, t', ins)
/* Checks cd's and extends D'(a) by t |--> t' if ok.

if CHCD(D', a, t, t')
   D'(a)t = t'
   ins = TRUE
else
   Output('CD violation, extend function failed')
   ins = FALSE  /* assigned for emphasis, ins must be false here
/* end of EXTEND_FUNCTION

Algorithm CHCD(D', a, t, t') : Boolean
/* Returns TRUE if the definition D'(a)t = t' respects commutative diagrams.
/* We check separately CD's with both paths beginning a, since D'(a)t not yet defined
/* (such paths would be checked twice, but second if is false).

  CHCD = TRUE
  for each CD = (pa, p'a)      /* p, p' are paths; the cd paths both begin with a
       if not ( D'(p)t' = D'(p')t' )
          /* Execute if test for comm fails
          CHCD = FALSE
  for each CD = (pa, p') or (p', pa)   /* p, p' are paths
       if not ( D'(p')t = undef ) and not ( D'(p)t' = D'(p')t )
          /* Execute if we can test for comm and the test fails.
          /* If we can't yet test (D'(p')t undef) we'll test this when defining D'(p')t.
          CHCD = FALSE

/* end of CHCD
```

Figure 1. An insert update algorithm for linear EA sketches

Proof. $LD(E') = \mathrm{colim}(J/E' \longrightarrow \widetilde{\mathbb{E}} \longrightarrow \mathbf{Set}_0)$ where the first arrow is the projection from the comma category, and the second is the database state D. Since J is bijective on objects, each object E' of $\widetilde{\mathbb{E}}[x_E]$ is of the form JF for some F. Consider J/JF, whose objects are of the form $(A, \alpha : JA \longrightarrow JF)$. Such α are either Ja for $a : A \longrightarrow F$, or $\beta x : J1 \longrightarrow JF$ with $\beta = Jb$, some b. Notice that there are no morphisms in J/JF to or from objects of the second kind (because there are no commmutative diagrams in $\mathbb{E}[x_E]$ involving x) and that $D(1) = 1$. Furthermore, the identity on JF is terminal among the other objects in J/JF. Thus the colimit is the sum as claimed. ■

We write $X_{E'}$ for the set $\{\beta : E \longrightarrow E'\}$ when E' is not E, and let $X_E = 1$. Then for any entity F we have $LD(F) = D(F) + X_F$. The unit of the adjunction has as components the injections $D(F) \rightarrowtail LD(F)$.

Proposition 13 *If the algorithm* INS *returns* TRUE, *then there exists an epi* $\phi : LD \longrightarrow D'$ *in* $\mathrm{Mod}(\mathbb{E}[x_e])$ *such that in* $\mathrm{Mod}(\mathbb{E})$, *with* m *the evident inclusion,*

$$m : D \rightarrowtail UD' \longleftarrow ULD : U\phi$$

and $(U\phi)\eta_D = m$.

Proof. We will construct ϕ explicitly.

We construct the components $\phi'_E : DE' + X'_E \longrightarrow D'E'$ in two cases.

Case 1: Suppose E' is not equal to E whence $D'E' = DE'$. Let the first component be the identity on DE'. The second component, $X'_E \longrightarrow D'E'$, is defined by $\beta \mapsto D'(\beta)t$.

Case 2: $E' = E$ whence $D'E' = DE + \{t\}$. Let the first component be the inclusion of $DE' = DE$ into $DE + \{t\} = D'E'$. The second component is the constant at t mapping $X'_E \longrightarrow DE + \{t\}$.

As constructed ϕ is a natural transformation. Notice also that each component is epi (in case 2, X'_E is non-empty).

Finally, let $m : D \rightarrowtail UD'$ be given by the evident inclusions and notice $(U\phi)\eta_D = m$. ■

Conversely, consider unrestricted insertions in a database state D — unrestricted in the sense that they may cascade, and result in further inserts. The next proposition says that our algorithm is powerful enough to produce all of these — it just needs to be iterated appropriately. In other words, the algorithm can be used to effect all insertions (as defined in Definition 8).

Proposition 14 *Suppose* D'' *is a database state,* $m : D \rightarrowtail UD''$ *a monomorphic natural transformation, and* $\phi : LD \longrightarrow D''$ *an epimorphic natural transformation, such that* $(U\phi)\eta_D = m$, *then* D'' *can be obtained from* D *by finitely iterated application of the algorithm in Figure 1.*

Proof. Choose an E such that $D''(E)$ is not equal to $D(E)$ but from which all arrows $a : E \longrightarrow E'$, satisfy $D''(E') = D(E')$ Choose also a $t \in D''(E) - D(E)$.

`INS(t, D, E, ins, D')` will ask for the values of $D'(a)t$ and we reply with $U\phi(ULD(a)((U\phi)^{-1}(t)))$ (independent of choice of representative in ϕ^{-1} by naturality of ϕ). Any diagrams that are required to commute will do so since they do in D'', so the `INS` will terminate with `ins = TRUE`.

Now repeat from the choice of E and t above. This terminates by finiteness and acyclicity of the sketch and because the database states are finite set valued, and the resulting $D' = D''$ as required. ■

Now, what about an $\mathbb{E}$ state X such that $m : D \rightarrowtail X \longleftarrow ULD : e$ with $e\eta_D = m$? In fact, all $\mathbb{E}$ states thus 'between' D and ULD can be obtained by iterated insertions.

Corollary 15 *Suppose* X *is an* $\mathbb{E}$ *database state,* $m : D \rightarrowtail X$ *a monomorphism of database states, and* $e : ULD \longrightarrow X$ *an epimorphism of database states such that* $e\eta_D = m$. *Then* X *can be obtained from* D *by finitely iterated application of the algorithm in Figure 1.*

Proof. X must be of the form UD' since $\mathbb{E}[x_E]$ states are just $\mathbb{E}$ states, together with an extra constant of type E, and $e : ULD \longrightarrow X$ determines that constant as $e(x_E(1))$. Furthermore e must then be of the form $U\phi$ for $\phi : LD \longrightarrow D'$ since, with J being bijective on objects, to give a natural transformation in $\mathrm{Mod}(\mathbb{E})$ is to give a natural transformations in $\mathrm{Mod}(\mathbb{E}[x_E])$, provided only that it is also natural with respect to x_E. Thus we have satisfied the premises of the proposition. ■

5.2 The finite-limit case

We now extend the insert algorithm to deal with finite limit EA sketches. This case covers most real applications since, although the full specification requires limits and coproducts, the latter are usually restricted to attribute domains and these are never updated.

The algorithm (Figure 2) presupposes that the initial insert entity `E` is not the vertex of a limit cone. This is without loss of generality since inserting into the vertex requires inserting into one of the base entities, and (using acyclicity) the appropriate base insert can be done instead. Furthermore, this is the appropriate way to achieve such inserts in practice.

The algorithm is closely related to the previous algorithm (Figure 1). The only significant difference arises when `E` is "minimal" in a cone base. *Minimal* means simply that there is no arrow in the cone base with `E` as its codomain. Inserting into non-minimal cone base entities will not alter the limit, and so can be dealt with by the previous insert algorithm. Inserting into minimal cone base entities results in recursive calls to `INS`. These calls may insert in the cone vertex since the relevant base instances are already being

```
/* INSERT ALGORITHM FOR FINITE LIMIT EA SKETCHES

Algorithm INS(t, D, E, ins, D')
/*  pre: D a database state, E an object of D, t not in D(E).
/* post: if ins is FALSE, D is unchanged, and D' is garbage;
/*       if ins is TRUE t is in D'(E), and D' is a database state for
/*       the extended sketch E[x].

D' = D                /* D' will be the new database state
D'(E) = D'(E) + { t } /* this is sum
D'(x_E)1 = t
ins = TRUE

/* The new element needs to have its image given for each arrow out of E as usual.
/* Then if E is a minimal cone base entity, it will determine
/* new elements for vertex.
/* To get them test for compatibility with every possible other tuple.
/* If compatible define all projection values for new vertex element
/* then call insert to add the element to vertex.

for each a : E --> E'
  /* loop completion means all D'(a)t are defined
  if D'(a)t is defined
     next a
  else
     INTER_ASSIGN(D', a, t, ins)
     /* Let user choose D'(a)t's value.
     /* Does not allow new element creation
  if not ins
     exit    /* on fail we return previous state and exit

for each cone C
  if E = C_h(M) for M minimal in base of cone C
  /* C_v, C_h : I --> G are graph homoms, I a finite graph,
  /* G the graph of the sketch, C_v constant,
  /* p : C_h --> C_v ( a nat transf - the projections)
    V = C_v(I)   /* the cone vertex
    for each x in product over B in I of D'(C_h(B))
       with x_{C_h(B)} = t  where C_h(B) = E
       /* an E'th coordinate of x has value t
      if CHECKCOMP(C, D', x)
        for each B in I
          D'(p_B)x = x_{C_h(B)}
        INS(x, D', V, ins, D')
        if not ins
           exit   /* on fail we return previous state and exit

/* end of INS

Algorithm CHECKCOMP(C, D, x) : boolean
/* this checks compatibility for possible new tuple
CHECKCOMP = TRUE
for each B in I
  for each B' in I
    for each Y in I
      for each path p : B --> Y in I
        for each path p' : B' --> Y in I
          if D'(C_h(p))(x_C_h(B)) <> D'(C_h(p'))(x_C_h(B')
             /* C_h extended to functor
             CHECKCOMP = FALSE
             exit
/* end of CHECKCOMP

/* The procedures INTER_ASSIGN, EXTEND_FUNCTION, and CHCD are shown in Figure 1.
```

Figure 2. An insert update algorithm for finite-limit EA sketches

```
/* DELETE ALGORITHM FOR FINITE LIMIT EA SKETCHES

Algorithm REMOVE(t, D, E)
/* removes t and all definitions D(a)t = t'
for a : E --> E'
   D(a)t = undefined      /* means (t,t') removed from the graph of D(a)
D(E) = D(E) - { t }       /* t might already be gone...
/* end of REMOVE


Algorithm DEL(t, D, E)

for each a : E' --> E, for each t' with D(a)t' = t
   DEL(t', D, E')

for each cone C with vertex E
   scan projections p:E-->E' of C for t' not in D(E') with D(p)t = t'
   if no such p exists        /* can there be more than one such p ?
         output(Choose a projection p : E --> E')
         input( p : E --> E' )      /* factor chosen to delete from
         REMOVE(D(p)t, D, E')
         DEL(D(p)t, D, E')

REMOVE(t, D, E)
/* end of DEL
```

Figure 3. A delete update algorithm for finite-limit EA sketches

inserted. The required inserts are determined by explicitly calculating the limit as a collection of tuples.

The delete algorithm for finite-limit EA sketches (Figure 3) is a recursive cascading delete. This is necessary since deleting from either a cone vertex or a cone base entity requires deletion from the other, and this in turn may recur to arbitrary (finite) depth. The order of `REMOVE` and `DEL` and the sequencing of partial functions "`scan the projections...`" is delicate and important.

6. Conclusions

One of the most interesting issues to arise from the application of categorical universal algebra to database systems is the nature of update algorithms in the presence of detailed categorical structure in the form of limits and colimits. This has not arisen previously because universal algebras are rarely updated (we rarely add or remove an element of a given type from an extant algebra), while database updates have only ever been applied to relatively flat systems (systems with many types, but very few limit types).

This paper provides both a precise definition of insert update and update algorithms for categorical sketches including those with finite limits. It makes explicit mathematically the link between the theory and the algorithms. This answers the question which motivated this paper and it will aid in the development of CASE tools needed for quickly developing interoperating sem-structured databases for design.

Although the application of the theory of the sketch data model would seem to depend on the development of the algorithms for mixed sketches, we can in fact make considerable progress using the algorithms for limit sketches since the vast majority of database systems can be designed so that the only colimits are non-updatable (being in fact attribute types).

Nevertheless, the authors have developed algorithms that deal with coproduct sketches, and they are investigating the possibilities for mixed sketches.

7. Acknowledgements

This research was partially supported by the Australian Research Council, the Canadian NSERC, and the Oxford Computing Laboratory. We are grateful to all these sponsors.

References

[1] M. Barr and C. Wells. *Category theory for computing science*. Prentice-Hall, second edition, 1995.

[2] M. Barr and C. Wells. *Toposes, Triples and Theories*. Grundlehren Math. Wiss. 278, Springer Verlag, 1985.

[3] F. Borceux. *Handbook of Categorical Algebra 3*. Cambridge University Press, 1994.

[4] Aurelio Carboni, Steven Lack, and R. F. C. Walters. Introduction to extensive and distributive categories. *Journal of Pure and Applied Algebra*, 84:145–158, 1993.

[5] P. P. -S. Chen. The Entity-Relationship Model—Toward a Unified View of Data. *ACM Transactions on Database Systems*, 2:9–36, 1976.

[6] E. F. Codd. A Relational Model of Data for Large Shared Data Banks. *Communications of the ACM*, 13:377–387, 1970.

[7] C. N. G. Dampney and Michael Johnson. TIME Compliant Corporate Data Model Validation. Unpublished report to Telecom Australia, 1991.

[8] C. N. G. Dampney and Michael Johnson. Fibrations and the DoH Data Model. Unpublished report to NSW Department of Health, 1999.

[9] C. N. G. Dampney and Michael Johnson. A formal method for enterprise interoperability: A case study in a major health informatics information system. Proceedings of the 13th International Conference on Software and Systems Engineering and their Applications, Paris, vol 3, 12-5, pp 1-6, 2000.

[10] C. N. G. Dampney, Michael Johnson and G. M. McGrath. Audit and Enhancement of the Caltex Information Strategy Planning (CISP) Project. Unpublished report to Caltex, 1993.

[11] C. N. G. Dampney, Michael Johnson, and G. P. Monro. An illustrated mathematical foundation for ERA. In *The unified computation laboratory*, pages 77–84, Oxford University Press, 1992.

[12] C. J. Date. *Introduction to Database Systems*. Addison-Wesley, sixth edition, 1995.

[13] C. J. Date. *Introduction to Database Systems, Volume 2*. Addison-Wesley, 1983.

[14] Zinovy Diskin and Boris Cadish. Algebraic graph-based approach to management of multidatabase systems. In *Proceedings of The Second International Workshop on Next Generation Information Technologies and Systems (NGITS '95)*, 1995.

[15] Zinovy Diskin and Boris Cadish. Variable set semantics for generalised sketches: Why ER is more object oriented than OO. In *Data and Knowledge Engineering*, to appear, 2001.

[16] Michael Johnson. The Tos, and the Logic of Queries. presented at CATS'94, UTS, 1994.

[17] Michael Johnson. Categorical Logic and Information System Specification. Manuscript, 2000. See also *Journal of the IGPL*, 4, OUP, 1996.

[18] Michael Johnson and C. N. G. Dampney. On the value of commutative diagrams in information modelling. In *Springer Workshops in Computing*, Springer-Verlag, 47–60, 1994.

[19] Michael Johnson and C. N. G. Dampney. Half-duplex interoperations for cooperating information systems. Accepted for CE2001, the Eighth International Conference on Concurrent Engineering, 2001.

[20] Michael Johnson and Robert Rosebrugh. Database interoperability through state based logical data independence. Proceedings of the Fifth International Conference on Computer Supported Cooperative Work in Design, 161–166, 2000.

[21] Michael Johnson, Robert Rosebrugh and C.N.G. Dampney. View updates in a semantic data modelling paradigm. *Database Technologies*. Proceedings ADC2001, 29–36, IEEE Computer Society, 2001.

[22] Michael Johnson and Robert Rosebrugh. View updatability based on the models of a formal specification. In press for the Proceedings of Formal Methods Europe (FME01), *Springer Lecture Notes in Computer Science* 2021, 534–549 2001.

[23] Michael Johnson, Robert Rosebrugh, and R. J. Wood. Entity-relationship models and sketches. Submitted to *Theory and Applications of Categories*, 2000.

[24] F.W. Lawvere. *Functorial semantics of algebraic theories*. PhD thesis, Columbia University, 1963.

[25] Saunders Mac Lane. *Categories for the Working Mathematician*. Graduate Texts in Mathematics 5, Springer Verlag, 1971.

[26] Frank Piessens. *Semantic data specifications: an analysis based on a categorical formulation*. PhD thesis, Katholieke Universiteit Leuven, 1996.

[27] Frank Piessens and Eric Steegmans. Categorical data specifications. *Theory and Applications of Categories*, 1:156–173, 1995.

[28] G. Southon, C. Sauer, and C. N. G. Dampney. Lessons from a failed information systems initiative: issues for complex organisations *International Journal of Medical Informatics*, Elsevier Science, 1999.

[29] R. F. C. Walters. *Categories and Computer Science*. Cambridge University Press, 1991.

Database Support for Evolving Data in Product Design

Dietmar Posselt* Gerd Hillebrand

IPD, Universität Karlsruhe, Germany

{posselt,ggh}@ira.uka.de

Abstract

We argue that database support for design processes is still inadequate, despite the many transaction models that have been put forth to fill this deficiency. In our opinion, substantial improvement is not to be had by introducing yet another transaction model, but by questioning the basic paradigm of transaction processing itself. Instead of the usual view of transactions as destructive operations replacing an outdated database state with a more current one, we propose to view design transactions as non-destructive operations importing additional knowledge about an essentially unchanging final design solution. This leads to a model of designing with constraints, and a natural way of concurrent design. We give a formal presentation of the model and then discuss implementation techniques for centralized and distributed constraint management.

1. Introduction

Product design, like most other business processes today, requires database support for all but the most simple undertakings. Yet, design processes differ from ordinary business processes in that they are not easily decomposable into a series of independent steps that would lend themselves to the ACID paradigm of classical database transaction processing. Instead, the various subtasks in a design task tend to be highly interrelated, and the design process is best viewed as a set of cooperative activities pursuing these subtasks simultaneously.

Consequently, research in database support for design processes has mostly focused on developing alternatives to the strict isolation property of classical database transactions, and a variety of mechanisms to define and control cooperative transactions has been proposed.

While these proposals have – sometimes considerably – departed from the notion of serializability as the classic correctness criterion, the basic paradigm of transaction processing has remained the same: transactions effect state transitions between consistent database states, which are defined as being complete and unambiguous descriptions of the current state of the miniworld under consideration.

From this point of view, the *amount* of information in a database is essentially constant: each consistent database state provides a complete description of exactly one state of the miniworld. Transactions switch between these states as dictated by the evolution of the miniworld; they do not, in general, increase the amount of information available about the same state. Thus, transactions are essentially *non-monotonous*: information about the old state is destroyed and replaced by information about a new state. From non-monotonicity follows non-commutativity, and therefore the importance of serializability.

In a design process, however, information is predominantly *added* to the information already available. Ideally, design means a continuous refinement of specifications until a unique artifact is singled out. In other words, the "state" of the world – the artifact that is the eventual result of design – never changes; instead, the amount of knowledge available about that state increases monotonically. Of course, real design processes are less straightforward; proposed designs may prove to be infeasible, in which case some specifications must be retracted. Nevertheless, the fundamental operation in design is narrowing the solution space, which is a *monotonous*, and hence commutative, operation.

We believe, therefore, that a database system useful for design processes should support the notion of *specification* (as a constraint on admissible artifacts) and *refinement* and *release* of specifications as primitives in its data model, and that the leeway offered by the commutativity of refinement should be exploited in its transaction model.

As a first step in that direction, we propose a constraint-based data model, thus adopting a perspective originally introduced by constraint logic programming and then taken up by constraint database systems. From this perspective, a database does not provide an explicit characterization of a single state of the world by assigning precise values to the state variables. Instead, a database contains

*Funded by the German Research Council (DFG) under project number SFB346.

a set of constraints over these variables, which delineate the admissible values and thereby the possible states of the world. Put in less abstract terms, a design database holds the requirements and design decisions gathered so far and thereby constrains the set of possible design solutions. As design proceeds, further constraints are added, until either a contradiction is reached and backtracking becomes necessary, or the specification progresses to the point that all information needed for eventual production of the artifact is available with sufficient precision.

The fundamental primitives in this data model are refinement, querying and release of constraints, and we demonstrate how these primitives play together in collaborative design. Given that the database always denotes an approximation to the design solution, the notion of consistency becomes rather simple: constraints must not be contradictory (at least not on a level the system can detect), and constraints introduced by one party must not be released by another party without proper authorization.

In the case of a single, centralized database, the data model and its associated consistency guarantees can be implemented in a straightforward way using techniques known from constraint databases. The situation becomes more difficult once data replication is taken into account, e.g., a sales representative loading a set of product specifications onto her laptop, making some changes offline at a customer meeting and then reintegrating the data into the main database. To maintain consistency in this case, changes to replicated data must, to a certain extent, be arranged in advance. We present a locking protocol designed to announce anticipated changes to replicated data and to prevent, in a pessimistic fashion, conflicts during reintegration. However, even if the protocol cannot be observed and unanticipated changes need to be made, conflicting specifications will be detected at reintegration time and can then be resolved manually.

To summarize, the contributions of this paper are as follows: we demonstrate that the notion of data as constraints lends itself in a natural way to concurrent design, and we present the protocols necessary to coordinate concurrent access and maintain consistency in both in a centralized and a distributed constraint store.

2. Related work

The view of design as refinement of specifications is fairly common in design theory (see, e.g., [?]). In the most general setting, specifications are given by arbitrary concepts (where a concept denotes a family of related artifacts), and the result of design is an artifact defined by the intersection of a sufficient number of such concepts [?]. However, there is no hope of treating specifications this general algorithmically. We limit ourselves to specifications that constrain the value of some attribute to a certain range. Such specifications do cover the most common cases in practice, and they admit efficient satisfiability tests.

As pointed out in the introduction, the constraint paradigm has been put to fruitful use in constraint database systems [?, ?, ?]. The primary motivation for this line of research was the desire to generalize ordinary relations (i.e., finite collections of tuples) to infinite relations while maintaining a finite description. Consequently, constraint database systems have become very popular for geographic and temporal applications, where regions or intervals containing an infinite number of points need to be manipulated. The transactions conducted by such applications, however, are rather conventional and in particular non-monotonous, so that constraint databases have not yet been able to realize the potential for increased parallelism inherent in the constraint framework.

Quite to the contrary, parallelism was the main focus in Saraswat's work on parallel constraint programming [?]. There, the notion of a shared constraint store is proposed as an elegant means of communication and synchronization between cooperating agents. Our data model is a rather simple application of the general framework laid out in this thesis. Saraswat, coming from a programming language background, was mostly concerned with a centralized constraint store; although he addresses distributed constraint systems, he does not discuss any mechanisms to maintain consistency in such a setting. This gap is filled by the protocol presented in Section ??.

Fuzzy sets [?] may be viewed as generalized constraints that allow varying degrees of satisfaction. Fuzzy database systems [?, ?] have been proposed to capture the uncertainty of specifications inherent in early design stages [?, ?]. Thus, fuzzy database systems share our view of design as refinement, or reduction of uncertainty. However, most research in applying fuzzy technology to design processes has focused on the modeling aspects and semantics of fuzzy sets, whereas we are mainly concerned with coordinating parallel access to design information. Yet, the data model and protocols presented below apply equally well to "crisp" and "fuzzy" constraints, provided that efficient consistency checking algorithms for the class of constraints under consideration are available (e.g., [?]).

Constraint and fuzzy database systems may be regarded as combinations of a non-standard data model and a standard transaction model. Conversely, there have also been many attempts to graft an advanced transaction model onto a conventional object-oriented or relational database system, seeking to provide better support for the long-running, cooperative transactions typical for design processes. Space limitations forbid a detailed discussion of this rather large body of work (see [?] for a survey). In general, these transaction models trade the strong consistency guarantees offered by serializable transaction processing against more flexible control over transaction isolation and notions of consistency. However, such flexibility comes at the price of more complex transaction management schemes and, worse perhaps, depending on each user to provide information about the desired semantics along with the transactions themselves. That, in our view, is a bad tradeoff; the consistency guarantees and semantics of a database system should be as simple as possible.

A much less drastic modification of the classic ACID paradigm is the notion of ϵ-serializability, introduced in [?] as a means of trading precision for concurrency. ϵ-serializability is applicable to data that possess a – numerically quantifiable – degree of uncertainty, and it extends ordinary serializability by permitting transaction histories whose results differ (according to a suitable metric) by no more than a quantity ϵ from a serializable history. Unfortunately, ϵ-serializability does not handle write transactions very well: in the absence of any knowledge about the internal workings of a write transaction, one must assume that small variations in the input can lead to arbitrarily large variations in the output, and hence accept the possibility of unbounded divergence of the database state from the result of a serial execution. [?] addresses this problem by suggesting that transactions that "went too far astray" be periodically undone, but in a design setting, with transactions easily representing a day's work, that may prove very costly.

3. A formal model of constraint-based design

We view design as the manipulation of specifications, where each specification poses a restriction on the admissible design solutions. Specifications imposed by external agents (i.e, design requirements) and specifications resulting from decisions internal to the design process are treated alike. We model the design space as a finite collection x_1, $x_2, \ldots, x_n$ of *design parameters* of types $t_1, t_2, \ldots, t_n$, such that every possible design outcome is uniquely characterized by an assignment of values to $x_1, x_2, \ldots, x_n$. The types $t_1, t_2, \ldots, t_n$ are arbitrary; thus it is entirely conceivable that a design parameter describes, say, a complex geometry.

A *specification* in general is a syntactical characterization of a subset of $t_1 \times t_2 \times \cdots \times t_n$. As pointed out in Section **??**, this notion is too broad to be algorithmically tractable, so we limit ourselves to specifications that restrict the value of a single design parameter x_i to a range S, where S has a syntactical characterization in some language L. Such one-dimensional specifications will henceforth be called *conditions*. The exact language L used to express ranges is left unspecified; for example, S might be defined by explicit enumeration, numerical intervals, logical formulas, etc. We do make certain assumptions about L, though:

1. If ranges $S_1, S_2, \ldots, S_m$ are expressible in L, then their intersection $\bigcap_{i=1}^{m} S_i$ is expressible in L as well, and there exists a (reasonably efficient) procedure for computing the corresponding expression.

2. There exists a (reasonably efficient) procedure for deciding whether the intersection of a collection S_1, $S_2, \ldots, S_m$ of ranges expressible in L is empty.

3. The ranges expressible in L are "convex" in the following sense: if $S, S_1, S_2, \ldots, S_m$ are expressible in L and $\bigcap_{i=1}^{m} S_i \neq \emptyset$ and $S \cap S_i \neq \emptyset$ for $1 \leq i \leq m$, then $\bigcap_{i=1}^{m} (S_i \cap S) \neq \emptyset$. (This is a technical and not particularly essential condition having to do with the *constrain* operator introduced below.)

These assumptions are obviously true for ranges defined by numerical intervals, but also for more powerful logical frameworks (cf. [?] for some examples).

Associated with each condition is a *principal*, i.e., an abstract representation of the agent that imposed the condition. This could be an individual designer, a design team, a manager, a customer or any other entity authorized to participate in the design process. The combination of condition and principal is called *constraint*.

A *design state*, or *design database*, is given by a collection of constraints, which together describe the values considered admissible for $x_1, x_2, \ldots, x_n$ in that particular state. A design state is called *consistent* if there is at least one assignment of values to $x_1, x_2, \ldots, x_n$ satisfying all constraints.

This notion of consistency is admittedly naive, because it does not take interactions between design parameters due to physical laws or technological limitations into account. For example, in microprocessor design the conditions "clock rate $\geq$ 1000MHz" and "power dissipation $\leq$ 1W" are formally consistent (because they refer to different design parameters), but currently not simultaneously achievable. If desired, such dependencies between design parameters could be introduced into the model as multi-dimensional conditions, at the expense of increased overhead for consistency checking. We believe, however, that even the simple-minded consistency checks defined here will prove beneficial in practice.

Design proceeds by means of three basic operations. Each operation is carried out atomically.

constrain(x, S) Adds a new constraint with condition $x \in S$ and the invoking agent as principal to the design state. This operation succeeds only if the new state is consistent. If it fails, the set of constraints conflicting with $x \in S$ is returned (this set is well defined because of the convexity property assumed above).

release(p, x, S) Removes the specified constraint from the design state, if it exists. (Note that this does not affect other constraints on x.) This operation succeeds only if the invoking agent is authorized to remove constraints introduced by principal p. The specific scheme whereby such authorization is obtained is left open.

peek(x) Returns a condition describing the set of values admissible for x in the current state (i.e., the intersection of all current constraints on x).

Let us briefly comment on these operations and the invariants they imply.

constrain is a monotone operation that is used to establish new design requirements, new design decisions or – this case will be discussed in the next section – a stake in existing decisions. *release* is the inverse operation, used to retract requirements or decisions that did not come to fruition. *peek* serves to query the current state, typically at the beginning of a design task. From the definition of these operations, it is clear that the following two properties hold:

Constraint Consistency (CC): The design state is always consistent. In particular, the result of a *peek* operation never denotes the empty set.

Constraint Durability (CD): Once a constraint has been successfully established, that constraint will be satisfied by any design solution unless explicitly released by a properly authorized agent (presumably with notification of the constraint's principal).

It is worth pointing out that the notion of durability described here is quite different from the one offered by ACID transactions: the latter refers to protection from system or media failures, but not the effects of other transactions. In fact, values written by ACID transactions are safe from other transactions only while they are *not* yet committed; after being committed, they may be overwritten at any time. This is contrary to the semantics of committed design decisions, which are supposed to be stable unless explicitly retracted.

4. Concurrent design with constraints

We will now illustrate how the constraint manipulation primitives play together in concurrent design.

The design process will usually begin by determining the major design requirements and entering corresponding constraints into the design database. As soon as a sufficient set of requirements has been established to allow a meaningful subdivision of the design task, design subtasks may be formulated and assigned. Meanwhile, the acquisition of requirements continues, resulting in further constraints being imposed.

A designer charged with a certain design task will initially use the *peek* operation to obtain the current constraints on the relevant design parameters. Three outcomes are possible:

1. The constraints are such that no solution is possible. In this case, the design process has reached a dead end, and some constraints need to be released.

2. The problem is underspecified to the extent that nothing useful can be done. In this case, the design task must be postponed until more information is forthcoming.

3. The constraints admit one or more solutions. In this case, a solution is determined and constraints describing the solution are added to the database, *together with constraints that describe the assumptions on which the solution was based.* In the simplest case, these additional constraints are simply copies of the constraints obtained from the initial *peek* operations, and thus redundant. More frequently, though, a design solution requires stronger assumptions than just the design requirements, so that the additional constraints describing the design assumptions do actually narrow the solution space.

In this way, many design tasks can proceed in parallel, while – by virtue of constraint durability – each designer can rest assured that the assumptions and results of her design remain valid throughout the entire process, unless explicitly retracted.

Of course, this also means that an attempt to establish a constraint may fail, because it conflicts with the assumptions or results of a another design activity. No attempt is made to resolve a conflict automatically, because resolution typically requires domain-specific expertise. The only recourse in a conflict is to contact the principal of the offending constraint (which is returned by an unsuccessful *constrain*) and to negotiate a solution.

Note that the consistency and durability guarantees offered by the constraint store do not require any locking protocols (besides executing each primitive operation atomically), but follow simply from the definition and the monotonicity of *constrain.*

5. Implementing a centralized constraint store

While the current prototype of our system is a stand-alone implementation using a main-memory database system, a layered implementation on top of a conventional relational database system can be envisioned in a rather straightforward way.

First, the design parameters describing the design space need to be identified. It is hardly feasible to submit every single dimension occurring in an engineering drawing or every line of code in a program to the constraint paradigm. Therefore, one would try to identify a set of parameters representing information that truly needs to be shared among concurrent design activities, and then use the transaction management facilities of the underlying database system to couple other, non-shared parameters to their "governing" shared parameters. For example, if two design groups working on a mechanical assembly needed to exchange only bounding box information for their respective components, then just the coordinates of these bounding boxes would be considered design parameters, and access to all other dimensions would be wrapped in ACID transactions that ensure that the bounding box data remain consistent with the true geometry data.

Second, a domain and a condition language need to be chosen for every design parameter. In mechanical design, most information is numerical, and lower and upper bounds usually suffice as conditions. On the other hand, software engineering usually deals with free-text specifications, and for these some sort of symbolic representation must be found, e.g., by picking the most salient keywords and defining conditions by explicitly enumerating the required keywords.

Third, the design parameters and their associated sets of constraints have to be represented in the relational model. For example, a numerical design parameter with constraints defined by intervals might be represented by a three-column table having a tuple (principal, lower_bound, upper_bound) for each constraint on the design parameter.

Finally, the operations *constrain*, *release* and *peek* need to be implemented as ACID transaction procedures, where *constrain* and *peek* would presumably invoke custom procedures for satisfiability testing and intersection, respectively.

Note that with this scheme, applications can continue to use ordinary ACID transactions. In fact, if "wrapper" transactions are written for access to non-shared parameters, transparently inserting calls to *constrain* and *release* their governing shared parameters as outlined above, then applications need not be aware of the constraint framework at all.

6. Handling distribution and replication

Our goal is now to extend the (CC) and (CD) guarantees to a distributed and replicated constraint system. We assume that replicas may operate temporarily in disconnected mode, without being able to exchange operations with other replicas. In this case, two incompatible *constrain* operations may be executed at different replicas, raising a conflict when the replicas are eventually merged. If (CC) and (CD) are to be maintained under these assumptions, no such conflicts can be allowed. Hence, *constrain* operations must be executed pessimistically. *release* and *peek* operations are less critical, because they can never raise conflicts on a merge.

6.1. Intention locks

We introduce intention locks as a means to protect disconnected replicas from incompatible *constrain* operations. Intention locks are acquired while the replicas are still connected, and held until reconnection occurs.

The idea behind intention locks is to announce changes anticipated for the period of disconnected operation. Obviously, some advance knowledge of these changes is necessary to acquire the proper locks, but this is not an unreasonable assumption. If a sales representative copies data to her notebook and meets with a customer to discuss a design, she probably has a rough idea of which data may change.

Formally, intention locks are conditions in the sense of Section **??**, i.e., one-dimensional specifications expressed in some language L. Intention locks are always associated with a replica and, unlike constraints, do not have a principal. Also, there are two kinds of intention locks, and their semantics is somewhat different from conditions.

The two kind of intention locks correspond to classical shared and exclusive locks and reflect the two different attitudes a designer may exhibit towards a design parameter.

Firstly, a design parameter x may be viewed as input to a design task. In this case, the designer will presumably use *peek* to obtain the current range of x and then use *constrain* to re-impose this range (or perhaps a range somewhat narrower) as a design assumption. The semantics of this constraint is that she is prepared to accept any outcome of the design process as long as her design assumption is satisfied. In particular, she is prepared to accept arbitrary constraints imposed on x by other principals, perhaps on other replicas, as long as these other constraints have nonempty intersection with her own. To express this attitude when about to enter disconnected mode, she would request a *shared intention lock* on x with a range equal to the minimum range acceptable to her. A shared intention lock on x with range S held by a replica prohibits other replicas from imposing constraints on x with a range disjoint from S. As such it is very similar to a constraint, except that shared intention locks are immediately and globally published, whereas constraints are only eventually published (see the section on protocol implementation below).

Second, a design parameter x may be viewed as a (yet unknown) output of a design task. In this case, a designer may foresee that she needs to impose certain constraints on x, but does not yet know what these constraints will be.

In order to keep other designers from restricting her freedom of choice, she will want to impose a lock that prohibits other principals from introducing any constraints on x that eliminate values she considers potentially interesting. This attitude is expressed by requesting an *exclusive intention lock* on x with a range that is the union of all constraints the designer might want to introduce later. An exclusive intention lock on x with range S held by a replica prohibits other replicas from imposing constraints on x with a range that does not include S.

The entire locking protocol is determined by the following rules. The first two rules govern the acquisition of locks, and the last rule governs the execution of *constrain* operations (*release* and *peek* operations are always permitted).

Shared lock acquisition A request for a shared intention lock on design parameter x with range S will be granted iff the following conditions are met:

1. If another replica holds an exclusive intention lock on x with range X, then $S \supseteq X$.
2. If another replica holds a shared intention lock on x with range S', then $S \cap S' \neq \emptyset$.

Exclusive lock acquisition A request for an exclusive intention lock on design parameter x with range X will be granted iff the following conditions are met:

1. No other replica holds an exclusive intention lock on x.
2. If another replica holds a shared intention lock on x with range S, then $X \subseteq S$.

Constraint introduction A request *constrain*(x, S) executed on some replica will succeed iff the following conditions are met:

1. The replica holds a shared intention lock on x with a range $S' \subseteq S$ or an exclusive intention lock on x with a range $X \cap S \neq \emptyset$.
2. The intersection of S and all other constraints currently imposed on x at that replica is nonempty and, if the replica holds an exclusive intention lock on x, meets the range of that lock.

Note that the admissibility of a *constrain* operation can be decided locally, whereas the acquisition of locks requires global communication. How exactly this global communication is implemented is discussed next.

6.2. Protocol implementation

The protocol for the acquisition of intention locks may seem simple at first, but the details are somewhat complex. We note first that *constrain* operations can be executed locally at a replica without communication. The resulting constraints can be propagated eventually to the other replicas, along with the release of the lock covering the operation. The difficult part is setting and releasing the intention locks. We call these operations *global*, because they must be propagated to all replicas. Global operations can be executed asynchronously as long as all replicas execute them in the same order. In particular, a disconnected replica will see and execute global operations originating from other replicas only at reconnection time, but as long as their ordering is preserved, the protocol will remain correct.

The set of replicas will usually be divided into a set of connected partitions. It is easy to show that only one partition can be permitted to execute global operations. Now the problem is to decide which partition is allowed to initiate global operations. Usually this is done using a quorum consensus algorithm. However, in our scenario the majority of replicas might be offline, so that there is no partition big enough to achieve a quorum. In this case the requirements for a quorum have to be modified accordingly, e.g. a quorum could be any partition that contains more than half of the members of the last quorum.

Last, but not least, even inside a quorum only one replica can initiate a global operation. No other replica can initiate a global operation until the preceding global operation has been completed by all other replicas within the quorum.

7. Conclusions and open problems

We started from the observation that design, unlike most other business processes, is a process of successive refinement and monotonous information accretion. This led us to adopt constraints as the basic design objects. We then investigated how concurrent design can be achieved within the constraint framework, and showed how the basic guarantees of the constraint model, combined with the explication of design assumptions, lead to a very natural coordination model that does not require any additional transaction management. Finally, we presented implementation mechanism for centralized and distributed constraint systems, thus demonstrating the basic feasibility of the approach.

The most interesting continuation of this work is certainly the handling of multidimensional constraints, where a constraint can specify dependencies between several design parameters. Unfortunately, the time needed to compute intersection and satisfiability of such multidimensional constraints tends to grow rather fast with the number of dimensions. Further work is necessary to identify constraint languages and implementation techniques that can handle multidimensional constraints with reasonable performance.

References

[1] P. Bosc and J. Kacprzyk. *Fuzziness in Database Management Systems*. Physica Verlag Heidelberg (ISBN 3-7908-0858-X), 1995.

[2] D. Braha and O. Maimon. *A Mathematical Theory of Design: Foundations, Algorithms and Applications*. Kluwer Academic Publishers (ISBN 0-7923-5079-0), 1998.

[3] G. Chen. *Fuzzy Logic in Data Modeling. Semantics, Constraints and Database Design*. Kluwer Academic Publishers (ISBN 0-7923-8253-6), 1998.

[4] P. Drew and C. Pu. *Asynchronous consistency restoration under epsilon serializability*. Technical report, Oregon Graduate Institute, 1993.

[5] A. K. Elmagarmid. *Database Transaction Models*. Morgan Kaufmann (ISBN 1-55860-214-3), 1993.

[6] R. E. Giachetti, R. E. Young, A. Roggatz, W. Eversheim, and G. Perrone. A methodology for the reduction of imprecision in the engineering design process. *European Journal of Operations Research*, 100(2):277–292, 1996.

[7] P. C. Kanellakis, P. Z. Revesz, and G. M. Kuper. Constraint Query Languages. In *Proceedings of the 9th ACM SIGACT-SIGMOD-SIGART Symposium on Principles of Database Systems (PODS)*, pages 299–313. ACM Press, 1990.

[8] G. Kuper, L. Libkin, and J. Paredaens, editors. *Constraint Databases*. Springer Verlag (ISBN 3-540-66151-4), 2000.

[9] C. Pu. Relaxing the limitations of serializable transactions in distributed systems. In *Proceedings of the 5th ACM SIGOPS European Workshop*, August 1992.

[10] P. Z. Revesz. Constraint databases: A survey. In L. Libkin and B. Thalheim, editors, *Semantics in Databases*, pages 209–246. Springer LNCS 1358, 1998.

[11] R. Saraswat. *Concurrent Constraint Programming*. MIT Press (ISBN 0-262-19297-7), 1993.

[12] H.-J. Sebastian and E. K. Antonsson, editors. *Fuzzy Sets in Engineering Design and Configuration*. Kluwer, 1996.

[13] H. Yoshikawa. General design theory and a CAD system. In T. Sata and E. A. Warman, editors, *Man-Machine Communication in CAD/CAM, Proceedings of the IFIP Working Group 5.2 Working Conference*, pages 35–38, 1981.

[14] L. A. Zadeh. Fuzzy Sets. *Information and Control*, 8:338–353, 1965.

[15] H.-J. Zimmermann, editor. *Practical Applications of Fuzzy Technologies*. Kluwer Academic Publishers, 1999.

On Processing Cycle Query's Strategy for Semistructure Data Model and Its Role in Design

Jiajin Le
Dept. of Computer Science
Donghua University
Shanghai 200051, P. R. China
Lejiajin@dhu.edu.cn

Xuemin Lin
School of Com Sci & Eng
University of New South Wales
Sydney, NSW 2052, Australia
lxue@cse.unsw.edu.au

Abstract

There is no obvious schema information for the characteristic of self-describing semi-structure data, since the data structure is dynamic but not regular nor complete. The data mode often adopted is a directed graph with its root, in which it often appears a circular-like structure. In this paper, we address the issue of querying semi-structure data. Particularly, we are interested in detecting circular-like structure, so that infinitude circulating can be avoided for the query on OEM model-based. As an application, the approach (the Processing Cycle Query's Strategy) proposed in this paper has been successfully implemented in the design of www. Ego365.com E-business network station (built by Authors' team).

1. Introduction

With rapid developing of the internet technology, the amount of electronic data has been set to explode. The types of data are different to other original data from language or graphic files, and especially to those data, which are from conventional computer system with strict schema constraints. Some of electronic data have non-structure, while others may be in loose structure, or in cryptic structure, or in certain structure often to be ignored. These electronic data are called semi-structure data, which are generally in existence of electric data source.

The characteristic of semi-structure data is that its data-structure is non-regulation, non-completion as well as dynamically changeful. It is mainly demonstrated for those data in non-regular mode, while its structure is often to be cryptic, its schema contains a number of information, and the schema are varieties for its flexible, and the schema can combine to its data to restore, and so forth.

The characteristic of semi-structure data is illustrated as:

1) The structure is non-regular. For the potion isomerous data are often to be presented in variety type or in differed structure, through these data are identical.
2) The structure is cryptic. Such as electric documentation SGML Format.
3) The structure is non-complete. There is no structure at all for some portion of data, or only to be in rough structure for some portion of data occasionally.
4) The structure is instructive. Instructive structure is the non-accurate description for the data structure, it can be acceptance for all new data only through modifying the structure frequently as its costs.
5) Post-existing schema conduct. While it is on existing data, the semi-structure can bring on its structure data through concluding current data.
6) The schema can be modified frequently. Homologous schema will be modified frequently since the data in database are to be changed frequently.
7) Borderline between data and schema is illegibility. Discriminate between data and schema is increasing the difficulty to make.

As the characteristic of semi-structure data is of non-regulation, non-completion, and dynamic changeful, the semi-structure data is hard to be described by strict and fixed data model (relational-model or object-oriented model). In consequence, semi-structure data is customarily described by Directing connected graph data schema with its marker[1,2, 3]. The self-described function is the advantage of this chart, the Object Exchange Model (OEM) is the typicality of it .

The main source of semistructure data:

1) It is coming from semistructure data source directly, such as Web page, various electric documents, E-mails, etc.
2) It is creating data when the isomerous data are integrated.

With the internet become prevalence, to meet the requirements of integrating the isomerous data source system the semistructure data have been grown up dramatically; and research on semistructure data is attracting people's attention more and more. Many researching results for semistructure data schema[1,2,5], query language[3,6, 7], query process[4,5], as well as store and presentation[1,3] have been published.

2. Semistructure Data Schema

As the characteristics of semistructure data are non-regulation and non-forecasted, it could be in handy use for the simple and convenient data model because of its simple, flexible, and less restrains characteristics etc.

The OEM model (Object Exchange Model) discussed in the work[1,5] is adopted in this paper as a foundation of our research. The data presented by OEM is self-described, and it is of a quite flexibility for its presentation, it can be with the simplest structure to present the key characteristic that is hidden in data.

The most difference between OEM and traditional database model is the OEM can easily contain a various non-regulation and non-completion, as seen in Figure 1:

1) The identical information can be in different types or different strictures potentially. For example, the NAME can be presented in single name, or in composite name (separated as first name and last name), can be formal name, and can also be shortening name or alias. Further, the STAFF can be presented by a value of a simple character string type and also can be presented by a complex structure.
2) An object can own one attribute, or more than one attributes, or nothing of it.
3) Some identical attribute can be on different level of same model, e.g. the name can be a attribute of University, or Department's attribute, or even the staff's attribute.
4) The object's constituting (reference) relationship can be in a various way, such as from up to down, or from left to right, or from right to left, e.g. the relationship for the reference University to University, the reference University to department.

On all accounts, the OEM makes the Database as a Connected Graph that consists of root, Direction, Node, as well as Mark-edge etc. The chart's note is presented as object or sub-object, and direction edge as an aggregation relationship (represented as constituting (reference) relationships among objects). The Mark-edge represents the description of such relationships. As the notes (objects) can be referred each other, and therefore the circularity-like structure may be potentially occurred in such directional chart. As in Figure 1:

&1→&5→&16→&11

&1→&5→&16→&11.........

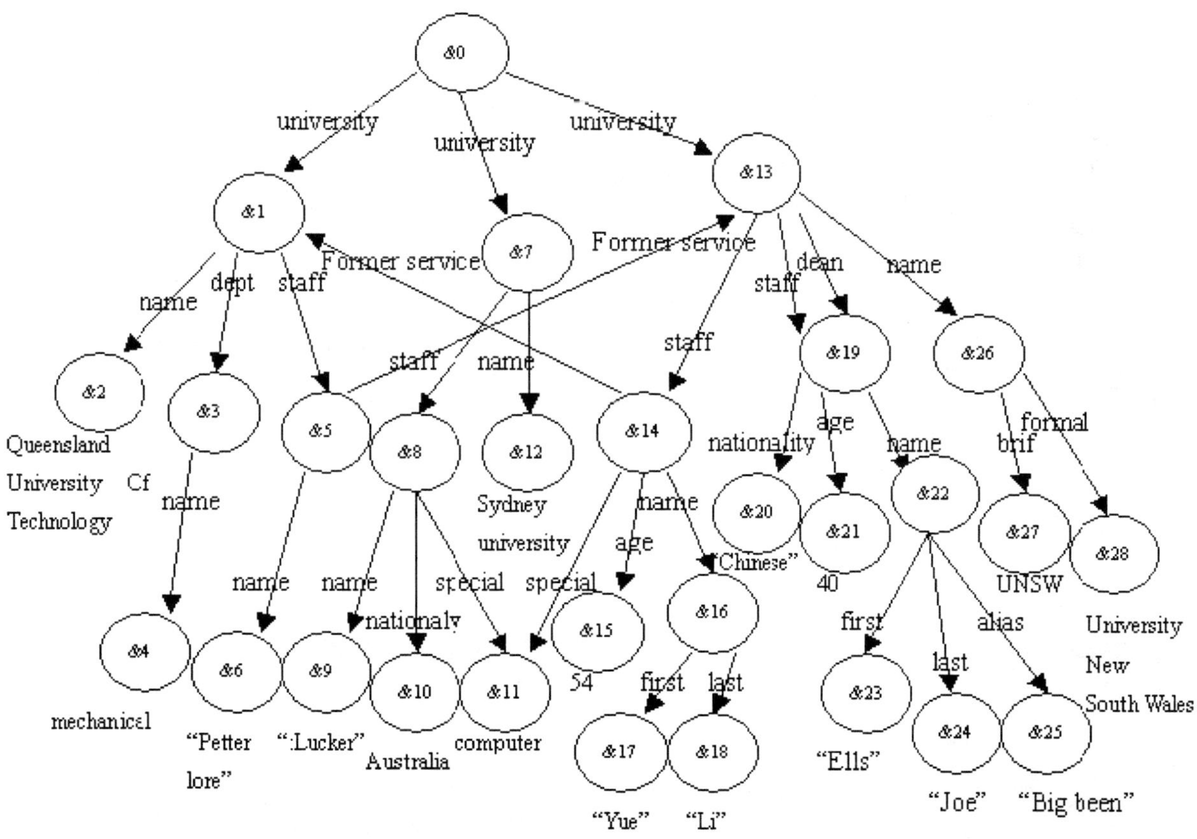

Figure 1. A portion of OEM database (Basic information of some University in Australia)

3. Detecting Cyclic Query Process Strategy

The following definitions that we have given for our efficient discussion about semistructure data model.

Definition 3.1 Key Vertex: Given G=(V, E) is non-directing connected graph. Supposing there are three different vertex: v, w, a, and v, w, a ∈ V, any path from "v" to "w" must be via vertex "a", the vertex "a" is called as the key vertex within G.

Definition 3.2 Strong Connection: Provided no key vertex in non-direction graph, such graph is called as Strong Connection.

Definition 3.3 Strong Connection Component: Within chart G=(V, E), provided both e_1 and e_2 are on the same return path, then the equivalence of e_1 and e_2 , taking this equivalence relationship plot out in K numbers: E1, E2, ..., Ek (k≥1), and all these E are not intersected each other. If and only if the equivalence of two edges, they are in the same subset. Gi=(Vi, Ei) , the sub-graph of G, is called G's Strong Connection Component, and Vi is the vertex set abut against Ei edge.

Definition 3.4 Self-loop: The self-loop is a loop with the 0 in length of edge.

Definition 3.5 Cyclic Query: Cyclic query refers to the looping query in the query graph. So-called looping is as the self-loop and Strong Connection Component in which the vertex that its number≥3 in the query graph.

3.1 Identifying the loop

To identify the loop means to find the Strong Connection Component in which the vertex in number≥3 and is no any loop made by these vertex.

Algorithm3.1 The algorithmfor identifying double connection component. [6]

Input: The adjacency table L[v] of each vertex within a non-directing connected graph.

Output: The edge set of Strong Connection Component within such non-directing connected graph.

Procedure:

```
Procedure searchb (v)
Begin change the mark v into "old";
    denumber[v]=count;
    count=count+1;
    low[v]= denumber[v];
    For L[v] each vertex w do
      If the mark w is "new" then
       Begin 1
         stack←(v, w); //press(v, w) into stack//
         point(v, w) ←1;
         father[w]←v;
         searchb(w);
         If low[w]≥ denumber[v]      // find a Strong Connection
                    Component//
         Then Begin 2
         pop-up the edge of stack top and print it;
         While the printed edge is not (v, w)
       Do pop-up the edge of stack top and print out a special mark *
       // print out a edge set of  Strong Connection Component//
            End 2
            low[v] ←min(low[v] ,low[w])
           End 1
      Else
        If w is not father[v] and vertex(v, w)=0
        Then Begin 3
           stack←(v, w);
           point(v, w)←1;
           low[v] ←min(low[v] , denumber[w])
         End 3
  End.
```

3.2 Identifying loop component

In query object graph, it is able to make up of another loop between the objects matched with a loop in query object graph, which refers to object loop. The intention of Identifying loop component is for finding out all object-oriented loop in object graph.

1) For the self-loop in query graph, the processing approach combined the right direction searches with nesting loop retrieve are adopted.
2) For the Strong Connection Component in query graph, we can consult both processing approaches presented in the document[14]: the processing approach combined the right direction searches with nesting loop retrieve, and the processing approach combined the reverse direction searches with nesting loop retrieve, both of these approaches can only adapt to such two situations as follows:

Given Vc as a vertex set of Strong Connection Component;

A) Each vertex of Strong Connection Component is connection with other two vertexes
B) Only one vertex's degree is greater than 2, or each degree of vertex is 2 identically, and there is only one goal class.

3.3 Process algorithm for Strong Connection Component

After the analysis on various Strong Connection Components in query graph, we read the main ideal in document[13,14] with prudence. We write following Process algorithm that has general sense for Strong Connection Component:

Algorithm3.2 Process algorithm for Strong Connection Component

(Given n as the vertex number; e as edge number)

Step 1. Using the algorithm that take out the results for non-directing connected graphic's generating tree within graphical theory[11], we can find out the generating tree that is to take a vertex $V_i(i = 1, \ldots, n)$ of Strong Connection Component as a root during the time of $O^{(n+e)}$.

Step 2. Within query object graph that is corresponding to the Strong Connection Component, each object of root class transmits its own cycling mark through other all class's object that is reachable for the generating tree, and this process can be achieved within the time of $O(\sum_{i=1}^{n} Numi)$, the Numi is for the number of object in the class represented by vertex Vi(i = 1, … , n).

Step 3. Award the mark "eligibility" to the relationships among all objects within query object graph that is corresponding to the double connection component.

Step 4. For i = 1 to n / * process each class in turn * /

Do Begin

For j = 1 to Numi / * process in turn each object in the class

represented by vertex Vi(i = 1, … , n) * /

Do Begin

"Eligible" Objects Oj represented

by vertex Vi in the class

are processed as follows:

1). Other class 's eligible Objects Ok associated in gear with

each "eligible" Objects Oj should be judged as follows:

If Tset _ Oj $\cap$ Tset _ Ok $\neq\{\}$

Then remain the mark "eligibility" of relationship

between Oj and Ok;

Else cancel the mark "eligibility" of relationship

between Oj and Ok;

2). If existing "eligibility" of relationship between Oj and it where is at least one "eligible" Objects in each border upon class

Then remain the mark "eligibility" of Oj ;

Else cancel the mark "eligibility" of Oj and cancel mark

"eligibility" of all relationship related to Oj

End

End

/ * this process can be finished within the time of O(Numi) * /

Step 5. If the mark "eligibility" of a object or relationship has been

canceled in Step 4

Then go to Step 4

Else Begin

(supposing the SET OC_Set being used for storage of

object loop)

1).OC_Set = ;

2). Following processes are to be on the "eligible" object and "eligible" relationship;

(supposing Q_Num_i (i = 1, … , n) as the number of

"eligible" objects contained by each class within the

Strong Connection Component)

(i) take a eligible object Oi (i = 1, … , n) out form each class

Ci (i = 1, … , n), of which the several eligible objects

can build a tuple in n, it can totally be $\prod_{i=1}^{n} Q\_Numi$

various tuples in n.

(ii) For each tuple in n should be judged as follows:

if several eligible objects, and the eligible relationships among

them are fully matched to the loop from query graph

Then several eligible objects and the eligible relationship among

them can be formed a structure of object loop(O1, O2, …, On);

End

/ * this process can be finished within the time of $O(\prod_{i=1}^{n} Q\_Num_i)$ * /

3.4 Testifying the validity of this algorithm

Theorem 3.1 All object loop in query graph can be fond by Strong Connection Component processing algorithm.

Testify: Supposing OC_Set as the set of object loop got from this algorithm, a set of AOC_Set is formed by all object loop in query object graph.

At first, to testify the validity of OC_Set $\subseteq$ AOC_Set. OC $\in$ OC_Set, OC must be satisfied with the terms in Step 4, i.e. the relationships among objects from OC are entirely matched to the loop from query graph, therefore, OC is an object loop, that is OC_Set $\subseteq$ AOC_Set, so that OC $\in$ OC_Set.

Then we can testify the validity of AOC $\subseteq$ OC_Set. A root class object ro is embodied in OC, supposing its cycling mark is as t_ro, on account of the other objects from OC are associated with ro in direct or indirect path. And in virtue of algorithm step2 along generating tree the ro transmits the t_ro to other objects in OC. In algorithm

step3, we award mark "eligibility" to the relationships among objects from OC. In algorithm step4, we detect objects from OC as well as their relationships among them. Supposing we take first detected object as O1, forasmuch the fact that all objects from OC have cycling mark t_ro, the intersection between set of O1 cycling mark and set of cycling mark of any object associated with O1 can not be empty. Thus we remain mark "eligibility" to the relationships between O1 and other objects from OC. Supposing A_Num is as the number of border upon class. In virtue of the attribute of object loop, it can reach to the number of A_Num for the objects associated with O1, and these objects in number of A_Num belong to border upon class of O1 respectively, and their relationships are to be of mark "eligibility"(as testified in the above subsection). So that within O1 and each border upon class, there must be at least one exists that "eligible" relationship among "eligible" objects. Thereby we remain mark "eligibility" of O1. Up to here, we have finished the task of detection. In this process, no any mark "eligibility" of object or relationship is to be effected or changed. Taking this as an example to do the detection for the others in turn, as to do that for the other objects from OC is rather than any effect on mark "eligibility" of object or relationship. It can be put in another way, no matter how much times the Step 4. to be executed, within OC no any mark "eligibility" of object or relationship is to be effected or changed. Step 5. of algorithm do the process on the "eligible" object and eligible relationship, it is obvious that OC matched entirely to loop in query graphic, so we should take OC into OC_Set, i.e. OC ∈ OC_Set. As a results AOC_Set = OC_Set. It is that Strong Connection Component processing algorithm can find out all object loops in query object graph.
That is end for this testifying.

4. The process after loop query

After finding out the loop from object query graph, the cyclic components can be taken as a special class, attributes of the special class are made by the attributes from various classes in cycling components. This object of special class refers to object loop that matches to cycling component. In virtue of algorithm 3.1, we can get the object of special class, in such way, the cycling query can transform into non-cycling query.

After that we can use relevant algorithm of non-cycling query (including optimizing algorithm or executing algorithm) to process the transmitted query, and get the final results of the query.

5. Summary

Along with rapid development of the Database technology on Internet Web, the semistructure query technologies are frequently used. The algorithm to be used to detect circular-like structure query has been discussed in this paper, and authors demonstrated its validity. In virtue of this algorithm, it can make the query based on OEM model successfully to avoid infinite query cycling. In our project of www.ego365.com network station, we wanted to do the cooperative work in design; and found this approach is helpful.

6. References

[1] The TSIMMIS Approach to Mediation: Data Modeels and Languages Hector Garcia-Molina, Yannis Papakonstantinou, DallanQuass et al, Stanford University, Stanford CA 94305 USA

[2] Extracting Schema from Semistructured Data Svetlozar Nestorov Serge Abiteboul Rajeev Motwani

[3] Proximity Search in Databases Roy Goldman, Narayanan Shivakumar,Suresh Venkatasubramanian, Hector Garcia-Molina Stanford University

[4] DataGuides: Enabling query Formulation and OptimiZation in Semistructured databases Roy Gold man Stanford University Jennifer Widom Stanford University

[5] Semistructured data: The Tsimmis Experience Joachim Hammer, Jason McHugh, and Hector Garcia-Molina department of Computer Science Stanford university CA 94305-9040

[6] NoDoSE- A Tool for Semi-Automatically Exatracting Structured and Semistrictured Date from Text Documents. Brad Adelberg

[7] From Data Independence to Knowledge Independence: An on-going Story Laurent Vieille

[8] Representative Objects: concise Representations of Semistructured, Hierarchical Data Svetlozar Nestorov, Jeffrey Ullman, Jnat, Janet Wiener, Sudarshan Chawathe department of computer Science Stanford University,CA 94305-9040, USA

[9] Rajaraman Anand, Norvig Peter, Virtual Database Tecnology: Transforming the Internet into a Database, IEEE Internet Computing, 1998, 1998 7/8, 55-58

[10]V.s.Subrahmanian et al, HERMES:A heterogeneous reasoning and mediator system, 1996, http://www.cs.umd.edu/projects/hermes/overwiew/paper

[11]Buneman O.P, Davidson S.B, Watters A., Querying IndependentDatabase,1992, Information Sciences, 52,12,1-34

[12]Kim K C, Kim W, Dale A. Cyclic query processing in object-oriented databases. In Codd E F, Harder T eds. Proc of the 5th int Conf on Data Engineering.Wahington,DC,1989.564~571

[13]Renee J.Miller,Using Schematically Heterogeneous Structures,SIGMOD,98.6,189---199

[14]S.Bergamaschi S.Castano M.Vincini, Semantic Integration of Semistructured Data Sources, SIGMOD,1999,VOL 28 NO.1

A Main Memory Database Method Supporting Synchronous Cooperative Application[1]

Qiujing Li, Yuchai Guo, Zongkai Lin and Shouxun Lin
CAD Lab., Institute of Computing Technology, Chinese Academy of Sciences
liqiujing@163.net

Abstract

There is no good solution of data management for CSCW applications in synchronous work mode . The choice now is no management or duplicate. In the former case, it is common that phone tool discards audio data that is worth to store. In the latter, the performance is still not high, especially when there involves large amount of multimedia data. We take advantages of main memory database technology to support synchronous cooperative work. After analyzes the cooperative design process based an cooperative concept design instance, the paper provides a main memory-based distributed data management model. The model schedules memory according to the features of synchronous cooperative application, greatly improving the performance of data access.

1. Background

In synchronous cooperative applications, there exist interactive operations among group members that need to be processed rapidly. Data management is essential for cooperative applications to share data. However, There is no good solution of data management for CSCW applications in synchronous work mode. The choice now is no management or duplicate. In the former case, it is common that phone tool discards audio data that is worth to store. In the latter, COAST[1] is an example. It duplicate all data in cooperative sites, transactions execute locally coordinating by a transaction manager. Full duplicate is not so good, especially when there involves large amount of multimedia data.

DBMS is important for cooperative applications to share data. However, database access is time consuming, which affects the rapid reaction of the applications. In general, DBMS is in client/server architecture. Cooperative applications and DBMS run in different hosts. Cooperative application submits a request to DBMS via network and waits for the result. After receiving the request, DBMS reads data from disk and calculates query according to the request. The query result is returned to the cooperative application via network. In many cases, database access is the bottleneck of the performance of cooperative applications. It is necessary to improve the efficiency of database access to meet the requirement of synchronous cooperative applications. It is well known that both bandwidth of network and processing capacity of computer are increasing, leading to the improvement of database access performance. However, it does not mean that the problem above is solved. On the contrary, the bottleneck of database is more obvious. The answer of the question is fully exploiting the capacity of computer.

Computer main memory is greatly increased in recent years. It is common that a server owns several gigabyte of main memory and a pc hundreds of megabyte. It is reasonable for us to take advantages of large amount of main memory to accelerate the operations on database. The data required for a client is somewhat limited in a period of time, leading to the high possibility of putting it in main memory of the client. Main memory database is not a new concept. The key to make it useful is a good schedule algorithm. There exists no universal good algorithm. Different schedule algorithm is needed for different application [2]. After analyzes the cooperative design process based an cooperative concept design instance, the paper provides a main memory-based distributed data management model. The model schedules memory according to the features of synchronous cooperative application, greatly improving the performance of data access. The model not only utilizes the main memory of server but also that of clients, which is seldom seen in other systems.

The paper is organized as following. First the paper introduces an application instance. After analyzing it, we can get some constructive hints. Then a model supporting synchronous cooperative application is presented, following the implementation. Lastly, the paper gives the conclusion.

---

[1] * The paper is supported by National Ninth Five-year Key Project 98-780-01-06-01 and National 863 Hi-tech Project 863-511-946-011, 863-511-942-016.

2. Application instance

Now we introduce a motorcycle cooperative concept design instance based on cooperative concept design. In the scenario, a group of engineers are discussing a concept design scheme of motorcycle. The data of design scheme includes designer, objectives, drawings of the motorcycle, technique requested or referred and etc. There may be several schemes for discussing. All the data is stored in database. DBMS is a client/server traditional database, which is not designed specially for synchronous cooperative applications. The engineers are discussing under the help of some cooperative tools such as application cooperation tool[3] and phone tool[4] developed by our research group. The drawings of the motorcycle is the focus of the debate. We should notice the size of drawings is of several hundred bytes or less, rather small compared with the size of PC's memory. The declaration is true for a large amount of files in concept design. The application cooperation tool reads drawing files from the database after it starts up. In the course of the discussing, some engineer may be interested in the technique referred of the motorcycle. Then the data concerned will also be fetched from database for him. All the update to the schemes will be write into the database. Some information such as who participate, their opinions and their operations is valuable for later use. So the related data will be stored in database too. In the case of no agreement with what is the best of all, different schemes need to be given repeated comparison. Phone tool helps to enforce communication. However, voice data is discarded as it will effect the rapid reaction of the application if we store it in database. Some engineer will participate the discussion in later time. He will synchronize with others after he get enough information from database.

There are some hints behind the scenario that should be noticed. First of all, each application cooperation tool should read the drawings from database. Some engineers may get the data at the beginning of the discussion. Some may join in at the middle time. In either of the situations, DBMS will read the drawings from disk and return to the client. In fact, application cooperation tool knows it clearly that some data such as the drawings should be used by all the participants. It is a pity DBMS does not know that, otherwise DBMS would utilize it to make data access more quickly. During the period of choosing the best scheme, engineers compare all the candidate schemes repeatedly. Some scheme denied , not interesting any more may be the focus of the talk in some later time, which can be due to the dynamism of cooperation. Now the client will ask for the data from DBMS and DBMS will read data from disk once more and return it. So Re-read is commonly happened in cooperative applications while it is rare in non cooperative ones. Data update is somewhat unusual in cooperative application too. During discussion, all the participants can modifying a map of drawing concurrently. Suppose there is a scenario in which Engineer A and Engineer B read a drawing file from database and begun to modify it. After some time, Engineer A write the update to the database. They continue modifying. Then Engineer B try to write the modification to the database. He will fail however. Because DBMS declare that Engineer B read some data which is updated by others, which is "dirty data". In fact, the drawing data shared by both of them is synchronous by the application. It is reasonable to let Engineer B finish "write" operation successfully.

From the analysis above we can see that there are large amount of read/write request for database as it is the base of sharing for cooperative applications. If there is no consideration for synchronous cooperative application in DBMS, it is difficult for cooperative applications to make rapid reaction in the interaction. On the other hand, there are repeated read/write requests for database in cooperative applications because a group of people are focused on something interesting. Furthermore, the applications can predict it sometime. We can utilize the feature to design DBMS. Lastly, traditional transaction model based on serializability is too strict for cooperative applications. Some adaptation is needed.

3. Main memory based distributed data management model

Main memory based distributed data management model is presented to improve the performance of DBMS. Because there is large amount of repeated read/write in synchronous cooperative applications, it is rewarded greatly if data needed are kept in the main memory all the time. The model is somewhat different from other main memory systems in which some part of data is shipped to client main memory for later query operation. The structure of data access is in figure.1 in which all processes included in an ellipse run in the same machine.

DMS(Data Management Server) maintains primary data copy and is responsible for consistency of data in the process of transactions. It keeps most frequently used data in main memory in order to access data quickly. LDM(Local Data Manager) is responsible for the data request from local site. When there is request from applications, LDM checks if it can get the data from local main memory to fulfill the request. If the answer is yes, LDM returns data to the application and at the same time send some tags to the DMS. Those tags are used for the maintenance of consistency of data. If there is no data in local site, LDM asks for DMS. DMS calculate the query and return the result. For some cases, DMS may pre-read data and send them to LDM.

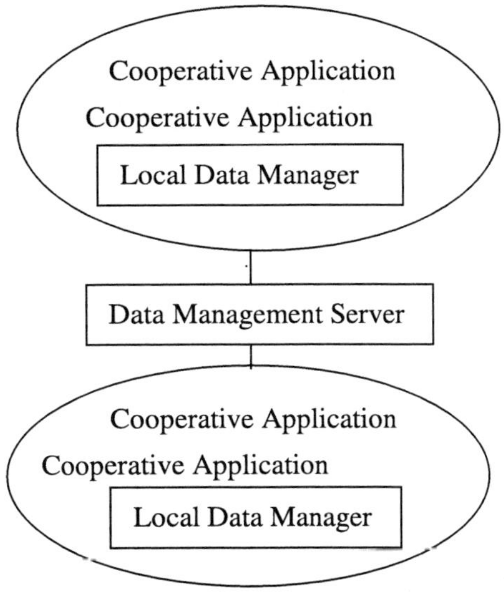

Figure 1. Structure of the Model

How can DMS or LDM decide which data is needed frequently? In fact, cooperative application can give some hints sometimes. Data can be divided into three category according to its possibility of re- read/write. Class 1 is high possibility. In the scenario we talked above, the drawing file is the focus of discussion. It is natural for all the participant applications to read it from DBMS. So we can say the reusability is high. Class 2 is that applications can not decide the possibility. Data concerned technique referred in the scenario is an example of Class 2. Class 3 is low possibility. Voice data stored in database just now is not likely to be read in the discussion.

We can define weight for data reusability. Heavy weight of data means it is worth to be kept in main memory.

Function1 WL(RR,RL,O)
WL: Weight of data in Local site
RR: Reusability of data for Remote sites
RR can be one of RH□High possiblity□, RU (Undecided) and RLow(Low posssiblity)
RL: Reusability of data for Local site
RL can be one of LH(High possibility), LU(Undecided), LLow(Low posssiblity)
O: current request Operation
O can be one of read, update and insert

It is natural that local site keeps the data that will most possibly reused by local site. So RL is the most decisive factors to calculate WL. Some other consideration for WL is data read tends to be more likely reused than data written because of the observation that re-read is commonly happened in cooperative applications and data is usually first read then updated. See table 1 for the detailed value of WL.

Table 1. WL(RR,RL,O)

	LH		
	RH	RU	RLow
read	23	22	19
update	23	21	18
insert	23	20	17

Table 1. WL(RR,RL,O) (Continues)

	LU		
	RH	RU	RLow
read	16	15	12
update	16	14	11
insert	16	13	10

Table 1. WL(RR,RL,O) (Continues)

	LLow		
	RH	RU	RLow
read	9	6	3
update	8	5	2
insert	7	4	1

Function 2 WS(RR,RL,O)
WS: Weight of Server site
RR, RL, O: the same as in WL.

WS is calculated in the server. The consideration for WL is still true for WR. However It is a little different that RL and RR is treated equally in the point view of server because both of they are the clients. See table 2 for detail.

Table 2. WS(RR,RL,O)

	RH		
	LH	LU	LLow
read	18	15	12
update	17	14	11
insert	16	13	10

Table 2. WS(RR,RL,O) (Continues)

	RU		
	LH	LU	LLow
read	15	9	6
update	14	8	5
insert	13	7	4

Table 2. WS(RR,RL,O)(Continues)

	RLow		
	LH	LU	LLow
read	12	6	3
update	11	5	2
insert	10	4	1

When cooperative application query data from database, it must declare the value of RR and RL in order that LDM and DMS can calculate WS and WL. If both RL and RS is zero, then WL and WS is zero too. It means the application is not a synchronous cooperative application.

4. Implementation

We discuss the implementation of the model based on relational database. The technique is applicable to object-relational database or OODB. Data types is divided into common type and file type. Common data type includes traditional types such as char, integer, string, float and etc. File data type is used to store large block of binary data. The storage of common type data and file type data is different. A field of file type contains some description information and pointer to content of a file which is of size less than a hundred bytes. The content of the file stores in file system.

There are three problems that need to be answered in the implementation. The first one is which data need to keep in the main memory. The second one is how to choose swap-out data when there is no enough memory for current query. The third one is how to maintain the data consistency between DMS and LDMs. DMS and LDM take different responsibilities. The consideration for their implementation is different too. LDM is expected to be simple and occupy least resource so as it reaches the objectives. Because the platform running LDM varies from low configuration to high configuration, the design of LDM should have good scalability. The design of DMS focus on the global performance and the consistence of data.

Concerning to the first problem, the answer is once there is free memory, data processed in the query can be kept in main memory. However common type data and file type should be treated somewhat a little differently. In the case of common type data, the detailed process is listed below under the suppose that main memory is large enough to hold the data.

1 read operation

If LDM has enough data to process the query, it calculates the query and returns the result. Otherwise LDM forwards the request along with WL and memory status of LDM to the DMS. After receiving the request, DMS calculate the query. Query is optimized to get better performance. However the query process is much simple in LDM. Most query is done quickly by ID hash in LDM. Complex query(such as join operation with more than three tables) is always forwarded to DMS. DMS returns the result once it gets it. Then DMS decides whether it send the tables concerned to the LDM. The factors related include WL, memory status of LDM, size of the tables and the operation type such as if it is complex query. The operation type is considered because it is mostly likely that the next operation on the tables is similar to this one.

2 update operation

When some data need to be updated, it is most likely that the data has been read, which suggests that data has most likely been in the main memory. If it is not, LDM just get data from DMS. The process is similar to read operation and simpler perhaps because there is no complex query.

3 insert operation

When a record need to be inserted into a table, the table is not necessary in the local site. But information such as the definition of the table, the current biggest ID and etc is required to check data type and get ID of the newly inserted record. The new ID should be verified in DMS. Because the information above is fetched from DMS, it is in the main memory of DMS. So the verification will be processed without reading from disk..

Now let's see how to process query related with file type data. Read and update operations means read/write file in its common meaning. Insert operation refers to the first check in the file in database. People use common type data and file type data in different way. Common type data can be processed in batches such as join several tables. However files in file type data are opened separately. One will not open too many files at the same time, especially in the cooperative work. When we read/write a file, the table concerned is not necessarily loaded into memory. But as insert is concerned, the table is needed because the information related with the file should be write in the corresponding field in the table. Because video and audio file occupy too much room, how to put them in memory should be more carefully planned. We stipulate that if the reusability of a audio/video is high and the size of it is less than 25% of free memory, it will be kept in the memory. When a audio/video file is to be inserted, LDM put it in disk not in memory. We will see it is reasonable later.

Memory will not be always large enough to hold all the data. Some of them must be swapped out. WS and WL is only one of factors to determine the priority of keeping data. The others factors include the tables related with uncommitted transactions, the operation of the data, most recent access time, access frequency, data type and etc. When we choose data to swap out, the first question is to ask if the data is related with uncommitted transactions. If yes, then we'd better keep it in memory because in the later commit phrase, the data will be needed. The second important factor is WL and WS. High weight means high priority to stay in memory. The third factor is operation enforced on data. Data updated or inserted has more reason to stay in memory than data read only. Now comes the recent access time. Data not access for longest time should give the room first. It is proper that data with high access frequency still stays there. Lastly, we consider data type. Audio/video file should not occupy too much memory. They should be swapped out when there is no free memory. All the consideration is true for LDM or DMS.

Since data is distributed in different site, it is important to keep the consistency of data. We utilize a method which is similar with AACC[5,6]. LDM applies for lock when it reads/writes data. However it does not wait for the apply for the request of lock. When the transaction is to be commit, all the lock should be confirmed before the commitment is finished. Data updated or inserted is written back in the server after commitment. However, it is not true for file type data, which differentiate the method here from AACC. After commitment, file type data is still in the local site (the table concerned is written back to DMS if there is modification however). Once there is request for the file from other clients or the local application is to be ended(in fact a synchronous request is sent to LDM), the file will be sent back to the server. We can see it more clearly as we think of the voice data store in phone tool. Because the size of voice data is large, not transmitting it to the server can speed the transaction completion. Now it is much easier for phone tool to store voice data in database for it is not troubled by the long delay when store the data in other DBMS. In the instance we mentioned above, the application can define WR as LU, then voice data is written to the disk immediately instead of staying in memory. We can see the mechanism is rather appropriate.

Transaction model guaranteeing ACID[7] based on serializability is too strict for cooperative applications. We adapt it so that the read-write or write-write conflict is treated differently for persons in the same group or persons from different group. The conflict in the same group is handled much more relaxed. Perhaps sometimes it should not be called conflict at all.

5. Analysis

When we evaluate the response time of a synchronous cooperative application, we distinguish two concepts. One is local response time, which means the response time of local site incurred by an local input. The other is remote response time which means the response time of remote site incurred by the local input. It is apparent that we expect short local response time and remote response time. Furthermore, the remote response time of different sites should be in balance which means the time of one site is nearer to the time of another site, the better.

As the application instance mentioned above concerned, the local response time as long as remote response time is greatly decreased after we adopt the method in chapter 4. The data in local site which make much network transmission unnecessary. In memory data access also greatly improves performance. When multimedia data is concerned, the efficiency is great as the network transmission time is delayed when application user is not waiting for a reply.

6. Conclusion

No management or duplicate is not good solution of data management for CSCW applications in synchronous work mode. We take advantages of main memory database technology to support synchronous cooperative work.. The paper provides a main memory-based distributed data management model. The model schedules memory according to the features of synchronous cooperative application, greatly improving the performance of data access. It is a great help for synchronous cooperative application, especially in cooperative concept design.

7. References

[1] Christian Schuckmann, Lutz Kirchner, and etc, "Designing object-oriented synchronous groupware with COAST", *Computer Supported Cooperative Work'96*, Cambridge MA USA, 1996, pp. 30-38

[2] Stankovic, John; Son, Sang H.; Hansson, Jorgen, "Misconceptions about real-time databases", August 13, 1998. ftp://ftp.cs.virginia.edu/pub/techreports

[3] Xu Baomin, technique document "Design of Application Cooperation Tool", CAD Lab., Institute of Computing Technology, Chinese Academy of Sciences, 2000

[4] Gao Qiang, Ph.D. Dissertation "The Research on Technique of Voice Tool in Enterprise Environment based on ISDN", CAD Lab., Institute of Computing Technology, Chinese Academy of Sciences, 2000

[5] Michael J. Frankly, and etc, "Transactional Client-Server Cache Consistency: Alternatives and Performance", *ACM Transactions on Database Systems*, Vol.22, No.3, September 1997, pp. 315-363

[6] M. Tamer & sut, Kaladhar Voruganti, and etc, "An Asynchronous Avoidance-Based Cache Consistency Algorithm", *Proceedings of 24th VLDB Conference*, New York, USA, 1998, pp. 440-451

[7] Ullman, J., *Principles of Database Systems,* Computer Science Press, 1982.

Part VI: Workflow Management Systems

Product-driven Workflow Design

W.M.P. van der Aalst[1,2], H.A. Reijers[2,3], and S. Limam[1]
[1] *Department of Technology and Management, Eindhoven University of Technology, PO Box 513, NL-5600 MB, Eindhoven, The Netherlands.*
[2] *Department of Mathematics and Computing Science, Eindhoven University of Technology, PO Box 513, NL-5600 MB, Eindhoven, The Netherlands.*
[3] *Deloitte & Touche Bakkenist, PO Box 23103, NL-1100 DP, Amsterdam, The Netherlands.*
E-mail: w.m.p.v.d.aalst@tm.tue.nl, hreijers@win.tue.nl, s.limam@tm.tue.nl,

Abstract

In manufacturing the interaction between the design of a product and the process to manufacture this product is studied in detail. Consider for example Material Requirements Planning (MRP), which is mainly driven by the Bill-Of-Material (BOM). For information-intensive products such as insurance, loans, permits, and many other services, the relationship with the supporting workflow process is often neglected. Typically, the workflow processes are designed without careful consideration of structure and characteristics of the product. In this paper, we present a method for designing efficient and effective workflows based on the products generated by the process rather than subjective interpretations of managers, consultants, and IT experts.

1. Introduction

From a logistical point of view, there are many similarities between administrative processes and production processes (cf. Platier [13]). Both kinds of processes, focus on the routing of work (workflow) and the allocation of work to resources. In a production system, the products are physical objects and the principal resources are machines, robots, humans, conveyor belts and trucks. In an administrative process the products are often informational (e.g. documents) and most of the resources are human. Although there are many similarities, there are also some logistical aspects in which an administrative process differs from a typical manufacturing process [1]:

- *Making a copy is easy and cheap.* In contrast to making a copy of a product like a car, it is relatively easy to copy a piece of information (especially if it is in electronic form).
- *There are no real limitations with respect to the in-process inventory.* Informational products do not require much space and are easy to access (especially if they are stored in a database).
- *There are less requirements with respect to the order in which tasks are executed.* Human resources are flexible and there are few technical constraints.
- *Quality is difficult to measure.* What is the quality of the decision to accept an insurance claim?
- *Quality of end-products may vary.* A manufacturer of cars has a minimal quality level that any product should satisfy. However, in an administrative process it might be attractive to skip certain checks to reduce the workload.
- *Transportation of electronic data is timeless.* In a network information travels at the speed of light.
- *Production to stock is seldom possible.* Every product is unique, therefore it is difficult to produce in advance. It is not possible to process an insurance claim before it arrives.

Nevertheless, the two types of processes have a lot in common. Consider for example performance indicators such as throughput time, waiting time, service level and utilization. These performance indicators play a prominent part in both domains.

In manufacturing, the Bill-Of-Material (BOM) is used to drive the production process [12]. Consider for example Material Requirements Planning, often referred to as MRP-I, which determines the production schedule based on the ordered quantifies, current stock, and the composition of product as specified in the BOM. Contemporary Enterprise Resource Planning (ERP) systems such as SAP also take resource availability into account and use more refined algorithms. Nevertheless, production is driven by the structure of the product.

Administrative processes generate information-intensive products such as mortgage loans, driving permits, customs declarations, salary payments, etc. Typically, support for the process execution is provided by workflow management systems [9][10][14]. In contrast to manufacturing, the relation between the product and the process is seldom made explicit in administrative processes. Clearly, the interaction between the product and the process is in the back of everyone's mind.

However, there is not a standardized way to describe the product and the workflow process is usually designed without proper consideration of the product. Consider for example the processing of insurance claims. The product is basically a decision: Either the claim is accepted (followed by a payment) or the claim is rejected. All kinds of data elements may play a role in making this decision. One can think of these data elements as raw materials or subassemblies. The workflow process should manufacture the decision while taking criteria such as average flow time, service level, handling costs, and product quality into account. As mentioned, the latter may be hard to measure.

Driven by management principles such as Business Process Reengineering (BPR) [7][11], the focus has shifted from data and organization to processes. The problem is not this focus but the way processes are designed [15]. In many BPR projects, the existing process is taken as a starting point. In workshops, small groups of consultants, managers and specialists try and improve such a process into a new design. As a result of this approach, the processes are usually sub-optimal, incomplete, subjective, and at a too abstract level ("The devil is in the details"). A careful analysis of the product in terms of data elements and a design which is driven by the structure of the product and the characteristics of the data elements, can solve most of these problems. Therefore, we propose a method for *product-driven* workflow design.

2. Product/data structures for workflow processes

The BOM used in manufacturing is a tree-like structure with the end product as root and raw materials and purchased products as leafs. In the resulting graph, the nodes correspond to products, i.e., end-products, raw materials, purchased products, and subassemblies. The edges are used to specify composition relations (i.e., *is-part-of* relations). The edges have a cardinality to indicate the number products needed. Figure 1 shows the simplified BOM of a car which is composed of an engine and a subassembly. The subassembly is composed of four wheels and one chassis.

For information-intensive process, the traditional BOM is not very useful. As was indicated in the introduction there are several differences between informational products and physical products. First, making a copy of information in electronic form is trivial from the process perspective since it takes hardly any time or resources to do this. Therefore, cardinalities make no sense. Second, the same piece of information may be used to manufacture various kinds of new information. Therefore, also non-tree-like structures are possible. For example, the age of an applicant for life insurance may be used to estimate the health risks and the risks of work related accidents. Finally, there are no physical constraints and therefore there are typically multiple ways to derive a piece of information. For example, health risks may be estimated using a questionnaire or a full medical examination. These observations lead to the following product/data model.

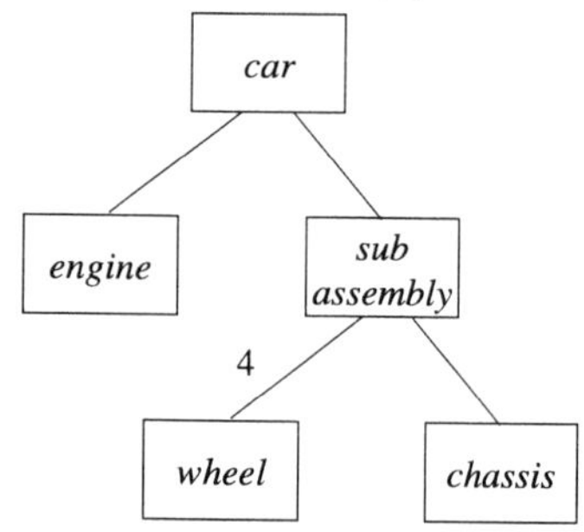

Figure 1. The BOM of a car.

Product/data model

A product/data model is a tuple (D, C, *pre*, *F*, *constr*, *cst*, *flow*, *prob*):

- D: set of data elements, $\underline{top} \in$ D,
- C: set of constraints, *true* $\in$ C,
- $pre : D \to \mathcal{P}(\mathcal{P}(D))$,
    - $R = \{(p, c) \in D \times D \mid c \in \bigcup_{es \in pre(p)} es\}$ is connected and acyclic,
    - $\forall (p,c) \in R : c \neq \underline{top}$,
    - $\forall e \in D : \varnothing \in pre(e) \Rightarrow pre(e) = \{\varnothing\}$,
- F : set of production rules, $F = \{ (p, cs) \in D \times \mathcal{P}(D) \mid cs \in pre(p) \}$,
- $constr : D \times \mathcal{P}(D) \not\to C$,
    - $dom(constr) = F$,
    - $\forall e \in D : pre(e) = \{\varnothing\} \Rightarrow constr(e, \{\varnothing\}) = true$,
- $cst : D \times \mathcal{P}(D) \not\to \mathbb{N}$,
    - $dom(cst) = F$,
- $flow : D \times \mathcal{P}(D) \not\to \mathbb{N}$,
    - $dom(flow) = F$,
- $prob : D \times \mathcal{P}(D) \not\to (0..1]$
    - $dom(prob) = F$,
    - $\forall e \in D : pre(e) = \{\varnothing\} \Rightarrow prob(e, \{\varnothing\}) = 1$,
    - $\forall (p, cs) \in F : constr(p, cs) = true \Rightarrow prob(p, cs) = 1$.

The product/data model expresses relations between data elements. These relations can be used to produce a value for the data element $\underline{top}$. The *pre* function yields for each data element *d* zero or more ways to determine a value for *d*. If we suppose for data-elements $d, e, f \in$ D, that $\{e, f\}$

∈ *pre*(*d*), then a value of *d* may be determined on basis of values of *e* and *f*. We say that (*d*, {*e*, *f*}) is a *production rule* for *d*. The expected probability that this production rule yields a result for *d* is given by *prob*(*d*,{*e*, *f*}). Furthermore, the rule can only be applied if *constr*(*d*,{*e*, *f* }) evaluates to *true*. The cost of producing a value for *d* with this rule is specified by *cst*(*d*,{*e*, *f*}). Likewise, its flow time (throughput time) is specified by *flow* (*d*, {*e*, *f* }). Note that a data element *d* for which holds that *pre*(*d*) = {∅} is special; it is called a *leaf*. No other data elements are required to determine the value of a leaf. There are also no constraints to determine the value of a leaf; the probability to determine a leaf's value is 1. Note that there may be costs associated with obtaining the value of a leaf – just as is the case for any other data element. Also note that for any data element the probability that a value can be determined is 1 if there are no constraints to determine its value (i.e., *true*).

Note that the probabilities given in a product data model are assumed to be independent. Because values can be less than 1, it is generally not ensured that the data element top can be determined for a given set of data element values. For example, suppose in a real-life situation that there are two alternative data elements that can be used to construct a value for some top, both with a probability of 0.9. Even if the values of both elements are available there is still a (1-0.9)·(1-0.9) = 0.01 probability that no value for the top element can be determined.

3. An example

An example of a product/data model is depicted in Figure 2. All nodes in this figure correspond to data elements. Arcs are used to express the *pre* relation that is in use to decide whether a candidate is suitable to become a helicopter pilot in the Dutch Airforce. Unlike the BOM in Figure 1 there are no cardinalities in effect. On the other hand, constraints, cost, flow time, and probabilities are associated with production rules. The meaning of the data elements is as follows:

- *a*: suitability to become a helicopter pilot,
- *b*: psychological fitness,
- *c*: physical fitness,
- *d*: latest result of suitability test in the previous two years,
- *e*: quality of reflexes,
- *f*: quality of eye-sight.

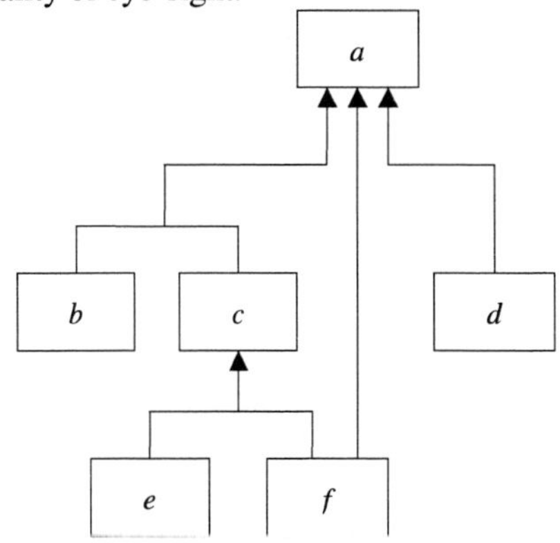

Figure 2. Helicopter pilot product/data model.

One of the things that follows from the figure is that there are three different production rules for the top element *a*. The suitability of a candidate can be determined on basis of:

- The results of the psychological test (*b*) and the physical test (*c*),
- The result of a previous test (*d*), or
- The candidate's eye-sight quality (*f*).

The relations *constr*, *cst*, *flow*, and *prob* for this example are as shown in Table 1. If *a* is a data element, the value of *a* is denoted with **a*.

From this table it follows that obtaining values for leafs is much more time-consuming in this example than other values. This represents a common phenomenon that actions that involve communication with external parties take more flow time than internal actions.

Furthermore, it can be concluded that if a candidate's eyes are worse than –3.0 or +3.0 dioptries this information can be used as a direct *knock-out* [3] for the test result. The probability that this will happen for an arbitrary case is 0.4. Note that each production rule for the top element can be a knock-out for a specific case.

Table 1 : Relations *constr*, *cst*, *flow*, and *prob* for testing a helicopter pilot candidate.

X	*constr*(*x*)	*cst*(*x*)	*flow*(*x*)	*prob*(*x*)
(*a*,{ *b*, *c* })	*True*	80	1	1.0
(*a*,{ *d* })	**d* ∈ {suitable, not suitable}	10	1	0.1
(*a*,{*f* })	**f* < -3.0 or **f* > +3.0	5	1	0.4
(*b*,{ ∅ })	*True*	150	48	1.0
(*c*,{ *e*,*f* })	*True*	50	1	1.0
(*d*,{ ∅ })	*True*	10	16	1.0
(*e*,{ ∅ })	*True*	60	4	1.0
(*f*,{ ∅ })	*True*	60	4	1.0

4. Conformance

Up to this point, no specifications are given of the production rules, other than their signature. For each element of the *pre* relation such a specification should be available to produce working process models. However, to consider the optimality of process models the exact specifications are not relevant. What is relevant is that a given process model conforms with the product/data model under consideration.

Process model

A process model PM on a product/data model (D, C, *pre*, *F*, *constr*, *cst*, *prob*) is defined by (T, *prod*, Ω) where:

- T is a set of tasks,
- *prod*: T → *F*, the production rule applied in the task,
- Ω is a set of firing sequences.

An execution sequence $r \in \Omega$ is a sequence $t_1 t_2 \ldots t_k$ of tasks with $k \in \mathbb{N}\backslash\{0\}$, such that for any $1 \leq i \leq j \leq k$ holds that $t_i, t_j \in$ T and $t_i = t_j \Rightarrow i = j$ (single executions of each task).

Conformance

A process model (T, *prod*, Ω) conforms to the product/data model (D, C, *pre*, *F*, *constr*, *cst*, *flow*, *prob*) if and only if for each sequence $r = t_1 t_2 \ldots t_k \in \Omega$, $k \in \mathbb{N}\backslash\{0\}$, holds that:

1. $\forall 1 \leq i \leq k, (p, cs) \in F : [prod(t_i) = (p, cs) \Rightarrow$
   $\forall c \in cs : \exists 1 \leq j < i, ds \in \mathrm{P}(D) : prod(t_j) = (c, ds)]$,
2. $\exists 1 \leq i \leq k, cs \in \mathrm{P}(D) : prod(t_i) = (\underline{top}, cs)$.

The first requirement ensures a proper order of the application of production rules. The last requirement guarantees that the top element *may* be produced.

5. Design criteria

Given a product/data model there are many conformant process models. Different subsets of tasks executed in some order may lead to the production of the top element. When designing product-driven workflows, the following three design criteria need to be taken into account: (1) *quality*, (2) *costs*, and (3) *time*. Costs and time are defined according to the functions *cst* and *flow*. In this paper we use a rather narrow definition of quality. Quality is defined as the probability that the value of the top element can be determined. Note that quality depends on the structure of the graph (i.e., function *pre*) and the probability that a production rule leads to a value. To allow for a formal definition of these design criteria we introduce the notion of a *plan*.

Plan

Let (*D*, *C*, *pre*, *F*, *constr*, *cst*, *prob*) be a product/data model. Any subset *S* of *D* is called a plan.

One can think of a plan as a sub-graph of the graph denoting the product/data model. The elements of *S* are the data elements that should be produced. The set $\{a, d\}$ is a plan corresponding to the product/data model shown in Figure 2. In this plan the production rules $(d, \{\varnothing\})$ and $(a, \{d\})$ are executed in some order. The set $\{a, e\}$ is also a plan although this plan will never lead to a value for data element *a*. For any given plan, we can determine the probability that a value for the top element is determined.

Quality of a plan

Let (*D*, *C*, *pre*, *F*, *constr*, *cst*, *prob*) be a product/data model. The quality of a plan $S \subseteq D$ is defined as $p\_quality(S) = q_{top}$ for all $d \in S$:

$$q_d = 1 - \prod_{(d,cs) \in F} \left(1 - \left(prob(d, cs) \cdot \prod_{e \in cs} q_e \cdot \delta(e)\right)\right), \quad \text{(i)}$$

where $\delta(e) = \begin{cases} 0, & e \notin S \\ 1, & e \in S \cup \{\varnothing\}. \end{cases}$

The quality of a plan is the probability that the value of the top element can be determined successfully assuming that all production rules only referring to elements in *S* are executed. Note that for any production rule $(p, cs) \in F$ holds that all elements in *cs* should be part of the plan in order to contribute to q_p.

Consider the product/data model shown in Figure 2 and three plans $S_1 = \{a, d\}$, $S_2 = \{a, b, c, e, f\}$ and $S_3 = \{a, e\}$. For plan S_1 holds that the quality of this plan is $p\_quality(S_1) = q_{top} = q_a$. According to formula (i), $q_a = 1-(1-prob(a,\{d\}).q_d.\delta(d))$ with $q_d = 1-(1-prob(d, \{\varnothing\}) \cdot q_\varnothing \cdot \delta(\varnothing)) = 1$. So, $p\_quality(S_1) = q_a = 0.1$. Similarly, for plan S_2, $p\_quality(S_2) = 1$ and for plan S_3, $p\_quality(S_3) = 0$.

Costs of a plan

Let $S \subseteq D$ be a plan. The costs of *S* are:

$$p\_csts(S) = \sum_{(p,cs) \in F} cst(p, cs) \cdot \delta(p) \cdot \prod_{e \in cs} \delta(e) \quad \text{(ii)}$$

The costs of a plan are simply given by the sum of all production rules costs relevant for the plan. Note that again it is assumed that production rule (*p*, *cs*) is executed if $\{p\} \cup cs$ is a subset of plan *S*. These costs can be interpreted as the maximum costs that are associated with the execution of a plan.

The costs of plans $S_1 = \{a, d\}$, $S_2 = \{a, b, c, e, f\}$ and $S_3 = \{a, e\}$ are as follows. For plan S_1 the only production rules relevant are $(a, \{d\})$ and $(d, \{\varnothing\})$. So, according to equation (ii), $p\_csts(S_1) = cst(a, \{d\}) \cdot \delta(a) \cdot \delta(d) + cst(d, \{\varnothing\}) \cdot \delta(d) \cdot \delta(\varnothing) = 20$ (see Table 1). Similarly, $p\_csts(S_2) = 405$ and $p\_csts(S_3) = 60$.

Flow time of a plan

The actual time required to produce all data elements of a plan depends on the order in which the production rules are executed. In a worst-case scenario where all production rules of the plan are executed sequentially, the total flow time is:

$$\sum_{(p,cs)\in F} flow(p,cs)\cdot\delta(p)\cdot\prod_{e\in cs}\delta(e) \quad \text{(iii)}$$

By executing some of the production rules of the plan in parallel, the total flow time can be reduced.

Consider plan $S_4 = \{a,b,c,d,e,f\}$. Assume that this plan is executed in the following order: $(d,\{\emptyset\})$, $(a,\{d\})$, $(f,\{\emptyset\})$, $(a,\{f\})$, $(e,\{\emptyset\})$, $(c,\{e,f\})$, $(b,\{\emptyset\})$, $(a,\{b,c\})$. Then the average worst case $flow\_time(S_4) = flow(a, \{b, c\})\cdot\delta(a)\cdot\delta(b)\cdot\delta(c) + flow(a, \{f\})\cdot\delta(a)\cdot\delta(f) + flow(a, \{d\})\cdot\delta(a)\cdot\delta(d) + flow(b, \{\emptyset\})\cdot\delta(b)\cdot\delta(\emptyset) + flow(c, \{e, f\})\cdot\delta(c)\cdot\delta(e)\cdot\delta(f) + flow(f, \{\emptyset\})\cdot\delta(f)\cdot\delta(\emptyset) + flow(e, \{\emptyset\})\cdot\delta(e)\cdot\delta(\emptyset) + flow(d, \{\emptyset\})\cdot\delta(d)\cdot\delta(\emptyset) = 76$ time units. Now suppose that the production rule $(a, \{d\})$ leads to a value for a, then the $flow\_time(S_4) = flow(a, \{d\})\cdot\delta(a)\cdot\delta(d) + flow(d, \{\emptyset\})\cdot\delta(d)\cdot\delta(\emptyset) = 17$ time units only. So, the average flow time of a plan may be much smaller because a value for a data element can be obtained before all elements of the plan are derived.

6. Product-driven reengineering rules

In the previous section we introduced the notion of a plan. Given a plan S, it is easy to calculate its quality $p\_quality(S)$ and the associated costs $p\_costs(S)$. To calculate the total flow time of a plan is more complex, because one has to make assumptions about the order in which the production rules are executed. Note that a plan is not a process model: it is merely a subset of data elements. However, the notion of a plan and the criteria $p\_quality(S)$ and $p\_costs(S)$ can be used for a heuristic approach towards product-driven workflow design. The heuristic uses the fact that $p\_quality(S)$ and $p\_costs(S)$ allow for the definition of a cost optimal plan given a quality level.

Cost optimal plan

Let $(D, C, pre, F, constr, cst, prob)$ be a product/data model and $q \in [0, 1]$ be a quality level. Plan $S \subseteq D$ is cost optimal if and only if

1. $p\_quality(S) \geq q$, and
2. $\forall S' \subseteq D$: $p\_quality(S') \geq q \Rightarrow p\_csts(S') \geq p\_csts(S)$.

Consider R the set of plans that can be derived from the product/data model of Figure 2. $R = \{S_1, S_2, S_3, S_4, S_5\}$ where $S_5 = \{a, f\}$. Assume $q = 0.8$. We already know the quality level of plans S_1, S_2 and S_3: $p\_quality(S_1) = 0.1$, $p\_quality(S_2) = 1$, and $p\_quality(S_3) = 0$. It is easy to calculate the quality level of plans S_4 and S_5: $p\_quality(S_4) = 1$ and $p\_quality(S_5) = 0.4$. Only plans S_2 and S_4 fulfill condition (1). For those plans, costs are $p\_csts(S_2) = 405$ and $p\_csts(S_4) = 425$. According to the definition of cost optimality, it appears that plan S_2 is the cost optimal plan.

A cost optimal plan gives the least costly subset of data elements that needs to be calculated to obtain a given quality level. Note that the costs associated to such a plan are the maximal costs, i.e., the costs that are made if all corresponding production rules need to be calculated. However, a knock-out may result in a value for the top element before the whole plan is executed. Therefore, it is important to order the production rules. The ordering of the production rules can be based on the time criterion or on the cost criterion mentioned in the previous section. The two extremes are:

1. *Breadth-first.* Start with the leaf nodes in the plan and execute as many production rules in parallel as possible.
2. *Depth-first.* Start with the part of the plan which has the best quality/cost ratio, i.e., execute the production rules sequentially and start with the most promising branches first.

Assuming sufficient capacity the breadth-first approach optimizes the process with respect to flow time but at high costs (in principle all production rules associated to the plan are executed). The depth-first minimizes the average costs but may result in substantial longer flow times. For the breadth-first approach there is no need to order the activities executing the production rules: The graph structure is used to maximize parallelism. For the depth-first approach these activities need to be ordered sequentially. To decide on the ordering of activities within a given plan S, we introduce the notion of a cost optimal chain of plans.

Cost optimal chain of plans

Let $(D, C, pre, F, constr, cst, prob)$ be a product/data model and $q \in [0, 1]$ be the required quality level. Let $S \subseteq D$ be cost optimal with respect to q and $1 \leq K \leq |S|$. S_1, S_2, ..., S_K is a *cost optimal chain of plans* if and only if

1. $S = S_K \supseteq S_{K-1} \supseteq S_{K-2} \supseteq ... \supseteq S_1$
2. $\forall\ 1 \leq i \leq K$, $S' \subseteq S$: $p\_quality(S') \geq p\_quality(S_i) \Rightarrow p\_csts(S') \geq p\_csts(S_i)$

A cost optimal chain gives the order in which the production rules should be executed. First execute the activities associated to S_1 in parallel, then the activities associated to S_2 but not S_1 in parallel, etc. The number K gives the number of steps in this sequential process. If K equals the number of elements of the set S, then the plan is executed sequentially, i.e., a pure depth-first strategy. If K=1, a pure breadth-first strategy is used.

If we consider the example of Figure 2 and a quality level $q = 0.8$, the cost optimal plan was $S_2 = \{a, b, c, e, f\}$. A breadth-first strategy leads to executing production rules $(e, \{\varnothing\})$, $(f, \{\varnothing\})$ and $(b, \{\varnothing\})$ in parallel. Then execute production rule $(c, \{e, f\})$ and finally $(a, \{b, c\})$. A depth-first strategy leads to executing production rules $(b, \{\varnothing\})$ then $(e, \{\varnothing\})$ and $(f, \{\varnothing\})$ in parallel, then $(c, \{e, f\})$ and finally $(a, \{b, c\})$.

A cost optimal plan identifies the class of process models that are best suited to implement a given product specification. Depending on the preferred ordering of production rules, a specific process can be derived.

7. Conclusion

In this paper, we presented a new way of looking at workflow process design. By taking the product as a starting point many of the problems mentioned in the introduction can be avoided. A formal model, a corresponding notion of conformance and a quantification of design criteria have been given. Finally, a heuristic approach towards product driven workflow design has been presented.

By now, Deloitte & Touche Bakkenist applied the method presented in this paper four times in client engagements in 2000 and 2001. Business processes of a large Dutch bank and a national social security agency have been reengineered. In each occasion, at least a 50% cost reduction and a 30% flow time reduction has been achieved, while maintaining the same quality level.

The method of product-driven workflow design is a promising new way of executing BPR initiatives. A practical boost would come from the development of tools to model product/data models and derive process models. Further research can lead to the incorporation of other relevant design criteria, techniques, and algorithms.

In Section 2, we compared the classical BOM with our product/data model for workflow processes. Current ERP systems allow for so-called *generic/variant* BOM's [6][8]. This way, it is possible to deal with large product families having millions of variants [16]. As is shown in [2], this concept can be translated to workflow processes facing ad-hoc and structural changes. It is interesting to extend the approach presented in this paper to deal with product families. Another topic for future research is the issue of variable cardinalities. In many cases, multiple instances of the same type of data element are required, e.g., multiple eyewitness reports for an insurance claim. The product/data model presented in Section 2 does not take these cardinalities into account.

8. References

[1] W.M.P. van der Aalst. On the automatic generation of workflow processes based on product structures. *Computers in Industry*, 39:97-111, 1999.

[2] W.M.P. van der Aalst. Generic Workflow Models: How to Handle Dynamic Change and Capture Management Information. In M. Lenzerini and U. Dayal, editors, Proceedings of *CoopIS'99*. IEEE Computer Society Press, pages 115-126, 1999.

[3] W.M.P. van der Aalst. Reengineering Knock-out Processes. *Decision Support Systems*, 30(4):451-468, 2001.

[4] W.M.P. van der Aalst, J. Desel, and A. Oberweis, editors. *Business Process Management: Models, Techniques, and Empirical Studies*, volume 1806 of Lecture Notes in Computer Science. Springer-Verlag, Berlin, 2000.

[5] J.A. Buzacott. Commonalities in Reengineered Business Processes: Models and Issues. *Management Science*, 42(5):768-782, 1996.

[6] F. Erens, A. MacKay, and R. Sulonen. Product modelling using multiple levels of abstraction - instances as types. *Computers in Industry*, 24(1):17-28, 1994.

[7] M. Hammer and J. Champy. Reengineering the corporation. Nicolas Brealey Publishing, London, 1993.

[8] H.M.H. Hegge. *Intelligent Product Family Descriptions for Business Applications*. PhD thesis, Eindhoven University of Technology, Eindhoven, 1995.

[9] S. Jablonski and C. Bussler. *Workflow Management: Modeling Concepts, Architecture, and Implementation.* International Thomson Computer Press, 1996.

[10] P. Lawrence, editor. *Workflow Handbook 1997, Workflow Management Coalition*. John Wiley and Sons, New York, 1997.

[11] R.L. Manganelli and M.K. Klein. The Reengineering Handbook: A Step-by-step Guide to Business Transformation. Amacom, 1996.

[12] A. Orlicky. Structuring the bill of materials for MRP. *Production and Inventory Management*, pages 19-42, Dec 1972.

[13] E.A.H. Platier. *A logistical view on business processes: BPR and WFM concepts (in Dutch)*. PhD thesis, Eindhoven University of Technology, Eindhoven, 1996.

[14] G. Poyssick and S. Hannaford. Workflow Reengineering. Adobe Press, Mountain View, CA, 1996.

[15] H.A. Reijers and K. Voorhoeve. On the Optimal Design of Processes and Information Systems (in Dutch). Informatie, 42:50-57, December, 2000.

[16] E.A. van Veen and J.C. Wortmann. Generative bill of material processing systems. *Production Planning and Control*, 3(3):314-326, 1992.

Workflow Interoperability – Enabling E-Business*

Yan Zhang[1], Meilin Shi[1], Chunyu Miao[1], Zixin Zhuang[2], Xinxiang Chen[2]
[1] *Department of Computer Science and Technology*
Tsinghua University, Beijing 100084, P.R. China
[2] *China Information Highway Corporation, Beijing 100032, P.R. China*
[zyan, shi, mcyu]@csnet4.cs.tsinghua.edu.cn; [zhuangzx, xxchen]@capinfo.com.cn

Abstract

Workflow interoperability is such a significant problem that can almost decide whether E-Business can be conducted successfully. In this paper, we first describe a typical, simplified E-Business flow to analyze the collaborative work in it and introduce the workflow interoperability problem. We define the collaborative model of E-Business as a hierarchical structure, which can also be regarded as the sketch map of the workflow interoperability in E-Business. Then two typical approaches to implement workflow interoperability are discussed. Our opinion is to combine the two methods and apply them to the different hierarchy. Workflow interoperability standards are presented in section 4.

1. Introduction

Propelled by the explosive growth of information technology, E-Business has been turned into the strategic imperative of today. And it has caught the attention of the whole world in unprecedented ways.

Unlike the traditional commerce, E-Business is to conduct business electronically, whether it is over the Internet or through other networks between businesses or between consumers and businesses. E-Business needs the deployment of business processes electronically and automatically for which workflow is obviously a key supporting technology. As the entities involved in an E-Business flow need to work cooperatively, interaction among different partners in an enterprise or cross enterprises have to be dealt with. Consequently, workflow interoperability is such a significant problem that can almost decide whether the E-Business flow can be conducted successfully.

In this paper, we first introduce the workflow interoperability problem by analyzing the collaborative work in a typical simplified E-Business flow. Then workflow interoperability related problems including the collaborative model of E-Business, approaches to implement interoperability are discussed in section 3. Finally, workflow interoperability standards are presented in section 4.

2. A typical simplified E-Business flow

In the past, manufacturers were the drivers of the market, which controlled the pace of production and distribution of goods. Today, customers are much more demanding and have taken ownership of the market in many industries. In such a competitive climate, enterprises must constantly seek ways to improve their business processes to meet the rising customer's demands for customization, high efficiency, and even the ability to follow the progress of order fulfillment.

2.1. Business flow analysis

The following is a typical simplified business flow, which has been used in various forms as the basis of a number of interoperability experiments and demonstrations by WfMC member companies. It describes the life history of an order fulfillment for a product (let's say a new car) from a retailer (lets say a mail order company), which places orders with manufacturers without stock. The retailer must therefore make arrangements with the manufacturer to produce the goods (a new car) and with a transport company to deliver the goods to the customer. If the retailer, manufacturer, and the transport company belong to the same virtual enterprise, the cooperative work among them is in an autonomy domain. Thus, E-Business flow will be accomplished in a virtual enterprise. Contrarily, if they are not the partners of one virtual enterprise, E-Business needs interoperability across virtual enterprise.

* The paper is supported by China National Science Foundation. (No.60073011).

Each entity in the business flow uses a workflow management system to manage business processes, which separately operates to deliver its contribution towards order fulfillment. It is a system that completely defines, manages and executes workflow processes through the execution of software whose order of execution is driven by a computer representation of the workflow process logic. To fulfill the business flow mentioned above, the following activities are to be accomplished: (as shown in Figure 1)

1. Retailer: place an order
2. Manufacturer: confirm the order
3. Retailer: pay for the car
4. Manufacture: give the invoice
5. Manufacturer: produce the car
6. Manufacturer: schedule delivery
7. Transport Company: confirm the delivery
8. Transport Company: transport the car
9. Retailer: pick up the car

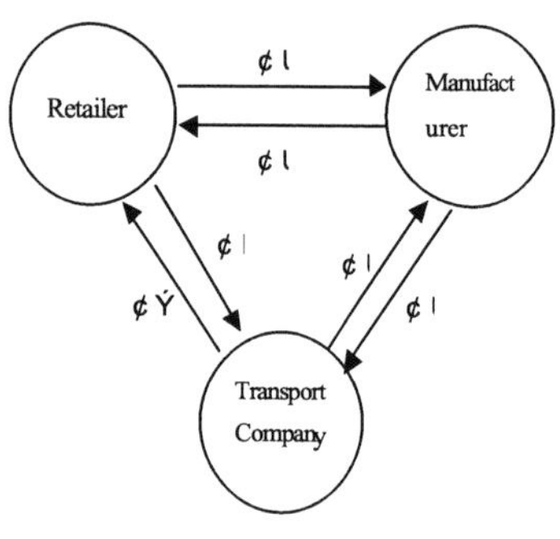

¢ Place an order ¢ Confirm the order ¢ Schedule delivery
¢ Confirm the delivery ¢ Transport ¢ Pick up

Figure 1. A typical simplified business flow.

When the retailer decides to raise an order, the first activity of its predefined process definition is to send a request to the workflow engine of the manufacturer, in which the requirements of the product, such as the color of the car, the material of the seat, acoustics in the car etc. are detailed. Once the manufacturer receives the message, its system will check whether it possesses the manufacturing capability. If it can meet the requirements of the retailer, it will send a response to confirm the order and its order fulfillment process is to be invoked. If it cannot, the manufacturer can only say sorry if the retailer would not change its requirements. When the retailer receives the order confirmation, the next step is negotiation, during which price, assurance, transport and other clauses are to be discussed and signed in a contract. Here, we ignore other factors and only focus on transport (suppose they decide that the manufacturer should be in charge of transport.). Then the system of the retailer will print the cheque and send it to the manufacturer. When the manufacturer receives the cheque, its system will give an invoice to the retailer automatically. During the production process, the manufacturer will contact with the transport company. When it is done, bill of lading will be sent to the retailer, which is the voucher for picking up the goods. All of these processes are accomplished by the workflow management system according to the predefined process definition.

When the workflow engine of the retailer instantiates the process of placing an order, it will trigger a sub-process instance on the workflow engine of the manufacturer. The manufacturer will check its resources to determine whether it possesses the capability to meet the requirements of the retailer. The workflow of the retailer must wait for the termination of the sub-process instance enacted on the workflow engine of the manufacturer before carrying on with its own process enactment. This kind of interoperability is an application of a nested sub-process model.

If the manufacturer is capable to manufacture the ordered car, an order confirmation will be sent to the retailer and then both of them will prepare to negotiate. They may simultaneously enact process instances like some preparing works and their workflow engines will need some interchange at the negotiation point in the definition of each process before continuing the enactment of their respective process instances. They must wait for each other until enactment of both process instances have achieved the respective rendezvous points. This process of interoperability can be modeled as a parallel- synchronized model.

The chained processes model of interoperability is also embodied in the order fulfillment process. For example, when the workflow of the retailer prints the cheque and sends it to the manufacturer, it will continue with other activities without waiting for the invoice sent by the manufacturer.

The analysis of the business flow shows that workflow interoperability is closely related to the cooperation of the participants of E-Business and the collaboration in the complex business flow can be decomposed to the three basic models of workflow interoperability, namely nested sub-process model, chained processes model and parallel-synchronized model.

2.2. Relevant data and operations

To implement the order fulfillment process, relevant data must be passed between the workflow management systems of the participants. As for the retailer and the manufacturer, the following relevant data should be transferred between their workflow systems.

- Order number

- Name of the retailer
- Name of the manufacturer
- Contract number
- Product code
- Requirements of the car, such as the color, the material of the seats, and acoustics etc.
- Delivery date
- Delivery address

If the retailer wants to be able to track the production process, data of production state are also need to be transferred.

And the basic operation primitives are as follows:

Workflow Engine of the retailer:
Create Process Instance
Workflow Engine of the manufacturer:
Response to Create Process Instance
Workflow Engine of the retailer:
Set Process Instance Attributes
Start Process Enactment
Workflow Engine of the manufacturer:
Response to Set Process Instance Attributes
Response to Start Process Enactment
Workflow Engine of the manufacturer:
Notify Process Instance Attribute Changed
Workflow Engine of the retailer:
Response to Notify Process Instance Attribute Changed
Workflow Engine of the retailer:
Get Process Instance Attributes
Workflow Engine of the manufacturer:
Response to Get Process Instance Attributes
Workflow Engine of the retailer:
Relinquish Process Instance
Workflow Engine of the manufacturer:
Response to Relinquish Process Instance

3. Workflow interoperability in E-Business

With the rapid development of the economy and society, the environments in which millions of enterprises struggle to find a foothold become very variable and the voices of customers are no longer identified. To meet the various and customized needs of the markets, the traditional enterprises have to change their business model to a brand-new one, which is the virtual enterprise [1]. It is a business coalition in which manufacturer combines with suppliers, channel members and customers to develop and sale customized products as quickly as possible. Virtual enterprises may use different hardwares and softwares to make the whole business flow automatic, which must support the cooperation among the virtual enterprise partners. And interoperation across virtual enterprises is also necessary. Each virtual enterprise can be regarded as an autonomy domain.

3.1. Collaboration model in E-Business

According to the type of participant, E-Business can be categorized to two types: business to business (B2B) and business to customer (B2C). (Here, government purchasing is ignored.) In B2B, virtual enterprises or their partners are the main participants, which may act as the buyer, vendor or some other roles. To fulfill any task in E-Business, technologies must be provided to support collaboration across virtual enterprise and virtual organization. In other words, collaboration across autonomy domain is so necessary that related technologies must be provided to support cross-organizational interaction. Similarly, partners inside a virtual enterprise also need to cooperate to fulfill tasks. That is also the case to the parts in a virtual enterprise partner.

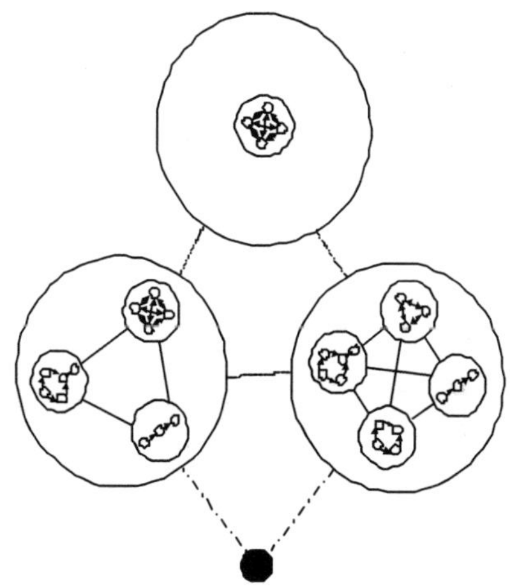

Figure 2. Collaboration Model in E-Business.

As shown in Figure 2, collaboration in E-Business is modeled as a hierarchical structure. The biggest circle represents autonomy domain. The middle one is the virtual enterprise partner. The smallest circle represents the basic unit of the virtual enterprise partner. In B2C mode, a node is used to represent the customer. The hierarchical structure is not static but dynamic, which means the partners on a hierarchy can be freely grouped according to the actual demand. Even partners on different hierarchy can exchange their positions.

Different circles represent different hierarchies and each hierarchy needs its entities to work cooperatively. As a result, the collaboration model can also be regarded as a workflow interoperability sketch map. Entities involved in each circle are coupled by various modes, which can partially decide the level of collaboration and the ways of interoperability among them. For example, inside a virtual enterprise, the manufacturer may need to activate and control the processes of the suppliers and the latter may need to return some messages about the process execution. Such collaborative level is higher than the one that only needs information sharing, so their implementation approaches will also be different.

3.2. Available workflow interoperability solutions

Common protocols and standard APIs are the basis to implement workflow interoperability. According to the function modules of WfMS, WfMC has established several standards on workflow interoperability (see section 4). On the basis of existent standards, there are two typical solutions to make workflow interoperate.

3.2.1. Agent based/enhanced workflow interoperability

Agent based workflow systems have been developed by a number of teams. In agent based workflow, the software agents take full responsibility for process provisioning, execution and compensation with each agent managing and controlling a given activity or set of activities. However, agent based workflow is likely to incur huge reengineering costs and training costs, and the process logic cannot be represented explicitly outside the agents.

In order to overcome the shortcomings of agent based workflow, agent enhanced workflow is proposed by Shepherdson [2]. Agent enhanced workflow is achieved by combining a layer of agents with a commercial workflow system. The agent layer is responsible for business process management, whilst the underlying workflow system handles process execution. Thus, workflow information is obtained by a Zeus agent and passed to the broker, which converts it to process interchange format (PIF). The information is transmitted to other brokers using Foundation for Intelligent Physical Agents (FIPA) Agent Communication Language (ACL), and converted to one or more Zeus goals. As each goal is asserted, a plan is formed and executed by Zeus agents, resulting in some change to the state of the agency's underlying workflow.

3.2.2. Contract-based workflow interoperability

This approach is based on the idea that the basis for cooperation in virtual enterprise is the contract, in which the encapsulated service and cooperation support services can be completely specified. The most typical contract-based system that has implemented workflow interoperability is the CrossFlow [3].

In CrossFlow, Partially Filled Contract (PFC) is used by service provider to advertise the service he can provide and the contract search is used by service consumer to search the service he needs. By matching the service demand and supply, the consumer may use the services offered by various external providers to outsource parts of its business process. XML is chosen as the language basis for contract specification and IES, which is a key factor to implement workflow interoperability in CrossFlow. Qos, Level of Control and Flexible Change Control are three areas of advanced cooperation support functionality to complement basic workflow interoperability.

3.3. Our opinions and future work

The methods mentioned above to implement workflow interoperability are both on the basis of the existent workflow management systems. There is another indirect way to achieve workflow interaction that is to establish a common workflow management platform.

As many existent enterprises have selected various workflow products to make their business process automatic and have invested a lot to make the workflow management system suitable for their own environments, enterprises would not change their workflow management systems easily due to the big shift cost. Consequently, the plan to establish a common workflow management platform for all enterprises is unpractical.

Although both of them can provide cross-organizational collaboration, the agent based/enhanced workflow and contract-based workflow still have some differences. In our opinions, these two methods suit different cross-organizational cooperation. Agent based workflow interoperability is appropriate for loose-coupled enterprises and the latter is suitable for tight-coupled enterprises.

In the collaboration model shown in Figure 2, outermost hierarchy is often coupled relatively loosely and partners inside the virtual enterprise are often coupled more tightly. So we would like to combine these two approaches and apply them to different hierarchy. In other words, we choose software agents as the basic tool to enable interoperation across virtual enterprise and adopt contract as the basis to implement workflow interoperation among the partners inside a virtual enterprise.

We are to design a self-organization architecture to further analyze the hierarchical structure of the collaboration model of E-Business. Then we will develop a mechanism to solve the workflow interoperability problem on different hierarchy by combining the two methods mentioned above.

4. Workflow interoperability standards

In the emerging networked economy, there is a wide range of potential benefits motivating today's enterprises to change their traditional business model to E-Business. It not only needs the efforts of numerous companies but also support from related organizations, government and even the whole society. To make the enterprises be able to cooperate with each other harmoniously, workflow interoperability standards must be enacted and abided by workflow product vendors.

The WfMC, an organization comprising of vendors, users, consultants, and researchers, has become the primary standardization organization in the workflow management system arena. Since the definition of Workflow Reference Model in 1994, it has defined several workflow interoperability standards.

4.1. Workflow standard – interoperability abstract specification

This document (Version 1.0) is published on Oct.20, 1996 by the WfMC [4]. It is an abstract specification that defines the functionality required to support interoperability between different workflow engines. It aims to provide a basis for early experimentation by workflow product vendors and help them to understand the principles of how interoperability between workflow engines is effected using the WfMC standards. Required mechanisms are defined for the workflow product vendors in order that one workflow engine may make requests of another workflow engine to effect selection, instantiation, and enactment of known process definitions by that other engine. The requesting workflow engine should (optionally) be able to receive back status information and the results of the enactment of the process definition and get the status data from the requested workflow engine.

In the interoperability abstract specification, levels and models of interoperability are introduced and the specification of operators that effects interoperation is detailed. Implementation issues and evaluation criteria are also discussed. It deals with the most basic problems in workflow interoperability, but it is still far from the requirements to instruct practical implementation.

4.2. Workflow standard – interoperability Internet e-mail MIME Binding

To effect interoperability as defined in interoperability abstract specification [WfMC1012], this document is enacted to provide a binding that gives concrete type definitions and message formats mapped onto a simple text interface, using Internet e-mail with MIME encoding as the transport mechanism [5]. From the first version published in October 1996, three versions have been enacted and the latest one is the version 2.0 issued on January 7 2000, which corrects a number of issues, errors and omissions in the version 1.0 document.

This MIME binding of standard specifies the protocol for exchanging interoperability requests and responses via e-mail. Several workflow management system vendors have applied the MIME-interoperability standard in prototype workflow engines by passing a set of test cases for proving conformance with the standard. The version2.0 of the MIME Binding only supports implementation of simple chains and nested sub-process interoperability scenarios. More complex interoperability scenarios may be achievable, but are not defined here, as they are the subjects of ongoing work by the Workflow Management Coalition.

4.3. Workflow standard – interoperability Wf-XML binding

This document represents a specification for an XML language designed to model the data transfer requirements set forth in the WfMC's Interoperability Abstract specification. It is the intention of this specification to describe a language that can be used to achieve the two basic types of interoperability defined in the abstract specification, namely simple chained workflows and nested workflows [6].

The Wf-XML specification enhances some of its predecessors' capabilities, such as:

- Providing a structured and well-formed XML body protocol that consists of message structures containing headers and data;
- Supporting these two types of interchange both synchronously and asynchronously;
- Being independent of any particular implementation mechanism, such as programming language, data transport mechanism, platform, hardware and so on;
- Easy extensibility through the use of XML and dynamic workflow context data.

WfMC' workflow interoperability standards have provided powerful support for collaborative work in E-Business. And there are some other interoperability related standards, such as jointFlow, SWAP (Simple Workflow Access Protocol) etc. As no single standard can address all aspects of interactions between business partners in an E-Business scenario, a variety of standardization efforts that deal with specific facets of the overall problem are ongoing.

5. Summary

As most enterprises have selected and will select different workflow management system to manage their business flow, workflow interoperability becomes quite an important problem to enable E-Business. Obviously, the emerging markets have put forward this demand in front of the workflow product vendors.

In this paper, the collaborative model of E-Business is defined as a hierarchical structure and each hierarchy needs its entities work cooperatively. In other word, E-Business needs interoperation on different hierarchy and workflow products should correspondingly provide interoperability functionality on different hierarchy.

The WfMC, acting as the core standardization organization has been endeavoring to enact workflow interoperability standards to provide a basis for workflow management system vendors. How to enable workflow products successfully support E-Business is also what they are focused on. However, establishment of the standards also depends on the further analysis of the functionality of workflow and the special needs of E-Business.

E-Business, which is pervasive in the future networked economy, has brought huge commercial opportunities to workflow products, but workflow interoperability still has a lot of problems need to resolve.

6. References

[1] Meilin Shi, Yan Zhang, "Virtual Corporation: A New Business Model for the 21st Century", *Journal of China Institute of Communications*, 1999, 20(9): 99-104.

[2] Shepherdson J.W., Thompson S. and Odgers B, "Cross Organizational Workflow Co-ordinated by Software Agents", *Work Activity Co-ordination and Collaboration Workshop paper*, February 1999.

[3] http://www.crossflow.org/

[4] WfMC-TC-1012, *Workflow Standard – Interoperability Abstract Specification, Version 1.0*, Workflow Management Coalition, 20-October-1996.

[5] WFMC-TC-1018, *Workflow Standard – Interoperability Internet e-mail MIME Binding, Version 1.2,* Workflow Management Coalition, 7-January-2000.

[6] WFMC-TC-1023, *Workflow Standard – Interoperability Wf-XML Binding, Draft 1.0 (Beta Status)*, Workflow Management Coalition, 11-January-2000.

An Object-Oriented Modeling Method for Workflow Applications

Jingzhou Li
Institute of Computing Technology
Chinese Academy of Science
Beijing 100080, P.R. China
ljz@cs.sdu.edu.cn

Haiyang Wang
School of Computer Science and Technology
Shandong University
Jinan, 250100, P.R. China
why@cs.sdu.edu.cn

Abstract

Based on the analysis of the workflow applications, the modeling process of workflow applications is divided into four levels in this paper. They are Application level, Structure and information level, Execution control level, and Implementation level. An object-oriented model corresponding to the four-level modeling process is also given, expressed in UML. Event mechanism is employed in this paper to realize both the control and distribution mechanism in the logical modeling of workflow applications, illustrated by an event meta-model consisting of four basic event classes. The event definition and notification mechanisms are separated from the event reaction objects, and also the definition of functions and the triggering mechanism of the functions are separated. This may facilitate the dynamic modification of the workflow definitions and also facilitate the description of the distribution of workflow applications.

1. Introduction

Workflow, as defined by WfMC (Workflow Management Coalition)[1], is the automation of a business process, in whole or part, during which documents, information or tasks are passed from one participant to another for action, according to a set of procedural rules. A Workflow Management System (WFMS) is a system that defines, creates and manages the execution of workflows through the use of software, running on one or more workflow engines, which is able to interpret the process definition, interact with workflow participants and, where required, invoke the use of IT tools and applications.

The modeling and the enacting of workflows are the two main functionality of a WFMS. The modeling phase is concerned with the specification of a workflow. The enactment phase is responsible for scheduling, executing, and monitoring activities of a workflow to ensure correct workflow execution. However, the workflow modeling is fundamental to workflow applications. This paper will focus on the modeling aspect of workflow applications.

There are many workflow modeling techniques, currently, one of which is the use of Object-Oriented (OO) technique [2][3][4]. In [2], three models are used, Organizational Model, Core Workflow Model, and Execution Model. The OO concepts are used mainly to describe the meta-models, using UML. In [3], The OO concepts are used to describe the modeling and enacting of processes. While in [4], the WFMS is combined with an object-oriented database management system which bring in many OO concepts. Although many modeling methods use OO concepts to model workflow systems, however, different methods use the OO concepts for modeling different aspects and elements of workflow systems, resulting in confusion and un-uniform of the usage of OO concepts in modeling of workflow.

We will present a simplified OO model of workflow applications in this paper. This model uses fewer constructs than those of the aforementioned methods. And we also propose an event-based mechanism for modeling both the control and distribution aspects of workflow applications.

The remained sections are organized as follows. Section 2 discusses the basic requirements of the modeling of workflows. Section 3 presents our OO model. Section 4 discusses the event mechanism of the model. Finally, section 5 gives the conclusions.

2. Requirements of workflow modeling

The workflow modeling actually maps the application (or a problem) to a certain representation level at which the workflow can be implemented by a kind of WFMS. Therefore, which information has to be definable using a workflow model depends on the requirements stemmed from the application domain. Based on the discussions in [2][4], we obtain the following basic requirements:

1) Functional: **which** workflows (or functions) have to be performed and how they are structured.

2) Informational: **what** data is produced and consumed by workflows and how data flows among workflows (data flow).
3) Behavioral: **when** the workflows have to be performed and how they are controlled (control flow).
4) Organizational: **who** is responsible of carrying out the activities of workflows (roles).

Therefore, we divided the modeling process accordingly but not exactly correspondingly into four levels:

1) **Application level:** the application domain (or a problem, task) should be recognized and specified at this level. This is usually the user view of the main functions a workflow should have, including roles and their responsibilities (or processing entities in [2]).
2) **Structure and information level**: the structure of or the relationships between the functions, as well as the data produced or/and consumed by these functions, should be specified. This is a process to specify the input/output of the functions and the dataflow between the functions while decompose the functions recognized at the application level. The structure of the functions may be hierarchical, representing the decomposition of functions.
3) **Execution control level**: how the above recognized functions are organized together to accomplish the corresponding application is determined by the execution control. Functions are related together in a certain execution or control order by proper mechanisms such as messages or ECA (Event, Condition, Action) rules.
4) **Implementation level**: the above three levels are more conceptual in that they all describe the application itself. However, how to get the application implemented remains a problem. The purpose of this level is to make the above model of an application to be carried out in a specific kind of WFMS. So, the jobs at this level are WFMS-specific.

It can be seen that the four-level modeling of workflow is a stepwise and hierarchical process. As a result, the modeling process of a workflow application becomes more operable and manageable, hence simple, than those suggested in [2][3][4]. Because, those modeling methods either give no explicit modeling process or have a very complex modeling process with too many model constructs. The major difference is in that the modeling process presented here is a four-level hierarchy.

According to whether a workflow model is implementation-specific or not, the workflow models used in the four-level modeling process can be sorted into two categories: physical or logical. Models at the Implementation Level and the initial models (may be a description of the current implemented application from which the new goal-model stems) at the Application Level can be treated as physical models, others logical. Therefore, the logical models segregate those problem-specific elements from implementation-specific elements, so that the modelers can concentrate on the workflow application logics other than the implementation details.

3. An object-oriented workflow model

According to the four-level modeling process, a workflow model should have the ability to represent the objects at different levels. Meanwhile, the model can also help to refine stepwise each result of every level. This is one of the main goals of the model presented here.

For the sake of simplicity, only the main part of the Object-Oriented Workflow Model (OOWM) is presented in this paper, using the Unified Modeling Language (UML), as shown in Figure 1 (a) and (b).

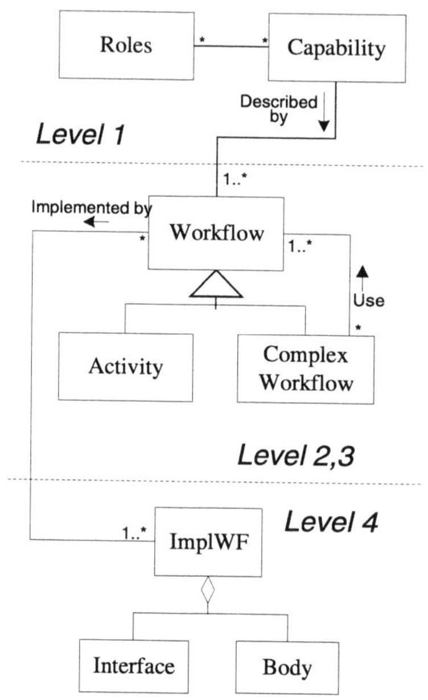

(a) Workflow Model

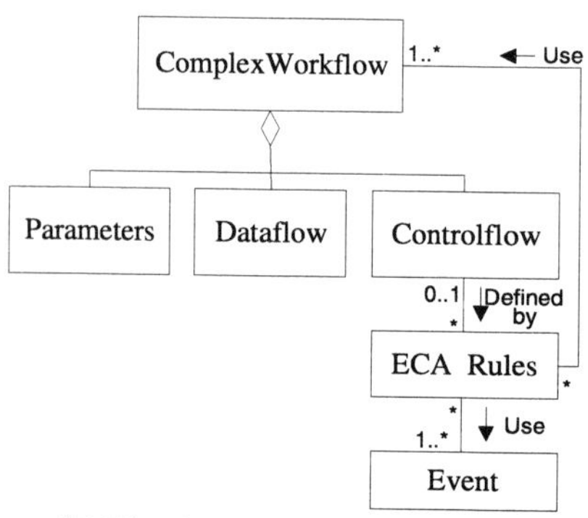

(b) The Complex-Workflow Definition

Figure 1. The Workflow Model.

It can be seen from Figure 1 that Level 1 and Level 4 are very simple while Level 2 and Level 3 are merged. For Level 1, *Roles* and their corresponding *Capabilities* and their relationships are identified. These are the organizational as well as the functional aspects of a workflow application. The *Capability* representing the functions of the application can be then used to define the workflow.

The model constructs for Level 2 and Level 3 are mixed together. The informational and execution control levels are actually the two aspects of a workflow, consequently they need not be staged. However, one can identify and define them in a sequence. The events and ECA rules used to define the control flow are also described as objects, which will facilitate the transition from a workflow model to a CORBA-based implementation. The model for these two levels is actually the logical model of the workflow applications. Once the logical model is determined it will be easy to transform it into an implementation level model. Therefore, this is the key part of the modeling of a workflow application.

For Level 4, any constructs of the workflow model of the last levels are transformed into an *ImplWorkflow* object which consists of only *Interface* and *Body*. So far, the workflow model becomes very easy to be implemented by a specific kind of, say CORBA-based, WFMS. It is because that all the objects of the workflow model are easy to be converted into CORBA objects.

The next section will focus on the modeling of event and ECA rules on which both the control and distribution mechanisms are based in the logical model of the workflow applications.

4. The event modeling mechanism

4.1. Related works

The HAD (Heterogeneous, Autonomous, Distribution) properties are the most important and difficult design goals of the workflow modeling and enacting, among those others [3]. As for heterogeneous and autonomous, the object-oriented technology has been the best solution. Based on DOM (Distributed Object Management), e.g. CORBA, the distribution aspect seems also easy to handle with. However, the distribution problem is often discussed in workflow enacting part, and most workflow modeling methods didn't touch upon it.

Two typical mechanism supporting distribution of the workflow enactment are event-based [5] and message-based [6]. In [5], a multi-server/multi-client architecture was proposed to support the distributed execution of workflows. Similar but different to CORBA, the concepts of Broker and Adaptor are also employed. An event server of EVE(Event Engine) is defined to manage the events. There can be multi-EVE-servers distributed on different sites to work together, accomplishing the notification of events, thus realizing the distribution of a workflow enactment. In [6], persistent message queues are employed to store and propagate messages related to the execution of workflow. These persistent queues are distributed on different sites. Therefore through passing messages from one site to another realizing the distribution of the execution of workflow. However, these kinds of distribution mechanisms are only for the workflow enactment.

We will present in next section an event-based method to cope with the distribution requirements as well as the flow control mechanism in workflow modeling.

4.2. The modeling of events and ECA rules

4.2.1 The Event Meta-model

As shown in Figure 2, there are four types of class in the event meta-model: the Event Class, the Reactive Event Class, the Event Detection Class and the Event Reference Class. These four types of class are called the basic event classes of the event model. As for the Application Class, it represents the functions of a workflow application. The functions, which may be either computer software systems or human operations, can be called by the workflow control system. Both the Event Class and the Reactive Event Class should be defined in the application part, while the Event Detection Class and the Event Reference Class are for the WFMS.

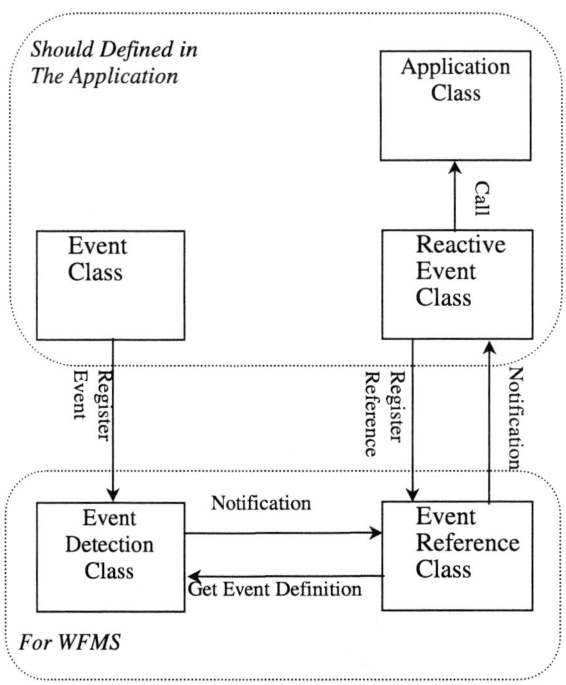

Figure 2. Event Meta-model.

4.2.2 The Event Class and the Reactive Event Class

The Event Class defines the primitive events. The Reactive Event Class defines the objects that are able to react to the occurred events with proper actions. The following is the simplified interface definition of these two classes:

Interface WFEvent{ //Event Class
attribute String EventName;
attribute String EventType;
attribute String SourceObjID; //The event source
attribute UTCtime TimeStamp;
//The time point of the event occurrence
string Get_EventName(in string EventObjID);
string Get_EventType(in string EventObjID);
string Get_Event_SourceObjID(in string EventObjID);
UTCtime Get_Event_TimeStamp(in string EventObjID);
}.
Interface WFReactEvent{ //The reactive event class
void RegisterEventReference(in string EventName);
void ReactToEvent(in string EventName);
}.

The ReactToEvent method in the Reactive Event Class is responsible for the actions should be taken after related events occurred, which means the <Event, Condition, Action> pairs should be implemented in this method. This is in fact the realization of ECA rules. However, the E and CA of ECA rules are defined separately here, which bring us some benefits. One is that treating the definition, generation and notification of events in a separate way may facilitate the implementation of event mechanism with existing event and message mechanisms such as CORBA. The second one is defining CA in the Reactive Event Class also separate the definition of the functions from the triggering mechanism of functions, which may facilitate the dynamic modification of workflow definitions. Another one is in that it may give help to the describing of distribution applications. This is because that the generation of events and the execution of corresponding functions are probably taken place in different places, therefore the event notification mechanism could transmit the occurred events to the places where the Reactive event Class reside.

4.2.3 The Registration of Event

All the primitive events are registered to the Event Detection Class which will detect the occurrences of the registered event types. Once it detects an occurred event the Event Detection Class will notify the event to the Event Reference Class.

4.2.4 The Event Reference Registration and Event Notification

The relationship between the event and its reactive objects, expressed by <Event, ReactObj> and called the Event Reference here, is defined in the Event Reference Class. When there is a new instance of Reactive Event Class created, the object ID together with all the events this instance can be react to are registered to the Event Reference Class. The registered events may also be composite events.

The Event Detection Class will notify the event occurrence to the Event Reference Class when it detects the generation of events. The Event Reference Class in turn notifies the event occurrence to all the instances of Reactive Event Class that are registered for the events according the event reference relationships. The instance of the Reactive Event Class then triggers the corresponding functions according to the CA pairs defined in ReactToEvent method. The whole process from event definition, occurrence, and notification to triggering the functions is completed hitherto.

4.3. The transition from logical model to implementing model

Different to [5], we based our event mechanism on CORBA Event Service[7][8] which is a world-wide standard and there are many compatible products of it. Consequently, what we need to do left is to transform the logical event mechanism into implementing event mechanism(i.e, the CORBA events).

This transformation is directly in that all the event classes in the meta-model shown as Figure 2 can be translated into CORBA objects first. Then, the notification between these objects can be implemented by the CORBA Event Service which provides a fundamental communication channel between objects.

Another way to implement the logical model is to employ the active database systems as the base of the whole workflow system. The Reactive Event Class directly corresponds to the ECA rules. Other classes may also be implemented by the management system of the active database system.

5. Conclusions

We presented a four-level workflow modeling process and a corresponding object-oriented workflow model in this paper. Consequently workflow objects can be defined in an object-oriented manner. The basic units of the workflow are object definitions ideally abstracted from real-world structures. This provides a clean, conceptual one-to-one mapping between the real-world objects and the model objects maintained within the system. Object-orientation, by using information hiding and abstraction, makes the workflow easier to understand by end-users and maintain by developers. This approach allows for a consistent work model across human executed and computer executed activities even as the definition and understanding of these tasks evolve over time.

The event mechanism is employed to realize the distribution mechanism as well as the control mechanism. Different to the ECA rules in most active database systems and the events of EVE, we separate the event definition and notification mechanisms from the event reaction objects and also separate the definition of functions from the triggering mechanism of the functions. This may facilitate the dynamic modification of the workflow definitions and also the description of the distribution of workflow applications.

6. References

[1] WfMC, "Workflow Management Coalition Terminology & Glossary", WfMC-TC-1011, Feb 1999.

[2] Markus Kradolfer, et al, "Modeling Concepts for Workflow Specification", Technical report of the University of Zurich, May 1997.

[3] Gregor Joeris, et al, "Towards Object-Oriented Modeling and Enacting of Processes", TZI-Report 07/98, Center for Computing Technologies, University of Bremen, 1998.

[4] G. Kappel, et al, "Workflow Management Based on Objects, Rules, and Roles", Bulletin of the Technical Committee on Data Engineering, Vol. 18 No. 1, March 1995, PP10-17.

[5] A. Geppert, et al, "Event-based Distributed Workflow Execution With EVE", Technical Report of the University of Zurich, May 1996.

[6] G. Alonso, et al, "Exotica/FMQM: A Persistent Message-based Architecture for Distributed Workflow Management", Proc IFIP Working Conf on Info Sys for Decentralized Organizations, Trondheim, Aug 1995.

[7] Douglas C. Schmidt, "An Overview of OMG CORBA Event Services", http://www.omg.org.

[8] IONA Technologies, "A CORBA-based Notification Service", http://www.omg.org.

CovaFlow: a General Workflow System by Cova

Jinlei Jiang, Yan Wu, Meilin Shi
Department of Computer Science, Tsinghua University, Beijing, 100084, P.R. China
[jjlei, wyan, shi]@csnet4.cs.tsinghua.edu.cn

Abstract

Although being exploited by businesses in a variety of industries, the current workflow system is considered to be too rigid. However, it is difficult to construct a new workflow system from scratch due to its complexity. On the other hand, meta-groupware is attracting more and more attention, in that it can be used to compose sophisticated systems. In this paper, we show how to implement workflow management system by Cova, a meta-groupware developed at Tsinghua University. Problems, including process design, run-time synchronous support and reliability, are addressed. It is efficient as well as easy to construct Workflow Management System (WfMS) based on meta-groupware. In addition, a system developed in this way presents many desired features such as flexibility, reliability, and so on.

1. Introduction

In modern organizations, especially those involving design and product manufacturing, close cooperation between co-workers is often needed in order to get the work done. New tools and applications have been developed to enable such cooperation. Examples of these are groupware[3] and workflow systems [13]. Groupware is a broad designation covering a range of technologies that support tasks that people carry out in groups that vary in size and composition, which reflects a change in emphasis from using the computer to solve problems to facilitate human interaction [3]. Workflow system enables a group of people to work together to achieve the common goal, which belongs to asynchronous groupware according to [3]. Nowadays workflow management technology is increasingly being exploited by businesses in a variety of industries. Although there are many workflow products available in the market, the current workflow system is considered to be too rigid to be applied to commercial environments. However, it is difficult to construct a new one from scratch.

Recently it has been highly demanded and widely accepted that a software development environment capable of providing some forms of general-purpose services that ease developers' endeavors should be developed [8]. Many toolkits and platforms [6,9,10,12] have emerged till now and among them is Cova[14]. We term such toolkits/platforms as meta-groupware [4].

The power of meta-groupware lies in that complex systems can be constructed based on them. In this paper, we show how to implement workflow management system by Cova. The paper is organized as follows. At first, workflow management system is introduced in section 2. Then section 3 gives an overview of Cova. In section 4, how to implement workflow management system is explained including system architecture, process design, synchronous support and reliability. In the end, we conclude our work in section 5.

2. Workflow Management System

Workflow management systems (WfMSs) are used to coordinate and streamline business process. According to Workflow Management Coalition (WfMC), a business process is "a procedure where documents, information or tasks are passed between participants according to a defined set of rules to achieve, or contribute to, an overall business goal"[13]. A workflow is a representation of the business process in a machine-readable format. Hence, a workflow management system, WfMS, is "a system that completely defines, manages and executes workflows through the execution of software whose order of execution is driven by a computer representation of the workflow logic" [13].

Workflow process execution conducted by a WfMS is based on the formal definitions of business processes. The formal definitions, which are called workflow models, are quite different in different WfMSs. In this paper, we use the model named "conditional directed graph"(CDG)[15]. A CDG is a directed graph where a node represents a work activity in a business process, while a directed edge represents the control flow and data flow between two work activities. The components described below follow the meta-model proposed by WfMC. These are described based on CDG model, depicted in Figure 1:

- **Process**, a description of the sequence of steps to be completed to accomplish some goal.
- **Activity**, each step in a process. An activity may include a manual activity and/or an automated workflow activity. Each activity has an input and an output.

- **Flow of Control**, specified by arrow line between activities, is the order in which activities are executed. This corresponds to the transition conditions of the reference model.
- **Flow of Data**, also specified by arrow line between activities, is a series of mapping between output data and input data to allow activities to exchange information.
- **Conditions,** which specify the circumstances under which certain events will happen. There are three basic types of conditions, i.e. transition conditions, start conditions and exit conditions. Start conditions specify when an activity will be started and exit conditions specify when an activity is considered to have terminated.

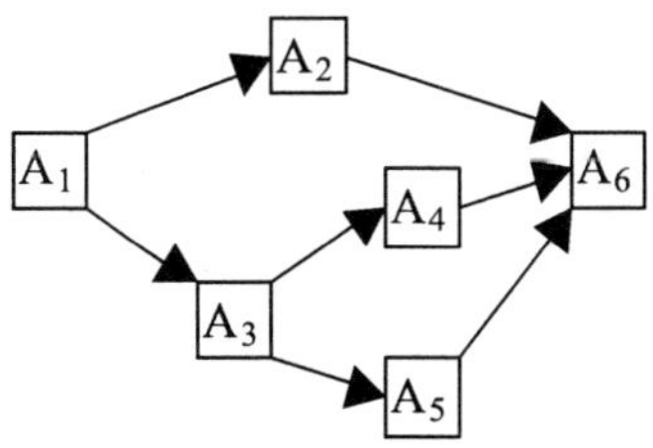

Figure 1. CDG Model for a Workflow Process.

New trends in workflow include transactional workflow, exception handling and recovery.

3. An Overview of Cova

In this section, we will examine Cova developed by CSCW Lab at Department of Computer Sciences, Tsinghua University.

3.1. Why Cova?

With increasing complexities of CSCW systems, it becomes too difficult to build a large-scale cooperative system from scratch, with the direct support of traditional distributed facilities and services. It is urgently demanded for a platform providing the general-purpose service and programming environment to ease the endeavor of the developers. It's the motivation of Cova (the word Cova comes from the combination of the four bold letters of word "***co***operati***v***e ***a***pplications").

Cova is a meta-groupware that provides groupware developers with a set of general-purpose cooperative computing services based on a cooperation model and the programming facilities to utilize common services. Services provided by Cova can be divided into two parts, i.e. core service set and extended service set. Core service set includes:

- Activity Control
- Replication Control
- Concurrency Control

While extended service set contains:

- Resource Directory
- Access Control
- Version Control
- Transaction Support

The desiring features of Cova are listed as follows:

- **Cooperation integration**: Different modes-cooperation will be modeled in a uniform way.
- **Dynamic Modification**: The new activities can be created by existing activities as if they were predefined. At the same time, controlling rules can also be modified or added dynamically.
- **Reusability**: The results of the process can be reused for the description of other processes.
- **Simplicity:** It is natural and easy for the developers without extra heavy burden.

3.2. The Cova Model and Languages

Based on the separation of computation and cooperation, we provide two kinds of mechanism for describing cooperative process: One is used for expressing data structure and semantics of the result in an activity. The other is used for specifying the relationship among the activities. Therefore, the Cova model consists of two parts: the object model and the cooperation model. Corresponding to them, two languages, i.e. **CODL**(Cova Object Description Language) and **CCDL**(Cova Cooperative Description Language) are designed. Cova object model abides by specification of ODMG and supports *multiple inheritance*, *polymorphism*, *encapsulation*, and *nested* or *local class definition*. For limit of space, we will not explain them here. Instead, we focus on the Cova cooperation model.

The Cova cooperation model consists of process and activity similar to those in workflow. A Cova cooperation process is modeled as a set of interrelated *activities*, each of which can be regarded as a step or a stage of cooperation process. An activity is modeled as a sextuplet $<n, c, e, R, A, T>$, where:

- ***n*** is the name of the activity.
- ***c*** is the name of the class of the data object (which we call it *activity object*) maintained and manipulated during the lifecycle of this activity.
- ***e*** is the *enable condition* for this activity. It will be evaluated whenever the internal state of the activity changes, e.g. the activity object is modified. When it evaluates to *true*, the activity will be enabled, which makes the activity object visible to users or applications.
- ***R*** is the *receiver list* that contains all the users and/or applications that can participate in an activity to access its data object ***o***. An element in ***R*** can be a single user or a group of users with different semantics.
- ***A*** is the *access list* which defines how the attributes and operations of ***o*** can be accessed, e.g. *read*, *write*

or *execute,* by each element in ***R***.

- ***T*** is the *trigger list* that specifies what actions should be triggered upon the occurrence of specific events. Each element t of ***T*** has the form as $\langle e, c, a\rangle$, which is similar to the well-known *event-condition-action* rule [7] developed for active DBMSs, where c is a boolean expression defined on the activity object. An action a is an invocation to a method defined on a *destination object*. When the event e occurs, if the condition c is evaluated to *true*, the action a will be executed on the specified destination object. The mechanism of trigger makes it possible for control and data to flow among different activities, which is critical for supporting asynchronous coordination, such as cooperation process defined in workflow management system. When a collection of activities "run" together, they cooperate with one another via transition relations defined in the trigger list *T*.

To describe the cooperation model, we design and implement **CCDL**, which is object-oriented in that process definitions are inheritable. Other important object-oriented concepts such as polymorphism, encapsulation are also introduced into Cova to promote the modularity and reusability.

4. Workflow by Cova

In this section, we will see several aspects about how to construct workflow system by Cova. Sharing some common features (e.g. similar model), the construction is straightforward.

4.1. System Architecture

Usually workflow management systems are implemented using client server mode, so do with us. We will depict the system architecture from two perspectives: build-time and run-time scenario.

The build-time function of a WfMS concerns with process definition, form definition (for participants to handle their tasks) and so on. The build-time system architecture looks like Figure 2, where CovaModeler is process definition tool. From this figure, we can see that the process definition is distributed. Indeed, CovaModeler itself is a collaborative tool. Here CovaServer acts as a role of database and corresponds to the process manager in workflow system. In this scenario, extended services are utilized.

The run-time function concerns with process execution. It is the key scenario involved cooperation. The run-time system architecture is similar to that of build-time except CovaModeler is replaced by CovaClient, which is a visual tool provided for users to handle their tasks. In this scenario, CovaServer functions as workflow enactor and core services are applied. According to the real load, various numbers of servers can be adopted. For example, if the load is light, one server is sufficient. However, more servers are needed when the load is heavy.

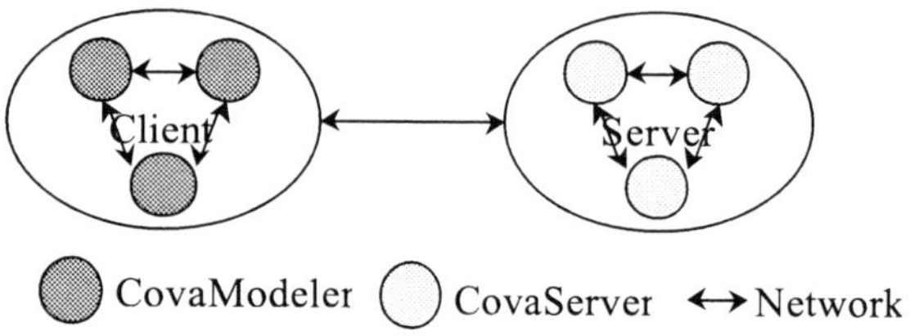

Figure 2 Build-time System Architecture.

4.2. Process Design

In our system, CovaModeler is supplied as process definition tool. As stated above, CovaModeler itself is a collaborative tool. As we all know, modeling collaborative processes is a complex task because it may involve many people with different backgrounds, each having his own focuses and interests. To get a perfect process description, a mechanism must be provided to support the cooperation among these people. It is in this sense that we design CovaModeler as a collaborative tool. CovaModeler provides workspace for process designer and can support task division, synthesis and code self-generation. At first, a task is divided into several sub-tasks and each sub-task is assigned to one or more experts. If the sub-task is still very complex, it can be divided further till it is simple enough. This phase is called meeting phase[5]. After task dividing, process design can be performed. This phase is called design phase[5]. During design phase, the corresponding Cova process definition (see Figure 3) is generated. In this way, the workload is reduced. For more details about CovaModeler, please refer to [5].

Even if working collaboratively, it is a time-consuming thing to define a process, especially a complex one. On the other hand, companies live in a varying environment. The once correct commercial rules may get unacceptable as time goes on. Therefore, process design becomes a heavy burden when the companies rely more and more work on workflow management system. So process reuse is badly needed to ease this burden. We meet this requirement by inheritance. As mentioned above, Cova exploits object-oriented technologies to describe cooperative process and can reuse existing process easily. An example is given in Figure 3, which illustrates a trip reservation process with three activities, i.e. RRequest, ReserveCar and ReserveHotel.

In the given program, the boldface fonts are keywords of Cova, where **extends** is used to indicate the inheritance relation. In this example, the process TripApplication is inherited from PApplication. Compile the program and we get the workflow process constructed. Roughly speaking, activities within the process are specified by the program. The participants of an activity are indicated by **role** and the start condition is specified by **startswhen**. The control flow can be inferred from the activity complete event. For an instance, method *Accept* of activity ReserveHotel is triggered when activity ReserveCar completes. Therefore, there is an arrow line from ReserveCar to ReserveHotel in the corresponding CDG model. The action *Accept* also specifics the data flow (as method parameters) between them. Due to space limitation, we will not explain other words here.

In fact, we can also invoke another process in an activity thus subtasks are well supported and flexible process composition becomes possible. By process reuse, the burden of companies on process design is eased.

```
public process TripApplication extends  PApplication startsat RRequest
{
  activity RRequest handles CForm //start activity of the process
  {
    role  Customer; //define the participants
    trigger ReserveCar.Accept(this) //action
        when after complete where true; //event & condition
    ...
  };
  activity ReserveCar handles CCar startswhen (this.GetRequest() !=
null) //start condition of the activity
  {
    role ServiceProvider;
    trigger EMailAgent.SendMail (GetCustomer(), false)
        when after abort where true; //execution failed & inform user
    trigger ReserveHotel.Accept(this)
        when after complete where true; //succeed & start next activity
    ...
  };
  activity ReserveHotel handles  CHotel startswhen (this.GetRequest()
!= null)
  {
    role ServiceProvider;
    trigger EMailAgent.SendMail (GetCustomer(), false)
        when after abort where true; //process fails
    trigger EMailAgent.SendMail (GetCustomer(), true)
        when after complete where true; //process succeeds
    ...
  }
}
```

Figure 3. A Trip Reservation Process in Cova.

4.3. Process Execution

Once the process definition is finished, the generated Cova code is compiled by CCDL compiler. The compiled result is then saved to the CovaServer and ready for execution.

To execute a process, a user sends the request to the server along with it the process name. The server then finds the class definition and constructs an instance of it. Each instance is assigned a unique name through which the user interacts with the server. The run-time/virtual machine maintains a list of running process instances and works in event-driven mode. Cova events can be classified as the following categories:

- **Time-related events** which occurs *at* a specific time point, *after* a specific interval, or occurs periodically with a specific interval.
- **Method invocation events** which are triggered exactly before or after the invocation to a specific method defined on a Cova object.
- **State-related events** which are triggered whenever the state of a process or an activity or a participant changes, e.g. when they are activated, inactivated, completed, or terminated, or when an activity object is opened, closed, or saved, or when a participant logs in or logs off.

The introduction of state-related events makes the communication (e.g. flow control, data exchange) between different activities possible. This is the key to support asynchronous activity. For an asynchronous activity, the participants need not let the operation results be known instantly by others. Instead, they open the activity object and do some operations on it and then submit the results to server. What the server need do is to combine the results from different participants according to the predefined rules. It isn't so with synchronous activities. We will explain it in the next section.

Using triggers, we can implement routing easily. Cova activity can have multiple inputs and/or outputs. When an activity is submitted, a completed event will occur in the server. The process enactor captures this event and starts to calculate the corresponding condition. Once it is met, the triggered methods are called. Therefore, the work is routed on. If there are more than one methods in the trigger list, split(OR) routing is achieved. Join(AND) routing can also obtained by setting related conditions. For example, in the process given in Figure 3, if we change the triggered method list of activity RRequest to

ReserveCar.Accept(this), ReserverHotel.Accept(this), then ReserveCar and ReserverHotel can run in parallel.

4.4. Synchronous Support

In the previous statements, we mention that Cova can support synchronous and asynchronous activity uniformly. Here we will show how it is implemented. In synchronous applications, concurrency control is very important, so we will also address it here.

To reduce the response time, we employ data replication as most real-time systems do. When an object is opened, it is replicated to every site involved. The concurrency control algorithm adopted by Cova is similar to operation transform introduced in [2,11]. That is, operations executed in one site will eventually be sent to other related sites, where the incoming concurrent operation is transformed against the local state and then executed. For each basic object type (e.g. *list*, *set*), primitive operations are defined and transformation function is specified. User defined operations are decomposed into the primitive ones on the basic types and then executed. Replications of data and concurrency control are both implemented in Cova virtual/run-time machine (CVM) as core services.

We can use concurrency control provided by Cova to support synchronous activity easily. This is shown in Figure 4. CVM is integrated with CovaClient. CVMs of different sites are connected by computer networks. When user from one site changes the value through interface presented by CovaClient, the underlying method (e.g. Update) is called to take the action into effect. The operation is then multicast to other sites belonging to the same activity and executed there.

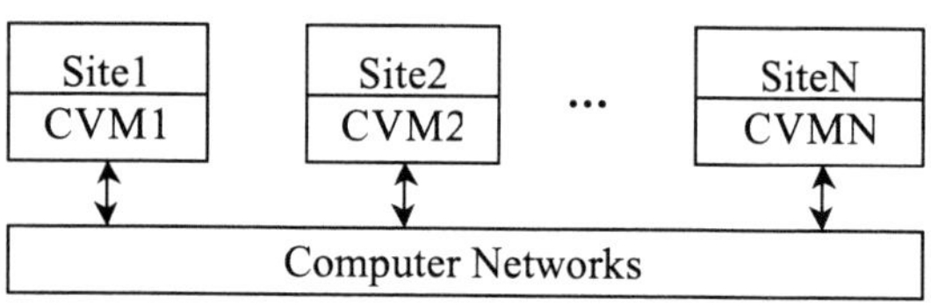

Figure 4 Synchronous Support Mechanism.

4.5. Reliability

As the organizations evolve, it is widely accepted that effectiveness or performance is no longer sufficient for them to achieve their goals. Reliability has been shown to be crucial, especially for mission-critical applications. During the long duration cooperative process, exceptions and failures are inevitable. How to survive upon these circumstances? How to guarantee data integrity when many activities run in parallel? Such problems have been solved successfully in traditional database system by the concept of transaction model. However, it is not applicable to workflow management system. Though many advanced transaction models have been proposed, they are too database-centric and still lack some features of workflow, such as user or program support [1]. In many aspects workflow models are a superset of advanced transaction models [1].

In Cova, a new transaction model named CovaTM is designed to support cooperative process. In this model, we view one execution of a process instance as a transaction with its sub-transactions corresponding to activities. It can be used for reliable and flexible process instance enactment, rollback and compensation. This is the key to make Cova be capable of supporting mission-critical applications. For limit of space, we will give a brief description of it below.

A CovaTM transaction T is a 6-tuple (**ST, O, PP, TP, AC, AD**) where

- **ST** is the sub-transaction set of T
- **O** is the partial order on **ST**
- **PP** is the set of all precedence predicates of **ST**
- **TP** is the set of all temporal predicates of **ST**
- **AC** is the set of all acceptable states of T
- **AD** is the administrator of T

Let **X** be the set of all possible execution states, which is a combination of sub-transactions' states. Thus, a precedence predicate pp_i for a sub-transaction t_i is a boolean function defined on **X**, where

$$pp_i : \mathbf{X} \rightarrow \{\text{true, false}\}$$

While a temporal predicate tp_i for a subtransaction t_i is a time requirement of t_i. It has the following format:

Operator Hour:Minute:Month:Day:Year

where Operator can be *before*, *after* and their combination,denoted by *between*. A transaction administrator is an active entity, such as a person or a program that is responsible for the execution of the transaction, i.e. it is up to the administrator to determine what to do when exceptions occur.

In CovaTM model, we can specify more than one acceptable state, thus different execution paths (strategies) are allowed. Characteristics of this model include:

- Integrated exception handling and recovery as well as access control.
- Sophisticated control of flow, which makes it possible to describe the wide spectrum of cooperative activities from structured to non-structured.
- Time requirements of activity are reflected. This can be used to depict time-critical and none time-critical activities uniformly.
- Mixed sub-transaction type. Compensable and none-compensable sub-transactions are specified as well as compound one. This makes it flexible.

By introducing transaction concept into workflow management systems, the integrity of process is guaranteed. Moreover, the flexibility is also promoted.

5. Conclusions and Future Work

We have addressed the problem of implementing WfMS by Cova, a meta-groupware developed at CSCW Lab of Tsinghua University. Meta-groupware provides common services needed by various applications and new services can also be added when required, thus it can meet a wide spectrum of CSCW applications. Constructing WfMS by Cova has the following advantages:

- Flexible According to the real requirement, new services can be added and/or existing service rules can be altered. Therefore, a WfMS constructed in this way can adapt to the varying environment well.
- Scalable Communication support is well established in groupware, thus the system can be extended from single server to multi-server easily.
- Powerful Using the object-oriented features provided by Cova, we could reuse the existing process definition, which eases the burden faced by enterprises.
- Efficient Utilizing the services supplied by meta-groupware, we can spare a lot of work and focus on other problems than concurrency control, data management and so on.

Now we can conclude that constructing workflow management system by Cova is easy but effective and efficient. It has most outstanding features mentioned above. Indeed, other cooperative systems can also be established besides workflow.

Currently we mainly focus on the centered business process execution. However, the system architecture makes it scalable. In the future, we will do some work on load assignment and balance among multi-servers. Issues like inter-domain workflow management and security will also be considered.

6. Acknowledgement

This paper was finished at Bell-Labs Research China. It is co-supported by National Natural Science Foundation of China under grant No. 60073011 and 985 Project of Tsinghua University. The previous work on workflow management system is co-sponsored by China National Science Foundation under grant No. 69773029 and China National Advanced Science & Technology (863) Plan under Contract No. 863-306-ZD-10-2B and 863-306-ZD-02-03-01.

The authors would like to thank Guangxin Yang, Yufeng Hsieh and Hong Yan for their work on the system.

7. References

[1] G. Alonso, D. Agrawal, A. E. Abbadi, M. Kamath, R. Gunthor and C. Mohan, Advanced Transaction Models in Workflow Context, IBM Research Report RJ9970, IBM Almaden Research Center

[2] C. A. Ellis and S. J. Gibbs, Concurrency Control in Groupware Systems, In proc. of ACM Conf on Management of Data, Portland, 1989, 399-407

[3] C. A. Ellis, S. J. Gibbs and G. L. Rein, Groupware: Some Issues and Experiences, Communication of the ACM, Jan. 1991, 34(1), 39-59

[4] C. A. Ellis and J. Wainer, A Conceptual Model of Groupware, In proc. of ACM Conf on Computer Supported Cooperative Work, Chapel Hill, 1994, 79-88

[5] Y. HSIEH, G. Yang and M. Shi, CovaModeler: a Multi-user Tool for Modeling Collaborative Processes, In proc. of CSCW in Design, HongKong, 2000

[6] D. Li and R. Muntz, COCA: Collaborative Objects Coordination Architecture. In proc. of ACM Conf on Computer Supported Cooperative Work, Seattle, 1998, 179-188

[7] D. R. McCarthy and U. Dayal, The Architecture of an Active Database Management System. In proc. of ACM Conf on Management of Data, Portland, 1989, 215-224

[8] J. F. Patterson, Comparing the Programming Demands of Single-user and Multi-user Applications, In proc. of ACM symposium on User Interface Software and Technology, Hilton Head, 1991, 87-95

[9] J. F. Patterson, M. Day and J. Kucan, Notification servers for synchronous groupware. In proc. of ACM Conf on Computer Supported Cooperative Work, Cambridge, 1996, 122 - 129

[10] C. Schuckmann L. Kirchner, J. Schummer and J. M. Haake, Designing Object-oriented Synchronous Groupware with COAST. In proc. of ACM Conf on Computer Supported Cooperative Work, Cambridge, 1996, 30-38

[11] M. Suleiman, M. Cart and J. Ferrie, Serialization of Concurrent Operations in a Distributed Collaborative Environment, in proc. of ACM SIGGROUP Conf on Supporting Group Work, Phoenix, 1997, 435-445

[12] J. Trevor, T. Rodden and G. Blair, COLA: A Lightweight Platform for CSCW. In proc. of European Conf on Computer Supported Cooperative Work, Milan, 1993, 15-30

[13] Workflow Management Coalition, The Workflow Reference Model, 1995

[14] G. Yang, Research on Meta-groupware—the Cova Programming Language and System, [PhD Dissertation], Beijing: Tsinghua University, 2000

[15] G. Yang, M. Shi, Y. Xiang and S. Wu, Wowww!: Managing Workflow on the World Wide Web, In proc. of IEEE Intl Conf. on Communication Technology, Beijing, Oct. 1998, S33-03-01~S33-03-05.

Integration of Workflow and Agent Technology for Business Process Management

Yuhong Yan
Enterprise Integration Lab
University of Toronto
Toronto, Canada
yuhong.yan@utoronto.ca

Zakaria Maamar
Systems Engineering Group
College of Information Systems
Zayed University, Dubai, U.A.E
zakaria.maamar@zu.ac.ae

Weiming Shen
IMTI
National Research Council
London, Ontario, Canada
weiming.shen@nrc.ca

Abstract

Both workflow and agent technology have recently been applied to business process management. The integration of these two technologies will definitely provide solutions to the problems not to be solved by any of them individually. This paper summarises the capabilities of these two technologies and discusses the forms and their benefits of integrating them for business process management. Generally, agent enhanced workflow management and agent-based workflow management are the main forms for applications of intelligent agents to workflow systems. Some research issues in each form are discussed. A conclusion with discussions of future research directions is also given.

1. Introduction

Many organizations have realized that although they have adapted the Information Technology (IT) to improve their working efficiency, the business processes within their organizations, and between themselves and their partners, have not been clearly described and streamlined. During the execution of business processes, there are not enough techniques and methods to follow-up and control the processes. This leads to the misunderstanding of responsibilities, blocks in coordination, and slow reaction to the changing market.

Workflow management technology is among the ones under development to overcome these shortcomings. It promises to provide an efficient way to model and control the complex business processes within and between organizations. Although workflow management technology has emerged for years, it is only in the last few years that it has become very popular in the commercial as well as the research world. The benefits of workflow management technology include explicit process definition, quick reaction to changing environments, and easy tracking of operations. While workflow management focuses on managing the process logic, it needs to integrate other technologies so to fully control a business process, such as activity assignment, and resource allocation. Compared to the formal workflow management methodology, software agent technology provides flexible, distributed, and intelligent solutions for business process management. A combination of these two technologies will definitely provide solutions to the problems not to be solved by any of them individually.

Integration of workflow and agent technology has recently attracted a lot of attention of researchers in the related areas. This paper provides an overview of this exciting research area. The paper is organized as follows: Section 2 discusses the advantages and shortcomings of the workflow technology for managing business processes; Section 3 introduces the recent research work on applications of agent technology to business process management; Section 4 overviews the integration of these two technologies; Section 5 gives a brief conclusion with some discussions of future research directions in this area.

2. Workflow Management for Business Process Management

Workflow management comes from the office automation area, where all kinds of documents need to be digitalized and transferred among co-workers. Nowadays, the workflow management attracts a lot of attention due to its ability in modeling, executing, and monitoring processes. The processes can be not only the business processes, but also any procedures that need to be presented and managed. But the motivation for its booming is its promising usage in managing business processes.

One widely accepted definition of workflow comes from the Workflow Management Coalition [17]:

> *"Workflow is the computerized facilitation or automation of a business process, in whole or part."*

The benefits of applying workflow technology to business process management are as follows:

- The business processes are explicitly defined, so that the responsibilities and the coordination relations are clearly determined.
- It is easy to optimize the business process because of their explicit definitions.
- Business processes are modularized and these modules can be reorganized by the WFMS to form new business processes, so as to react quickly to unexpected changing business needs and conditions.
- WFMS can track daily operations.
- WFMS integrates applications on different platforms into a business process.
- WFMS provides personal workplaces.
- WFMS stem separates business logic of a process from the tasks themselves in the process. Therefore, the user in the workflow system does not have to deal with the route of the business process, but only concentrates on the task itself.

From a technical perspective, WFMSs bring together principles, methodologies and technologies from various areas of computer science and management science. For example, workflow techniques involve database management, client-server computing, heterogeneous distributed computing, graphical user interfaces, application and subsystem integration, messaging, document management, simulation, and business practices and re-engineering. However, the current generation of WFMS has some shortcomings:

- Relying on one central control: WFMSs have a central workflow server that defines and controls all business processes. While in distributed enterprises or large companies, the business processes have no way to be managed by a central point. The independent companies make their decision on how to do their pieces of business.
- Lack of automation: WFMSs only determine the process logic, but most of the activities are still fulfilled by human. WFMSs can't even start a workflow without a human's intervention.
- Lack of reactivity [13, 18]: WFMSs require a pre-defined representation of a business process and all potential deviations from that process.
- Lack of resources management [13, 18]: WFMSs do not control the resources of a business process, and so rely on a business process being dimensioned beforehand.
- Lack of semantics [13, 18]: WFMSs lack an appreciation of the content of a business process and do not make decisions based on the nature of the information generated by a business process.
- Lack of generic interfaces: WFMSs need to exchange data between activities or interface to other applications. Currently, these operations depend on API calls. There should be some generic interfaces to eliminate the effort to develop interfaces between WFMSs and other applications.
- Lack of interoperation [13]: independent vendors now develop WFMSs. Though WfMC has presented its work to enable the interoperation, this does not help much.

One solution to these shortcomings is to "embody" WFMSs with mechanisms from other technologies. For example, integrating project management for resource management and scheduling, integrating groupware for message passing and personal working environment. Software agent technology is among the ones that can benefit more the workflow technology. We will discuss more on this in the following sections.

3. Software Agent Technology for Business Process Management

For the notion of agent and autonomy used in this paper, we comply with Jennings and Wooldridge's definition [10]: "An agent is a computer system situated in some environment, and that is capable of autonomous action in this environment in order to meet its design objectives." An autonomous agent should be able to act without the direct intervention of humans (or other agents), and should have control over its own actions and internal state. An "agent based system" means one in which the key abstraction used is that of an agent [10].

Agent technology provides an extension and alternative to business process management problems. Among many of the properties software have, the properties of being autonomous, collaborative, and intelligent are those most interested for distributed business process management systems. The benefits of applying agent technology to business process management include:

- Distributed system architecture: For the scenario concerning multiple workflow systems, agent technology provides loose coupled distributed system structure for integrating distributed business process management systems.
- Automation: The inherent autonomy of software agents can fulfill activities as human substitution. Moreover, agents can start a workflow based on event triggers or more complex reactions to environment changes.
- Interaction: Software agents enable organizations to interact with each other, normally through semantic message exchanging.
- Resource management: Agents can represent resources. Task assignments and resource allocations are done through negotiation among these agents.

- Reactivity: Agents react to changing circumstances and have the ability to generate alternative execution paths. This ability normally involves agent's intelligent features, such as learning.
- Interoperation among heterogeneous systems: Agents can be heterogeneous. The interactions rely on semantic messages for exchanging plans and service definitions. That makes interoperation more feasible than API calls.
- Intelligent decision-making: Some high-level features of agents, such as learning, are also very helpful in workflow management, though they are not matured techniques nowadays.

However, implementation of business process management systems using only agent technology has the following problems:

- A coordination mechanism is usually missing, which could make the systems unstable and unreliable.
- Business process optimization is difficult due to the lack of explicit definitions and representations of the business processes.
- It is not easy to track the daily operations as it is easily done in workflow management systems.

4. Integration of Workflow and Agent Technologies

Some workflow management systems vendors have declaimed that they have implemented software agents in their commercial workflow products. However, these agents are usually very basic, and are used to implement autonomous activities. More and more researchers believe that agent technology can provide system architectures for integrating multiple workflow systems. From the system point view, most workflow management systems implemented with agents are multi-agent systems. Applications of agents to workflow management systems can be classified into two forms: agent-enhanced workflow management and agent-based workflow management.

4.1. Agent-Enhanced Workflow Management

Agent-enhanced workflow management is the basic form for application of agents to workflow management (cf. Figure 1). There is one central workflow engine, which controls all the activities. Agents are invocated during the execution of one work item to implement certain tasks. Workflow system controls the generation and elimination of the agents. This form is how current commercial WFMSs use the concept of agents. There are several things agents can achieve in this scenario:

- Human interface: it is part of the workplace environment the workflow system provides to its user. The agent acts as a personal assistant. Some typical usages are sorting emails, replying email, reminding events, acquiring work items. Example is A1 in figure 1.
- Autonomous activity: it implements the tasks autonomously without human's interruption. Example is A2 in figure 1.
- Interface to other applications: agent can provide interface to other applications. Instead of defining API for application interoperation, semantic messages can be defined to exchange high-level information. This is also a direction for designing generic interface to applications.

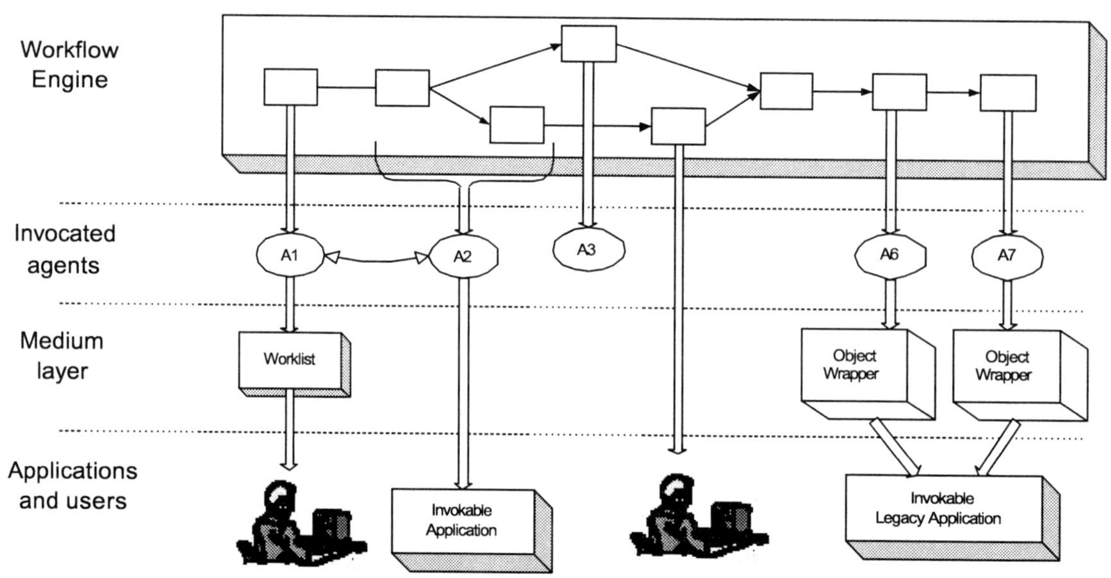

Figure 1 Agent Enhanced Workflow Management System.

In this scenario, agents are like services provided by the workflow management system. The aim to using agents here is to increase automation for workflow system. As substitution for humans, agents can carry out many tasks without human involvement. The main beneficial features of agents include autonomy, communication, and reactivity. From the system view, agents do not necessarily interact with each other. In fact, the workflow engine controls their actions. The information exchange, if there is any, is through workflow engine. The workflow engine is responsible to create and eliminate the agents. Second, the agents in this scenario are not necessarily "intelligent", which means, it might not have the strong notion of agents. In most commercial WFMS products, the agent is more like a piece of ordinary software, e.g. IBM MQSeries Workflow [6] and InConcert [8].

Some research projects on this schema try to fully use the agent abilities in every dimension. One example is the project in BT Lab by Shepherdson and colleagues [11, 15]. The agents correspond to activities one by one. Judy et al. [11] illustrates the ability to route work items at the or-split point in a workflow definition. The following activities are some identical activity, which means the work item can go through either path. In current WFMSs, path is chosen probably or wait until the user represents on activity acquires this work item. Now, by using agents to represent activities, the work item is routed by the result of negotiation. The contract net [4] is the supporting negotiation technique. The work item is reward to the agent that posts the lowest cost. [15] mentioned other benefits, such as provisioning, interoperability, supporting for visualization and verification services, and protecting investment in workflow technology. Zeus agent toolkit developed in BT is chosen as the agent development environment.

Another example is TRP support environment (TSE) by Chang and Scott [3] in Andersen Consulting. Their work has a central workflow agent (as workflow engine) and a set of agents acting as a personal assistant, communication facilitator, database connection manager, etc. TRP Agent Communication Language is defined for communication. This system also has a WWW interface.

4.2. Agent-Based Workflow Management

An Agent-based workflow system is a distributed system consisting of multiple agents (cf. Figure 2). These agents are independent to each other and each is responsible for process execution. In this scenario, the whole business process is formed by the pieces of sub-networks within those agents. The process logic is embedded in the agents, rather than being explicitly represented elsewhere.

This scenario is quite interesting in real world; the business process is across several units in a company or even spreads across several companies. Thus a central workflow engine is unable to get all the information of the whole business process in order to control it. A more likely solution is to have one workflow engine residing in each organization or unit. Through interactions among the multiple workflow engines the whole business process is fulfilled.

Agents in this scenario take full responsibilities of a workflow management system, which means that agents have all the means to analyze, automate, integrate, and inspect workflows. More important, agents should have means to communicate and interact with each other. Third, agents' high-level ability, such as learning, and negotiation, can add values to workflow system.

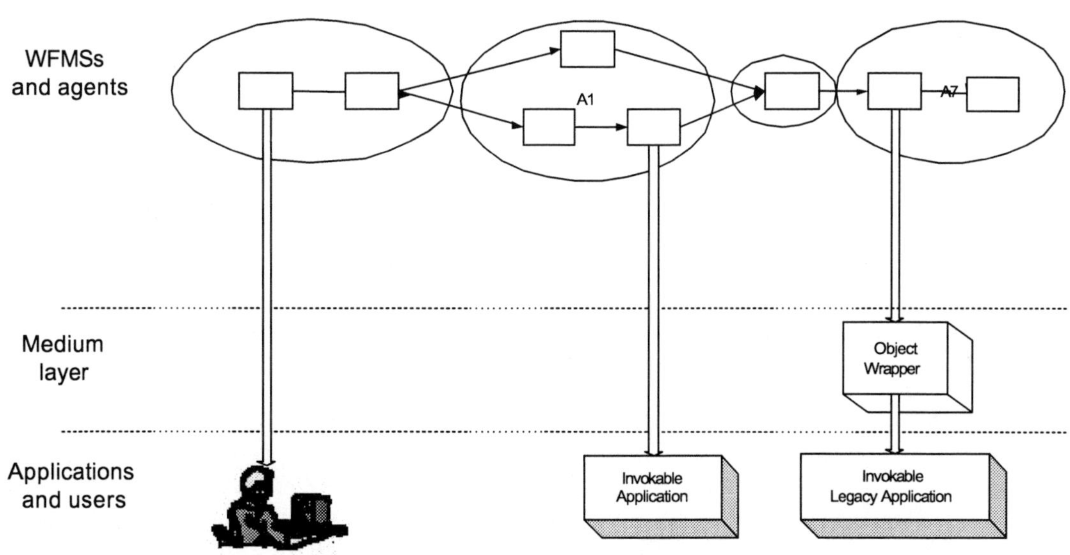

Figure 2. Agent-based Workflow Management System.

Agents in this scenario should be much more complex than the first one. The main features include: autonomous, communication, self-consistent, goal-oriented, and react to environment. Other high-level features such as learning, and negotiation are also benefit. But since these techniques are not matured, these features are just promising ones.

The usage of agents in this scenario benefits workflow technology in the following ways:

- *Providing distributed system architecture:* There are several system architectures in multi-agent systems [14], which can be used in a distributed system for implementing workflow management systems.
- *Providing communication methods:* Agent communication languages are studied a lot. There are several communication languages based on semantic messages, such as KQML. Theses languages can enable interoperation of workflow systems, especially for the heterogeneous ones.
- *Providing automation behavior:* Agent has the ability to execute tasks on its own without human involvement. Agent also has some decision power according to its goal.
- *Reacting to environment:* Agent can adjust itself, for example, forming new activity and new routing.
- *Benefits of high level features:* These high level features include learning, negotiation, and planning. Though these features are not fully implemented for industrial applications, they are what promising.

Nowadays, no commercial product falls in this category. But, several prototype systems exist in research areas, namely ADEPT [9] and FireFlow System [19]. This is an area attracting many research attentions. Some important research issues are:

(a) *System architecture:* various system architectures are presented. Normally an agent represents a functional team/department, sample systems are ADEPT [9]. Debenham [5] presents an ad-hoc workflow system for processing applications received by a university department from potential research students. In his proposal, agents act as application agent, supervisor agent, control agent, and administration agent. These agents delegate the functions of correspondent users. Another routing agent is to mediate the communication between agents. Another kind of system architecture uses mobile agents [2]. Mobile agent represents work item and the mobile agent decides which path the agent should go. On each node a server agent resides, which accepts mobile agents, interface to user, invokes the function condition for each work item agent, and prevents the work item idle in one node. Inomoto [7] presents another ad-hoc system to deal with internal document delivery. He uses an e-form agent as a mobile agent to deliver documents. A manager agent dispatches the e-form agent and has the ability of learning.

(b) *Negotiation:* negotiation is an important topic in multi-agent systems. Same negotiation methods and protocols can be used in agent-based business process management systems. The methods are categorized into theoretical ones, such as Game Theories [16], and contract nets [4], and non-theoretical ones, such as knowledge-based, and model-based methods [9]. Communication language is a relevant topic of negotiation. Although many proposals are related to communication languages, some protocols especially useful in exchanging business process information are presented. Jennings et al. [10] presents SDL (Service Description Language). Provided by an agent, a service is described by a name, its inputs, its outputs and its body.

(c) *Resource management:* via negotiation. Normally, agents represent resources, through negotiation to commit to certain tasks [9].

(d) *Scheduling:* via negotiation, decide start time [9].

(e) *Modeling:* all kinds of ways to analyze and model the workflow [1, 7, 12].

(f) *Learning:* decide new processes; communication is necessary.

(g) *Agent architectures:* Many architectures proposed and developed for multi-agent systems [14] can be used to implement agent-based workflow management systems.

4.3. Remarks

In the above two sub-sections, we discussed the two forms of integration of workflow management technology with agent technology. Here we will summarize the features of their combination.

If we consider the life cycle of a business process management, according to O'Brain, and Wiegrand [13], it can be divided into three phases: creation, provisioning, and enactment (cf. Figure 3).

The *creation stage* involves the analysis modeling and definition of the business processes. The business process can be represented as a diagram or a verbal format, for example, PIF. Some ad-hoc methods are presented, such as State-transition table [7]. The main aim of this stage is to present the logic of the activities in business process. Currently, this is mainly done at design time and in a manual way. The *provisioning stage* involves the assignment of resources, including people, equipment, and computing time. This requires negotiating, planning, and scheduling, to ensure that there are sufficient resources to handle expected throughput of work for a business process. This takes place at the run time. The *enactment stage* involves the management activities required to ensure that each instance of a business process is effectively executed. This includes routing of work, passing of information, activation of automated activities, and the handling of work list. This is also done at run time.

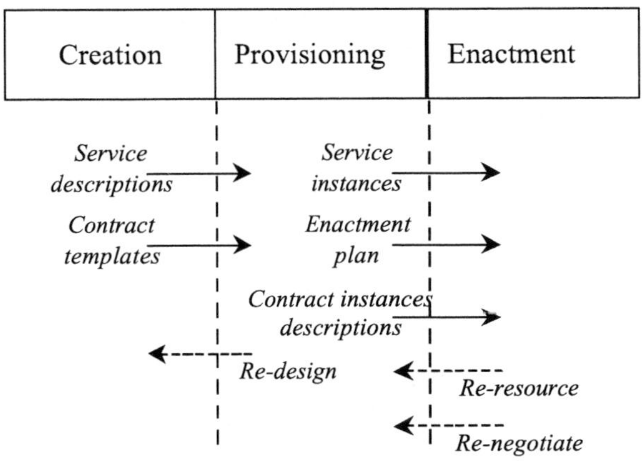

Figure 3. Life Cycle of Business Process Management (adapted from O'Brain, and Wiegrand, 1998).

Current workflow technology is good at the enactment stage. It uses all means to control and supervise the execution of business processes. Current workflow management systems also provide tools for user to define business process in workflow templates. These process-editing tools are vender dependent, and normally are visual graphic tools with/without description languages. But these methods are not mature, for they depend on human user's vision and lack of mathematic support. For the provision stage, current workflow technology has not provided support yet.

Agent technology, on the contrary, is good at providing the provisioning functions, for example, negotiation, scheduling, and resource allocation. Some cases can be found in above two sub-sections. But agent can also take part in the other two stages. For enactment stage, one important usage of agent is to provide automation mechanism for activities. Agent can also present work items, and through negotiation, decide the route that these work items will follow. If there is change in business process, agent can react to it. For creation stage, agents can form new business processes through negotiation.

Workflow technology combined with agent technology provides potential perspectives in business process management. Some currently successful areas for intelligent agents to apply in workflow management are:

(1) *Automation activity:* Agent helps to fulfill activity.

(2) *Integration with other applications:* Agent is a good way to provide a wrapper of other applications.

(3) *Communication:* Agent technology helps to set up protocols to exchange semantic information.

(4) *Interaction:* Agent technology helps to set up schema for the interactions among different workflow systems.

Something that agents promise to do, but are not mature enough yet includes:

(1) *Negotiation:* Although this study has plenty results. Real world is still more complex to be handled by current techniques.

(2) *Learning:* That would be a very interesting topic, especially when business process not well structured.

(3) *Personal working environment:* By integrating Groupware technology and smart agent technology, it is a promising direction.

Combination of workflow with agent technology also has some drawbacks: more complex systems; too many issues joined together; some agent-based methods are not mature yet. Besides the combination with agent technology, workflow management can benefit from integration with other technologies, such as project management, resource management, and intelligent scheduling.

5. Conclusion and Research Opportunities

Agent-enhanced workflow systems are widely adopted by commercial products, while the agent-based workflow systems are still in prototypes in research laboratories. However, the agent-based workflow systems are attracting more and more attention for studying the interaction of multiple workflow systems, which are the requirement of next generation of workflow systems.

Agents can help to automate the management of business processes to a much greater degree than has been possible before. The potential benefits include a reduced need for human involvement, which impacts the overall cost of managing business processes; a reduced response time to the changing working environment; and improved interoperability among heterogeneous systems.

The following research topics are identified as future research opportunities:

- *Modeling*: Business processes need to be well described. We need modeling methods especially for the processes across organization boundaries. We also need verification methods, based on mathematic analysis.
- *Communication:* Efficient communication languages for exchanging service definitions.

- *Negotiation*: Negotiation methods and protocols compliant to the features of the workflow management process.
- *Personal working environment*: Good user interfaces by using interface agents and personal assistant techniques.
- *Integrated with other technologies*: Integration with other technologies, such as project management, intelligent scheduling, and optimization, etc.
- *Learning*: Creation of new business process logic during run time through learning.

6. References

[1] S. Aknine, S. and S. Pinson, "Agent Oriented Conceptual Modeling of Parallel Workflow Systems", In Lecture Notes in Computer Science 1611, pp. 500-509, Springer, 1999.

[2] Budimac, Z., Ivanovic, M., and Popovic, A. Workflow Management System Using Mobile Agents, In Lecture Notes in Computer Science 1691, pp. 168-198, Springer,1999.

[3] J.W. Chang, and C.T. Scott, "Agent-based Workflow TRP Support Environment (TSE), Computer Networks and ISDN Systems, Vol. 28, issue 7-11, pp 1501, 1996.

[4] R. Davis and R.G. Smith. "Negotiation as a metaphor for distributed problem solving". *Artificial Intelligence*, 20:63-109, 1983.

[5] J. Debenham, "Constructing an Intelligent Multi-agent Workflow System", in Lecture Notes in Computer Sciences 1502, Springer-Verlag, pp. 119-130, 1998.

[6] IBM, "MQSeries Workflow", White Paper, 2000.

[7] Inamoto, "Agent Oriented System Approach for Workflow Automation", *International Journal of Production Economics*, 60-61:327-335, Elsevier, 1999.

[8] InConcert, "How is Infostrada able to offer innovative, customized services to a rapidly growing customer base in real-time?", www.tibco.com/customers/infostrada.html, 2000.

[9] N.R. Jennings, P. Faratin, T.J. Norman, etc., "ADEPT: Managing Business Processes Using Intelligent Agent", *Proceedings of BC Expert Systems 96 Conference*. 1996.

[10] N. R. Jennings, M. J. Wooldridge. Applications of Intelligent Agents. In N. R. Jennings, M. J. Wooldridge, (eds.), *Agent Technology: Foundations, Applications, and Markets*. Springer, 1998, pp. 3-28.

[11] D.W. Judy, B.R. Odger, J.W. Shepherdson, and Z. Cui, "Agent Enhanced Workflow", BT Technical Notes, Vol. 16, No. 3, p79-85, 1998.

[12] Z. Maamar and W. Shen, "An Approach Based on Perspectives for Modeling Workflows", in Proceedings of CSCWD 2000, Hong Kong, 2000.

[13] P.D. O'Brien and M.E. Wiegand, "Agent based process management: applying intelligent agents to workflow", the Knowledge Engineering Review, Vol. 13/2. 161-174, Cambridge University Press, 1998.

[14] W. Shen, D.H. Norrie and J.P. Barthes, "Multi-Agent Systems for Concurrent Intelligent Design and Manufacturing", Taylor and Francis, London, UK, 2000.

[15] J.W. Shepherdson, S.G. Thompson, and B. R. Odgers, "Cross-Organizational Workflow Management and Co-ordinated by Software Agent", in Proceedings of WACC'99 (International Joint Conference on Work Activities Coordination and Collaboration) Workshop on Cross-Organizational Workflow Management and Co-ordination February 22nd 1999, San Francisco, CA.

[16] Y. Shoham, "Agent Oriented Programming", Artificial Intelligence, 60(1), 51-92, 1993.

[17] WFMC, "The Workflow Management Coalition Specification, Workflow Reference Model", *Document Number TC00-1003*, Nov. 1994.

[18] K. Trammel, "Workflow without fear", *Byte*, April 1996.

[19] Y. Yan, "*Agent-based Process and Workflow Management*", PhD Thesis, Tsinghua University, P.R. China, 1999.

Verification of XRL: An XML-based Workflow Language

W.M.P. van der Aalst[1], H.M.W. Verbeek[1], and A. Kumar[2]
[1] *Faculty of Technology and Management, Eindhoven University of Technology, PO Box 513, NL-5600 MB, Eindhoven, The Netherlands.*
[2] *Database Systems Research Department, Bell Laboratories, 600 Mountain Ave., 2A-406, Murray Hill, NJ 07974, USA.*
w.m.p.v.d.aalst@tm.tue.nl, h.m.w.verbeek@tm.tue.nl, akhil@acm.org

Abstract

XRL (eXchangeable Routing Language) is an instance-based workflow language that uses XML for the representation of process definitions and Petri nets for its semantics. Since XRL is instance-based, workflow definitions can be changed on the fly and sent across organizational boundaries. These features are vital for today's dynamic and networked economy. However, the features also enable subtle, but highly disruptive, cross-organizational errors. On-the-fly changes and one-of-a-kind processes are destined to result in errors. Moreover, errors of a cross-organizational nature are difficult to repair. In this paper, we show soundness properties of XRL constructs by using a novel, constructive approach. We also describe a software tool based on XML and Petri-net technologies for verifying XRL workflows.

1. Introduction

Recent years have seen the proliferation of workflow management systems developed for different types of workflows and based on different paradigms [9]. Despite the abundance of such tools, the critical issue of workflow verification is virtually neglected [1]. Few tools provide any form of verification support. The tools Woflan [11] and Flowmake [10] are two noteworthy exceptions.

To complicate matters, more and more workflow management systems are used to support inter-organizational business processes, e.g., in the context of Business-To-Business (B2B) E-commerce. Especially for *open* E-commerce (i.e., doing business among parties having no prior trading relationship), the workflow support should be *trustworthy* in the sense that trading partners who do not know each other, and may even come from different countries and cultures, may conduct business with the assurance that their interests will be protected in the event that "things go wrong", whether by accident, negligence, or intentional fraud. One of the prerequisites for this is the guarantee that the workflow process definitions do not contain any logical errors.

Another requirement for *open* E-commerce is that one cannot make any arbitrary assumptions about the workflow processes implemented in the participating organizations. For instance, in the context of inter-organizational workflow it is unrealistic to assume that the different organizations share a common process model. Therefore, we developed the *eXchangeable Routing Language* (XRL), which describes processes at an instance level [4]. Traditional workflow modeling languages describe processes at a class or type level [7]. An XRL routing schema describes the partial ordering of tasks for one *specific* instance. The advantages of doing so are that: (1) the workflow schema can be exchanged more easily, (2) the schema can be changed without causing any problems for other instances, and (3) the expressive power is increased (workflow modeling languages typically have problems handling a variable number of parallel or alternative branches) [4].

The semantics of XRL are expressed in terms of Petri nets [4]. Such formal semantics allow for powerful analysis techniques, an efficient and compact implementation, interfaces to many existing tools, and, last but not least, an unambiguous understanding of XRL.

We have developed a workflow management system, named XRL/flower [4], to support XRL. XRL/flower benefits from the fact that it is based on both XML and Petri nets. Standard XML tools can be deployed to parse, check, and handle XRL documents. The Petri net representation allows for a straightforward and succinct implementation of the workflow engine. XRL constructs are automatically translated into Petri net constructs. On the one hand, this allows for an efficient implementation. On the other hand, the system is easy to extend: For supporting a new routing primitive, only the translation to the Petri net engine needs to be added and the engine itself does not need to change. Last, but not least, the Petri net representation can be analyzed using state-of-the-art analysis techniques and tools.

In this paper, we present a verification tool that can analyze workflows specified in terms of XRL. The tool is built on top of Woflan [11] and detects design errors resulting in deadlocks, livelocks, etc. We also give some analytical results which show that for a subset of XRL correctness is guaranteed if the design is consistent (i.e., a well-formed and valid XML file) with the XRL Document Type Definition (DTD).

2. XRL and XRL/flower

The focus of this paper is on verification. Therefore, we limit ourselves to only a brief introduction to XRL, the translation of XRL to Petri nets, and the workflow management system XRL/flower. The syntax of XRL is completely specified by the DTD given in the appendix. A routing element is an important building block of XRL and it can be any one of the following: *task* (a step to be performed), *sequence* (a set of tasks to be done in a specific order), *any_sequence* (a set of tasks to be done in any order), *choice* (any one task out of a set of tasks), *condition* (test a condition and determine next step based on result of the test), *parallel_sync* (create multiple parallel routing elements and later join them), *parallel_no_sync* (create multiple parallel routing elements which do not have to join), *parallel_part_sync* (create multiple parallel routing elements, some of which must join), *wait_all* (insert a wait step to wait for the completion of a group of events), *wait_any* (insert a wait step to wait for the completion of any one of a group of events), *while_do* (enable repetition of a task while a condition is true), *stop* (end the execution of this particular path of the workflow instance), *terminate* (end this workflow instance).

An example of a consistent XRL file is shown in Figure 1. This corresponds to the Petri net shown in Figure 2. It describes a workflow where a customer complaint is *registered* and then a questionnaire is sent to the customer in parallel with the complaint being sent to a manager for *evaluation*. If the customer does not fill the form within 3 days the event *receipt* occurs any way and puts a token in the *receipt* place. This event and the *evaluate* activity enable the loop in which *process* and *check* steps occur. After the result of *check* step is okay, the *archive* step is enabled.

This translation from XRL to Petri net is done according to rules given in [4]. For the sake of readability, we removed superfluous transitions and places that were generated by the rules. Note that task transitions are represented by squares and routing transitions by bars.

```
<?xml version="1.0" encoding="UTF-8"?>
<route name="example">
  <sequence>
    <task name="register"/>
    <parallel_sync>
      <sequence>
        <task name="send"/>
        <wait_all>
          <event_ref name="receipt"/>
          <timeout time="3 days">
            <task name="timeout">
              <event name="receipt"/>
            </task>
          </timeout>
        </wait_all>
      </sequence>
      <sequence>
        <task name="evaluate"/>
        <condition>
          <true>
            <sequence>
              <wait_all>
                <event_ref name="receipt"/>
              </wait_all>
              <while_do condition="okay">
                <sequence>
                  <task name="process"/>
                  <task name="check"/>
                </sequence>
              </while_do>
            </sequence>
          </true>
        </condition>
      </sequence>
    </parallel_sync>
    <task name="archive"/>
  </sequence>
</route>
```

Figure 1. An example XRL file.

In the remainder, we will use the term *XRL route* to refer to a consistent XRL routing schema such as the one shown in Figure 1. A consistent sub-workflow made from these constructs is also called a routing element, e.g., `re_1`, `re_2`, etc. Please note that, since an XRL route specifies the life cycle of a particular workflow instance (i.e., work case), any instance can be modified without reference to some underlying workflow schema type.

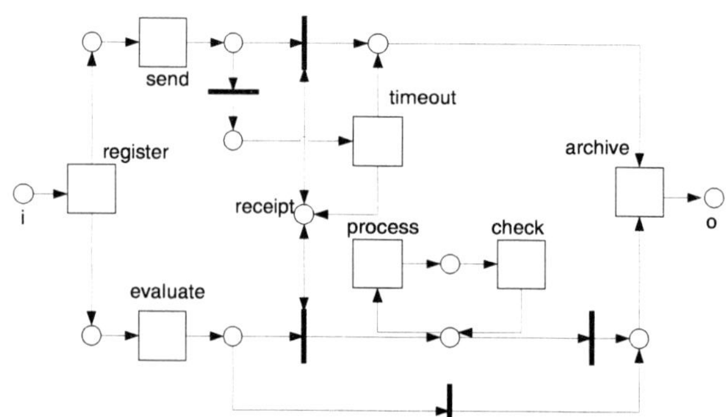

Figure 2. Petri net of example XRL file.

Based on XRL, we have developed a workflow management system named XRL/flower. XRL/flower. XRL/flower can handle XRL files arriving through e-mail or ftp. An incoming XRL file, i.e., workflow instance, is parsed and translated into a Petri net. The Petri net description drives the workflow engine, which calculates enabled tasks. The enabled tasks are offered to the proper workers through role-based worklists. Whenever a task is executed, the engine calculates newly enabled tasks. The engine or an authorized user can also decide to migrate a running instance to another workflow engine. For migration, an XRL file is created with entries for the current workflow state and shipped through e-mail or ftp.

3. Verification

XRL is aimed towards application domains where workflows cross organizational boundaries and change over time. In these applications domains, the correctness issues are particularly relevant because the distributed and dynamic nature of the workflow is a potential source of errors. Unfortunately, today's workflow management systems do not support advanced techniques to verify the correctness of workflow process definitions. These systems typically restrict themselves to a number of simple syntactic checks. Therefore, erroneous conditions such as deadlocks and livelocks may remain undetected. This means that an erroneous workflow may go into production, thus causing dramatic problems for the organization. An erroneous workflow may lead to extra work, legal problems, angry customers, managerial problems, and depressed employees. Therefore, it is important to verify the correctness of a workflow process definition before it becomes operational. In fact, for inter-organizational workflows, the costs of putting an erroneous workflow process definition into production are enormous because of the efforts required to repair errors crossing organizational boundaries.

The *soundness property*, defined in [1], relates to the dynamics of the workflow process definition expressed in terms of a so-called *workflow net*. A workflow net is a Petri net with one unique source place (for the initial state) and one unique sink place (for the final state). A workflow net is *sound* if the following requirements are satisfied: (1) termination is guaranteed, (2) upon termination, no dangling references (tokens) are left behind, and (3) there are no dead tasks, i.e., it should be possible to execute an arbitrary task by following the appropriate route. Soundness is the minimal property any workflow net should satisfy. Note that soundness implies the absence of livelocks and deadlocks. Soundness can be verified using Petri net techniques. In fact, we have developed a workflow verification tool, named *Woflan*, to decide soundness [12]. For a given workflow net, Woflan is able to decide whether it is sound. For this purpose, Woflan uses an interesting relation between soundness on the one hand, and liveness and boundedness on the other. A workflow net is sound, if and only if, the net obtained by connecting the sink place to the source place via an additional transition t^* is live and bounded. This relationship enables the use of efficient analysis techniques and the deployment of powerful software packages.

In a workflow net there is one unique source place and one unique sink place. For a straightforward translation of XRL to Petri nets, this requirement is too restrictive. Therefore, we introduce *extended workflow nets*. An extended workflow net may have multiple source places and multiple sink places. Furthermore, the requirement that only source places can indicate the arrival of a case and only sink places can indicate completion of a case is too restrictive. For this reason, we introduce so-called start- and end places. A workflow instance (i.e., case) is started by marking *one* of the start places, and it is completed if only end places are marked. Typically, a start place is a source place and an end place is a sink place. However, we also allow for instance the situation with end places that are no sink places (i.e., with outgoing arcs). We define start and end places as follows. A place is a *start place* if and only if it is a source place, i.e., a place without any ingoing arcs. A place p is an *end place* if every transition that consumes tokens from p also produces at least one token for p. Formally, a place is an end place if and only if it is a trap [6]. Note that any sink node is a trap, i.e., an end place. Any node of the extended workflow net should be on a path from some start place to some end place. Finally, we drop the assumption that every transition corresponds to some task. Transitions that do correspond to tasks are called task transitions.

Definition I. Extended workflow net
A Petri net is called an extended workflow net if and only if (1) it contains at least one start place, (2) it contains at least one end place, and (3) every task transition is on a path from some start place to some end place.

The Petri net shown in Figure 2 is an extended workflow net, There is one start place (i.e., place i) and two end places: *receipt* and o. Note that *receipt* is an end place but not a sink place. A state is called a *start state* if precisely one of the start places is marked. The set of *reachable states* of an extended workflow net are those states that are reachable from any start state. A reachable state is called a *final state* if and only if only end places are marked.

Definition II. Soundness
An extended workflow net is *sound* if and only if (1) from any reachable state it is possible to reach a final state, (2) no transitions are enabled in any reachable final state, and (3) there are no dead task transitions, i.e., it should be possible to execute an arbitrary task by following the appropriate route.

Recall that the soundness property is important because it implies absence of deadlocks and livelocks. It is easy to verify that the extended workflow net shown in Figure 2 is indeed sound.

4. Modeling XRL constructs

Although a (non-extended) workflow net is too restrictive in general, it is sufficient if we limit ourselves to use only certain XRL constructs. These constructs typically can be modeled as a workflow net and preserve soundness. We show this in a constructive manner. Afterwards, we discuss constructs for which workflow nets are too restrictive.

The Petri net corresponding to a *task* is a workflow net, provided it does not generate any *events*. It is straightforward to show that this net is sound.

For all constructs, we assume that each embedded routing element corresponds to a Petri net that is a sound workflow net.

For the *sequence* construct, it is straightforward to see that its corresponding Petri net is a workflow net. This workflow net is sound under the assumption mentioned above, which is easy to verify.

In a similar way, the constructs *any_sequence*, *parallel_sync*, *choice*, *condition*, and *while_do* have corresponding Petri nets that are workflow nets and are sound under the assumption as mentioned above. For sake of completeness, we mention that the semantics as given in [4] for the *condition* construct assumes that exactly two routing elements are embedded: one if the condition evaluates to true and one if it evaluates to false. In general, the semantics of the *condition* construct can be described as in Figure 3.

Result 1: If an XRL route is restricted to the constructs *sequence*, *any_sequence*, *choice*, *condition*, *while_do*, and *parallel_sync,* and if no *events* are produced by the *tasks*, then the corresponding Petri net representation is a sound workflow net.

The use of *tasks* producing *events* and the constructs *parallel_no_sync*, *parallel_part_sync*, *wait_all*, and *wait_any* results in additional end places. For the *parallel_no_sync*, an output place is added to each embedded routing element. Such an output place is an end place. For the *parallel_part_sync*, a rather complex network is used to make sure that the remaining control threads are detached, i.e., superfluous tokens are removed. This network contains an additional end place to signal completion. Furthermore, each event place is also an end place. As a result, if the constructs *stop* and *terminate* are not used, the Petri net corresponding to an XRL route is an *extended* workflow net. This net is sound, which is straightforward to prove, if the following conditions hold: (1) for each *parallel_part_sync* the number of branches k to synchronize is valid, i.e., at least 1 and at most the total number of branches n, (2) every *wait_all* and *wait_any* construct has a *timeout* defined. From now on we will assume that all *parallel_part_sync* constructs are valid, which seems reasonable.

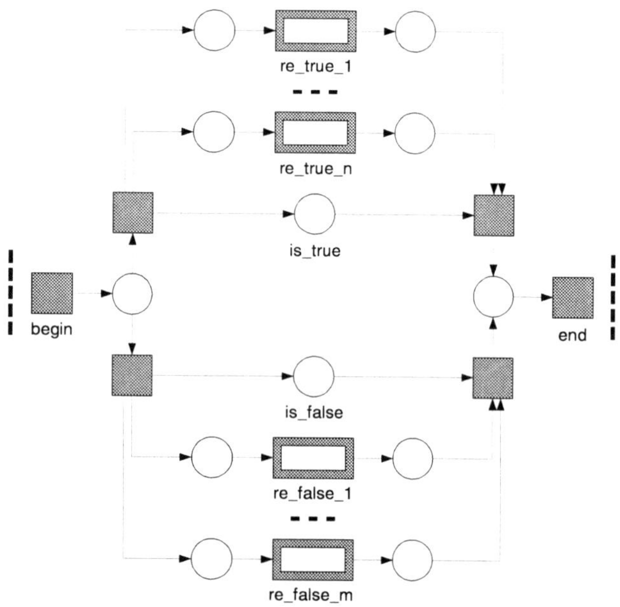

Figure 3. Construct *condition*.

Result 2: If an XRL route does not contain *stop* or *terminate* constructs, and if every *wait_all* and *wait_any* construct contains a *timeout*, then the corresponding Petri net representation is a sound extended workflow net.

This result can be applied in many situations and requires only checks at the syntactical level. However, whenever *stop* and *terminate* constructs are used and/or *wait_all* and *wait_any* constructs without a *timeout* are used, a more detailed analysis is required.

Assume the *stop* construct is put in the middle of a sequence, e.g., in-between tasks *t1* and *t2*. Task *t2* can never be executed, thus violating the third requirement of soundness. Therefore, the *stop* construct should only be used at the end of a sequence. Moreover, when a sequence is embedded in, for instance, a *parallel_sync* construct, then no *stop* construct is allowed in it.

Similarly, the *terminate* construct can invalidate soundness. Moreover, the *terminate* construct can also prevent tasks executed in parallel from being executed.

By enforcing syntactical requirements it is possible to extend Result 2 to nets using *stop* and *terminate* constructs. For example, in any sequence a *stop* should not be followed by a task. However, if some *wait_all* or *wait_any* does not contain a *timeout*, it is not possible to decide whether or not the net is sound using only syntactical criteria. To deal with this situation, we have developed a link between XRL and our workflow verification tool Woflan [11].

5. Verification tool: XRL/Woflan

In this section we describe an automatic translation from XRL to Woflan. This way any workflow management based on XRL can benefit from state-of-the-art verification software.

Woflan (http://www.tm.tue.nl/it/woflan) is designed as a WFMS-independent analysis tool. In principle it can interface with many workflow management systems. At present, Woflan can interface with the workflow products COSA (Thiel Logistic AG/Software Ley), METEOR (LSDIS), and Staffware (Staffware), and the BPR-tool Protos (Pallas Athena).

Woflan can read Petri Net Markup Language (PNML) files. PNML is a Petri net file format based on XML [8]. Therefore, it is obvious to use XSL to automatically translate an XRL route into a PNML representation that can be diagnosed using Woflan. This translation can be done in a rather straightforward way; the only difficulty is the generation of unique names for the transitions and places used in the various constructs. Figure 4 shows a snippet of the PNML file corresponding to the example XRL file:

```
<?xml version="1.0" encoding="UTF-8"?>
<net id="example">
  <name>
    <value>example</value>
  </name>
  <transition id="register_begin">
    <name>
      <value>"register begin"</value>
    </name>
  <place id="register_executing">
    <name>
      <value>"register executing"</value>
    </name>
  </place>
  <arc id="register_begin_executing"
       source="register_begin"
       target="register_executing"/>
  ...
</net>
```

Figure 4. The resulting PNML file (snippet)

Note that it is only necessary to use Woflan if Result 2 does not apply, i.e., for verifying *wait_all* and *wait_any* constructs that do not contain *timeouts* sub-elements or nasty *stop* and *terminate* constructs. We tested Woflan on the example XRL file of Figure 1. In this case there is one *wait_all* without *timeout*. Therefore, we use Woflan to check whether this *wait_all* can deadlock. At first, Woflan concludes that a transition corresponding to the other *wait_all* is dead. But for this *wait_all* a *timeout* is defined, so this situation is not problematic. After removing the dead transition from the net, Woflan concludes that the net is a sound extended workflow net.

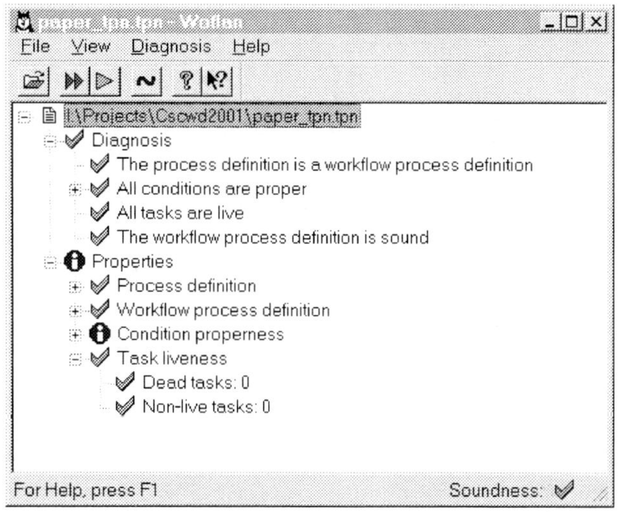

Figure 5. Woflan's results.

6. Conclusion

We presented a way to verify XRL routes. Depending on the constructs used in an XRL route, we can (1) claim soundness, (2) decide soundness by checking the XRL route's structure, or (3) decide soundness by using Woflan as a verification tool. We only need to use Woflan if *wait_all* or *wait_any* construct are used that do not contain a *timeout*. If those constructs are not used, soundness depends only on the positioning of the *stop* and *terminate* constructs.

Because Woflan only could check soundness on workflow nets, we have extended Woflan. Woflan can now also check soundness on extended workflow.

For translating XRL route to Woflan, we have build an XSL file that translates the XRL route to a PNML file. Woflan can read PNML files.

7. References

[1] W.M.P. van der Aalst. The Application of Petri Nets to Workflow Management. *The Journal of Circuits, Systems and Computers*, 8(1):21-66, 1998.

[2] W.M.P. van der Aalst. Process-oriented Architectures for Electronic Commerce and Interorganizational Workflow. *Information Systems*, 25(1):43-69, 2000.

[3] W.M.P. van der Aalst, J. Desel, and A. Oberweis, editors. *Business Process Management: Models, Techniques, and Empirical Studies*, volume 1806 of Lecture Notes in Computer Science. Springer-Verlag, Berlin, 2000.

[4] W.M.P. van der Aalst and A. Kumar. XML Based Schema Definition for Support of Inter-organizational Workflow. Technical report Bell-Labs. 2001.
[5] T. Bray, J. Paoli, C.M. Sperberg-McQueen, and E. Maler, "eXtensible Markup Language (XML) 1.0 (Second Edition)," http://www.w3.org/TR/REC-xml, World Wide Web Consortium (W3C), October 2000.
[6] J. Desel and J. Esparza. *Free choice Petri nets*, volume 40 of Cambridge Tracts in Theoretical Computer Science. Cambridge University Press, Cambridge, 1995.
[7] S. Jablonski and C. Bussler. *Workflow Management: Modeling Concepts, Architecture, and Implementation* International Thomson Computer Press, 1996.
[8] M. Jüngel, E. Kindler and M. Weber. The Petri Net Markup Language. In S. Philippi (ed.), AWPN, Koblenz, 2000.
[9] P. Lawrence, editor. *Workflow Handbook 1997, Workflow Management Coalition.* John Wiley and Sons, New York, 1997.
[10] W. Sadiq and M.E. Orlowska. Analyzing Process Models using Graph Reduction Techniques. *Information Systems*, 25(2): 117-134, 2000.
[11] H.M.W. Verbeek and W.M.P. van der Aalst. Woflan 2.0: A Petri-net-based Workflow Diagnosis Tool. In M. Nielsen and D. Simpson (editors), Application and Theory of Petri Nets 2000, volume 1825 of Lecture Notes in Computer Science, pages 475-484. Springer-Verlag, Berlin, 2000.
[12] H.M.W. Verbeek and T. Basten and W.M.P. van der Aalst. Diagnosing Workflow Processes Using Woflan, BETA Working Paper Series, WP 48. Eindhoven University of Technology, Eindhoven, 2000.
[13] Workflow Management Coalition, Workflow Standard – Interoperability Wf-XML Binding, Document Number WFMC-TC-1023, Version 1.0, http://www.wfmc.org, May 2000.

Appendix: XRL Document Type Definition

For the semantics of the following DTD, we refer to [4].

```
<!ENTITY % routing_element
"task|sequence|any_sequence|choice|
condition|parallel_sync|parallel_no_sync|
parallel_part_sync|wait_all|wait_any|
while_do|stop|terminate">
<!ELEMENT route (%routing_element;)>
<!ATTLIST route name ID #REQUIRED
  created_by CDATA #IMPLIED
  date CDATA #IMPLIED>
<!ELEMENT task (event*)>
<!ATTLIST task name ID #REQUIRED
  address CDATA #REQUIRED
  role CDATA #IMPLIED
  doc_read NMTOKENS #IMPLIED
  doc_update NMTOKENS #IMPLIED
  doc_create NMTOKENS #IMPLIED
  result CDATA #IMPLIED
  status (ready|running|enabled|disabled|
aborted|null) #IMPLIED
  start_time NMTOKEN #IMPLIED
  end_time NMTOKEN #IMPLIED
  notify CDATA #IMPLIED>
<!ELEMENT event EMPTY>
<!ATTLIST event name ID #REQUIRED>
<!ELEMENT sequence
((%routing_element;|state)+)>
<!ELEMENT any_sequence
((%routing_element;)+)>
<!ELEMENT choice ((%routing_element;)+)>
<!ELEMENT condition (true|false)*>
<!ATTLIST condition condition CDATA
#REQUIRED>
<!ELEMENT true (%routing_element;)>
<!ELEMENT false (%routing_element;)>
<!ELEMENT parallel_sync
((%routing_element;)+)>
<!ELEMENT parallel_no_sync
((%routing_element;)+)>
<!ELEMENT parallel_part_sync
((%routing_element;)+)>
<!ATTLIST parallel_part_sync number NMTOKEN
#REQUIRED>
<!ELEMENT wait_all (event_ref|timeout)+>
<!ELEMENT wait_any (event_ref|timeout)+>
<!ELEMENT event_ref EMPTY>
<!ATTLIST event_ref name IDREF #REQUIRED>
<!ELEMENT timeout (%routing_element;?)>
<!ATTLIST timeout time CDATA #REQUIRED
type (relative|s_relative|absolute)
"absolute">
<!ELEMENT while_do (%routing_element;)>
<!ATTLIST while_do condition CDATA
#REQUIRED>
<!ELEMENT stop EMPTY>
<!ELEMENT terminate EMPTY>
<!ELEMENT state (event+)>
```

Modeling Workflow within Distributed Systems

Yuhong Yan
Flydragon Computer Consulting, NY, USA
flydragon960@yahoo.com

Alex Bejan
ANCA Tech. RI, USA

Abstract

Workflow management techniques are aiming at supporting business process across organization boundaries. Current techniques are lacking of the formalism tools to model and analyze workflows in a large scale. And current WFMSs do not have the ability to react to the response of another WFMS dynamically. Using Petri Nets as the modeling tools, we present the concept of Standard Workflow Structure and a set of standard workflow blocks are designed. We prove that modeling a workflow process in Standard Workflow Structures can guarantee the soundness of a workflow network defined by Aalst [1]. We also presented the decomposing method of a large scaled workflow network into sub-networks. From our analysis and prove, we draw a very good conclusion that the workflow network composed by self-loop connected sub-networks maintains soundness under proper structure conditions. The methods presented are not only a design language for the specification of complex workflows, but also powerful analysis techniques to verify the workflow procedures.

1. Introduction

The main purpose of a workflow management system is the support of the definition, execution, registration and control of processes. These processes can be any kind, but most stimulation comes from its promising usage in managing business processes.

Modeling business processes is the start point of the workflow management techniques. Workflow techniques promise to provide the efficient way to model the complex business processes, so as to optimize them, reuse them and control them. But the fact is though many workflow management systems have modeling tools, these tools are paid much less effort than the workflow engine itself. These modeling tools normally use icons and links to represent the activities and the sequential order between the activities (e.g. ActionWorks, IBM FlowMark, Filenet, etc)[2][3][4]. Associated with these visual tools, description languages (e.g. ActionWorks, FlowMark)[2][3] are designed to allow linguistically programming and manipulating these process instances. However, current techniques in workflow modeling have many drawbacks. First, nearly all the methods developed so far are limited on the level of graphical description. No mathematical skills are used to analyze the properties of the processes. The correctness of the modeling is based on human check and considerations. Second, the current methods are not suitable in large scaled analysis, where the processes are more complex and across organizational boundaries. Not one workflow system can define and execute theses processes. Third, the diversity of current methods prevents the interoperation among WFMS from different vendors. Although WFMC has set up a reference model [5] to provide basic conceptions and enhance interoperation among heterogeneous workflow management systems, the situation is not improved.

We consider the Petri net is a natural model for accomplishing the need of workflow modeling. The main reasons for using Petri nets for workflow modeling [1] include: Formal semantics, a workflow process specified in terms of a Petri net has a clear and precise definition, because the semantics of the classical Petri net and several enhancement (color, time, hierarchy) have been defined formally; Graphical nature, Petri nets are a graphical language and easy to learn. This feature made Petri nets to be the choice to be used in workflow definition Tools, designed in WFMC Workflow Reference Model [5]. Expressiveness, Petri nets support all the primitives needed to model a workflow process. All the routing constructs present in today's workflow management systems can be modeled. Analysis methods, this is a great asset in favor of the use of Petri nets for workflow modeling. The analysis techniques can be used to prove properties and to calculate performance measures. Vendor independent, Petri nets are not based on a software package of a specific vendor and do not cease to exist if a new version is released.

Using Petri Nets as the modeling tools, we present the concept of Standard Workflow Structure and a set of standard workflow blocks. We prove in this paper that modeling a workflow process in Standard Workflow Structures can guarantee the soundness of a workflow network defined by Aalst (1998). We also present the method of decomposing a large scaled workflow network into sub-networks. From our analysis and prove, we draw a very good conclusion that the workflow network composed by self-loop connected sub-networks maintains soundness under proper structure conditions.

2. Workflow Modeling Based on Petri Nets: State of Art

Petri net is the promising tool of workflow modeling. Here is its definition.

Definition 1 (Petri nets) A PN is a three-tuple (P, T, F) [6], where
1. P = {p1, p2, p3, ..., pn} is a set of places
2. T = {t1, t2, t3, ..., tn} is a set of transitions
3. $F \subseteq S\times T \cup T\times S$ is a set of acres linking the places and the transitions
4. $P \cup T \neq \Phi$, $P \cap T = \Phi$

The early work in workflow modeling lays in Zisman's work in his thesis [7]. His modeling modified the usually Petri net firing rules by adding other rules to act as guards for transition firing. Li [Li, 1990] also employed a Petri net variant for representing office work. While both of these models are essentially high-level Petri nets, their control flow is specified using the basic Petri net firing rules. [8] presents the Information Control Net (ICN) model which is intended to represent control flow and data flow in office procedures. ICNs identify firing per transition by associating a token with each transition, then using semantics similar to color Petri nets. The recent work down by Aalst [1] presents the formal definition of workflow network based on the concept of Petri Net [1]:

Definition2 (WF-net) A Petri net PN= (P, T, F) is a WF-net (Workflow net) if and only if

(1) PN has two special places: i and o. Place i is a source place $\bullet i=\phi$. Place o is a sink place: $o\bullet=\varnothing$.

(2) If we add a transition t* to PN which connects place o with i (i.e. $\bullet t^*=\{o\}$ and $t^*\bullet=\{i\}$), then the resulting Petri net is strongly connected.

Aalst also discusses the correctness of the model. He considers a correctly modeled workflow should transfer the token from the start place to the end place and when there is a token at the end place, there is no other token in any other places. Second condition is that any transition in the workflow net can be triggered in a path through out the network. He gives this property a name of soundness.

Definition 3(Sound) A procedure modeled by a WF-net PN = (P, T, F) is sound if and only if

(1) For every state M reachable from state i, o is reachable. Formally:

$\forall M$, if $(i[\sigma>M))$, then $\exists f$, $M[f>o$

where σ is transition series

(2) State o is the only state reachable from state i with at least one token in place o. Formally:

$\forall M$, if $i[\sigma>M \wedge M=o$ then $M=0$

(3) There are no dead transitions in (PN, i). Formally:

$\forall t\in T$, if $\exists M$, M' $i[\sigma>M$, then $\exists t$ $M[t>M'$

The soundness of a WF-net implies the correctness of the model. Our following analysis based mostly the work of Aalst.

3. The Definition of Standard Workflow Network

The motivation of defining the standard workflow network comes from the observation of the illness structures a workflow net concept can conduct. See the two workflow nets show in figure 1.

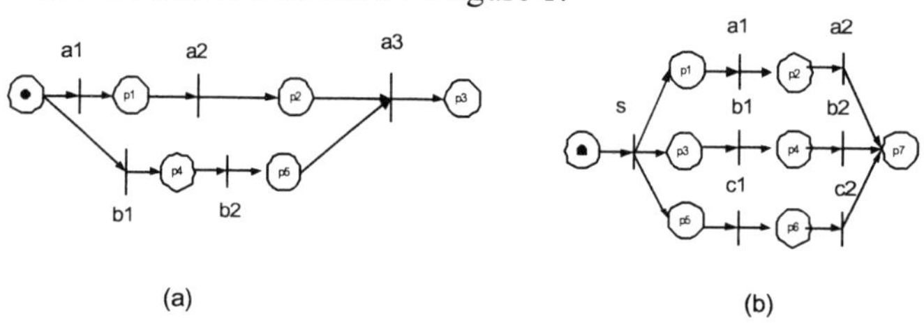

Figure 1. Two Samples of Illness Structures

Figure 1 summarizes two kinds of ill structures. In (a), following the source place are two Or-branches, so only one branch can be traced. But no matter which branch is executed, the token is blocked in p2 or p5, because the transition a3 can never be triggered. The reason of this illness comes from the structure itself. Two Or-branches after an Or-split point are synchronized by one transition. This causes this transition dead. A symmetric case is in (b), where the And-branches are synchronized by one Or-join point. This causes the problem of resource competition and this structure cannot be ended properly. This means that the structure of a workflow network should have more restrict limitation, so that the Or-split point and And-split point cannot appear as a pair. In the following section, we introduce the definition of Standard Workflow to avoid this problem.

Definition 4 (the Basic Sequential Branch) the procedure formed by sequential linked transitions and places. Noted by Seq_Branch(P, T, I, O). Formally:

Seq_Branch(P, T, I, O)=[P0][tkPk]*,

where palces: P={p0, p1,..., pn}, and transitions T={ t0, t1,..., tn }. $\forall i=0,1,...n-1$, I(pi,tt+1)=1; $\forall i=1,2,..,n$, O(pi,ti)=1; other I(pi,tj)=0, O(pi,tj)=0. P0(t1) is called the start place (transition)of the branch, pn(tn) is called the end place (transition) of the branch. See figure 2 (a).

Definition 5 (the Basic Conditional Branch) a procedure formed by several basic sequential branches which share the common start place and the common end place. Note by Cond_Branch(P, T, I, O). Formally

$$[p_s][\bigcup_k^m\{t_{k,0}Seq\_Branch(P_k,T_k,I_k,O_k)t_{k,nk}\}][p_e]$$

ps(pe) is called the start (end) place of each branch, tk,0(tk,nk)is the start (end) transition of the kth branch. See figure 2 (b).

Definition 6 (the Basic Parallel Branch) a procedure formed by several basic sequential branches which share the common start transition and the common end transition. Note by Par_Branch(P, T, I, O), formally:

$$[p_s][t_s][\bigcup_k^m\{t_{k,0}Seq\_Branch(P_k,T_k,I_k,O_k)t_{k,nk}\}][t_e][p_e]$$

ts(te) is called the start (end) transition of each branch, pk,0(pk,nk)is the start (end) place of the kth branch. See figure 2(c).

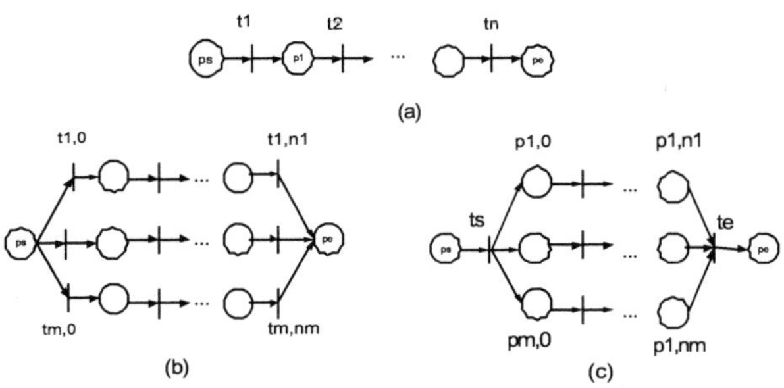

Figure 2. The basic structure of workflow network

Definition 7 (the Complex Sequential Structure) the procedure formed by serializing the basic

$$\{[Seq\_Branch(P_s,T_s,I_s,O_s)]$$
$$*[Cond\_Struct(P_c,T_c,I_c,O_c]$$
$$*[Par\_Struct(P_p,T_p,I_p,O_p)]*\}*$$

sequential branches, the complex conditional structures and the complex parallel structures. Noted by Seq_Struct(P,T,I,O), formally:

Definition 8 (the Complex Conditional Branch) the procedure formed by paralleling several complex sequential structures with the common start place and the end place. Noted by Cond_Struct(P,T,I,O), formally:

$$[p_s][\bigcup_s^m\{t_{s,0}Seq\_Struct(P_s,T_s,I_s,O_s)t_{s,ns}\}][p_e]$$

Definition 9 (the Complex Parallel Branch) the procedure formed by paralleling several complex sequential structures with the common start transition and the common end transition. Noted by Par_Struct(P,T,I,O), formally:

$$[p_s][t_s][\bigcup_s^m\{t_{s,0}Seq\_Struct(P_s,T_s,I_s,O_s)t_{s,ns}\}][t_s][p_e]$$

Definition 10 (the Standard Workflow Network (SWF-net)) the network formed by the above complex structures with shared start place and end place.

As a stronger definition than the WF-net, SWF-net inherits the properties of the WF-net (i.e. the common start place and the end place, the strong connection). In real work, we found its logic covers all kinds of the business processes, so that it can be used in most applications. The SWF-net benefits at least too things. First, it can be used in designing workflow engine. The execution algorithms are designed according to these standard blocks. Second, the SWF-net can be used in large scaled system analysis. As we will discuss in section 3.2 the large scaled network can be divided into several sub-networks while maintaining its properties, which also found the theoretical base to link several distributed networks together.

A SWF-net has the following properties:

Theorem 1. A SWF-net (N, i) is bounded.
Theorem 2. A SWF-net (N, i) is live
Theorem 3. A SWF-net is sound.

According to the definition of SWF-net, the first two theorems are easy to prove. The third theorem is based on the Theorem 1 in [1]. The three theorems provide the fundamental for workflow decomposition.

4. Workflow Network Decomposition

4.1 Workflow Network Decomposition

A large scaled workflow process normally involves several companies or several departments in one company. Thus a workflow network is composed of several sub-networks that are modeling independently and executed by several separated workflow management systems. A sample process of order processing is shown in figure 7. The specific transitions and places are listed in table 1 in Appendix.

The sales department (CA) accepts orders from customers, and then passes the order to production department (SA). The production department therefore makes the production plan. In order to gather enough information about inventory, the production department must check the inventory management department (PMA). Thus SA responses to CA after PMA sends it answers. And CA responses to customer request according to SA's report. Figure 3 is the whole Petri Net model.

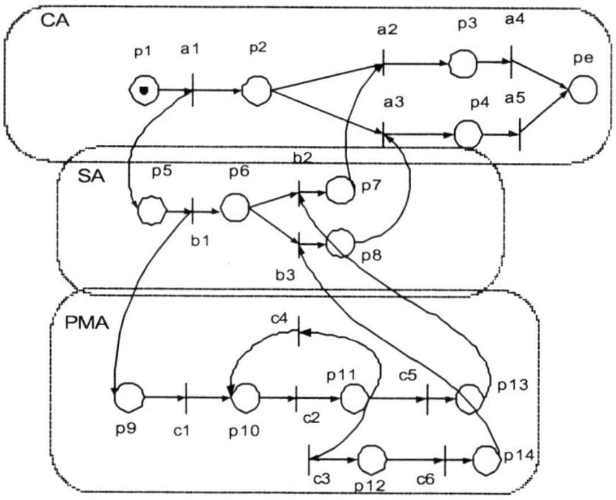

Figure 3 order processing

In the real work, the whole process is managed by three departments. This makes us to consider how to divide the whole network into three sub-networks. To decompose a network, we must consider three problems: 1) how to divide the whole network, 2) how to build the synchronous relation between sub-networks, 3) the properties of the sub-networks.

For the first problem, one nature division is to divide the network by nature boundaries of organizations, for normally organization will have an independent workflow engine. For the case in figure 3, the network can be divided into three sub-networks and each sub-network corresponds to the tasks of each department. In order to make sub-network compliant to the definition of SWF-net, a start place and an end place normally need to be added into the sub-networks. For the case in figure 3, the sub-network of SA and PMA must be added with a start place and an end place respectively.

The adding of the new start places and end places is reasonable, for sub-networks are controlled by different WFMSs and each should have starting and ending conditions. Since the starting conditions are not only related to the state the local sub-network, but also related to the other sub-networks, the definition of the added start places and the added end places are different.

The second problem is the synchronization of the sub-networks. From figure 3, man can see the CA controls the start of the SA at the transition a1, while SA controls which branch of the two Or-connected branches in CA. SA and PMA have similar relations.

Taking out CA and SA from figure 3, we get two sub-networks in figure 4. Note that the new start place and the new end place are added into figure 3. The arc (p1,b1), (b1,p1) form a loop. This loop has no influence on CA, while SA is controlled by CA. Similarly, (a3, p7)(p7, a3) and (a4, p7)(p7, a4) form two other loops, but SA controls CA. This loop is controlled as self-loop in Petri net [9].

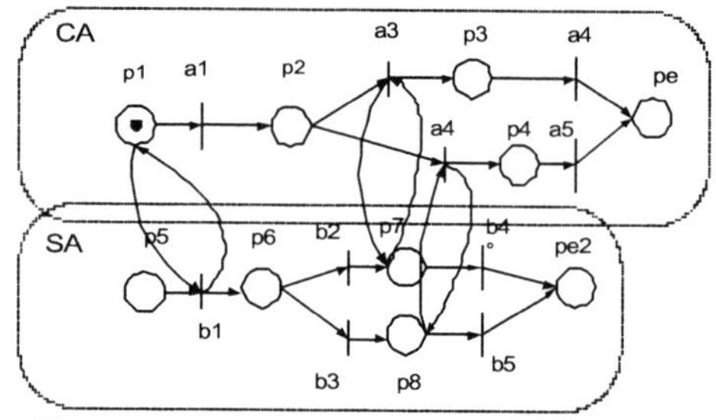

Figure 4. Sub-network Decomposition

Definition 10 (Self-loop) two sub-network $N1=(P1,T1,I1,O1)$, $N2=(P2,T2,I2,O2)$, $T1\cap T2=\varnothing$, $P1\cap P2=\varnothing$, and two initial state is $M1$, $M2$. If $\exists p^*\in P1$, $t^*\in T2$, $I(p^*,t^*)=O(p^*,t^*)=1$, a self loop is formed from N1 to N2. p^* is called connection place in this self loop, t^* is called connection transition in this self loop. N1 is the active network and N2 is the passive network. If whole network is noted by $N=(P,T,I,O)$,then $P=P1\cup P2$, $T=T1\cup T2$, $I=I1\cup I2\cup I(p^*,t^*)$, $O=O1\cup O2\cup O(p^*,t^*)$, $M=[M1, M2]$.

Until now, we get one conclusion:

A standard workflow can be divided into several self-loop connected sub-networks.

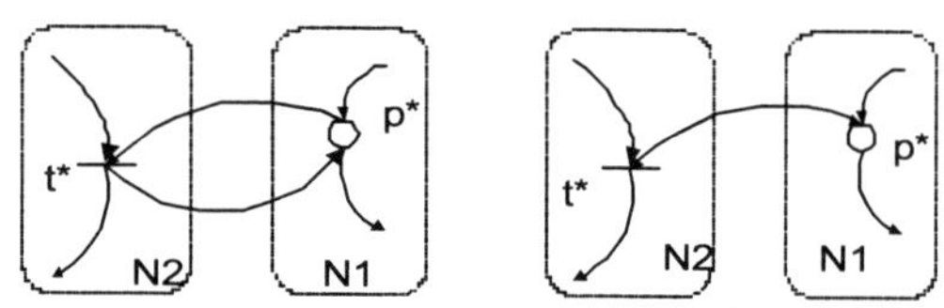

(b)

Figure 5. Self-loop connection

In real work, we are facing a reversed process. A business process across organization bounded normally is modeled and managed by several independent WFMSs and actually none knows the whole picture of the entire process. Thus, we are facing the third problem—if each WFMS models its work in the proper way (i.e. guarantee the soundness of its local network), how are the properties of the network made up of the sub-networks?

4.2 The property maintenance of sub-networks

In this section, we discuss the conditions of property maintenance of sub-networks. More actually, the question is if the sub-networks are sound, under what condition can the joint network keep the property? After several lemmas, we present the sufficient and necessary conditions for property maintenance. Some of the following work has link to [10], but our work is based on definitions for workflow network.

Lemma 1. Suppose N1 and N2 are two SWF-nets, N1 is the active net and N2 is the passive net. N1 and N2 make up a joint network N, note the reachable set under state M is R(N,M), then for $\forall M^*=[M1^*, M2^*] \in R(N,M), M1^* \in R(N1,M1), M2^* \in R(N2, M2)$.

Prove: see appendix.

Lemma 2. Suppose N1 and N2 are two SWF-nets, N1 is the active net and N2 is the passive net. N1 and N2 make up a joint network N, then (1) N is bounded; (2) the necessary and sufficient condition of N1 and N2 and N are live.

Prove: see appendix.

Lemma 3. Suppose N1 and N2 are two SWF-nets, N1 and N2 make up a joint network N. There are two self-loops. One of the self-loops is from N1 to N2 , and the other is from N2 to N1. Then (1) N is bounded; (2) the joint workflow N is live iff (a) if pi* and ti* are on the same sequential branch, then $\forall i$ (i=1,2), pi* is the preset of ti*; (b) if ti* and pi* are not on the same branch, then $\forall i$ (i=1,2) ti* is at the start transition of the conditional structure; or (c) if ti* is on a branch of a parallel structure, pi* is the preset of the end transition of this structure.

Prove: see appendix.

We build the extended network of the joint workflow network in figure 6 in order to prove the condition of sound for a joint workflow network.

Theorem 4. If the two standard workflow network N1 and N2 are sound, then their joint workflow network N keeps sound under the conditions of Lemma 3.

Prove: removed for context limitation.

Theorem 4 is the most important result of this paper. If we satisfy the three conditions in Lemma 3, we can safely link the multiple workflow networks and assert the joint workflow is sound.

It is very useful in real work. Normally we can guarantee a single workflow network is sound. With the above theoretical analysis, the workflow network can keep sound if certain conditions are satisfied. But we do not emphasize the three conditions must be satisfied, because there is no picture of the whole network. The meaning of theorem 4 is not only on providing the necessary and sufficient conditions, but also on providing the criteria when the exception handling activities should be added.

5. Conclusion

This paper presents a set of methods for modeling large-scaled workflow network. Using the concept of Standard Workflow Net presented in this paper, the soundness of single workflow net and the joint workflow net can be guaranteed. The theorem 2 can be inferred to multiple self-loop cases or multiple sub networks (>2). This discuss is not donc in this paper.

This method concerns only the determined relationship between activities. It is not suitable in analyzing the un-determined relationships, e.g. several activities can be executed in un-determined sequence. For some more complex situations, high Petri Nets (colored, timed) might help. But their verification is much more difficult. Another reason for not using high Petri Nets is that those complex methods are unlikely to be used in commercial products for training cost is too high and unfeasible. If the user can change the description of the problem a little bit and concentrate only to the final goal, the classic Petri Nets would be able to describe any relations, at least not worse than the block diagram provided by WfMC.

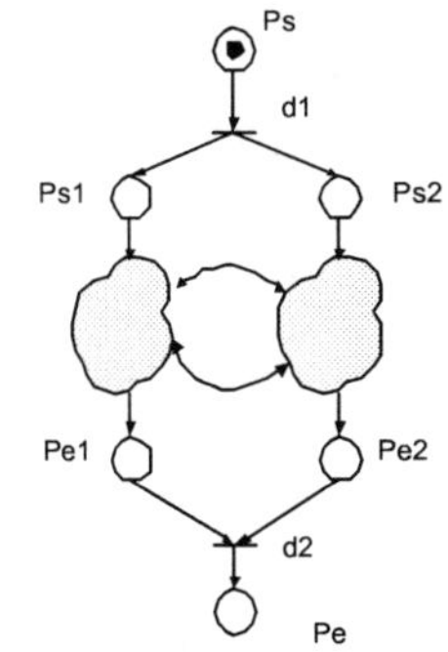

Figure 6. Joint Workflow

Self-loop connection might be only one control relation between sub networks. We are going to investigate other control relations.

References:

[1] W.M.P. van der Aalst, "The Application of Petri Nets to Workflow Management", *The Journal of Circuits, Systems and Computers,* 8(1), p21-66, 1998.

[2] Action Technology Inc. "ActionWorks Developer's Guide Version 5", *White paper*, 1999.

[3] IBM, "MQSeries Workflow", *White Paper*, 2000.

[4] FileNet, "FileNET's eProcess Offering: The Future of Web-Based Business Process Management", *White Paper*, 2000.

[5] WFMC, "The Workflow Management Coalition Specification, Workflow Reference Model", *Document Number TC00-1003*, Nov. 1994.

[6] Zhou, M. C. and DiCesare, F., *Petri Net Synthesis for Discrete Event Control of Manufacturing Systems*. Boston: Kluwer Academic Publishers, 1993.

[7] ZISMAN, M. D., "Representation, specification and automation of office procedures", *Ph.D. dissertation*, Univ. of Pennsylvania, Wharton School of Business, Philadelphia, Pa. 1977.

[8] Ellis, Clarence and Nutt, Gary, "Modeling and Enactment of Workflow Systems", *Advances in Petri Nets*, pp. 1-16, June 1993, Springer Verlag.
[9] Ferrarini, L., Narduzzi, M. and Tassan-Solet, M., " A new Approach to Modular Liveness Analysis Conceived for Large Logic Controllers' Design", *IEEE Transactions on Robotics and Automation,* Vol. 10, No.2, p169-184, April,1994
[10] Zhibin Xu, "Performance Analysis of Petri Net for a class of DEDS", Master Thesis, Tsinghua University, China, 1995.

Appendix:

1. Table.1 the definition of places and transitions in figure 2.

Transactions	Description	Places	Description
a1	send new order to SA	p1	start
a2	get Commit message	p2	wait
a3	get Reject message	p3	success
a4	report Commit	p4	fail
a5	report Fail	pe	end
b1	send new order to PMA	p5	get_order
b2	send Commit message to CA	p6	wait
b3	send Reject message to CA	p7	success
c1	divide order into parts	p8	fail
c2	send production schedule to work unit	p9	get_order
c3	get Reject message from work unit	p10	planned
c4	get Commit message from work unit	p11	wait
c5	report Commit to SA	p12	get_refuse
c6	send Reject message to SA	p13	success
		p14	fail

2. Prove of Lemma1:

Lemma 1. Suppose N1 and N2 are two SWF-nets, N1 is the active net and N2 is the passive net. N1 and N2 make up a joint network N, note the reachable set under state M R(N,M), then for $\forall M^*=[M1^*, M2^*]\in R(N,M)$, $M1^*\in R(N1,M1)$, $M2^*\in R(N2, M2)$

Prove: (1) from the definition of reachable set, has $\forall M^*=[M1^*, M2^*]\in R(N,M)$, $\exists f$, $M[f>M^*$. Consider different cases of t. Before t happens, $M^*=[M1^*, M2^*]$, $M1^*\in R[N1, M1]$, $M2^*\in R[N2, M2]$; after t happens $M''=[M1'', M2'']$, $M''(p)=M^*(p)+O(p,t)-I(p,t)$. Suppose T^* is the self-loop transition set and $T^*\subseteq T2$, P^* is the self-loop place set and $P^*\subseteq P1$, M1', M2' are the marks when t happens in each individual net without considering the link of self-loop.

$t\in T1$:

when $p\in P1$, $M''(p)=M^*(p)+O(p,t)-I(p,t)=M1^*(p)+O(p,t)-I(p,t)=M1'(p)$;

when $p\in P2$, $M''(p)=M^*(p)+O(p,t)-I(p,t)=M2^*(p)+0-0=M2^*$

$t\in T2\backslash T^*$

when $p\in P1$, $M''(p)=M^*(p)+O(p,t)-I(p,t)=M1^*(p)+0-0=M1^*(p)$;

when $p\in P2$, $M''(p)=M^*(p)+O(p,t)-I(p,t)=M2^*(p)+O(p,t)-I(p,t)=M2'(p)$;

$t\in T^*$

when $p\in P1\backslash P^*$, $M''(p)=M^*(p)+O(p,t)-I(p,t)=M1^*(p)+0-0=M1^*(p)$;

when $p\in P^*$, $M''(p)=M^*(p)+O(p,t)-I(p,t)=M1^*(p)+O(p,t)-I(p,t)=M1^*(p)$;

when $p\in P2$, $M''(p)=M^*(p)+O(p,t)-I(p,t)=M2^*(p)+O(p,t)-I(p,t)=M2^*(p)$;

From above, after t happens, $M1''\in R(N1,M1)$, $M2''\in R(N2,M2)$. From the initial mark M=[M1, M2], let f=t1t2...tn, then $M[f>M^*=M[t1> M' [t2> M2''[...[tn>M^*$, it can be deduced $M1'\in R[N1, M1]$, $M2'\in R[N2, M2]$, $M1''\in R[N1, M1]$, $M2''\in R[N2, M2]$, ..., $M1^*\in R[N1, M1]$, $M2^*\in R[N2, M2]$, so we can get $M^*\in R(N1,M1)\times R(N2,M2)$.

(2)if transition serial σ has only transitions in T1, from (1), get $[M1, M2][\sigma>[M1^*,M2]$, then for the joint network $M=[M1^*, M2]$, $M1^*$ is sub-vector of M.

(3) as (2), if σ has only transitions in T2, then proved. If $t^*\in\sigma$, then because N1 is live, t* can be triggered, after it is triggered, $[M1, M2][\sigma>[M1,M2^*]$. So the final result is proved.

3. Prove of Lemma 2.

Lemma 2. Suppose N1 and N2 are two SWF-nets, N1 is the active net and N2 is the passive net. N1 and N2 make up a joint network N, then (1) N is bounded; (2) the necessary and sufficient condition of N1 and N2 and N are live.

Prove: (1) From lemma 1,∀ M*=[M1*, M2*] , M1*∈ R(N1,M1), M2*∈ R(N2,M2). And N1 and N2 is bounded, M1*<C1, M2*<C2, then M*=[M1*, M2*]<[C1,C2], so N is bounded.

(2) prove sufficient condition first, if t∈T1, because N1 is live, ∀t∈ T1, ∃M1", M1'∈R[N1, M1], i1[σ>M1", M[t> M1'. For the joint network, M=[M1,M2], f is a transition serial made up of transitions in N1, then [M1,M2][f>[M1", M2][t>[M1', M2], i.e. t is live.

When t∈T*, suppose from N1 to N2, there is k self-loops. Because N1 is live, M1(pi*)∈R[N1, M1], i.e. ∃fi, M1(pi*)=1; to joint network M=[M1,M2], ∃σ1 has only transitions in T1, fi∈σ1, [M1,M2][σ1>[M1' (pi*), M2]; to N2, because N2 is live, i.e.∀t∈ T2, ∃M2", M2'⊂R[N2, M2], i2 [σ2>M2", M2"[t>M2', so ∃M2˜, M2˜∈R(N2, M2') ti*∈EN(N2, M2˜). Thus for M=[M1,M2], [M1,M2][σ1>[M1' (pi*), M2][σ2>[M1' (pi*), M2"], ∃ M2˜, [M1' (pi*), M2˜]∈R(N, [M1' (pi*), M2"]), ti* ∈ EN(N,[M1'(pi*), M2˜]), i.e. ti* is live. If t∈T2\T*, consider only N2, there are two cases, if ∃σ, ti*∉σ, and M2[σ>M2', t∈EN(N, M2'], then proved. If ∃σ, ti*∈σ, and M2[σ>M2', t∈EN(N, M2'], split σ to sections, σ=σ1ti*σ2, M2[σ1>M2˜, because t* is live, ∃λ, λ is made up of transitions in T1 and M[λ>[M1˜, M2], then [M1, M2][λ>[M1˜, M2][σ1>[M1˜, M2˜][ti*σ2>[M1˜, M2'], t∈EN(N, [M1˜, M2']), proved.

4. Prove of Lemma 3

Lemma 3. Suppose N1 and N2 are two SWF-nets, N1 and N2 make up a joint network N. There are two self-loops. One of the self-loops is from N1 to N2 , and the other is from N2 to N1. Then (1) N is bounded; (2) the joint workflow N is live iff (a) if pi* and ti* are on the same sequential branch, then ∀i (i=1,2),pi* is the preset of ti*; (b) if ti* and pi* are not on the same branch, then ∀i (i=1,2) ti* is at the start transition of the conditional structure; or (c) if ti* is on a branch of a parallel structure, pi* is the preset of the end transition of this structure.

Prove: (1) From lemma 1,∀ M*=[M1*, M2*] , M1*∈ R(N1,M1), M2*∈ R(N2,M2). And N1 and N2 is bounded, M1*<C1, M2*<C2, then M*=[M1*, M2*]<[C1,C2], so N is bounded.

(2) sufficient condition: when condition (a) is satisfied, if t=t1*, because N1 is live, then ∃σ1, t∉σ1, M1*[σ1>M1', t∈EN(N1,M1'). Because N2 is live, then ∃σ2,t∉σ2, M2*[σ2>M2', M2'(p2*)>0. Now if t2*∉σ2, then M*[σ1σ2>M', t∈EN(N,M'). If t2*∈σ2 , note σ2=f1t2*f2, from t1* is enabled, and pi*∈PRE(ti*), get M1'(p1*)>0, M1'[t2*>M1', then [M1*,M2*][σ1> [M1' ,M2*][f1 t2*> [M1' , M2*'] [f2 > [M1' ,M2'], t∈EN(N,M'). Similarly, it can be proved when t=t2*. If t∈T1\t1*, because N1 is live, if ∃σ1, t1*∉σ1, M1*[σ1>M1', t∈EN(N1,M1'),then proved. If t1*∈σ1, σ1=g1t1*g2, because N2 is live, get ∃σ2, t∉σ2, M2*[σ2>M2', M2'(p2*)>0. Now if t2*∉σ2, then M*[g1t1*σ2 g2>M', t∈EN(N,M'). If t2*∈σ2 , note σ2=f1t2*f2, because t1* is enabled and pi*∈PRE(ti*), get M1'(p1*)>0, M1'[t2*>M1', then M*[g1t1* f1t2*f2g2>M', t∈EN(N,M'). Similarly it can be proved when t∈T2\t2*.

When condition (b) is satisfied, suppose t1* is the start transition of the branch L1. L1 is in the condition structure of N1. Because N1 is live, token can reach every place which are not on L1, including p1*, then it can be deduced that transition t2* in N2 is not controlled by t1*. So all transitions in N are live.

When condition (c) are satisfied, from pi*∉ Li , we know branch pi* has a token making M(pi*)>0, then ti* is enabled. Thus the end place of the parallel structure can be enabled. So all the transitions in N are live.

Necessary conditions: suppose (a)(b)(c) are not satisfied, then (1) if on a sequential branch, suppose pi* is behind ti*, then ti*∈EN(N, M(pj*)), tj*∈ EN(N, M(pi*)), but to reach M(pj*), ti* must be triggered, this causes a dead-lock. In another case, suppose pi* is before ti*, but pi*∉PRE(ti*),suppose PRE(t1*)=M(p1'), we get to know, the condition to trigger t1* is that M(p1')>0 and M(p 2*)>0. Similarly, the condition to trigger t2* is that M(p2')>0, M(p 1*)>0,so when t 1* is enabled, t2* cannot be triggered, so N cannot be live.

(2) suppose pi*, ti* are on the same branch of a parallel structure, it is like condition (a), which is proved. Suppose pi*, ti* are not on the same branch of a parallel structure, and ti* is not the start place of the branch, then a transition series exists, which transfers the token to the precedent place of ti*. At this moment, M(pi*)=0, dead lock appears. (3) suppose pi* and ti* on the same branch of a parallel structure, this is like condition (a), and is proved. Suppose pi*, ti* are not on the same branch, and pi* is not in the present of the end transition of this parallel structure. Then we can construct a transition series fi, which transfers the token on the branch where pi* is on to the end place of this branch. At this moment, if ti* is enabled, M(pi*)>0 must be satisfied. And because pi* is not the end place, so M(pi*)=0. Thus if pi* is not the end place, then N is not live.

A Workflow System for Supporting Group Activities of an Enterprise

Feng Li, Zongkai Lin, Yuchai Guo, Jintao Li
Cooperative Design Group, CAD Laboratory, Institute of Computing Technology
Chinese Academy of Sciences, Beijing, 100080, P.R. China
Lifeng_public@sina.com.cn

Abstract

Workflow technology provides a good way to mediate the responsibility flow in a business process from person to person and from task to task. However, most current workflow systems are not able to do this very well. The main reasons are: most activities of a workflow are executed asynchronously and respectively according to their workflow model, and cannot be executed synchronously; External GroupWare tools cannot be integrated very well, which makes the system not be best of supporting human-human synchronous interactions. CFlow (Cooperative workFlow) has been developed by our group recently to overcome the above shortcomings. In this paper, the main implementation techniques of CFlow are described; a process model visually describing method based on constraints is introduced; and an approach to integrate GroupWare tools is presented.

1. Introduction

There are division of labor and cooperative work among members of a group participating in an enterprise group activity. In general, the division of labor is called individual work, and the cooperative work is called cooperative work. So, an enterprise group activity consists of individual work and cooperative work. In cooperative work, the collaborative mode can be divided into synchronous collaboration and asynchronous collaboration in time dimension.

Now, in most workflow systems, all activities are executed independently. The collaborative mode often is "end-start" mode [3]. Which is called simple asynchronous collaboration. That means that an activity can be started only when it's preceding activity has completed, and there is no synchronous collaboration between the two activities. Even among the parallel activities, all activities are executed respectively, there is no collaboration among them. This will reduce the collaborative ability of a workflow, and cut down the efficiency and flexibility of workflow systems. But in many enterprise group activities, such as co-design and co-edit, there are requirements of synchronous collaboration among activities, such as the activities "design" and "review" of a product design flow that is illustrated in Figure 1 can be executed synchronously. And they two activities can share the middle result each other. In order to meet these requirements of cooperative work, we have developed a workflow prototype system CFlow recently. In CFlow, a new visual describing approach of process model based on constraints is introduced. It softens the dependency between two sequential activities, so that the two sequential activities can be executed synchronously or asynchronously. And the successor can use the middle result of its preceding activity. In order to make two parallel activities executed in synchronous collaboration mode, a synchronization dependency is introduced between them.

Figure 1. A product design flow.

Most workflow management systems put their focus to manage business processes of an enterprise, and support their execution in a WFMS. Moreover they concern with how to integrate relevant persons, software tools and data objects. However, the software tools concerned often are individual software tools, such as AutoCAD. External GroupWare tools cannot be integrated very well. Because human-to-human collaboration and coordination is critical in GroupWare tools [2], particularly synchronous human-human interaction, and the work mode can be divided into synchronous work and asynchronous work in time dimension. So, in order to enhance the ability of synchronous human-human interaction of workflow systems, it is natural to explore how to synergies GroupWare and workflow. And the very simple and preferred approach is to integrate GroupWare tools into workflow systems. Which will enhance the collaborative ability, particularly synchronous collaborative ability. But, it is more complex to integrate GroupWare tools into workflow systems than to integrate individual software tools. Many issues must be handled, including how to make up a group, which cooperative work mode will be employed, how to describe the dynamic change of persons, how to save cooperative record and how to use the results of cooperative work. In CFlow, a mechanism of integrating GroupWare tools is present.

The rest of this paper is organized as follows: Section 2 introduces a novel workflow model, a three level workflow model. Section 3 discusses the process model visual describing approach of CFlow; on this basic, an example of describing co-design process is introduced in Section 4. Section 5 presents our approach to integrate groupware tools; and Section 6 summarizes the paper.

2. Workflow Model

A workflow model is an abstract description of an enterprise business process and a key point of a workflow system successful application. Primarily, there are four components in a workflow model: data objects, tasks, performers, and constraint [1]. And how to describe these components and their dependencies is a key question that need to solve in a workflow model. Now, many workflow-modeling methods have proposed, but most of them are focused on process description [4,5]. In this paper, we present a novel workflow model, a three level workflow model. Which includes process level, task level and resource level. And these levels and their relationship are illustrated in Figure 2.

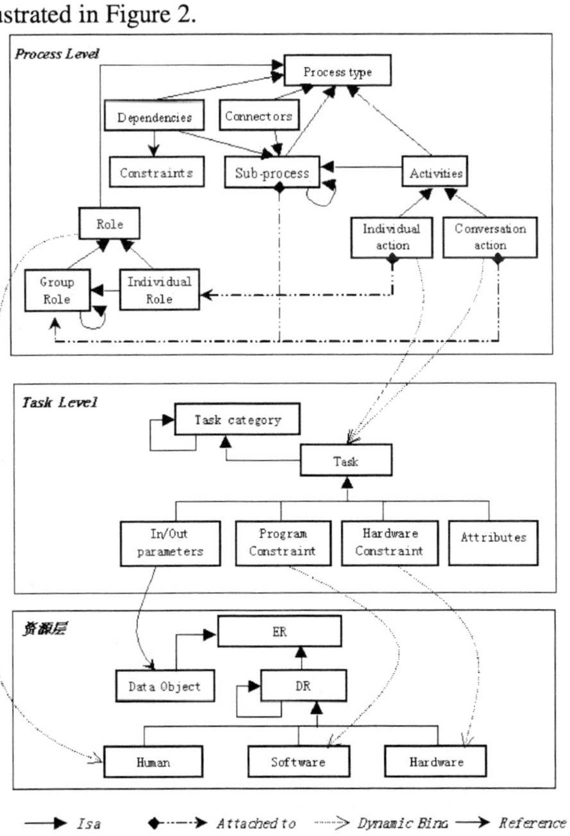

Figure 2. A three level workflow model.

In this three level workflow model, the IT resources, such as human, software and hardware, and data objects are described in resource level. And a resource management tools and a resource proxy are implemented in order to manage these IT resources and data objects. In order to manage the IT resources in different places or different departments, they can attach to different DR according their distribution feature.

In task level, we first organize all tasks into two categories: common tasks and temporary tasks. And they are saved in task-base and temporary-task-base respectively. Of course, these two categories can be divided into some sub-categories, such as individual tasks and conversation tasks. Every task contains context-free attributes and context-dependent attributes. And these context-dependent attributes are defined by parameters, such as input/output parameters. Most relationships between task and resources are defined by constraints, such as program constraint. On this basic, it can be implemented to dynamically bind relevant resources. In addition, every task can have its compensative task that will undo the effects of executing the task.

In process level, the process model is represented with hierarchical DAG. And activities are represented with nodes, and dependencies among activities are described with directed arcs. Every activity can refer to a relevant task in the task level. And these context-dependent attributes of the task must be defined in the activity, such as input/output data objects.

3. A visual describing approach of process model

Now, most process models are often described based on directed graph in most workflow systems [5]. In this modeling approach, activities or states are represented with nodes, and dependencies among activities are described with directed arcs. And this modeling approach is more visual and more apt to understand. So, this modeling approach is called advanced modeling method [7]. In CFlow, a hierarchical process model based on DAG is presented. Which can make it easy to reuse sub-processes, handle failure, and will improve the flexibility and adaptability of the workflow system. The hierarchical process model can be described as PF=(N, E)□and N is a set of nodes, which represents activities, sub-processes and so on, E is a set of directed arcs, which represent dependencies among activities. In CFlow, dependencies are divided into sequence, condition, and synchronization three types in order to facilitate to describe collaboration between two activities flexibly. Constraints can be set in a sequence or synchronization dependency in order to represent the collaborative relationship between two activities. And many constraint parameters can be set in the nodes that represent activities or sub-processes, such as role-constraint parameter and resource-constraint parameter. The strict dependency relationship between two activities is softened by constraints of their dependency, and the flexibility of activity execution is enhanced with constraint parameters of a node. For example, participants' dynamic binding can be implemented with role-constraint parameter of a node.

3.1 Constraints of a dependency

The dependency between two sequential activities often is "end-start" collaborative dependency, which is called simple asynchronous dependency, illustrated as Figure 3. It means that the activity B will be start when the activity A is completed and C will be start when B is completed. And the activity B can read the result data of activity A as its input data, and activity C can read the result data of activity B as its input data. So, there is no middle data sharing among these activities, and there is no synchronous collaboration.

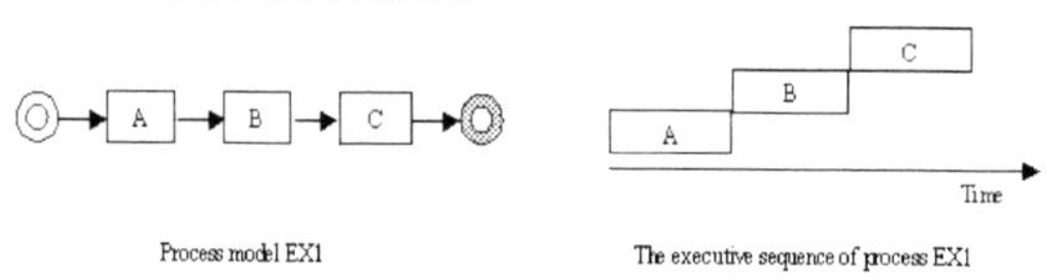

Figure 3. A process model including sequential activities.

However, the synchronous collaboration between two sequential activities is required in some business processes [3], such as in co-design process, illustrated in Figure 1, activity "review" could read the middle result of activity "design", and the activity "review" could start anticipant execution before the activity "design" completed. Which can reduce the number of feedback and improve working efficiency. But, how dose this synchronous collaboration be represented?

In Figure 3, we can suppose that there are constraints of "B.Starttime=A.Endtime" and "B.Starttime= A.Endtime" between activities A and B and B and C respectively. These constraints are too strict to represent synchronous collaboration. If some soft constraints are set between two sequential activities, they will be executed in synchronous collaborative mode. Such as, if a constraint of "B.Starttime>=A.Starttime AND B.Endtime >= A.Endtime" is set between activities A and B, activities A and B will can be executed synchronously. If a constraint of "C.Starttime >= B.Starttime AND C.Endtime >= B.Endtime" is set between activities B and C, activities B and C will can be executed synchronously, illustrated in Figure 4.

So, some constraints can be set for a sequence dependency in order to soften the dependency relationship between two sequential activities in CFlow.

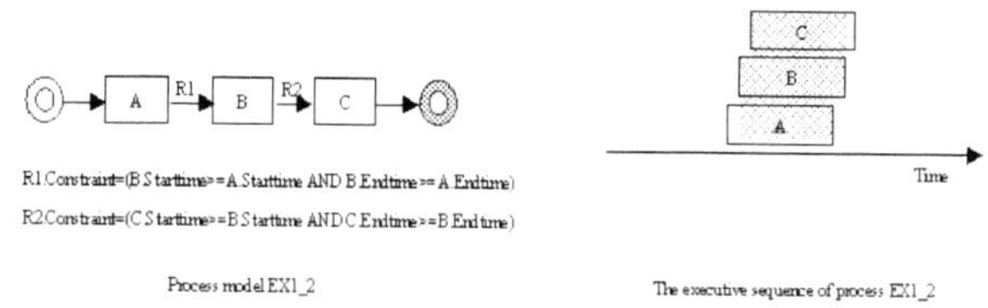

Figure 4. Implementation of synchronous collaboration in sequential activities.

3.2 Synchronization dependency

Parallel activities can be represented in a process model based-on directed graph, illustrated in Figure 5. However, in most workflow systems, these parallel activities are executed respectively, such as activity D1 and D2, there is no collaboration among them. This will reduce the collaborative ability of a workflow, and cut down the efficiency and flexibility of a workflow.

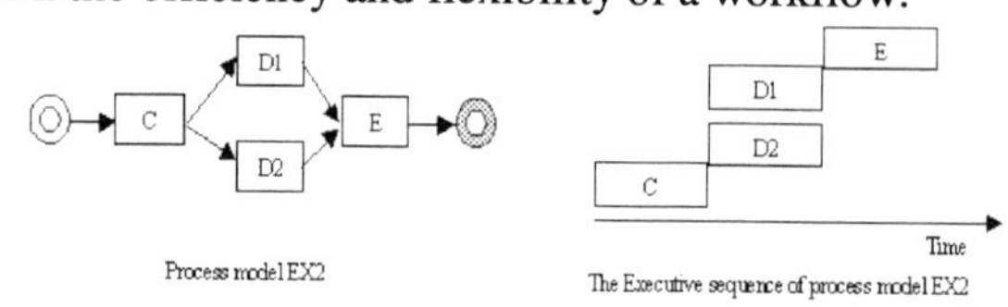

Figure 5. A process model including parallel activities.

However, in some case, there need some collaboration between parallel activities D1 and D2. In order to meet this requirement, a new synchronization dependency is introduced in CFlow. For example, a synchronization dependency can be added between activities D1 and D2. Which is illustrated in Figure 6. Therefor, activities D1 and D2 can be share middle results each other.

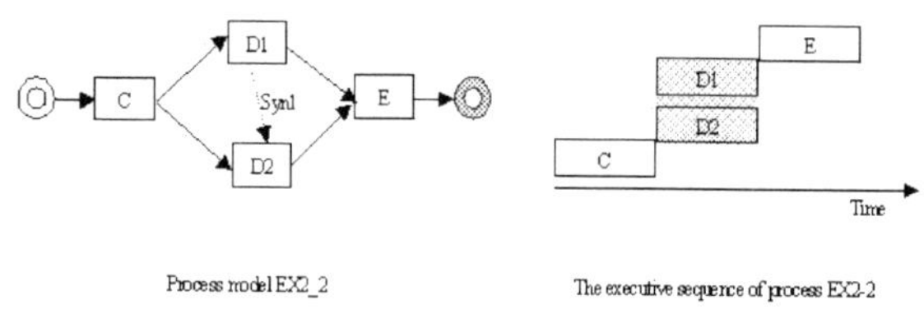

Figure 6. Implementation of synchronous collaboration between parallel activities.

3.3 Constraints of a node

In CFlow, Constraints are introduced for a node that represents an activity or a sub-process yet. And these constraints can be divided into preceding constraint (Pre_constraint) and executive constraint (Exe_constraint) in time dimension. In addition, these constraints can be divided into role constraint, resource constraint and so on according objects constrained. Pre_constraints will be checked before its activity starts. For example, dynamic binding of participants can be implemented with Pre_constraint of a role, and dynamic binding of GroupWare tools can be implemented with Pre_constraint of resources. The executive behaviors of activities are constrained by their Exe_constraints.

4. Description of a co-design process

To further illustrate the use of the visual describing approach of a process model, we present a description of a co-design process. There are six activities in the process, illustrated as Figure 7. Dependency R1 between activities "Layout" and "Review" has a constraint, which means that activities "Layout" and "Review" can be executed synchronously or asynchronously. And there is a synchronous dependency between activities "Part1 Design" and "Part2 Design", so these two activities can be executed synchronously, and they can share middle results each other. In activity "Meeting", there is a Pre_constraint of role, so its participants can be dynamically bound according the Pre_constraint of role. Other constraints are not discussed in this paper.

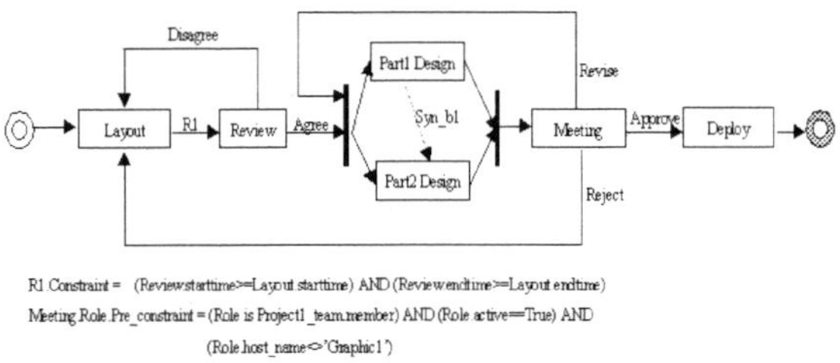

Figure 7. An example of a co-design process.

5. A method of integrating GroupWare tools

In order to enhance the ability of synchronous human-human interaction of workflow systems, the preferred approach is to integrate GroupWare tools into workflow systems. Which will enhance the collaborative ability, particularly synchronous collaborative ability. In CFlow, there are two types of GroupWare tools, one are GroupWare tools that have developed by our group, including white broad, voice tool, and application collaboration [6], the other are external GroupWare tools. In order to synergy these GroupWare tools with CFlow, we introduce the two different approaches according their type.

A new node type is introduced in process model in order to integrate the GroupWare tools developed by our group. We call this node type as a meeting node. The role of a meeting node is similar with electronic meeting tools. A virtual meeting can be started with a meeting node according its constraints. There are white broad, voice tool and application collaboration components [6] in a meeting node.

In order to integrate external GroupWare tools, we provide a method based-on proxy, illustrated in Figure 8. The correlative parameters and constraints are mapped to a proxy by CFlow engine. Then the proxy starts a corresponding GroupWare tool. And the proxy monitors and records the executive history of the GroupWare tool. Finally, the proxy will return the result to CFlow engine.

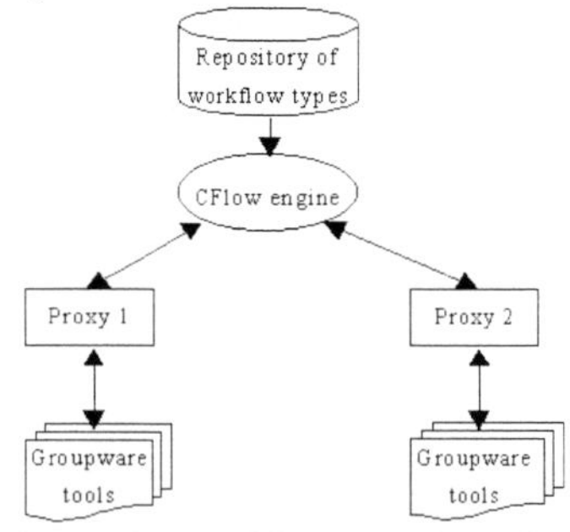

Figure 8. A method of integrating GroupWare tools based-on proxy.

6. Conclusion

It is a significant means for improving group work efficiency to enhance the synchronous collaboration between task and task or between human and human. But most workflow management systems are very limited to support synchronous collaboration. In CFlow, the ability of synchronous collaboration has been strengthened. And these methods have been applied into the concept design system. In order to make CFlow more practical, there is much work to do in the future. In this paper, we only discuss the representation of synchronous collaboration in process model layer. And the correlative executive mechanism will be discussed in other papers.

7. References

[1] Wasim Sadiq and Maria E. Orlowska, "Modeling and Verification of Workflow Graphs", technical report, department of computer science, the university of queensland, 1996.

[2] Israel Z. Ben-Shaul and Gail E. Kaiser, "Integrating Groupware Activities into Workflow Management Systems", http://citeseer.nj.nec.com/338725.html

[3] Danicia Grigori, Hala Skaf-Molli and Francois Charoy, "Adding Flexibility in a Cooperative Workflow Execution Engine", Http://www.loria.fa/~skaf/paper/hcpnoo/flexlncs.htm

[4] Ellis C A, Nutt G J. "Modeling and enactment of workflow systems", In Marsan A ed. Application and Theory of Petri Nets, Lecture Notes in Computer Science 691, Berlin: Springer-Verlag, 1993. 1-16

[5] LUO Hai-bin, FAN Yu-shun, Wu Cheng, "Overview of workflow Technology", Journal of Software, 2000,11(7):899-907

[6] Xu Baomin, "Research on Application Cooperation Technique", Ph.D thesis, Institute of Computing Technology, Chinese Academy of Sciences, 2000

[7] Gregor joeris and Otthein Herzog, "Towards Flexible and High-Level Modeling and Enacting of processes", CaiSE'99, Heidelberg, Germany, 1999.

Collaborative Modeling and Simulation for Order Management in Virtual Enterprise

Xiaoming Zhang*, Ziqiong Deng, Hongze Ma
Narvik Institute of Technology, Norway
{xiz, zd, hm}@hin.no

Abstract

Collaborative modeling and simulation takes strong advantages of resource sharing across the Internet for the extended enterprise or virtual enterprise. Collaborative graphical modeling shows more attractiveness and difficulties when applied in web. One solution to integrate 2D-graphical modeling and web online analyzer is brought up and a prototype is realized through campus network. An order management model is designed collaboratively using modeling tool of EXTEND to map the operation in scheduling, production and transportation of a geographically distributed iron ore enterprise. The distributed managers can submit orders and then get the simulation results online from a web online analyzer which realized by apache web server, PHP and mySQL tools. The prototype shows effective experimental results. It will be studied more in creating complete model base for simulation server and graphical user interface for browsers.

1. Introduction

With the rapid development of network communication technology, more and more market opportunity can be found for the extended enterprise or virtual enterprise by effective collaboration and coordination across the network. The research of CSCW (Computer Supported Collaboration Work) becomes more important and urgently. Collaborative environment and technologies such as platform creation, middleware, distributed product design and distributed decision making are studied and applied largely from the literature. Deng [1] proposes an architecture for electronic networking for remote collaboration and coordination of supply chain. There exist three layers of communication, collaboration, and application in the architecture. To run an inter-enterprise collaborative net-tasking, six functions of application sharing, white boarding, file transfer, text chat, audio multimedia talking and video multimedia seeing each other in the collaboration layer are basically required. Many users can collaborate interactively by these functions. The famous supplemental tool for collaboration across the network is Windows NetMeeting with strong functions including video, audio, chat, white board and file transferring.

For application sharing, web-based technologies are popularly developed for these years, and most effects are focused on Java-based distributed modeling and simulation. Text-based and dialog-based simulation can be easily realized for real-time collaboration feature. Pooley [2] describes a prototype distributed simulation environment in Java for decision-making. Now, graphical modeling and simulation becomes more attractive. Icon-based, 3D-based and VRML-based models can be created and simulated through the network. Rezayat [3] shows inter-linked VRML and HTML files to access to all MCAD data in E-Web. A virtual reality environment will allow complete user immersion in the interaction. It will show the same effectiveness just like what a 3D-game player obtained.

For discrete-event system (DEVS) simulation, Sarjoughian [4] presents a collaborative distributed network system and a middleware called Collaborative DEVS modeller. The modeler shows simple icon-based object model such as aircraft model comprised of atomic and coupled models [5]. Similarly, Ammerlanhn [6] develops a geographically distributed enterprise simulation system for management of the nuclear stockpile in the US Department of Energy's Nuclear Weapons Complex. Java-based technologies are adopted in the three solutions of synchronized human participation within the simulation, distributed ownership and security, and integration of legacy models. The realizations of user interface and parallel discrete-event simulation system are also in dialog and text-based presentation. Lin [7] developed an extended enterprise supply-chain analysis tool, the Asset Management Tool (AMT) in IBM Company to help users visualize the supply-chain activities or orders, goods, and trucks in animation as browser/server structure. The AMT is designed in Java to

* Xiaoming Zhang is a visiting scholar in Narvik Institute of Technology at present. He is from Beijing Institute of Petrochemical Technology.

show a graphical user interface to the browser. AMT benefits include over $750 million in material costs and price-protection expenses saved in 1998 [7].

In the field of scheduling management, Mehandjiev [8] suggested an agent-supported visual interactive system-SAMBA for distributed team building and working. In addition to the use of NLPP (non-linear process plan) for flexible load balancing and reactive rescheduling, Jan Kemperaers [9] present a new architecture for collaborative process planning and scheduling system in the ESPRIT project, COMPLAN. Lopez [10] presented a methodology with top-down approach and bottom-up approach to implement cooperative scheduling. Besides, Interrante [11] uses active rescheduling in the development of a multi-agent architecture for scheduling to promote collaboration in a dynamic manufacturing environment.

From the above stated, the following success have been obtained:

- Complete graphical product design can be provided for collaboration in intranet. This is called product-alliance oriented collaboration [1]. However, it is unclear whether it can be useful in Internet or not.
- Collaborative text-based modelling and simulation can be easily realized across the web.
- Collaborative 3D-graphical modelling and simulation will be difficult to do through Internet. In practice, because of the limitation of network bandwidth, icon-based models can be designed by collaboration through intranet in the enterprise.

Today, the web has been a unified collaborative environment for thousands of people. Moreover, discrete-event method is popularly used for performance analysis in the extended enterprise planning and construction. Through graphical modelling and simulation, enterprise models can be more easily understood and accepted.

Therefore, how to integrate the web technology with graphical modelling and simulation for performance analysis in virtual enterprise or extended enterprise? Next, the authors will provide one solution for this.

2. Principle of collaborative graphical modelling and simulation

In order to realize graphical modeling and simulation, a collaborative environment is created with web technology ,as shown in Figure 1. The features of the environment are designed as three parts:

- Enterprise model design by a modeling team;
- Manager request submiting and process through web online analyzer;
- Simulation server to realize simulating when simulation event comes.

Enterprise models can be created by one person if it is a small enterprise. However, for an extended enterprise or virtual enterprise which are distributed geographically, models must be designed by a modeling team. This group may be organized in a local network to accept modeling tasks. Each member will have his own ability to design one or more modules for a big enterprise model. At least one manager in the team is needed to coordinate and control the dispatch of sub-tasks and integrate all modules. The modeling team can share enterprise data from the Data base and developing tools from the Tool base. Actually, the Tool base can be also provided by collaborative members from all over the world to form a virtual consultant company. Then, an integrated enterprise model can be accepted by simulation and

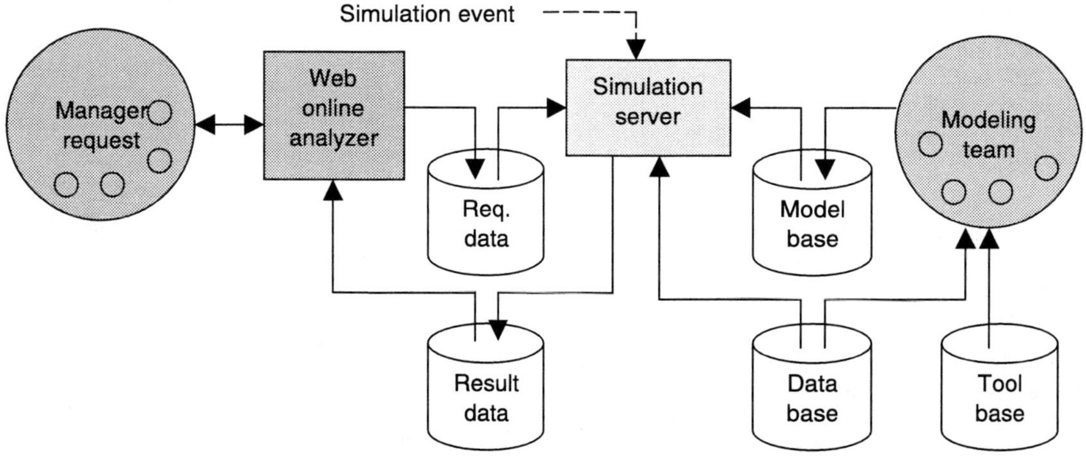

Figure 1. Integration of collaborative modeling with web technology.

entered into the Model base. The Model base will be used for Simulation server. Because the modeling environment is mainly constructed in a local network, graphical models can be easily designed.

For the second part on the left side in Figure 1, a browser/server structure is considered to process Manager request online. Basically, the Managers are the distributed managers of the extended enterprise or virtual enterprise. What kinds of requests can be processed across the web ? It depends on the network opeating mechanism in the enterprise. For such an enterprise like IBM Company, its manufacturing sites are linked with tens of thousands of suppliers and distribution channels all over the world. In the AMT [7], the modeling engines reside on a server computer (eServer). The AMT modeling database contains such supply chain data as bills of materials, demand forecasts, lead times, costs, inventory polices, and customer-service requiements.

The Web online analyzer in Figure 1 accepts requests from Managers and put them into Request database. This Request data are collected from all the Manager before a time or one command for simualtion. When the simulation event comes, the Simulation server input these Request data to analysize the task. Then, it inputs some necessary data from the enterprise Data base,and selects a suitable model from the Model base to begin simulation. The simulation results will be stored into the Result data base. When a manager in the browser is querying the answer of his request, the results are selected by the Web online analyzer from the Result database and displayed for the manager. This finishes the complete collaborative process through the network.

3. Collaborative design of order management model: a case study

A collaboration prototype of modelling and simulation is designed for order management. EXTEND, an excellent modeling and simulation tool made by the Imagine That, Inc., is installed in a server which enables each person of the campus network in Narvik Institute of Technology to access and execute it. The model base can be accessed completely for the collaborative team in the intranet. It will be fast to operate the modelling tool in local computer.

The order management model is studied for an iron ore enterprise, which owns many geographically distributed units in marketing, production and product delivering across countries.

Referring to Figure 1, the orders enter the Simulation server simultaneously when simulation event comes. In the experiment, the simulation event is set as a departure time of work in a workday. For example, the time can be at 17:00 o'clock. It means that the distributed managers must submit their orders before the time. When the moment of the time arrives, the Simulation server begins to work. The clients can view their own results in network before 8:00 o'clock the next day. That will be helpful for the geographically distributed managers in decision making of marketing.

Before collaborative design of order process model, some basic objects are defined to describe the ordering, scheduling, production, transportation, inventory and delivery. Modules are defined as three parts of scheduling, production and transportation. Firstly, the collaborative manager in the team has designed these modules for requirement of input and output interface. This helps each collaborative member to have a unified and explicit modeling space. Some of the defined objects are list in Table 1.

Table 1. Some defined objects.

Objects	Properties	Meaning
Order	*OrderT*	Order time, due date
	OrderQ	Ordering quantity
Prod.	*ProdC*	Production capacity
	ProdQ	Production quantity
Trans.	*TransT*	Transportation time
	TransC	Transportation capacity

3.1 Design of scheduling module

In the scheduling module, other than the estimation of order processing time, scheduling rule is very important for order management. In order to control due date of orders from customers, many job-shop scheduling approaches are brought up. From the view of scheduling rules, there are at least FCFS (First-come-first-service), SPT (shortest processing time), EDD (earliest due date) and heuristic rules. Researchers also study some artificial intelligent methods such as searching, neural network and fuzzy logic scheduling rules for many years. SPT method is only based on the estimated processing time of an order, while EDD method is reflected only on due date of an order. In fact, *OrderQ* and *OrderT* defined in Table 1 are all affect the scheduling process. In order to consider the affection of due date and quantity of an order, the smallest ratio of due date to quantity called STQ rule is presented by authors as the following:

$$Tq = \frac{OrderT}{OrderQ} \qquad (1)$$

Where, Tq represents a priority of an order. This means that an order with a short due date and a large ordering quantity will be in priority. That is STQ rule: an order with min (Tq_i) will be first out of the queue of buffer. The min (Tq_i) is the minimum of Tq in the buffer.

Here, FCFS, SPT, EDD and STQ in the scheduling module can be selected when simulation, as shown in Figure 2.

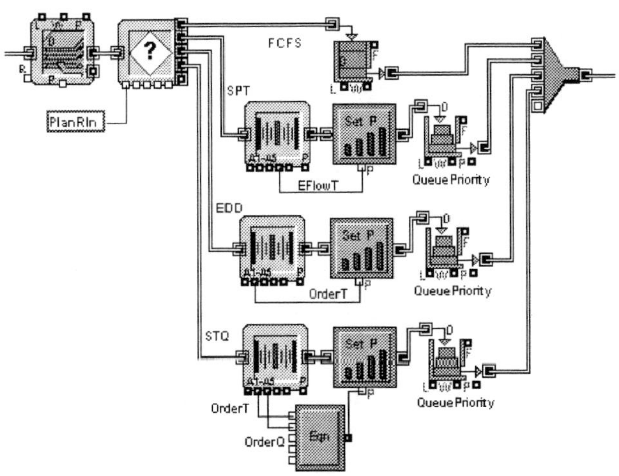

Figure 2. Selection of scheduling rules.

3.2 Modules in production and transportation

The two modules are designed independently. Several production modules are needed to simulate corresponding geographically distributed production units. The production module is shown in Figure 3.

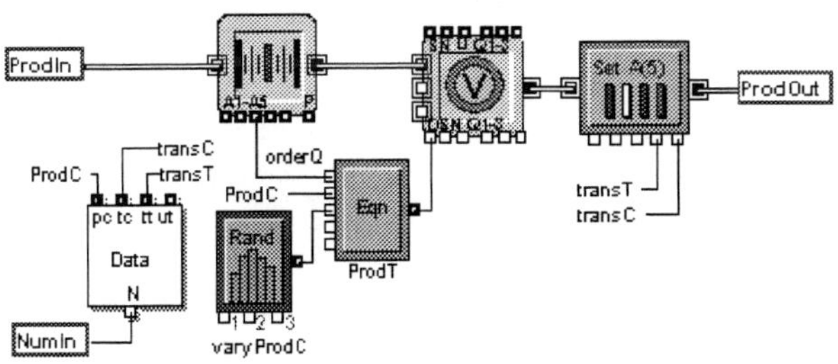

Figure 3. Production module.

3.3 Data filtering module

The modelling tool, EXTEND, has strong ability of programming to support modeling and simulation. Programs are easily developed to make a new block or embedded into an existed block owing to its programming language, ModL, is similar to C language.

When simulating on designed model, data can be stored into FileOut Block as step number or by run number. However, if step number is selected, data created in each simulation step will be written into a file. This will contain many invalid data. If run number is selected, only the last simulating data can be stored in a file. It means that only simulating result of the last scheduled order can be put into a file, while results of other orders have been thrown away. Because each order should have its simulation result, the two data writing modes are not suitable for our requirement.

Then, a data filtering technology is added in FileOut Block by programming. Only effective data can be stored in a file when selecting mode of step number. Figure 4 shows data filtering module for file storing.

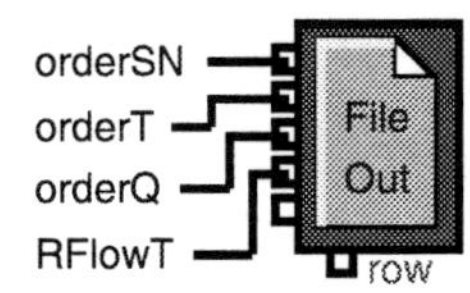

Figure 4. Data filtering module.

4. Web online Ordering

A simple prototype has been realized for web online ordering for a single product. The web online analyser is designed with apache server of version 1.3.19 for win32, php4 for code design and mySQL for data management under Windows NT 4.0. The text files from the finish of simulation will be converted into database in mySQL to get strong control of network data access. Figure 5 shows user interface for manager to fill in order information after a successful login. The simulation results for managers will show the orders valid or not.

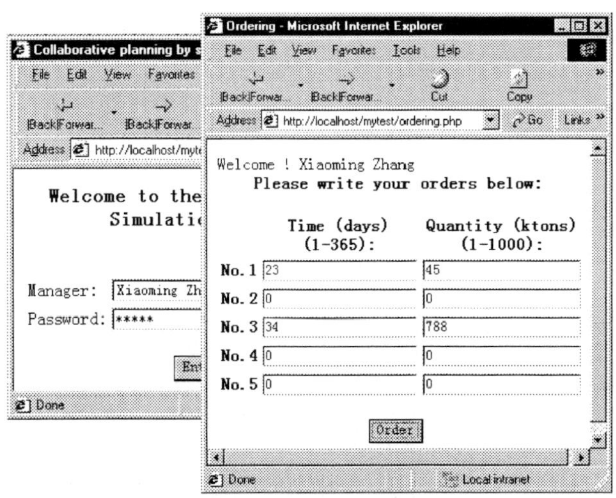

Figure 5. Order input for distributed manager.

5. Conclusion

Collaborative graphical modeling and simulation takes advantages of resources sharing. It will be more practical and effective to integrate collaborative work for humans with web technology. The modeling and simulation tools such as EXTEND can be used in global market for the extended enterprise or virtual enterprise when they own features of collaborative design capability across the Internet. The Simulation server in Figure 1 act as a role of middleware to link collaborative modeling with web online analyzer. It will be a further research to make the prototype practical with a complete model base and more graphical user interface for distributed managers.

6. Acknowledgement

This research is supported by The Research Council of Norway.

7. References

[1] Z. Deng, B. Solvang, and W.D. Solvang, "Collaborative and Co-ordination of Virtual Supply Chain via Electronic Networking", *Proceedings of 16th IFIP World Computer Congress*, Beijing Hope, August 21-25, 2000.

[2] R. Pooley, and P. Wilcox, "Distributing Decision Making Using Java Simulation across the World Wide Web", *Journal of the Operational Research Society*, 2000, 51:395-404.

[3] M. Rezayat, "The Enterprise-Web Portal for Life-Cycle Support", *Computer-Aided Design*, ELSEVIER, 2000, 32:85-96.

[4] H.S. Sarjoughian, B.P. Zeigler, and S. Park, "Collaborative Distributed Network System: a Lightweight Middleware Supporting Collaborative DEVS Modeling", *Future Generation Computer Systems*, ELSEVIER, 2000, 17:89-105.

[5] H.S. Sarjoughian, B.P. Zeigler, "The Role of Collaborative DEVS Modeler in Federation Development", *Proceedings of SIW September 1999*, Orlando, Florida. Also in http://www.acims.arizona.edu.

[6] H.R. Ammerlahn, M.E. Goldsby, M.M. Johnson, and D.M. Nicol, "A Geographically Distributed Enterprise Simulation System", *Future Generation Computer Systems*, ELSEVIER, 2000, 17:135-146.

[7] G. Lin, et al., "Extended-Enterprise Supply-Chain Management at IBM Personal Systems Group and Other Divisions", *INTERFACES 30*: 1 January-February 2000, pp.7-25.

[8] N. Mehandjiev, B.R. Odgers, "SAMBA-agent-supported visual interactive control for distributed team building and empowerment", *Journal of BT Technol*, 1999, 17(4): 72-77.

[9] Jan Kempenaers, Jos Pinte, Jan Detand, and J.P. Kruth, "A collaborative process planning and scheduling system", *Advances in Engineering Software*, 1996, 25: 3-8.

[10] P. Lopez, P. Esquirol, L. Haudot, and M. Sicard. "Cooperative system design in scheduling", *INT. J. PROD. RES.*, 1998, 36(1): 211-230.

[11] L.D. Interrante, and D.M. Rochowiak, "Active Rescheduling and Collaboration in Dynamic Manufacturing Systems", *Concurrent Engineering: Research and Applications*, 1994, 2: 97-105.

Part VII: E-Business/E-Commerce and Virtual Enterprises

An Agent-based Life Cycle Management for Dynamic Virtual Enterprises

Vaggelis K. Ouzounis, Volker Tschammer
ECCO – Electronic Commerce Center of Competence, GMD-Fokus
Kaiserin-Augusta-Allee 31, D-10589, Berlin, Germany
[ouzounis, tschammer]@fokus.gmd.de

Abstract

Virtual Enterprises (VEs) enable the deployment of distributed business processes among different partners to support shorten development and manufacturing cycles, reduced time to market, increased customer satisfaction, and global operation and reach. Dynamic Virtual Enterprises are an emerging category of VEs, where the different partners are being selected dynamically during business process execution based on market-driven criteria and negotiation. In this paper, we discuss concepts and technologies that are considered to satisfy key requirements of dynamic virtual enterprises, and propose DIVE, a framework for the specification, execution and management of shared business processes in dynamic virtual enterprises.

1. Introduction

In the digital economy, companies are continuously seeking for new ways of business in order to cope with the increasing competitive pressure generated by the global market place. The need for global scale and reach, short development and manufacturing cycles, reduced time-to-market and operational costs, increased customer satisfaction, and rapid adaptation to new market changes has led companies to intensify automation, collaboration, and distribution [15][16]. The Internet is increasingly becoming the general basis for such purposes together with its advanced services, including electronic market places and the World Wide Web. New organisational forms develop, like collaborative commerce and virtual enterprises. In such virtual, collaborative organisations, partners, particularly small and medium enterprises (SMEs), can concentrate on their own core-competence and exploit partners' resources and capacities.

Distribution of tasks and co-operation has been a central aspect of trade and business since centuries. Manifold relationships between enterprises and parts of enterprises, often invisible for the customer have contributed to the goal of satisfying the customer's needs for timely and high quality services. In the digital economy, where many of those relationships are realised via the Internet, new forms of business must develop which support communication and co-operation between distributed and automated business processes. Such Internet-based, collaborative organisations are usually called virtual organisations or virtual enterprises.

Some of the common characteristics of virtual enterprises are distributed execution and control of business processes across organisations barriers, autonomous execution of processes within administrative domains, access control and authorisation of remote business process requests, and dynamic selection of partners and automated negotiations.. Virtual enterprises are not a new concept in management studies [2]. Some of the big manufacturing companies, and particularly car manufacturers, have already business relationships with their suppliers and customers based on electronic procedures and protocols, such as EDI [6]. However, the level of integration and the usage characteristics of information and communication technologies vary and are often not optimal. Most of the activities are still performed manually and the cost of implementing and integrating electronic solutions and the time required to deploy virtual enterprise concepts is high.

Based on the common features of virtual enterprises described above, two well-defined categories can be defined, namely static virtual enterprises and dynamic virtual enterprises (DVEs).

In the case of static virtual enterprises the set of business partners is linked together in a static way, i.e. the relationships between partners are fixed and the shared business processes are tightly integrated. The business process interfaces are pre-defined, tightly coupled, and customised according to the partners' environments. Changes during the execution of business processes are not foreseen, except for critical situations, like failures or loss of partners.

In dynamic virtual enterprises the business partners are found and linked dynamically, on-demand, and according to the requirements of the customers with the help of the services provided by a virtual market place. The business relationships in a dynamic virtual enterprise can change continuously based on market-driven criteria. The virtual market place provides services for the registration of business process offerings based on generic, well-known process templates [8]. Business domains that want to engage in virtual enterprise relationships can register offers at the market place in relation to the process templates. Whenever a business domain wants to use a particular process, it is visiting the virtual market place and locates all the potential partners that can provide the required process. As soon as the market place search has resulted in a list of candidate partners, the partner selection process starts. Included in the process are negotiations about capacities and services required and contracting about roles and tasks of each partner. The negotiation process can be performed manually, or automatically. The result of it is usually a contract, which regulates the business relationships established. With the support of the virtual market place, the number and role of partners can easily be adapted to the requirements of customers and the currently available capacities and abilities of the partners. This is a significant evolutionary step, which lets virtual enterprises take advantage of the demand and supply regulated by an open, internet-based market.

Consequently, from a business point of view, dynamic virtual enterprises appear to be the most promising approach. However, from the technical point of view, the required solutions and systems are more complex and more sophisticated. Therefore, evaluating and integrating actual information and communication technologies is an important step towards an efficient, flexible, and easy-to-use support environment for dynamic virtual enterprises.

In this paper we first describe virtual enterprise concepts and we introduce the dynamic virtual enterprises, which are considered to be a decisive step towards collaborative e-business and, therefore, are the subject of intensive research and development. The next section describes and analyses technologies, which are considered to be essential for the development and management of virtual enterprises. These technologies are the basis of our approach towards a support environment for dynamic virtual enterprises. The proposed DIVE Framework is described in section four, including a business model, an architecture, a life-cycle model, a set of agents and a set of ontologies. The paper concludes with a discussion of design decisions made and results achieved related to the technologies used and the implementation environment available. Future plans for the continuation and improvement of the developments are outlined.

2. Technologies for Dynamic Virtual Enterprises

Several technologies can be used for the development of dynamic virtual enterprises concepts, including those supporting the exchange of information in loosely coupled inter-organisational environments, the co-operation of automated business components, the control of work processes across organisational boundaries, and the dynamic mediation of business process providers and users.

Systems supporting the information exchange between the loosely coupled systems of dynamic virtual enterprises must enable asynchronous transactions across organisational boundaries, support application domain-specific types and formats of information, and allow for autonomous behaviour of the communicating and co-operating business processes. Given the requirements for asynchronous and loosely-coupled communication among the different co-operating partners, solutions based on messaging techniques are in favour. Such technologies include EDI, Web Interface Definition Language, Commerce XML, Common Business Library, EbXML, BizTalk Framework, and Commerce XML [17]. Most of these technologies are industry driven standard formats for exchanging business data and not business process execution and management. Although such transactions are performed asynchronously, the currently provided format of the messages is static and cannot be easily extended for dynamic virtual enterprises. The scope and context of such documents is limited and rather impossible to change. Therefore, it is difficult to use such standards as the basis for general-purpose inter-domain business process execution and management.

Systems supporting the co-operation of automated business components must enable distribution, intelligent behaviour and high degree of automation. Several technologies have been proposed like Object-Oriented Components, like Enterprise Java Beans, and Intelligent Mobile Agents. Object-Oriented Components (OOC) are widely used as a framework for distributed computing and information processing due to the ease of integration and deployment, high degree of distribution, and standard underlying distributed protocols, like CORBA-IIOP and RMI. In these technologies, back-end systems and clients integrate themselves with the distributed framework using the application programming interfaces (APIs) and object models exposed by the underlying levels of the architecture. While clients are insulated from the APIs of the back-end systems, they are tightly bound to the APIs provided by the framework [18]. This design choice has two implications. First, by using object binding as the interaction technique, as opposed to document exchange used in EDI, OOC applications must be adopted at once by all participants in the cross-organisation relationship. Upgrades to back-end systems, the component framework, and the business application must be co-ordinated across all participants. Second, because of the tight binding, security issues are a major factor. For communication and co-operation, objects in different business domains must adopt the same object model. This poses a significant barrier to interworking in cross-organisational environments. Additionally, the OOC frameworks do not provide a complete solution, but instead serve as the starting point for developers to build applications. These choices make the OOC frameworks most appropriate for deployment inside a single company that needs to link multiple distributed divisions or sites [19].

Intelligent mobile agents combine many of the benefits offered by messaging systems and OOCs. Agent systems are loosely coupled and communicate asynchronously [1]. Messages exchanged are well defined through the FIPA agent communication language standard [7]. Agents realise the concept of ontologies, which makes them more flexible and autonomous and, agent systems are deployed within a distributed object-oriented framework, like CORBA or Java, and thus can access any type of standard business component. Through these combined benefits, agent systems support autonomy and flexibility, scalability, and adaptability. Flexibility is supported by the distinct communication and co-operation models which have been developed for agent-based systems, scalability comes from the migration capabilities of agents, and adaptability relies on the inherent intelligence of agents [3]. Although agent technology is a very good candidate for the support of virtual enterprises, it has some problems as well. One of the key issues currently is the requirement for a mobile agent platform for the provision of agent life cycle and migration management services. Several mobile, intelligent agent platforms have been developed so far, which, however, are mostly incompatible between each other. Agents sitting on different platforms have difficulties in communication and co-operation. Therefore, several standardisation activities have emerged, like OMG-MASIF and FIPA, which deal with these problems.

In the area of control of Business Processes across organisational boundaries, workflow management systems are used to specify, execute, manage, and streamline business processes. Workflow management systems in general provide several useful functions to be used for business process execution and control across the organisational boundaries, which exist within a virtual enterprise. However, conventional workflow management systems have certain drawbacks in relation to the virtual enterprise concept. One of the main problems is their limited autonomy and flexibility. So far there are no extensions to the existing business process specification languages towards the support of cross-organisational business processes. Furthermore, remote invocation of business processes provided by different business domains should be preceded by access control, authorisation, and contract checks. Finally, in current workflow systems, shared business processes are structured inflexibly with respect to sub-processes that have to be executed remotely, i.e. the virtual enterprise partners which are to provide those processes are specified statically.

In dynamic virtual enterprises a workflow management system must deal with the fact that for the same business process specification different instances can exist. For every instance a set of different partners may be selected according to the needs and requirements of the process. Standards, currently proposed, are not directly dealing with market-based cross-organisational business process execution and management. Critical open issues, like inter-domain workflow execution and management, business process specification languages for inter-domain business processes, and dynamic selection of workflow providers during process execution are not effectively discussed.

Agent-based workflow management systems are to solve several of these problems. For example, control of business processes shared between multiple business domains can be assigned to agents which are either deployed directly in each of the domains involved or can migrate from domain to domain as required by the processes structure. However, most of the issues related to such agent-based workflow management systems, cross-organisational business process execution, and dynamic selection of partners are still under investigation and stable, well-defined solutions and concepts are missing.

Finally, Virtual market places are central parts of dynamic virtual enterprises. They increase flexibility and scalability through their search and mediation functions and support the selection of business partners during the establishment and reconfiguration of dynamic virtual enterprises. A market place can administer offerings for business processes made by potential providers and can perform searches for partners satisfying required quality attributes. The search and matchmaking services can be further complemented by advanced services, like automated negotiation and electronic contracting. Intelligent mobile agents can be a profitable technical implementation of such functionality due to their inherent autonomy, adaptability, and learning characteristics [11]. However, most current implementations of agent-based virtual market places do not consider emerging agent standards, like FIPA-ACL and FIPA protocols. Furthermore, most of the negotiation approaches, techniques, and models have basically concentrated on the area of business-to-consumer and consumer-to-consumer electronic commerce and are not addressing the needs of business-to-business market places and the needs of dynamic virtual enterprises in particular. Although some of the above techniques can be extended for the dynamic selection of partners in virtual enterprises based on service templates, this area is considered new and further research is needed.

3. The DIVE Framework

In the following, we introduce DIVE, a life cycle management approach for dynamic virtual enterprises. It is to realise dynamic virtual enterprise concepts based on agent technology and virtual market places. The framework is composed of a business model, an architecture, a life-cycle model, a set of agents that perform specific operations and a set of ontologies that the agents are using to communicate and interoperate.

3.1. The DIVE Business Model

The DIVE business model is specified as a set of different administrative domains having specific relationships. The business model for the DIVE framework consists of the following administrative domains:

- **Customer**, is the domain that has subscribed to the services provided by the VE and is allowed to use them,
- **VE representative**, is the domain that represents the VE to the outside world, i.e. to the customers, and provides the VE services to them. The VE representative is the responsible domain where the end-users are logged in, deploy and manage the provided services, i.e. the business processes,
- **VE Candidate/Partner** is the domain that registers its service offerings into the marketplace and negotiates with other VE partners on-demand to establish business relationships. When a successful negotiation has been achieved between a VE partner and a VE candidate partner, then this domain acquires the status of the VE partner. The VE partner authorises and authenticates business process requests, based on the negotiated contracts, and executes and manages the business processes on behalf of other domains,
- **Virtual Marketplace**, is the domain that provides registration and selection services for VE candidate partners. This domain is responsible for administrating service types and service offers and for managing the daily operations of the marketplace.

The logical relationships among these domains are depicted in the following picture.

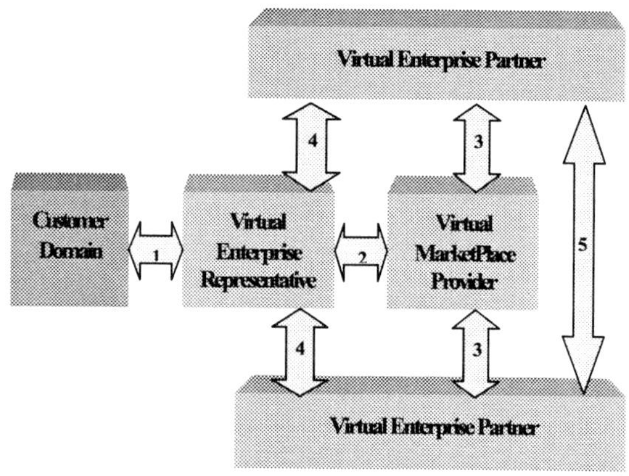

Figure 1: DIVE Business Model.

It should be noted that the VE representative domain and the VE Candidate/Partner provide similar services to each other and they deploy the same mechanisms. The main different is that the VE representative actually represents the VE to the external world, i.e. to the customers. Otherwise, the services, roles, responsibilities and internal components of both domains are identical from technical point of view.

Based on the previous business model, the following logical relationships among the domains can be specified:

- **Customer - VE representative (1):** The customer can log in into the system using a standard web browser and can start a business process, get a status report about a running processes, and manage existing processes, i.e. suspend, resume or terminate a process. Finally, when a running process completes, the result of the process is returned to the customer.
- **VE representative and Partner - Virtual marketplace (2,3):** the VE domain uses the marketplace to register local business processes and to search for potential VE candidate partners that can provide a service. More specifically, the VE representative can register, de-register or modify an existing business process offer stored in the virtual marketplaces. Additionally, the VE representative can search the virtual marketplace based on some constraints and get a list of potential VE candidate partners that can provide a specific service. The requirements imposed by this relationship are reflected into the specifications of the *virtual marketplace ontology*.
- **VE Representative – VE Candidate/Partner (4,5):** the VE representative starts a negotiation process with a VE candidate partner. This negotiation process results into an electronic contract that regulates the co-operation among the domains. As soon as the contract has been established, the VE representative can start the agreed remote process, resume, suspend, or terminate it upon request of a customer. The requirements imposed by this relationship are reflected

into the specifications of the *inter-domain* and *negotiation ontology*.

Having specifying the key business domains of the platform and the key relationships that they have, the individual human roles that exist in every domain can be specified.

In the VE representative and VE Candidate/Partner domain the role of the *Business Process Analyst* exists. This person is responsible for the specification of the business processes of this domain by deploying the business process definition language. The analyst also specifies which processes will be provided by this domain, i.e. local processes and which processes will be deployed remotely and dynamically by other partners, i.e. remote processes. Additionally, the analyst specifies the terms and conditions concerning the offering of local processes to potential partners. Based on these terms and conditions, the registration of local business processes to the virtual marketplace is done. The terms and conditions are actually logical constraints that relate process properties with certain min or max values. In the Virtual Marketplace domain the role of the *administrator* exists. This person is responsible for specifying and managing the service types that have been created in the virtual marketplace. The main operations that this role performs are creation, modification, list, and deletion of a service type. Finally, in the Customer domain the role of the *end-user* exists. This person initially subscribes to the services provided by this domain and gains access to them. This person has no particular responsibility except to log in and use the provided services.

3.2. The DIVE Architecture

The layered architecture of the DIVE Framework for the management of dynamic virtual enterprises consists of three respective layers. These are:

- **Agent-based Business Process Specification, Registration and Management System** and **Agent-based Virtual Marketplace System** that provide the basic operations for the specification of inter-domain business processes, the registration of them in the virtual marketplace, the selection and negotiation of partners, the access control and authorisation of process requests, and the execution and management of business processes.
- **Mobile Agent Platform (MAP) and Supporting Services** that provide the basic agent life-cycle services, migration services, messaging services, and access to services provided by the underlying distributed processing environment like XML parsers, legacy systems, etc.
- **Distributed Processing Environment (DPE)**: that supports the key operations for object life-cycle management and distributed services like Remote Method Invocation, access to persistent repositories, deployment of existing legacy systems, enabling services like vectors, etc. The distributed processing environment that will be deployed is that of the Java Framework with support of CORBA middleware services in order to enable interoperable access to distributed objects and components.

The last two layers are the basic infrastructure where the agent-based platform for the management of VE has been developed. The different services and components provided by these two layers will be directly used by the upper layer, i.e. by both the agent-based business process specification, registration and management system and the agent-based virtual marketplace. This architecture is presented in the following Figure 2.

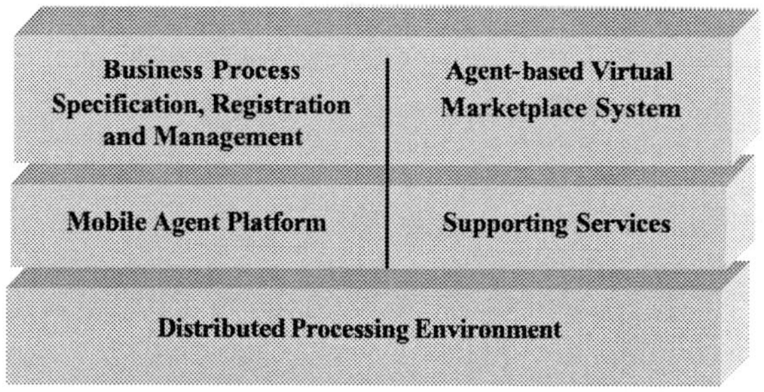

Figure 2: System Reference Architecture

Additionally, the first layer of the architecture is split into the business process specification, registration and management sub-layer and the virtual marketplace sub-layer. However, every administrative domain deploys this architecture in a rather different way. More specifically, the virtual marketplace domain uses only the agent-based virtual marketplace sub-layer, the VE representative and Candidate/Partner domain uses the business process specification, registration and management, while the customer domain deploys only a standard web browser for accessing the business processes. In both cases the two lowest layers deployed are the same, i.e. the Distributed Processing Environment and the Mobile Agent Platform and Supporting Services.

3.3. The DIVE Life-Cycle Model

In addition to the DIVE business model, a life-cycle model for the establishment and management of dynamic virtual enterprises has been developed. In general, a life-cycle model describes the key phases and activities existing during the life-time of an entity. According to ISO (ISO 91), a life cycle can be defined as "the finite steps a system may go through over its entire life history. The different life cycle phases define types of activities which are pertinent during the life cycle of the entity".

In our case, the virtual enterprise life-cycle model consists of two key phases, which are to be followed for the establishment, execution and management of a virtual enterprise. These phases are the Business Process Specification and Registration Phase and Business Process Execution and Management Phase.

During the Business Process Specification Phase a virtual enterprise candidate partner specifies his or her business processes. Such a business process is a hierarchical structure consisting of the root process and a tree of sub-processes. Sub-processes that cannot be further sub-divided are called tasks. Tasks are the actually activities performed by a several business objects, such as a software component, a device, or a robot. The specification of business processes is done using the DIVE business process definition language. For every business process, the input parameters, the output parameters, the sub-processes, the tasks, and the conditions among the sub-processes and tasks are specified. Additionally, every sub-process and task is specified as a local or remote entity. Local entities are those, which can be entirely performed by the offering domain, while remote entities are those, which must be provided by remote domains. Furthermore, for every specified task the associated business objects, that will be deployed, are specified.

Within the agent-based approach of DIVE, each business object is encapsulated and represented by a specific agent, called Resource Provider Agent (RPA). In this way, autonomous agents can handle legacy services provided by existing distributed objects that are physically located at different network locations within the domain. The result of a business process specification in DIVE is an XML-encoded document, which can be interpreted for execution during the Business Process Execution and Management Phase.

During the Process Registration Phase, a business domain, which wants to participate in dynamic virtual enterprise relationships, registers its processes at the virtual market place. The market place administrator registers the type of the process offer together with attributes, which are to be taken into account during the partner selection and negotiation phase.

After successful registration, the offering business domain becomes a virtual enterprise candidate and the business process can be executed upon request from a customer or from a higher-level business process. The execution phase is supported by DIVE process management facilities. When a request arrives at the management facility, the request is directed to a virtual enterprise representative, which creates an instance of the process. First the process description is retrieved from the business process repository. Then it is interpreted and the execution of the process is started. The virtual enterprise representative is realised in DIVE by a Personal User Agent (PUA). Each request is administered by a dedicated PUA. The PUA forwards the request to the DIVE agent-based business process management system, which instantiates, interprets, and executes the business process by means of a set of co-operating agents, called Workflow Provider Agents (WPAs).

Communication and co-operation between the WPA and RPA agents, if they all belong to the same domain, are based upon an intra-domain ontology, which comprises the set of messages, which are defined within the domain for the exchange between WPA and RPA agents. If a sub-process has been specified as remote, then the corresponding WPA creates a specialised negotiation agent, called Requestor Negotiation Agent (RNA), and sends it to the virtual market place in order to locate potential virtual enterprise candidate partners for the required remote sub-process. Upon request, the virtual market place informs the RNA about all the registered domains that can provide the process, i.e. all the potential virtual enterprise candidate partners. After that, the RNA starts the negotiation process by contacting the negotiation agents located in each candidate partner. These agents are called Provider Negotiation Agents (PNA). The negotiation process is performed among the RNA and PNAs by using a negotiation protocol and a negotiation ontology. The negotiation protocol used is the FIPA Contract Net protocol, which is based on a modified version of the original Contract Net protocol. The result of this negotiation process is the selection and contracting of the best virtual enterprise candidate domain according to the constraint given by both sides and the compromise negotiated.

As soon as a virtual enterprise partner for a particular remote process has been selected and contracted, the PNA returns back to his original location and informs the corresponding WPA agent about the selected partner. Then, the WPA contacts that partner domain and requests the execution of the business process by referring to the contract id that has been fixed during the negotiation process. The virtual enterprise partner domain checks the list of existing contracts and starts the execution of the requested process if a legitimate contract is found. The agent responsible for the access control and authorisation is called Domain Representative (DR) agent.

A customer that has requested the execution of a business process can influence and manage the process during execution time. The main operations that can be performed are suspension, resumption, and termination of the process execution. Every request of the customer is served initially by the virtual enterprise representative domain. All the WPA and RPA agents, related to the execution of the business process instance, are suspended, resumed, or terminated according to the customer's request. In addition to the activities within the representative domain, all the remote sub-processes and tasks, which have been started in the context of the business process, must be suspended, resumed, or terminated, too. Therefore, similar requests are issued and sent from the virtual enterprise representative to the corresponding virtual enterprise partners. Whenever the DR agent of a virtual enterprise partner gets a request to suspend, resume, or terminate an existing local business process, it checks the contract id, and if correct orders the related internal WPA and RPA agents to resume, suspend, or terminate. In that way, unauthorised requests for process suspension, resumption, or termination are not served. The communication and co-ordination between the different agents responsible for remote business processes are regulated by an inter-domain ontology. The inter-domain ontology is actually the set of messages that the agents are allowed to exchange during the execution and management of remote business processes. In addition to the execution control actions provided, the customer can request information about the current status of the business process. Similar to the information flow for execution control, the virtual enterprise representative requests from all agents associated with this process, local or remote, to declare their current status. When a process finally completes its operation, the DR agent of the virtual enterprise representative partner is informed. In the sequel, the DR agent informs the customer by posting to him or her the output results of the process and other status information, like the time of completion. If a fatal problem occurs during the execution of a process, the WPA agent responsible for the process instance informs the DR agent that the execution of this process could not be continued and consequently, the WPA agent aborts itself. The DR agent informs on demand, either the customer or the associated virtual enterprise partner, about this event and stops the execution of the business process.

The virtual market place plays the role of a third party administrative domain within the virtual enterprise. It provides matchmaking services to the virtual enterprise partners. The virtual market place administers a set of so-called service types, which represent the business process types defined and describe in a consistent way the interface of business processes The market place enables virtual enterprise candidate partners to register and administer offers in relation to defined service types and virtual enterprise representatives to search for potential partners that can provide particular business processes associated with existing service types. For every service type, the name of the type and a set of named properties are specified. The name of the service type is the name of the business process, while the input and output parameters of the process are named properties of the service type. Additionally, extra properties, related with the negotiation phase, are also included into the service type. For every property a (name, value) pair is associated. Service types are managed by the virtual market place administrator. The service type management includes creation, deletion, modification, and retrieval of service types.

Virtual enterprise candidate partners that want to register their process offerings at the market place must match each offer with an existing service type or must initiate the creation of a new type. For a common understanding between the partners making an offer and the domains looking for matching offers, it is necessary that every service offer is an instance of a given service type. Each offer assigns values to the properties defined by the type. Each candidate domain is responsible for its service offers. Actions, which a candidate domain can perform on its offers, include the registration of an offer, the withdrawal, the listing of offers and the modification of an offer.

Virtual enterprise representatives that want to find suitable partners, which can provide a particular process, retrieve from the market place all the registered offers that satisfy given constraints. The search and matchmaking functions required, are provided by the market place. Therefore, the basic functions, which the market place must provide, include service type management, service offer management, and service offer retrieval management. In DIVE, each of these functions are implemented by a dedicated agent, namely:

- **Service Type Agent (STA),** which is responsible for the management of service types, including the registration, removal, listing, and modification of service types.
- **Service Offer Agent (SOA),** which is responsible for the management of service offers including the registration, withdrawal, description, and modification of service offers.
- **Service Offer Retrieval Agent (SOR),** which is responsible for the retrieval of offers associated with a service type based on constraints given by the requestor.

These three agents are FIPA-compliant agents, i.e. they communicate with other agents by exchanging standard FIPA ACL/XML messages. The content of these messages is described in XML and it follows the DIVE virtual market place ontology. The virtual market place ontology describes the set of input and output messages that the market place agents can exchange with other agents. Agents that want to communicate with the STA, SOA, or SOR agents must formulate and understand messages based on this ontology. The key concepts of the DIVE virtual market place are very close to those of the standard OMG Trader. In fact, the STA, SOA, and SOR agents implement the functionality of an OMG-Trader in a FIPA compliant way.

In the case of business process registration by a candidate partner, a service type check is to be made first. For that purpose, an instance of the Provider Negotiation Agent (PNA) migrates to the market place, composes a FIPA ACL/XML request, and sends it to the SOA agent. The communication protocol among the PNA and SOA agents is based on the standard FIPA request-response protocol. The SOA receives the message, parses it from the ACL and XML parser, retrieves the action type from the message, and performs the requested action on the service offer repository. After action completion, the SOA composes an ACL/XML inform message, and sends the result back to the PNA. The PNA receives the message, parses it first from the ACL and XML parser, checks the reply of the SOA, and migrates to the virtual enterprise candidate partner to inform its domain about the results achieved.

If the business process registration requires a new type to be created, an instance of the Provider Negotiation Agent (PNA) migrates to the virtual marketplace, composes a FIPA ACL/XML request, and sends it to the STA agent. The communication protocol used between the PNA and the STA is again based on the standard FIPA request-response protocol. The STA receives the request, parses it from the ACL and XML parser, checks the type of the request and decides which action is required. Then, the STA performs the requested action together with the Service Type Repository, generates an ACL/XML inform message and sends it to the PNA. The PNA, upon receipt of the message, parses it first from the ACL and XML parser, checks the response and migrates back home in order to inform the virtual enterprise candidate domain about the result achieved.

4. Implementation, Validation and Conclusions

The DIVE framework has been completely implemented and tested, using the open, interoperable and standard technologies described above [16][17]. The different DIVE agents have been developed in Java on top of the Grasshopper [9] agent platform with OMG-MASIF and FIPA extensions. The agent communication protocols obey related FIPA standards, including FIPA request-response, request-query and Contract-Net.

The virtual marketplace agents implement standard OMG-Trader functionality, including the Service Type Repository, the Service Offer Repository and the OMG Constraint Language Parser objects.

For the persistent storage of local and remote business process offers, contracts, and business process specifications, dedicated repositories have been developed. The Offer, Contract, and Business Process Repositories are XML-based persistent modules. The customer's interface to the specification, execution and management of business process is Web-based. For that reason, special mechanisms based on TCP/IP and HTTP protocols and the standard Java Servlets technology have been developed.

The validation of the proposed system has been done by the development of a set of characteristic scenarios that demonstrate the features of the system. The validation scenarios have been developed in the context of the ACTS/MIAMI [13] and the EURESCOM P815 project [5].

Based on the development, testing, and validation of the DIVE framework, the following conclusions can be drawn according to the selection of technologies and the design decisions during implementation:

- **Openness** has been achieved due to flexible, XML-based ontologies and the use of open, interoperable standard technologies like XML, FIPA, FIPA-ACL, and Java.
- **Dynamicity, flexibility and evolution** are due to the dynamic selection of virtual enterprise partners and the automated negotiation during business process execution and management.
- **Asynchronous and loosely coupled communication** has been achieved through communication mechanisms supported by FIPA platform.
- **Distribution and scalability** due to the autonomous and distributed execution and management of shared business processes by means of intelligent, autonomous agents located in different administrative domains.
- **Autonomy** has been achieved due to the asynchronous and loosely coupled communication of agents during the execution and management of business processes.

Apart from those benefits, a key drawback has been identified. This drawback is the lack of performance, which is the result of certain design decisions. The performance limitations mostly originate from parsing the messages, the asynchronous message transportation imposed from the FIPA communication mechanisms, and the underlying agent platform.

Although the presented work intended to provide a coherent solution for the management of dynamic intelligent virtual enterprises, certain issues are subject to further improvement and research. These are the negotiation strategy algorithms for automated negotiation and partner selection, fault tolerance and exception handling during the execution of shared business processes, and secure inter-domain communication during inter-domain process execution. These and other open issues are addressed by two international projects, recently initiated, namely COVE [4], CREATIVEThink and E-Organisations.

5. References

[1] Breugst, M., Hagen, L., Magedanz; T. (1998a) "Impacts of Mobile Agent Technology on Mobile Communication System Evolution", pp.56-69, IEEE Personal Communications Magazine, Vol. 5, No. 4

[2] Camarinha-Matos, L. M., Afsarmanesh, H., (eds) (1999a), Infrastructures for Virtual Enterrises. Networking Industrial Enterprises. IFIP TC5 WG5.3 / PRODNET Working Conference for Virtual Enterprises (PRO-VE'99), Porto, Portugal, 27-28 October 1999, Boston: Kluwer Academic Publishers

[3] Choy, S., Breugst, M.,Magedanz, T., (1999), "Beyond Mobile Agents with CORBA - Towards Mobile CORBA Objects", 6th ACTS Conference on Intelligence in Services and Networks (IS&N), pp. 168-180, H. Zuidweg et.al (Eds.), IS&N 99, LNCS 1597, ISBN: 3-540-65895-5, Springer-Verlag, 1999.

[4] COVE Project, http://www.uninova.pt/~cove/coveproject.htm

[5] EURESCOM P815 Project http://www.eurescom.de/

[6] Filos, E. and V. Ouzounis, (2000) "Virtual Organisations: Technologies, Trends, Standards and the Contribution of the European RTD Programmes" International Journal of Computer Applications in Technology, Special Issue: "Applications in Industry of Product and Process Modeling Using Standards"Virtual-organisation.net, "Newsletter", Vol. 1, No. 3-4, 1997

[7] FIPA (1998) http://www.fipa.org/spec/FIPA98.html.

[8] Geppert, A. Kradolfer, M. Tombros, D. (1998) "Market-Based Workflow Management" Int'l IFIP Conf. on Distributed Systems for Electronic Commerce, Hamburg, Germany, June 1998

[9] IKV ++, Grasshopper, http://www.grasshopper.de/

[10] ISO/IEC (1991) 9596, Information Technology, Open Systems Interconnection, Common Management Information Protocol (CMIP) – Part 1: Specification, Geneva, Switzerland, 1991.

[11] Magedanz, T. (Ed.), (1999) Special Issue on "Mobile Agents in Intelligent Networks and Mobile Communication Systems", in Computer Networks Journal, ELSEVIER Publisher, Netherlands, vol. 31, No. 10, July 1999

[12] Martesson, N., Mackay, R., Björgvinsson, S., (eds) (1998), "Changing the Ways We Work. Shaping the ICT-Solutions for the Next Century" Proceedings of the Conference on Integration in Manufacturing, Göteborg, Sweden, 6-8 October 1998, Amsterdam: IOS Press.

[13] MIAMI project1998-2000, http://www.fokus.gmd.de/cc/ecco/

[14] Miller J, Sheth A, Kochut K, and Palaniswami D. (1998) "The Future of Web-Based Workflows" International Workshop on Research Directions in Process Technology, Nancy, France, International Workshop on Research Directions in Process Technology, position paper, Nancy, France

[15] Ouzounis, V. (1998) "Electronic Commerce and New Ways of Work – An R&D RoadMap", European Commission – Directoral General III

[16] Ouzounis, V. Tschammer V. (1999) "A Framework for Virtual Enterprise Support Services", 32nd International Conference on Systems and Sciences (HICSS32) Maoui Hawaii, 3-5 January 1999

[17] Ouzounis, V. Tschammer V. (2000) "Einsatz mobiler intelligenter Agenten zur Unterstützung virtueller Organisationen" in "Virtuelle Organisationen in Zeitalter von E-Business und E-Government", pp. 177-198, Springer Verlang, ISBN 3-540-41171-2

[18] Orfali, R. "The Essential Distributed Objects Survival Guide ", John Wiley and Sons, ISBN 0-471-12993-3 1996

[19] Stricker, C. Riboni, S. , Kradolfer M. and Taylor. J. (2000) "Market-based Workflow Management for Supply Chains of Services." In Proc. 33rd Hawaii Int'l Conference on System Sciences (HICSS-33), Maui, Hawaii, January 2000.

[20] Tombros, D. Geppert, A.(2000) "Building Extensible Workflow Systems using an Event-Based Infrastructure", Proc. 12th Conf. on Advanced Information Systems Engineering, Stockholm, Sweden, June 2000.

Research and Development of Electronic Commerce System—ESHOPDIY

Xiufen Fu[1, 2], Yong Tang[2], Jing Wang[2]
[1] *Huazhong University of Science & Technology, Wuhan, P.R. China*
[2] *Faculty of Computer Science and Engineering, Guangdong University of Technology Guangzhou, 510090, P.R. China*
xffu@gdut.edu.cn, ytang@gdut.edu.cn

Abstract

As an open standard for data representation, XML breathes new energy into Web applications and Electronic Commerce. In this paper, the authors elucidate the concept and connotation of XML; introduce the system structure and function model of the Electronic Commerce System—ESHOPDIY which we developed integrating three kinds of Electronic Commerce sales styles including B to C, B to B and C to C; deal with developing technology and realization of ESHOPDIY; and finally discuss its features.

1. Introduction

With the fast development of the Internet and its wide application, especially with the development of the Electronic Commerce based on Internet, great amount of information need to be treated. It has become more and more important how to organize the information on Internet to ensure its structure and search the information semantically. Nowadays developers have found the limitation of HTML in information presentation. Since XML— an optimizing subset of SGML (Standard Generalized Markup Language) is an open criterion of information presentation [1]. Its characters such as concise, open, extensible, flexible and self-describing and so on have supplied a gap of HTML and it has a wide application in information management, data exchange, Web application and Electronic Commerce, etc [2]. The effective Electronic Commerce system—ESHOPDIY has just been developed based on XML technology. This system is a set of Electronic Commerce system with great functions of business managing, sales counting and orders managing, etc. It has integrated three electronic business styles including B to C (Business to Consumer), B to B (Business to Business) and C to C (Consumer to Consumer). It is a synthetic Electronic Commerce Internet site as an Internet electronic shop of B to C, a business communication center of B to B and secondhand trade markets of C to C. On the base of description of XML technology, the paper introduces the structure and function of ESHOPDIY system, describes its key developing technology and discusses its characters in the end.

2. XML technology

XML is an optimizing subset of SGML [4]. SGML is a specified, structured, extensible markup language standard published in 1986 by ISO and can be used to make large information library. An optimizing subset of SGML, XML has appeared because the complexity of SGML has made it difficult to be used on Internet. XML has got rid of the complex part and the part used scarcely in SGML so that it can be used in Web easily. XML has a simple syntax structure, places its importance on file contents and is good at describing hierarchical data, making information orderly and combining structured data of different source to make search more meaningful. By this way Web application systems can analyze the file structure quickly and process corresponding structure objects. XML mainly has three parts. It including DTD (Document Type Definition) used to specify logic structure of files and define Web page elements and their attributes, XSL (Extensible Style Language) used to specify styles of XML files and XLL (Extensible Link Language) used to support simple links and extensible links in Web [5].

The application of XML has supplied many gaps of HTML. XML can be used by Internet users to exchange information among different databases, can process data mainly downloaded from Internet servers in client side and can make the same data have different interfaces to different users. Compared to HTML, XML is more powerful to describe data, can design its own file types, define new application and also is more powerful in hypertext linking. Because XML has none business with specified software and hardware platforms the data presented by it has a more widely application. The XML data is structured and is separate in data and its presentation so it has promoted information opening and reusable ability greatly. At the same time the validity check of XML files has excluded the disturbing of rubbish data to the system and promoted the system greatly in robust and performance.

In a word, XML is an open standard of data presentation, it will become an important part of Web source, especially have important applications in the integration of isomeric data resource, native computing, multiple presentation of data and interoperations of Web applications and so on. It will bring evolutionary change to all kinds of Web applications [3].

3. System structure and developing environment

Electronic Commerce is business activities in Internet [6]. The present EC system can be divided into three EC modes including B to C, B to B, C to C according to both sides in dealing and its virtual contents.

Electronic Commerce provides enterprises with virtual worldwide dealing environment, promotes the proficiency and service quality greatly and so strengthens their competition. The comprehensive EC web site —ESHOPDIY (Electronic Shop Do It Myself) we developed is fit for Chinese own situations, perfect in business modes, security and dealing management and so on and compatible with B to C, B to B, C to C. The structure of the comprehensive electronic business shop is shown in Figure 1.

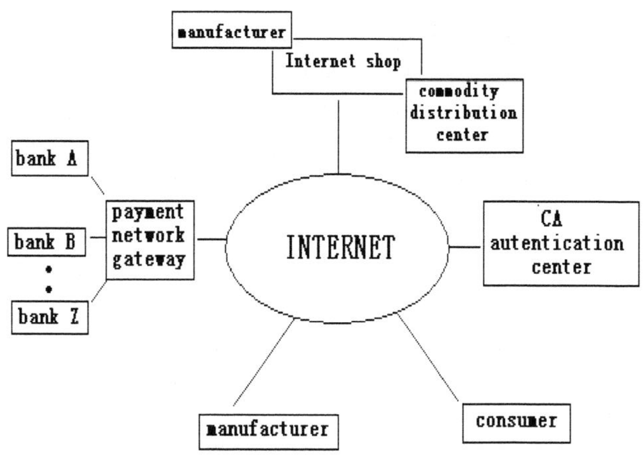

Figure 1. Structure of the electronic business shop.

3.1 The structure and function of system modules

There are six main modules in ESHOPDIY. They are B to C Electronic Commerce module (see Figure 2), B to B business communication center module (see Figure 3), and C to C secondhand trade market module (see Figure 4), system management module, shop financial manager module, and customer personality setting module.

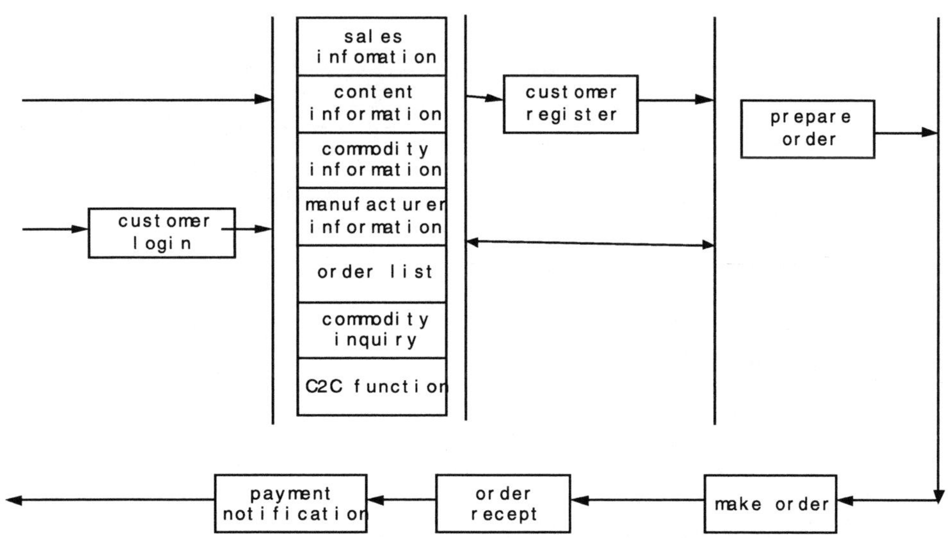

Figure 2. Shopping flow chart of B to C system.

B to C E-business module: It is composed of shopping sub-module and shop manage sub-module based on the concept of Electronic Commerce shop. In shopping sub-module, It provides member register and enroll function, at the same time it includes several functions such as search engine of goods, the realization of shopping car, make order for goods and so on. Shop management module is the top of the Web site. It provided a lot of powerful functions such as order management, goods management, sales promotion, stock management, shop setting, material modification and key modification. In these functions, shop setting is the distinguished feature of

ESHOPDIY. The shopkeeper can edit his own Electronic Commerce shop web page as he likes. In this module, It provides strong editing function for page in order to realize the personality of sale.

B to B business communication center module: It originated from B to B E-business module. In this module, we design register information and searching information sub-module. In register information sub-module, the shopkeeper and producer can public some information with their own needs. The procedure where is running in backstage will delete useless information automatically. In searching information sub-module, It provide a lot of searching module on the basis of category, area, synthesizing condition. The searching module is vague in file. It provides use information for customer easily.

C to C second hand trade market module: It is composed of goods material management sub-module, trade searching sub-module, goods bid sub-module.

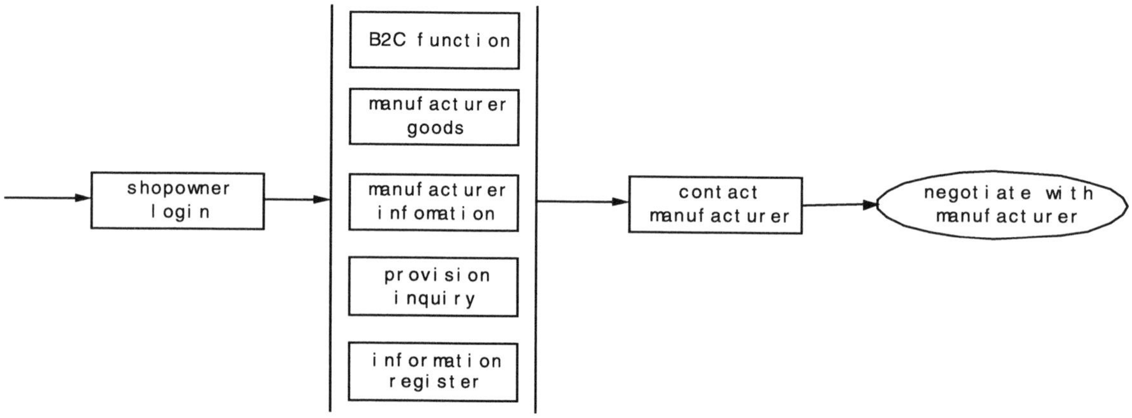

Figure 3. Dealing system flow chart for B to B shop-owners.

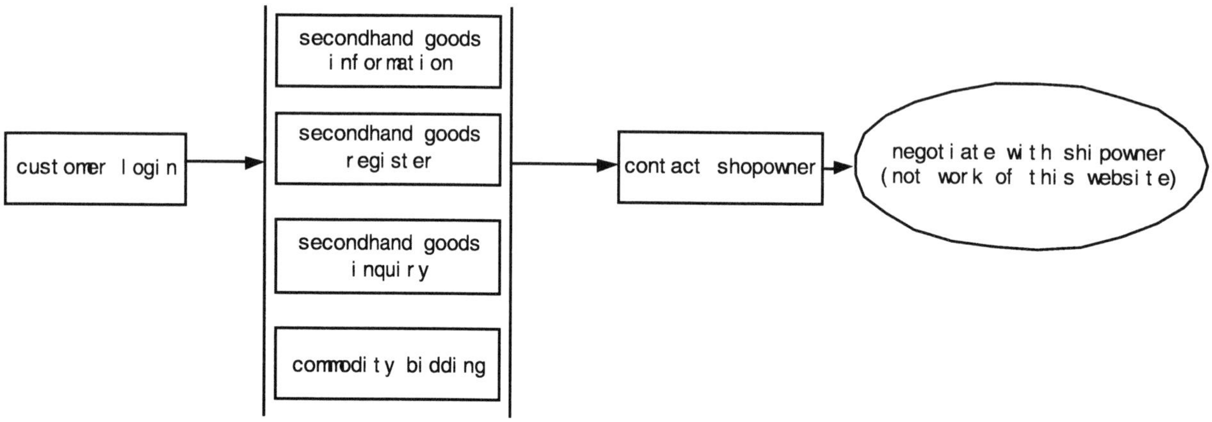

Figure 4. Secondhand dealing flow chart for C to C system.

System management module: The web management system which is based on web server is used in ESHOPDIY. The management part includes three modules: member data management, information updating management and merchandise data management.

Shop financial management module of the Web site: This module uses data resource recorded in databases to list clearly sales condition of a month, kinds of sales amount for the shopkeeper and also make personal financial management for the shopkeeper including the automatic building of profit and loss table in accounting, dynamic building of the sales lists and vivid representation of the histogram and so on so that it have provided a practical auxiliary tool for shopkeepers to managing their Web shops. This module is made up of the profit and loss table building sub-module, sales lists building sub-module and histogram building sub-module.

Customer personality setting module of the Web site:

This module provides customers with multi-personality service to make the Web site more convenient for customers to do shopping and more easy to select what they like. we set personality for customers according to the point of view in their shopping, the view in the sales promotion activities, their last shopping records and shopping guides they provide, etc.

3.2 System developing platform and developing tools

Developing platform:

When selecting Web developing platform we place the importance on its support to the system, technology requirement and safety. The Web site uses ASP (Active Server Pages) technology, which is convenient to operate databases, included in IIS3.0 and 4.0, and can only run with support of Windows NT operating system. And also for the reason of great amount of data the Web site has it is quick and convenient to manage databases using large databases managing system such as SQL Server 7.0 as well as ODBC data source. SQL Server 7.0 can also run with support of Windows NT operating system. So we have selected the newest edition of Windows NT, Windows 2000, as the developing platform. At the same time the following seven steps have been taken to protect Windows NT Server: use NTFS not FAT as the file managing system, rename the system manager account, track the audit system Abolish NetBIOS of TCP/IP. Close unnecessary inward TCP/IP interfaces. Cancel the convenience of Access from Network. Don't publish information rashly

Key developing technologies of the system include: ASP technology, MS SQL Server database managing system, ODBC technology, XML technology, C++ Builder programming technology, Flash technology of dynamic vector diagram and technology of combining Java and Jscript, etc.

Developing tools: Members of the Visual Studio 6.0 family: Visual InterDev, Macromedia, Drumbeat2000, MS SQL Server 7.0, C++ Builder 5.0, MS FrontPage 2000, Macromedia Dreamweaver 3.0, Adobe PhotoShop 5.5, Macromedia Flash 4.0, XML styler, Msxml 3.0 Release Beta (July 2000)

4. Key technologies in the designing

4.1 Design of the search engine

At present the Internet search engine is mainly aimed at pages of the Web site, not its databases, and the databases cannot be visited freely so the biggest and also the most important resource of Internet has been wasted. The search engine technology of DSHOPDIY system is based on XML, Visual Basic programming technology, COM technology and DOM (Document Object Model) technology, uses Microsoft parser (Msxml.dll) to visit DOM, maintain and search XML files. This search engine first makes a request to the Web site, gets the right data from its databases and encapsulate data to XML files, then saves XML files in client computers and in the end operates on XML files using the client program. In the search engine algorithm we present a kind of blur-searching algorithm in which by calling several mf_docase () functions to make the exclusive SQL sentence without collision dynamically according to all kinds of input conditions when they are more than one.

4.2 Images uploading/showing

The process engine must be made ourselves because uploaded files should be encoded in the form of multipart/form-data. At present there are three ways in IIS to get uploaded files: use CGI/ISAPI; use uploaded groupware; and write ASP codes. We use ASP to write process engine. Access of binary files is not supported in ASP so images must be stored in databases. In database operations we use a field of image type (blob type) to record images. This kind of field can't be accessed directly by object Recordset and must be accessed by GetChunk/AppendChunk, a method of object Field. Types of images are stored in field PhotoType. Also we take some safety measures to keep hackers from invading client browsers by uploading ActiveX components.

4.3 Use Web page-templates

In the page-editing module of the shop we use templates, in which the system provides users basic system functions and users can build their own shops using our templates of different styles selectively. We have used XML technology so each kind of template can define its own attribute freely and some attributes can be modified. Each kind of template is written as a sub-procedure and the whole template will display once the sub-procedure is called. So our templates have great extensibility. We can make a new template if only we write a new sub-procedure. Styles of templates are different to different template types. There are three template types including search engine template, merchandise classifying template and merchandise lists template. Among them the merchandise lists template is the most complex. We split codes of the merchandise lists template into universal functions and sub-procedures to make codes reusable and thus to satisfy different users.

4.4 Encrypt certification

In the system the encrypt certification is in form of Challenge/Response in which passwords are stored in databases and also random encryption is adopted in the transportation to ensure transported passwords can't be intercepted and used. The system takes TEA as the encryption algorithm that is fast to encode, powerful to stop difference invading, simple to carry out and is up to the information summary algorithm. Our encryption program uses the user password as the password and use the ASCII code of the string "ESHOPDIY " as the initial evident text to make the once-encrypted user password and store it to databases. When the user asks for password verification a random number will be produced by the server and written to client pages as a hidden input. After the user inputs the password the program also uses "ESHOPDIY " as the initial evident text and the once-encrypted result as the password to make the second encryption in which the result will be sent to the server to be verified as well as the user name. Then when the server gets the encrypted password also the produced random number is taken as the initial evident text and the password in the database as the password to make compare and judgment, thus to finish the verification procedure.

4.5 Make the sales histogram

According to the trade condition of the shop the system automatically makes contrastive histograms of the shop sales including the month contrast, the quarter contrast and the year contrast. The contrast contents include sales earning, sales cost and merchandise. The shopkeeper can see intuitionistically through histograms the earning condition of the shop and the welcome level of the merchandise. Only the shopkeeper can see the histogram. In the identity verification the uniform identity certification page of the Web site is called to get high maintenance and ensure data encryption. Data of the histogram come from the table Statistics that records the trade sum of each shopkeeper in each month. Histogram display as images whose width is adjusted according to the contrastive proportion.

5. System characters

In the designing of the Web site we make a detailed analysis to the present condition of Electronic Commerce in China, know some existing problems in its developing style, summarize its developing trend and then have developed an popular Electronic Commerce system aimed at China's own condition. The Web site has the following designing characters:

5.1 High safety

The first question of Electronic Commerce is the safety of Internet trade so this Web site has paid great attention to safety. It has taken steps to protect the server operating system, tested modules of the program over and over, ensure the communicating information safe to reduce the possibility of error and ensure high system safety.

5.2 Integration of three Electronic Commerce styles

The Web site has tried to integrate three styles of Electronic Commerce: B to C, B to B and C to C, which have played good supplementary roles to each other. To shopkeepers the whole service from stocking, sale to managing has made a good sale environment, to customers they not only can buy new merchandise they like but also can sell out old ones, to manufacturers they can promote sales with so many shopkeepers and customers.

5.3 The strong function of editing shop Web page

In this Web site, in order to stand out personality of shop we provide the strong function of editing shop Web pages. The shopkeeper can choose all kinds of styles for template freely and in each template several edited modules are provided. In each module, the shopkeeper can insert some information such as edited photo, word, table, at the same time, the shopkeeper also can freely choose function objects provided by the system. For example, search engine, all kinds of goods list, vote system and so on. So each shopkeeper can edit personal and powerful shop.

5.4 Advanced XML program technology

In the editing function of the shop Web page, we use XML program technology. Flexible markup functions allow people define their own marks according to different needs so they are fit for different jobs and different areas. Thus, the template we develop has strong extensibility.

5.5 The strong function of manage shop

To help the shopkeeper manage the shop, the system provides the powerful management functions, which contains not only the basic management functions, such as: merchandise management, and orders management, but also some advanced management functions, for example: sales promotion management, and stock management. What is more important is that we provide finance management function. It is convenient for the shopkeeper to query a lot of financial data with this function, such as: finance status, sales profits and sales tax.

5.6 Personal setting of shopping

At present, almost all the Company Web sites provide the personal setting. But only a few of the Electronic Commerce Web sites provide the same setting. For this trend, the Web site provides the personal setting for the customers. Through the shopping records and settings, the shopkeeper can conjecture the shopping intent of the customers, so the shop can help the customers get more information of the merchandise, and the customers can select the favorite goods more easily than before.

5.7 System management program

If we design the system management program as Web pages, the data will pass through several middle transactions, so the security problem arises. But the system management program is the kernel of system, which can't bear the existing of system security holes. Considering the security of the system, we design and run the system management program on the server. And some of the redundant data can be deleted periodically, which means system resources will not be wasted.

6. Conclusion

With the fast development of Internet and applications of Electronic Commerce based on Internet wider and wider, the Electronic Commerce is rapidly becoming a new kind of commercial pattern. The synthetic Electronic Commerce system we developed integrating three kinds of Electronic Commerce sales styles including B to C, B to B and C to C has a wide application scene. We also look forward to the development of the XML information-processing platform and developing methods, technology of XML applications.

7. Acknowledgement

This work is supported by the Natural Science Foundation of Education Department of Guangdong Province and the Guangdong Provincial Natural Science Foundation (Team project).

8. References

[1] R.khare, A.Tifkin, XML:A Door to Automated Web Applications, IEEE Internet Computing, July/Aug, 1997, Vol. 1, No.4, pp.78--87

[2] Frank Manola, Technologies for a Web Object Model, IEEE Internet Computing, Jan/Feb.1999, Vol.3, No.1, pp.38--47.

[3] Robert J.Glushko, Jay M.Tenenbaum, Bart Meltzer.an, XML framework for Agent-based E-commerce, Communications Of The ACM, 1999; Vol.42, No.3, pp.106-114.

[4] T.Bray, J.Paoli and C.M.Sperberg-McQueen, eds., Extensible Markup Language (XML)1.0, W3C Recommendation, World Wide Web Consortium, 1998; available online at http://www.w3c.org/TR/REC-xml.

[5] Qu Yuzhong, Zhanfeng, A Survey of XML and Related Technologies, Computer Engineering, 2000, Vol 26, No12, pp.124-127

[6] Liang Zhengyou , Research on Integration of E-commerce and ERP, Proceedings of The 7th Joint International Computer Conference, Oct. 2000, pp.1144-1148.

Virtual Supply Chain Management: Information Framework and Agreement Network Model

Charu Chandra
Industrial and Manufacturing Systems Engineering Department
College of Engineering and Computer Science, University of Michigan - Dearborn
4901 Evergreen Road, Dearborn, Michigan, 48128-1491, U.S.A.
charu@umich.edu

Alexander V. Smirnov
St. Petersburg Institute for Informatics and Automation of the Russian Academy of Sciences
39, 14th Line, St. Petersburg, 199178, Russia
smir@mail.iias.spb.su

Abstract

The interest in virtual supply chain decision-making problem is growing fast. The purpose of this type of enterprise is to transform incomplete information about customer orders and available resources into coordinated plans for production and replenishment of goods and services in the temporal network formed by collaborating units. This paper discusses a framework that integrates intelligent information support group-decision making and agreement modeling for a supply chain network. It offers the platform for design and modeling of diverse implementation strategies related to the type of agreement, optimization policies, decision-making strategies, organization structures, information sharing strategies and mechanisms, and business policies for the virtual supply chain.

1. Introduction

The concept of supply chain (SC) is about managing coordinated information and material flows, plant operations, and logistics [1]. The fundamental premise of this philosophy is synchronization among multiple autonomous business entities represented in it. That is, improved coordination *within* and *between* various supply chain members is achieved within the framework of commitments made by Members. Increased coordination can lead to reduction in lead times and costs, alignment of interdependent decision-making processes, and improvement in the overall performance of each Member, as well as the supply chain network (Group) [2, 3].

Enterprises deploy Internet technologies that impact the market substantially by driving costs down through standardized networking technology and creation of entirely new enterprises and relationships by real-time interconnection of companies with their customers. As a result, enterprises mutate into virtual SCs (VSCs).

Management of these technologies requires capabilities for real-time decision support. The challenge is to accommodate interaction between various units of an enterprise with competing objectives as they share information in the course of transacting with each other and provide decision-making capabilities towards achieving their shared goals. This paper offers insights into a supply chain problem-solving approach that enables implementing integration capabilities for intelligent information support, and group decision-making utilizing a common agreement model.

2. General information framework of a virtual supply chain

The information support architecture for a VSC builds on an agent classification implementation of decision-support technologies, namely custom-ordered management, configuration management, constraint satisfaction and propagation, and data and knowledge management [4]. These are described below.

- *Problem solving agent* (PSA) is oriented for a unique class of SC management problem applied for problem solving of different processes (sourcing, inventory management, etc.).
- *System information agent* (SIA) is oriented for consistency checking, controls of constraints for

integration between units of a VSC and its management and maintenance.

- *Facilitator agent* (FA) is oriented for distribution of knowledge and identification of access level in sharing data and knowledge base. It has three major components - (i) conventional reusable knowledge management technique [5], (ii) common knowledge map that describes distribution of knowledge sources for a VSC, and (iii) ontology management.
- *Unit information agent* (UIA) is oriented towards application of reusable methods, techniques and solutions for problem solving at the unit level. It also includes the *user information agent* (UsIA).
- *Object - relational adapter* that maps relational models to object-oriented data models and vice-versa.

In order to solve a problem, the PSA through the FA deploys reusable knowledge management to communicate with shared object-oriented data model (based on template libraries and repositories) in SIA and ontology-based / user-oriented knowledge domain models of UIA. Unit knowledge domain model describes the environment at the unit level, such as objectives, constraints and resources [6]. FA utilizes the shared object-oriented data model for consistency checks of unit's data and knowledge for problem solving. FA communicates through object-relational adapter with Internet-based distributed relational databases that describe the dynamic environment of a VSC.

3. Structure of negotiation environment for a supply chain

The structure of VSC negotiation environment supported by a group decision support system (GDSS) consists of three types of agents [7], namely (i) VSC manager agent, (ii) VSC unit agent, and (iii) VSC coordinator agent. Depending on the information exchanged, different agents perform different functions. Using agent classification [8], Unit and Coordinator agents may be classified as regular agents, while the Manager Agent is a planning agent. Figure 1 depicts general structure of the decision-making approach applied to the GDSS. It has following steps, based on which unit and group decisions are formed:

1. The task is ready to be processed as soon as the manager agent prepares a list of attributes for evaluation.
2. The coordinator agent receives incoming message (M_{inp}) about task readiness for processing. It prepares information for the unit's local database. Using internal set of function (F), it prepares messages for unit agent.

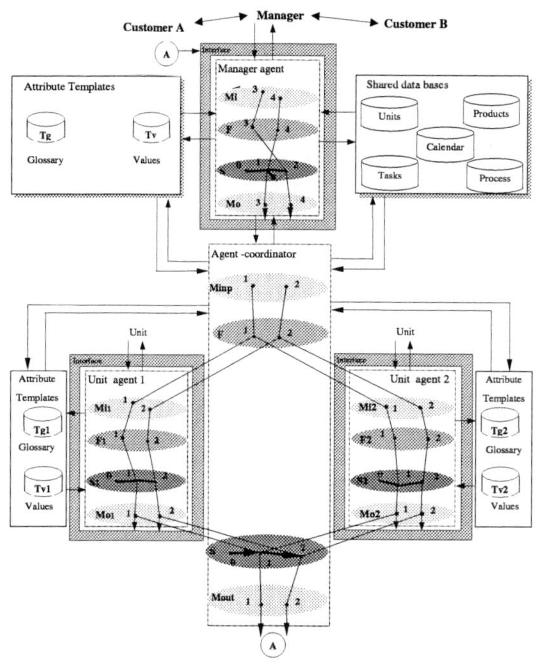

Figure 1. A general structure of interacting agents during negotiation.

3. The unit agent checks existing tasks (analyses incoming messages M_{ij}, where j is a unique number of the unit and i is a number of the process step) and develops a set of actions F_j.
4. Unit agent increments readiness degree of the task on the local level when the unit finishes (transition S from readiness degree N into readiness degree $N+1$). It sends a message (M_{oj}) to coordinator agent.
5. The coordinator agent analyses whether similar messages have been received on readiness from all other members of the group. If someone is not ready, there is a wait on the entire group readiness. The readiness of the task is incremented when everybody is ready (transition S from readiness degree N into readiness degree $N+1$). As a rule, after a stage of opinions making by each unit, a stage of conformation follows.
6. The coordinator agent sends the message (M_{out}) to the manager agent that a stage of the conformation follows.
7. Depending on the coordination type, that is (a) automatic control by the coordinator agent, or (b) personal control by the VSC manager, there are two possible outcomes, (i) an automatic check of conformation is received, or, (ii) the task is stopped and the VSC manager analyses whether it is necessary to advance the task readiness or to return back a step.
8. Appropriate decision is sent to the coordinator agent who either advances the task per unit further, or returns the unit agent on to the previous stage.

Unit agents use local data which are representation of data prepared by the manager agent, while each agent at local and group levels works with its own internal representations. In order to obtain conforming agents' decisions, two levels of abstraction are distinguished: unit's decisions and manager agent's level aggregating unit opinions. A feedback is applied to different stages of an algorithm: group decision aggregation, initial data representation or agent's role. In order to obtain conformity, decision agents change utility functions iteratively, or group estimation is involved as an additional attribute. A wholesome function is used to select a decision. The following algorithm depicted in Figure 2 and based on a unit's preference changes according to an aggregate estimation is applied. Quantitative attributes are estimated using limit function values in terms of direct and reverse relations for maximizing and minimizing factors respectively, assuming each factor to be linear at the considered interval. Let I be a number of attribute and j - a number of alternative, and k - a number of a unit: $j = 1,..,m,\quad i = 1,...,n\quad k = 1,...,l.$ Then an aggregate matrix R can be obtained according to following relations:

$$r_{ij} = \frac{x_{ij} - x_i^0}{x_i^* - x_i^0} - \text{direct},\quad r_{ij} = \frac{x_i^* - x_{ij}}{x_i^* - x_i^0} - \text{reverse},$$

where r_{ij} is a weighted unit's estimation, x_{ij} - an attribute estimation, x_i^0 - low limit of an attribute estimation, x_i^* - high limit of an attribute estimation.

Calculation of the R matrix is performed as an aggregation of alternative estimations for quantitative and qualitative attributes. Attribute weights matrix W is calculated according to the analytic hierarchy process [9]. A decision vector is calculated based on an agent wholesome function:

$$\varphi_k(m) = \sum_i w_{ik} r'_{ij},\quad \sum_i w_{ik} = 1.$$

This function permits representation of a group preference in the form of additive weighing of local wholesome functions. Each unit of a VSC group has aggregate ranking of other unit as follows:

$$\varphi_g(m) = \sum_{l \neq k} h_l \sum w_{il} r'_{ij},$$

where h_l - is a weight for *i-th* unit.

In the case where ranking for alternatives φ_g and φ_k are similar for all agents, a totally conforming decision is obtained. But usually decision-makers are interested not in one and the only, but in a number of partially conforming decisions.

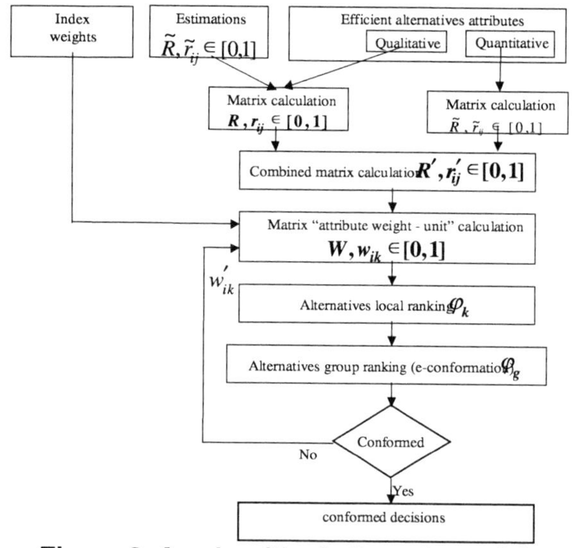

Figure 2. An algorithmic flow chart of the decision making process associated to each agent.

In the case where ranking for alternatives φ_g and φ_k are similar for all agents, a totally conforming decision is obtained. But usually decision-makers are interested not in one and the only, but in a number of partially conforming decisions.

To take group decision into consideration, "a VSC group estimation" attribute is introduced as follows:

$$\varphi'_k(m) = \sum_i w_{ik} r'_{ij} + w_g \varphi_g,$$

where w_g is the importance of a VSC group opinion for an agent.

To make the conformation process convergent, this VSC group estimation is used for scaling of previous estimations as follows:

$$w'_{ik} = w_{ik}(1 - w_g),\quad \sum_i w_{ik} = 1.$$

In the event, conforming decision is not obtained after a number of iterations, the conforming procedure leads to partially conforming decision. Multiple criteria choice methods, such as minimum/maximum criterion for preliminary ranking and Hurwitz method, using index weights for final ranking are applied.

4. Agreement model for a virtual supply chain

As evident in discussions in previous sections, adopting VSC management techniques to deal with integration issues requires a common understanding of goals, objectives, and policies by members belonging to the network. In order to implement these, however, a common agreement among VSC members is required. It is based on following guiding principles for the VSC. These principles encapsulate trends in production and logistics management that a VSC arrangement is usually designed to capture [2]:

- Supply chain is a Co-operative System.
- Supply chain exists on Group Dynamics of its Members.
- Negotiation and compromise are norms of operation in a supply chain.
- Supply chain system solutions are Pareto-optimal (satisficing), not optimizing.
- Integration in supply chain is achieved through synchronization.

Figure 3 captures the essence of these guiding principles for a supply chain. It depicts a generic approach to decision-making in the supply chain network. Local decisions are made at the member level. A type of local decision made *within* a member network is forecasting inventory at Member_A based on the historic demand. Another type of local decision made *between* a member network is planning production at Member_B for forecasted demand projected by Member_A. Global decisions are made at the group level. A type of global decision made at the Group level is how the supply chain will collectively support capacities for planned production by the network.

Implementation of the supply chain philosophy is facilitated through *agreement(s)* entered into by members. The main elements of an agreement are -- type of agreement, optimization policies, decision-making strategies, organization structures, information sharing strategies and mechanisms, and business policies. These are briefly described below.

Type of agreement signifies the scope and focus of commitment made *within* and *between* members of the supply chain. Local or member level agreement relates to commitments on costs and lead times. Global or group level agreement relates to commitments on price and manufacturing capacity.

Optimization policies for a VSC can be modeled to manage any or all of the following market conditions:

1. Manufacturer optimizes its own cost and imposes its policies on suppliers. In this case, no cooperation among members is forthcoming.
2. Global cost of the supply chain is minimized by cooperation among members at various levels.
3. Each level optimizes its local cost objective that may or may not optimize the global cost objective of the supply chain. In this case, cooperation among members is selective.

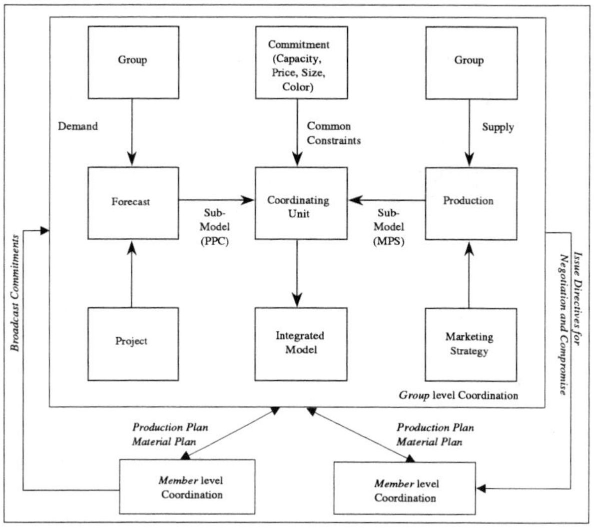

Figure 3. A generic agreement modeling representation in a virtual supply chain

Decision making strategies for a supply chain fall in following general categories:

1. Centralized, with *all* decisions made at the Group level and implemented down to a Member level.
2. Decentralized, with decision-making for:
   - *commonly* shared resources or policies at the Group level, and
   - *independent* resources or policies at the Member level.
3. Autonomy, with decisions made autonomously by each Member at the Member level and some loose form of agreement on price, cost, and capacity sharing, etc.
4. A combination of the above, where some members may opt for centralized decision-making while others may choose decentralized, or autonomous form.

Organization structure for the supply chain network can assume various forms, such as:

1. M-form organization, which prescribes a multi-divisional structure with each Member, set up as a division of an organization and adapting a specific, or a combination of above noted decision-making strategies.
2. U-form organization, which prescribes a functional structure with each Member, set up as a function of the organization and adapting a specific, or a combination of above noted decision-making strategies.

Information sharing strategies for the supply chain are designed so that, (a) at the local or member level, information is shared with regard to demand forecasts, inventory replenishment schedules, inventory costs, etc., and (b) at the global or group level, information sharing pertains to market plans, production plans, pricing strategies, etc. As depicted in Figure 3, members broadcast their intentions for making commitments on capacity sharing, cost structures, lead times, etc. throughout the supply chain network. The level of commitment is negotiated between members of the supply chain and final agreements concluded. These agreements are then issued as directives for VSC members to follow in meeting common VSC objectives.

Business policies pertain to how VSC conducts its business with regard to inventory management, production, etc. For example, it may follow a push, pull, or synchronous strategy to manage inventory; and a product-to-stock or produce-to-order strategy to manufacture products. For illustrative purposes, a *Generic Supply Chain Network Agreement Optimization Model* is described in the Appendix. It combines various elements of a supply chain network agreement described above. As depicted in the Appendix, coordination schemes for supply chain network optimization fall in three main categories:

1. Price-directed approach [10] solves the problem via pricing resources in the network. The coordinating unit announces tentative prices for common resources to divisions. At the same time, divisions respond with tentative production plans, with associated profits to the coordinating unit. The procedure is repeated until an optimal solution is obtained.
2. In the resource-directed approach [11, 12], the coordinating unit announces a tentative portion of common resources to divisions. Divisions report maximum profits of given resources and valuations of allocated resources to the coordinating unit. The procedure is repeated until an optimal solution is obtained.
3. A mixed price and resource directed approach blends the above two strategies as a problem-solving technique.

5. Conclusions

A Virtual Supply Chain Network approach offers various alternatives for enterprise integration through shared information infrastructure for decision-making. The motivation in proposing this approach is to incorporate elements of scalability, reconfigurability and reusability of enterprise components. A *General Framework of Supply Chain Management* that integrates intelligent information support, group-decision making, and agreements modeling for a virtual supply chain network has been proposed. It provides the mechanism to experiment various types of behavior patterns that may emerge through interaction of virtual enterprise members and apply lessons learned in developing robust supply chain management models. Implementation of this methodology would fundamentally change global business environment by enabling constructive collaboration among its network units to achieve shared objectives. This approach satisfies the need for increasingly complex business relationships and underlying technology infrastructure that companies are implementing to support their global strategies.

6. References

[1] H.L. Lee, and C. Billington, "Material management in decentralized supply chains", *Operations Research*, Vol. 41, No. 5, 1993, pp. 835-847.

[2] C. Chandra, "Enterprise architectural framework for supply-chain integration", Proceedings of the 6th Annual Industrial Engineering Research Conference, May 17-18, 1997, Miami Beach, Florida, Institute of Industrial Engineers, Norcross, Georgia, pp. 873-878.

[3] P. Sousa, T. Heikkila, M. Kollingbaum, and P. Valckenaers, "Aspects of co-operation in Distributed Manufacturing Systems", Proceedings of the Second International Workshop on Intelligent Manufacturing Systems, September 1999, Leuven, Belgium, pp. 685-717.

[4] A. Smirnov, and C. Chandra, "E-Management of Scalable Supply Chain Configurations: Concept and Reusable Knowledge Management", Proceedings of the 6th International Conference on Concurrent Enterprising. Toulouse, France, 2000, pp. 91-99.

[5] R. Fikes, and A. Farquhar, "Distributed Repositories of Highly Expressive Reusable Ontologies", *IEEE Intelligent Systems*, 1999, pp. 73-79.

[6] A. Smirnov, and C. Chandra, "Ontology-based Knowledge Management for Co-operative Supply Chain Configuration", Proceedings of the 2000 AAAI Spring Symposium "Bringing knowledge to business Processes", March 20-22, 2000 Stanford, California, AAAI Press, pp. 85-92.

[7] M. Pashkin, A. Smirnov, and I. Rakhmanova, "Multiexpert: an Internet bases support environment for solution evaluation", Proceedings of the Special Day on pan-European Co-operation and Technology Transfer, Zakopane, Poland, 1996.

[8] K. Fischer, J.P. Müller, H. Heimig, and A.-W. Scheer, "Intelligent Agents in Virtual Enterprises", Proceedings of the First International Conference and Exhibition on the Practical Application of Intelligent Agents and Multi-Agent Technology, The Westminister Central Hall, London, UK, 1996, pp. 205-223.

[9] T.L. Saaty, "*The Analytic Hierarchy Process*", McGraw Hill, 1980, pp. 287

[10] G.B. Dantzig, and P. Wolfe, "Decomposition principles for linear programs", *Operations Research*, Vol. 8, 1960, pp.101.

[11] P.S. Bender, W.D. Northup, and J.F. Shapiro, "Practical modelling for resource management", *Harvard Business Review*, March-April, 1981, pp. 163-173.

[12] A.T. Kate, "Decomposition of linear programs by direct distribution", *Econometrica*, Vol. 40, No. 5, 1972, pp. 883-890.

Appendix

A Generic Supply Chain Network Agreement Optimization Model

Model Notations:

- R - technology matrix of the total enterprise.
- A_j - technology matrix of the jth Member representing common constraints.
- b_0 - resource vector for the jth Member with respect to common constraints.
- D_j - technology matrix of the jth Member representing independent constraints.
- b_j - resource vector for the jth Member with respect to independent constraints.
- B - vector of constraints for the total enterprise.
- C_j - vector of objective function coefficients for the jth Member.
- X_j - vector of decision variables corresponding to j^{th} Member.
- t - iteration index.
- j - Member index.
- u - dual price computed on the Member level for their part of common constraints.
- U - dual price of common constraints computed at the Group level.
- I_j - identity matrix of the jth Member.
- Π_j - dual price for the upper bound of Member j's variable in the aggregate problem.
- F_j^t - local bound contribution of Member to the Group.

A M-form (Multi-divisional structure) supply chain network:

Model Characteristics:
Division (Members) share common resources

Problem Formulation:
Maximize, $z = C_1X_1 + C_2X_2 + \ldots + C_nX_n$ *Common Objective*

subject to (constraints):

$$
\begin{array}{llllll}
A_1X_1 + & A_2X_2 + \ldots & A_nX_n & \le b_0 & \textit{Common Constraints} \\
D_1X_1 + & & & \le b_1 & \\
& D_2X_2 + & & \le b_2 & \\
& \ldots & & & \\
& & D_nX_n & \le b_n & \textit{Independent Constraints}
\end{array}
$$

$X_j \ge 0, \forall j$

A U-form (functional structure) supply chain network:

Model Characteristics:
Production and sales are coordinated

Problem Formulation:
Maximize, $z = C_1X_1 + C_2X_2 + \ldots + C_nX_n$ *Common Objective*

subject to (constraints):

$$
\begin{array}{llllll}
I_1X_1 + & I_2X_2 + \ldots & I_nX_n & = 0 & \\
D_1X_1 + & & & \le b_1 & \\
& D_2X_2 + & & \le b_2 & \\
& \ldots & & & \\
& & D_nX_n & \le b_n & \textit{Independent Constraints}
\end{array}
$$

$X_j \ge 0, \forall j$

Coordination schemes for supply chain network optimization is depicted in Figure 4 below:

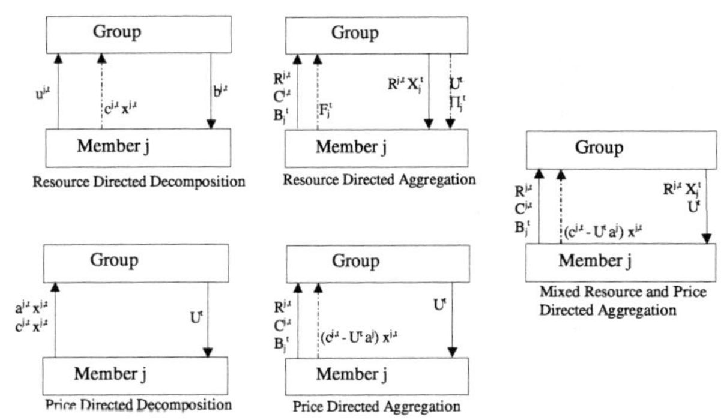

Figure 4. A supply chain coordination scheme.

Business to Business Standard and Supply Chain System Framework in Virtual Enterprises

Deren Chen
Dept. of Computer Science & Engineering, Zhejiang University, Hangzhou, 310027, P.R. China
drchen@cs.zju.edu.cn

Abstract

Electronic Commerce provides a revolutionary way of doing business. It is changing the traditional commerce activities and forming a new pattern of international commerce running by establishing dynamic virtual enterprises. Due to its potential advantages, specifically in business to business fields, EC has become an active area recently, with many standards and solutions defined, proposed, and emerged. This paper introduces concepts of electronic commerce and virtual enterprise with the trade circle, then discusses recently proposed, developed and applied Internet B2B standards, including OBI, ICE, RNIF&PIP. It focuses on the relation of supply chain with enterprises and partners. It also presents an application framework of supply chains in trade partners of virtual enterprise.

1. Electronic Commerce and Virtual Enterprise

Electronic Commerce (EC and Electronic Business or EB) is the process of linking businesses electronically with their suppliers, distributors, manufacturers, and customers to facilitate, create, or support the whole or part of the business cycle. For a long time, traditional commerce activities developed around seller market and concentrate the main work into promoting the sales of products. The EC is changing the traditional commerce activities and forming a new pattern of international commerce running by establishing dynamic virtual enterprises. Virtual Enterprise (EV) is set up for quick response to the faster change competition environment and a foreseen market chance. It is a common profit organization for all its member companies and partners, and is a way to approach the global agile manufacturing and transacting.

EC is the application of technology toward the automation of trade transactions and workflow. The application of EC runs through the whole trade process from pre-trade, the trade to post-trade and forming a trade circle [1]. During the pre-trade, people needs prepare to publish/update catalog, prices, to browse catalog, or to negotiate terms. In the middle of trade, people place order, make payment, do billing and inventory. In the post-trade, people manage inventory with suppliers, coordinate with shipping company, resolve issues and maintain relationship.

According the Figure 1 presented, EC establishes both the integration between modules and chairs (buyer and seller). It can reach dynamic modification of flow and feedback loop, and is relate to transaction, contractual and partnerships as well as industry specific requirements. Special important, the model and flow can help corporations perform efficient business activities between companies and partners (business to business or B2B).

2. Business to Business Standards

The EB activities should be uniform across all of the different system platforms and business practices. There are two types relative standards in EB. One of standards dedicate the foundation of EB, For example meta-language standard SGML and HTML, common document exchange EDI, XML and ebXML and SET and SSL which in the EB security protocol fields. The other types are about application of EB in B2B or B2C. Ongoing industrial standards mainly include Open Trading Protocol (OTP) for on-line shopping procedures[8]; Open Buying on the Internet (OBI) for B2B non-production materials purchasing; Information, Content Exchange Protocol (ICE) for another B2B, and RNIF&PIP in RosettaNet.

2.1 OBI

OBI standard is an open, flexible framework for B2B internet commerce solution. The purpose of the OBI is to provide a standard framework for secure and interoperable B2B internet commerce with an initial focus on automating high-volume, low-dollar transactions between trading partners. There are four essentials entities involved in an OBI system. The Buying Organization procures items as part of its daily business operations. The

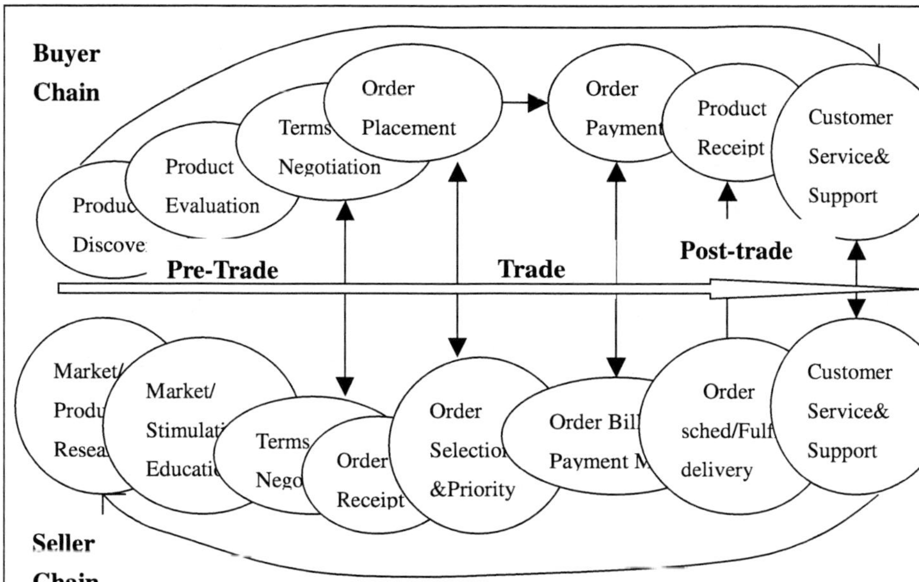

Figure 1: Electronic Commerce Lifecycle

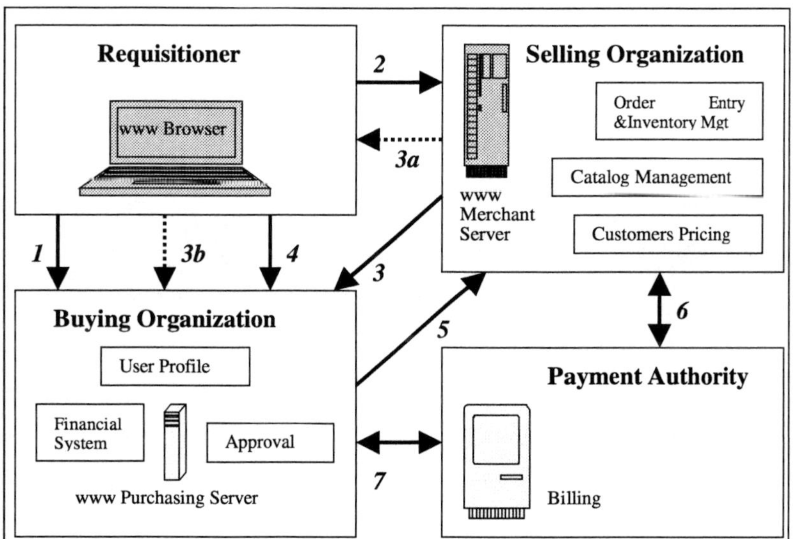

Figure 2: OBI Architecture

Requisitioner, a member of the buying organization, is interested in procuring certain items as part of the nonmission critical process of the organization within his/her command. The Selling Organization supplies goods and services to other businesses. Finally, the Payment Authority, which may not exist in an OBI scenarios, as a clearing-house for all payment and settlement activities between the selling and buying organizations. All the aforementioned entities should be connecting to the Internet and have digital certificates that uniquely and securely establish their identities. The whole OBI architecture (Figure 2) is based on the following model of B2B commerce [4]:

(1) A requisitioner, using a Web browser, connects to a local purchasing server located at the Buying Organization and selects a hyperlink to a Selling Organization's merchant server containing an on-line catalog of goods and services.

(2) The Selling Organization's server authenticates the requisitioner's identity and organizational affiliation based on information presented in the requisitioner's digital certificate. Authentication information is used, in conjunction with profile information optionally presented

by the requisitioner's browser, to uniquely identify the requisitioner and to construct a specialized catalog view. The requisitioner browses the catalog, select items, and "checks out."

(3) The content of the requisitioner's "shopping basket" and identity of the requisitioner is mapped into an order request (EDI-compatible). A digital signature is calculated (optionally); the order request (and digital signature if used) is encapsulated in an OBI object which is encoded and transmitted securely to the Buying Organization over the Internet using HTTP and SSL. There are two alternative methods for transmitting an encoded OBI object containing an order request over the Internet using HTTP. These are referred to as the server-to-server method (step 3 in figure) and the server-browser-server method (step 3a-3b in figure). The Buying Organization server receives the encoded OBI object, decodes it, extracts the order request, verifies the signature (if appropriate) and translates the order request into an internal format for processing.

(4) Administrative information (including payment type) is added to the order request at the Buying Organization (automatically from a profile database and/or manually by the Requisitioner), and the order is processed internally either automatically or through a workflow-based process.

(5) The completed and apprved order is formatted as an OBI order (EDI-compatible) and a digital signature is calculated if desired. The order (and digital signature if appropriate) is encapsulated in an OBI object which is encoded for transport and transmitted securely from Buying Organization server to Selling Organization server via the Internet using HTTP over SSL. The Selling Organization receives the encoded OBI object, decodes it, extracts the order, verifies the signature (if appropriate), and translates the order into its internal format.

(6) The Selling Organization obtains credit authorization, if necessary, and begins order fulfillment.

(7) The payment authority issues an invoice and receives payment.

2.2 ICE

The Information and Content Exchange (ICE) Protocol is a B2B one for exchanging document between content provider and content distributor. The goal of ICE is reducing the cost of doing business online and increases the value of B2B relationships[9]. ICE facilitates the controlled exchange and management of electronic assets between networked partners and affiliates.

ICE covers four general types of operations: Subscription Establishment and Management, Data Delivery, Event Logs, and Miscellaneous. There are two entities involved in forming a business relationship where ICE is used: a Syndicator and a Subscriber. A relationship between a syndicator and a subscriber starts off with some form of subscription establishment. In ICE, the subscriber typically begins by obtaining a catalog of possible subscriptions offers from the syndicator. The subscriber then subscribes to particular subscriptions, possibly engaging in protocol parameter negotiation to arrive at mutually agreeable delivery methods and schedules. The relationship then moves on to the steady state, where the primary message exchanges center on data delivery. ICE uses a package concept as a container mechanism for generic data items. It defines a sequenced package model allowing syndicators to support both incremental and full update models. ICE also defines push and pull data transfer models for server-to-server protocol.

2.3 RNIF & PIP

RosettaNet's mission is to facilitate electronic exchange of standard business documents between trading partners, adhering to the Partner Interface Processes (PIPs) specified and standardized by RosettaNet. Fundamental to this are the RosettaNet Implementation Framework (RNIF), the PIP specifications, and the business and technical dictionaries[10]. Figure 3 shows the relationship of the RosettaNet Implementation Framework and its constituent parts to the rest of the RosettaNet specifications in a trading partner implementation.

An organization's business processes can be divided into two broad categories. The business processes that are internal to the organization are called "private processes," while the business processes that involve interactions with trading partners are known as "public processes."The public processes are business processes through which partners conduct EB. Within the context of RosettaNet, these are the partner interface processes that are visible between trading partners. Public processes implement the RosettaNet PIP specifications to exchange standard business documents over standard Internet transfer protocols, as specified by the RosettaNet Implementation Framework. Within trading partner enterprises, private processes interface with public processes and with back-end business systems as needed to facilitate EB exchanges between trading partner organizations.

A major part of RosettaNet's standardization effort is alignment of business processes between trading partners in a given supply chain. RosettaNet specifies these as Partner Interface Process (PIP) specifications. RosettaNet divides the entire e-business supply chain domain for which PIPs are specified into broad classifications called "clusters." Each cluster is further sub-divided into two or more "segments." Each segment comprises several PIPs. PIPs contain one or more Activities, and Activities in turn specify Actions (see table 1).

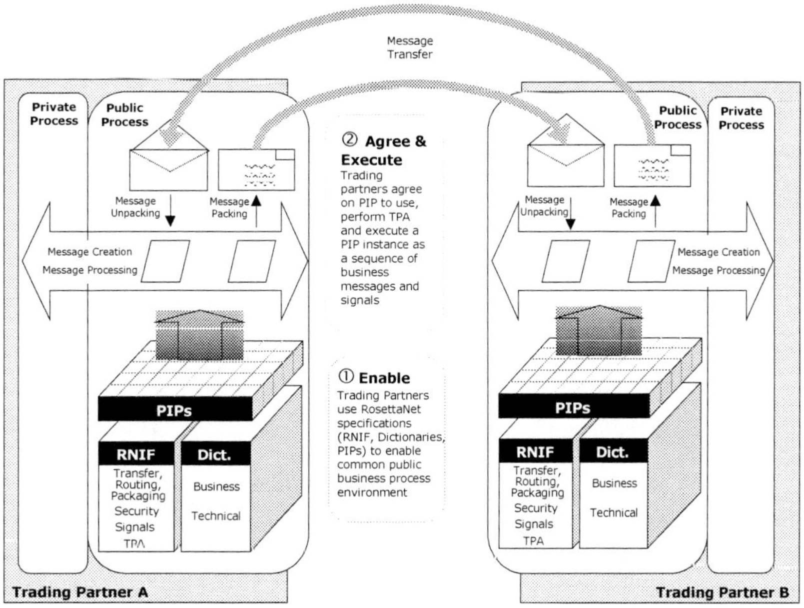

Figure 3. RosettaNet Specifications in a Trading Partner Implementation

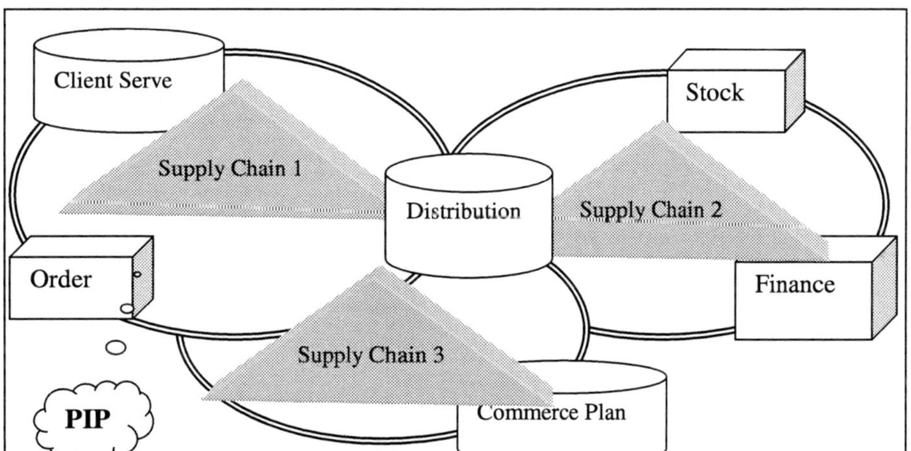

Figure 4: Agile Configuration of Supply Chain Systems

PIPs include specification of partner business roles (Buyer, Seller etc.); business activities involved between the roles; and type, content, and sequence of business documents exchanged by the role-interactions while performing these activities. They also specify the time, security, authentication, and performance constraints of these interactions. Structure and content of the business documents exchanged is specified through XML Document Type Definitions (DTDs) and associated Message Guidelines.

Trading partners that participate in the PIP exchange business documents that conform to the DTDs and Message Guidelines in the subject PIP specification, using network protocols that are specified and supported by the RosettaNet Implementation Framework.

3. Supply Chain System Framework

There are many virtual enterprises in today's global electronic marketplaces and there are some different required supply chain systems in one or more virtual enterprises. Every enterprise or partner runs in these supply chain system by standard B2B proposal framework. It is necessary to do agile configuration for each supply chain system such as Figure 4. There are relative management or agent affair according to standard has been described for every module in each supply chain system. Table 2 lists the entire PIPs order management from segment A to C of cluster 3.

Table 1. The entire EB supply chain domain in RosettaNet

Cluster	Segment	PIPs Listing number
0: RosettaNet Support	0A: Administrative	0A1
1: Partner Profile Management	1A: Partner Review	1A1~ 1A3
	1B: Product and Service Review	1B1 ~ 1B3
2: Product Information	2A: Preparation for Distribution	2A1 ~ 2A9
	2B: Product Change Notification	2B1 ~ 2B11
	2C: Product Design Information	2C1~ 2C6
	2D: Collaborative Design	
3: Order Management	3A: Quote and Order Entry	3A1 ~ 3A9
	3B: Transportation and Distribution	3B1 ~ 3B10
	3C: Returns and Finance	3C1 ~ 3C6
	3D: Product Configuration	3D1 ~ 3D16
4: Inventory Management	4A: Collaborative Forecasting	4A1 ~ 4A5
	4B: Inventory Allocation	4B1 ~ 4B4
	4C: Inventory Reporting	4C1 ~ 4C4
	4D: Inventory Replenishment	4D1
	4E: Sales Reporting	4E1 ~ 4E6
	4F: Price Protection	4F1 ~ 4F5
5: Marketing Information Management	5A: Lead Opportunity Management	5A1 ~ 5A3
	5B: Marketing Campaign Management	5B1 ~ 5B6
	5C: Design Win Management (Electronic Components)	5C1 ~ 5C5
	5D: Ship from Stock and Debit (Electronic Components)	5D1 ~ 5D6
6: Service and Support	6A: Provide and Administer Warranties, Service Packages, and Contract Services	6A1
	6B: Provide and Administer Asset Management (Merged with 6A)	
	6C: Technical Support and Service Management	6C1 ~ 6C6
7:Manufacturing	7A: Design Transfer	
	7B: Manage Manufacturing Work Orders and WIP	
	7C: Distribute Manufacturing Information	

4. Summary

The OBI, ICE, RNIF&PIP standards provide relative general system architecture and message protocol to enable smooth B2B EC transactions to be conducted on the Internet. There are a lot of middle-small enterprises and partners for manufacturing in China and they are combining each types virtual enterprises and organizing each type supply chain framework. It is necessary to study and develop standardized supply chain system.

There are several evolutionary trends of EB such as: (i) from buy-side and sell-side centric to B2B marketplace and exchange, (ii) from fixed price model to the dynamic pricing including auctions, reverse auction and dynamically negotiated pricing, (iii) from standard catalog buying to build to order and design to order, (iv) from one to one interaction to many to many interactions, (v) from the pre-planed model to the on-line decision support and deep computing, (vi) from the static HTML to dynamic HTML then to the XML based JSP, and (vii) the integrated supply chain management, collaborative and mobile commerce.

Table 2. All of Order Management Activities of Cluster 3 from segment A to C

3A1: Request Quote	3A2: Query Price and Availability
3A3: Transfer Shopping Cart	3A4: Manage Purchase Order
3A5: Query Order Status	3A6: Distribute Order Status
3A7: Notify of Purchase Order Acceptance	3A8: Change Purchase Order
3A9: Cancel Purchase Order	
3B1: Distribute Transportation Projection	3B2: Notify of Advance Shipment
3B3: Distribute Shipment Status	3B4: Query Shipment Status
3B5: Change Shipment	3B6: Notify of Shipper's Manifest
3B7: Create Delivery Appointment	3B8: Notify of Transportation Claim
3B9: Notify of Delivery Exception Appointment	3B10: Notify of Cancel Delivery
3C1: Return Product	3C2: Obtain Financing Approval
3C3: Notify of Invoice	3C4: Notify of Invoice Reject
3C5: Notify of Billing Statement	3C6: Notify of Remittance Advice

5. References

[1] S. Fu, J.-Y. Chung, W. Dietrich, V. Gottemukkala, M. Cohen, and S. Chenetc: "A Practical Approach to Web-based Internet EDI", Proceedings of 19th IEEE International Conference on Distributed Computing System (ICDCS) Workshops on Electronic Commerce, pp.53-58, May 1999

[2] Deren Chen & Jen-Yao Chung: "Internet Based Electronic Business Framework, Application and Business to Business Standards", The 2nd International Symposium on Electronic Commerce (ISEC2001), Hong Kong, April 2001

[3] Shen-sheng Zhang & Guo-jun Gao: "Virtual Enterprise and Agile Supply Chain, Computer Integrated Manufacturing System", Vol. 5 No. 2 pp.1-5, April 1999

[4] The OBI Consortium, "Open Buying on the Internet(OBI) Technical Specifications, Release V2.1, November 1999

[5] The ebXML Technical Requirements Project Team:, "ebXML Requirements Specification Version 1.0", May 2000

[6] Robert S. Sutor, "IBM XMML Industry Standards Liaison", IBM Japan XML Day, August 1999

[7] XMLSolutions Corporation, XML and EDI: Peaceful Co-Existence, 1999, http://www.xml.org

[8] The Open Trading Protocol Consortium, "Internet Open Trading Protocal Version 0.9", January 1998, http://www.otp.org

[9] ICE Authoring Group, The Information and Content Exchange (ICE) Protocol AC Review Version 1.1 (May 2000), Revision R, http://www.icestandard.org.

[10] RosettaNet, RosettaNet Implementation Framework: Core Specification, Release 3, Jan 2001, http://www.rosettanet.org

Part VIII: Distance Learning

Cooperative Mechanism for Real Time Long-distance Teaching System*

Junzhou Luo, Jianxiang Wei and Huasong Xu
Department of Computer Science and Engineering, Southeast University,
Nanjing 210096, P. R. China
jluo@seu.edu.cn

Abstract

Cooperative mechanism is the most important portion in the study of CSCW. As for a virtual background of teaching, this paper, by analyzing the cooperative relation between teacher and students in real time long-distance teaching system, constructed a Petri net model based on cooperative mechanism within CSCW. Thereafter, a cooperative controller to be realized in software is designed, and applications of the controller are discussed.

1. Introduction

Computer Supported Cooperative Work (CSCW) focuses on the cooperative work among humans with computers, which makes it possible for persons at different places to form a virtual group and work on the same task at the same time. The application of corresponding techniques is called groupware. A groupware mainly consists of two parts. One is called shared space. The other is cooperative mechanism, which allows many customers to access the shared space at the same time and fulfill the task efficiently.

The long-distance education is one of the most important CSCW application areas[1]. The long-distance teaching is a key activity in the long-distance education, which is known as a way to allow the teacher and students who are in different places to talk by face to face based on network. Now the most of the existing long-distance teaching systems are broadcasting modes, with the many defects, known as "one-way teaching". In those systems, it is impossible for teachers to adapt and improve their teaching in time because they cannot get feedback information from the students. The same thing happens to the students, they are not allowed to solve their puzzles immediately. Such long-distance teaching system can hardly pledge a good teaching quality and scarcely can it satisfy the requirements of long-distance education. Therefore, it is greatly meaningful to study an interactive long-distance teaching system[2].

In an interactive system, the teacher and students can see each other, hear each other as in the same classroom although they are at different places. However, the teaching happens on network background, and the teacher should be the center of such activities, in other word, the teacher should be authorized to allow the student in and out, to ask questions, to allow and forbid the students to speak etc, while the students should be allowed to ask questions anytime. That is to say, all the activities take place are random, uncertainly and synchronous. The cooperative mechanism should be improved to meet all the requirements[3,4,5]. This paper described, analyzed and realized such cooperative mechanism.

2. Cooperative relation

Virtual classrooms are designed to fulfill all functions of the traditional classrooms by setting up a virtual interactive teaching environment within network, which is shared by the teacher and students online. It allows the teacher and students to see each other, hear each other, and realize their immediate communication and cooperation[6]. We can divide the virtual class into sub-ones by the student's characteristics, such as amount and area etc.

In this virtual classroom, the teacher and students play the same roles as in traditional classroom. Nevertheless, there are a few differences between them as follows:

- □ Differences in place, usually in virtual classroom, the teacher and the students are at different places, even the students are at different places.
- □ Difference in sight, the teacher can hardly see all the students in virtual classroom. The teacher can merely choose a sub-classroom to observe. When long-distance learning is being carried out, there would be a great number of students, which make it impossible for the teacher to see everyone.

The teaching becomes much complicated because of such differences. In order to describe the cooperative

* This work is supported by National "973" Fundamental Research Programme of China

relation of the teacher and students during the teaching courses correctly, we need to make out all the statuses of the teacher and students in details. During the courses, the teacher's statuses may be described as listening and speaking, while the students may be described as listening, speaking, pausing speaking, continuing speaking, pausing listening, continuing listening, as described in figure 1.

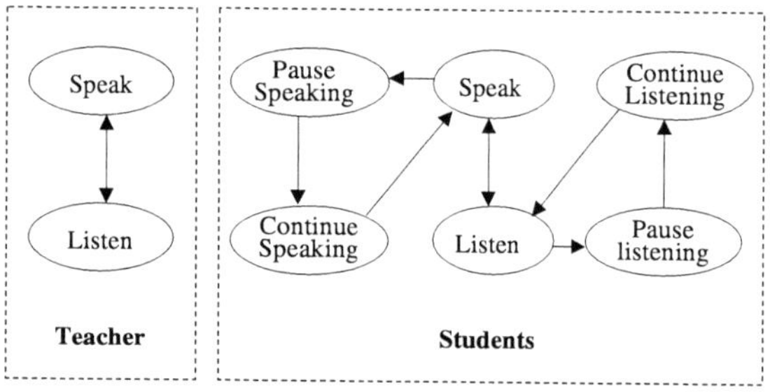

Figure 1. State transformation figure

We can describe cooperative relation between the teacher and students as following according to the variety of statuses. We can have only one speaker, this is to say, either the teacher, or the student, during the teaching courses. If the teacher is speaking, then the students should be listening, on the other side, if the teacher asks a question or a student applies for speaking and is allowed, then the student speaks, while the teacher listens. Such statuses may be transformed when needed. In order to present a good order of the teaching, the transformation of students' state should be controlled by the teacher. The teacher should be authorized to forbid or allow the student to speak, to forbid the students who was considered not welcomed to listen, and re-allow the students to listen etc. Within network background, all the status transformations will be unforeseen and synchronous activities. For example, the students may ask question any time, many students will apply for speaking at the same time. In order to solve all the troubles that may emerge, we need to work out a cooperative mechanism to make it possible to control all the synchronous activities.

3. Specification and analysis of cooperative mechanism

3.1. Petri net model

According to the cooperative relation of the teacher and students in teaching, it is feasible to describe the synchronous activities by Petri net. We've got a Petri net model of cooperative mechanism in figure 2. Here we suppose that there are n students and a teacher in the virtual classroom. The figure includes 6 transitions, 6 places, 22 arcs and n+3 token. N students are represented by "1" to "n" respectively and the teacher by "m". In addition, "p" is used for a token providing mutual exclusion of potential speakers. The label <x> is free variable and can be any of the token <1>...<n> and <m>.The six transitions represent six different processes. The transition "stol" represents the student or the teacher from speak to listen, and the "ltos" represents a person from listen to speak. The transition "dell" represents deleting a student is not welcomed temporarily, and the "addll" represents adding a deleted student to listen. The transition "pauss" represents pausing a student speaking, and the "cons" represents let a student continue speaking.

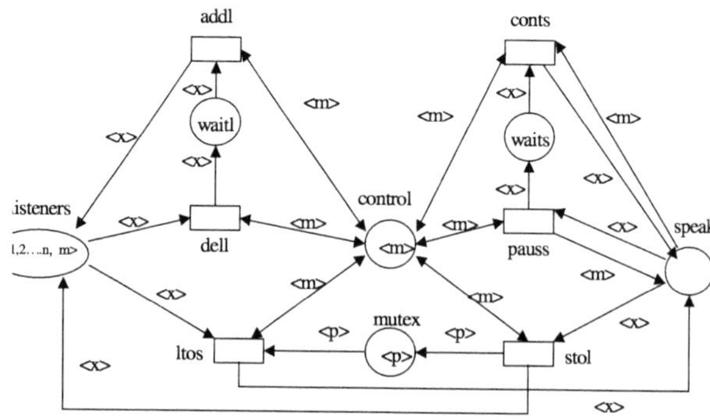

Figure 2. Petri net model of cooperative mechanism

3.2. Model Specification

Figure 2 shows the initial status of the system. The tokens represent teacher and students are located in "listeners" place. When the teacher begins teaching, the transition "lots" fires and the token "m" flows into the "speaker" place. At the same time, the token "p" disappears. Now the transition "ltos" has no authority to fire. That is to say, when a person is speaking, the other cannot speak. While the teacher answer a question or a student wants to speak, the transition "stol" fires and the token "m" flows from "speaker" to "listeners". The teacher begins to listen. At the same time the "mutex" place gets a token "p". The transition "ltos" has authority to fire. One of the token <1>...<n> flows into "speaker" place when it happens. A student begins to speak. During speaking of a student, the teacher may interrupt him by firing the transition "pauss". The token <x> flows into place "waits" and the token <m> into place "speaker" from place "control". The teacher begins to speak. The student can continue to speak when the teacher agrees. The transition "cons" fires. When the student completes his speaking, the transition "stol" fires and after the transition "ltos" fired, the teacher or another will begin to speak. In order to keep an organizational discipline of the teaching, the teacher can pause a student listening when he is considered unwelcome, where transition "dell" works and the transition "addl" may let him return to class.

4. Implementation Policy

Through the analysis of the Petri model, we can see there are mainly eight synchronous activities in teaching. They are consistent with the six transitions in the figure 2. In order to put the mechanism into practice, we present the concept of cooperative controller. It acts as a kind of application software that is designed to control all the synchronous activities in a cooperative system. The figure 3 is a cooperative controller, which is able to control all activities that occur in a real time long-distance teaching system.

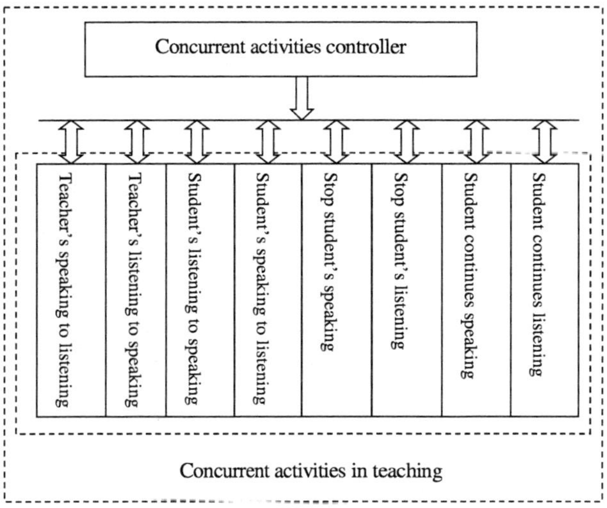

Figure 3. Cooperative controller

In order to realize the cooperative controller's actions, we define the eight synchronous activities as eight process functions. They are: tstol( ), tltos( ), sstol( ), sltos( ), addl( ), dell( ), pauss( ) and conts( ). A main procedure includes the eight functions above. Two notation "s" and "t" represent the mutex mechanism between the processes. Their initial values are 1, while P and V operations relies on them.

P(s) is defined as follow (the same as t):

(1) s=s-1,

(2) if s>=0,then the process continue,

(3) if s<0, then it will be blocked, and inserted into blocked queue of s, system recall another process to work.

V(s) is defined as follow(the same as t):

(1) s=s+1,

(2) if s>0,then the process continue,

(3) if s<=0, it implies that there are processes blocked, then system will recall a process from blocked queue of s and make it ready. This process is going on.

The following is the description of application system:

```
main( )
{        s:=1; t:=1;
//the following processes are unorderly
        tstol( )
        {P(s);
         teacher: from speaking to listening
         V(s)  }
        ......................................
        tltos( )
        {P(s)
         teacher: from listening to speaking
         V(s)  }
        ......................................
        sstol( )
        {P(s);
         student: from speaking to listening
         V(s)  }
        ......................................
        sltos( )
        {P(s)
         student: from listening to speaking
         V(s)  }
        ......................................
        addl( )
        {P(t)
         add a student to listening
         V(t)  }
        ......................................
        dell( )
        {P(t)
         delete a listening student
         V(t)  }
        ......................................
        pauss( )
        {P(t)
         pause a student's speaking
         V(t)  }
        ......................................
        conts( )
        {P(t)
         let a student continue speaking
         V(t)  }
        ......................................
}
```

5. Further Considerations

In practical teaching, the students are located in few centers and connected with each other by high-speed local network inside these centers. Every two centers are connected by wide area network. Usually the teacher is at one local network of a certain network. The data mainly flow from teacher to student because the teacher is speaking and the students are listening in most time. So when we design a long-distance teaching system, we choose the centralized C/S architecture[7]. The server of the network, where the teacher located, acts as main server, while the servers of other centers act as secondary server. Figure 4 gives a specific design.

As most data flows from main server to secondary server, the delay of the teacher's speaking may reduce to the least. On the other hand, the delay of the student's speaking will become long. It will hardly affect the system's performance because the student has a little chance to speak. In this design, all the operations aren't performed locally, but should be transferred to the main server. Then the main server transfers it to all secondary servers so that all the users keep the same interface. So the cooperative controller may be installed on the main server.

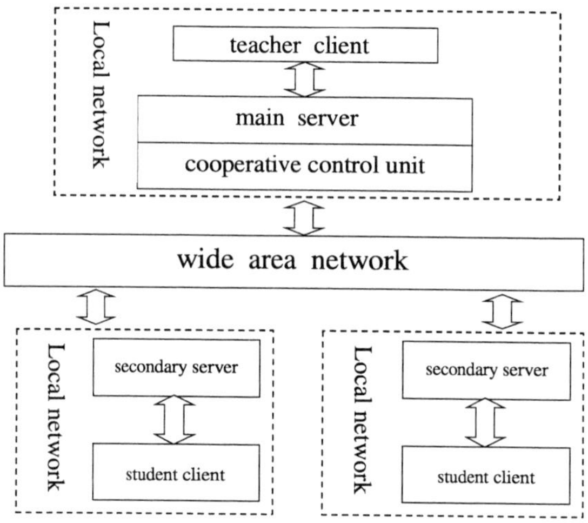

Figure 4. Real time long-distance teaching system architecture

6. Conclusions

The research on the cooperative mechanism of long-distance teaching system is the foundation for us to develop a practical system[8,9]. This paper constructed a Petri net model to be realized by software, after describing and analyzing the possible activities in a teaching course and the cooperative mechanism that we need to adapt such activities, and its application as well. We should say the description and analysis on the cooperative mechanism in long-distance teaching system is exhaustive, but it is still pretty complicated and new problems may emerge anytime during our design. This article may be regarded as a reference for the research in cooperative mechanism in other fields, while it focus on long-distance teaching system here.

7. References

[1] Chen Qingzhang, Hong Ning, "Research on CSCW Cooperative Mechanism through a Computer supported Cooperative Learning System", *Computer engineering and Application, 2000, 6*: 93-96

[2] Zhao Jianying, Wang Zhengmao, "Study on One Application of CSCW System, The Real-time Cooperative Edit System of Newspapers", *Computer Engineering and Application, 2000, 11*: 148-150

[3] Shao Huairong, Liu Bin, "The Model and Implementation of A High Performance CSCW Supporting Environment", *Computer Research and Development, Vol.37, No.5, 2000*: 583-588

[4] Huang Liwen, Li Chenkai, "The Model of the User Cooperation Layer in LCSDM", *Computer Science.Vol.27, No.3, 2000*: 22-25

[5] Grief I, et al, "A Case Study Of CES: A Distributed Collaborative Editing System Implemented In Argus", *IEEE TSE, 1992(9)*: 827~839

[6] Pendergast M O.A, "Comparative Analysis of Groupware Application Protocols", *Computer Communication Review, 1998, 28(1)*: 28-41

[7] Li Chengkai, Mao Bing, Xie li, "LCSDM: A layered model for cooperative systems development", *Proc. The Fifth Intel. Conf. for Young computer Scientists (ICYCS'99), Nanjing, China, August 17-20, 1999*: 789-794

[8] Grudin J, "Computer-supported cooperative work: history and focus", *Computer, 1994, 27(5)*: 19-26

[9] Ted Selker (1994).coach, "A teaching agent that learns", *Communication of the ACM, 1994, 37(7)*: 92-99

CSCL: An Internet Based Education Model

Shengjun Xue, Ran Tan, Hua Huang
Institute Computer science and Technology, Wuhan University of Technology
sjxue@china.com

Abstract

This paper provides a new educational model on the Internet — CSCL (Computer Supported Cooperative Learning). It introduces the current models of education, overviews the whole system, elaborates the process of realization, discusses its implementation using the Web technology, and finally, it provides some prospective of the CSCL model.

1. Introduction

With the development of the Internet in the world, the number of users has exceeded 1,000,000,000, covering 143 countries and districts. The population increases at the rate of 1,000,000 every month. The Internet will spread to everywhere in the world, because the Internet has thc advantages of being convenient, low-priced, interactive, everyone will use the Internet very easily. The Internet will become the primary channel by which the people attach the knowledge.

2. Current models of education

A number of types of education are commonly today. They range from the traditional classroom-training course to the use of stand alone CBT course. There are advantages and disadvantages in both of these. We will consider each in turn and then outline a new model which we term CSCL which is based on CSCW (Computer Supported Cooperative Work). In the remainder of the paper we will outline the architecture and functionality of the CSCL system and illustrate its use by considering a number of case studies.

2.1 Traditional classroom based training

The classroom-based course has a number of advantages.

1) The learner is directly supported in their training by either the lecturer of tutor.
2) Perceived difficulties can be addressed during the course.
3) The learner can ask questions. There are a number of problems with this approach, which we note below.
4) Courses tend to be expensive.
5) One must book well in advance so JIT learning is not really supported.
6) You are away from your desk for a few days which may simply not be economically viable for a small company.
7) Support rarely extends beyond the duration of the course.
8) Many such courses are generic and it is unlikely that they will be tailored to meet the individual needs of the learner in their context.

Perhaps the most worrying aspect is the scarcity of trainers, which will mean either larger class sizes or reduced availability of places on such courses.

2.2 The CBT approach

In this approach the learning material is delivered to the learner via a computer. In almost all cases the courses is intended to be self contained and the learner is free to proceed through the material at their own pace.

The advantages would seem to be

- You do not have to travel to a particular place to take the course: you can do so from home or at work.
- The courses are usually a lot less expensive than traditional courses.
- The learner can use the material at their own pace.

But a number of problems become clear, which may have a detrimental effect on the learning experience.

- Lack of contact with other learners.
- Easy to lose motivation.
- Quality of material may not be good.
- No expert on hand to ask questions.
- Material not tailored to the individuals needs.

A number of companies are currently favoring CBT on economic grounds. However, if staff fail to get the required benefits from training delivered this way then it may turn out to be a false economy in the longer term. We need to balance the training requirements of the learners with the economic requirements of the company that are purchasing the course.

2.3 The CSCL model

The CSCL model seeks to combine the favorable aspects of both models. The training material, which is delivered, is similar to the CBT course. It can be delivered to the desktop. However, it can quickly and easily be tailored to meet the needs of each individual student.

Unlike the CBT model, in CSCL his or her owns tutor can directly support each student. The tutor can monitor the learner's progress, suggest additional material that may be relevant to them and help them overcome difficulties, which they encounter.

In addition, the learner can be linked to other learners taking the course and thus a virtual classroom environment can be created. Classes and their tutors can be geographically separated without this interfering with the learning process.

We can maximize the use of scarce resources such as trainers and tutors. Courses can be significantly cheaper than in the traditional model and as CSCL potentially opens up a much larger market for CBT manufacturers the costs of courses may decrease even further.

3. An overview of the CSCL system

The CSCL system sits on top of a standard World Wide Web server and is accessible over the Internet. The CSCL server software is available on Linux and Windows NT/2000 versions. On the client side CSCL is accessed using a standard Web browser such as Netscape Navigator or IE. Thus the user is free to access the system from a range of different client platforms. Access to the system is restricted to registered users who have Ids and passwords. A number of databases are maintained on the server. These contain profiles of the users, the actual learning material and various submissions and messages, which have been sent between users.

Broadly speaking the CSCL system contains functionality, which supports 4 distinct roles in the education and training enterprise.

- The learner
- The tutor
- The author
- The administrator

3.1 The learner

A student or learner using the system will have access to their individual course material. They can proceed through it at their own pace and submit exercises or assignments to their tutor. When their tutor has graded these assignments they can retrieve them and read their tutor's comments and suggestions.

They have access to a class discussion list, which is essentially a bulletin board, which allows communication with other learners taking the course. At any time during using the system they can contact their tutor via a messaging system which is part of CSCL.

At the same time they can discuss the problems with the remote tutor or their classmates using the whiteboard in the CSCL system. They can obtain the answer by the pattern of word or graph. Using the whiteboard, they know the answer very quickly and very simple and very clear.

They can participate in the exam not going to the school by themselves. They just have access to the Examination in the CSCL and login. The paper will be transported to the front of them.

3.2 The tutor

Each learner can have a tutor assigned to him or her. The role of the tutor is to monitor and support the learners as they proceed through the course material. A tutor can communicate with their learners using the built-in message system. They can access and contribute to the class discussion list.

When learners submit course assignments the tutor can access them, grade and add comments. Using the whiteboard, the tutor can answer the problem of the learners by word or graph. In addition to this, the tutor can choose to alter a given learner's course by adding additional or remedial learning material. This may come from a pre-existing store of such material or may be authored on the fly by the tutor.

3.3 The author

Support is provided to enable an author to structure and write courses or learning material. In particular, there is extensive support for developing exercises and assignments. The author can set Multiple-Choice Question exercises, which will be automatically graded by the system. They can also specify actions to happen depending on the grade of the exercises. These include automatically grafting additional material for the student or alerting the tutor.

3.4 The administrator

The administrator is provided with support to allow registration of new users on the system. They can manage the information about the learners, the tutor and the system manager. They can install or remove courses and information. They can create new classes and assign course material to these classes.

4. The design and realization of the CSCL model

4.1 The two realization patterns of the CSCL

Now there are two patterns in the research of the CSCL. One is the pattern of software realization based on the Web. In this pattern, if the learner has a computer linked to the Internet, he will attain the knowledge by the remote education. It needs not the special hardware. In this pattern, the learners can study, test, discuss in the real-time, ask the question, and submit the exercises and so on. The learners and the tutor discuss by the word, graph, sound, and even the image. The other pattern is based on the video meeting system. Except the software of the first pattern, this pattern needs many special hardware, which transport and compress/decompress the real-time information of the sounds and images. The learners and tutor can hear and see each other at the same time.

The following will discusses the first pattern.

4.2 The logic structure and function of the CSCL

The CSCL system includes three parts: remote learning, remote exchanging, remote examining. The following is the logic structure picture (Figure 1).

Remote learning

This is the basic function of the CSCL system. In this part, the learners study the courses in the Learning by themselves. The courses exist in the way of media. A media includes the word, picture, sound, and animation of the course. They are combined organizationally. After the learners study, they can check their degree of the course learning by the problem in Testing.

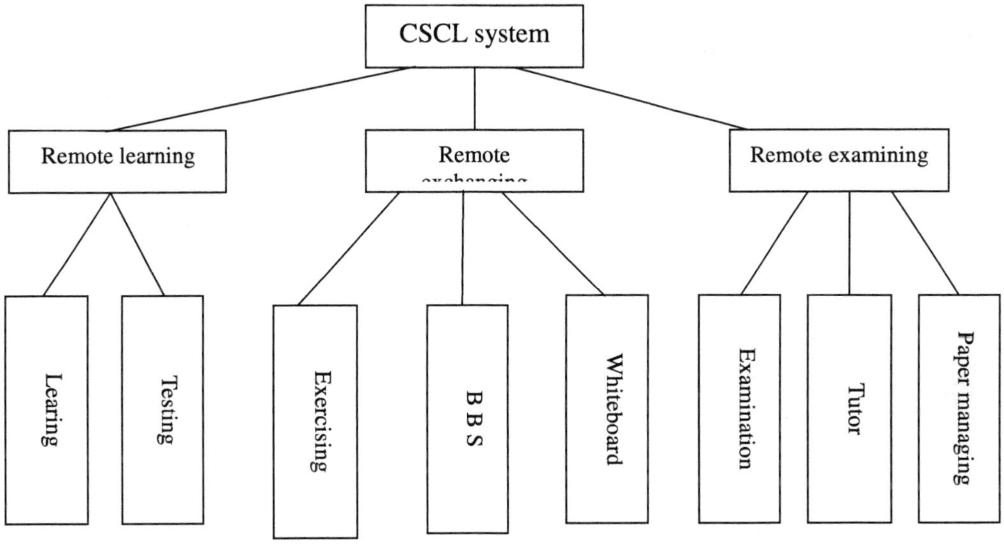

Figure 1. The logic structure of the CSCL system.

Remote exchanging

The remote exchanging system includes a server side and many client sides. The client sides can all be run in the standard browser. The tutor decides the speed of the progress. The learners can discuss the problems with the tutor by the way of word, sound, graph, using the whiteboard, BBS, email and exercising.

Remote examining

This part includes the Examination, Tutor, and Paper managing. It imitates the function of the traditional examination. It has the functions such as the paper's managing, producing, marking and the score's statistics.

Only if the browser of the examiners has linked to the server, they can participate in a remote examination, testing and looking over their score. The tutor can correct the learners' paper, analyze the score of the examination.

4.3 The logic structure and function of the CSCL

The following is the net structure of the CSCL system (Figure 2).

In the picture 2, the computers, which the learners use to take part in the remote education, are called "client". The computers, which are run in the backstage, are called "server". The Web server publishes the homepage of the CSCL. The whiteboard server manages the use of the whiteboard. The Internet links the client side and the server side. The program of the database links the server and the database. The database includes the paper, the information of learners and tutor, and the media of the courses and so on. The whole CSCL system is run by the way of the homepage.

The following will introduce the work process of the CSCL system.

The learners provide their request to the server according to their learning case. For example, the studying of the course. The courses in the server are a series of interrelated homepages, include the word, sounds, graph. After the server receives the request of the learners, it will take the relevant homepage to the learners' browser. By the browser, the learners will start there studying of the courses. For example, a learner discusses with the remote learners or the tutor. The first, he accesses the homepage of the whiteboard in the CSCL. If the whiteboard server is opened, he cans login

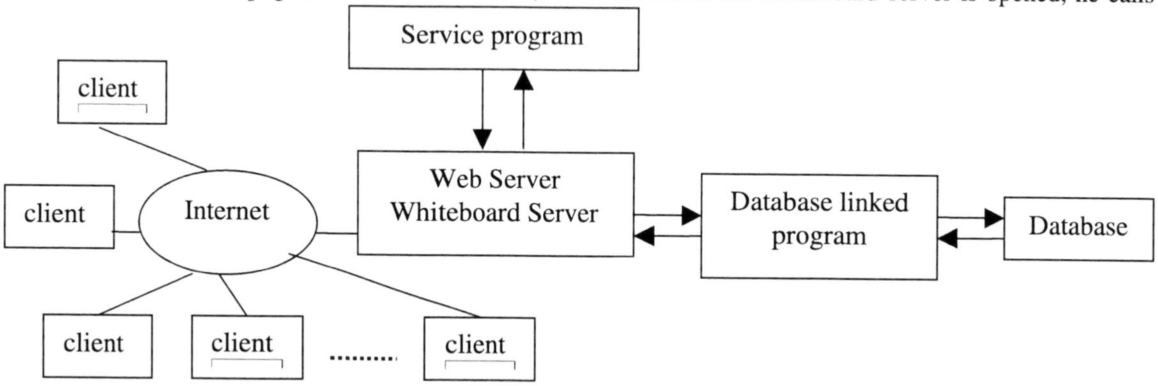

Figure 2. The net structure of the CSCL system.

the system. The whiteboard server provides the Java applet of discussing to the browser of the users. The Java applet needs not install like the common software. It can directly run in the browser. It responds the users'request and manages the discussing among the users. After a user login to the whiteboard, he will see the others in the system. These people may be distributed in the different class. The users can still choose a class. The way of the users discussing can be the word, or graph. If the users meet something, which can not be expressed by the word, they can draw the graphs.

4.3 The realization technology of the CSCL

The realization of the CSCL uses much technology. Among these, the Web technology and the Internet programming have the key effect. The following will introduce this technology.

HTML technology

HTML is the base of the Web technology. Thus it is the basic unit of the CSCL.

VBScript / JavaScript technology

They are the simple and easy used script languages. The object, which is provided by VBscript/JavaScript language and browser, can embed with the program's source code in the HTML document. Thus the Web pages of realizing the dynamic interactivation is the basic programming tools which make the courses' context vivid.

ASP technology

The full name of ASP is the active server pages. Its using must be combined with the IIS of the Microsoft. If it is combined with the script language of the server side and the script language of the client, it can realize the real-time dynamic HTML. Using this technology, we can realize the part function of the CSCL, such as the learners'register, the degree's statistics, the context's searching, BBS and so on.

Java technology

Java is a language and developing environment, which is not interrelated with the plat. Java can develop both the independent program and the Java applet embed in the HTML. Java can realize the interactive and dynamic pages. The developing of the whiteboard used this technology. After the Java applet is downloaded to the browser of the client, the users can see the page. In this point, it can not compare with ASP. But in the platless, ASP can not compare with it.

Web database technology

In the Web database, there are some important informations about the system, users. Because the database has the feature of data sharing, consistence, complement, dependable. The design of the whiteboard must face to the database. The Web database usually uses SQL Server, Oracle, or mySQL.

Web multimedia technology

In the developing of the remote learning, the Web multimedia technology is the key of realizing the transportation of the sound and the video. Because the sound and video have the feature of the real-time, large capacity, the Web multimedia may be the key technology, which affects the system models good or not.

5. Conclusion and further direction

The CSCL system has been built to address a particular set of perceived problems with delivery and support of Training. It has been made possible by timely developments in technology, principally the World Wide Web.

Because it is based upon the Web it means that new developments such as the provision of streamed audio and video over the Internet can quickly be incorporated into the training material which is being delivered. This will continue to be the case in the future.

We believe that in time, when a sufficient body of learning material has been developed for use via CSCL, that the system will be able to make a significant contribution towards facilitating less expensive and more effective training.

6. References

[1] Xue Shengjun,Gao Xiaohong,Tan Ran. Multi-Agent System Architecture Research. CSCWID'98.

[2] Xue Shengjun,Gao Xiaohong,Jiang Jingjue. The Mechanism of Multimedia Communication Synchronization in the CSCW Environment. CSCW'97.

[3] Tan Ran Xue Shengjun. An Autonomous Agent and Recognition Based on Self-organizing Neural Architecture. Proceedings of International Workshop on CSCW in Design

[4] Lin Zongkai,Jean-Paul Barthes. Proceedings of International Workshop on CSCW in Design.

[5] David A.Robinson and Callum R.Lester and Neil M.Hamilton, Computer Networks and ISDN Systems 30(1998)301-307

[6] Cornell.G and Horstmann,C.S, Core Java.SunSoft Press(Prentice-Hall), California, 1.1996, pp.12-16,177

[7] Soloway E, Technology in education. Special Issue Comman ACM 36.No.5(1999)

[8] Henry MeLoughlin, WEST,Computer Networks and ISDN Systems 28(1999)1887-1890

Integrated Process Design for E-Learning: A Case Study

Hartini Ahmad
School of Management, Universiti Utara Malaysia, Kedah, Malaysia
hartini@uum.edu.my

Zulkifli Mohamed Udin
School of Information Technology, Universiti Utara Malaysia, Kedah, Malaysia
zulkifli@uum.edu.my

Rushami Zien Yusoff
Quality Management Institute, Universiti Utara Malaysia, Kedah, Malaysia
rzy278@uum.edu.my

Abstract

The higher education (HE) institutions are expanding fast. Universities are becoming genuine learning organisations and providers of a wide variety of opportunities for e-learning. Despite all efforts, the processes carried out do not meet learner's expectations, and often lead to ineffectiveness and inefficiency. Clearly, we need to develop a better understanding of the processes in e-learning in order to satisfy learner requirements. This research examines the nature of learner's demand in e-learning process such as in registration process. In particular, it examines how the integration in process design could support e-learning. The first part of the paper will raise issues and questions about the meaning of e-learning and about the challenges which members of HE institutions have to face and cope. Section two deals with the application integrated process design to enhance the competitiveness of e-learning. This study identified one HE institution as a case study. The study found the current initiatives on e-learning, which the transformation of the entire processes were considered in order to accommodate the new realities, such as cost effective, reliable and effective learning tools, co-operatives technologies, and easy to operate. In addition, the process design must incorporate learner requirement information at all stages of e-learning process.

1. Introduction

It has been accepted that to get education is a human's right. This is extremely true in the era of knowledge awareness. Malaysia has allocated millions to build the education infrastructure, in order to ensure that it's people get adequate education. Although, in reality this amount of money is still not enough to fulfil the increasing demand in student admissions particularly in HE institutions each year. That's why the Government of Malaysia has encouraged local universities to offer distance learning so called to utilise e-learning in their education systems. HE institutions will seek to be effective on a global basis through the efficient use of new communications technology. This is true that HE institutions should be innovative and proactive, which means that they should be sensitive to any changes and new demands.

The HE institutions today compete within and beyond national boundaries, thus, to compete in the global environment. The advancement in technologies specifically related to information technology (IT) creates new challenges and opportunities. The strategy in utilising the information and knowledge will determine on the competitiveness of organisation. This is mainly due to advancement of information technology and the reforms in international business, which called for much more free and liberal business and trade arrangements [1,2]. The Ministry of Education also stresses on 'Malaysia-Regional Centre of Excellence' which means to build up excellent institutions of learning, to liberalise and globalise the education. The technological changes are paralleled, however, by changes in demand from learners.

IT is a new tool of information delivery such as using e-learning. E-learning challenges the top management of HE institutions to rethink their traditional methods of doing work and commit to a customer-focused process [5,6,7]. The HE institutions are looking for new and innovative ways of meeting learners expectation. The current development of academic industry such as e-learning and globalisation have become the catalyst for a rethinking and reengineering of the HE institutions.

E-learning will be required to expand flexibility and innovation in learning and teaching. E-learning becomes a tool implemented by HE institutions (especially private HE institutions) and multinational companies. The common understanding is that e-learning is synonymous to online learning, which utilise computer and web site. In addition it is also known as distance learning through internet. In this context, e-learning refers to the utilisation of any one of on-line systems in term of administrative management, interactive materials, and collaborative tools.

For the HE institutions that implement e-learning in their organisations, they should realise that e-learning offers opportunities for flexible and accessible study. It should be clearly stated that e-learning should try to meet the need of learners that are changing rapidly and continually from time to time. E-learning should not be implemented as just a matter to change the operational processes from manual to computer-based system, but it should innovate the processes that are involved in it. This means an integrated process design for many HE institutions must be reshaped and re-conceptualised even their manual work activities. The process redesign is needed in term of dramatic change basis or drastic change approach, and no more incremental improvement to ensure the competitive advantage [3,4]. In other words, the HE institutions need to constantly venture into new and innovative methods, in the shortest time possible.

Thus, e-learning will give benefits to the organisations, in terms of reduction in costs such as documentation, processing, staff, classroom, etc. Furthermore, through the integrated process design it could improve greater accuracy and speed, less reworking, fewer mistakes and errors which lead to an improvement in the cost base, higher customer satisfaction and improved employee satisfaction [4,8,9].

2. Background of e-learning in Malaysia

Educational institutions have long been very much closed systems. To compete in the global environment, HE institutions should interact in different ways and in various directions by using multi-dimensional ICTs. Specifically, HE institutions should be able to provide learning that can take place anywhere and everywhere for the learners. In other words, HE institutions should not be forever constrained by traditional boundaries which failed to meet the modern industralised societies needs.

E-learning in Malaysia has been pioneered by Universiti Tunku Abdul Rahman (UNITAR) and the Multimedia University (MMU). The use of e-learning from registration process to learning process including lecture notes distributions, the information on the courses offered, course contents, academic discussions, online quizes etc. To ensure that this system could operate smoothly, every student has been given the identification number and password to log in. The universities have progressively begun to drive institutional responsiveness in educations towards customers priorities. Increasingly, these universities must satisfy learners demands. In addition, they need to meet the 'quality' imperative. Ensuring the courses are delivered, monitored and evaluated with a course on high quality provision.

3. Research objectives

This research intends to determine how process design should be integrated in order to implement e-learning to meet learner's expectation. This study attempts to chart some of the processes and demands involved in these changes.

4. Research method

This study identified one HE institution as a case study. Semi-structured interviews with the head of departments and deans, lecturers and administrators provided the data used in this study. The data were collected over a three-week period. The questionnaires were divided into three parts. Part 1, was for the organisation/ respondents profile. Part 2, was utilised in order to investigate some actual processes that have been carried out specifically in terms of collaborative and integrative approach. Part 3, consisted of the questions to get information on the advantages and disadvantages of utilising e-learning, particularly this article will report on the issues related to the processes involved. The study looked in-depth on how the processes were being carried out before and after implementing e-learning. In specific, the rank order from '1' to '10' was used, which '1' for 'the most importance' and '10' for 'the least importance'. Moreover, the Likert Scale was utilised and modified to four-point scale, which '1' indicated 'strongly disagree', '2' indicated 'disagree', '3' indicated 'agree' and '4' indicated 'strongly agree'.

5. Analysing the results

This article will just highlight some issues in Table 1 that shows the analysis on respondents' perspective on the features for e-learning.

Table 1. Ranking for e-learning features

Features	Rank order
Co-ordination	1
Fast response	2
Update work activity	3
Well defined process	4
Intranet and internet	5
Paperless	6

Co-ordination through co-operative work (IT) found as a major contribution for members in the university. Thus, they found that it is different from the manually co-ordinated work activities which are less efficient in terms of management process. Furthermore, the use of e-learning should enable the staff to respond quickly to learners online rather than depend on bureaucracy-oriented. In addition, the work activities could be easily updated rather than be rigid on working procedures. The respondents also hoped to practise a well defined process to all the learners, instead of doing different process to different learners. Moreover, the staffs prefer to use intranet and internet to avoid 'sneaker-net'. The study found that the respondents prefer to go practice paperless rather than getting involved with many documentation or using many papers such as application forms, acknowledgement cards, referee letters, etc.

However, it was found that after implementing e-learning, the members of the university are still faced with the problems in the processes performance. The study found that the problems faced by the members are as shown in Table 2.

Table 2. Problems in process

Issues	Mean
Process overlap	3.42
Inadequate infrastructure	3.31
Network breakdown	2.94
Vague of the process	2.53
Maintenance costs	2.47
Other	1.68

The study found that although the university implemented e-learning but some of the task procedures remained the same. Instead of doing the processes manually, they utilised computer-based systems, which operated on the same processes. These changes demand new responses from lecturers, an integration process redesign is called for which is more creative to increase personal satisfactions for learners. Knowledge is being constructed by learners in ever more open systems, rather than being assimilated through existing subjects and modes of delivery in conventional educational settings.

This research investigated the process of the course registration, which showed in the process map that had been done manually before implementing e-learning (Figure 1). The flowchart of the course registration process using the Internet also shows (Figure 2). This study then recommends a new integrated process design in that particular process (Figure 3). The illustration on the integrated process design should be seen in this simple diagram (Figure 4), which demonstrates how IT plays an important role connecting the nodes of processes [10].

Table 3. Reasons implementing e-learning

Attributes	Mean
Cost effective	3.94
Reliable and effective	3.41
Co-operative	3.19
Easy to operate	3.18
Other	3.11

In addition, it was found that the university implementing e-learning is demanding cost effective, reliable and effective learning tools, co-operative, and easy to operate (Table 3). Other factors considered by staffs in term of reputation, open system, delivering style and trend. The cost efficiencies and savings are made in terms of money spent on learning and revisiting learning, class facilities, and course maintenance.

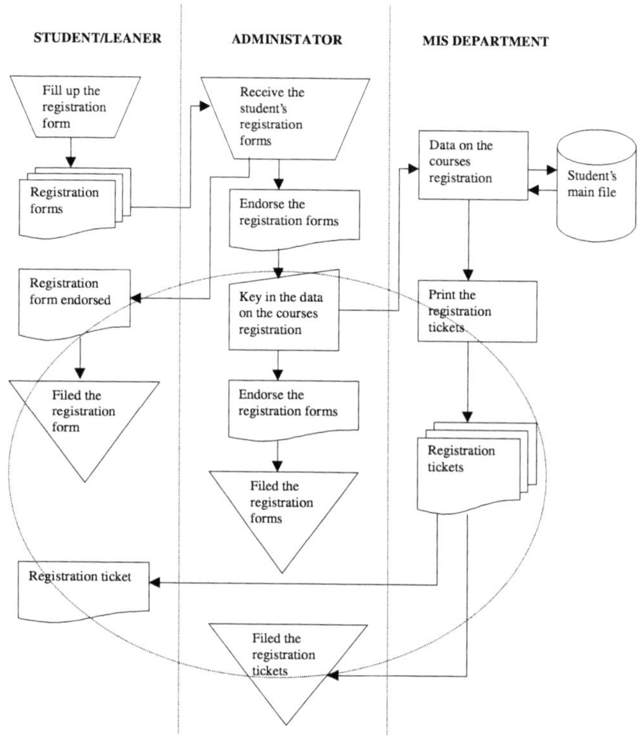

Figure 1. The course registration process (manual).

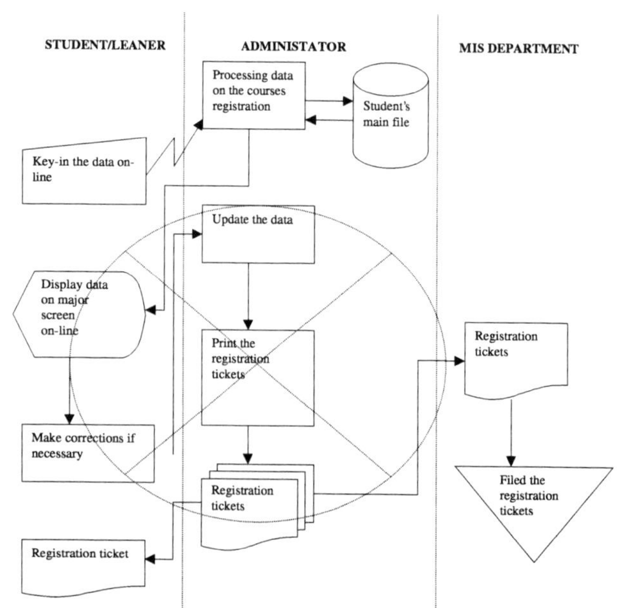

Figure 2. The course registration process (on-line).

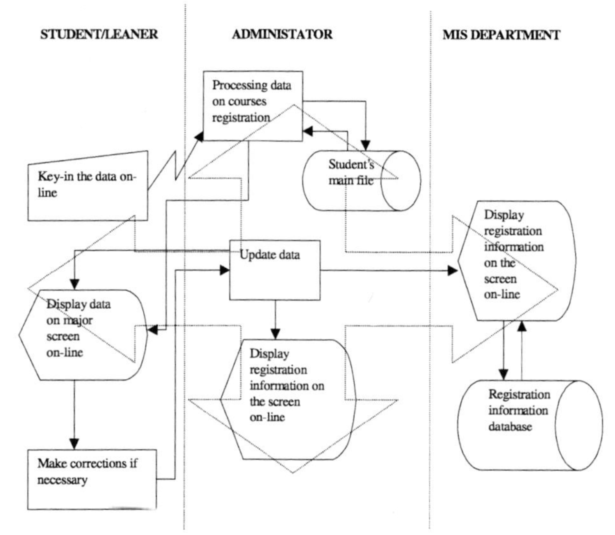

Figure 3. Integrated process design for the courses registration.

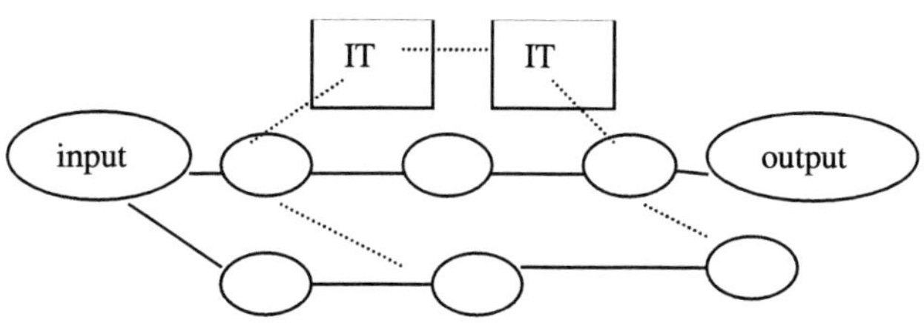

Figure 4. Operational perspective on the role of IT.

6. Conclusions

In this article we provide the examples that there is need for a better integrated process design in HE institutions that implemented e-learning to incorporate learner's expectation into the desired process. Chart 3, we propose a basic design of the courses registration process that could be integrated on-line. It is recommended that the integration of productive processes from registration process to learning process should be created to ensure learners' expectations are consistently met. This is supported by many studies, which stated the connection between IT and process redesign [11,12,13]. In addition, this innovation in process could ensure the organisations to keep a high level of performance whether nationally or internationally and globally.

7. References

[1] Dean, E.B. (1996). Business process reengineering from the perspective of competitive advantage.

[2] Dale, M.(1994). The re-engineering route to business transformation. *Journal of Strategic Change*.

[3] McAdam, R. & Donaghy, J. (1999). Business process re-engineering in the public sector. *Journal of Business Process Management*, 5(1), 33-49.

[4] Childe, S.J., Maull, R.S. & Bennett, J. (1994). Frameworks for understanding business process re-engineering. *International Journal of Operations & Production Management*, 14(12), 22-34.

[5] Zinser,S., Baumgartner, A. & Walliser, F. (1998). Best practice in reengineering : a successful example of the Porsche research and development center. *Journal of Business Process Management*, 4(2), 1-9.

[6] Hvam, L., & Have,U. (1998). Re-engineering the specification process. *Journal of Business Process Management*, 4(1), 1-15.

[7] Giannini, P., Grupe, F., & Saholsky, R.(1997). Reengineering through simulation modeling: optimizing a telephone ordering system at GPO. *Information Systems Management*, Summer, 61-66.

[8] Dale,B. & McQuater, R. (1998). Managing business improvement & quality: implementing key tools & techniques. *Journal of Business Process Management*, 4(2), 1-3.

[9] Pereira, Z.L. & Anspinwall, E. (1997). Total quality management versus business process re-engineering. *Journal of Total Quality Management*, 8(1), 33-39.

[10] Tinnila, M. (1995). Strategic perspective to business process redesign. *Management Decision*, 33(3), 25-34.

[11] Martinsons, M.G. (1995). Radical process innovation using information technology: the theory, the practice and future of reengineering. Journal of Information Management, 15(4), 253-269.

[12] Saxena, K.B.C. & Ammaly (1995). Information technology support for reengineering public administration: A conceptual framework. *Journal of Information Management*, 15(4), 271-293.

[13] Zairi, Mohamed & Siclair, D. (1995). Business process re-engineering and process management : a survey of current practice and future trends in integrated management. *Journal of Business Process Re-engineering & Management*, 1(1), 8-30.

Extracting WEB Table Information in Cooperative Learning Activities Based on Abstract Semantic Model[1]

Ning Gu, Guowen Wu , Xiaoyuan Wu and Baile Shi

Department of Computer Science and Engineering, Fudan University, Shanghai 200433,P.R. China

ninggu@fudan.edu.cn

Abstract

A great deal of Web table information exists in cooperative learning activities. This paper presents a new method that extracts information from tables of Web document. Using tabled abstract semantic model to describe complicated tables and understand tables from the point of view of semantics, this method reduces the dependence for the design difference of table constructions in the extraction process. At the same time it utilizes the characteristics of HTML and the technique of natural language process to design some heuristics rules, and thus aids the identification of table items. On the above basis, we design a prototype "EXTable" and then gain a better result according to the experimentation.

1. Introduction

Teachers and students need a great deal of Web table information from Internet to compose some knowledge points in cooperative learning activities. In practice, some Web links are usually obtained through general search engines, and all the web pages have to be opened and searched one by one if retrieving some contents. This indirect method above is inconvenient and fallible, so it is urgent for us to research on obtaining web information directly and exactly, i.e. retrieving information from web documents.

Now researches focus on web text documents, because of comparative easiness to research on and better references for the retrieval of other media. But even for text documents, they present quite complex forms in Web, such as some containing general "flat" texts, given structures, tables, etc. So we need to adopt corresponding methods according to different document contents.

There exist a great number of web data extraction methods, which are summarized and analyzed respectively as follows:

- Constructing data extractor based on HTML tags in web pages[1,2,3,4]. Extractor first analyzes special HTML tags in web pages and extracts special information according to extraction rules, then stores them in some given forms. This method largely depends on web pages' layout, format etc, so extractor needs to be rewritten, once pages' layout or format changed. This case often happens in web document, thus this method sometimes doesn't fit table extraction.
- Manual construction of data extractor is very arduous. Therefore, people construct extractor by means of automatization and semi-automatization, such as studying by examples [5,6]. Though heavy tasks of extractor construction are lightened, it doesn't change extractor's dependencies on page's format.
- Making use of natural language process (NLP)[7]. In the area of artificial intelligence, researchers have put great emphasis on natural language process and given birth to many comparatively mature methods. But a large number of examples are needed for training, and the process is so slow. Due to research on table in natural language process, table extraction can also adopt NLP.
- Method based on semantic model [8]. Given semantic model, it describes contents of extracted pages, and it designs extraction rules according to characteristics of domain and web. The method is independent of the format of extracted web pages, and even when domain is changed, only semantic model should be modified. Of course how to design semantic model is a relatively difficult task. Generally, we apply context in identifying the concept of semantic model, but there exist fewer relationships between the contexts of labels in table so that it is more difficult for us to design the semantic model and former methods are not suitable.

[1] This work was supported by the National Key Natural Science Foundation of China under Grant No. 69933010

Based on the analysis above, we can see that table makes the exaction fall into mess because of its specialty itself although table is only a part of web document. Table in the web is similar to the table in the relational database, and extraction seems to process the table line by line only. In fact, even the tables of the same contents, are greatly different in their formats, and extracted data need to be adjusted and stored in given forms. Therefore, it'll involve much manual intervention. We will discuss the intrinsic characteristics of the table deeply from the point of view of semantics [9]. So that extracted contents can be naturally turned into the data in relational database. Our paper will focus on information extraction of the table in web pages.

2. Abstract semantic model of table

To extract table information automatically from web document, we must put aside diverse formats of the table, and understand table from point of view of semantics. Therefore, we will discuss abstract semantic model of table.

Table1 is a common statistics table of student marks, which is composed of a series of cells and data items. We name these cells and data items as items, and items may be figures, texts, symbols, graphs, mathematic formula and even tables. We can divide these items into two kinds: one is called entity, which represents basic data in table; the other is title, which is used to locate basic data. Title can be divided into distinct classes according to its distinct levels; table can be divided into four parts: column title, row title, row title head, table body according to its forms. The set of quadrate cells is named block. We will explain a series of terms involved in table process as follows:

Title: label that locates basic data, may be any character, string, and even null string, such as "test" and "Average mark" in table1. These titles can be divided into different kinds based on their respective semantics. For example, titles in table1 can be divided into three kinds: year, term and mark.

Title set: the set of titles, which includes not less than one title, is expressed in ordered pairs (label, set).

Title domain: it is defined by the following recursive method:

1. A null title set (L, φ) is a title domain.
2. A title set of title domain is a title domain.
3. Only the title domains obtained from rules 1,2 are legal ones.

In title domain D=(l, s), we use lbl (D) to express title *l* and set (D) to express *s*.

Title Domain: (mark, {(Homework, {(first time, φ), (second time, φ), (third time, φ)}), (test, {(Midsemester, φ), (final term, φ)}) (Average mark, φ)})

Title Tree: title domain is usually described with title tree. One node in title tree expresses one title domain, and the title domain expressed by every outer node comprises one null title set.

Figure 1 illustrates the relation between title domain and title tree.

Title Sequence: We link the titles of the title domains in dots, which possess inclusive hierarchy relation and name them title sequence. Usually title sequences are used to express sub-domain in title domains.

For example, in title domain D= (*l*, s) , *l* is a title sequence that identifies D. By analogy, if the title domain that *l* identifies contains title domain set $\{ (l_1,s_1) , \ldots, (l_r,s_r) \}$, l. l_i is a title sequence that expresses title subdomain (l_i,s_l).

Suppose title sequence is *l* and Ω(*l*) expresses a title domain identified by *l*, And lbl(*l*) and set(*l*) express the identifier and the set of Ω(*l*) respectively.

Table 1. Table illustration.

year	Term	Mark					
		Homework			Test		Average mark (0.3A+0.3M+0.4F)
		First time	Second time	Third time	Midsemester	Final term exam	
1999	Spring	85	80	75	69	75	75
	Autumn	80	65	75	60	70	70
2000	Spring	80	85	75	55	80	75
	Autumn	85	80	70	70	75	75

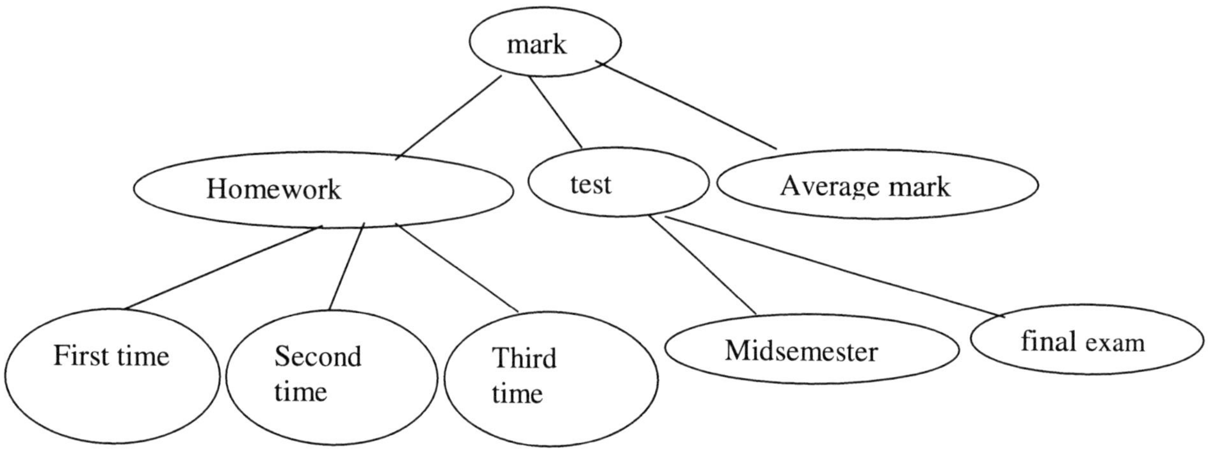

Figure 1. Title domain and title tree.

Boundary Title Sequence: if title domain identified by title sequence includes null set, the title sequence is named as Boundary title sequence.

The boundary of title domain, fr(D), is the set of all boundary sequences in D. If C is a title domain set, then fr(C)={fr(D)|D∈C}.

The relation between fr(D) and D is one-to-one. If title sequence set S meets the following conditions, it will be one-to-one with title domain:

- All the titles in S are common in the first one.
- S is non-prefixal, i.e. if x. y∈ S, x∉ S.

We can construct a title domain D from S, which satisfies fr(D)=S. We call such S coherent.

Suppose there are $n \geq 1$ disintersectant sets A_1, A_2,..., A_n, whose unordered Cartesian product $A_1 \otimes A_2 \otimes \ldots \otimes A_n$ is a set A, and every element in A is a set, which consists of A_i(i□n). For n disintersectant title domains $D_1, D_2, \ldots, D_n$, unordered Cartesian product of Boundary Title Sequences is $fr(D_1) \otimes fr(D_2) \otimes \ldots \otimes fr(D_n)$. Also, suppose the set of title domains $D_1, D_2, \ldots, D_n$ is C, we represent $fr(D_1) \otimes fr(D_2) \otimes k \otimes fr(D_n)$ with $\otimes fr(C)$.

Table T: we adopt the following function to express:

Δ: $\otimes fr(C) \rightarrow V$, in which C is a finite set of title domains and V is all the values of entity.

We know that function is easy to be transformed into relation model. For table T above, suppose $\{f_1, f_2, \ldots, f_n\}$□$fr(D_1) \otimes fr(D_2) \otimes \ldots \otimes fr(D_n)$ and $v \in V$, we can also describe table T as $\{f_1, f_2, \ldots, f_n, v\}$.

If $V = \phi$ in function δ, we also call table T framework. Suppose the dimension of table T dim(T) expresses the size of C, i.e. the number of classes in C. For example, the dimension of table 1 is 3, and size of table size(T) is that of $\otimes fr(C)$, i.e. the number of entity in T. The size of table 1 is 24.

3. Extraction of table

During the process of extraction, we suppose that all tables belong to the same domain, which describes congeneric information, i.e., their title classes are coherent, but the number of titles and title domains in title classes can be different. Under the assumption, the range that suits for table extraction is comparatively great, The order of the titles in the same title domain can be variant, and title domain and the number of titles in title classes can be different. In general extraction method, we generate description file of the table from the its layout, and then process extraction . The method depends on the layout of the table greatly.

3.1 Table Recognition in HTML

The tags of table are explicit in HTML files. For example, Figure 2 is one part of the description in HTML form of table 1:

```
<table border="1" width="100%">
 <th>
   <td width="12%" rowspan="3">year</td>
   <td width="12%" rowspan="3">term</td>
   <td width="76%" colspan="6">mark</td>
 </th>
 <th>
   <td width="37%" colspan="3"> Homework </td>
   <td width="26%" colspan="2">test</td>
   <td width="13%" rowspan="2"> Average mark </td>
 </th>
```

```
<th>
   <td width="12%">first time</td>
   <td width="12%">second time</td>
   <td width="13%">third time</td>
   <td width="13%"> Midsemester </td>
   <td width="13%">final term</td>
</th>
<tr>
   <td width="12%" rowspan="2">1999</td>
    <td width="12%"> spring </td>
   <td width="12%" bordercolor="#FFFFFF"
        bordercolorlight="#FFFFFF">85</td>
   <td width="12%" bordercolor="#FFFFFF"
      bordercolorlight="#FFFFFF">80</td>
   <td width="13%" bordercolor="#FFFFFF"
        bordercolorlight="#FFFFFF">75</td>
   <td width="13%" bordercolor="#FFFFFF"
        bordercolorlight="#FFFFFF">60</td>
  <td width="13%" bordercolor="#FFFFFF"
        bordercolorlight="#FFFFFF">75</td>
  <td width="13%" bordercolor="#FFFFFF"
      bordercolorlight="#FFFFFF">75</td>
</tr>
...
</table>
```

Figure 2 Part of HTML description.

According to HTML tags of table, we can find some rules from Figure 2. By further analysis, we can educe the following heuristic identification rules:

Table Recognition: the whole table is embodied by < TABLE >...</TABLE>.

Column and Row Recognition: every row is embodied by <TR>...</TR> , whose column is embodied by <TD >...</TD >.

Separation and Combination: if rowSpan=i included, the item must repeat in the next i-1 rows. Analogously, if colSpan=i included, the item must repeat in the next i-1 columns.

By means of above rules, we can design extraction algorithms to locate tables in web pages and extract felicitously and construct tabled abstract semantic model.

3.2 Title recognition

To create abstract semantic model of table, the first step is to identify titles in table. Because title classes are given, sum of title's rows and columns are decided. But we don't know the layout of rows and columns. And if the number of row titles is known, the number of column titles is also known. Vice versa. By analyzing the design of the table in HTML form, we generate the following rules to apply in title recognition.

Font size detection: Check the attributions of <TR> and <TD> , and if <font size...>...</font> item exists and the values of size attribution are different, rows (or columns) with big font are title rows (or columns).

Font color detection: in the rows and columns of table, if font color change, the rows and columns with the same color in the top and left of table are titles. Rows (or columns) including <STRONG>...</STRONG> attribution are title rows and columns.

Identification detection: rows embodied by <TH>...</TH> are title rows of table.

Like Figure 2, its title includes <TH>...</TH>, so we can check that the title is the first row.

After recognizing titles, we form titles of table into title tree or title domain boundary in term of different domains, and then form into abstract semantic model of table with the method introduced at above section.

3.3 Special problems

During the period of process, we use NLP method and standardize the titles' description to the best of ability. For example, "year" in table 1 may be described as "time", etc in many tables, and we standardize them with "year". Another example, its substrate title "1999" may be described as "99", and we unify them into four-digit description. these means have no effect on the extraction , but reduce the maintenance of the generated database . No more about these techniques will be discussed in the paper. Please refer to [7] and etc, if you are interested in them.

In addition, not all tables, in fact, are as normative as table 1 and there may exist some "reasonable" problems, which trouble us in the extraction. So tables are standardized before the extraction. The following questions are met frequently:

Data items grouping: if the values of data items are equal in some block of the table's body , these items, such as the first column in Figure 3, can be combined into one cell .

Title omission: if all the data in some row or column of the table are self-interpreted, the title is often omitted. Please refer to the third column in Figure 3.

Data item congregation: data items that some titles correspond to centralize data, which correspond to data items that other titles correspond. Such as the second column in Figure 3.

Cement	Standard Mixes	
Ordinary	ST1,ST2,ST3	BS 12
	ST4,ST5	BS 146
Blastfurbace	ST1,ST2,ST3	BS 6588
	ST4,ST5	BS 146

Figure 3. Simple illustration of problem table.

3.4 Extraction and its example

Table extraction is complicated and with the technique introduced above we have designed a prototype -- EXTable, which mainly implements the following functions:

- Identify tables, tables' titles and data items from HTML.
- Standardize tables.
- Form abstract semantic model of table.
- Converse tabled abstract semantic model into the table of relational database , and optimize it.

In the following, we converse table 1 into tabled abstract semantic model with our system. The HTML code of table1 acts as the input of the prototype, and the result outputs as follows:

Table titles in chart are divided into three classes and Their respective title domains are:

(year, {(1999, φ))(2000, φ)})

(term, {(spring, φ))(autumn, φ)})

(mark, {(Homework, {(first time, φ))(second time, φ))(third time, φ)}), (test, {(Midsemester, φ))(final term, φ)})(average mark, φ)})

Table1 use relative model mode to express as following:

{year.1999, term. spring, mark. Homework. first time, 85}
{year.1999, term. spring, mark. Homework. second time, 80}
{year.1999, term. spring, mark. Homework. Third time, 75}
{year.1999, term. spring, mark. Homework. test. Midsemester, 60}
{year.1999, term. spring, mark. Homework. test. final term, 75}
{year.1999, term. spring, mark. Homework. average mark, 75}
{year.1999, term. autumn, mark. Homework. first time, 80}
{year.1999, term. autumn, mark. Homework. Second time, 65}
{year.1999, term. autumn, mark. Homework. Third time, 75}
{year.1999, term. autumn, mark. Homework. test. Midsemester, 60}
{year.1999, term. autumn, mark. Homework. test.. final term, 70}
{year.1999, term. autumn, mark. Homework. average mark, 70}
{year.2000, term. spring, mark. Homework. first time, 80}
{year.2000, term. spring, mark. Homework. second time, 85}
{year.2000, term. spring, mark. Homework. Third time, 75}
{year.2000, term. spring, mark. Homework. test. Midsemester, 55}
{year.2000, term. spring, mark. Homework. test.. final term, 80}
{year.2000, term. spring, mark. Homework average mark, 75}
{year.2000, term. autumn, mark. Homework. first time, 85}
{year.2000, term. autumn, mark. Homework. second time, 80}
{year.2000, term. autumn, mark. Homework. Third time, 70}
{year.2000, term. autumn, mark. Homework. test. Midsemester, 70}
{year.2000, term. autumn, mark. Homework. test.. final term, 75}
{year.2000, term. autumn, mark. Homework. average mark, 75}

Then, we transform above result to relational database, and optimize it. Here, We do not discuss the optimization of relational database.

4. Test result

Using EXTable, we extract 344 homepages, which are divided into three classes: tables of the performance target of the machines, statistics tables of the students' marks and the lists of the raincoats' models and prices . We figure out the rate of recall and precision respectively with the following formula:

$$(1)\ R = \frac{CON}{TN}$$

$$(2)\ P = \frac{CON}{OTN}$$

CON represents the number of relative table obtained correctly.

TN represents the total of relative table.

OTN represents the total of obtained relative table.

From the statistics of the results, R counts34%,38%,32% respectively, and P 72%,65%,71% respectively. The results are about 5% and 8% higher than those generated by other usual methods, namely, the extraction from the description files of the tables. Moreover, it is burdensome to manually describe the layout of the table. Especially for more complex tables, even though some kind of tables are described completely, there are so many elements involved that as long as one is changed, the whole description file must be modified too.

5. Conclusion

This paper describes complicated tables with abstract semantic model and understand them from the point of view of semantics, in order to supporting participators in cooperative activities to obtain and share Web information conveniently. This method reduces the dependence on the diversity of the structures at the design phase. During the period of table extraction, the characteristics of HTML identification are full utilized and some heuristics rules are designed to aid recognizing elements. In addition, some techniques of the natural language process are applied in the design of the prototype. EXTable generates satisfactory results in many tests, but there is still much room to improve on. In the next future we will add some learning functions in the equivalent conversion of the tables and NLP modules to enhance the extraction ability of our system.

6. References

[1] J. Hammer, H.Garcia-Molina, J.Cho, R.Aranha, and A.Crespo *Extracting Semistructured Information from the Web.* 1997, SIGOD Record, 26(2):18-25.

[2] Huang Yu-qing, Qi Guang-zhi, Zhang Fu-yan *Constructing Semistructured Information extractor from the Web document.* 2000, Journal of Software 11(1):73-78.

[3] Naveen Ashish and Craig Knoblock *Wrapper Generation for Semi-structured Internet Sources.* 1997 SIGMOD Record, 26(4):8--15.

[4] Dan Smith and Mauricio Lopez *Information extraction for semi-structured documents.* 1997. In Proceedings of the Workshop on Management of Semistructured Data, in conjunction with PODS/SIGMOD, Tucson, AZ, USA, May, 12

[5] William W. Cohen, Wei Fan *Learning Page-Independent Heuristics for Extracting Data from Web Pages* 1999 In AAAI Spring Symposium on Intelligent Agents in Cyberspace

[6] Mark Craven, Dan DiPasquo, Dayne Freitag, Andrew McCallum *Learning to Extract Symbolic Knowledge from the World Wide Web.* 1998 In Proceedings of the 15th National Conference on Artificial Intelligence, pages 509--516.

[7] Cowie, J., and Lehnert, W. *Information Extraction.* 1996, Communications of the ACM 39,1(1996),80-91.

[8] D.W. Embley, D.M. Campbell, Y.S. Jiang, S.W. Liddle, D.W. Lonsdale, Y.-K. Ng, R.D. Smith *Conceptual-model-based data extraction from multiple-record Web pages* 1999 Data & Knowledge Engineering 31 227-251

[9] Xinxin Wang. *Tabular Abstraction, Editing, and Formatting.* PhD thesis, University of Waterloo, Waterloo, Ontario, Canada, 1996.

[10] Douglas, S., Hurst, M., & Quinn, D. 1995, *Using Natural Language Process for Identifying and Interpreting Tables in Plain Text.* In Proceedings of the Fourth Annual Symposium on Document Analysis and Information Retrieval, pp.535-546.

A Cooperative System for Multimedia Courseware Evaluation

Ming Zuo, Xiaotian Zhu
China University of Mining and Technology, Xuzhuo, Jiangsu, China
mingz@cumt.edu.cn, zxt@cumt.edu.cn

Abstract

This paper presents a new multimedia courseware evaluation system. It is based on the internet-supported cooperative work. After we build the logical model upon workflow of the evaluation and management system, we have developed the cooperative system for the evaluation of multimedia courseware. Expert's work is distance to use the evaluation system on Internet; they would reduce expert's working hours and enhance management's working efficiency to a great extent.

1. Introduction

CSCW is rising in the multi-interdisciplinary area, and they are remarkable from a potential research and a very bright future. We should take the application of the system to management and evaluation for multimedia courseware into account. New we should like to bring forward the cooperative work model. At the same time, we will integrate them into one that is management technology of multimedia courseware, Internet communication and synchronization control technology. In this way, we shall promote CSCW to research and development (J-C. & L. 1999).

We will make use of CERNET(China Education and Research Network)to build the environment about evaluation and management of multimedia courseware, we shall develop application research for cooperative system. It shall improved expert the work efficiency and work quality in a way.

At present, multimedia courseware have come into use in the distance learning, but then we think to evaluate quality about multimedia courseware, it has still invited experts by administrator. Then they shall come from every department to attend conference. Because this would process repetitive work and more long working cycle, it's not to take any advantage of Internet communication ability and cooperative work advantage (Huang X. et al.2000).

In this system, we could make use of Internet and CSCW technology at the full, and they were cooperative work with such as administrator, multimedia Courseware applicant and evaluation expert.

2. Evaluation cooperate model

Evaluation cooperate model was the basis for multimedia courseware evaluation cooperative system. The model target is when prior condition is time and space separate how we describe, and how we coordinated with every multimedia courseware and evaluation experts as well.

First, we will discuss cooperate model characteristics. Then we will carry on analyses and define basic element for the evaluation cooperate system. Finally we are based on basic element and we will give a completed structure model to system.

2.1 Basic characteristic for evaluation cooperate model

The evaluation cooperate model express form to us, it's under simulating natural work environment, for work object can be shared, we organize multi-user operation in accord with habit and logic.

Characteristic of evaluation cooperate model give expressions for interaction, distribution, role support, perception and visualization and so on.

2.2 Basic element for evaluation cooperative system

We abstract some basic element from multimedia courseware evaluation cooperative system. and describe as below:

2.2.1 System user

<system user object>::< user class><user attribute>

<user class>::<administrator>|<multimedia courseware>|<expert>.

<user attribute>::<account, password, courseware title, abstract>.

2.2.2 Courseware object

<courseware object>::<courseware form><courseware content>.

<courseware form>::<courseware title, courseware

type...>.
<courseware content>::<chapter, section, segment...>.

2.2.3 Evaluation object

<evaluation object>::<evaluation form><evaluation page>.
<evaluation form>::<courseware form><evaluation content>.
<evaluation page>::<courseware content><evaluation page content>.
<evaluation page content>::{<evaluation unit>}.
<evaluation unit>::<evaluation unit position >< evaluation unit content>.
<evaluation unit position>::<begin on position, end of position>.

2.2.4 Information object

<information object>::<information class><correlative content>.
<information class>::<announce>|<edit>|<invite> |<agreement>|<veto>.

2.3 Role mechanism

In evaluation system, every roles have a very different from responsibility and claim, that is social roles. In this system, we define three class role model, that is administrator, multimedia courseware and evaluation expert. Administrator perform item's management, he is competent for announce which multimedia courseware that take part evaluation, invite expert to view and evaluation multimedia courseware through Internet. Administrator may rely on expert's opinion to proceeded evaluation conclusion. Final he should ask that multimedia courseware author over again edit, or he announced which had been ratification or veto (Sandhu & Edward et al 1999).

Author was an editor of the multimedia courseware, after administrator announced that apply to the multimedia courseware for evaluation, author must fill in application for the multimedia courseware, then they must supply the multimedia courseware to evaluation system through internet. The experts are taken up with evaluation for the multimedia Courseware. After administrator invited expert to review the multimedia courseware, the evaluation system qualify expert to evaluate the multimedia courseware.

2.4 Process model for evaluation cooperative system

We divide the process model into four phases. First phase was administrator announced basic terms and requirements. Second phase was author submit that multimedia courseware and application to the evaluation system. Third phase was administrator referred to every expert opinion, and asked author for re-modify content. Last phase was administrator should make the decision. The process model is shown in Figure 1.

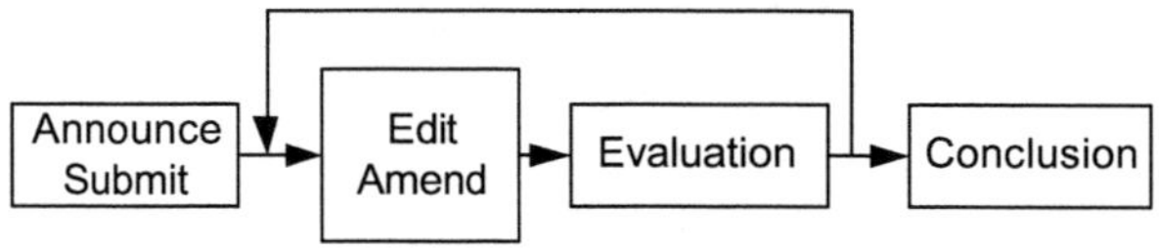

Figure 1. Process model for evaluation cooperative system.

2.5 Integration model for the evaluation cooperative system

We have described the role mechanism and the process model before. Now we shall get on with discussion about the integration model (Figure 2). First we defined four basic element space of system:

Apply courseware object space: Each object opposed a space to the multimedia courseware's apply, author should submit the multimedia courseware and application to an appointed space, and author corrected the multimedia courseware in there.

Experts object space: It's for workspace of the evaluation experts. The evaluation system asked expert to register entry and view. They should fill evaluation content in the evaluation template and they should finish work in the space (Wang et al. 1996).

Administrator space: In this space, administrator should proceed to announce message and manage experts to evaluate the multimedia courseware.

Information object space: It's a space make contact with administrator, author and expert (Zuo. M et al. 1999).

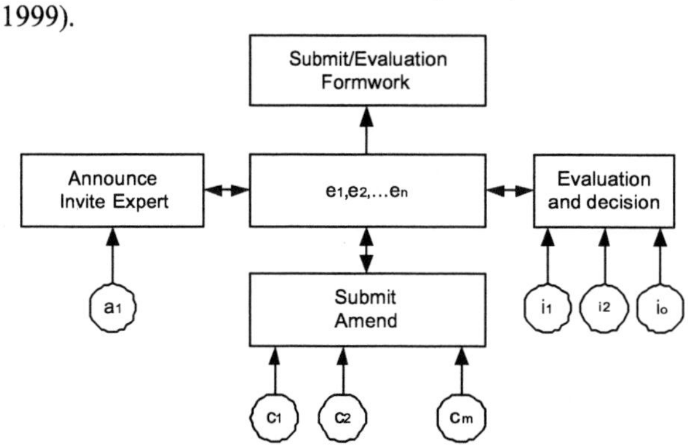

Figure 2. Integration model of the system.

They rely on one another relation above four system element space, we should the model definition as follows:

Suppose:

$A = \{a_1\}$ the gather for administrator.

$C = \{c_1, c_2, ... c_m\}$ the gather for submit courseware.
$I = \{i_1, i_2, ... i_o\}$ the gather for the experts.
$E = \{e_1, e_2, ... e_n\}$ the gather for the information object.
...

3. Systemic key technology design

The evaluation system is designed for the multimedia courseware, the key technology is that how to realize organization, cooperate control and multi-user interface of the multimedia courseware.

3.1 The structure of the organization

We designed the structure of the organization that was composed of management contents, expert library, courseware contents and evaluation contents.

Administrator created management contents, it contain all content of the multimedia courseware, application and register information.

Administrator obtained the information for the expert library, and he should decide to invite which experts attend the evaluation system.

Courseware contents contain all content of chapter, section, segment, and also it contain homework and answer system.

Evaluation contents deposited expert's opinion in there.

3.2 The cooperative control

The evaluation system managed to submit, re-modify and respond, it must observe the asynchronous mode. Before the courseware has not submitted, administrator was not able to see the multimedia courseware. Likewise, before the evaluation expert has not submitted the evaluative report, administrator and author were not able to see the evaluative report too (Gao Xu et al.2000).

In addition, the jurisdiction of the each role must is close control. For example, author and expert were not able to modify the evaluative requirement. Expert was not able to modify the courseware content and he only was to evaluate correlative the courseware. Administrator was not able to modify the courseware content and the evaluative report content. They would not able to modify the multimedia courseware and the evaluative report when they have been submitted. Author was only able to see oneself correlative content.

3.3 Information mechanism

How we announce the multimedia courseware, How administrator invite expert to review the multimedia courseware? In this system, we are adopted the information mechanism (Guohua Bai, 1999).

When administrator claimed that author submitted the multimedia courseware to the evaluative system, administrator shall only send information to the author of the multimedia courseware. When administrator shall invite expert to attend the evaluation system. Likewise, administrator shall only send information to expert.

Only when the expert or the author had received the information, they were able to copy correlative content to their computer system and that should delete correlative information at the same time. So that, we bring information mechanism into use the evaluation system, it should bring down network flow in a way, and we should make the most of the service resource.

4. Conclusion

We have developed the evaluative system that is in general use visual C++ 6.0 and B/C/S structure. It consist of the system administrator module, the courseware module, the submit module and the expert module. The system administrator applied oneself to build, edit and delete the system user in the servers (Makris L. 1998).

The evaluative system has regard the administrator, applicant and expert as client. We supplied different server and interface according to different user.

We have applied the evaluative system to teaching evaluation management system of CUMT on trial. We also need to improve function as follows:

We should build up that take control of the data security and conformableness.

We should consider how the evaluative content had been displayed in different system.

5. Acknowledgement

I would like to thank Pro. ZongKai Lin, Dr. QianPing Wang and XiaoLing Guo for giving the support to make application in the specific field of the cooperative system for the evaluation of multimedia courseware at China University of Mining and Technology, and for their many helpful and suggestions remarks during the research.

6. References

[1] J-C. Leon, L. Fine, "A Design Review Environment for Complex Products", *CSCWID'1999 Proceedings*, Compiegne, France, Sept. 1999, pp199-208.

[2] Sand-hu R. and Edward J. et al, "Role-Based Access Control Models ", *IEEE Computer, Vol.29 (2)*, Feb. 1996, pp 38-47.

[3] Wang Q.P. et al., "An Engineering Information Base Management System", *In Proceedings of the First International Workshop on CSCW in Design*, Peiking, China, May. 1996, pp218-222.

[4] Zuo M. et al., "Coordination Editor of Dispatch Geographical Information", *In Proceedings of the Fourth International Workshop on CSCW in Design*, Compiegne,

France, Sept. 1999, pp29-33.

[5] Makris L., “A CSCW Application for Remote Medical Diagnosis Support and Teleconsultation”, *IEEE Transactions on Information Technology in Biomedicine, Vol.2 (2)*, Feb. 1998, pp62-73.

[6] Guohua Bai, “Communication Management in Information Systems Co-Design Activities”, *ICYCS'99*, Nanjing, China, Aug. 1999, pp295-300.

[7] Gao Xu et al., “A New Floor Control Approach for Distance Education”, *Computer Engineering, Vol.26*, Shanghai, China, Oct.2000, pp636-640.

[8] Huang X. et al., “Research of Online Multimedia Courseware's Development”, *Computer Engineering, Vol.26*, Shanghai, China, Oct.2000, pp650-654.

A Teaching Support Program for Distance Learning

Xiaotian Zhu, Zhijian Huang, Yanchun Ma, Ming Zuo
China University of Mining and Technology, Xuzhou, Jiangsu, China
{zxt, zhijianh, yancunm, mingz}@cumt.edu.cn

Abstract

A description of a teaching support system that has been put into use in CUMT is given in this paper. The system is actually an Internet based program that provides a variety of multimedia courses for CUMT's students both in and out of the campus. Besides that, it also provided a friendly environment for the better communication between students and teachers. By using it, many teaching and learning tasks like consultations, discussions and seminars, can be fulfilled. In addition, yet the program has a useful function that allows a large number of experienced teachers in different departments to work co-coordinately. This working method, which distributed teachers' work as collaborators, makes the instant replacement of information possible. With it's remarkable multimedia potential, the system lets students enjoy the desired discussion with their favorable teachers as much as they like. The modification plan and the prospect of the system are also given in this paper.

1. Introduction

Though distance learning is not a new term in today's educational vocabulary and has gone through several stages like correspondence courses, TV-based classroom, and email-based education, it seems not to be changed much about its appearance as a passive method clinging to give the information of certain courses. (TAN Kun.et al. 1999.) Actually, getting education is not a matter of acquiring a certain mount of knowledge and knowing certain principles in some specific fields. (Crozat & Trigano, 1999) It has a variety of pedagogical aspects such as thinking, sharing, appropriation and application of knowledge, which cannot be easily obtained just by reading books or listening to lectures. (Mivet, 1996) To get education needs an educational environment (Lebeau, 1998, Moore & Rocklin, 1998) in which it is possible for students to attend lectures to get vivid ideas, exchange their opinion with classmates to mature their understanding of life, discuss question even have arguments with teachers to sharp their way of thinking. But how these real educational activities can be fulfilled in the case of distance learning is the question, which needs to answer and is going to be discussed in this paper. Thankfully, rapid development of Internet technology has made the establishment of a virtual educational environment possible. (Huang, Li, & Ding, 1999) Web-site-based distance learning performs much better in simulating an educational environment than it's previous one does. (Scgulz, Reuding, & Erd, 1998) Despite a web-site-based educational support system can not yet provide a virtual environment in which students feel as same as in a real campus, it's potential interactive functions do make a breakthrough for distance learning in many aspects in the sense of educational activities.

In this paper, we are going to introduce such a web-site-based educational support system. Its main idea is to establish a virtual environment for fulfilling all sorts of learning activities and especially trying to make the contact between students and teachers as easier as possible.

The aim to use this support system is mainly to provide education for the students who engaged in CUMT's distance learning program. But the students who take classical courses within the campus are also allowed to access to the system on certain courses as long as they officially registered. At present, there are eight different courses provided on the system. Of course, for normal students within the campus, it seems to be a complementary devise since other supports like human intervention teaching activities, paper document and so on are still available.

2. Description

According to its functional structure, this support system can be divided as following three parts:

- On-line electronic version courses
- A virtual platform for distributed teaching groups
- Several ways for students to take virtual learning activities

2.1 Electronic courses

Since this web-site-based educational support system has been designed for the students in all the majors in CUMT, the courses involved have been considered among more common one like general Mathematics, Physics, etc, which are the compulsory courses for nearly every student in the University. While putting those common courses into internet to bring a result in increasing the efficiency and lowing the cost, it also bring a risk of facing the question about whether the students would get the reasonable results through this new learning method. Of course, it is not a matter for distance learning students, because they have no classroom lecture to attend anyway. But it is a matter for the students in the campus. So only a few classes have been chosen to take part in the experiment at the beginning and also a parallel teaching scheme has been available at present experimental stage, so guaranteeing that the students on the experiment would get similar learning support as others do. This means that the students chosen in parallel scheme not only can learn certain course from web-site-based numeric material, but also have the opportunity to go to a number of multimedia classrooms in where the same material would be lectured by teachers in a way somehow like a traditional class.

After logged in the support system, students can get into their desired course easily through the indication menu on the screen. Normally what you meet in first place is the chapter, which is synchronizing with the one lectured in the multimedia classrooms and the other two chapters they are before and after current one. Of course, the completed electronic versions of all the eight courses are provided in a common database area for students to review.

Each of eight on-line courses was carefully prepared by a team of teachers supervised by one professor in that field. For instance, a professor from Mathematics department would be the supervisor for the Mathematics course group. Other team members are distributed teachers. One member could be a young teacher from computer science department or even someone from art department in order to make the numeric course more attractive.

2.2 The virtual collaborated platform for group teaching

After being published on the web site the main content of the electronic courses surely needn't to be changed often at least within a period of several semesters. Some aspects of the courses like summary, review class and seminar, however, can not be just left there without any change one term after another. They should be revised instantly according to how the students doing in this semester and what have happened recently in the related research field. So there needs a collaborated platform in which distributed teachers could work as collaborators in writing, editing and revising easily.

Based on the idea above the support system set up a collaborated platform for group teachers to modify the teaching content that need to be revised for some reason. It includes a common dynamic database and a friendly interface that is available for both teachers and students to access. Of course, only teachers have the right to write.

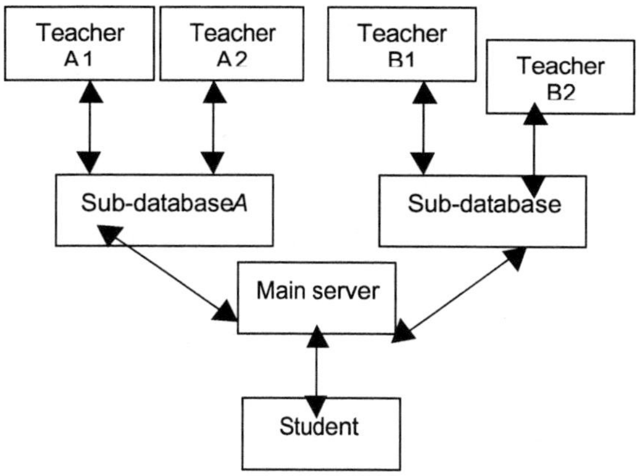

Figure 1. Structure of collaborated teaching.

Each teaching group also has its sub-interface and sub-dbase. If any part of the content needs revising distributed group members could work collaborated on the matter they concern in different terminals. Some discussion should be taken and some arrangements also need to be reached before anything actually has been done. During the modification stage anyone's work can be seen by other cooperative teammates at same time or they can simply work in a way of completed synchronization. New works replace the old one in the dynamic database as soon as the submission is carried out and then the students can see the update material. The structure of this collaborated work was shown in Figure 1.

2.3 Several ways for students to take virtual learning activities

One of the most remarkable advantages for a web-site-based distance-learning package is its interactive function. By pressing those friendly buttons nearly all the educational activities can be fulfilled. The communication between students and teachers in a geographical apart, which was the main obstacle in the distance learning, become so easy that the students no longer need to worry about how to get in touch with the teacher who they want to ask some unsolved questions.

In this support system, there are four major programs students could use for helping study or communicating with teachers and each other.

2.3.1 The common discussion area. Since eight on-line courses are given in the support system currently so same number of discussion areas should be provided under the supervision of the related teaching groups. The students can raise questions or write articles in the discussion area any time. Anyone who know the answer, no matter whether he is a teacher or not, can have the opportunity to write an article in response. Unlike those students who answer the question may just because they are interested in it the certain teaching group is required to respond every question as long as it relates to the content of the course. And, of course, they are also required to delete those that are far more beyond the range of academic study. Good questions and articles will be selected by teachers and discussed further at the on-line seminar. This is a real-technique-based live class in which the students in different geographic parts can receive both video and audio live signal through the internet and can talk to the teacher directly with the multimedia interactive facilities.

In the common area, the questions are not just to be presented by words. They can also be illustrated by graphics or even by pictures as long as they have a reasonable size for data transmission. By using picture function in the common area make the presentation of the question, especially for the subjects as Physics and electronic circuit, for it is very difficult to try making question understandable without using sketch or diagrams, more efficient.

2.3.2 On-line discussion. This is a regular arrangement for students to ask questions to certain teacher. At the beginning of the new term, times table that students can find who will be the on-line teacher and when he or she will be there is announced in the system's questions and answers section. Normally one teacher of the certain course group gives a two hour on-line class in fixed time, for example 3:00—5:00 p.m. Monday, once a week. During this time, discussions can be undertaken in an environment somehow like a chat room enforced by a sharing board in which both sides could exactly see what the other side is doing on the screen, an ideal tool for writing formula and drawing sketch. Questions can be raised and answered without any difficulty as long as you could type fast enough and use mouse skillfully to draw sketches.

Having a microphone and a pair of earphone the students then can easily use the audio facility to ask question orally .The ability of directed talk between students and teachers in addition of the sharing board makes the discussion much more efficient than ever before.

The procedure of the audio communication is that the student find out teacher's call number (IP address) first and make a on-line request then after being accepted by teacher he could join in. The teacher acts as a chairman and decides when or which student he would speak to at the moment. Usually the choice is made by a waiting queue. All the accepted students could hear the conversation in progress. So one thing the teacher needs to do when he or she is on duty is to announce his or her call number on the entry page to let students know where to call up in case they want to use on-line audio communication. Of course, if someone never thinks to change his terminal during on-line answering he needn't to announce the call number each time. He can simply put his call number along with his name on the timetable.

2.3.3 Virtual classroom. A virtual classroom actually is an ordinary studio in that plus Internet function based on the real technology. A diagram of the virtual classroom is indicated in Figure 2.

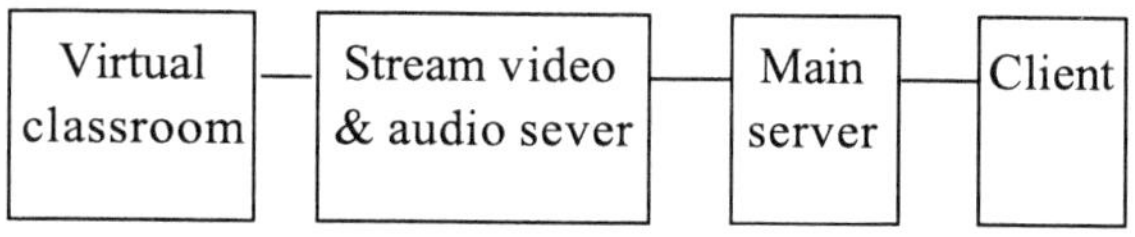

Figure 2. Virtual classroom.

The studio could be an actual classroom or a laboratory as long as they equipped with some video camera and microphones. The teaching activities in the virtual classroom can be televised and published on the Internet. The students could clearly know what happened in the room by watching the motion images and listening the sound that are transmitted live as they access to the support system. While watching and listening the students can still raise questions to the spot room. The interactive function is currently limited on sending message. Multimedia means of interactive function is hopefully developed in short future.

This virtual classroom is very useful in some teaching areas such as physical principle demonstration, Chemical experiment demonstration, summary class, seminar and so on. Experimental part is a part not as successful as other parts among the teaching activities in web-site-based distance learning. We do add some real format videos about experimental demonstration and simulation of the experiment into the on-line courses. But the students certainly feel less interesting compared with attending a virtual classroom in which they can choose one desired experiment and technicians either show the experiment step by step or do what the students asked. Though virtual experiment can only be held once or two times a semester they still draw a lot of attention on it. Other less complicated virtual classes like seminar are held as many as students hoped.

2.3.4 On-line exercise. Accordance with each chapter in electronic course, there are sets of exercise. Most of them are simple questions with multiple-choice-answer to choose. Each student has an electric exercise book. After the exercise being finished the correct answer will be given and the result will be recorded in the systems exercise database. The students are allowed to review their previous exercises and it is also possible for them to do any of these exercises again. Because the type of exercise forms is limited the sole use of on-line exercises cannot meet the teaching requirements for the subjects. At present, paper work is still playing the main role in terms of doing exercises. Although remote students may send their assignments in emails, the type of exercises and the way of correction are still the same as the conventional one.

3. Consideration of improvement

From educational point of view, however, a web-site-based learning system is only a new powerful measure among others such as classroom lecture, paper documents, oral presentation, face-to-face consultation, etc. Also it's a method just for supplementing not for replacing existing ones. In the sense of education, it is the teacher not the machine that students want to learn from. A lot of work to increase interactive and communication being done at current system in fulfilling that purpose, but how to make the learning process more personalized is still our major target in future. The modification of system is already under the consideration.

Firstly audio presentation for each numeric course is in the preparation stage, which will let students concentrate on what they are reading and more significantly emphasize what is imprint in the materials when it is needed.

Secondly elaborated evaluations for exercise correction are undertaken. They include more personalized corrections and a variety of question types. In order to prevent students from choose answer at random a combination of choices is considered. But for the some scientific courses like Physics and Mechanics a well-designed evaluation is still to be coming up with.

Thirdly multimedia means is another important way to reinforce system's personalized feature. A demand-on-video system is on the construction. It would consist of a great mount of video materials compressed either in MPEG format for students download or in real format for students on-line watching. In addition, one-way video on-line consultation and more virtual classroom are also considered in short future. And an optical fiber cable net work has been distributed on campus thoroughly so as to meet the heavy broad-width consumption brought by multimedia data requirements.

4. Conclusion

After having been put into place for one semester this web-site support system has shown a very bright prosperity in self- learning especially in long distance learning area. It's been proved as a potential teaching method with efficiency, less labor consumption and no confinement of time and place. Though being a trail version, with the coverage limitation both in students and courses, it gave a strong impact on University's convention teaching. Among trial courses, teaching efficiency increased dramatically at last semester. The students also showed their enthusiasm on this new learning approach by raising all sorts of questions in common discussion area and attending on-line consolation. In order to provide more vivid and easier numeric courses the teachers have to think a new measure for presentation of the concept and principles, which will hopefully be able to lead to a further understanding about subjects concerned. Besides these advantages, however, some reasonable declination of teaching results among those trial students have been found as expected. Even though it is well believed that this promising method would eventually lead to a historical change in shaping and oriented current teaching and learning device.

5. Reference

[1] Huang Xinyuan, Li Chang and Ding Yuzhan, "An Application of CSCW to Atatistic and Analysis in LMIS", *Proceedings of Fourth International Workshop on CSCW in Design,* Compiegne, France, 1999.

[2] Scgulz M, Reuding T. and Erd T, "Analyzing engineering Simulations in a Virtual Environment", *IEEE computer Graphics and Applications*, December 1998, pp46-52.

[3] Lebeau, R., "Cognitive tools in a clinical encounter in medicine: Supporting empathy and expertise in distributed systems", *Educational Psychology Review*, 10,1998, pp3-24.

[4] Moor, J. L., And T. R. Rocklin, "The distribution of distributed cognition: Multiple interpretations and uses", *Educational Psychology Review*, 10, 1998, pp97-113.

[5] Vivet Martal, "Evaluation educational technologies: Evaluation of teaching material versus evaluation of learning? ", *CALISCE'96,* San Sebastian, juillet 1996.

[6] Crozat S., Trigano P., "First Results from Distance Learning Experiment in Algorithmic", *Proceedings of Fourth International Workshop on CSCW in Design,* Compiegne, France, 1999.

Cultivating a Community of Practice: Designing The TeleLearning Exchange

Griff Richards
British Columbia Institute of Technology
richards@techbc.ca

Tom Calvert
Technical University of British Columbia
calvert@techbc.ca

Aude Dufresne
Université de Montréal
dufresne@COM.UMontreal.ca

Lyn Bartram
Simon Fraser University
lyn@cs.sfu.ca

Marek Hatala
Technical University of British Columbia
hatala@techbc.ca

Abstract

CSCW is one mechanism by which the TeleLearning Exchange Project promotes collaboration in the interdisciplinary research of the TeleLearning Network of Centres of Excellence (TL-NCE). Affiliiated researchers are encouraged to share informal knowledge of their projects via three major strategies: site visits by an integration officer, a community web repository (http://teleport.techbc.ca) and collaborative visits by students. An informal goal of the TeleLearning Exchange is to scaffold the next generation of collaborative researchers by actively involving students affiliated with the TL-NCE as agents of interdisciplinary cooperation and technical integration. This paper describes the design and development issues surrounding a web repository tool to build a community of practice.

1. Introduction

Since 1995, Canada's TeleLearning Network of Centres of Excellence (TL-NCE) has brought together approximately 120 public and private sector partners to explore emerging network technologies for impact in education and training.[1] This collaborative research organization has been fortuitous in its timing. The global spread of the Internet has lead to an unprecedented acceptance of e-learning as a favored mechanism for distance education. Given the pressing demands for R and D in information technology, and the difficulties of interdisciplinary collaboration especially across Canada's space and time obstacles, the Systems Models Theme of TL-NCE sought additional mechanisms to create and reinforce a community of collaborative research. Complementing the traditional integration methods of site visits and conferences, The TeleLearning Exchange [2] was launched as a community web repository that encourages researchers affiliated with TL-NCE to contribute informal project knowledge, and the explicit strategy of employing graduate students as agents of collaboration and integration. This paper discusses the TeleLearning Exchange as an example of computer supported cooperative work, with particular emphasis on its functional design and the complementary strategies employed to spread its acceptance and use.

2. Communities of practice

Although present before the Guilds, the benefits of a community of practice have been most recently documented by Seely-Brown and Duguid [3]. Formal and informal networks of work and study provide mechanisms for sharing common practice, disseminating technical information, and seeking social support. In a vast country such as Canada, networks play an important role in linking researchers together to form a critical mass of scientists to tackle large technical and social problems. In this regard, Canada's Network of Centres of Excellence (NCEs) are unique partnerships among industry, universities and government designed to connect excellent research with industrial know-how and practical investment [4]. The advantages of networks are strongest if the researchers can be kept in constant communication with each other, unfortunately this would leave little time to pursue local initiatives, or fulfill other academic obligations.

Communication, and technical integration are essential elements of collaborative research, and while they are of enormous value, they also require effort. It takes time to write formal papers, attend conferences and to travel to distant research sites to share ideas with the community of practice. The cost in time and resources is such that not all members of a research group may attend so opportunities to participate in the formal and informal rites of community building are not shared equally. Indeed, the venerable members of the community of practice are the best networked and the most active participants, while those with fewer personal connections (typically new researchers and students) have reduced opportunity to participate.

There is a difference between collaboration and sharing. It is the authors' belief that collaboration is easier to promote early in a project through exchanges of informal information. Formal sharing of information is effected through publication. This requires waiting for a longer period and it usually only becomes available at a stage where the main research direction is set and collaboration is less likely to have significant impact.

3. Promise of network technology

Internet technology provides the opportunity to attenuate the isolation of time and space. Applications such as e-mail, Internet telephony, and group work can reduce the overhead in distance collaboration. In particular, web sites provide a low cost publishing mechanism for disseminating research results. In the domain of research and new technologies, the pace of development is fast and is more in pace with Internet publication than paper exchanges. Even for reviewed journals, more and more journals are now developing an electronic version which can be diffused faster, wider and at a lower cost.

A by-product of the ease of publication has been a relaxation in the formality of publication. Unlike a paper which is locked into its hard copy the moment the press is run, web sites are malleable. Web content can be changed relatively easily, so authors are encouraged to post less formal information: information that seeks input from its audience, and engages discussion in the community of practice. Web sites also are also more easily referenced: URLs can be shared with collaborators, formally attached to curriculum vitae, offered as reference in courses or simply referenced to from other web documents.

4. Repository sites for integration and collaboration

A web repository is a centralized collection of information placed on-line by members of a community. These "information objects" may be works in progress or final. A repository provides fast and wide-reaching dissemination, but of more interest to our goals of community building is the potential for feedback and iterative development of the work with other members of the community.

An early lesson in the creation of community web sites was that all authors are not blessed with equal knowledge of HTML and web mastery. In 1995 Richards [5] created SEEDS, a focal site for a community of practice in educational technology. While this early repository received much initial recognition, the awkwardness of the HTML editing process dissuaded long-term perpetuation of the site. In essence, the cost of contributing new information was too high, and interest died off as the initial content became stagnant. For a site to self-perpetuate it must facilitate the activities of its community while bringing some savings in time, costs or energy to do tasks that would normally be done by some other means.

Several repositories now exist where contributors can post their work without running the gamut of a formal editorial authority, in the expectation that others will add beneficial comment and suggestions for improvement. The objects deposited may be textual, graphic, multimedia, or interactive in nature. For example, MERLOT [6] (www.merlot.org) is a repository for learning objects sponsored by a consortium of North American universities and colleges. Members contribute java applets, and multimedia demonstrations that can be used in a variety of teaching scenarios. Other members of the community can view the learning objects, and post comments and lesson plans for their use.

Repositories are also used as tools for knowledge management. Hatala [7] developed CEDAR – Conceptually Enriched Document Archive – a tool developed at The Knowledge Media Institute, UK for organizational learning systems [8]. CEDAR facilitated the depositing of documents which could then be annotated or commented on in an appended discussion forum. Readers within the organization could freely add related links to other sites or demonstrations, or upload related documents. As distributed authoring renders impractical strict hierarchical organization, cataloguing, meta-tagging and search engines take on new importance in management and usability of the content. As is the case with any repository, to have up-to-date content, the addition of new information reflecting the development in each particular area must be assured and supported. This work is carried out, drawing on the work of Gantt and Nardi [9] by local developers (similar to our "integration agents") who are members of the research teams, but have a greater familiarity with tools and methodology.

5. Design guidelines

5.1 Efficiency

The repository must load quickly and be easy to use. The interface in our case also needed to be bilingual (English/ French)

5.2 Structural Coherence

The hierarchy must reflect the formal project structure of the TL-NCE. In addition, it must support additional structures which arise over time as research interests cross the vertical boundaries and as community members bring in new interests and new members from outside their immediate project. It is expected that distributed authoring will result in a variety of structures. Good navigational aids and search engines can reduce the complexity.

5.3 Extensibility

Extensibility reflects the values of the community in terms of its membership and the scope of the site. While some repositories e.g. MERLOT, allow anyone to join the community, we required introductions. Any member of the community is able to extend membership to newcomers to the community. Members are also welcome to add additional pages or branch on new research clusters

5.4 Community Responsibility

This is a hierarchy of responsibility that allows content or authors deemed to be unacceptable to be removed. We used the inherent tree-structure of our database to create spheres of responsibility. Every member is responsible for the behavior and content of those in their sphere of responsibility. The site administrator has the ultimate responsibility to investigate complaints and in consultation with the community, take corrective action.

5.5 Security and Privacy

Only documents of a certain formats and size will be accepted to the main repository. Executable code demonstrations are allowed to be linked to but not to reside in the repository. Authors may designate their pages to be public or private (password protected). This allows for distant workgroups to shelter early work that is not ready for dissemination.

5.6 Bidirectional Information Flow

In addition to pulling information from the community, monthly updates or urgent messages can be pushed to the community. A regular scan of the database identifies new or updated materials, which can be scanned by the site administrator for inclusion in an electronic newsletter to the community. Some consideration has been given to automating parts of this process. This enlarges the possible impact of contributions in the repository, making it accessible, not only to a core more active researchers but also to those who follow the progress of the community.

5.7 Participation Incentives

A repository requires ongoing incentives to contribute and participate. The TeleLearning Exchange is developing rewards program is planned whereby authors of significant contributions can be provide travel opportunities for visits to research sites, exchanges and to attend conferences – activities which provide even closer collaboration, networking opportunities. Interdisciplinary exchange may also counter the impact of competition among themes, and encourage faster reporting of results.

5.8 Architectural Flexibility

The exchange needs to be open to suggestion from its community and easy to modify as the needs or structure of the community change.Architecturally it is import to emphasize a database approach, separating the elements of style and structure from the content itself.

5.9 Openness

The Exchange encourages the diffusion of international information in the domain in general, so researchers are motivated to visit on a more regular basis. Their contribution on the state of the art is precious and serves as a shared "Technological watch", where the most recent advances can be confronted.

These broad guidelines were developed through a collaborative design process. We started with a basic knowledge of the user community and let the TeleLearning Exchange grow by an iterative design process. NCE members were invited to use the site and report on its strengths and weaknesses. In many cases, doing so allowed them to refine their own experience and expectations of what such a service could support, influencing both the next design version and their commitment to collaborative research. While the essential guidelines remain the same, this ongoing process has continued to shed light on which aspects of the TeleLearning Exchange are most functionally useful and where the gaps are.

6. Experience to date

Technically, the Exchange builds on Hatala's CEDAR. After initial prototyping in Lisp, the system was re-coded to use Active Server Pages with a SQL7 database. Students affiliated with TL-NCE Theme 3 projects provided content and links to project home pages. Significant consultation and formative evaluation of the user interface was provided by Dufresne and her students at the Université de Montréal. The project was subsequently demonstrated at the TeleLearning 2000 Conference in Toronto, where it garnered the People's Choice Award in the Demonstration Category. Based on feedback and observations from users, the design has been streamlined. We are pruning some features because they are simply not used – for example instead of allowing the public to add demonstrations and links, this privilege is restricted to registered authors. At time of writing (February, 2001) the interface conversion to French has been completed and the use of the site is expanding from Theme 3 to the rest of the TeleLearning-NCE. Figure 1. illustrates one view of a project in the repository.

The quest for content remains a challenge. While the use of graduate students as integration agents remains viable in principle, the current economic demand for information technologists has diminished the number of available graduate students with strong communication skills in English or French. This is expected to be a temporary problem and does not appear to extend to the education and communications themes of the TL-NCE.

7. Future Plans

Coupling the repositories with other collaboration strategies will be key to the promotion of research collaboration. One strategy under consideration is for TL-NCE is to use the exchange as the submission channel for an electronic journal. Submissions could be screened on a monthly basis and key contributions would be summarized and pushed back to the TL-NCE community. Authors would be encouraged not only to report on their research activities, but also to identify potential collaborators within the network. An incentive program of physical site visits to enable face-to-face collaboration is being considered for selected papers.Given our NCE mandate to prepare highly qualified people (HQP) many of the strategies reinforce the participation of graduate students, but always in the context of their affiliation with the Principal Investigators of the TL-NCE. Consideration is also given to creating a resume drop for affiliated students and researchers as a method to acknowledge their contributions, and to demonstrate the HQP output of the NCE.An unexpected outcome of the TeleLearning

Exchange has been the growth of an extended project team of developers, evaluators, content contributors and reviewers. The development of the site has been itself an exercise in community building. It has brought closer collaboration with other TL-NCE community building activities such as the Global Educators Network, the TL2000 conference, and the main TL-NCE web site. We find we cannot preach collaboration without collaborating ourselves.

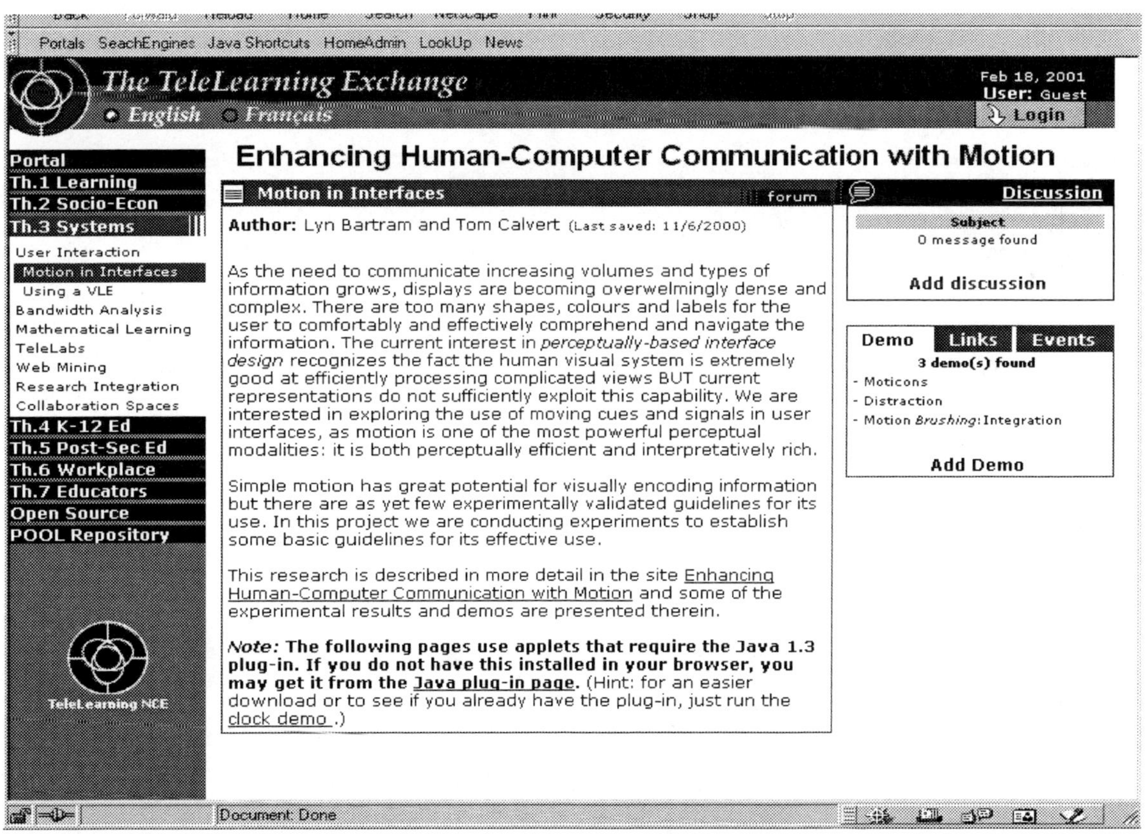

Figure 1. A view of a one project in the repository.

8. Summary

In this paper we have explained the context and development of The TeleLearning Exchange. We have outlined our design guidelines for the creation of community building respositories, and briefly explained our experiences and intentions. Our initial findings lead us to believe that such repositories can be effective tools for disseminating results, stimulating collaboration in research, and networking the next generation of collaborative researchers. However, web repositories need to reflect the needs of their communities and provide utility to their users. As such they need to be employed in the context of wider strategies for the integration of research.

9. Acknowledgements

The authors wish to acknowledge the valuable input of Alain Breuleux, TL-NCE's Director of Integration, and Gordon Yip of Simon Fraser University who programmed the ASP version during his cooperative work term. This project is supported by the TeleLearning Network of Centres of Excellence.

10. References

[1] TeleLearning Network of Centres of Excellence http://www.telelearn.ca

[2] The TeleLearning Exchange http://teleport.techbc.ca

[3] Seely Brown, J. & P.Duguid, *The Social Life of Information*, Boston : Harvard Business School Press, 2000.

[4] Networks of Centres of Excellence http://nce.nserc.ca/en/abouteng.htm

[5] Richards, G. (1995) SCOET Explores Electronic Delivery Systems (SEEDS) (no longer online).

[6] MERLOT (Multimedia Educational Resource for Learning and On-Line Delivery) http://www.merlot.org

[7] M.Hatala: Contextually-enriched Documents: Publishing of Organisational Learning. *The Open University Technical Report KMI-TR-85*, Milton Keynes, UK. January 2000.

[8] P.Mulholland, Z.Zdrahal, J.Domingue and M.Hatala: Integrating Working and Learning: a Document Enrichment Approach. *Journal of Behaviour and Information Technology,* 19(3), 2000, p. 171-180

[9] Gantt, M. and Nardi, B. 1992, Gardeners and Gurus: patterns of cooperating among CAD users. In P.Bauersfeld, J.Bennet and G.Lynch (eds.) *Proceedings of ACM CHI'92 Conference on Human Factors in Computing Systems* (May 3-7), Monterey.

An Interactive Multimedia-Board*

Junzhou Luo, Huasong Xu, Jianxiang Wei
Department of Computer Science and Engineering, Southeast University, Nanjing, 210096, P. R. China
jluo@seu.edu.cn

Abstract

In order to meet requirements of large-scale users' cooperation with real time multimedia in the IP network, this paper presents the concept of multimedia-board and the definition of multimedia-board system, describes the conversation model with Petri net and provides a concurrent control mechanism and algorithm which may be applied to multimedia-board and other CSCW systems. At the end of the paper, we give a design of multimedia-board with B/S structure and its perspective in practice.

1. Introduction

Multimedia synchronization of real time interaction is one of the key technologies of Computer Supported Cooperative Work (CSCW) in long-distance education[1,2]. It utilizes a lot of kinds of information networks to realize a virtual classroom, where the education scale can be expanded and the teaching of famous teachers can be shared. People can be out of limitation of time and locality to enjoy the excellent education resources. Therefore, the technology is attracting more and more attention of the departments of education, science research and industry all over the world.

Nowadays some countries have realized the long-distance education through three following ways[3]: One is to transmit by satellites, which is one direction broadcast and there is no interaction between teachers and students. The second way is based on H.320. Although it can support dual-direction interaction, MCU must be needed because the technology is based on circuit exchange, which is expensive, may not be merged into Internet to support a lot of users and cannot be used widely. The last one is based on IP networks. This approach can be used in dual-direction interaction. Depending on the available Internet, it is scalable, low-cost and easy to be built etc. Although there is limitation of bandwidth now, with the development of wide band networks, the long-distance education based on IP will have a bright future.

To design a multimedia cooperation system of the long-distance education that can support scalable users on IP networks, we must solve the interaction problem between the teacher and students. The quantity of data may be very large and all kinds of type, including text, image, video, animation, audio and so on. Therefore, we raise the multimedia-board concept based on electronic white-board, which can allows the cooperation between teacher and students. According to the prevailing publishing and sharing method on IP network now, we choose HTML documents to describe the information. It is because that HTML has the welcome document format, strong ability to present, more editors and being supported by third-party vendor widely.

2. Definition of multimedia-board

The main goal of CSCW is to help members to complete the same task together. To create a shared space is the first step. In fact, multimedia-board is a virtual shared working space through which teacher and students can carry out teaching and discussion of special topics. The screen of every attendee represents a virtual shared working space. Each attendee can add, delete or modify it by token. After operation, the space will be refreshed and client PC can be refreshed too. So the system must support the following basics: 1.texts and graphics, used to interact and dynamically note and so on; 2.multimedia electronic document, used to exchange the electronic data when teaching and discussing; 3.Video/Audio, interaction between teacher and students; 4.Operation feedback, to refresh the virtual space and clients data; 5.Other functions, to get the authority of speaking and judge the operation attributes of a virtual space.

Multimedia-board can display the teaching materials prepared by teachers with third-party tools and allow teachers to create dynamic notes. Logically, the display space of the multimedia-board can be considered two levels. One is used to show materials, named document

* This work is supported by National "973" Fundamental Research Programme of China

level. The other is to show dynamic notes, named overlay level because it is always over the document level.

The information data of the system is transmitted based on TCP protocol. Video/Audio data are sent on RTSP protocol. Multimedia transmission adopts multicast methods to reduce data amount. It uses less bandwidth and reduces the delay of transmission. The following figure presents inter-layer relation of the system:

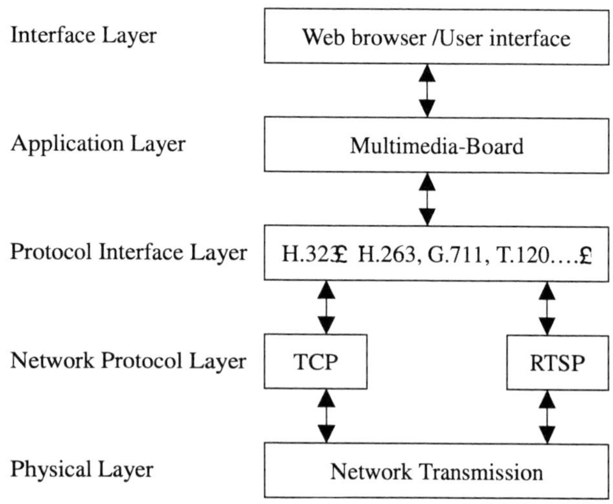

Figure 1. Inter-layers relation.

3. System model and algorithm of concurrent control

In order to design a CSCW system, the characteristics of cooperative work should be analyzed and known at first, and the model of group work should be set up. This is the foundation of development of a system, too. At the same time, the essential concurrent control mechanism should be made out according to the group work model. Then we can realize the goal of the CSCW system.

In a CSCW system, the behaviors of participant are unpredictable, especially all kinds of concurrent actions. The theory of Petri net is developed from the concept of concurrency. It describes a real system by simple graphics form. It can express the uncertainty in the system succinctly, and can be used to express various kinds of abstract concepts. Now, we can set up synchronous conversation model of the multimedia-board using Petri net at first[4,5], then put forward a kind of concurrent control mechanism and algorithm which is suitable for the multimedia system.

3.1. Petri net model of the system

Multimedia-board system maintains the virtual shared space, in which all participants can operate the multimedia-board system in coordination with each other simultaneously. Now, we describe the synchronous conversation course of multimedia-board system that is defined above in Petri net.

For simplicity, figure 2 gives a simultaneous operation example, which contains two participants and one shared resource. The place Ss stands for the very beginning place. This model begins with transition Ts. If T11 fires, the first participant wins the token of operation. After that, every place could create a copy of itself. Through the copy, the conversation course would end at any time by the transition T1e or T2e with the participation of S1e or S2e. Fig 3 gives the example in which the first participant wins the token of operation.

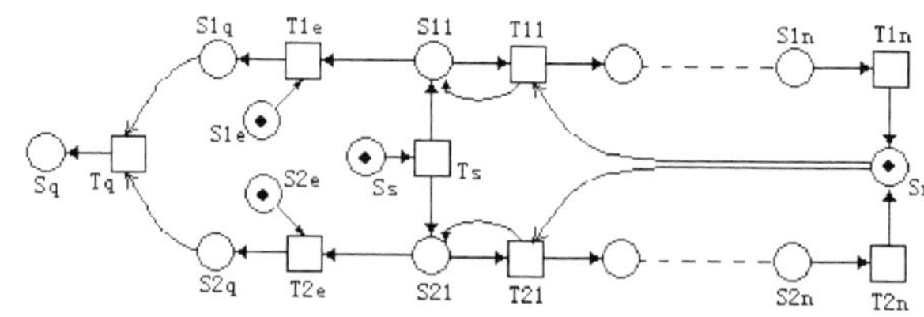

Figure 2. Petri net model.

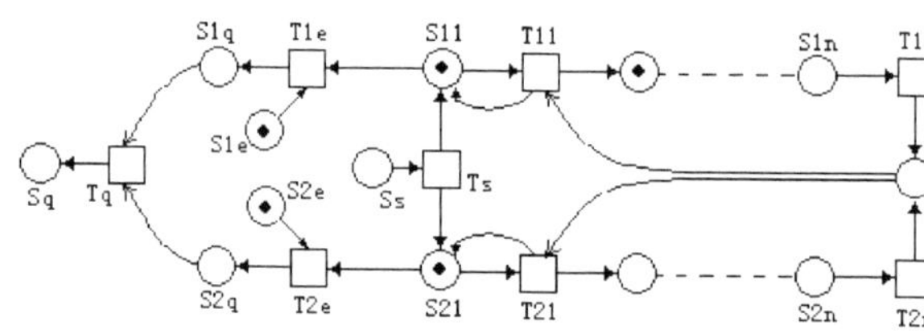

Figure 3. The first participant wins the right of operation.

The actual application situation in this model is: While the cooperative work begins, the virtual workspace stores all objects, which exchange with each other simultaneously. Each object is regarded as a shared resource and each participant can operate on one certain object to express its own opinion or different views. But two participants cannot use the same resource at the same time.

The special case of the model is to deal with the virtual space as a whole shared resource.(In Figure 2, Sr stands for this kind of whole resource.) Obviously, when one gets the token of operation, he will occupy the virtual workspace. Other persons can only receive information from him until he gives up or loses right to speak. Such treatment is very suitable for real time teaching system, because the teacher is the coordinator in the classroom, and the student just could make a speech at a certain moment. The flow direction of the data is asymmetric. In other words, at a certain moment during the course of teaching, only one person is speaking in all the time. As a result, other participants keep in audience's status. (When one student is making a speech, the teacher is a audience, too.) It is not hard to imagine what the result would be if students can alter the content of the multimedia-board when the teacher is giving lessons.

This model describes the starting, concurrent operation and exiting process in the conversation of the multimedia-board. It reveals the characteristics of a CSCW system, such as sharing space, concurrent operation, and so on. Generally speaking, there may be a lot of shared resources and concurrent process in a multimedia-board system.

3.2. Concurrent control

The cooperation mechanism in CSCW system is the key to realize sharing. And concurrent control is the main problem of the mechanism. Its aim is to solve the conflicts, which occur when the user visits the shared space. The concurrent control in CSCW system has essential difference from that of traditional multi-user system. The latter tries hard to make each user think that oneself is the only user of the whole system. Just in the opposite, to sense each other's status is one of the purposes that should be achieved in concurrent control. We can give a kind of concurrent control mechanism and algorithm suitable for the multimedia-board and other CSCW application system.

Because many users operate on the virtual shared workspace at the same time, different user's virtual workspace might be inconsistent, which might lead to errors. A concurrent control mechanism should be provided to solve the conflicts. According to the application characteristic of the multimedia-board, the centralized concurrent control should be adopted. Figure 4 shows the system structure of the concentrative control mechanism.

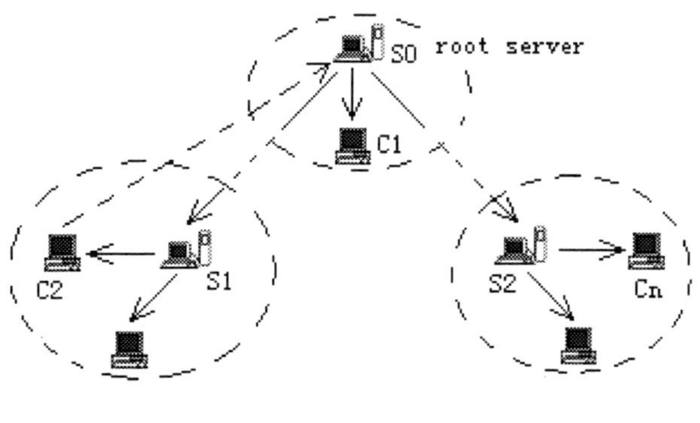

Figure 4. Centralized control mechanism.

Figure 4 describes a long-distance system consists of three LANs. In this system, the multimedia data is transmitted in multicast way. If the client c2 wants to speak, all operations that this website sends out will be delivered to root server through the up-link way at first, then they will be multicast to all the users by the root server.

This structure has mixed the characteristics of data concentrated type and application duplicated type. The concentrative control programs in the application server will be in charge of interaction of semantics and keep the consistency of semantics. At the same time, they clone the semantics onto the distributed sites. The root server takes charge of dealing with all user's input and output events.

In order to solve the problem of concurrent control, token is the simplest and the most common method. Some experts including *C. A. Eills* had introduced some algorithms of concurrent control in their papers. Here, we give a concurrent algorithm, which adopts the method of token. During cooperative work, only one participator owns the token of each shared resource, and only he can access the virtual shared workspace. The following is the algorithm.

Step 1: Allocate each shared resource a token.

Step 2: In a certain client C1, if the participant needs to use a certain resource, he chooses the resource at first, then he asks the application server for the token of the resource R1.

Step 3: The application server receives the request.

Step 4: If the source R1 is being used, the server refuses this request.

Step 5: If the source R1 is free and more than one participants ask for it at the same time, the application server will allocate the token of the resource R1 to one of the participants according to the principle of First Come First Served or the priority settled in advance. Then he sends the refusal messages to other participants. If there is no competition, the application server allocates the token to the applicant.

Step 6: Only a participant owns the token of a certain resource can he operate on it.

Step 7: At any time, only one participant possesses the token of a certain resource and the resource will be marked being used. The operation of this resource will be sent to other participants through the application server.

Step 8: After using the resource R1, the participant reports to the application server that the token has been released.

This algorithm guarantees the consistency of shared workspace. It assigns every participant operation privilege of some shared resource and gets it back after a certain period of time. The problem of right assigning has been solved by the control of token-ring. In order to realize it, we provide four primitives to realize this mechanism: ask, assign, refuse and release.

4. System Design

According to the centralized control system structure that has been introduced above, the program of multimedia-board consists of two parts: the server and the client. The server monitors the link request from the clients, and establishes connection with it, accepts the uploaded data from it and then transmits them to other clients. The task of the client is to sent the data to the server, and at the same time accepts other clients' data from the server. Each client only communicates with the server, not establishing direct connection with other clients.

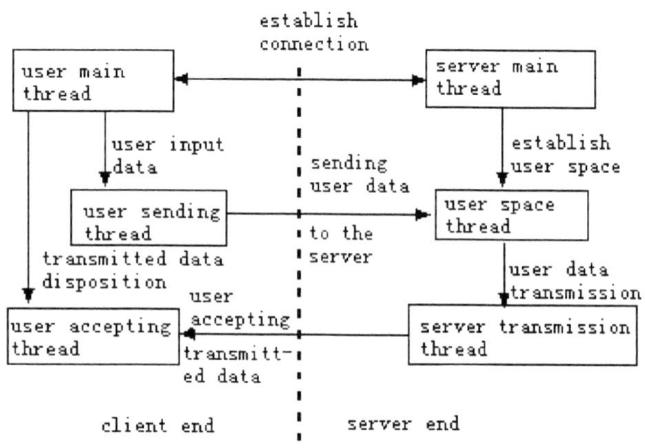

Figure 5. Program model of the multimedia-board.

Figure 5 shows the program model of the multimedia-board. The client will send link request to a certain port of the server immediately when running. After the connection is established, the user's sending thread will send the input data to the server. The main thread on the server is responsible for monitoring a certain port, and setting up a user space thread for each user who sends request. This thread is responsible for accepting and recording the user's information, updating user list table, accepting the data transmitted by the user's sending thread and sending the data to the server's transmitting thread for transmission. According to the user list table, the server's transmitting thread sends the data one by one to every user's accepting thread. For the client, the user's accepting thread accepts the data from the server and sends them to the user's main thread. After disposed by the main thread, the data will display.

As the system is mainly used to create network connection and transmit data and there are different data types and many threads in the system that needs to communicate with each other, and the heterogeneity of IP network, we may choose Java language for programming[6]. This is because Java is a kind of object-oriented network programming language, and it supports multi-thread. Moreover, its platform-independency and security is unsurpassed by other programming languages.

Creating protocol set of the application layer is the key problem of program realization. The communication protocol set can be created according to the functions that the multimedia-board needs. It may include log on, text transmission, file transmission, graph transmission, video/audio transmission, refresh workgroup member list, join workgroup, exit workgroup and pause/resume communication, etc.

TCP/IP protocol has been used to ensure the validity of data transmission, so on the premise of ensuring the functions, the communication protocol set should be designed as compact as possible to improve the efficiency of the bandwidth.

The following is an example of the transmission protocol set. The meaning of the symbols included is: (S) means being sent by the server and accepted by the client; (C) means being sent by the client and accepted by the server; (short) means 2-byte short integer; (int) means 4-byte standard integer.

Log on (C): log→user name string.

Graph transmission (C): draw→(int)color value→(short)line starting point X-coordinate→(short)line starting point Y-coordinate→(short)line ending point X-coordinate→(short)line ending point Y-coordinate.

Graph transmission (S): draw→(int) color value→(short) line starting point X-coordinate→(short) line starting point Y-coordinate→(short) line ending point X-coordinate→(short) line ending point Y-coordinate.

Refresh workgroup member list(S): refresh→user 1 name string→user 2 name string→...→complete→ok.

Join workgroup(C): join→applicant's name string.

Exit workgroup(C): quit.

Pause communication(C): pause.

Resume communication(C): continue.

Due to space limitation, only parts of the communication protocol set followed by the multimedia-board software is given above as an example. The expansibility of this protocol set is very strong. The protocol set can be expanded at any time according to improved functions. For example, round pattern needs to be transmitted. Then the following protocol can be added to the protocol set: circle→(int)color value→(short)circle center x-coordinates→(short)circle center y-coordinates→(short)circle semi-diameter.

When designing a specific program, two issues should be noted:

- □ Some OS thread's operation mechanism is not wholesome enough, and when the thread is canceled, the resources occupied by them can not be released completely, as a result, multimedia - board server will use up the system resources gradually while it is running. Therefore, when the server is initialized, user service thread should be built according to the maximum permitted simultaneous on-line user number, waiting for connection of client. When the user exits, the thread is not canceled. It will delete the user's data and prepare for serving the next user.

- Every protocol of the protocol set needs to establish a synchronized code block. All the operations of the protocol should be accomplished in this block, so as to ensure that the thread can execute the protocol completely and correctly. However, because synchronized operation can reduce the efficiency, the synchronized code's length should be shortened as much as possible on the premise of executing the protocol perfectly.

5. Application areas

Multimedia-board, which can make the long-distance classroom and real-time answering question become true, is an application of CSCW and a vital technology of long-distance education.

5.1. Long-distance classroom

By the long-distance classroom, students can attend a class in different places. Multimedia-board provides a shared space, on which the teacher can display the teaching material beforehand prepared and the teacher or the students can create dynamic notes. During teaching, multimedia-board can present to every body synchronously and each modification in it can be sent to every one immediately. Not only static page can be placed in the multimedia-board, but also stream data, such as video, audio and stream-media courseware. These stream-medias keep the synchronization relation with the teacher's video and audio, which makes the teacher can use a picture to speak.

In order to absorb the students' concentration, the teacher can use a cursor to point out an area in the multimedia-board. Maybe he can use a light tool or other remark tools to create dynamic notes. These notes can be overlay on the teaching material. It is common to establish some simple characters or geometric figure for explanatory note, or to embed some brief flow media etc. The teaching material that the teacher has got ready can be shown according to an order that be prepared in advance. For example, according to page number order, the teacher can use " next page" and" last page" to browse all pages. Maybe the teacher can visit a certain page faster according to the file's index.

5.2. The real-time multimedia answering of question

Answering of question is a part of long-distance education. Its basic goal lies in answering various kinds of queries of the student. Answering system divides mainly into two modes of functions. The common one answers question through the E-mail and belongs to the asynchronous mode. Using the multimedia-board the teacher can answer questions and explain immediately and synchronously. The answer can be written down in HTML's file format and also the teacher can quote picture, sound or the video in the material storehouse to make the answer better.

6. Conclusions

This paper starts from a system design of CSCW to discuss the multimedia-board system, conversation model, and concurrent controlling mechanism and algorithm and the thoughts of systematical design. The multimedia-board is a CSCW application in the long-distance education and the key technology to realize long-distance classroom and real time multimedia answering system. We believe, it will become an important learning tool for people in the near future to use.

It's necessary to talk about some more issues of IP multicast. At present, IP multicast is mostly used in LAN and more extensive IP multicast is still being studied. It is less mature than unicast TCP. In fact, most routers have not support multicast protocol. All issues must be solved. One feasible method is to introduce a SMCU realized in software for every sub-network . We can see it in figure 6.

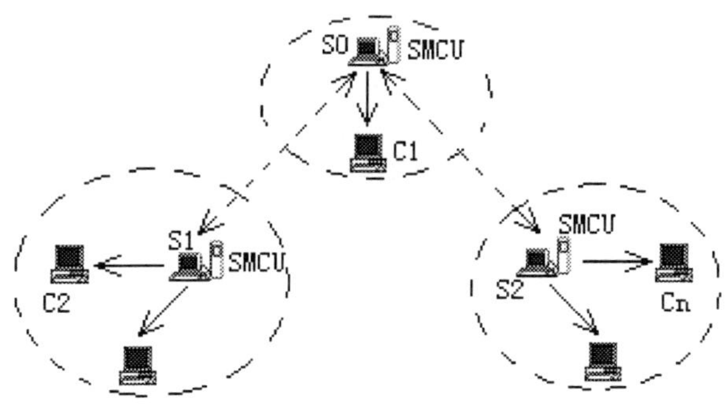

Figure 6. The system structure of SMCU.

SMCU is responsible for data transmitting and routing within the network, which can solve transmission problem of multimedia data through sub-network and problem of synchronous transmission among many nodes. As figure 6 shows, the data comes from the root node's SOSMCU, then they are distributed to S1SMCU and S2SMCU to be transmitted to its sub-network separately. The above only presents some ideas of SMCU. We'll do further work about it.

7. References

[1] Dan Diaper and Colston Sanger (Ed.), "CSCW in Practice: An Introduction and Case Studies", 1993

[2] Sankt Augustin, "Proceedings of the ERCIM workshop on CSCW and the Web", Germany, Feb. 1996

[3] Zheng Qinghua, Liu Jun, "Multimedia Distance Learning System over IP Network", Proc. *The 5th International Conference on CSCW in Design, Hong Kong, Nov. 29-Dec. 1, 2000*: 198-203

[4] Luo Junzhou, Gu Guanqun, "Groupware Service Description and Specification", *In: Zongkai Lin eds. Proc. of 1st Workshop on CSCW in Design, Beijing: International Academic Publishers. May 1996*: 270-274

[5] Reisig W, "A Primer in Petri Net Design", *Berlin: Springer-Verlag, 1992*: p.120

[6] http://java.sun.com/docs/books/jls/index.html

Cooperative Work in Multimedia Network Teaching System

Shaozi Li, Tangqiu Li, Liying Wang
Department of Computer Science, Xiamen University,
Xiamen 361005, P. R. China
szlig@xmu.edu.cn

Abstract

This paper presents the basic architecture of CSCW (Computer Supported Cooperative Work), then some issues in CSCW systems such as concurrency controlling and synchronization is discussed, it's application in multimedia network teaching system such as whiteboard, chatting, etc is given last.

1. Introduction

Along with the development of the distributed computer system and high-speed network, computer systems have been developed from the traditional problem solving, some scientific calculation to helping user interaction and supporting cooperative work [1]. To support this new method, Computer Supported Cooperative Work (CSCW) was brought forward and developed rapidly in about recent 10 years.

A key issue in Computer Supported Cooperative Work (CSCW) application is how to provide participants with an effective cooperative support. CSCW application is such a system that under the computer supported environment, a group of participants cooperate with each other doing a common task and aiming at a same target [2]. Currently the main applications in Computer Supported Cooperative Work (CSCW) are collaborate compile, collaborate design, collaborate management, communication and knowledge management, etc[4].

Multimedia network teaching system (Netclass) is a multi-user collaborative system. In this system, there are one or more teachers and a number of students (40 to 80 students) to work together. How to work cooperatively and share the common resource between teacher and students or between student and student is a question. This paper will first present basic architecture of CSCW system, and then some issues about collaborative and synchronous work, and lastly, its application in multimedia network teaching system.

2. Basic Architecture of CSCW System

A basic CSCW application can be divided into four parts: Application Service (AS), Cooperative Service (CS), Synchronous Service (SS), Transmission Service (CTS) as shown in Figure 1.

AS is a user interface of CSCW application in a domain such as multimedia network teaching system. With this service, the system of CSCW can interact with user and one user can cooperate with others. There are some interactive utilities in this service, such as chatting, whiteboard, multimedia network cinema, etc. Otherwise, it can process the message from other user.

CS provides a mechanism that multi-users cooperate each other. In this part, there are two functions to be done. One is the access controlling to user, the other is concurrency controlling and coordination to sharing object.

A collaborative application (e.g., whiteboard) must provide a means of concurrency control to ensure that shared data held by each instance of the application be consistent throughout the session [3]. SS provides a mechanism that ensures the sharing object to be consistent.

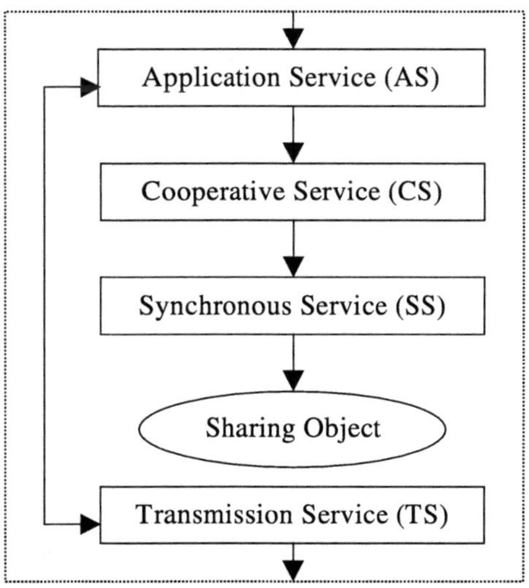

Figure 1. Basic architecture of CSCW.

TS is a foundation of the cooperative application. There are many events, data flow and multimedia flow to be transmitted between collaborative users. For a multi-point to multi-point multimedia interaction information and data information, by using datagram UDP socket and TCP socket, a multicast (for local participant) agent and a directional transfer (for remote participant) communication agent are created, and a reliable data communication is completed [2].

3. Collaborative and Synchronous Work in CSCW System

Collaboration and synchronization is the key issue in CSCW application. Collaboration can make the multi-users cooperate with each other, synchronization can ensure the sharing object to be consistent.

3.1 Collaborative Work

Collaboration provides a mechanism that multi-users cooperate each other. There are two functions to be done, one is the access controlling to user, the other is concurrency controlling and coordination to sharing object as shown in Fig 2.

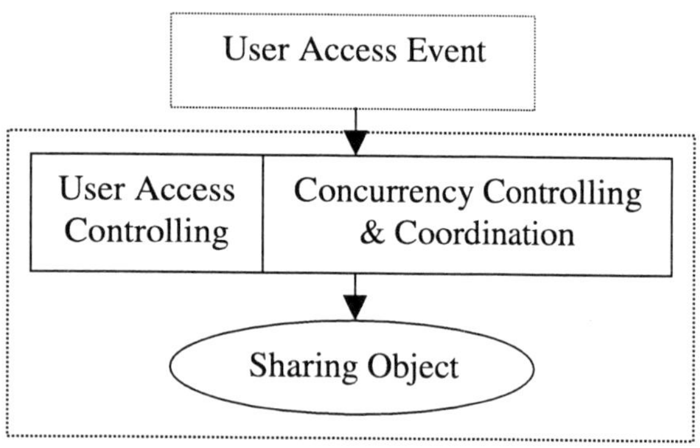

Figure 2. Collaborative mechanism.

When the system receives the access request to operate the shared object from user, the user access controlling will check that the request is valid or not. If it is valid, the access request will be sent to the concurrency controlling and coordination, else it will be refused.

When the participants cooperate with each other, it is possible that there are two or more participants to operate the sharing object concurrently. So, the system must provide the concurrency controlling and coordination to sharing object.

At that time that user wants to access the sharing object, the user access controlling checks whether it is valid or not at first. If it is a valid request, then system will let user access the sharing object according to the concurrent controlling algorithm (such as FIFO, Priority etc).

There are two concurrent controlling and coordination mechanisms in our system Netclass. One is that the user of sharing object is collected by teacher when the students want to talk to teacher. The other is FIFO, such as in function 'discussing free'.

3.2 Synchronous Work

Synchronization can provide a WYSIWIS (What You See Is What I See) environment. In Netclass system, we have used the event mechanism to make the participants synchronous. The events can be divided into two kinds: internal event and external event. The internal event is an operation from local participant, the external event is an operation from other cooperative participants.

For better synchronization, techniques of multiple sockets, time stamps and multiple buffers can be used. Socket ports are set up for different uses (Audio, Video, Whiteboard, Chatting, etc), then the thread knows what kind of data is by judging different socket ports. It is useful to put the time stamps over audio/video data packets and take use of multiple buffers, the streams will be played back and showed at receiving endpoint after analyzing the identifiable bits, then synchronization and ordering among different kinds of data sources can be guaranteed.

4. It's Application in Multimedia Teaching System

4.1 Some Issue in Network Transmission

In our system Netclass, there are two kinds of data streams: one is audio/video streams which can tolerate certain error rate, but strict with delay, the other is event messages and whiteboard operation messages which require credible transferring.

Our system uses two protocols. The audio video streams should be achieved with UDP protocol, UDP doesn't retransmit although error happening, which guarantees the real-time and smoothness of audio video streams. The event message and whiteboard operation messages should be realized with TCP protocol, it can assure interaction credibly.

4.2 Implementation

We have designed the multimedia network teaching system (NetClass) based on CSCW. This system works on peer to peer networks. The operating system is Windows 9X or NT Workstation, programming language is VC++, and transmit protocol is UDP or TCP/IP.

The shared whiteboard is one of the functions in this system as shown in Fig 3. It is provided for the discussion in the teaching. When a student wants to ask some questions to teacher or others, he can use the whiteboard (Fig 3 (a)) to express the questions, and at the same time the teacher can explain the questions through the whiteboard.

In the shared whiteboard, we have implemented the functions that teaching is needed in server (teacher's site) and client (student's site), such as load/save a image, input the text, change the text style, the color, line's style, erase a picture, draw any line, box, circle, ellipse, arrow, etc. With the help of the collaborative controlling and synchronous mechanism as described in Section 3, simultaneous operations on the same object are performed in the same order at all sites.

The chatting such as 'free discussion' is another function in Netclass system as shown in Fig 3(b). Every participant can chat with others through text or audio. When any participant wants to speak by voice or text, it must send a request to the system. Then system can select one from the participants who want to speak according to the concurrent controlling algorithm, such as FIFO.

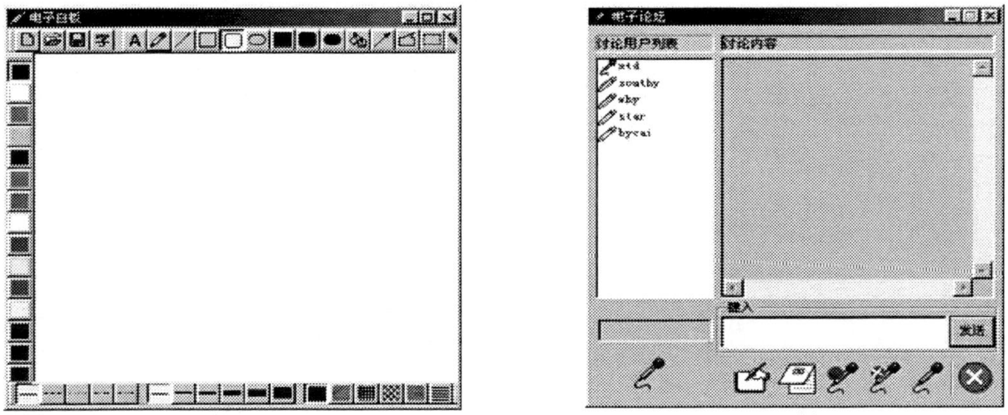

(a) (b)

Figure 3. The shared whiteboard of NetClass system.

5. Conclusion

In this paper, we have presented the basic architecture of CSCW system, some issues on collaborative controlling and synchronization, and its application in multimedia network teaching system (NetClass) such as in shared whiteboard, or chatting.

In future work about this system, we will take more research on the quality of service (QOS) to make the Netclass system more performance in real time.

6. Acknowledgements

A number of people contributed to the work described in this paper especially Mr. Tan Yongqiang, Mr. Cai Bingyue, Mr. Wang Yuxiang and Mr. Xu Shurei for the assistance in the implementation of Netclass system.

7. Reference

[1] Ellis C A etal. Groupware: Some issues and experiences. Comm. Of the ACM, 1991, 34(1):38-58

[2] Wang Kuisheng, Li Renhou, etal. Design and Implementation of Computer Supported Cooperative Design System. Proceddings of Fivth International Workshop on CSCW in Design. Nov. 29 ~ Dec. 1, 2000. Hong Kong. 75-80:

[3] Pei Yunzhang, Liu Yan, etal. Totally ordered reliable multicast for whiteboard application. Proceddings of Fourth International Workshop on CSCW in Design. Sep. 29 ~ Oct. 1, 1999. Compiegne, France. 391-395

[4] Hu XiaoFeng, Cao LiHua. Key technology of video transmission in CSCW. Proceddings of Fourth International Workshop on CSCW in Design. Sep. 29 ~ Oct. 1, 1999. Compiegne, France. 427-430.

MA-C: A Multimedia Instruction Model based on Multi-Agent and CSCW

Shenkang Wang , Qinming He, Chunyang Yu
Artificial Intelligence Institute, Zhejiang University, Hangzhou, PRC. 310027
wsk@mail.hz.zj.cn

Abstract

For the construction of a cooperative learning environment, the authors propose a multi-media instruction model based on multi-agent and CSCW. The structure of the model and some important technologies, such as the communication and control of agents, are discussed in this paper.

1. Introduction

The aim of traditional Computer-Assisted Instruction (CAI) system is to achieve or enhance individual instruction. However, from the view of pedagogy and learning theory, education is a social activity that is organization oriented, distributed, interactive and cooperative [1]. These features happen to have the same view with CSCW (Computer Supported Collaborative work) technique, appeared in the middle 1980th. Combining CSCW technique with CAI leads to cooperative learning, where instructor and students discuss and exchange their views of the topic in a cooperative way, and thus have a better understanding of the subject and gain advanced ability for cognition.

CSCW system must be based on technology that works in a way of group or team, multi-agent is the appropriate technology. An agent is a self-determined entity that runs independent in certain environment with its own persuasion, promise, obligation and intend. It is knowledge based, self-dependant, cooperative, and have the ability to learn. Agents are equal in a multi-agent based learning system, they cooperate with each other to achieve the aim, these learning characteristics have something in common with modern education. Furthermore, the intelligence of agent will play an important role to emphasize the individual's importance in a cooperative working environment. Since multi-agent system achieves CSCW successfully and provides with excellent solutions, the use of multi-agent has become a tendency in cooperative CAI [2][3].

2. MA-C model

Based on multi-agent and CSCW technology, we bring out a multi-media cooperative CAI model MA-C. Before coming to the MA-C model, let's investigate the use of Agent and CSCW in the model.

2.1 Agents

Multi-agent system consists of a group of agents that cooperate or compete with each other. When comes down to an agent system, the above all thing is to determine the main intelligence body based on the problem.

Role mechanism is indispensable in a cooperative system. Role is a group of rights and obligation attributes provided to entities in cooperative system. There are two obvious roles in the model: instructors and students. And two corresponding types of agents: instructor's agent for management on behalf of instructors, student's agent for assistant student learning.

To make the instructor's agent and student's agent work in line, each type of agent involves several specified functional sub-models. Student's agent consists of two application sub-models: one give support to student's learning activity and one to support self-testing of current status, mainly to show student's grade and weak points by interactive with the evaluation database.

Teacher's agent consists of four application sub-models. The first one is used to watch and inspect on student's learning progress and current status in order to choose the right navigation policy for students. It acts as the system resource monitor too. The second sub model interact with cooperate-rule-database and process the predefined cooperating request. The third sub model makes use of the collected information and evaluation database to evaluate the collective learning status, in order to improve the effect of teaching. The fourth sub model helps to communicate, discuss with students and give instructions.

2.2 Mode of Cooperative Work

The aim of CSCW system is to provide support for group work in computer environment. Generally speaking, there are four cooperative working modes according to the interactive method and distribution of entities. They are synchronous mode, distributed synchronous mode, asynchronous mode and distributed asynchronous mode.

As a cooperative instruction system, MA-C extends former CAI model by giving support to cooperative work.

MA-C system supports three kinds of cooperative work: individual tutorship, group instruction and free discussion.

Individual tutorship starts by the student's agent and send request to instructor's agent. Request waits for dispatch in instructor's request queue (the queue use a FIFO policy), once the request is dispatched, instructor and student start one-to-one communion.

When instructor's agent detects a number of similar request in the individual tutorship request queue, or when the monitor function detect a point in the knowledge base where mistake occurs for certain times, a group instruction is created and all individual tutorship request of this kind is canceled. Group instruction has a higher priority than individual tutorship, once group instruction is dispatched, instructor and students start a one to many communion.

Instructor's agent raises free discussion by sending request to cooperate queue, when he/she feel some topics worth discussing.

2.3 Architecture of MA-C

Integrating agent type and cooperate mode discussed above, we bring forward the model of MA-C system, depicted in Figure 1. The model describes architecture of cooperative learning environment for instructors and students in multi-media network. It advanced traditional CAI to a hyper-media based, distributed, cooperative educational environment.

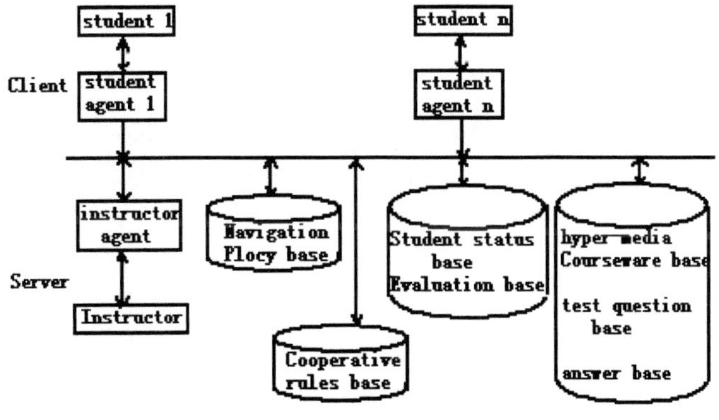

Figure 1. Model of MA-C system.

3. Key Points of MA-C System

3.1 Presentation of Agent

From the view of formalization, agent can be described in a 6-tuples, A_Depiction =<Aid, F, S, T, R, C>.

Aid: the descriptor of agent

F: agent's exterior characteristics

S: the state collection of agent

T: the transaction collection of agent

R: the resource collection of agent

C: the restriction defined on T and R.

Using object oriented technology, agent can be presented by classes that encapsulate special inside structure. Thus, agent in MA-C system is a 5-tuple element, A-Object=<Aid,A,M,C,I>.

Aid: the unique id of A-Object, correlated to Aid in A-Depiction.

A: the private data area of A-Object with certain data structure, act as the attributes collection, correlated to F, S and R in A-Depiction, e.g. agent object type.

M: the method collection of A-Object, defining the independent behavior characteristics needed to achieve its aim.

C: the operation collection of A-Object, defining the operations when cooperative request is send or received.

I: the agent's interactive operations with outside environment and other Object

The last three parts (M,C,I) makcup transaction management collection T in A-Depiction.

3.2 Communication of Agent in MA-C system

MA-C is an instruction system based on multi-agent and CSCW. In a cooperative system, communication is the base, cooperation is the form and coordination is the key. Accordingly, agent's framework presentation can be divided into three layers by the view of function: communication layer, custom layer and management layer.

Communication layer defines the network interface and communication interface and is in charge of the communication and interaction between agents, correspond to the first part of A-Object defined in 3.1.

Custom layer defines type, inside structure and task of agents, correspond to Aid, A and M in A-Object.

Management layer defines the cooperating protocol and starts specified sub-models of agents according to messages passed by the communication layer. It is implemented by the C element in A-Object.

In multi-agent system, cooperation is carried out by interaction between agents. Any kind of interaction is finally achieved by communication layer which makes communication layer the key part in a multi-agent system. Agents communicate in four ways:

- private-line, communication gains high-level of security in this way.
- blackboard, messages are placed in a broad available blackboard.
- Predefine storage location, agents store information to a predefined location to carry out the communication.
- a mixture of second and third way, agents are grouped and each group has its own blackboard.

Agents communicate in more than one ways in MA-C system. Usually go though private-line in individual tutorship and using a blackboard in group instruction and free discussion. Since KQML (Knowledge Query and Manipulate Language) has become the actual standard of syntax and semantics definition in agent's communication, we define a KQML compatible agent communication language in C++ Builder programming environment. The language has the form of a 4-tuples: (action, source agent, destination agent, contend).

Moreover the system is build on TCP/IP protocol and works in a Client/Server mode. TCP/IP provide us with a service oriented point-to-point service whereas we need a great amount of one to many interaction between instructor and students. To achieve multicast in Socket programming, the system is designed in a master-slave way to simulate multicast. Each instructor's agent and student's agent participated in the multicast have a custom state table. Instructor's agent manages the state of its member via a monitor, and thus provides support for broadcasting functions by maintaining the member state table dynamically. What a student's agent has to do is simply connecting to an instructor's agent and sending data, while instructor's agent transmits the package according to the member state table.

3.3 Agent Working Area in MA-C

The different role of instructors and students leads to different operation though their work areas have a uniform user interface. Both student's agent and instructor's agent has four sections in the work area: private work area, broadcasting bar, free discussion area and member state bar.

It is obvious that only the agent itself can use its private work area.

Since broadcasting bar is for group instruction and important notices it provides read-write rights to instructor's agent but read-only rights to student's agents.

Free discussion area needs a WYSIWIS (What You See Is What I See) user interface which is a basic requirement of CSCW system. All members taking part in the free discussion have the rights to operate and browse which brings forward the consistency problem of free discussion area. Adding a speaker control can solve this problem. There is at most one agent who has the right to speak, thus operate on free discussion area, at a time. The speaking priority dispatching use a queue mechanism, usually follows the first in first out rule but can be altered by instructor dynamically. Instructor's agent always has the highest priority. The dispatching use a preemption priority policy which means higher priority request will break the current job without delay. Cooperative function model started by instructor's agent takes charge of the speaking priority dispatching.

Member state bar can be operated only by instructor's agent, and provides browse rights to student's agent. Besides the difference in access authorization, student's agent can start member state bar only when he joins a free discussion while instructor's agent starts member state bar immediately after the agent starts itself. What's more, member state bar shows discussion member list and the current active speaker to student's agent, but list of all the student login and their learning progress to instructor's agent.

3.4 More Discussion

Taking advantage of Winsock programming, MA-C system provides a more convenient communion environment between instructors and students with e-mail tools. Winsock is a extended function lib based on Socket model and Windows event driven mechanism. Winsock supports two kinds of socket: stream socket and datagram socket. Since the model requires high reliability, we choose to use the stream socket.

Courseware in cooperative instruction must have powerful, well structured message model. The databases of the courseware in this system use hyper-media document based structure because it contains various elements (node, link) which can give expression to the internal relationship of knowledge and satisfy the demand for cooperative accessing.

4. Conclusion

The multi-agent and CSCW based multi-media instruction system supports the cooperative learning activity and communion between instructors and students, provides students with a vivid, socialized and personalized learning environment and greatly enhances the cooperation and communication quality in learning activity. It fulfills the education requirement of information society. Future works can be added to the following aspects: the navigation policy for hyper-media document, real time synchronous communication for multi-media data (including rapid response to feedback and feed-in, sequential synchronization and the synchronization for multi-media information's arrival to the agent). Agent's intelligence can be improved as well.

5. References

[1] Koch M., The Collaborative Multi-user Editor Project, Iris.TUM info XI, 1995-08.

[2] Lin Shouxun, Lin Zongkai and Guo Yuchai, MACE: a Multi-Agent Collaborative Work Environment, Chinese Journal of Computers, V21,N2,1998

[3] Qi Yinjian, Zhao Zhiyong and Qi Yinhui, Research on CSCW and System Model, The 1st National Confenrence on CSCW, Electronic Industry Press, 1996.

Part IX: Applications

Telematic Applications for Supporting Telework Related Activities

Joseph Barjis, Boris Shishkov
Department of Information Systems & Software Engineering,
Delft University of Technology, Delft, The Netherlands
J.Barjis@IS.TWI.TUDelft.NL, Shishkov@IS.TWI.TUDelft.NL

Abstract

This paper introduces further study towards telematic systems, especially in relation to their importance for telework. With respect to this issue, the paper considers the role of telematic applications in support of telework. In particular, it investigates telework support brokerage systems, which turn out to be crucial for the spread of telework among broad social circles. As a possible infrastructure for such systems, the paper considers telecenters. It is shown how such a brokerage system could be practically implemented, providing ideas about the system architecture. And finally, attention is paid also to some social issues, discussing the great importance of modern information and communication technologies (including telework as one of the best examples of successful mass application of these technologies with great benefits for society) for the human prosperity of tomorrow.

1. Introduction

Social prosperity depends dramatically on the rates of spreading Information and Communication Technologies (ICT) among people. In order to reach this purpose, it is crucial to create conditions for and motivate people in using ICT in their every day activities. For this reason, it is important to investigate telework (and some related issues), as a sphere of particular importance for the mass spread of these technologies.

This paper, as its title suggests, introduces further study towards telematic systems. In particular, it investigates the role of telematic applications in support of telework.

Although this direction of study is comparatively new, however it takes its origin in the last century, more precisely in the first half of the nineties. One of the first investigations of this type of work was introduced in [1]. In addition to basic concepts of telematics reported in his book, it also provides one of the first, but fundamental definitions of telematics. This definition says, "the new type of technologies (integrating computer, information and telecommunication technologies), as well as programming applications, supported by such technologies, is telematics". The definition was further developed in the context of telematic application [9]. Later on, telework turned out to be a high-priority sphere for the development of Information society. And even now, there is no doubt that by stimulating people to use ICT in their every-day activities, telework could definitely become an important milestone for the social progress. Information society could become reality only if every member of our human society has access to mutual information space, knowledge, data etc., using new and modern ICT. This has inspired the authors to work in this sphere. First results of our study were reported and discussed in [7]. Feedback and suggestions given from these discussions convinced us to continue the work further.

The main contribution to the domain and society of what is introduced in this paper could be briefly outlined in several issues:

- The paper considers one very important aspect of telework – its being effectively supported by telematic applications.
- An extended view over the domain is realized, discussing not only the mere application related telework matters, but also providing ideas about a suggested infrastructure, social issues, etc.
- An issue of great social significance, like telework, is related to modern application design principles, which could only be useful for the development in this sphere.

The following section of the paper overviews the telework concepts and considers the potentials of telematic applications in support of telework, outlining one special aspect – the brokerage functions, which could be performed by applications. Telecenters are suggested as a suitable communicational infrastructure for telework support systems. In section 3, an example is studied, based on a concrete application, investigated in [7] in order to demonstrate how a practical implementation could be carried out, e.g. in a system of telecenters. Section 4 considers some social aspects of the mass spread of telework.

2. Brief overview of telework and TSS

Before turning to telework support brokerage systems, investigated in this paper, it is essential to provide some clarification in relation to the concepts of telework as well as to discuss in general the possibilities, given by modern telematic applications in support of telework. This section treats also the problems, related to the necessity of an adequate communication infrastructure, on which the telework related applications should operate properly. Telecenters are suggested as a possible solution in this case.

2.1. Telework

The following is our definition of telework adapted from [7]:

Telework is a kind of work, in which all parties (worker-supervisor, customer-supplier, etc.) use the possibilities of the new ICT including modern software environments to exchange data (e.g. work results) over some distance, and thus – being relatively not dependable on time and space respects.

ICT create new opportunities, which could dramatically change our ways of living, and thus also working. The possibility to establish a full-value distant communication and to exchange all kinds of information determines the new reality in which one could work for his company from home, meeting his colleagues and granting the work results to his boss virtually. The collaboration technologies, which are based entirely on the new ICT even create some additional "centripetal forces", as defined in [3], for successful distant collaboration - e.g. telework, global teams, etc.

There is no doubt that modern society should benefit to a maximum degree from telework, because this could not only improve the people's standard of living but also save resources, travel time, office equipment expenses, and increase mobility and flexibility of users and business itself. From this point of view, telework would be crucial for the social prosperity of tomorrow.

Telework application area already covers vitally important spheres such as remote education, remote services (e.g. banking) etc. [2]. As for the use of telematic systems and telework in education, in the related literature it is referred as telelearning [5].

For more information about telework, interested readers are referred to [4].

2.2. Modern telematic applications in support of telework

Business systems could be more effective and efficient if exploiting fully the possibilities of the new ICT, relying on modern software applications and hardware platforms. Telematic applications defined in [9] as the type of applications that support distributed users using communication infrastructures and information technology, usually across the boundaries of organizations, are developed not only to facilitate the communication between distant business processes [8] but also to create a sound information system background supporting these processes.

The fact that some of the current business systems were designed before the modern technology came in, without the particular thought in mind that they should possess an IT background, created the necessity of re-engineering some of these systems, re-designing their business processes in order to make them easily supportable by telematic applications. As long as telework appeared as a result of the ICT spread, it doesn't possess any of these problems. Something more, telework is inherently open to any kinds of application support, because of the mere fact that it exists via modern software applications.

So, there is no doubt that telematic applications could support telework in many different ways, some of which are outlined below.

1) Telematic applications could perform brokerage functions: administrating data concerning people looking for telework positions, companies, offering telework positions, positions themselves; realizing matchmaking functionality, etc. Such applications are investigated in this paper.
2) Telematic applications could facilitate the mere telework process, creating an environment, suitable for a full-value employer-employee communication, data exchange, etc.
3) Some specific telematic applications could support the tele-workers in the process of executing their concrete work activities.

2.3. TSS

As stated already, the paper investigates in particular the telework support brokerage systems, which in this paper are called Telework Support Systems (TSS). In order to perform effective matchmaking functionality and operate over many locations, it is essential that TSS function within reliable distributed environments. This could be possible only if ICT are fully exploited in:

- establishing effective communication between distant locations (regardless of the regional telecommunication infrastructures);
- exchanging easily all possible types of data.

These are among the factors (according to the authors) for successful TSS development, some of which are outlined in Figure 1.

Analogous brokerage systems are applied also in other spheres of high priority for the development of the Information society, e.g. e-commerce. The importance of these systems for the business processes they serve results from their functioning as communicational centers for all the processes, in maintaining full-value "dialogues" among the parties involved. This particular role of the brokerage systems invokes the basic problem in their design and development: how to cover locations of different level of telecommunication infrastructural development (including areas, not even possessing such infrastructure) and provide reliable and easily maintainable distant communication and exchange of all types of data.

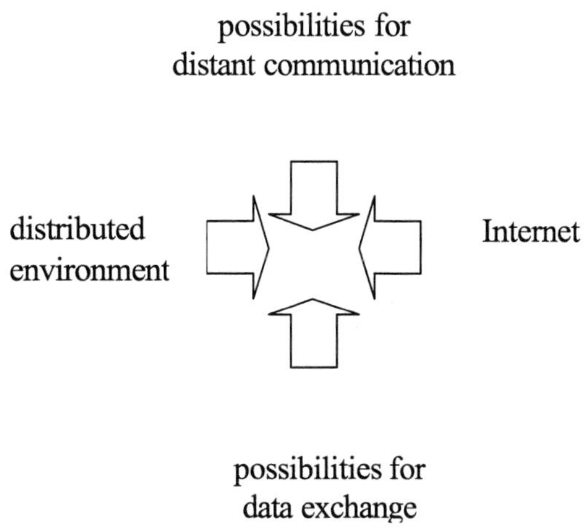

Figure 1. Factors for successful TSS development.

The solution comes from the latest ICT development. The alternative ways of communication, e.g. wireless communication, the Internet – providing powerful collaboration technologies, allowing many opportunities for data exchange, and the great social involvement and concern (producing already promising results, e.g. the mass appearance of telecenters as a way to overcome the lack of adequate telecommunicational infrastructure in some locations) – all this makes it possible to design powerful TSS, which could operate effectively, contributing to the development of telework.

2.4. Telecenters as a possible infrastructure for TSS

According to [7], **telecenters** are public facilities, situated in one building, and equipped with computer, telecommunication and office techniques, aiming at stimulating the development of Information society by spreading the new ICT. The exchange of any kind of information and the realization of full-value distant communication are crucial for the existence of telecenters.

Telecenters turn out to be important for spreading ICT among the human actors and in this particular role of theirs they should reach all the populated areas of a country, even those not supplied with any communication infrastructure, e.g. telephone connectivity, Internet, etc., using in this case some alternative ways of communication. So, in order to stimulate the development of Information society, it is essential to stimulate the spread of both telecenters and telework among maximum broad social circles. This is the first reason for suggesting telecenters as a communicational infrastructure for TSS (they should exist "everywhere" and it would be possible to be used by "everybody"). The second reason is that, equipped with modern telecommunications, hardware and software, telecenters would be a suitable environment for running modern telematic applications, including telework related ones. Adding to these facts some social aspects like the possibilities of telecenters, related to IT education of people, spreading the ideas of telework, and even using telecenters like telework outposts, there is no doubt for the authors that telecenters are at least a suitable infrastructure for telework related applications.

3. Example of a concrete TSS

A concrete TSS was introduced in [6] and later on further investigated by the authors in [7]. This TSS, operating in a system of telecenters is considered in this section to demonstrate how a TSS could be practically implemented.

The system of telecenters considered should consist of a main telecenter (the central bundle of the system) and a multitude of local ones, building up the distributed telecenter system.

The TSS could be based on the usage of the following systems:

- Microsoft IIS – processing the basic client queries;
- Oracle Application Server 4.07 for Windows NT – processing the system functions;
- Oracle 8.0.5. Database Server – storing the whole system data, as well as the data of the functions, realized with PL/SQL packets and procedures.

Two notions have been taken into account:

- a purely centralized database could not be sufficiently flexible to respond to the requirements of the system users;
- a completely distributed design (without any central database) would complicate the architecture and introduce additional overhead in terms of maintenance costs.

As a consequence, a hybrid approach consisting of a central database and several local slave servers was chosen. A typical schema of distributed system architecture is shown on Figure 2.

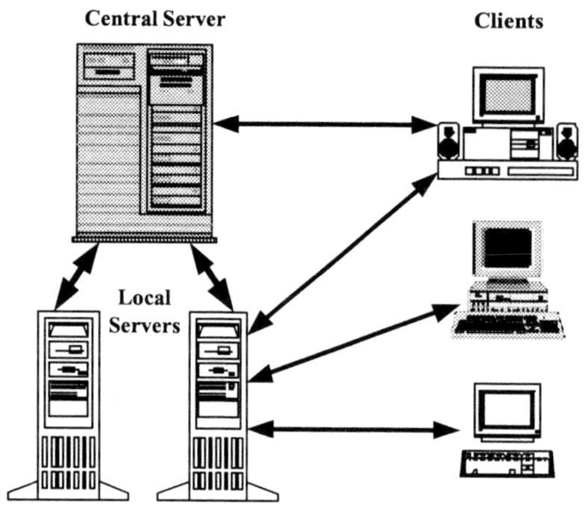

Figure 2. Distributed System Architecture.

Hosted on the **central server** will be a central telework web site and a central database. The server manages all data, used by the TSS; all data related to the proper functioning of the system must be stored on it. Insertion, modification, and deletion of data may occur at the central server only (locally or distantly).

The central server is to be situated in the main telecenter of the system, because of its fundamental role for the entire system of telecenters; the coordination activities concerning the entire system of telecenters are performed by this main telecenter.

The main entrance (gateway) to the TSS would be through the central web site. This site would contain links to the web sites of the local telecenters.

Each of the **local servers** would possess a replica of the central server's core database that is updated in regular intervals. This replica is read-only and is modified only by a well-defined updating process from the central server. Replication should occur as often as necessary but not too often with respect to the costs according to network bandwidth consumption. A nightly update seems to be feasible as a starting point.

All operations, which involve some storage of data on the database, will use the central database. For example, if a subscriber submits a demand request, using a local server, the record would be saved in the central database, not the local one.

This approach was chosen, because of the level of flexibility it allows. The local telecenters have enough "space" for offering local services, while at the same time – no need appears for a complicated architecture.

4. Social implications of modern telecommunications

The social aspects of the problems discussed are of great importance not only because society could benefit a lot from the wide spread of new information and communication services, but also because the social demand for such services is an additional factor, motivating the fast ICT development.

The positive social effect of applying advanced TSS (and of telework in general) could become reality only after providing people with:

- access to modern IT equipment;
- adequate knowledge about the possibilities of ICT and their usage.

With respect to this issue, distance education should be considered as an extremely important sphere, crucial for the development not only of telework but also of ICT. Telecenters are among the public facilities, which have proven their usefulness in providing people with both access to equipment and opportunities for acquiring IT knowledge.

There is no doubt that the social benefits from involving people in telework could be enormous. People would have much more convenience and free time, having less need to travel and opportunity to more effectively distribute their duties. Companies would concentrate their expenses on core business activities, instead of spending a lot on office equipment and organizational matters. Public authorities would benefit from an increased business effectiveness, decreased travel intensity, etc. And finally all this could save public resources, which would result in positive effects for the environment. This vision is outlined in Figure 3.

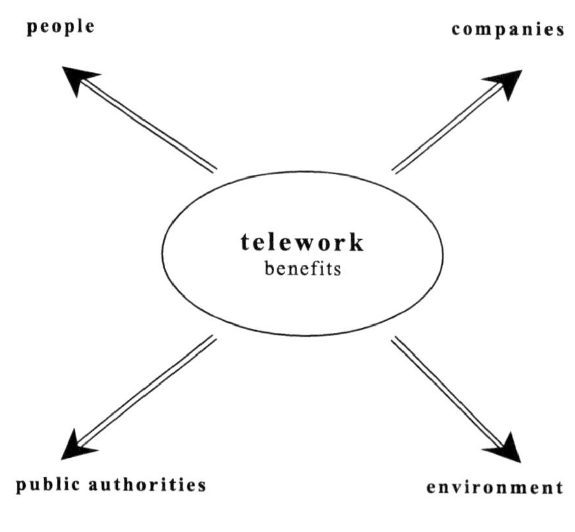

Figure 3. Social benefits, related to telework.

So, stimulating the development of telework is motivated by significant social factors. However, solutions pose new problems while solving others: the availability of powerful telework related applications turns out to be crucial for the success of telework. This encouraged the authors to study TSS in order to contribute to one aspect of the development of telework.

5. Conclusion

The purpose of the suggested paper, as it was stated in the introduction, is to investigate the role of telematic applications in support of telework. In particular, the paper considers some problems related to TSS, their practical implementation and social importance. A concrete example of such a system is studied in order to demonstrate how, according to the authors' ideas a TSS could be realized e.g. in a network of telecenters.

It was shown in this paper that TSS are not only an example of a successful implementation of the new ICT in practice, but also an important prerequisite for the development of telework. The new possibilities, offered by ICT, make it possible such systems to be realized. Something more, telework itself has appeared as a result of the dramatic development in the information technologies and telecommunications. So, the paper could contribute to the development of telework, investigating one particular implementation sphere of ICT.

As a future research, authors will study application of telematics and telework by example of a virtual university, where admission, enrolment, education and other interactions are carried out via telematic systems, using the Internet.

6. References

[1] Andriessen, J.H.E., Roe, R.A., 1994. Telematics and Work.

[2] Bonyuet, D.; Bagayas, L.H. The Internet in the New Millenium. In the proceedings of World Multiconference on Systemics, Cybernetics and Informatics (SCI 2000), July 23-26, 2000, Orlando, Florida, USA.

[3] Carmel, E., 1999. Global Software Teams.

[4] European Commission, 1999. Status Report on European Union Telework.

[5] Fezzani, D.; Desbiens, J.; Abdelhak, R.; Ghazel, A. Internet-based Solution for Telelearning: the PRICAT Experience. In the proceedings of World Multiconference on Systemics, Cybernetics and Informatics (SCI 2000), July 23-26, 2000, Orlando, Florida, USA.

[6] Shishkov, B., 1999. Design and Development of a System of Telecenters, supporting Telework and stimulating the development of Information Society, M.Sc. Thesis, Sofia University, Bulgaria.

[7] Shishkov, B.; Barjis, J. Analysis of a System of Telecenters as a Distributed Network. In the proceedings of the 14th International Conference on Systems for Automation of Engineering and Research (SAER'2000), September 2000, Varna, Bulgaria, pp. 108-112. ISBN: 954-8329-22-0.

[8] Stamper, D.A., 1994. Business Data Communications.

[9] Verbraeck, A., Saanen Y.A., 1998. Building blocks for Effective Telematics Application Development and Evaluation, Internal report, available at TU Delft.

Analysis and Modeling of Cooperative Processes in Dynamic and Complex Environments

Francine Merat-Bodin
Délégation Générale pour l'Armement (PARIS)
Armored Cavalry Major
Mission Chief in Ergonomics and Human Factors
f.meratt@infonie.fr

Tahar Hakim Benchekroun
Conservatoire National des Arts et Métiers (PARIS)
Researcher in Ergonomics
benchekroun@cnam.fr

Abstract

The theme of this research is assistance to team activities in a command situation. The framework is the introduction of a computer system into a team working in distributed mode (regimental command post, abbreviated CP). The system has not been subjected to a specific analysis as concerns horizontal cooperation within the CP. Data is transmitted to the CP in multiple modes and from multiple sources and is characterized by incompleteness, uncertainty, and requires both individual and group inferences to build up the tactical situation. We analyzed video tape recordings of simulation exercises and an operational control exercise in the field and attempted to model the proxemic cooperative process in this dynamic environment. The analysis proceeded in two phases: quantitative and qualitative. The qualitative analysis crystallized the idea of a proxemic multi-addressing zone of gradual involvement in a dynamic, complex environment which, in conceptual terms, prompted a number of different lines of thought.

1. Introduction

The theme of our work is "assistance to team activities in a command situation." It is of practical interest both from the viewpoint of understanding the collective decision-making process and from the viewpoint of assisting this process. The framework of investigation is the introduction of a computer system to assist decision-making in a group working in distributed mode in a dynamic environment. The platform used for this research is a regiment-level command post (abbreviated "CP"). The prescribed task was first analyzed by computer specialists in charge of developing the information system, on the basis of analysis of official documents (utilization regulations), and led to a representation of CP activity as a group activity of co-action among operators. Our purpose is to show that this collective activity cannot be represented simply as a set of juxtaposed individual activities, but has to be analyzed as a team activity of distributed cooperation. To this end, our approach was oriented toward the analysis and modeling of the proxemic cooperative processes involved in the collective construction of the tactical situation, and more especially of the enemy situation. We will attempt to show how the CP officers, who have to deal with partial, incomplete, or even uncertain data coming from sources that differ both in mode and hierarchic level, construct not only their individual inferences but also participate in the development of group inference. The expression "proxemic multi-addressing zone of gradual involvement" (PMAZGI) is intended to mean the process that allows the construction of this group inference. This group inference, which will be accepted and validated by all the CP member officers, will be presented as his own by the intelligence officer in charge of presenting the tactical situation to the CP commander, who will generate is maneuver decision on the basis of the tactical situation.

2. General context

2.1. Research Platform

The research platform for our study is a command post (CP) of a French army regiment or group of the contact type (armored cavalry or infantry). The CP is an entity consisting of officers, subalterns, and recruits who are either active professionals or reserve personnel, *i.e.*, having some other civilian activity and participating in CP activities only during their reserve duty periods ranging in length from a few days to a maximum hundred days per year. This reinforcement of staff is needed to maintain the round-the-clock rhythm of the maneuver, lasting several days. This particular structure results in different levels of expertise, not only in terms of active/reserve status but also rank and experience acquired through qualifications (specialization diplomas) and/or active foreign duty (Gulf war, ex-Yugoslavia).

This working team is part of a decisional command

chain with strong hierarchy and holonistic, pyramidal structure.

The CP consists of two technical cells and several operational cells, each directed by an officer who may be assisted by another officer, subaltern, and even one or two recruits when these are not occupied by CP logistics tasks.

There are five operational cells and even more, depending on the regiment's commitment context, and they work in distributed mode:

- ***MAN cell*** *(maneuver) is a special cell because it is directed by the CP commander, regiment corps chief, and/or his assistant. This cell is responsible for coordination and implementing the orders from the higher echelon.*
- ***INT cell*** *(intelligence) is in charge of tracking and interpreting information concerning the enemy for presentation to the CP commander for decision-making purposes, and writing the enemy situation paragraph of the regimental orders.*
- ***FIRE cell*** *(operations) has to keep the enemy situation up to date in order to be able to present coherent situations and conduct the maneuver in progress.*
- ***LOG cell*** *(logistics) ensures both internal and external logistic links (human and equipment support).*
- ***LIAI cell*** *(liaison detachment) Engineering, Artillery, or other corps that may be present depending on the commitment context, provides its competence to the regiment for the mission in progress.*

These cells have two radio networks (to higher and subordinate echelons) except for the INT cell, which has only one radio network with the brigade. In the cell, there is at least one 1:50,000 scale cell map and organization charts and manuals, depending on the officer.

2.2. Technological Context

It is planned to equip all CPs with a global information system in the coming years. This system is part of the principle of the digitized battlefield, which is already an old concept, the first premises of which can be found in the United States in the early eighties. Intelligence is considered to be the "fifth dimension" of the battlefield, after land, air, sea, and space. For military strategists, it is becoming a challenge that has to be met. Information technologies will be used to acquire, transfer, and analyze the pertinent numerical data so that each commander, each combatant, and each support unit can have a clear and precise view of his maneuver space.

The information system that will support the regimental level CP will be the SIR (Matra SI) and will be integrated in the C3I concept of command, consisting of three echelons (Figure 1)

This system should make it possible to get rid of hard copy maps and transmit information and orders from the different echelons by data transmission. It will be installed on command vehicles, which can thus function at different sites.

2.3. Research Objective

This research falls within the framework of the weapon program design scheme with preponderant software (PALP), which is broken down into phases in which one phase, once stabilized, is succeeded by another to improve upon it. More particularly, the work integrates into the framework of "upstream research" in the DGA/SH for writing recommendations as concerns "the consideration of the cooperative aspect of information and decision assistance systems".

3. Methodology

3.1. Data Collection Methodology

Data is collected in three main phases:

1) Understanding the research platform by regulatory and theoretical documents of the utilization rule type, and open observation of part of a simulation exercise at the CEPC (CP training center at MAILLY (10) France).

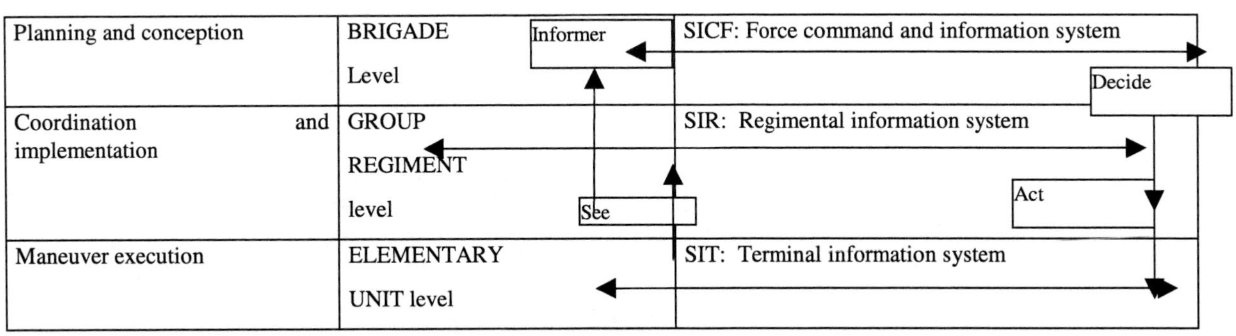

Figure 1.

2) Audiovisual recording of two simulation exercises at the CEPC and an operational control in the free field, in light of different combat phases.
3) Individual and limited-size group meetings during the exercises (during periods of low activity) or after the data collection phases.

3.2. Data Processing Methodology

To bring out the cooperative activity within the CP, we chose to process the collected data quantitatively and then qualitatively, by analyzing the exchanges among officers during the different phases of combat. It seemed pertinent to concentrate on the viewpoint of enemy situation construction in order to see the complexity of the interactions more clearly. To do so, we literally and fully re-transcribed the combat periods present in each exercise.

We break down these periods according to enemy status:

- ***Low intensity:*** *The regiment and all its friends are not yet in contact with the enemy, aside from a few reconnaissance elements.*
- ***Medium intensity:*** *The regiment is in contact with the enemy, but is not bearing the central line of effort.*
- ***High and very high intensity:*** *The regiment is in contact with the enemy and is also subjected to the main line of attack.*

4. Results

4.1. Quantitative Analysis: Effect of Combat Intensity

We observe a sharp variation in the number, content, and duration of exchanges with the variation in combat intensity, but also a shift of the exchanges among the cells (see Table 1).

The number of exchanges increases with combat intensity up to the point where the field units are in contact with the enemy, at which time the exchange intensity becomes inversely proportional to that of the combat. The exchange duration decreases in the same ratio, until such time as the exchanges reduce to a single verbal round. The increase in combat intensity prompts a switch from the generation of the shared representation of the tactical situation to the maintenance of it [3]. During the phases of very high intensity, the officers and especially the FIRE cell commander are monopolized with listening to their radio network and are no longer available for the group. The volume level of the various networks then increases (from 3 to 7 on a scale of 10) so that intelligence transmitted is audible to everyone. Moreover, in a strategy to simplify the problems encountered with the dynamism and volume of information to be processed in those phases of highest combat activity, the officers consider only the information concerning their specialty chain. So they are aware of only a limited number of parameters concerning the tactical situation, and especially very little about the second-echelon situation. It is then that the CP commander generates a directed review that allows a better shared generation of the tactical situation. However, these situation reviews are only narrow aspects of the group activity. We generally find them every six hours at the time of shift changes, or even sporadically, or one every three hours. We can thus affirm that the shared construction of the tactical situation representation works in another mode. We therefore adopted a qualitative approach to show this.

4.2. Qualitative Analysis: Importance of Multiple Addressing

By multiple addressing, or multiple-addressed communication, we mean that once a message is transmitted by one speaker it will undergo multiple and various processing by more than one person (information distribution, richness of the processing and group involvement, time gain). Many authors feel that this type of communication is one of the most powerful of those involved in information distribution and sharing, and the least costly on the cognitive level [8], [1].

Table 1. Data extract from CEPC simulation 2

	INT FIRE	INT ART	INT LOG	INT CHIEF	FIRE ART	FIRE LOG	FIRE CHIEF	ART CHIEF	LOG CHIEF
LOW INTENSITY	5	5	2	6	3	5	10	3	4
HIGH INTENSITY	20	6	2	15	3	4	12	4	2
VERY HIGH INTENSITY	2	6	0	4	3	0	7	3	0

The officers have to construct a pertinent diagnostic on the basis of available information or data, and do so in order to have indicators of use in comparing the current situation with the previously estimated situation. The information reaching the CP comes from two distinct hierarchic levels: first the subordinates, the field, and the field elements in contact with the enemy, and then the superiors, the brigade, which centralizes the information coming from all of the regiments and allied units. Consequently, we will consider that this information corresponds to two types of information, distinguished in terms of their completeness and certainty:

- **Type 1 information** is certain, but with a degree of imprecision. It stems from real information, not instrument-generated by sensors or inferred by information grouping. Operationals call this information.

- **Type 2 information** comes from the higher echelon or allied units of the same level. This is precise information, but is marred with uncertainty because it is inferred or estimated on the basis of knowledge coming from structural organizations constructed on the basis of intelligence generated before the conflict began.

Because of the distributed character of the CP, however, not every officer is in possession of all the available data. External multiple-addressing (radio) provides access to this data, but the sufficiency of this depends on the warfare intensity and officer availability. Internal multiple-addressing allows access to this data, but also makes it possible to bring out a pertinent piece of information that had not been put to use in the sum of information reaching the CP, by multiple modes and from multiple sources. The pertinent information is generally put in relation with the rule to which it can be attached, in

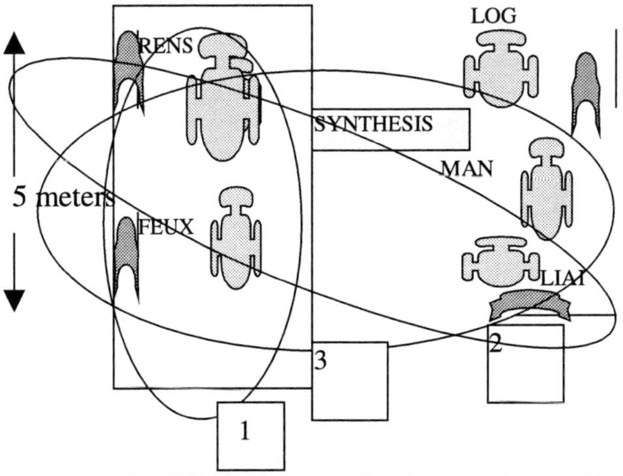

Figure 2. The PMAZGI dynamic also evolves with operator availability and their acquantanceship

order to strengthen its pertinence but also to share knowledge and know-how, the purpose being to construct a collective inference to achieve a shared representation of the tactical situation. This process is necessary in order to compensate the different levels of expertise of the various operators (reserves and actives) by validating an individual inference, but also in order to compensate the limitations of information dissemination by the different radio networks.

The proxemic multi-addressing zone of gradual involvement is the cooperative space in which the operators modify the context in which their communications are embedded [6]. A PMAZGI, however, cannot be reduced to the idea of a context as communications facilitator. It is the locus of officers' activities and interactions with the shared resources in this space. Moreover, a PMAZGI is not a fixed space but has its own dynamic and its organization evolves in accordance with informational requirements, each operator's inputs, both on the level of knowledge and that of know-how. The PMAZGI dynamic also evolves with operator availability and their acquantanceship in general. Proxemic multi-addressing zones of gradual involvement are thus created at the outset by an exchange between two operators, generally the FIRE and INT officers because of the radio network distribution, and then extend with the context to the other CP officers. This principle is in confirmation of the Goffman's works [4][5] concerning people who are located in a common perceptive space. In our work, we have only addressed and unaddressed listeners and perhaps chance listeners, but never any indiscreet ears. However, contrary to what Goffman affirms, the operators in a multi-addressing zone of gradual involvement do not need a "sustained look" to change status (unaddressed/addressed) as the fact of knowing how to interact [1] in a common area, with the possibility of hearing both the external radio network and voice speech is enough. The arrival of other officers in this proxemic multi-addressing zone may be due to a knowledge review concerning the situation, or to a reinforcement of context information from another source, or even to another mode than the one that originated the initial exchange. This proxemic cooperation is based on invariants that render one's intended commitment and involvement manifest to oneself and others at the time it is needed and how it is needed.

Generally, information from the first echelon arrives in the multi-addressing zone in the order FIRE-INT, then ART with INT the FIRE, possibly LOG with FIRE, and lastly it arrives at the CP commander. The variations encountered in this order of arrival stem from the initial source of information. A permutation between ART-INT and FIRE-INT generally concerns information concerning the second echelon. This is a real dynamic and not a successive arrival. In a PMAZGI, the question is to analyze the verbal initiative and circulation, but especially the information processing in order to coordinate actions and anticipate with respect to the enemy's strategy. Whatever the level of warfare intensity, we found at all times that, in the multi-addressing zone, exchanges aimed at reducing uncertainty and incompleteness occurred in three phases:

1) Information sharing and exchange
2) Knowledge and know-how sharing
3) Individual inferences and justification toward a collective inference

These three phases are present even in a situation of very high activity. The FIRE officer in charge of all of the first-echelon information, though submerged by the intensity and amount of information coming from the field, makes sure that these three phases remain clearly present. That is, he will naturally increase the volume of his loudspeaker, draw his colleagues' attention by a gesture, but will especially repeat all of the information coming through the network, then verbalize his reasoning aloud, while marking the transmitted or inferred information on his map. During this time, his colleagues (especially INT and ART) will approach the FIRE map, move back and forth between this and their own map, in order to keep listening to their radio network, but especially in order to repudiate or confirm the diagnostics and pre-alert the various echelons on their own networks, particularly the brigade.

We were able to bring out the idea of acquaintanceship (knowledge of others' state of availability) in the creation of the multi-addressing zone of gradual involvement. That is, the cooperative diagnostic catalyst effect can occur only if the various officers consider the acquaintanceship of the other CP members. This idea refers to that of cooperation developed by Schmidt [9], which is an augmentive cooperation at the level of information sharing, but also integrative cooperation as concerns the integration of each officer's competence in the various cells.

We found that spatialization was an important part of multiple-addressing quality. Spatialization makes it possible to:

- Listen to various radio networks by external multiple-addressing
- Listen to verbal accounts, by internal multiple-addressing
- Access what we call "open objects": cell map, synthetic map.

The CPs of the various regiments are not subject to the same spatial organization, especially when, in exercises, they have to submit to a structure that does not allow them to use the space as they need it. We twice witnessed the consequences of low-performance spatialization for the collective inference. The multi-addressing zone of gradual involvement can then no longer be constructed, and we found a loss of information-sharing, knowledge confirmation, and even of validation of individual inferences by the team.

All of these remarks were confirmed to us in individual and group interviews which were conducted after the combat phases.

5. Discussion and Research Perspectives

This first phase of our research showed that, although the regimental CP team is highly distributed in its dynamic environment, the constraint of organizational markers [1], *i.e.*, the elements that define and describe the spatial distribution of operators and stations, is not strong enough for the group activity to be considered a co-action activity. That is, the operators move in what we will call a proxemic cooperation space in which the dynamic warfare environment generates incomplete, uncertain data coming from multiple-mode sources. Proxemic multi-addressing zone of gradual involvement are thus created in this group, where the officers share and compare information, knowledge and know-how to arrive at a collective inference concerning the tactical situation diagnostic. This diagnostic construction dynamic should thus be taken into consideration in building the SIR information system.

Our study nonetheless shows a number of limits that should be pushed back. These concern essentially the role of maps in the inference construction. That is, we have to look into how a coherent representation is constructed and how, on the basis of these "open object" representations (support for inferences and working material for the personnel that consult them), the officers are able to detect the enemy's deception strategies and anticipate the enemy's action possibilities.

We thus show the importance of "team situation awareness", as Endsley [2] shows, but also the dynamic aspect of this construction, in the conception of construction awareness. It is important to consider the ideas of information uncertainty and incompleteness in the knowledge generation and sharing dynamics, while also considering the acquaintanceship of the other operators.

In this first portion of research, we have dealt only with the "information management strategy". At the level of "meta-strategy", it would seem necessary to determine at what point each cell officer transmits a diagnostic to his chain of specialty. In the framework of introducing the information system, this diagnostic will no longer be disseminated to the other regimental CPs, or even to the other subordinates. Will this lost "party line" have no consequences at the level of team inference?

6. References

[1] T. H. Benchekroun (2000) Les espaces de coopération proxémiques; in Le travail collectif: perspectives actuelles en ergonomie T.H Benchekroun and A. Weil-Fassina (Eds) Octares Editions

[2] M.Endsley (1995) Measurement of situation awareness in dynamic systems, Human Factors, 37 (1)

[3] J.M. Hoc, A. Loiselet,. and R. Amalberti(2000) From the management of interference to the elaboration of common frame of reference 9th Human Workshop, IMASSA, Brétigny sur orge 16 May 2000

[4] E. Goffman (1973) La mise en scène de la vie quotidienne. Les relations en public, Editions de Minuit, Paris

[5] E. Goffman. (1983) Façon de parler, Editions de Minuit, Paris

[6] J. Leplat (2000) La gestion des communications par le contexte, in Actes du XXXV° congrès de la SELF, "Communication et travail, Toulouse 20,21 & 22 September 2000 Octares Editions

[7] F. Merat-Bodin (2000) Analyse et modélisation de processus coopératifs proxémiques en environnement dynamique et complexe : la construction de la situation tactique au sein d'un poste de commandement de niveau régimentaire. Mémoire de diplôme études approfondies en ergonomie EPHE CNAM, Université de Toulouse 2 Le Mirail

[8] B. Pavard (1994) Systèmes coopératifs : de la modélisation à la conception Toulouse Octares

[9] K. Schmidt (1991) Cooperative work: a conceptual frame work. In J Rasmussen, B Brehmer and J. Leplat, Distributed decision making, cognitive models for cooperative work Wiley & Sons.

Paperless in the Electronic Era: Millennium Dream and Reality

Azmi Mohd Yusoff, Manjit Singh Sidhu
Computer Science and IT Department, Universiti Tenaga National
KM-7, Jalan Kajang-Puchong, 43009, Kajang,Selangor Darul Ehsan, Malaysia
azmi@uniten.edu.my; manjit@uniten.edu.my

Abstract

In this third millennium, the world is undergoing a revolution where technologies are being introduced almost daily. And until now, paperless has been a long dream of every organization and individual, which never became a reality. This paper will unveil all available technologies covering the communication, storage device, and retrieval methodologies toward paperless activities. We will see in the perspective of the provider and end user. By understanding the point view of both parties, it will help when deciding or concluding the real future of the approached technologies. We also cannot run away with legal issue, so we will look into any factors which affect or demoralized this idea of paperless and green world.

1. Introduction

The concept of the "paperless office" has been discussed for quite some time. The paperless concept is similar to predicting the end of the world, it is always being predicted but never happens. To date, it is still not 100% achievable. New tools are constantly being created to help achieve the paperless goal, but there are still many obstacles and considerations [15].

The concept of the paperless office was mooted in the 1970s and received with a great deal of optimism. The rapid rate of progress in the IT sector led many people at that time to believe that the paperless office was the inevitable way of the future. It sounded impressive - a system designed to facilitate filing, faxing, word processing, document retrieval, analysis and even basic communication, all done efficiently using computers without the need to use paper. Today, however, almost every office is still filled with piles of paper and rows of filing cabinets. What happened to being paperless?

The initial development of the paperless office was slow as only a handful of specialists at that time had the remotest idea of information storage and document retrieval [1]. The biggest idea at that time was the microfiche and portable microfiche viewer [2]. Then came the computer, and word processing done using the computer. Merging electronic data processing with word processing introduced electronic data storage, which became the backbone of the paperless office. Scanners and OCR software made it possible to convert existing paper documents to electronic documents. Computers then took a giant leap forward with their ability to connect everyone together with just a phone line and a modem. Electronically sending out documents using emails and fax-modems was seen as a way of eliminating the need to use postal service to send out documents [1].

Sad and surprisingly, the statistic shows that instead of a decrease in paper use in the office, there has actually been an *increase*. The American Forest and Paper Association estimates that use of printer and copy paper has increased by 30% over the past decade [3]. In most firms, hardcopy records are growing at about 7% annually [4]. The proliferation of paper has increased dramatically with the introduction of computers. Computers allow us to correct errors and make numerous revisions to documents very easily. The paperless office is supposed to be facilitated by more use of computers, but ironically every computer is hooked up to at least one printer, making it easier to generate more paper documents. In our quest for perfection, a document is printed out again simply because of a single typo, or because the formatting doesn't look so great. Documents are still printed out for reading even though they are available in electronic format. Currently, the average working person in Australia uses 30kg of paper every year [9] - hardly paperless!

1.1 What types of offices implementing paperless operation?

For some organization, paperless is no longer choice or something which is nice to have. Instead, paperless is a must issue and unavoidable. The companies with the following criteria can no longer survive with an old method of printing, faxing and photostats when transmitting the information. Those criteria are:

a. Companies which produces a large amount of document. It is stated that giant corporation like

Aetna has eliminated 87 million pages of paper.

b. Companies which required frequent access of data. This is applied when data retrieving or searching due to customer number of inquiries is too large.

1.2 Advantages or aims of paperless operation

With all the hard-work and sweat in order to achieve this goal, everyone is having common goals to achieve. Some of the goals are giving tangible benefits and some are giving intangible benefits. The three goals are:

a. Space saving.

United Services Automobile Association claimed that they had saved 35.000 square feet of office space that used to be taken up by paper files. Directly, hard money which required for renovation or warehousing are also saved.

b. Easy searching and retrieving.

Technology that enable electronic storing and retrieving are now available in many different forms.

c. Labor saving.

When end users can survive on their own for data retrieving, man power involved can be reduced. The same cased true ex. warehouse maintenance that no longer need so many people or man power.

1.3 Other issues to consider

Besides all the beauty and positive points that we like to hear, there is also a dark side of paperless in office automation. So we look at some negative points that eventually stop or slow this paperless idea to progress in the rapid phase. Two main issues that always hunting are security and legal issues. For some small transaction's company, another issue is costing. This third issue can be easily overcome once they seen the long effect or future direction of the companies. For this reason, cost and benefit playing a major role in convincing the necessity. It is interesting to note that there is a cases where this paperless is being implemented not because they want to do it, but due to enforcement by their customers. Basically the customer can choose to do business only with company that using latest form of communication that is paperless. This simple point dictate the possibility that paperless is no longer a choice but a way of life.

2. Deciding on the right tools

For any technologies or methods available, the application is decided by a lot of factors. Those factors are:

2.1 Read-only or Writable

a. Once the documents are stored can they be edited?

b. If editing is allowed, what type of security do you want to control who has the rights to edit the files?

c. Do you need an audit trail tracking who and when modifications were made?

2.2 Accessibility

a. Should all documents be accessible to everyone or do you require limited access?

b. Do all documents need to be accessible immediately or can they be retrieved from some type of archival media?

c. Are you using portable computers that need access to documents remotely?

2.3 Frequency

How often do you need to access the documents?

- several times a week
- once a month
- or seldom

2.4 Duration

- days
- months
- years

3. Technologies available

The rapid development of technology presents that make everyone to be excited again since the old long dream is now can be a reality. Surely more new technologies will be introduced. For now, we will only concentrate our discussion on technologies currently available.

3.1 Electronic Data Interchanged (EDI)

This activity involves exchanged of data between organization of transaction such as orders, confirmations, invoices, etc.

An EDI message contains a string of *data elements*, each of which represents a singular fact, such as a price, product model number, and so forth, separated by delimiter. The entire string is called a *data segment*. One or more data segments framed by a header and trailer form a *transaction set*, which is the EDI unit of transmission (equivalent to a *message*). A transaction set often consists of what would usually be contained in a typical business document or form. The parties who exchange EDI transmissions are referred to as *trading partners*.

General Motors(2000) orders via EDI about 90 percent of the production materials it uses in the manufacture of its automobiles. Likewise, Ford's Computer network links over 4800 of its dealer and over 3000 of its production material suppliers to its corporate data center. IBM is connecting 2000 of its largest suppliers worldwide via EDI network. Sears Roebuck's EDI network has 135,000 terminals in 7000 locations and over 800 suppliers.[13]

So all these big figure showed what is the actual progress toward paperless.

It is interesting to note that most of the paperless through EDI is initiated by one company. This one company or organization then imposed a new requirement to all the connected or trading partners to stop submitting or sending any printed material ; but through EDI.

What so significant about EDI? The answer is that EDI is indirectly forcing more than one parties toward paperless.

3.2 Optical Character Recognition (OCR)

OCR (optical character recognition) is the recognition of printed or written text character by a computer. This involves photoscanning of the text character-by-character, analysis of the scanned-in image, and then translation of the character image into character codes, such as ASCII, commonly used in data processing.

In OCR processing, the scanned-in image or bitmap is analyzed for light and dark areas in order to identify each alphabetic letter or numeric digit. When a character is recognized, it is converted into an ASCII code. Special circuit boards and computer chips designed expressly for OCR are used to speed up the recognition process.

OCR is being used by libraries to digitize and preserve their holdings. OCR is also used to process checks and credit card slips and sort the mail. Billions of magazines and letters are sorted every day by OCR machines, considerably speeding up mail delivery.

The OCR is not totally paperless because the technology is that to enable the device to read an information from printed material. However, OCR contributes toward paperless by reducing number of unnecessary copies when reading or detection can be done by device at a lightning speed.

3.3 Optical Mark Reader (OMR)

Contains a light beam that passes over the scan sheet and detects the filled-in ovals. This is commonly used when filling the given shape on answer sheet. The lead pencil used to fill the shape will stop the beam from getting through.

This is similar concept to OCR. Paper circulation for grading or marking can be eliminated once the existing technology can perform with a better result compared to human being. Again, the purpose of paperless operation is not only to reduce the printed material but also stop the physical circulation

3.4 Magnetic Ink Character Recognition(MICR)

This a technique which commonly used in banking environment where customer identification, bank information are preprinted with ion oxide based ink. The device which detect the ink will translate to the readable form.

An MICR reader translates these characters into digital form for the computer. [10]

3.5 Intelligent Character Recognition (ICR)

Software capable of reading handwriting. It adapts and learns according to the shapes of letters and configurations encountered. It is used by the United States Postal Service for scanning mail.

Through ICR, how paper can be saved? The answer is the scanning and recognizing capability. When the information can be stored and displayed on the electronic device, the target toward paperless is visible.

3.6 Audio/Video Conferences

The telecommunications system may involve only audio (sound) exchanges among the participants, as in a telephone conference, Audio-graphic conferences using microphone-speakers, a computer terminal, and electronic tablet, make it possible to hear and see written information at the same time. This method which looked simple is surely reducing the number of printed material.

3.7 Electronic Document Conferencing

Preparation of documents among geographically removed users. The originator displays the master document on a screen; it is also displayed on the screen of the other participants. The participants can annotate the document while discussing it.

This is clearly a good method. Instead of sending the printed material to all the attendees, document will be viewed electronically. Other benefits are clarification and attraction of the presented materials which should be animated.

3.8 Imaging / Scanning & Indexing

The conversion of documents from a physical medium (for example, paper) to a digital form where they can be manipulated by computers for easier access and retrieval. Most image management systems include workstations for viewing and manipulating images; storage devices; and servers to manage the image database, run application software and connect to host devices.

It is time consuming and exhausting to retrieve documents stored in filing cabinets. Statistics show that the average working person in a traditional paper-based office spends an average of 15% of his time retrieving documents [5]. Electronic data storage reduces the inconvenience and time often associated with document retrieval. Furthermore, flexibility is a strong point of electronic data storage. Documents can be sorted and stored according to the liking of the user, and its electronic nature facilitates document tracking and easy distribution of documents. Electronic documents can also be accessed simultaneously by different people, whereas a paper-based process forces serial work [6]. All this enhances office productivity.

Another excellent feature is indexing.. The name of the index corresponds to the name of the paper or report title. In the index software search engine, we can pick up through many different indexes or by entering a key word search. For example, if I were looking for a letter written by Mr. Jones and I remember it came sometime in December 1996, we would type the key words "Jones AND December AND 1996." The key word search uses the same nomenclature as LEXIS and the Internet, using the operators "and" "or" "w/_____" "not" etc. The indexing software would search the indexes and find any text files which meet the three key word criteria. The computer then lists all of the text files meeting the search criteria which may be a long list of files requiring a second level of search. Searching for documents using these key words is somewhat of an art form and requires practice.

Once you have found the corresponding text file, the computer will electronically link the text file to the graphic file which is the actual photocopy of the document. You then pop the electronic document on the screen, page through it to read, review, print, fax or attach electronic "sticky notes." This same document can be searched, found and viewed from a computer outside the office via modem and telephone lines.[14]

So, once document can be scanned, stored, and retrieved, then future need to reprint the material will can be reduced tremendously.

3.9 Personal Digital Assistant (PDA)

Personal digital assistant PDA (personal digital assistant) is a term for any small mobile hand-held device that provides computing and information storage and retrieval capabilities for personal or business use, often for keeping schedule calendars and address book information handy. The term handheld is a synonym. Many people use the name of one of the popular PDA products as a generic term. These include Hewlett-Packard's Palmtop and 3Com's PalmPilot.

Most PDAs have a small keyboard. Some PDAs have an electronically sensitive pad on which hand-writing can be received. Apple's Newton, which has been withdrawn from the market, was the first widely-sold PDA that accepted handwriting. Typical uses include schedule and address book storage and retrieval and note-entering. However, many applications have been written for PDAs. Increasingly, PDAs are combined with telephones and paging systems.

Some PDAs offer a variation of the Microsoft Windows operating system called Windows CE. Other products have their own or another operating system.

3.10 Screen Input System

Manufacturers have also recognized the importance of the need to write on the screen, and have responded quite admirably. While there is still no widely accepted screen input system for the desktop computer, hand-held electronic organizers that can be "written on" is already in the mainstream. 3Com's Palm has garnered incredible popularity thanks mostly to its intuitive pen-to-screen interface [7]. In many cases, it has completely replaced the office worker's diary. In its latest incarnation, it even offers a wireless Internet connection - an appetizing sign of things to come.

3.11 Document-Sharing Software

"Document-sharing" software has appeared on the market, attempting to solve the problem of incompatible document formats. Software such as Adobe's Acrobat serves to read documents of various formats and export them to a single standard format, which in this case would be the PDF format [8]. This format would then presumably be readable across the board. The idea behind this is ease of management and use. However, document-sharing software in its current state of infancy appears to be a crude hack at present, not having the power and finesse to truly implement complete document coherency and integration throughout the workflow. Still, it is a promising step in the right direction.

4. Legal Issues Involved

Among the various legal issues that may arise are the validity of electronic documents and digital signatures, defamation, professional negligence, cyber crimes and intellectual property issues. However, among these reasons, I would only touch on legality and crimes issues .[12]

4.1 Validity of Electronic Documents and Digital Signatures

In contrast to the paper-based documents, although EDI can satisfy the requirement of informative function, it may create potential legal problem relating to the evidential and symbolic function. Further, there are yet very disturbing apprehensions —would digital signatures secured by cryptography constitutes a signature when that is required and would the digital signature constitute proof of the signatory's responsibility for the correctness of the information on the document and to the genuineness of the source from which it is supposed to have been sent.[12]

4.2 Cyber Crimes

With the technological advancement and progress in information technology, a whole spectrum of activity – various types of abuse of computer system, coupled with the traditional crimes the penetration of which get facilitated, such as offence of theft, deception and fraud. Computer systems are particularly vulnerable to computer crimes because of a number of factors. Some of the important factors relate to the density of information and processes, open connectivity, system accessibility, and electronic technology and last but not the least the human factor. Computer data can be incidentally changed or erases with minimum chances of detection e.g. virus or logic bomb. Anybody can easily modify the files and then screen the evidence of the offence. Data can be duplicated on floppies/tapes without any audit trails. Due these concern, some organization reluctant to fully paperless.[12]

5. Other relevant issues

a. High Up-front Cost

Apart from computers, we need file storage system, scanner, high quality monitors, computer network, etc.... It costs a large amount of money for these installation. This is a very expensive start-up cost.

b. High Ongoing Cost

In order to have the best performance of computer system, we need a large amount of money for maintenance. Besides, companies need to employ network managers to supervise the establishment of computer network. Therefore, the ongoing cost is also very high.

6. Conclusion

In conclusion, we have seen a lot of improvement in new technologies, but human seem to think that the two problems stated – legality and security always overcome the benefit of paperless. Other than that, it is always been understood that human always comfortable with the old ways of life and always reluctant to accept changes. However, we have to admit there has been some progress made, there are substantial gaps to a totally paperless . The enthusiastic predictions made way back in the 1970s have not been fulfilled in its entirety - no doubt, we now have some offices that use very little paper, but a fully paper-free office environment remains an impracticality in most cases. The only real foothold that the paperless idea has taken in daily activities is the use of an Intranet for internal communication, but even that is mostly restricted to emails, memos, and the occasional document or two.

We believe that given enough motivation and technological advancement, we could very well have a completely paperless at some point in the future. But the question needs to be asked : Do we really want a paperless ? It all depends on how far the technology can be pushed. If we ever reach the point where a computer looks and behaves almost exactly like a piece of paper, then in all likelihood, yes, a paperless is a sound possibility. Otherwise, we suspect that the answer will be that we will still want paper for a lot of things, simply because it is so convenient and natural. [11]

One thing for sure that the major obstacle isn't technological, or even economic. Instead it's that people prefer paper.

7. Acknowledgement

The authors gratefully acknowledge the support provided by Universiti Tenaga National (UNITEN).

8. References

[1] The paths to the paperless office, June 30 1975

[2] T.Zito. World without paper, the first totally electronic office. *Washington Post*, May 3 1979.

[3] Robert O'Harrow Jr. Save to disk? yes and make a printout,too. *Washington Post*, page 21, November 3 1997.

[4] Sara Peeler. The paperless office, 1998

[5] Sandra Salmans. The debate over the electronic office. *New York Times*, page 132, Nov 14 1982.

[6] Erik Sherman. Paperless office still a goal, but reality remains elusive. *Macweek*, 10(24), June 17 1996

[7] Bad news for trees. The Economist, December 19 1998.

[8] Reid. Reviving the concept of the 'paperless office'. *Washington Business*, page 24, November 28 1994.

[9] Selina Mitchell. Whatever happened to the paperless office? *APC magazine*, August 1 1996.

[10] Kenneth C. Laudon., Jane P. Laudon, "Managing Information Systems".

[11] Khai Weyn Ong and Amy Yoon. The Paperless Office : Myth or Reality?, Jan 27 2000.

[12] Dr. Lalitha Sreenath & M.R. Sreenath. The Impact of Information technology on the Legal System, 21 Nov 2000

[13] A. Seetharamam, Balachandran & S Saravanan, EDI and Purchase Accounting : Problem and Prospects, 21 Nov 2000.

[14] Paperless,http://www.gate.net/~tennyson/paper.html,18 Jan 1998.

[15] Paperless,http://www.ahinc.com/paperles.html, 30 Oct 2000.

A New CSCW Prototype System for Color CRT CAD/CAM[1]

Yuxiu Cao, Chunxiao Xing, Long Tang, Lizhu Zhou
Department of computer science and technology, Tsinghua University
Bejjing 100084 P. R. China
yxcao@cad.cs.tsinghua.edu.cn

Abstract

A new Computer Supported Collaborative Work (CSCW) prototype system is developed to meet the need of color CRT CAD/CAM. This Paper discusses characteristics, and defines key elements of the CSCW prototype system. Cooperation relation and cooperation work flow of color CRT CAD/CAM are provided. Compared with the traditional color CRT CAD/CAM system, the new CSCW prototype system is more suitable for the rapid design of color CRT. The efficiency of the new CSCW prototype is validated. A general CSCW architecture to support integrated product/process design and development is presented. A case study and evaluation of the architecture is also discussed. The integrated architecture will facilitate information access, sharing, and analysis among design team members using the open World Wide Web platforms and resources to design CRT products.

1. Introduction

Color CRT consists of fluorescent screen, shadow mask, deflection yoke and electron gun. Engineers engaged in the field of color CRT is used to separately design the each part of CRT (i.e., fluorescent screen, shadow mask, deflection yoke and electron-gun), this results in badly match designing components of CRT which have to be devised over again. CSCW has been viewed as the key for reducing cycle times and improving product quality and reliability. CRT CAD/CAM is a systematic approach to product development that achieves a timely collaboration of necessary disciplines throughout the product life to satisfy customer needs better. Therefore, color CRT modern design has to be transferred into the CSCW environment now. CSCW is a set of software, hardware, language components, and procedures that support a group of people in a decision related meeting [1]. The primary goal of this research is to bridge the apparent gap between CSCW and CRT CAD. This research effort will allow potential industry users of CSCW technology to identify the requirements of the CSCW system for distributed collaboration. It will also examine some productivity metrics used to measure the effectiveness and productivity of the CSCW/CAD/CAM integration. The case study presented at the end of the paper will show how CSCW technology can be applied in a complex process design involving project management, data analysis, etc. Secondary goals of this research were to complement ongoing efforts to get a much clearer understanding of the application of CSCW in a CAD/CAM setting [2-4].

2. Color CRT CAD/CAM framework based on CSCW

Applying CSCW technology to CRT design and development process, color CRT CAD/CAM cooperative workflow is acquired so as to raise the efficiency of the design and manufacture of color CRT. CRT CAD/CAM is viewed as a distributed and iterative process that involved significant sharing of information in the design process [5]. The CSCW environment provides the ideal environment for real-time discussion of design ideas and sharing of design data and test results. The collaborative efforts among the design teams will ultimately determine the most effective methods and processes to design and manufacture the CRT product. For example, decisions made with respect to product geometry, materials and loads will affect the selection of a process layout. The CRT CAD/CAM process is illustrated (as Figure 1), here, electron gun design tool is GUN-3D software; deflection yoke design tool is a magnetic field analysis mould in ANSYS software; drawing design tool is AUTOCAD14; 3D Modeling design tool is PRO-ENGINEER software.

3. A CSCW Architecture for CRT

The CSCW field has emerged from a growing interest of software product developers in supporting networked

[1] This work is supported by the national foundation research project No. G1999032704.

groups and from a common interest among researchers in various disciplines in how the productivity of work groups can be improved [6]. A CSCW system for color CRT should encompass the following characteristics to ensure success:

- Interaction in synchronous and asynchronous mode among team members.
- Coordination of the various tasks performed by members of the team.
- Distribution to enable people to interact from remote places.
- Visualization and accessibility of data by team members.
- Sharing of data, engineering drawings, applications, etc. among participants.

The challenge of CSCW designers is to provide suitable tools and supported platform that will allow participants to organize and structure the product/process design data in an understandable and retrievable format [7]. The central workspace in the CSCW architecture and the effectiveness and productivity derived from this architecture will be discussed later (as Figure 2). The CSCW Architecture consists of WWW server, sharing engine, cooperation agent, system information service, Tool set of CSCW, cooperation workspace, KB, and DB.

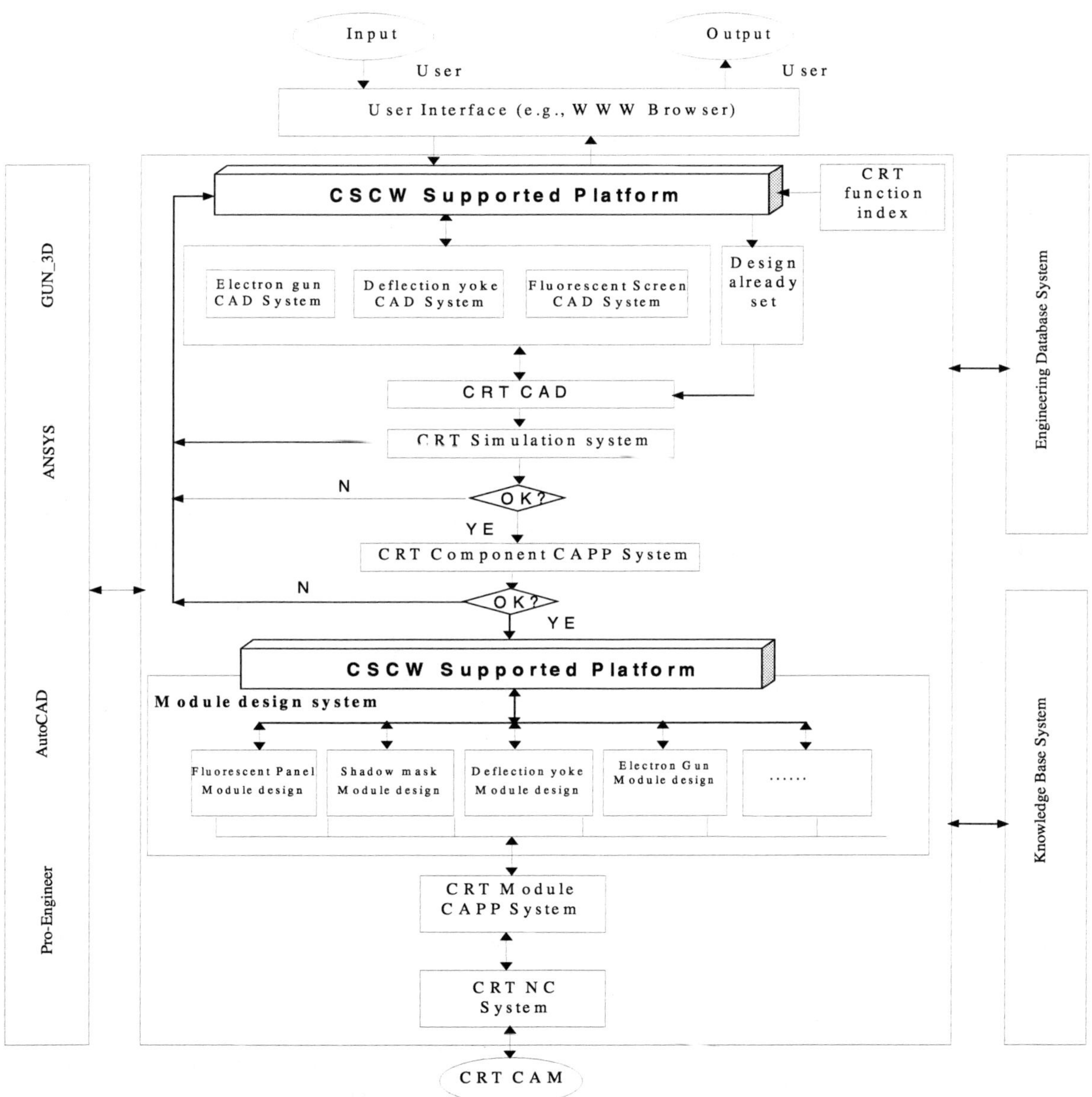

Figure 1. Color CRT CAD/CAM Cooperative Framework.

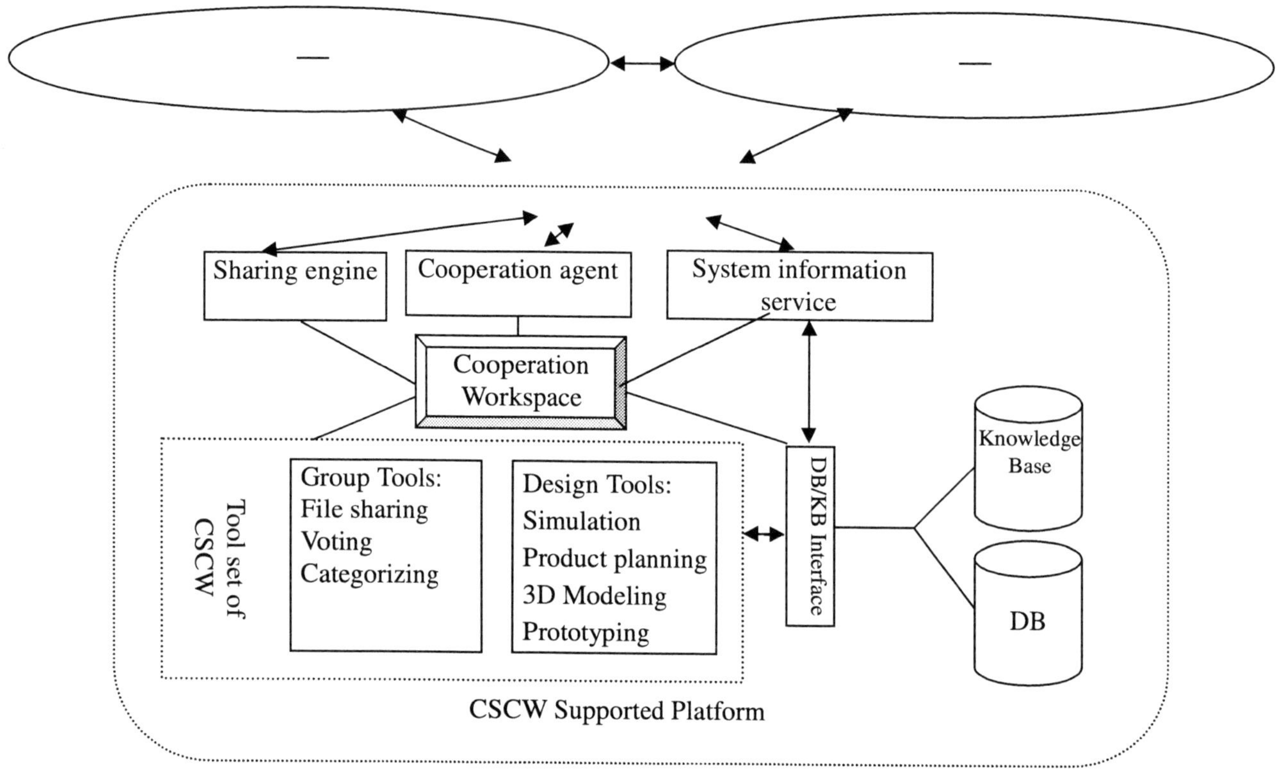

Figure 2. CSCW Architecture.

The shared WWW workspace consisted of a groupware-based kernel and a tool set containing a variety of engineering project management tools for CRT CAD. The role of the groupware-based tools is to foster collaboration among the work teams. It supported group activities by offering a collection of software tools such as information gathering, idea generation and organization, voting, preference aggregation, and consensus building. The WWW server/browser provides the interface to other project management tools such as spreadsheets; simulation programs, financial planning templates, project scheduling, computer-aided design tools, and shared whiteboard tools. Besides the shared environment, project teams can bring the results in the shared workspace environment for further analysis and discussion of CRT/CAD. The promised benefits of this integrated CSCW architecture include:

- Reduction of lead times, by changing from serial to a collaborative engineering approach.
- Getting designs right first time by having access to relevant data in a timely manner.
- Better communications between departments.
- More effective use of design teams in that they can locate, retrieve, distribute, and re-use data in a speedy and more efficient manner.
- Reduction in the time required for design teams to reach consensus on the product/process design.
- Reduction in the overall costs during the design and development during the design and development phases.
- Production of designs of a better quality.

The color CRT CAD/CAM system aims to provide a comprehensive set of inter-working tools for efficient multi-group collaboration using the WWW as an information-sharing platform. To serve the changing needs of CRT CAD/CAM teams, the system architecture must have the following capabilities:

- Access to information from a universal workspace.
- Internet-based remote access via a standard WWW browser.
- Database and knowledge base management,
- Multimedia interaction for easy deployment and administration.
- Low cost of implementation.

A shared workspace was developed on the WWW to ease the exchange and sharing of complex data including text documents, spreadsheets, project schedules, engineering drawings, images and 3D modeling graphics. Integration of design and management tools with the WWW in a shared environment is accomplished. The WWW offers an excellent opportunity to support collaboration within widely dispersed work teams. The CSCW architecture is shown in Figure 2.

4. Conclusion

Computer Supported Collaborative Work is increasingly being used by engineering design teams to reach a consensus on a range of design issues. In this paper, a new CSCW architecture has been developed to support CRT CAD/CAM product/process design and development. Test and analysis of the results shows that designing and evaluating of CRT CAD/CAM processes can obtain productivity enhancements, improve decision quality by using CSCW tools and environment. The CSCW technology used in this study incorporates distributed interactions but further work is necessary to support product development teams working in different places at different times across global platforms.

5. References

[1] Huber GP. Issues in the design of group decision support systems, MIS Quarterly, 1984, pp.195-204.

[2] Monplaisir LF, Benjamin CO, Riordan CA. Groupware for computer supported communications in quality teams. In: Proc. 46th Int. Industrial Eng. Conf. 1995: pp.500-509.

[3] Monplaisir LF, Innovative applications of groupware for solving engineering design problems. Eng. Management J. 1997, 9(1): pp. 11-16.

[4] Nof SY, Eberts RE, Papastavrou JD. Computer-based collaborative integration of distributed manufacturing engineering, Proc. 1995 NSF Design and manufacturing Grantees Conf. University of California, San Diego, 4-6 January, 1995, pp.283-284.

[5] Yuxiu Cao, Research on CAD Theory And Application of Color CRT Electron-Beam Landing Errors Correction [D], Northwestern Polytechnic University of China, 1999.

[6] Leslie Monplaisir. An integrated CSCW architecture for integrated product/process design and development. Robotics and Computer-Integrated Manufacturing 15 ,1999, pp.145-153

[7] Chunxiao Xing, A Study on Distributed Multimedia System and Its Application to Digital Library [D], Northwestern Polytechnic University of China, 1999.

Designing and Implementation of Tele-Health Care Information System based on Health Care Model

Tsutomu Matsumoto[1], Toshikatsu Mori[1], Hirofumi Ohtsuka[1],
Yasuyuki Shimada[1], Shigeyasu Kawaji[2]
[1] *Kumamoto National College of Technology*
2659-2 Suya Nishigoshi Kikuchi, Kumamoto, Japan
[2] *Graduate School of Science and Technology, Kumamoto University*
2-39-1 Kurokami Kumamoto-city Kumamoto Japan
{matumoto,mori,ohtsuka,shimada}@ec.knct.ac.jp
kawaji@cs.kumamoto-u.ac.jp

Abstract

This paper describes a concept and system implementation of tele-health care support system. Traditional medical consultation has carried out with face to face. The tele-health care support system should make up the physical distance between client and clinician. The medical consultation model is proposed based on analyzing task and information of medical consultation for creating a designing methodology to tele-health care system. The structure of proposed medical consultation model is created by idea of feed back control system. Clinician takes play of a controller and client is regarded with a controlled system in the model. How and which parts of the model could be supported by computer and communication technology are discussed. The designing and implementation of collaboration among client, clinician and co-medical including family and establishment of communication channel are explained as one actual example based on proposed medical consultation model.

1. Introduction

These guidelines include complete descriptions of the fonts, spacing, and related information for producing your proceedings manuscripts. Please follow them and if you have any questions, please contact the Conference Program Committee Chair.

The increasing population of senior citizens in the community requires the health care system to outreach to the people. However, it is not possible for the clinician and client's family to continuously monitor the client's everyday activities, over twenty-fours hours. There have been published papers related to this matters to improve this situation.

It is pointed out that the research to improve QOL (quality of life) of aged people has been a very important subject [1]. Ohta reported about the system, which measures person's moving among rooms at home using infrared sensors [2]. He proposed the visualization of person's moving among rooms at home and total movement distance. It is considered that the non-contact type sensor is useful for measuring person's action, because it does not create additional burden on a subject. Monitoring system as similar as Ohta's way is also reported [3]. It is explained that finding method of the observed person's whereabouts can be obtained by a very simple statistics. Above-mentioned reports focused on only the way of measuring vital signal or human behavior at home.

Papers related on Tele-medicine have been discussed on transferring medical image data to support diagnosis, medical counseling, sharing clinical information [4][5]. Above reports have not cleared to support which parts of medical consultation, treatment or health care to client. This paper proposed a health care model with analyzing task and information of traditional medical consultation that clinician performs to clients. The designing and implementation based on the health care model are described in this paper

2.Analyzing Task and Information at Medical Consultation

In this section task and information from client are described are described. And medical consultation model is proposed.

2.1 Task of Medical Consultation

In order to create a medical consultation model, first we discuss behavior of physician in medical consultation. Physician interviewees a new patient for acquiring various personal information, disease history of patient and family, life style, main complaint and so on. And physician does inquiry, inspection, and auscultation to find abnormal state of patient. Physician takes blood check and/or bio-chemical medical check as occasion demands.

Physician estimates state of patient with medical check up data and inspection, and identifies disease. We can say this is a diagnosis. Then physician makes a plan for treatment. This paper means giving some advice and/or remedy. Treatment has two types. One is to provide directly treatment to patient. The other one is to provide indirectly patient treatment for example through nursing staff. In this case physician gives an instruction to nursing staff then nurse staff carries out the instruction from physician.

There is a case that physician may give a description to patient. As occasion demands physician gives some advice to family. The family cares own patient. This situation can include a treatment in a broad sense. Some time later physician evaluates the effect of treatment given to patient and improves the treatment plan. Above-mentioned behavior of physician derives an automatic feedback control system, a controlled system as patient, healthy as command and manipulated variables as medical treatment with cybernetics.

2.2 Information for Medical Decision Making

Information (output) from patient is used to diagnose. We classify the information according to attribute [7]. Static information is not to change content of variable or if change; we can ignore time interval of changing. For example sex, age, etc. Dynamic information means variation with time and other causes. For example body temperature, auscultation sound, medical check up data. In this paper information from patient is classified as follows.

(1)Zero order information:
This is a personal information such as sex, age, disease history, career, disease history of family, life style (smoking, drinking, meal custom etc.

(2)First order information:
This type information could be directly obtained from patient or obtained by using simple equipment. For example there are main complaint, inspection, stethoscope, blood pressure, pulse count, body temperature etc.

(3) Secondary information:
This information is so called clinical laboratory data. For instance, there are blood check and bio-chemical check up data and information from analyzing available sampled data of living body.

(4) Tertiary information:
This type data could be obtained by information technology. For example, ECG, EMG,X-ray image, MRI,X-ray CT etc.

First, second and third order information is categorized in dynamic information. Tertiary information is more detailed medical information.

2.3 Health Care Model

Combination medical consultation model and structured information from patient derives a block diagram as shown Figure 1. Disturbance of figure1 means inputs for patient to take bad turn. Figure 2 is simplified with detector part and treatment part showing each one box. We name it Medical Consultation Model.

Physician plays role like a controller in the Medical Consultation Model. Physician diagnoses using output **y** of patient, makes a treatment plans and gives directly and/or indirectly medical treatment to patient for recovery. This medical treatment is recognized as same as operation in control system. At this time physician verifies whether or not treatment is given

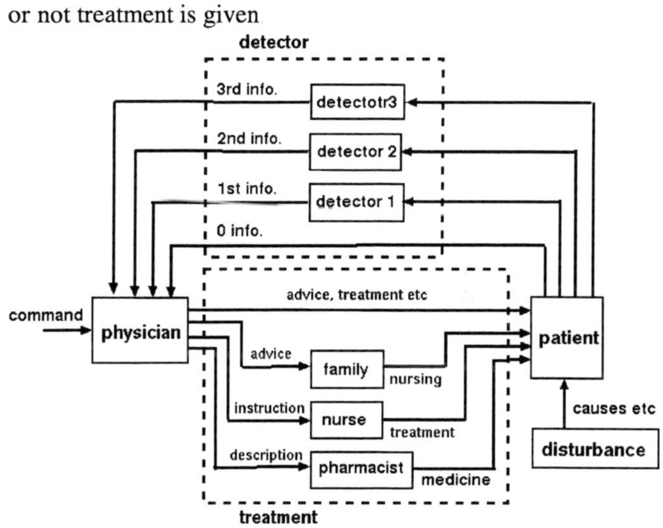

Figure 1. Task and Information at Medical Consultation.

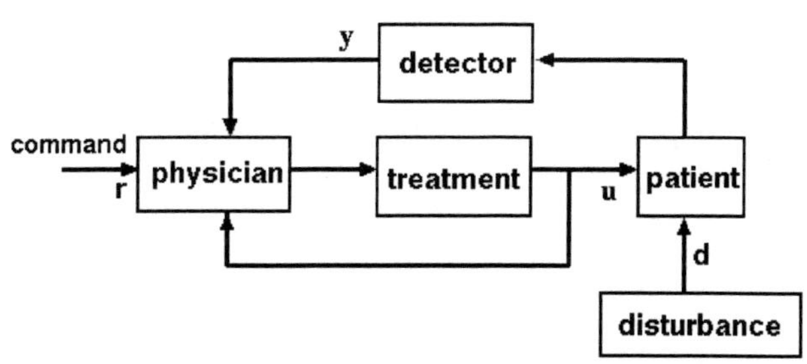

Figure 2. Health Care Model.

3. Tele-Health Care System

This section describes designing of tele-health care information system based upon above the health care model. This system focus client who lives alone at home.

3.1 System Design

Medical consultation includes keeping health condition. Consequently tele-health care is that feedback system is set up at the occasion demands and supports to share/transfer the information among client, clinician and co-medical. This paper defines the tele-health care information system as follows. When an irregular habitual life behavior and abnormal health condition is detected, the system confirms safety of client. And the system sends a message to clinician, nursing staff and/or neighbors depending on answer of client. In addition the system provides health advice from clinician to client.

Consequently the system has following specifications in order to perform above mentioned.

1. Obtaining the information of client
2. Transferring the information of client
3. Finding problem of client
4. Confirming client safety
5. Informing urgency to co-medical, clinician
6. Providing health care advice

The fourth and the fifth mean the collaboration among client, co-medical and clinician. Finally the tele-health care information system can be designed using layer structure shown by figure 3. The top view of figure 3 is as same as figure 3 and figure 5.

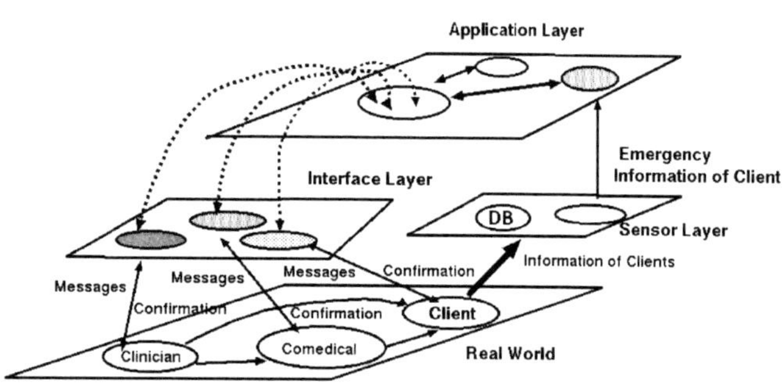

Figure 3. Layer Structure for Tele-Health Care.

3.2 Layer Structure

This section describes the relations among layers and functions of each layer.

3.2.1 Real World

The object of this system are clients, clinicians and co-medical including client's family and/or neighbors.

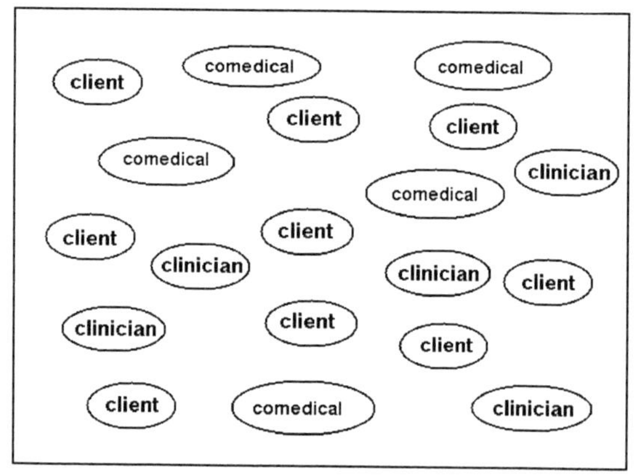

Figure 4. Real World.

3.2.2 Sensor Layer

The sensor layer is to gather information of client and to find urgency of client. If sensor layer finds an urgency of client, the sensor layer sends a message to application layer. The message has ID of client and urgency level that client needs clinician or not, needs nursing staff or not and so on. In addition the sensor layer provides health care record to clinician via application layer. This system focuses life information (Zero order), pulse and body temperature as the third order.

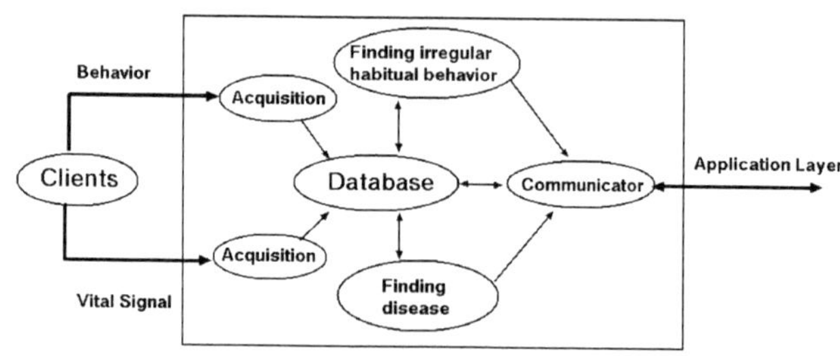

Figure 5. Sensor Layer.

3.2.3 Application Layer

Application layer searches contacts of family doctor, co-medical and/or client corresponding to urgency level for sending a message. With the search results application layer sends a message to interface layer. Application layer continues to send a message in order to the reply from receiver. If nobody replies, the system calls an ambulance.

The application layer requests information of client to sensor layer after receiving a message of "Request : providing information of client". The application layer transfers the data to appropriate interface layer.

3.2.4 User Interface

1. Clinician Interface
   Clinician interface informs urgency of client and transfers reply from clinician to the application layer. Clinician inputs the process into the system via the clinician interface. Clinician is able to retrieve health record of client.

2. Co-medical Interface
   Urgency of client is informed to registered co-medical via the co-medical interface. Co-medical who received a message has to inform the condition of client and/or result of treatment to the system via the co-,medical interface. Also co-medical retrieves health advice from the system that clinician has provided, and co-medical guides client to keep health condition.

3. Client Interface (Interactive Response System)
   Client interface sends a message to confirm safety. Client can reply "No problem", "need help", "need clinician" or no answer. The client interface informs content of reply to the system.

4. System Administrator Interface
   The system administrator interface provides functions of system administrating such as registration, removing, modifying user (client, clinician, nursing staff, neighbors).

4. Software Designing and Implementation

This section describes the software designing of tele-health care system based on UML (Unified Modeling Language).

4.1 Use Case and Sequence Diagram

The use case view shows functionality of the system as perceived by outside users, called actor. This system consists of three use cases.

(1) Sharing information
  Health care advice
  Retrieve information of client

(2) Emergency jobs
  Receiving urgency message from sensor system,
  Informing urgency of client to clinician, co-medical
  Confirming the safety of client

(3) Administration
  Adding/removing users
  Modifying users

Each use case shows the views by figure 6, Figure 7, figure 8.

Here two sequence diagrams for showing a set messages arranged in time sequence are shown as figure 9 and figure 10.

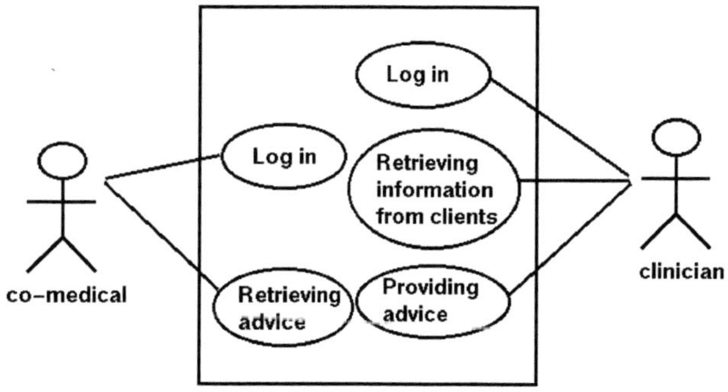

Figure 6. Use Case View: Sharing Information.

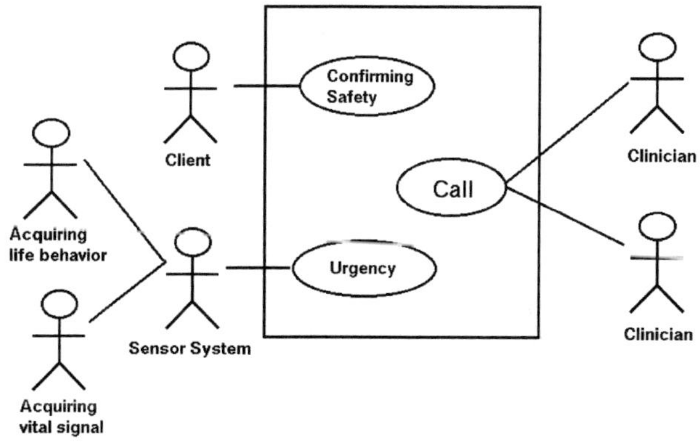

Figure 7. A Use Case View: Finding Emergency.

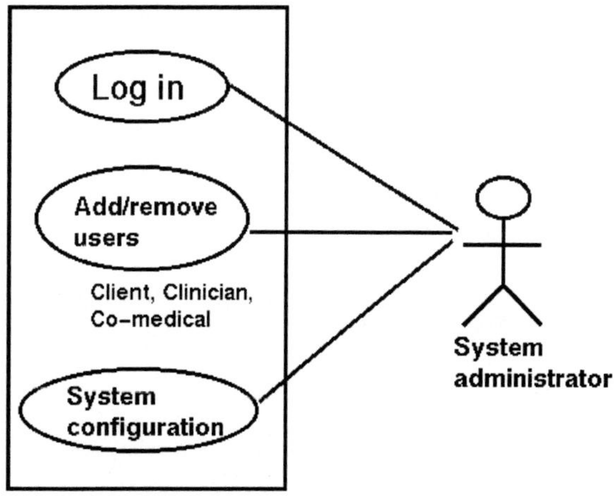

Figure 8. A Use Case View: Administration.

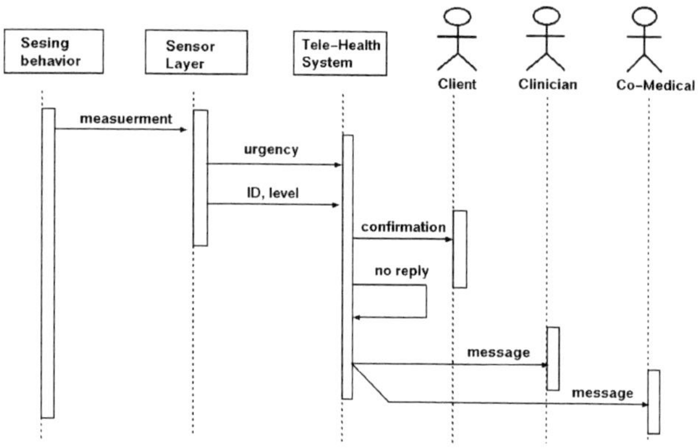

Figure 9. A Sequence Diagram: No Response of Client.

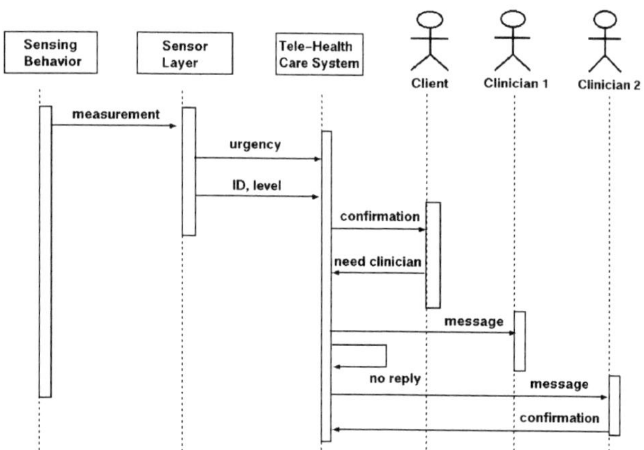

Figure 10. Sequence Diagram: No Response from a Clinician.

4.2 Implementation

The software system has been developed on the Linux System. All source programs have been programmed by Java using JDK1.3.PostgreSQL has used to create a database.

Right now the system consists of about sixty classes, each class is about 100 statements. The system has verified by behavior as same as designed scenario. User interface layer and sensor layer communicate the main system with TCP/IP. The classes in the main part communicate each another with sending/receiving message supported by Linux. Thus any layer can be located anywhere on the network.

5. Conclusion

This paper discussed the health care modeling and describes that tele-health support system should set up feedback loop between clinician, co-medical and client on demand. As an actual example, designing and implementation of the system which supports collaboration and sharing information among clinician, co-medical and client after finding the urgency from obtained life behavior and vital signal of client has reported in this paper. This paper does not talk how the sensor layer finds the urgency of client based on obtained life behavior data. We have already reported this matter [8] [9]. Gathering vital signal of client is carried out by simulation.

Proposed methodology in this paper is applicable to hospital. Because physician and nursing staff are not able to continuously care the clients over twenty-four hours.

6. References

[1] K. Ishikawa and H. Inada : An experimental study on home health care support information system construction, Proceeding of the World Congress on Medical Informatics (MEDINFO 95), pp.581-585 (1995)

[2] Shigeru Ohta: Health Monitoring System for the Aged, Information, Processing Society of Japan Magazine, Vol.41 No.6 pp.639-643 (2000)

[3] Tanikawa Tomohiro et al : In-house Movement Monitoring System for the Aged Living Alone, Japan Journal of Medical Informatics, Vol.20 No.2 pp.270-271 (1999)

[4] S-M Yoo, H Lee and S-Y Kim: Medical Counseling through Computer Communication Networks, Proceeding of the World Congress on Medical Informatics (MEDINFO 98), Vol.1, pp.337-340 (1998)

[5] R. Neame and M Olson: How can Sharing Clinician Information be Made to Work, Proceeding of the World Congress on Medical Informatics (MEDINFO 98), Vol.1, pp.315-318 (1998)

[6] H. Takeda and Y. Natsumura: Functional Evaluation of Telemedicine with Super High Definition Images and B-ISDN, Proceeding of the World Congress on Medical Informatics (MEDINFO 98), Vol.1 pp.311-314 (1998)

[7] T. Matsumoto, S. Kawaji, Weiming Shen and Douglas H. Norrie : Designing a Multi-Agent Information System for Tele-Health, Proceedings of the Second ICSC Symposium on Engineering of Intelligent, Systems (EIS2000), pp.545--548 (2000)

[8] T. Matsumoto, Y. Shimada, K. Shibasato, H. Ohtsuka and S. Kawaji : Creating behavior model of a senior solitary life and detecting an urgency, Proceedings of the International ICSC Congress on Intelligent System and Applications, pp.234-238 (2000)

[9] T. Matsumoto,Y. Shimada, K. Shibasato, H. Ohtsuka and S. Kawaji : Developing of Human Behavior Model at home and Its Application, World Congress on Medical Informatics 2001 (to be published).

New Multimedia Telecommunication Technologies and CSCW for Telemedicine

Wenhua Huang[1], Yuhua Ai[2], Shizhen Zhong[1], Qingan Zhu[1],
Weidong Zhao[1], Lin Yuan[1], Zhenyu Chen[3], Jingdong Yan[2], Zhexing Liu[2]
[1] *Institute of Anatomy, The First Military Medical University, Gaungzhou, 510515, P.R.China*
[2] *Information Center of Southern Hospital, Guangzhou, 510515, P.R.China*
[3] *Guangdong Technical College of Water and Electricity Power, Guangzhou, 510635 P.R.China,*
hwh@fimmu.edu.cn

Abstract

This paper introduces developing status of telemedicine. Telemedicine is based on the technology of telecommunication, and it is one of the applications of Computer Supported Cooperative Work (CSCW). Computer supported teleconferencing applications have been developed in the field of CSCW and are currently providing more effective support for telemedicine services. This can be achieved by a combination of high performance networks and can reduce data transmissions during the conference. A scheme of telemedicine system is proposed.

1. Introduction

In recent years, advances in information technology and telecommunications have acted as catalysts for significant developments in the healthcare sector. In the near future, healthcare institutions will make extensive use of computer networks, mass storage devices, and sophisticated workstations at which humans and machines will interact, assisted by advanced information processing tools and techniques of knowledge engineering, to achieve integration of multimodality multimedia diagnostic data and expert medical knowledge [1]. Computer networks also provide physical links to other hospital departments and patient wards in order to improve interdepartmental communications and patient monitoring procedures.

Telemedicine may be defined as the application of communication technology to facilitate the distance independent practice of medicine. Telemedicine came about with the introduction of the telephone. With the advent of computer networks communication of clinical information both within and between institutions has entered a new era. One thing is clear from even a cursory examination of the history of electronic communication in medicine. The trend is toward increasing volume and complexity of information and increasing demand for communication bandwidth.

Telemedicicine began with the application of analog video technology (close circuit television) in the early 1960s [4]. Recent advances in digital wide area networks have led to a resurgence in interest in expanding beyond analog video systems to encompass n-way digital video teleconferencing, remote physical examinations and the image.communication systems of teleradiology [5]. To date most telemedicine projects have focused on relatively low bandwidth communication channels and commercial off the shelf videoconference technology. It is becoming clear, however, that each medical specialty and subspecialty has unique requirements for application software and bandwidth.

Information Center of Southern Hospital in partnership with the Clinical Anatomy Institute of the First Military Medical University is developing a prototype Surgery Research Center to provide a comprehensive system for remote diagnosis, patient counseling, and integration of on-line multidisciplinary academic expert clinical teleconsultation with local treatment providers. This center will facilitate greater interaction of government facilities with academia, industry, consumer groups, and community outreach services in researching and implementing the most effective methods of screening and diagnosis, including CAD, CSCW, teleradiology, telerpathology, and remote consultation. The Surgery Research Center will provide a platform to enhance promising state-of-the-art technologies from the leading national academic, government, and industrial laboratories to patients, collaboration between the various medical cares and integration with local treatment providers.

2. Program goals

Our program goals are to support the overall goal of establishing a virtual clinical anatomy and to explore the utilization and human factors issues involved with teleproctoring and telepresence of surgical procedures done in high volume. This would allow generalists operating on the battlefield or in peacetime to receive real-time expertise form senior level surgeons at a different location. When technological advances over the next decade make this inexpensive, there will be improved quality of care for surgical procedures performed at distance sites, as well as cost savings by not having to transport every patient to a tertiary center.

We will take a rapid prototyping, object-oriented approach to software and hardware technologies to create a true plug and play system of components, allowing easy upgrades in hardware and software as new technologies are developed and become available. The proposed management approach may provide low cost, rapid solutions to clinical testing and implementation of promising new technologies to improve diagnosis and quality of care for patients.

The program produces: Distance Learning, Call the Expert, Customer Service, Visual Rehearsal, and Remote Diagnosis.

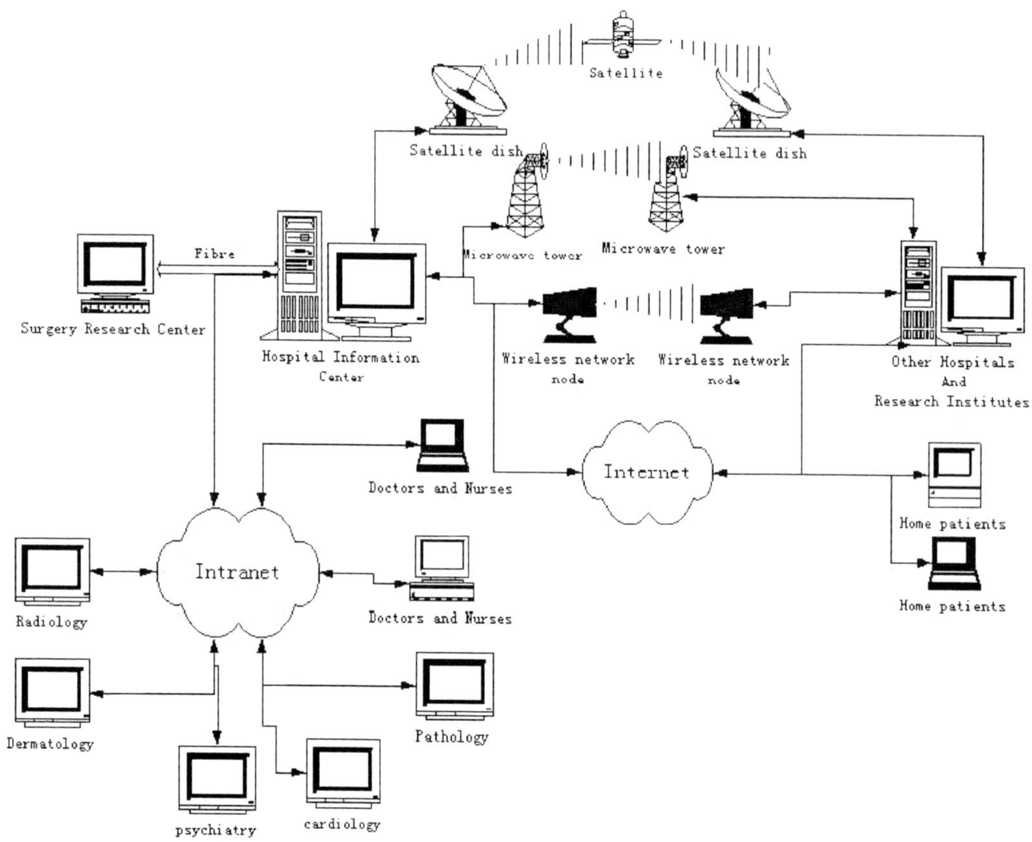

Figure 1. A simplified graphic representation of the proposed telemedicine system.

3. Technical approach

Following McGee and Tangalos [2] telemedicine encompasses three components: (1) diagnostic and consultative support to remote medical practitioners, (2) direct delivery of medical care to patients at distant sites, and (3) continuing medical education. The major focus of telemedicine research and pilot programs continues to be improving the quality of medical care in under-served rural areas [3].

All telemedicine services involve a mixture of multimedia data communication and Computer Supported Cooperative Work (CSCW), ranging from the most demanding real-time synchronous communication, to the simpler case of asynchronous communication and cooperation. The temporal characteristics of this information exchange and the type of data involved determine which applications are required to support a particular telemedicine service. When used in telemedicine, it takes the form of multimedia mail composed of diagnostic data (static or dynamic), medical and demographic patient data, as well as comments by physicians, written or spoken. The asynchronous mode of communication suggests that electronic mail does not require high transfer rates and can be implemented over any networking infrastructure.

Figure 1. is a simplified graphic representation of the key communication components underlying telemedicine. The purpose of this project is to build academic communications between the Surgery Research Center (SRC) and other hospitals and institutes, to provide medical consultation and service, and to provide technical support within the hospital for doctors, nurses, and relevant departments through the Information Center of Southern Hospital (ICSH).

SRC and ICSH will be connected with optical fiber. ICSH and departments can be connected via intranet using optical fiber between different buildings to acquire bandwidth over 1Gbps. Class 5 wire inside the buildings will provide bandwidth of 100M. Thus the clinical information, such as X-ray film, CT, MRI, ECT, can be shared and studied using multimedia technology. ICSH can be directly linked to remote medical organizations through satellite, microwave and wireless network to carry out telemedicine diagnosis and discussion. Or remote medical organizations and patients at home can get telemedicine consultation and emergency help from SRC through the Internet.

HIS has now been built in Southern Hospital. The PACS under construction will be integrated with HIS to provide X-ray films, CT, MRI, and ECT images in DICOM format [6].

4. Discussion

Potential benefits of telemedicine services include improved access to diagnostic imaging facilities, reduced costs, reduced isolation for both patients and medical personnel, improved quality of care through real-time diagnosis, advanced decision support tools, increased collaboration and continuous education, reduced loss of patient data, and direct access to remote computational facilities for advanced image processing and 3D visualization. The above benefits are likely to become evident after a transition period, during which users of telemedicine services receive proper training and learn to trust them, potential legal issues are resolved, and capital and operational costs are reduced, while service quality is improved with new technological developments.

The main problems of the project involve the image acquiring and format converting, the interface among HIS, Internet and intranet, the high efficient store, transfer, security and reliability of medical information in HIS.

5. References

[1] S.C. Orphanoudakis, E. Kaldoudi, and M. Tsiknakis, ,Technological Advances in Teleradiology, European Journal of Radiology, 22, 205-217, 1996.

[2] McGee,R.and E..G.Tangalos,1994."Delivery of Health Care to the Underserved: Potential Contributions of Telecommunications Technology." Mayo Clinic Proceedings, 59:1131-1136.

[3] Puskin, D, 1992. "Telecommunications in Rural America, Opportunities and Challenges for the Health care System." In: Extended Clinical Consulting by Hospital Computer Networks. D.F. parsons, C.M. Fleischer and R.A. Greenes (eds.) PP. 67-75. Annals of the New York Academy of Sciences, vol. 670, New York Academy of Sciences, New York, NY.

[4] Bashshur, R.L. P.A. Armstrong, and Z.L. Youssef (eds) 1975. Telemedicine: Explorations in the Use of Telecommunications in Health Care. Charles. C. Thomas, Spring field , IL.

[5] Goeringer, F. and R. Leckie, 1994, U.s. Military Medicine: PACS< Telemedicine & Personal Digital Assistant Devices. In: SCAR94 Computer Applications to Assist Radiology, J. Boehme , A Rowberg, N.Wolfman (eds.), PP. 297-285. Symposia Foundation, Carlsbed. CA.

[6] Hindel, R(ed.) 1994. Implementation of the DICOM 3.0 Standard, Radiological Society of North America, Oak Brook, IL.

CSCW Based Telemedicine on the Internet

Seok-Soo Kim
Dept. of Computer Engineering
Dong-yang University, Korea
sskim@phenix.dyu.ac.kr

Dae-Joon Hwang
School of Electrical and Computer Engineering,
Sung-kyun-kwan University, Korea
djhwang@yurim.skku.ac.kr

Abstract

This paper proposes the telemedicine system which is a CBM-based real-time interactive telemedicine system using personal computer on the internet. And this system presents an intelligence type remote medical examination system for both on-line and off-line mode to transcend time and space on the web being participated by anybody, which is cheap type to solve problems in existing remote medical examination system such as high price based on hardware, incompatibility, and so on.

And, this is a system where data processing of general member and charge member, doctor is differently performed by being divided into each class, certification of each member must follow by all means because service access right is endowed according to this, and remote medical system for western medicine of mutual conversation type between 3 parties(patient, doctor, pharmacist) are jointly operated ; and this presents the content regarding electronic medical examination chart and data processing for efficient and fast medical examination and treatment.

1. Introduction

Recently, various applied common ownership service such as remote education, remote medical examination, multimedia video conference, electronic approval system, etc. based on dispersed multimedia and network environment is being developed owing to high performance of PC and establishment of ultra-high speed information communication network of national level.

A remote medical examination system is a system to support bi-directional audio and video examination between doctor and patient using information communication system as much as possible using hardware and software technology of communication and video/audio according to development of modern equipment. However, establishment of data compression technology and ultra-high speed information[1].

Such modern remote medical examination must be accompanied by multimedia technology as well as high price hardware element and network environment. However, cyber doctor system proposed in this paper is a quick medical treatment system to transcend time and space among patient, doctor, and pharmacist in internet environment, whose service scope is different depending on charge or free service, and efficient medical examination, prescription, and medicine preparation according to separation of medical practice and pharmacy can be performed by this system, which is supposed to perform exact medical examination, quick medical care, and efficient data management by continuously searching connected data thereafter through management of treatment data by proper patient, doctor, and pharmacist [2,3].

2. Cyber doctor system

The study having been performed until now is limited to partial medical information, but this study establishes medical professional site of portal concept, so as to make new medical system as well as to bring mutual benefit among patient, doctor, pharmacist, and health related business with various type of service [4,5].

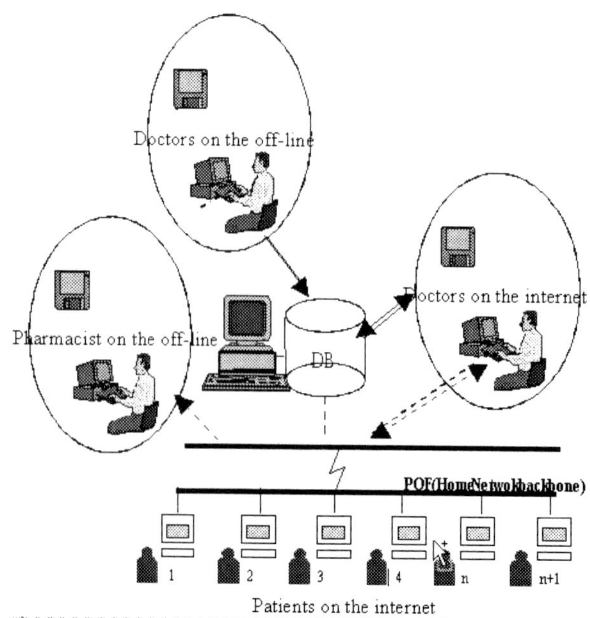

Figure 1. Network structure of cyber doctor system.

Table 1. Configuration of Cyber doctor.

Purpose / Member		Role/benefit of each member
Operating manager(OM)		1. Members using Cyber doctor site are connected to each doctor to receive medical examination on cyber, and are connected to doctor who is near home to receive medical examination. 2. Free member- Consultation only is possible in inquiry type on web site. Charge member- Direct consultation is possible with private doctor, and special benefit is provided when visiting hospital of private doctor. 3. Site is operated with member registration fee(charge number, doctor) and advertisement fee. Medical examination data of members are composed as D/B so as to be used anywhere in the world, and are preserved semi-eternally.
Patient	Free member	1. Private medical consultation is possible through bulletin on web site. 2. It is possible to use things other than menu limited on web site.
	Charge member	1. All rights which free member has are effective. 2. A medical prescription can be received from private doctor. (Periodic examination on cyber) 3. When it comes to the things impossible such as prescription, examination, preparation, etc. on internet, members can visit doctor directly for examination or prescription. 4. In case of visiting doctor directly, charge and free members can receive special benefit.
Medical doctor		1. Doctors can provide consultation directly for registered members, and it there is request from patient, the patient can visit hospital for examination. 2. In case of providing prescription of medicine, members can visit hospital directly, or can receive medicine by mail. 3. The doctor who lives at the nearest place from the place where charge member lives becomes private doctor.

2.1 Design of Web system and D/B

This enables medical examination and consultation on cyber space, and medical examination impossible on Internet is directly connected to doctor for medical examination. Personal (family) medical examination record can be referred to in all over the world, and can be preserved semi-eternally [6].

(1) Special member (private doctor) system and general member (free service) system

Customers who use center can receive most service freely as general member, and can be differentiated as special member in order to receive better quality service. This method enables periodic medical examination and reservation with private doctor of home doctor concept, and enables effective examination and service in case of using off-line hospital (medical examination, inspection, care).

(2) Management of personal case history

Personal medical examination data can be seen by patient, which enables self health control, and quick examination, error prevention, and effective care based on the past case history through this data in case of emergency or in case of receiving medical examination at other hospital institution.

(3) Connection with other medicine related site

This cyber medical examination center links many registered doctors (hospital and clinic), pharmacist (pharmacy), and various health related institution (health center, health institution, medical tool agent) in order to bring various cooperation relationship and mutual benefit. That is, patient can perform professional self control with regard to health by using various portal site specializing in medicine, and each doctor performed free consultation for patient, but such responsible doctors are visited in case of using off-line hospital, and it is possible to obtain various information through medical examination cooperated with other doctors. And, pharmacist can perform smooth commercial interchange through mutual interchange with registered doctors pursuant to separation of medicine and pharmacy, and can contribute greatly to national welfare pursuant to this. And, it is possible to obtain various commercial effect when using off-line and on-line(commercial transaction) while giving benefit to patient by providing various types of information for related medical tool enterprise or health related center.

2.2 Configuration

Configuration of cyber doctor presented in this paper is shown in the table 1. Operating manager(OM) is web master to manage site, and patient is divided into charged patient or free patient, whose service is differentiated. And, patient who performs consultation on cyber of patient is performing actual medical examination on off-line system as well. The data of medical examination on off-line mode are treated by electronic medical examination chart, whose content is transmitted to the pharmacist designated by patient in order to prepare medicine.

2.3 Scenario

1) Required information is provided for the users who need consultation with regard to health through internet.
2) Patient can be examined at hospital more conveniently in case of wanting to receive medical examination at near hospital through medical consultation provided on cyber.
3) Medical examination data of patients who are suffering from large operation or long chronic disease are connected to this site, and the stored data can be used anywhere.
4) Medical examination data is checked and medicine is prescribed at any pharmacy through personal ID.

3. Design of intelligence type database

3.1 Development process

Relation type database is realized by using WinNT 4.0 which is an operation system in order to operate opendoctor site, using IIS 4.0 which is web server program, using SQL and Access 97 for database of medical examination data with members during web operation, and using ASP language, VBScript, and JavaScript in order to handle DB.[7,8]

Data treatment is different between general member and charge member and between doctor and pharmacist, and service access right is endowed according to this, so certification of each members must follow. Login and certification treatment diagram of each member of cyber doctor system which is common field part of such members is supposed to include member ID, password, member name, sex, resident number, grade, home phone number, cell phone number, mail number, home address, e-mail, recommended ID, class, member check question, answer, point, etc., and difference between members is determined by class.

Charge member can request consultation to expert, and when requesting consultation, doctor is selected first and then the content of consultation is inputted to doctor. When doctor finishes consultation, charge member can see the content of consultation. Doctor receives the content of consultation first and then performs medical examination, and the medical examination data is opened to charge member. If this consultation data is presented to pharmacist for medicine, the pharmacist prepares medicine.

4. Realization and analysis

This cyber doctor system presents the content regarding electronic medical examination chart and data treatment for efficient medical examination and fast treatment ; database is established by interconnecting ASP and SQL on IIS 4.0 web server to enable system integration for efficient data handling of both on-line and off-line mode, consultation between patient and doctor, medical examination in off-line mode, transmission of prescription sheet to the pharmacist designated by patient, preparation of medicine, semi-eternal medical examination data storage owing to storage and check of medical examination data, more exact medical examination and prescription using this medical examination data of patient and doctor, etc. ; and figure 1 shows an example of history application where the content of description sheet is transmitted to the designated pharmacy and medicine is prepared after medical examination of patient is finished.

1) Log-in and certification of each member
2) Consultation treatment for patient member
3) Database processing for consultation treatment of patient member
4) Consultation treatment list for doctor member
5) Database processing for consultation treatment of doctor member
6) Prescription treatment of pharmacist
7) Database processing for Prescription treatment of pharmacist

Figure 2. A process for implementation of the cyber doctor system.

Especially, semi-eternal class between members according to efficient intelligence type data handling presented in this paper, management of differentiated medical examination data, fast medical examination through disease history, prevention of mal-examination, and effect of cooperated medical examination between doctors are brought.

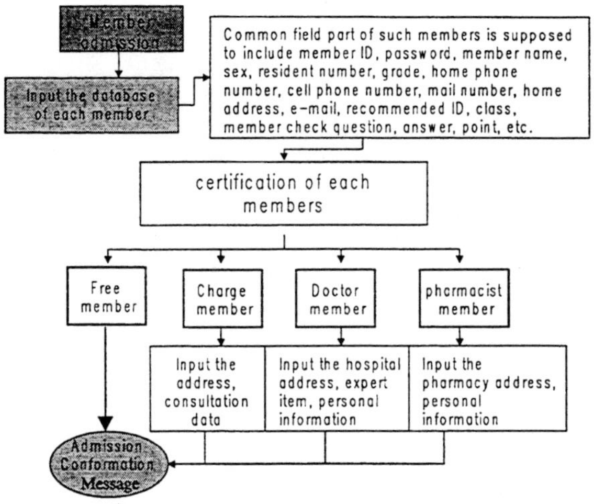

Figure 3. Login and certification treatment diagram.

5. Conclusion

In this paper, we are proposes the telemedicine system which is a CBM-based real-time interactive telemedicine system using personal computer on the internet.

This telemedicine system is an intelligence type for both on-line and off-line mode to transcend time and space on web where anyone can participate. And, this is a system where data processing of general member and charge member, doctor is differently performed by being divided into each class, certification of each member must follow by all means because service access right is endowed according to this, and remote medical system for western medicine of mutual conversation type between 3 parties(patient, doctor, pharmacist) are jointly operated ; and this presents the content regarding electronic medical examination chart and data processing for efficient and fast medical examination and treatment.

Also, database is composed by using intelligence type of object control which enables effect of treatment for heredity including check and mutual holding of medical examination data among family, fast treatment effect pursuant to situation of past medical examination through control of private disease history, cooperated medical examination among doctors, and so on.

6. References

[1] Medical Communications - The home of medical image distribution and teleradiology systems, *http://www.medicalcommunications.com*

[2] *ITU Draft Rec.* T.120, Data protocols for Multimedia conferencing, International Telecommunication Union, Feb. 1999.

[3]. "Telemedicine Coming of Age", The Telemedicine Research Center Home Page, *URL: http://trc.telemed.org/, 2000*

[4]. Kilgore C (Dec 1999) Patients take the wheel with internet health records. *Telehealth Magazine* 5(7) : 7-8.vol 5, no.7, 2000, Dec.

[5] Seok. S. Kim, Hae. M. Kwon, "Design of a Multimedia Telemedicine System for CBM Based Collaboration Environment : TELME, " *EDEN conference* '97 June 23-25. 1997

[6] D. J. Hwang, Seok. S. Kim "Development of Telemedicine which is a CBM based Collaborative Multimedia System on LAN Environment ", *The transactions of the korea information processing society* v.4,n.5 pp.1153-1161 May 1997.

[7] Alex Homer, *Active Server Page,* 1997.

[8] Scot Johnson, *Active Server Page Application*, 1998.

Design and Implementation of a Communication Mechanism for WEBGIS Based on a Distributed Component Model

Chaozhen Guo[1], Qingming Wang[2], Yuan Xu[1]
[1] *Department of Computer Science and Technology, Fuzhou University*
[2] *Institute of Globosity Information Science, Fuzhou University,*
Fuzhou, Fujian, 350002 P.R. China
gczwfm@pub5.fz.fj.cn

Abstract

This paper is an introduction to the design and implementation of a communication mechanism for WEBGIS (CGMGIS) based on the distributed component model. Since the format and method of the data transmission on the network have great influence on the performances of WEBGIS, great emphasis is placed on how to improve the transmission of the data and to create a new communication mechanism for WEBGIS by utilizing the advantages of both the DCOM and WebCGM technology and introducing a new WEBGIS data transmission and communication mechanism. Meanwhile, a new approach of real-time dynamic index is introduced focusing on the display of the spatial typographic information at the client end, in order to solve the problem of repeated transmission of the spatial typographic information so that the flow of the network data transmission can be cut down.

1. Introduction

WBGIS derives from the combination of the technology of Internet and GIS. WEBGIS is a geographic information system based on WWW (World Wide Web). Compared with the normal information systems based on the Web, WEBGIS's characterized by its implementation of the dynamic connection of graphs, geographic data, and property data under the spatial framework and the function of the visual query and spatial analysis.

The necessary steps for the operation of GIS include the construction of the geographic entity object, the organization and transmission of the data. However in the WEBGIS system, how to transmit the user's requests and the result data effectively has a great influence on the system's performances due to the constraint of the protocol of the network transmission, the quantity of the real-time visit, and the used bandwidth. So emphasis is given to how in this paper, CGMGIS system overcomes the constraining factors and forward a series of complete systematic modification plans from the sever end, the middle protocol layer and the man-machine client interface.

First, CGMGIS system introduces a new communication and transmission mechanism for the new spatial data based on the technology of the distributed component object in order to get rid of the demerit that normal WEBGIS application system is only capable of conducting the simple transmission of data.

Second, vector graph the main format of is spatial data transmission by the CGMGIS system. The ratio of the compression of vector graphs format is far better than the bitmap graphs, especially when the spatial geographic objects in the GIS system is mostly made up of simple graphic elements such as points, lines and areas etc. In addition, it is very convenient to analyze the file of vector graphs to get the relevant information[1]. CGMGIS utilizes the most potential format of web vector graph in order to be compatible with other products in other companies.

Finally, a new algorithm of real-time dynamic index for the spatial information query is introduced in order to cut down the workload in the net transmission and the machine at client's end and to improve the performances of the query of the CGMGIS system.

2. The Brief of the Distributed Component Object Model (DCOM) and WEBCGM Vector Graph Format

2.1 Distributed Component Object Model (DCOM)

The key to the design and development of WEBGIS is how to reduce the spatial dataflow transmitted on the web, especially when the bandwidth of current network is not desirable.

Normal solutions for the WEBGIS server end, such as CGI, API, ASP need the HTTP or WEB severs as a medium. The basic data exchanged between the objects of server and client is still HTML text. There is little room for the improvement of rate when it comes to the application of complicated client/server where a large amount of net communication is involved. So a web solution called object web based on the distributed object is introduced in June 1997. This kind of solution admits the client object invokes the object's method on the server for the bottleneck created by the CGI and other methods can be avoided. The distributed object platform supplies a flexible system structure from server to server so that the server object can run on different server machines to dynamically balance the workload as requested by clients, CGMGIS uses DCOM as basic protocol structure because the practical system is running on the Windows operation system.

COM (Component Object Model) is a kind of technology that makes it possible for objects to go beyond the limits of process and machine. To achieve the goal, it is stipulated in COM. The only way to operate the data and method in an object is the interface.

DCOM (Distributed Component Object Model) is the extension of COM in the distributed environment. The introduction of DCOM brings COM technology to a new stage. The application program objects can communicate and run via network such as the Internet, intranets, LAN, WAN. DCOM permits many objects that are part of an application program running on many different computers individually. The local application program can not only deal with the communication between different objects but also control the establishment and running of the remote application program objects. Most important, DCOM enables local application program to use the remote objects in much the same way as the local objects . It seems that these remote objects are local. The basis of the DCOM technology is RPC (Remote Procedure Call) technology[2]. DCOM enables component objects to communicate with each other via various network protocols including TCP/IP and IPX/SPX protocol. The employment of DCOM is capable of control over various aspects in the security mechanism, such as the control visits and area identification. In addition, DCOM can load a component object in another computer into the memory and run it remotely. These special characteristics of DCOM enable software developers to develop the real distributed application system without giving consideration to the performances of network, the capacity of the system or the integration of all kinds of components.

2.2 WEBCGM the Format of Vector Graph

Another important choice during the design of the WebGIS system is to choose the vector graph or raster graph at the client end. If the raster graph is chosen, there is only the x and y coordinate available for the mouse that can be read on the client end. If the operations of the map such as zoom-in, zoom-out, pan and select etc are needed, the x, y coordinate has to be sent to server for further handling, then the newly-made raster graph is sent back to client end for display. Because the information of raster graph is relatively great, the burden of the server and network transmission is increased. Some current WEBGIS sites use MapInfo Proserver and MapXtreme of MapInfo Company and IMS of Esri Company as development platform. The technology employed is to Commit all the operations to the server and transmit the result to the Client and in the form of the raster graph. As a result, the response time on the client end is long. On the other hand, if the vector graph is chosen, it is necessary to install plug-in on the client end in advance or download the JAVA applet or ActiveX control automatically in operation. The raster graph at Client end does not call for any extra programs. But with the help of plug-in, Java applet or ActiveX control, many operations can be performed locally such as zoom-in, zoom-out, pan, etc., which to a certain extent decreases flow of data transmitted over the network and the workload on the server end.

So far in the research field of WEBGIS, WEBCGM is considered as a hopeful vector graph format with great technical potential [3]. WEBCGM is a child set of the CGM□COMPUTER GRAPHIC METAFILE□graph format standard under specific constraints stipulated in the standard of ISO/IEC 8632:1992. But these specific constraints not only improve the interoperation of WEBCGM graph format in the network environment but also simplify the work of the interpreter tool of the WEBCGM graph and improve efficiency.

A WEBCGM graph is made up of one or more pictures, each includes CGM graphic elements and some special applied structures, for example, the layer structure stands for a group of graphic elements. Applied structure has properties as or features graphic elements do.

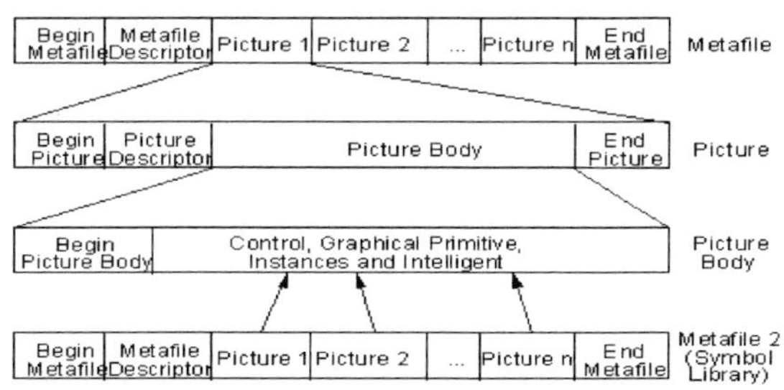

Figure 1. WebCGM File Structure.

In the Figure 1, we can see that WEBCGM has a typical nesting structure. The following is the textual description of a typical WEBCGM file:

```
Begin Metafile
      Begin Picture
            Begin PictureBody
                  Begin Layer1
                        Point1
                        Point2
                        ......
                  End Layer1
                  Begin Layer2
                        PolyLine1
                        PolyLine2
                        ......
                  EndLayer2
            End PictureBody
      End Picture
End Metafile
```

3. The Design of Communication Mechanism of CGMGIS

A new WEBGIS data transmission and communication mechanism is introduced by way of synthesizing and refining the advantages and merits both DCOM protocol and WEBCGM vector graph format. The following is the basic design principle of CGMGIS communication mechanism:

Definition: the communication mechanism of CGMGIS is a communication mechanism used for the transmission of the spatial data of WEBCGM vector graph format between the WEBGIS software at client end and server based on the distributed component object model.

Aim: to establish a new transmission and communication mechanism of the WEBGIS data to encapsulate the objects used in the WEBGIS system to enhance its practicality so the flow of the spatial data on the network can be lessened further and the speed of response of software at client end can be improved . So it's necessary to optimize the format and mechanism for the transmission of spatial data. The layer of protocol based on the communication mechanism of CGMGIS is shown in the Figure 2.

From Figure 2 we can draw the conclusion: through the communication mechanism of CGMGIS, the server software and client software can communicate with each other. Only by invoking the corresponding objects in the class library of the communication mechanism of CGMGIS without considering the details of data transmission format and its implementation on the lower layers.

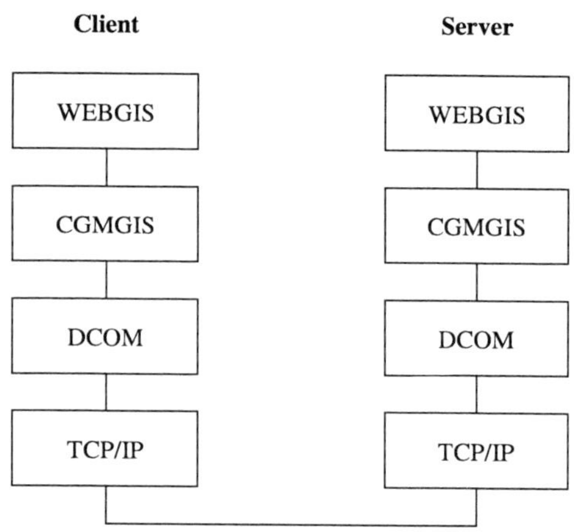

Figure 2. The layers of protocol of CGMGIS.

4. The Algorithm of Dynamic Index for the Design of GIS Interface at the Client End

In the process of the design of client interface for the WEBGIS applical system, there are two important factors of design, namely the effect of display and the response time. A common designing strategy is to download in advance the spatial information of the region that is several times the current region being shown from the database on the server according to the sizes of the region currently needs. When the user wants to pan the graph of the current region, the user feels that the new graph is displayed immediately without any flashes because the adjacent graph around the current graph have been partly downloaded and stored in a local machine. There is no need to get the information from the database on the server. However, this technique brings up some problems, for example, the flow of the data transmitted on the network is greater because each time a graph that is much bigger than the current graph should be downloaded to the local machine. Then the more flow of network data inevitably leads to the worse response time inevitably, thus, a negative impression of the whole system will eventual imprint on the user.

In this paper a method is employed using the dynamic index of the graph displayed at the client end. On the basis of this, relevant algorithms can be introduced to avoid the repeated transmission of the graph information that has been formerly transmitted so the whole flow of the network data can be lessened.

- **The first step: establish the dynamic index of the graph information**

First, we can divide the spatial graph into blocks. Then we should find a new way to guarantee that one block of the spatial graph is coded solely and is easy to be implemented,

The exclusiveness principle in the Cartesian coordinate system gives us the revelation. It is stimulated that a point in the Cartesian coordinate system can de determined by a planar coordinate with the format as (x, y). According to this principle, if each block containing the graph is seen as a point logically, a Cartesian coordinate can differentiate this point from other points on the same plane. This means each block containing the graph can be determined exclusively.

According to this idea, the first spatial graph displayed on the client's window region will be coded as the initial block with values of (0,0). And the other 8 blocks around the initial block will be coded with values from (-1, -1) to (1,1). In this way, an arbitrary central block with the code number of (x, y) has the surrounding 8 blocks with values from (x-1, y-1) to (x+1, y+1) as is shown in Figure 3. So each block can get a definite number index exclusively. And the code needn't be determined in the program in advance, because the initial block is arbitrarily determined. What's more, there's no need to store the table of code number index in the computer because they can be established dynamically while the program is running. Therefore, it is referred to as dynamic index.

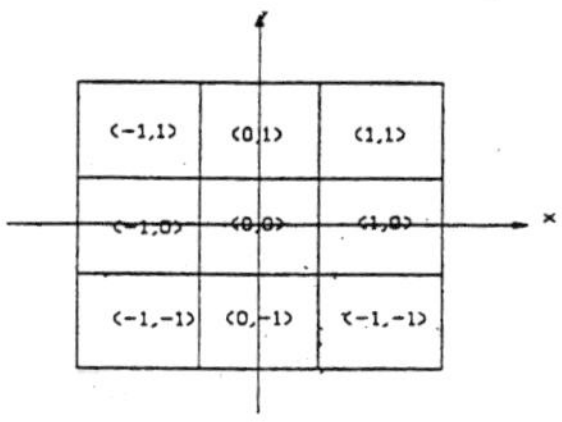

Figure 3. Dynamic index of blocks.

- **The second step:**

(1) Operation of get all the block's numbers that need to be judged whether they exists at the client end or not.

There are 8 possible directions of moving graph among which 4 on the diagonal line can be seen as the combination of move on both directions of x axis and y axis. At the same time, the moves on x axis and y axis are the same in nature. So we can assume the former position of the graph is on the NO 5 block, then the graph move to the shadow region , namely the 3*3 block group with the dashed line border centering around block number 5'as is shown in Figure 4.

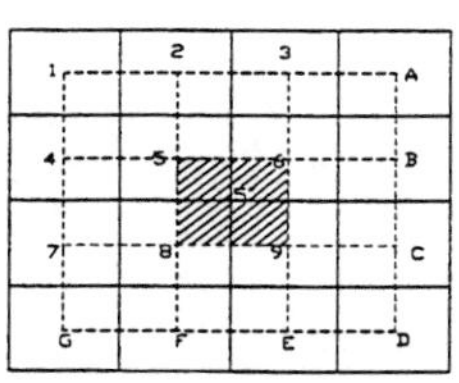

Figure 4. The moving of the graph.

Because the direction of the moving graph is random, and in the position of the shadow region 5 the adequate necessary condition to guarantee that there's no need to wait for the extra information downloading when the graph is moving is horizontally downloading in advance the graph whose size is nice times the size of current graph displayed. Therefore, in Figure 4, the 3*3 block group with the center shadow block numbered 5', namely the block group with the dashed border line, is the block group after moving. And the former block group with the central block numbered 5 including the blocks with numbers from 1 to 9 must have been downloaded at the client end in last downloading. At the same time, the unit of downloaded graph information is based on blocks. As a result, blocks A, B, C, D, E, F, G, namely those blocks covered by the graph region beyond the region of the former block group in the moving direction ,should be downloaded to client region. In implementation, we can figure out all the block index numbers necessary for the judgment as to whether they are at clients end or not by way of the region index numbers in which the summits of the shadow region block are situated.

(2) exclude the blocks at client end

A table of database should be produced to store the index code of those blocks that have been exist once at the client end, then a comparison between the block index numbers got in from first step with those already in existence will exclude blocks appeared at client end I enhancing WEBGIS performances and define the ones that should really be downloaded.

5. Conclusion

This paper is about the application of the CGMGIS communication mechanism. In the development and application of the WEBGIS system, the key to improving the performances and response time is the flow of the spatial data transmitted on the network. In order to solve this problem, CGMGIS is the combination of the technology of object-oriented, database the latest development on the net and former GIS technology with administration system for the planning of oceanic regions as the background for application and internet as the basis CGMGIS becomes a WEBGIS system in which users are capable of inter operation on line. However, there is still much to be discussed in the algorithm of the transmission speed of spatial data based on the Internet.

References

[1] Kong Yunhong and Lin Hui ,Intergrated Research on WWW For GIS System, Remote Sensing Journal, Vol. 2 No2.

[2] Markus Horstmann and Mary Kirtland, DCOM Architeeture, PDC 97 Conference Papers, July 23, 1997 .

[3] ISO/IEC 8632:1992 Standard.

Computer Supported Force Analysis and Layer Imagine for Masked Rapid Prototyping System

You-Min Huang , Jeng-Ywan Jeng, Cho-Pei Jiang, Jia-Chang Wang
Department of Mechanical Engineering, National Taiwan University of Science and Technology, Taipei, Taiwan
ymhuang@mail.ntust.edu.tw

Abstract

An innovative rapid prototyping (RP) system using direct mask photo curing (DMPC) is proposed in this paper. The underlying principle of RP system is to fabricate a three-dimensional part layer by layer. The design concept of the DMPC RP system is to use a dynamical photo mask for the layer profile. The image of the dynamic mask, generated by a thin film transistor liquid crystal displayer (TFT LCD) panel, is calculated from the slicing data of the CAD model. In the DMPC system, the photo-polymer is exposed and shielded by the LCD mask, and the light source to cure the photo-polymer is from the bottom of the transparent resin container. A residual force is formed between the layer of the cured photo-polymer and the silicon film on the container. This residual force is measured using a load-cell and its theoretical model is further developed and simulated using finite element method. Hence, the force measurement, force simulation, layer imagine calculation, LCD control, position control and process parameters are supported by a host computer. As a result, good quality part with low cost can be built by this proposed DMPC RP system.

1. Introduction

RP technology has been widely used in most industries to fabricate models or prototypes. The key technologies of RP process can be categorized as laser technology and nozzle technology [2]. The photopolymer fabrication process is the most important process for the laser technology. Stereolithography (SL) is the most popular photopolymer fabrication process to cure successive thin layer of photopolymer until a 3D part is fabricated. It uses an ultraviolet laser to cure the associated photopolymer [1,8]. However, using UV laser in the SL system is expensive in terms of equipment and maintenance. Lower-priced RP systems are in high demand in the RP industry [10], because the high capital and maintenance costs of the SL system limit wider application and popularity of the RP system. On the other hand, the curing speed of photopolymer using a laser point is slow. This relatively low curing speed of laser point sacrifices curing strength in part. Hence, a post curing process using higher power UV light is used to fully cure the photopolymer. However, the uncured and partially cured resins trapped within the photopolymer result in inhomogeneity of curing in the post curing, causing shrinkage and distortion [4].

One of the solutions to the low laser point curing speed is to use an alternative method of irradiation, a masked lamp. In this method, a patterned mask is prepared and laid in the path of light from a lamp. The mask allows light to pass through to cure the parts of the surface that form a single cross section of the desired object, but blocks light from reaching those regions which are intended to remain liquid [1]. Although the masked-lamp curing process accelerates the fabrication speed, the preparation of masks is another implementation problem, which makes the Solid Ground Curing (SGC from Cubital) process complex. Hence, the capital and maintenance cost of the Solider is still very high.

Sensitive photopolymer with a small shrinkage coefficient will improve the accuracy of the prototyped profile. Ullett [5] investigated the novel liquid crystal resin for stereolithography process parameters and mechanical analysis to gain the accuracy. Furthermore, several studies have been conducted on smoothing the differences of each lamination with lasers vertically and obliquely by Narahara [9] and by Hagiwara [6], respectively. However, no work has been done on reducing equipment cost and on-line monitoring by computer supported. For this reason, one of most important technical problems is to shorten the expensive processing time. In order to reduce the RP system cost, UV laser and galvanometer-scanning mirror have to be replaced.

In this paper, a new masked rapid prototyping system has been developed, named direct mask photo-curing system. To shorten the processing time, the liquid crystal display (LCD) used for generating photo mask for each layer. To stabilize the bounded layers, computer supported force analysis and on-line monitoring of stabilizing applied for keeping an eye on the stability. The friendly user interface and fundamental theory of pulling force is also introduced.

2. The Principle of DMPC System

There are two major innovative designs of the DMPC system. Instead of laser scanning technique, a metal halide 575Watt lamp used as the light source and a dynamic photo mask of a large TFT panel are used to produce a 2D layer. The exposure system is different from the upper surface curing of SL system, because a bottom illumination is used to reduce the complexity of resin level control. The photosensitive resin is exposed by the light shielded thorough the LCD mask. Optical system, as shown in Figure 1, is used to generate parallel light to cure the photo polymer from the bottom of the resin tank. The TFT LCD is not only used to generate the mask for exposure, but also used for shuttering. When the current layer has been cured, the LCD turns to black color and blocks all the light as a shutter until the next layer is ready to be cured. Figure 2 illustrates the construction of the DMPC system. A thin and soft silicone film is adhered at the top of the contacted window to reduce the pulling force during the pulling up process. In the fabrication process, a load cell is used to monitor the pulling force of every layer and a joint connection is designed to eliminate the bending moment of pulling force.

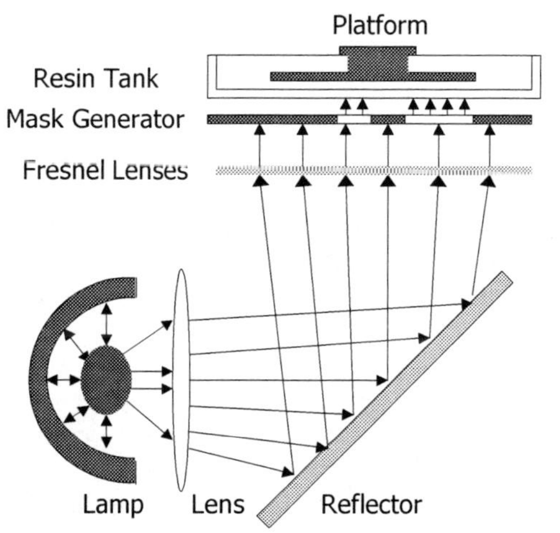

Figure 1. The diagram illustrates the optical system of DMPC.

In order to unify the light distribution, the optical energy passing through the LCD mask is measured by a photo-detector, and then a compensated image can be calculated. The distribution of the optical energy after LCD mask is shown in Figure 3. As shown in Figure 3, the energy distribution at the board of the working area is not as uniform as at the center. Hence, only the area at the center is employed for the part fabrication.

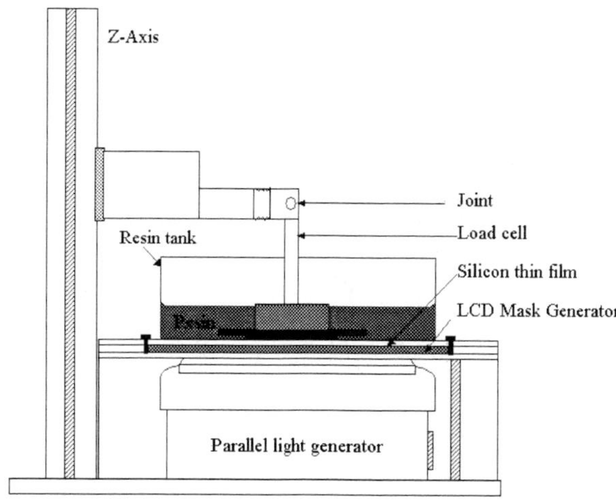

Figure 2. The diagram illustrates the construction of the DMPC system.

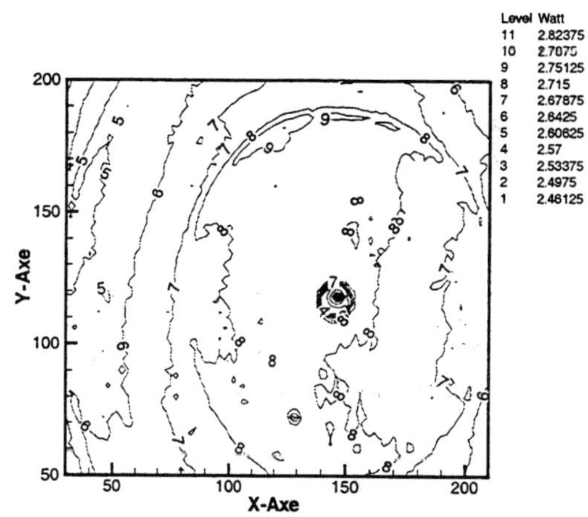

Figure 3. The diagram illustrates the energy distribution after passing through LCD mask (unit Watt/cm^2).

3. Principle of Computer Supported Force Analysis and On-line Monitoring

One of the advantages of the DMPC system is to use a surface constrain method to fabricate prototype, because the layer is cured at the bottom of the resin container. This fabrication design eliminates the problem of uneven surface cured at a free surface of SLA process and saves the usage of photo-polymer. However, the layer cured on the contacted window results another problem of separation between the cured layer and window, because the photo-polymer cured and stuck on the window during curing process. After the current layer has been fabricated, the platform is pulled up to separate the part and the window, and at the same time the resin fills into the gap between the part and window automatically. The new cured layer or part may not be pulled up successfully, because the stuck force between the cured layer and the window may be larger than that of part strength or bonding strength of layers. In order to solve the problem of pulling up, the computer supported force analysis and on-line monitoring is developed and employed in DMPC system.

The fabrication flow of the DMPC system is shown in Figure 4. As shown in Figure 4, the 3D CAD model of part must be constructed for further slicing calculation and layer image generation. Once the layer image has been generated, the layer image is transferred to the LCD as the shutter and mask to fabricate the part layer. Also, the computer calculates the required pulling force according to the force analysis model and layer geometry information. Then, the parallel light passes through the LCD mask and cures the photo-polymer. Once the photo-polymer has been cured, Z-axis moves up to pull up the part and separate the part from the contacted window. Figure 5 illustrates the operational window of the DMPC system. As shown in the right-up side of Figure 5, the control of Z-axis movement can be categorized as several steps according to force analysis model. Also, a load cell is employed to monitor the pulling force in-situ as shown in Figure 5 to avoid the failure of part pulling up. That whether the pulling up is successful can be determined by the force measurement of load cell and force analysis model. On the other hand, if the strength of the current layer is too low to pull up the part, the computer will make the current layer to be exposure longer to increase the part strength. This fabrication process repeats until all the layers have been cured.

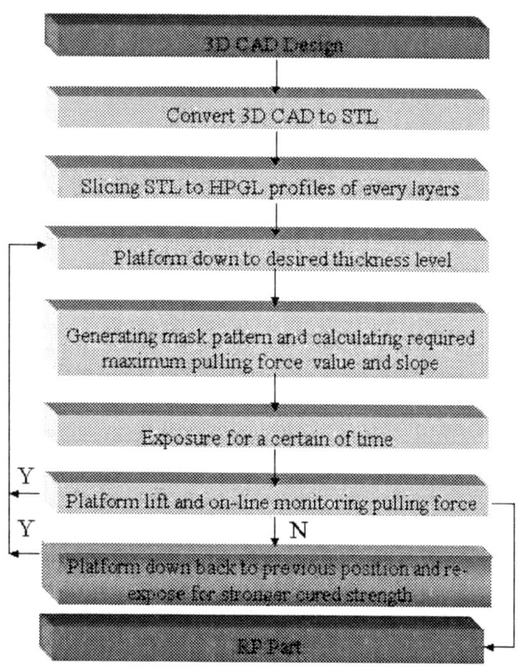

Figure 4. The diagram illustrates the fabrication flow chart of the DMPC system.

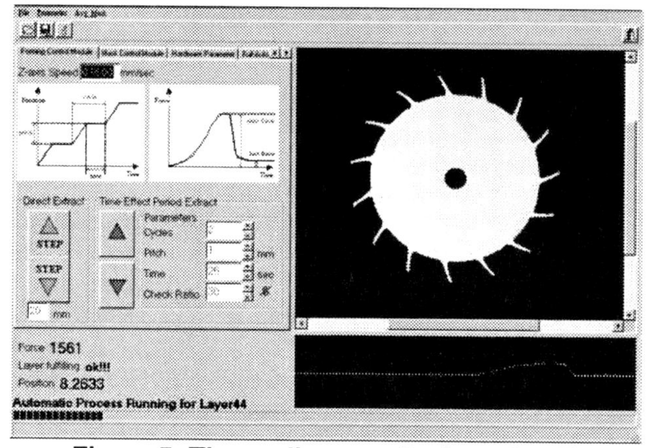

Figure 5. The on-line monitoring of pulling force value and slope.

4. The Effect of Exposure Time on Curing Depth

The absorbable wavelength and reaction rates are two important characteristics of photo-polymer used in the photo-polymerization type of RP systems. The wavelength and power of laser or optical system used in the RP system determine the curing depth and reaction time of photo-polymer. Most of the lasers or optical system used in the RP system is in the range of UV light. The photo-polymer of DMPC system used in this study is NAF202 from Nippon Kayaku Company. A special photo-initiator has been added into NAF 202 to initiate the photo-polymerization in the range of visible light. It was designed for the EDARTS RP system [3], which uses a red diode laser with the wavelength of 650nm and power of 40mW. The optical system of DMPC system provides a visible light in wavelength of wider range and higher power to cure the photopolymer. In order to control the process parameters of the DMPC system to fabricate a 3D part, the effect of the exposure time on the curing depth must be evaluate for further force analysis.

In order to evaluate the effect of the exposure time on the curing depth, an experiment is designed to make a part with $20x20mm^2$ rectangular 2D shape using several different periods. The exposure time can then be converted into exposure energy density and the curing depth can be measured. The experimental results are shown in Figure 6. As shown in Figure 6.(a), the curing depth is affected by the exposure time. The longer the exposure time, the higher the curing depth. Hence, the relationship between exposure time and curing depth can be approximated as Equation (1), where E is the exposure energy density in J/cm^2.

$$CuringDepth(E) = 0.0007E^4 - 0.0236E^3 + 0.2693E^2 - 1.0575E + 1.4093 \quad (1)$$

In particular, the curing depth is almost proportional to the exposure time at the curing depth between 0mm to

0.5mm as shown in Figure 6.(b). Hence, the relationship between exposure time and curing depth can be approximated as Equation (2).

$$CuringDepth(mm) = 0.1329E - 0.3198 \quad (2)$$

As shown in Figure 6.(b), it indicates that the critical exposure energy density for the photo-polymerization must be higher than 2.4 J/cm^2. The reason for this critical energy is the required energy for the photo-initiation. Also, the reaction rate becomes moderate when the exposure energy density is higher than 10 J/cm^2. The reason for this slower reaction rate is that some of the cured material blocks the optical light.

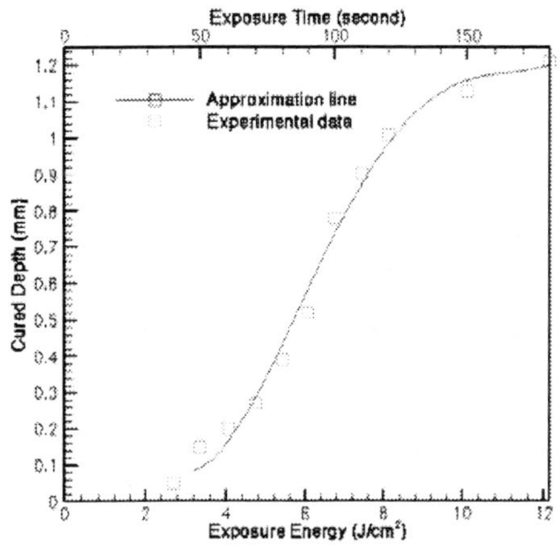

(a) The relationship between the Exposure Energy and Cured depth

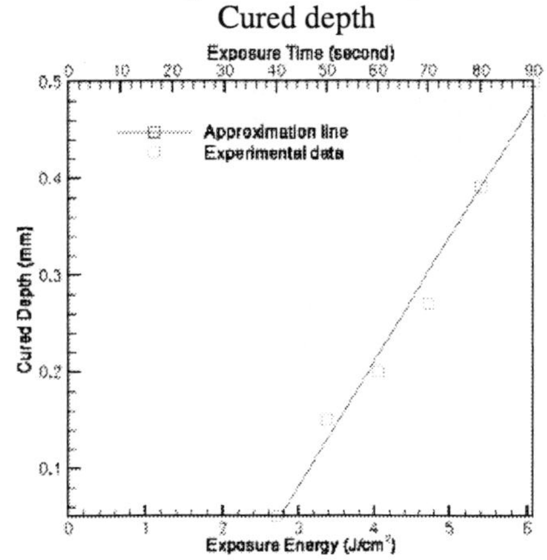

(b) The proportional segment of exposure energy and cured depth.

Figure 6. The diagram illustrates the effect of the exposure time on the curing.

The minimum energy, required to cure the resin with 0.2mm thickness, can be calculated according to the above equation. It is 3.9J/cm^2 for the optical energy and 55 seconds for the exposure time. The design of the DMPC system is a bottom exposure system. When a new layer has been cured, it will bound with the previous layer to provide a stronger bounding force to pull up the fabricated part, which is stuck on the contacted window during the curing process. Hence, the layer over-curing is required for the successful pulling up process. Therefore, the optical energy and exposure time for the layer thickness of 0.2mm must be higher than that of the calculated value. However, the increase of optical energy and exposure time sacrifices the part accuracy, though it solves the problem of pulling up. This is the practical problem of the bottom exposure system. It also limits the wider application of this type RP system, though this type RP system owns the saving advantage of resin usage.

The DMPC system provides a solution for the problem of overcuring, because it employs a TFT LCD for the image mask. Actually, the cross section of the current cured layer affected the part strength is different from that affected part accuracy. The cross section of the current cured layer affected part strength is at the union area of the current layer and the previous layer. The curing depth of this union area does not increase any more even at the longer exposure time or higher optical energy, but it does increase the bounding strength of the fabricated part. On the other hand, the difference area of the current layer with the previous layer dramatically affects the part accuracy. Therefore, the dynamic control of the LCD mask makes the DMPC system very flexible and enables the union area and difference area of the current lay and previous layer with various exposure time. Therefore, the problems of layer over-curing and accuracy sacrifice can be solved using the direct LCD mask. This is the most innovative solution for the bottom exposure RP system. However, the degree of over-curing is still very difficult to be determined, because it is affected by the stuck force between the current cured layer and the contacted window.

5. The Fundamental Theory of Pulling Force

Although the employment of LCD mask solves the problem of overcuring and accuracy sacrifice, it is not easy to determine the degree of overcuring, because it is affected by the stuck force between the current cured layer and the contacted window. In order to reduce the stuck force, a soft silicon film is used as a soft interface between the tank and the current cured layer to make the separation of fabricated part and contacted window easily. In this section, the modeling of the pulling force and the thickness of the silicon film has been developed. Also, the fundamental theory of the pulling force is reported.

In order to measure the stuck force and pulling force during fabrication, a load cell is employed and assembled on the pulling axis (Z-axis). Also, in order to eliminate the effect of bending moment, a joint connection between the load cell and the over hang cantilever beam is used to monitor the normal force as shown in Figure 7. When the Z-axis moves up, the load cell is stressed and then the pulling force is measured. Hence, the relationship between the pulling force and the thickness of thin silicone film can be constructed and modeled.

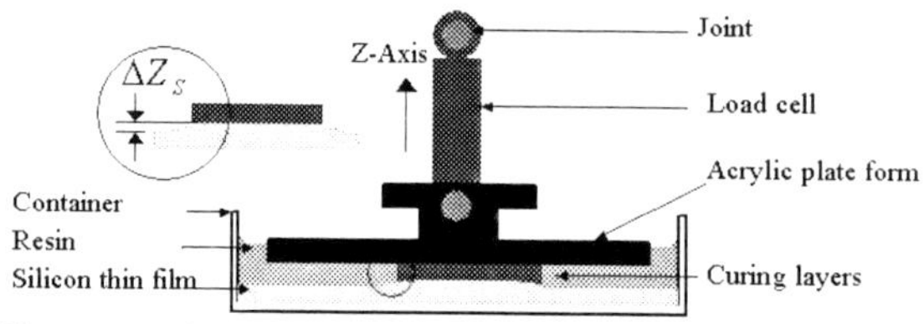

Figure 7. The structure and deformation of the pulling force measurement.

The free body diagram is shown in Figure 8. The silicon film has stiffness coefficient K_s and the deformation ΔZ_s. The stiffness coefficient of the load cell is K_l and the deformation is ΔZ_l. During the pulling, the results of the relationship between the pulling force and the level of Z-axis are shown in Figure 9. The pulling process can divide into two zones: I. the elastic deformation and II. the plastic deformation. Within the elastic deformation zone, the interface between the bottom of the curing surface and the silicon film will generate a large bounding force f_B. If the weight of the cured photopolymer has been ignored, the measuring force from the load cell will be the elastic deformation force of the silicon film.

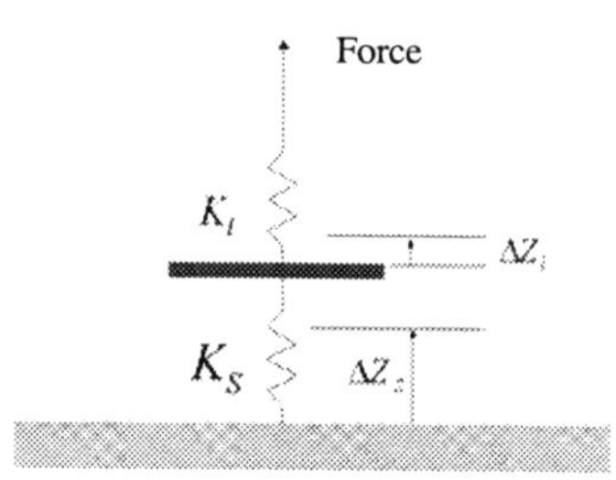

Figure 8. The free-body diagram of the pulling force measured by load cell.

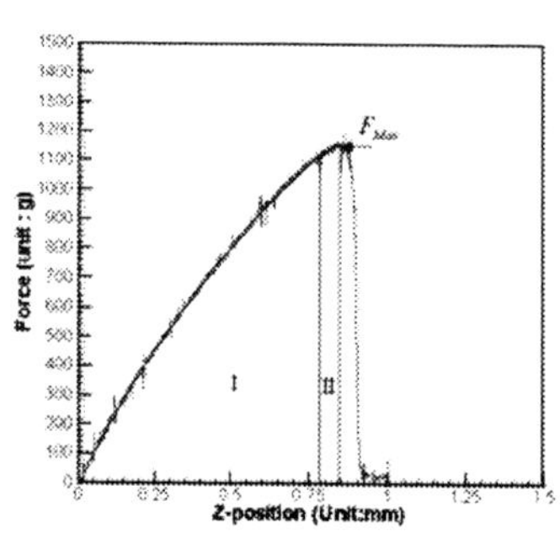

Figure 9. The relationship between the pulling force and the level change (position) of Z-axis.

The force equilibrium equation as Equation (3),

$$F = K_l \Delta Z_l = F_E + F_P = K_S \Delta Z_S + K_{IC} \tag{3}$$

where F_E is the elastic deformation force and FP is the plastic deformation force. According to the fracture mechanics, the creaking can be concluded as the plan strain creak initiation toughness K_{IC} [7] and the elastic force $K_S\Delta Z_S$. In the elastic deformation zone, K_{IC} is zero, besides F_E is the same as the f_B. The bounding force is a function of the interface bounding coefficient and the size of the bounding area. When the bounding force is larger than the elastic deformation force of the silicon film, it reaches the plastic deformation zone. The value of bi-material toughness K_{IC} between the cured layer and the silicon film will be launched. When the ultimate tensile stress F_{MAX} reached, a fatigue point generated and the de-bounding between cured photopolymer and the silicon film continues. The value of K_{IC} is dependent upon the thickness of silicone film. The required departing force decreases as initiation toughness decreased.

Further testing has been carried out to evaluate the K_{IC} value as well as the elastic force $K_S\Delta Z_S$. The pulling force of rectangular shape with different area size and different thickness of the silicon film has been monitored. The results of the maximum pulling force is shown in Figure10. It represents the bi-material initiation toughness of the photo-polymer and the silicon film decreased as the thickness of silicon film increased. The maximum elastic force $K_S\Delta Z_S$ is a function of the de-bounding coefficient α_s and the area size A as Equation (4). The de-bounding coefficient α_s is approximately to 0.0897.

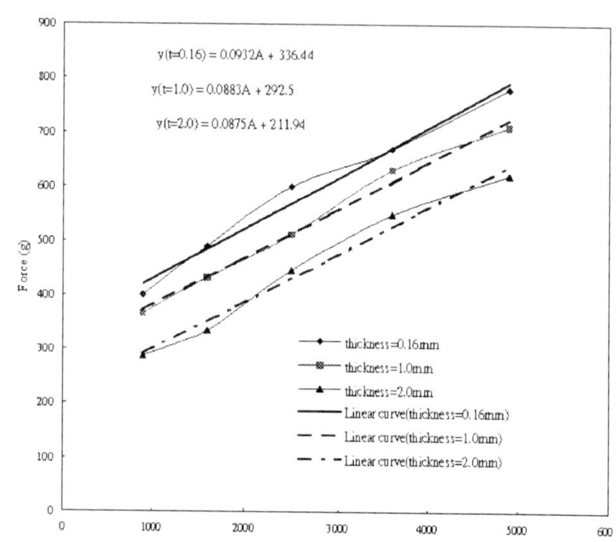

Figure 10. The relationship between the maximum pulling force and the area size.

$$F_E = K_S \Delta Z_S = \alpha_S A \tag{4}$$

According to the results, the de-bounding force can be predicted before lifting. The pulling force profile can be used to monitor the building state during the fabricating of a part. If the maximum pulling force is far less than the expected de-bounding force, it means the bounding force is not enough and the rate of the over-curing has to be increased. If there is no swiftly drop of the pulling force, there is no de-bounding and further rising of the platform is necessary.

6. Performance Evaluation of the DMPC RP system

The process parameters can be determined from the analysis of the pulling force and the curing depth. The silicon film used in this DMPC system is 2mm to take the advantage of decreasing the plan strain creak initiation toughness. The layer thickness is 0.2mm and the exposure time is 80 seconds to achieve 100% over curing. A 3D turbine blade with 365 layers as shown in Figure 12 has been fabricated using the process parameters. In comparison with the part built by SLP-4000 Desktop RP system [11], the amount of over curing has been significantly reduced. Also, the fabrication time using SLP system is 53 hours and it is down to 11 hours using the DMPC RP system.

Figure 11. (a) The turbine blade built by DMPC

Figure 11. (b) The turbine blade built by SLP-4000.

7. Conclusion

The usage of silicon film and the control of the over-curing depth make it feasible to fabricate a part with better accuracy and surface finish in the DMPC RP system. This low cost and user-friendly interface DMPC RP system is well suitable for desktop concept model applications. The computer supported force analysis and dynamic control of LCD mask has been successfully applied to the DMPC RP system to solve the problem of pulling force and the accuracy sacrifice. A 3D turbine blade benchmark has been fabricated by the DMPC and SLP-4000 RP system respectively to evaluate the performance of the DMPC RP system. It demonstrates that the accuracy of the evaluated part has been significantly improved and the cost of the DMPC RP system is much lower than that of all the available commercial RP systems.

8. Acknowledgement

The authors acknowledge supported from National Science Council through research grant NSC89-2212-E011-024. The authors would like to thank Prof. M.J.Tsai for his help with system automatically controlling.

9. References

[1] Burns, M. (1993), Automated fabrication- Improving productivity in manufacturing, PTR Prentice-Hall, Inc.

[2] Carrion, A. (1997), "Technology forecast on ink-jet head technology applications in rapid prototyping" Rapid Prototyping Journal, Vol.3, No.3, pp.99-115.

[3] E-Darts Solid Printer (1999), Product Brochure, Autostrade Corporation, Japan.

[4] Fuh, F.Y.H., Lu, L., Tan, C.C., Shen, Z.X., and Chew, S. (1999), "Curing characteristics of acrylic photopolymer used in stereolithography process" Rapid Prototyping Journal, Vol.5, No.1, pp.27-34.

[5] F. S. Ullett, F. W. Schultz and R. P. Chartoff (2000), "Novel liquid crystal resin for Stereolithography-processing parameters and mechanical analysis", Rapid Prototyping Journal, Vol. 6, Num. 1, pp. 8-17

[6] Hagiwara et al. (1996), "Laser Lithography for Masking Jewelry", Journal of JSPE, Vol. 62, No. 11, pp. 1587-1591

[7] James W. Dally (1991), Experimental Stress Analysis, McGraw-Hill, pp.91-121.

[8] Johnson, J.L., Principles of Computer Automated Fabrication, Palatino Press, Irvine, (1993).

[9] Narahara et al. (1996), "Study on improvement of surface roughness of complex model", Journal of JSPE, Vol. 66, No. 11, pp. 233-237

[10] Kai, C.C. and Fai, L.K. (1997), Rapid prototyping: principles & applications in manufacturing, John Wiley & Sons, Inc., Singapore.

[11] SLP-4000 Solid Laser diode Plotter (1997), Product Brochure, Denken Corporation, Japan.

Intelligent Assembly Modeling Based on Semantics Knowledge in Virtual Environment

Jianrong Tan, Zhenyu Liu, Shuyou Zhang
State Key Laboratory of CAD&CG
Zhejiang University, Hangzhou, P. R. China
egi@zju.edu.cn

Abstract

This paper proposes a semantics knowledge modeling approach for product assembly, which is suitable for virtual reality interaction. Assembly semantics knowledge is used to express design conception, constrain the relationship between parts, and encapsulate design knowledge in semantics knowledge modeling. Using assembly semantics, the designer is able to express design intention in engineer's language, which is natural and convenient for designer. In addition, the intelligence of the system can be promoted through the inference of implicit knowledge of semantics. These methods are implemented in the development of VIRDAS (VIrtual Reality Design and Assembly System) and several assembly examples are also given in this paper.

1. Introduction

Virtual assembly is a typical application of virtual reality (VR). With VR, the designer can manipulate the scene objects in a natural and intuitive way and receives immediate feedback (visual, aural and haptic feeling) about the model changes in the virtual environment. The designer examines various aspects of an assembly design by performing virtual assembly operation on virtual parts just as real parts are being put together through real assembly operations into a real assembly. Thus, not only the costly physical prototype can be eliminated, but also the quality and speed of assembly design can be promoted. VR is particularly helpful towards the assembly of large and complex product.

At the present time, researchers have addressed the application of VR technology to product assembly. Jayaram S. et al. proposed a virtual assembly system named VADE, which is a VR-based engineering application that allow engineers to evaluate, analyze and plan the assembly of mechanical product[1].VADE focuses on utilizing an immersive virtual environment tightly coupled with commercial CAD systems. Dewar R G researched assembly planning based on VR technology[2]. Jung B. et al. presented a virtual assembly system named CODY, which is a knowledge-based system that enables the interactive assembly of 3D visualized mechanical parts to complex and novel aggregates in virtual environment [3].

This paper proposes a semantics modeling approach for product assembly, which is suitable for virtual reality interaction. Assembly semantics is used to express designer's conception, constrain the relationship between parts, and encapsulate design knowledge. Using assembly semantics, the designer is able to express design intention in engineer's language, which is nature and convenient for designer. In addition, the intelligence of the system can be promoted through the inference of implicit knowledge of semantics.

2. Semantics modeling in virtual assembly

2.1 Assembly semantics

Assembly semantic[4] is the abstract description of assembly relationships, which implies the constraint between parts, assembly rule, assembly knowledge and assembly action. Assembly semantics, that is more suitable for the engineer to express design intention, has the abstract character in expression and is dependent of application domain.

We built a semantics dictionary that is composed of semantics in common use according to the joint relation, position relation and transmission relation of the assembly. For example, the joint semantics includes "bolt nut connection", "cylindrical pin connection", "taper pin connection", "key connection" (Figure 1), "multiple key connection", "shape connection" and "river connection" etc. The transmission semantics includes "straight gear mesh", "skew gear mesh", "belt transmission", "chain transmission" and "worm wheel gear mesh" etc. The position semantics for shaft includes "shaft shoulder – locked collar mate", "shaft shoulder –elastic collar mate", "shaft shoulder – sleeve mate", "shaft shoulder –round nut mate" and "shaft shoulder –end collar mate" etc.

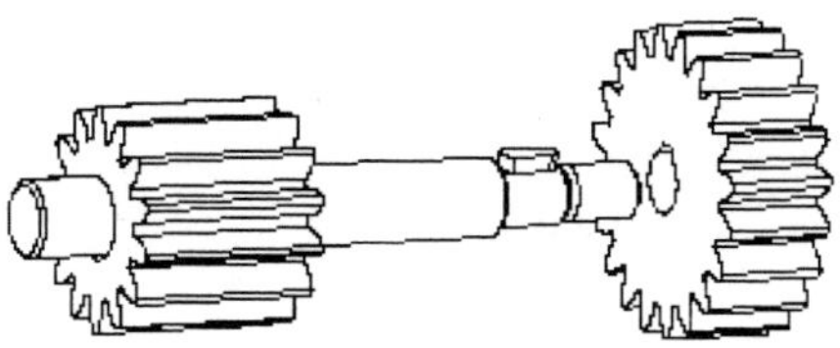

Figure 1. "Key connection" semantics.

The semantics model is represented as a graph, in which the node represents the mechanical part and the arc between two nodes represents the constraint relationship of mechanical parts. A direct arc connects the parts, which co-related to the same semantics. The direction of the arc implies the assembly sequence of the parts. A direct semantic loop is formed by connecting the first part and the last part in the semantics. Thus, a semantics model can be expressed as SCG =(H, E, S), where H means the set of mechanical parts, E means the set of constraint relationship, S means the set of assembly semantics. Figure 2 shows an example of semantics model.

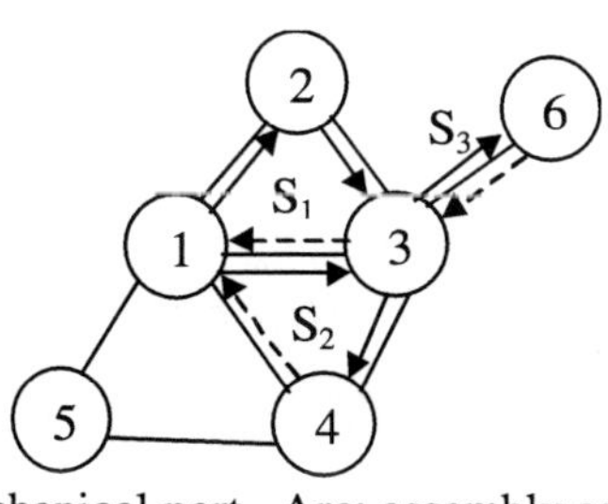

Node: mechanical part , Arc: assembly constraint
Direct Loop: assembly semantics
S1□"key connection"
S2□"shaft shoulder – sleeve mate"
S3□"straight gear mesh"

Figure 2. Assembly semantics model.

2.2. The benifit of assembly semantics

1) Simplify interaction and capture design intention.

As we all known, product design is a process from abstract to concrete. In the early stage of product design, the assembly relationship and parameter are also abstract and uncertain. Semantics can express the abstract design concept perfectly. Using assembly semantic as the basic primitive of interaction, not only the designer can express the design idea more freely and concisely, but also the system can capture the design intention more completely.

2) Express rich assembly constraints.

Assembly semantics is the abstract expression of assembly constraint that implies rich constraint information including position constraints and engineering constraints. The position constraint constrains the related position of parts while the engineering constraint includes the dimensional constraint, kinematics constraint, and dynamic constraint. For example, the position constraints implied in "bolt nut connection" include: □ alignment of the axes of bolt and nut; □ alignment of the hole axes of two connection parts;□ alignment of the hole axis and the bolt axis; □ mate of the end faces of two connection parts. The engineering constraints implied in "bolt nut connection" include: □ The diameter of two holes equals; □ Length of bolt > (Thickness of connection part1 + Thickness of connection part2+ Thickness of nut).

The assembly semantics can express the assembly relationship just as geometry constrain does. But the semantics describes the assembly relation from a higher level that is more abstract and concise. In addition, the information implied in assembly semantics is richer than geometry constraint.

3) Imply assembly knowledge.

Assembly semantics is the basic unit to store assembly knowledge and rules. The assembly semantics implies the assembly precedence relationship. For example, "key connection" implies that the key should be assembled to the shaft before the gear to be assembled (Figure 3). That means the assembly precedence relationship between shaft, key and gear is "shaft > key > gear". The precedence relationship implicated in the semantics will result in much lower computation times for the assembly sequence planning.

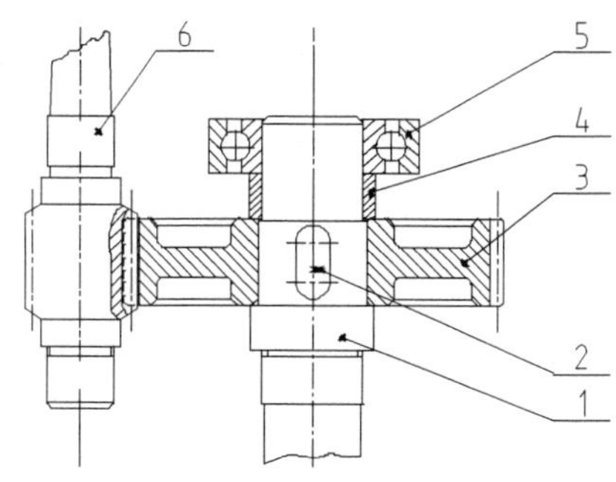

1.shaft 2.key 3.gear 4.sleeve
5.bearing 6.gear shaft

Figure 3. Gear transmission assembly.

3. Algorithm in semantics modeling

Algorithm 1: The Capture of assembly semantics

Let p_j be the part to be assembled, Ψ be the set of unassembled parts, P be the set of assembled parts.

- step1: Select part p_j from Ψ. (The user is able to grasp or issue speech command to select p_j directly)
- step2: Position the part p_j related to P roughly through moving the p_j by hand
- step3: If the user expressed the assembly semantics S_i by gesture or speech command explicitly, go to step 6
- step4: Recognize the assembly semantics according to the attributes of p_j and P. Take the recognized result as implicit semantics between p_j and P, go to step 6
- step5: Recognize the assembly constraints between p_j and P according to the position and orientation of p_j and P
- step6: Position p_j precisely according to the semantics S_i or constraint C.

Algorithm 2: Conversion of assembly semantics

Let Si be the semantics to be conversed, $H=\{h_l, i=1$ to $j\}$ be the set of mechanical parts which are connected by semantics S_i, H_l $(H_l \subset H)$ be the set of parts which have added to semantics model. Now add h_{j+1} to the semantics model.

- step1: Convert semantics S_i to constraint set $C=\{C_l, i=1$ to n $\}$
- step2: if the element C_i of C describes the constraint relationship between h_{j+1} and H_l,add C_i to $\varphi=\{\varphi_i \mid i=k \quad to \quad r\}$
- step3: Judge the conflicts between φ and the constraints that have existed in semantics model. If there is no conflict happes, goto step5
- step4: Report the conflict message to the designer, return
- step5: If node h_{j+1} does not exist in the semantics model, then add node h_{j+1} to the semantics model
- step6: Add constraint information to semantics model according to φ
- step7: Connect h_j to h_{j+1} in direct arc. If $(j+1)=m$, then connect h_{j+1} to h_l to close the semantics loop.

4. Implementation

We have developed a virtual reality design and assembly prototype system named VIRDAS. A head-mount display (Virtual i_glasses™) was used to provide the stereo images in VIRDAS. The hand gestures and hand motions input were implemented through the 5th-Glove™ and the Ascension Flock of Birds™. IBM viavoice™ was used to recognize the speech command input. Thus, the designer can express design semantics via voice, gesture or the combination of them.

VIRDAS imports the part models through neutral files, which are exported from CAD system including Pro/Engineer™ and Autodesk MDT™. The designer can grasp and assemble the mechanical part directly. And the design interactive intention can be captured through the recognition of assembly semantics.

Figure 4. shows the designer performs assembly modeling in VIRDAS. Several products were used as the experiment to verify our methods. Figure 5. is the assembling process of gear pump. Figure 6. is the final assembly of piston and connecting rod component of automotive engine. The experiment shows that semantics modeling can promote the assembly design efficiency rapidly.

5. Conclusion

Virtual reality brings a completely new interface to CAD system with which the designers can design and assembly the product in a more natural and intuitive way. Semantics modeling is an efficiency approach to express design conception, constrain the relationship between parts, and encapsulate design knowledge in virtual environment. Using assembly semantics, the designer is able to express design intention in engineer's language. In addition, the intelligence of the system can be promoted through the inference of implicit knowledge of semantics. The methods are implemented in the development of VIRDAS (VIrtual Reality Design and Assembly System) and several assembly examples are also given in this paper.

6. References

[1] Jayaram S, Jayaram U et al. VADE: a Virtual Assembly Design Environment, IEEE Computer Graphics and Applications ,1999, 19(6): 44 –50

[2] Dewar R G et al. Assembly planning in a virtual environment. In: Proceedings of Portland International Conference on Management and Technology (PICMET '97), Portland, 1997, 664 -667

[3] Jung B et al. Knowledge-based assembly simulation for virtual prototype modeling. In: IECON '98,Proceedings of the 24th Annual Conference of the IEEE Industrial Electronics Society, Germany, 1998 ,2152 -2157

[4] Liu Zhen-Yu, Tan Jian-Rong et al. Design semantics of assembly modeling in virtual environment: expression, transfer and transformation, Chinese Journal of Computers, 2000,23(11):1208-1214

Figure 4. Designer performs the assembly design and analysis in virtual environment.

Figure 5. Assembling gear pump in virtual environment.

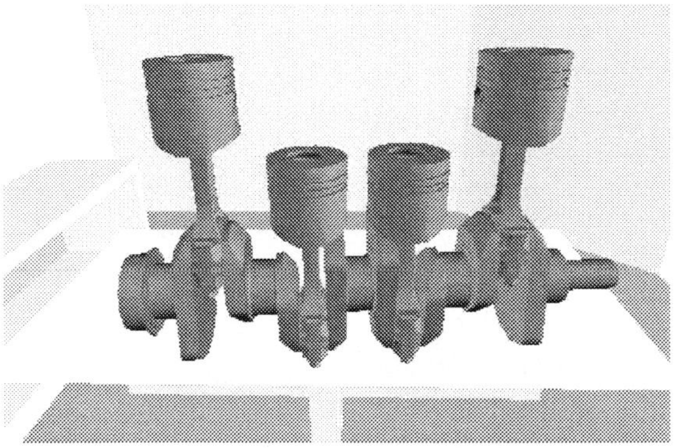

Figure 6. Final assembly of piston and connecting rod component of automotive engine

Development of a Web-Based Academic Time Management System

Zulkifli Mohamed Udin, Ahmad Suki Che Mohamed Arif, Azmi Md Saman
School of Information Technology, Universiti Utara Malaysia, MALAYSIA
[zulkifli, suki1207, azmi1326]@uum.edu.my

Abstract

The web-based application system is being widely used in managing organisation. This research paper reports on how web-based technology is used in developing an academic time management system. This prototype system, which is currently used in the School of Information Technology of Universiti Utara Malaysia, has showed significant improvement in time management. Through this web-based application, users could manage their time for traditional meetings, student consultations and lectures. Furthermore, this application would enable users to discuss things collaboratively, design projects and conduct meetings through the virtual environment. This research also discovered some constraints and limitations of the system. Finally, an avenue for further research is also identified.

1. Introduction

In the emergence era of Information and Communication Technology (ICT), the Internet has become a technology that is widely used as a tool to communicate between various locations. The used of web-based application which will enable users from any kind of machine and operating system to access the information has shown the tremendous improvement in the organisation [1]. Therefore, the capabilities to develop a web-based application are extremely valuable to the developers and organisations. The web-based application has several advantages compared to the conventional client-server system such as a capability to access, communicate and share data regardless the differences in machine platform, architecture, protocol and operating system [5].

There is an urgent need for the Universiti Utara Malaysia to find a better application in the time management for the academic staffs. Thus, this research is considered as a catalyst for other development in the web-based applications. Previously, it is hard for the administrators, students and top management to identify the available time of the academic staffs for some occasion such as for monthly academic meetings, committee meetings and student consultations. The times for these events are difficult to arrange due to the differences in the staff timetable such as their lecture times, consultation projects and administrative works. These factors affected the performance and quality of works in this department. The activities that should involved all the staffs sometimes cannot be implemented due to these problems.

2. Research method

The objective of this study was to develop a prototype of web-based academic time management system, which can be used to improve the productivity and quality of works in the School of Information Technology (IT school), Universiti Utara Malaysia. This system has the capabilities in providing the best available time for all staffs in the specific time and date. Furthermore, this system would aid the management and secretary of academic committees to manage the information that related to the school or committees such as the announcement of events, events scheduling, virtual discussion and knowledge management.

The method that used in developing this prototype was a combination of two methods including Waterfall Method and Rapid Prototyping. These methods were appropriate for the development, which involved a systematic phase development [2]. Furthermore, both of the methods were chose because this system development involves many users with different needs and perspectives [3]. In addition, the time constraints were also taken into consideration while selecting these methods. Before the final decision was made, the other methods that have been investigated were DOD-STD-2167A, Spiral and Parallel. Moreover, Waterfall method and Rapid Prototyping will help to minimise the risks in choosing appropriate interfaces. In the programming part, Visual Basic and HTML were chosen as a programming language to develop the application due to its stability and ease to use [4].

3. Results and Discussions

After spending about four months in designing and developing, this research had produced a model and prototype of web-based academic time management systems. This application system is currently being used

573in the IT school and after two months evaluation periods, this system has shown the significant improvements in productivity and quality of works. In this evaluation period, researchers also discovered some limitations in this system and its have been identified as an opportunity for the future research to be carried out. Among the limitations were related to the issues of documents security, systems stability and connectivity.

Specifically, this system comprised of three main modules, which were Academician Information System (SMK-STM), Committee System (SJK) and Available Time System (SML). These three modules have its own sub-modules including the virtual meeting, time searching, event scheduling, managing knowledge, announcement of events and on-line staffs timetable.

3.1 Academician Information System (SMK-STM)

SMK-STM is a module where all the information that related to the academic staffs is organised. For example, the information on staff personal details, timetable, online lecture notes and research project are organised in this module. There are 2 main functions in this module, firstly, the function that related in managing the staff knowledge and secondly, the set up of virtual discussion rooms. In managing the knowledge of the academic staffs, this function has the abilities to search, collect and select the information based on appropriate conditions and requirements. In addition, the process of updating the knowledge could be done through the web browser. Users could input the relevance information to their profiles. Each user was assigned their own username and password to protect the data. Other users could access the related information or knowledge by selecting the appropriate staff name in the pull-down menu or by clicking the photo of the staff. The interface of this module is shown in Figure 1.

Figure 1. SMK-STM Query Interface.

This module also capable in providing the virtual discussion rooms that could be conducted among the staffs. The initiator could invite the participants either through pull-down menu or by clicking the photo of the staffs. An acknowledgement message, which stated the discussion date and time, will be sent to the participants through emails. In this discussion room, the discussion members could share, access and work collaboratively. In addition, each participant was granted the permission by initiator to modify or update the related documents or resources. This function is widely used among the staffs specifically for research and development projects.

3.2 Committee System (SJK)

SJK or Committee System is a subsystem that used for school committee activities. This subsystem was developed to fulfil the requirements of users who involved in the school committees. The purpose of this subsystem was to manage the committee activities, set up the virtual meetings, allocate the suitable time for activities and manage the committee information. Figure 2 is an interface in SJK that used for time allocation in which each user is requested to fill up their time for a particular semester.

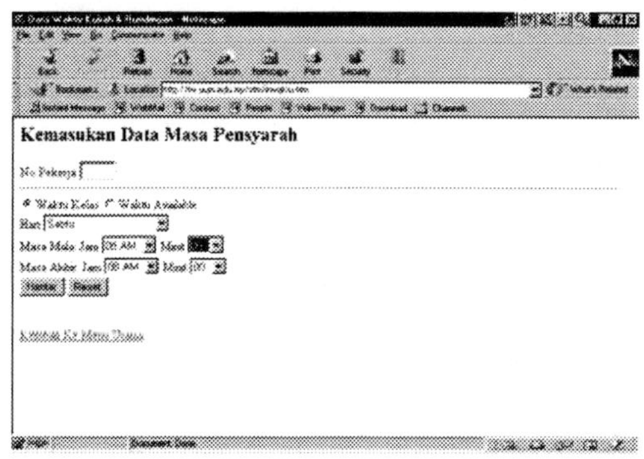

Figure 2. SJK Time Allocation Interface.

This recorded time will be used when the committee plans to schedule their activities. By utilizing the search engine in this subsystem, it will suggest the best time for any specific activity. Furthermore, this subsystem allows users to work collaboratively among committee members, particularly when they need to share the same resources and plan to provide the common decision. By utilizing this subsystem, the school committee could work together and this would minimise the procedures in making the decisions for the school.

3.3 Available Time System (SML)

The functions that available in this subsystem are used for accessing, arranging and providing an available time for any activities whether it was organised by the school management, school committee or staffs. Users could query the staff available time through browser by selecting the interface as shown in Figure 3.

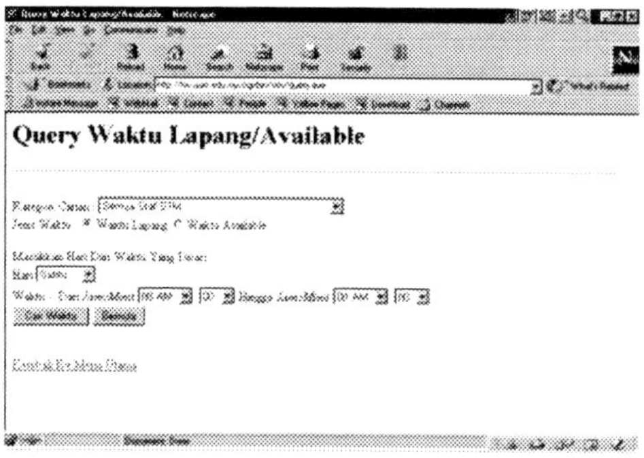

Figure 3. SML Query Interface.

In the process of making query, users could make selection whether to specify the condition or not. The engine in this subsystem will look into the condition before providing any solution. Based on the observation during the evaluation period most of the users provide the conditions in order to receive the best solution from the system. In addition, about 90% of school activities will include all academic staffs that available on that time. This scenario showed the significant differences before and after implementing this system.

Generally, the process of this system is based on the model that shown in Figure 4. This model explains the basis on how the system works through six activities or steps. These activities would enable users to access the information in the data sources through the web browser and these data will be appeared based on the user preferences. Users could also personalise and customised the appearance of the data that accessed from the data sources. Furthermore, this system enables the users to retrieve the historical data to view the time management trends for academic staff in this school.

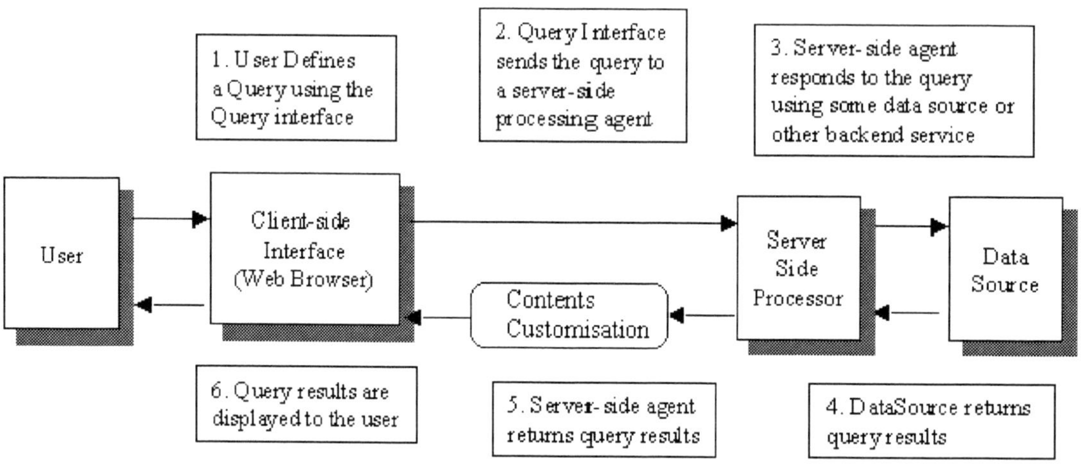

Figure 4. Web-Based Application Model.

4. Conclusions

This system still needs improvements to overcome the limitations in order to optimise the functionality of the system. The aspects or issues related to security, stability and system connectivity should be a domain for the next research. In addition, the utilisation of other technologies such as XML, Portal, and Intelligent Agents should be also considered as avenue for further research to improve the performance and functionality of this system. Even tough this system has some limitations, researchers believe this system will be a stepping stone for other web-based applications specially for e-university project in Universiti Utara Malaysia. We would recommend that this system could be used as a basis to develop other collaborative applications on the web technology. Moreover, it also could be used as a model to other researchers in developing a web-based collaborative system for higher academic institutions, which has the same processes and activities. Researchers also believe, in order to get an effective web-based application system, the processes of activities should be taken into consideration. In other words, the process innovation or process redesign should be implemented before the web-based application development process embarks, in order to gain the competitive advantages in term of cost effective and efficiencies.

5. References

[1] Maurer, H. (1996). Hyper-G Now Hyperwave, The Next Generations web Solutions, Singapore: Addison Wesley Longman.

[2] Whitten, B. (1994). System Analysis and Design: An Applied Approach, Singapore: John Wiley & Sons Inc.

[3] Smith, M.F. (1991). Software Prototyping Adoption, Practice and Management, Singapore : MCGraw-Hill Inc.

[4] LaOr, O. (1997). Hands-On Web Development, CGI Programming with Visual Basic 5, Singapore: McGraw-Hill Inc.

[5] Lu, M. & Yeung, W. (1998). A Framework for Effective Commercial Web Application Development. Electronic Networking Applications and Policy. 8(2): 166-173.

Transformation of Binary Relations

Jamal Said
Katholieke Univesiteit Leuven
Department of Computer Science
Celestijnenlaan 200 A
3001 Heverlee – Belgium
jamal.said@cs.kuleuven.ac.be

Eric Steegmans
Katholieke Univesiteit Leuven
Department of Computer Science
Celestijnenlaan 200 A
3001 Heverlee – Belgium
eric.steegmans@cs.kuleuven.ac.be

Abstract

In the object-oriented paradigm, as the complexity rises, the cost to develop and to maintain software systems grows exponentially. This complexity emerges from the continuous evolution in the software systems to cope with changing requirements. This crucial problem can be dealt with by performing an active transformation of the elements (i.e. classes and relations) of the conceptual model to produce an optimum design model. In this paper we will show transformations for a simple conceptual model consisting of three inter-related classes. Each of these transformations satisfies particular software quality factor(s), from which the software engineer can choose the one that matches the intended non-functional requirements. The added value of this approach is that less manual optimization is required and high maintenance is achieved.

1. Introduction

In the mid-nineties the idea of *design patterns* started to attract considerable attention in the area of object-oriented software development. However, the underlying representation of a design pattern and of its application, and the binding between these two levels is not exactly defined and thus can be interpreted in different ways [7]. Other researchers such as Fowler made promising work towards the use of refactoring [5] where they improve the design of existing code. However, refactoring is risky. It requires changes to working code that can introduce subtle bugs. Other researchers [3] have followed the *qualitative design trend*s, which lead to designs that exhibit a desirable quality and trigger a movement from bad design models to good design models. Although these approaches are of interest to software engineers, they lack the ability to keep traceability and maintainability between analysis and design models.

The analysis phase usually ends up with a *conceptual model*, in which the external world is described in terms of classes, relations and other high-level concepts. Once the conceptual model fully describes the functional requirements for the software system, *design* is started aiming at a model satisfying, in addition to functional requirement, the non-functional requirements. The transition from the conceptual model to the design model is often an iterative process; thus it s crucial to be able to develop a *framework* that performs a reliable and convenient transition between the two models. Currently, software developers try to accomplish some actions manually, which in most cases leads to a big distinction between experienced and inexperienced developers and increases the cost of the software system due to maintenance. Given the fact that software engineering is aiming at building robust and reliable software systems, an approach that supports modeling and provides insights into understanding the software requirements and the software design is crucial. This approach should not restrict the software engineer to a particular phase of the software life cycle. It must maintain a link between the early phases (analysis and design).

Our approach transforms elements of the conceptual model based on object-oriented concepts and the software quality factors. Throughout this paper we will work out some transformations of a simple conceptual model involving three classes involving a binary relation. The structure and the semantics of this pattern (i.e. binary relation) will be described in the next section.

2. Binary relations at the level of analysis

Relations at the level of the conceptual model must express *existential dependency*; a relation is said to refine objects of a given class, this refinement expressing that objects cannot exist without being related to objects of the classes participating in the given relation [8, 9]. In our example (Figure 1), a relation will be introduced refining the objects of the class of accounts, in order to express that no accounts can exist without being related to a person on the one hand, and to a bank on the other hand. Such kind of relations is called binary relations. They involve two participating classes and one refined class.

The definition of a binary relation identifies the classes participating in the given relation, each of which includes a series of constraints determining the way their objects can participate in the given relation. Among those constraints are the multiplicity and mutability. The representation of binary relation in UML (Figure 1) illustrates the multiplicity to insure that a given person can hold many accounts at a given bank and that a bank can grant many accounts to different persons. The constraint of mutability is introduced to allow a transfer of accounts from one person to another.

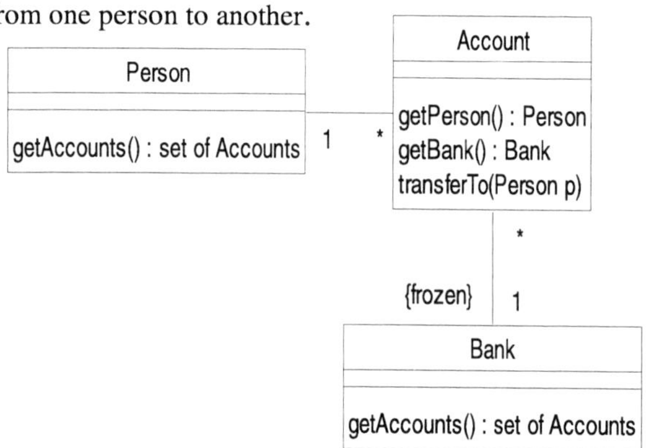

Figure 1. Structural and behavioral aspects of three classes involved in a binary relation (UML class diagram).

A relation is complemented with a specification of operations for manipulating the objects involved in it. For classes refined by a binary relation, at least a constructor, destructor and two queries must be introduced. The constructor will initialise the bindings of the new refined object with the given objects of the two participating classes; the inspectors will return the objects of the two participating classes involved in the refined relation. Furthermore, the refined class may introduce mutators for changing the binding of refined objects to some other objects of the participating classes. For example constructing a new account requires specifying the person that will hold this account and the bank that will grant this account. Furthermore, a destructor for closing a given account, queries for retrieving the owner (getPerson) and the grantor (getBank) of a given account, and a mutator (e.g. transferTo) for transferring accounts from one person to another are required.

3. Transformation of binary relations

During analysis, the software engineer focuses on the issue of specifying the objects relevant for the system, and on appropriate relationships involving these objects. In other words, the software developer is only interested in which objects are needed. The feasibilty of those defined objects are the subject matter of the desin process.

This paper presents a *transformational approach* to object-oriented design. Basically, a design model is obtained by transforming patterns (i.e. classes and their relations), as they can be observed in conceptual models. Because a single pattern can be designed in many different ways, the designer chooses the most appropriate one, based on quality factors for the ultimate system.

This paper discusses transformations for a pattern defining the refinement of a class by means of binary relation. For a pattern consisting of binary relation (Fig 2), there exist different alternatives to transform it to design. In this paper we will focus on transformations involving integration.

4. Integration

Integration in general is defined as integrating some of the classes in the conceptual model into a single class in the design model. In other words, the integration requires objects of the integrated class to be hidden inside objects of the integrating class. In the limit, and under the assumption that no other characteristics are defined for the integrated class, objects of that class can then be represented by simple attribute in the integrating class at the level of design. For the pattern of the binary relation, integration transformation can be classified into two categories. The first category is full integration where two of the involved classes are integrated in the third class. The second category is partial integration where only one class is integrated in another, while leaving the third with an association relationship with the integrating class.

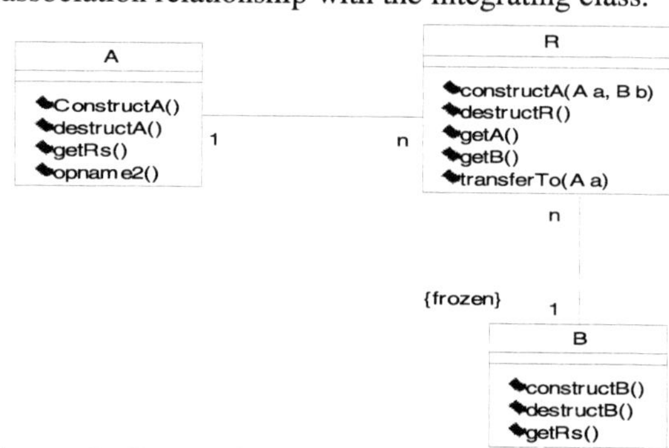

Figure 2. General pattern of classes involved in binary relation; structure and behavior.

4.1. Full Integration

Full integration of classes involved in a binary relation is defined as integrating two of the involved classes to become attributes of the other class. Full integration can be performed by either integrating two classes in one of the participating classes or integrating the participating classes into the refined class. Both types of integration carry different semantics and syntactical strategies. Only the former approach will be discussed in this paper. The resulting structure at the design level is shown in Figure 3.

Transformation using full integration is accomplished by defining two attributes, both of which will give information about objects of the integrated classes. For example, in the banking application (Fig. 1) both classes Account and Bank can be integrated in the class of Person by defining two attributes that will tell how many accounts this person holds and how many banks this person is registered in.

As we mentioned before once the integration transformation is performed the structure of the classes involved will be changed. However, not only the structure is changed but also the behaviour of these classes will change as well. As can be seen in Figure 2, classes involved in a binary relation have certain functionalities which have to be transformed and should be kept transparent after the transformation is performed.

- The constructor for objects of the refined class R as it is defined at the level of analysis, will be transformed into the method constructR (shown in Figure 3 and implemented in implementation 1), applicable to objects of class A at the level of design. As can be seen in the implementation below, this method will increment the number of R-objects integrated in the given A-object. Since the construction of a refined object requires an object of both participating classes, the arguments defined at the level of analysis (see Figure 2) are converted into a boolean at the level of design. This boolean will insure whether the newly created R-object requires us to increment the number of B-objects that are currently integrated in the given A-object or not. In the same manner, the construction of a B-object is transformed into the method constructB applicable to objects of class A at the level of design.
- The destruction of objects of the refined class R has been transformed into the method destructR applicable to A-objects at the level of design. This method will decrement the number of R-objects for the given A-object. Notices that during the destruction of R-objects at the level of design the user has to tell whether the number of B-objects involved must also be decremented or not. Therefore, a boolean flag is added as a formal argument for the destructor. In the case of the participating class B, the method destructB defined at the level of analysis will be transformed into the method destructB applicable to A-objects at the level of design. In this case, the destructor is complemented with a counter, registering the number of refined objects must be destroyed at the same time.
- Both the inspectors getA and getB introduced at the level of the refined class R cannot be offered at the level of design since the objects of the refined class are hidden inside the objects of the participating class A, the same also applies for the inspector getRs defined at the level of the participating class B. However, the inspector getRs defined at the level of the participating class A will be transformed into a method getNbRs returning the number of refined objects integrated in the given A-object. Furthermore, another inspector getNbBs is defined to return the number of B-objects integrated in a single A-object.
- The mutation at the level of analysis is transformed into the method transferR at the level of design. This method when applied to some A-object will decrement the current number of R-objects integrated in it. At the same time, the number of R objects integrated in the other A-object involved in the transfer is incremented.

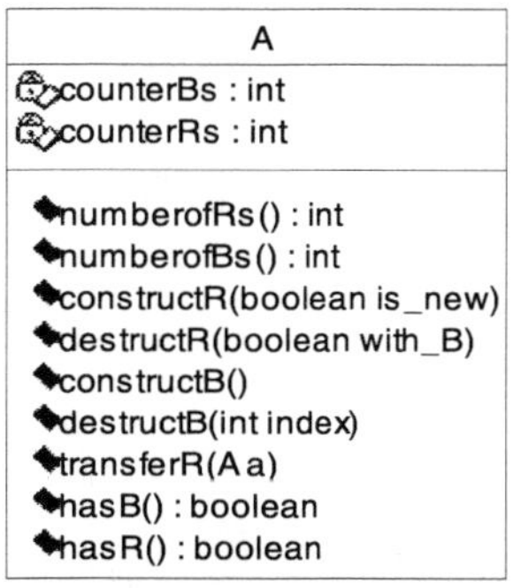

Figure 3. Resulted structure of the three classes inf Figure 2 after integration transformation of binary relation.

```
/**
 * Definition of the participating
 * class-A, integrating objects of the
 * refined class R and of the other
 * participating class B.
 */
class A {
/**
 * Initialise a new A-object with
 * neither R-objects nor B-objects
 * integrated in it.
 * @post | new.getNbBs() = 0 and
 * @post | new.getNbRs() = 0
 */
public A()
/**
 * Construct a new R-object integrated
 * in this A-object.
 * @post | new.getNbRs()= getNbRs() + 1
 * @post | if (newB) then new.getNbBs()=
 *        | getNbBs() + 1
```

```
 */
public void constructR(boolean newB)
/**
 * Destroy one of the R-objects
 * integrated in this A-object.
 * @pre  | getNbRs()> 0
 * @post | new.getNbRs()= getNbRs() - 1
 * @post | if (lastB) then new.getNbBs() =
 *         getNbBs() - 1
 */
public void destructR(boolean lastB)
/**
 * Destroy all R-objects integrated in
 * this A-object, and involving some
 * B-object.
 * @pre  | getNbRs()> 0
 * @post | new.getNbBs() = getNbBs() - 1
 * @post | new.getNbRs()= getNbRs()- amount
 */
public void destructB(int amount)
/**
 * Transfer one of the R-objects
 * integrated in this A-object to the
 * given A-object.
 * @pre  | a <> null
 * @pre  | getNbRs()> 0
 * @post | new.getNbRs() = getNbRs() - 1
 * @post |(new a).getNbRs()=a.getNbRs() + 1
 */
 public void transferR(A a)
}
```

Implementation 1: Implementing full integration of the three classes involved in a binary relation.

With this type of transformation, the software engineer chooses for the quality factors of efficiency and simplicity over flexibility (adaptability) and re-usability.

- Efficiency is emphasized first of all because memory requirements are low. Instead of assigning separate memory to three objects, two of these objects are represented by a simple integer attribute inside objects of the integrating class.
- The transformation is not a very flexible to withstand future modifications. As soon as objects of the integrated classes become more and more important, integration should be replaced by other types of transformations.
- Finally, chances for re-using a class in which other classes are integrated are not that high. Indeed, the integrating class can only be re-used in applications in which objects of the integrated classes are just of minor importance.

Notice that the transformation cannot be applied if the original queries getA and getB defined in the refined class R, as well as the query getRs defined in the participating class B shown in Figure 2, are of major importance.

4.2. Partial Integration

Integration transformation can be done partially by just integrating one of the three involved classes into either of the other two. In other words partial integration can be done either by integrating the refined class into one of the two participating classes or one of the participating classes into the refined class. In this paper only the former transformation will be worked. In that case, the original conceptual model shown in Figure 2 is transformed into the design model shown in Figure 4 with a many-to-many bi-directional association between the two participating classes A and B. Notice that, in general, this transformation leads to a bi-directional association with multiplicity many to many. For example in a banking application, a person can hold accounts at many banks and a bank can grant accounts to many persons. The behavioural aspects of this transformation are addressed below.

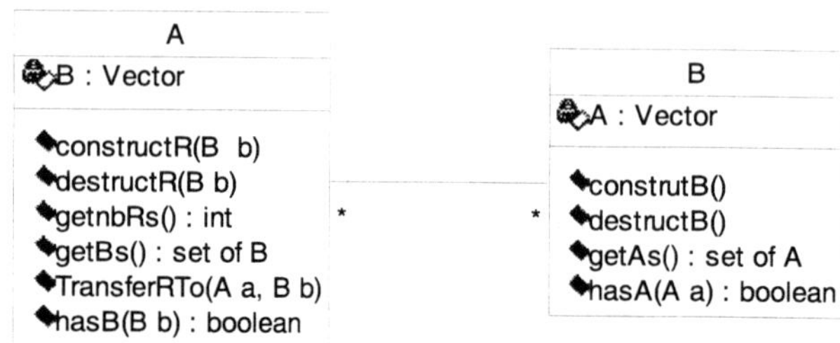

Figure 4. Partial integration where the refined class-R is integrated into the participating class-A.

Because objects of the refined class R are integrated into the participating class A, the constructor for creating new objects of the refined class is transformed into the method constructR applicable to objects of class A. At the level of analysis, the constructor involved an argument of both participating classes. Because the constructor is transferred to the participating class A, that argument becomes implicit. The method constructR, applicable to objects of the participating class A, is left with a single argument, identifying the object of the other participating class B, that is involved in the construction of the new R-object. As it is explained in the specification of the method, the number of refined objects integrated in the given A-object and involving the given B-object is incremented by 1. Because of the bi-directional binding, a similar clause is needed stating changes to the state of the given B-object. At the level of the implementation of class A, information (a data structure) is needed concerning the B-objects involved in the refinement of R-objects integrated in each A-object.

The principles applied in transforming the constructor for objects of the refined class R, equally apply to the destructor. As a result, the destructor is transformed into the method destructR (shown in Figure 4 and implemented in implementation 2) applicable to objects of class A. Because the actual R-object is hidden inside an A-object, information is needed concerning which of the hidden R-objects must be destroyed. For that purpose, the object of the other participating class B, refining the R-object to be destroyed, is passed as an argument. The specification of the destructor destructR is the inverse of the specification of the constructor constructR.

The inspector getRs() in the participating class-A defined at the level of the conceptual model in Figure 2, is transformed into the method getnbRs in the design model (illustrated in Figure 4) applicable to A-objects. Obviously, because R-objects are hidden inside A-objects, the refined objects themselves cannot be returned. When applied to an A-object, the method getnbRs returns how many R-objects are currently integrated in it.

Notice that the result of the inspector getnbRs() is specified in terms of another inspector getnbRs(B), returning the number of R-objects currently integrated in the A-object to which it is applied, and involving the given B-object. Another inspector getBs() is introduced to return a vector of B-objects currently involved in the refinement of R-objects integrated in a given A-object. At the level of analysis, these inspectors are not basic inspectors. As an example, the result of the inspector getnbRs(B) is obtained by first asking the set of all R-objects refined by the given A-object, followed by a projection over B-objects involved in the set of R-objects. Because the inspector getRs() cannot be offered, non-basic inspectors such as getnbRs(B) and getBs() must be introduced, in order to be able to retrieve as much information as possible from objects of both participating classes.

In view of the methods discussed so far, the definition of the participating class can be complemented with other mutators and inspectors. As a simple example, an inspector hasB(b) is defined. Notice that this inspector is easily specified (and implemented) in terms of the inspector getnbRs(b) defined above. A more complicated method for transferring a refined object to be integrated in another A-object, is also included in the definition of the class. This method corresponds to the mutator transfer, introduced at the level of the conceptual model. The specification (and implementation) of this method is basically obtained as a combination of the more basic methods constructR and destructR.

```
/**
 * Definition of the participating
 * class-A, integrating objects of the
 * refined class R.
 */
class A {
/**
 * Initialise a new A-object
 * integrating no refined R-objects.
 * @post| new.getnbRs()= 0
 */
public A()
/**
 * Terminate this A-object
 * @post|for each b in B:
 *       | not b.hasA(this)
 */
public void terminate()
/**
 * Construct a new refined R-object
 * integrated in this A-object.
 * @pre | b <> null
 * @post| new.getnbRs(b)= getnbRs(b) + 1
 * @post| (new b).getnbRs(this) =
 *      |   b.getnbRs(this) + 1
 */
public void constructR(B b)
/**
 * Destroy one of the refined R-objects
 * integrated this A-object.
 * @pre | hasB(b)
 * @post| new.getnbRs(b)= getnbRs(b) - 1
 * @post | (new b).getnbRs(this) =
 *        b.getnbRs(this) - 1
 */
public void destructR(B b)
/**
 * Transfer one of the refined R-
 * objects integrated in this A-object,
 * and involving the given B-object
 * <b>, to the given A-object.
 * @pre | (a <> null) and (a <> this)
 * @pre |  b <> null
 * @pre | hasB(b)
 * @post| new.getnbRs(b)= getnbRs(b) - 1
 * @post| (new b).getnbRs(this) =
 *      | b.getnbRs(this) - 1
 * @post|(new a).getnbRs(b)=a.getnbRs(b)+1
 * @post|(new b).getnbRs(a)=b.getnbRs(a)+1
 */
public void transferR(A a, B b)
/**
 * Return the total number of refined
 * R-objects integrated in this A-
 * object.
 * @return The sum of the number of R-
 *        objects integrated in this
 *        A-object,involving B-objects.
 *   | sum( {{ b in B : getNbRs(b) }} )
 */
public int getnbRs()
/**
 * Return the number of refined R-
 * objects integrated in this A-object,
 * and involving the given B-object <B>.
 * @return| if (getBs().contains(b))
 *        |then result > 0 else result = 0
 */
public int getnbRs(B b)
/**
 * Return a vector containing all the
 * B-objects involved in the refinement
 * of R-objects integrated in this A-
 * object.
```

```
  * @return |result <> null
  * @return|for each i in 0..result.size():
  *       | result.elementAt(i) instanceof B
  *@return|for each i,jin 0..result.size():
  *         |  if (i <> j)
  *         |   then result.elementAt(i) <>
  *         |     result.elementAt(j)
  */
public Vector getBs()
/**Check whether at least one refined
  * R-object is integrated in this
  * A-object, and involving the given B-
  * object <B>.
  * @param <b>
  * @result | getnbRs(b) > 0
  */
public boolean hasB(B b)
}//end of class-A
```

Implementation 2. Integration of the refined class R into the participating class A.

The specification of the other participating class B is not included in this paper. Next to a constructor and a destructor, it will introduce inspectors for retrieving information concerning the participation of B-objects in the refinement of objects of the integrated class R. As an example, the class B will introduce inspectors getAs() and getnbAs() to retrieve the objects of the other participating class to which a given B-object is associated, respectively the number of such objects.

With partial integration transformation, the software engineer gains some flexibility and re-usability over efficiency and simplicity.

- The resulting design is more flexible, in the sense that it can cope with more future extensions. In particular, additional characteristics ascribed to objects of both participating classes are easily added. In the same way, the design model must not be changed drastically, when objects of both participating classes can become involved in other associations with other classes.
- This transformation is also considered to be more re-usable than full integration since more of the structural and behavioral aspects of the classes specified at the level of analysis are kept at the level of design. However, because of the integration of the refined class in a participating class, chances for re-using the resulting class are still rather low.
- As far as efficiency is concerned this type of transformation is less efficient than full integration in terms of time and space requirements.
- Finally, the resulting design model is more complex compared to the one resulting from a full integration. Notice that particular inconsistencies may arise, if bi-directional associations are not designed and implemented with great care.

5. Conclusion

In this paper we have shown that designing convenient and transparent software systems is highly supported by performing an active transformation of the conceptual model's elements (i.e. classes and relations) to produce a design model that performs the system intended functionalities. We have seen that this approach offers the software engineer a series of transformations, each transformation typically emphasizing other quality factors. We have shown that full integration offers the software engineer efficiency and simplicity over adaptability and re-usability, whereas, partial integration offers more flexibility and re-usability, but less efficiency. As a result, this new technique doesn't require the software engineer to optimize the design model manually, which is lacking in current methodologies. Relatively speaking, this technique establishes a certain correspondence between conceptual models at the level of analysis and design models at the level of design, which awards high maintenance throughout the software system.

6. References

[1] E.Gamma, *Design Patterns: Elements of Reusable Object Oriented Software*, Addison-Wesley, 1995.

[2] Bertrand Meyer, *Object Oriented Software Construction*, Prentice Hall, 1997.

[3] Arthur, J. Riel, *Object Oriented Design Heuristics*, Addison Wesley, 1996.

[4] Rumbaugh, J. Jacobson, G.Booch, *The Unified Modeling Language Reference Manual*. Addison Wesley, 1999.

[5] Martin Fowler, *Refactoring: Improving the Design of Existing Code*. Addison Wesley, 1999.

[6]Charles Richter, *Software Engineering Series. Designing Flexible Object –Oriented System with UML*, Macmillan Technical, 1999.

[7] Gerson Sunye, Alain le Guennec, and Jean-marc Lezequuel, "*Design Patterns application in UML", ECOOP 2000, Sophia Antipolis and Cannes, France* , Springer – NY, June 2000 , *volume 1850 of lecture notes in Computer Science,pages 44–62.*

[8] S.Van Baelen, J. Lewi, and E. Steegmans,"*Constraints in Object-Oriented Analysis and Design*", *Technology of Object-Oriented Languages and Systems TOOLS 13*, eds. Magnusson, B., Meyer, B., Nerson, J.-M., and Perrot, J.-F., Prentice-Hall, Hertsfordshire, UK, 1994, pp. 185-199.

[9] S.Van Baelen, J. Lewi, E. Steegmans and B. Swennen, "*Constraints in Object-Oriented Analysis, Object Technologies for Advanced Software*", LectureNotes in Computer Science 742, eds. Nishio, S., and Yonezawa, A., Springer-Verlag, Berlin, D, 1993, pp. 393-407.

[10] J. Said, E. Steemans, "*Transformation of Unary Relations*", Proceeding ICSSEA 2000, 13th International Conference on Software & System Engineering and their Applications , Paris – France 2000, volume 4.

Author Index